Springer Series in Computational Mathematics

40

For further volumes:
http://www.springer.com/series/797

Sven Gross • Arnold Reusken

Numerical Methods for Two-phase Incompressible Flows

Springer

Sven Gross
Universität Bonn
Institut für Numerische
Simulation
Wegelerstrasse 6
53115 Bonn
Germany
sgross@ins.uni-bonn.de

Arnold Reusken
RWTH Aachen
Lehrstuhl für Numerische Mathematik
Templergraben 55
52056 Aachen
Germany
reusken@igpm.rwth-aachen.de

ISSN 0179-3632
ISBN 978-3-642-19685-0 e-ISBN 978-3-642-19686-7
DOI 10.1007/978-3-642-19686-7
Springer Heidelberg Dordrecht London New York

Library of Congress Control Number: 2011926983

Mathematics Subject Classification (2010): 35A35, 53A05, 76D05, 76D07, 65F08, 65F10, 65N22, 65N30, 65N75, 76M10, 76T10

Cover design: deblik, Berlin

Printed on acid-free paper

Springer is part of Springer Science+Business Media (www.springer.com)

Dedicated to Agnes and Monique

Preface

Numerical treatment of two-phase incompressible flow problems

In the past few decades there has been tremendous progress in the development and analysis of numerical methods for *one*-phase incompressible Stokes and Navier-Stokes equations. There is an extensive literature on space- and time discretization methods, iterative solvers and other numerical issues, e.g., implementation aspects, related to this problem class. This literature comprises a huge number of papers and many monographs. The research activities resulted in output ranging from new fundamental mathematical insights to software packages that can be used for the simulation of incompressible flow problems. Nowadays open source and commercial software packages are available that perform satisfactory when used as black or gray box solvers for a fairly large class of incompressible one-phase flow problems. Although very big progress has been made, there are still important topics which require further research. For example, in the field of development and analysis of numerical methods for the simulation of turbulent flows, non-Newtonian flows and flows coupled with chemistry the state of the art is not satisfactory, yet.

The work that has been done on numerical methods for one-phase incompressible Navier-Stokes equations forms a solid basis for an extension to the class of *two*-phase incompressible flow problems. In the past decade research on this topic has started. Until now most research results in this field have been published in the engineering literature. There are only few papers that have appeared in the numerical mathematics literature and address rigorous mathematical analysis of methods for two-phase flow problems. This book is meant to give an overview of, and introduction to this field of (analysis of) numerical methods for incompressible two-phase flow problems. We do not know of any other monograph devoted to this topic. In our opinion, time is ripe for substantial progress in the field of numerical analysis of methods for two-phase incompressible flows. There are several important issues relevant for the simulation of two-phase flows that are non-existent in one-phase incompressible flow problems. We briefly address a few of these:

Numerical treatment of the unknown interface. Even in the simplest case of immiscible fluids, i.e. no phase transition or evaporation phenomena, the numerical treatment of the unknown interface is a difficult task. Several numerical techniques are used, ranging from interface tracking, based on an explicit parametrization of the interface, to (VOF or level set) interface capturing methods, which are based on some indicator function. Until now many problems related to e.g. the coupling between the interface evolution and fluid dynamics, mass conservation, accuracy of discretizations and treatment of topological singularities (droplet collision) are largely unsolved. Only very few rigorous mathematical analyses related to these problems are known.
Numerical approximation of surface tension forces. The surface tension force is localized on the interface and in many two-phase systems it determines the flow behavior to a large extent. In case of topological singularities it is not obvious how this force should be modeled. An accurate numerical approximation of this force is often of major importance for a successful simulation, since an insufficient treatment leads to numerical oscillations at the interface (so-called *spurious velocities*). Only few approximation methods are known and analyses of these methods are very scarce.
Simulation of mass and heat transport from one phase into the other. The transport of a dissolved species from one phase into the other is usually modeled by convection-diffusion equations in the two phases that are coupled by a certain condition at the interface. If the species can attach to the interface this gives rise to (open) modeling problems. In general the concentration of the species is *dis*continuous across the interface. In that case one has to determine numerically a solution of a transport problem that is discontinuous across an evolving unknown interface. This topic has hardly been investigated in the literature. In certain systems it may be important to model a dependence of the surface tension on the concentration of the dissolved species or on the fluid temperature at the interface. If this is the case it results in a complicated strongly nonlinear coupling between the two-phase fluid dynamics and the mass or heat transport. The problem of how to handle numerically this coupling has hardly been addressed.
Simulation of surfactants, which are transported on the interface. It may happen that in the two-phase system there is a species (called tenside or surfactant) which adheres to the interface and is transported on the interface due to convection and molecular diffusion. An interesting modeling problem is how adsorption and desorption effects can be described. This surfactant transport results in a convection-diffusion equation on the interface only. As in the case of mass or heat transport discussed above there may be a dependence of the surface tension on the surfactant concentration. First studies of numerical methods for solving such surfactant transport equations, coupled with two-phase fluid dynamics, have appeared only recently.
Further interesting issues, which however will not be treated in this monograph, are the modeling and numerical treatment of evaporation, phase transition, topological singularities and reaction processes at the interface.

In this monograph we address topics, such as the four mentioned above, that are important in the numerical simulation of two-phase incompressible flow problems. We give a fairly complete treatment of such flow problems in the sense that we derive models, discuss appropriate weak formulations, introduce and analyze discretization methods, investigate iterative solvers and finally pay attention to implementation aspects and present results of numerical experiments. On the other hand we restricted ourselves to *in*compressible flows and do not consider important phenomena like phase transition and topological singularities. Also concerning the class of methods we made a severe restriction: we only treat discretizations based on finite elements. Within the problem and method classes considered in this book we tried to give a fairly complete overview. We do not present an overview of work that lies outside this problem class, e.g., flow problems with phase transition or compressible two-phase flows, or outside this method class, e.g., finite difference discretizations of two-phase flows.

Contents of this monograph

We start with an introductory chapter in which the basic models for one- and two-phase incompressible flows, for mass transport between the phases and for surfactant transport are derived. The book consists of five parts. We outline the main topics treated in these parts.

Part I. We first consider the incompressible Stokes and Navier-Stokes equations that model a *one*-phase flow. We treat numerical methods for these one-phase flows that are also used as basic building blocks in the simulation of *two*-phase flows, which is studied in Part II. The space discretizations that we consider are based on finite element methods. Therefore one needs suitable variational (weak) formulations. For this we collect some results on function spaces and variational formulations for (Navier-) Stokes equations known from the standard literature. We explain Hood-Taylor finite element discretization methods on multilevel tetrahedral triangulations and popular time discretization methods for Navier-Stokes equations. The topic of efficient iterative solvers is addressed. We give an introduction to multigrid methods and discuss certain Schur complement preconditioners for saddle point problems. In this part as well as in the other parts, for a specific method (or approach) we often address three aspects: 1. We try to give a clear description of the method. 2. A mathematical analysis of certain important aspects (e.g. discretization error, rate of convergence), often for a simplified model problem, is presented. 3. The method is implemented and results of numerical experiments, which illustrate certain phenomena, are presented.

Part II. We consider the fluid-dynamics in a two-phase incompressible flow problem with surface tension. The issue of interface representation is treated and a weak formulation of a two-phase Navier-Stokes equation with a localized surface tension force is given. Finite element discretization methods are developed and analyzed. In particular a special method for the discretization of the surface tension force and so-called extended finite element spaces (XFEM) for the pressure approximation are studied. Time discretization schemes are derived and finally iterative solvers are considered.

Part III. We address the numerical simulation of mass transport between the two phases. An appropriate weak formulation is derived. Based on this, finite element space discretization methods and time discretization schemes are discussed. We distinguish between problems with a stationary and with a non-stationary interface. An important issue is the numerical treatment of the discontinuity in the solution across the interface.

Part IV. In this part the convection-diffusion problem which models transport of surfactants is treated. Suitable weak formulations are discussed. Finite element methods based on both interface tracking and interface capturing techniques are presented.

Part V. This is an appendix consisting of two chapters. In the first chapter we collect some elementary results from differential geometry. In the second chapter we give some main results on variational problems in Hilbert spaces (e.g. Lax-Milgram lemma) and on Schur complement preconditioning of saddle point problems in Hilbert spaces.

Among the many numerical approaches treated in this monograph there are some that deserve special attention because these turned out to be particularly useful for the efficient simulation of our two-phase flow problems or we consider them to be promising for future applications. Therefore we emphasize these already in this preface:

- The finite element spaces that we use are based on *a hierarchy of nested tetrahedral triangulations*. The nested hierarchy allows very easy and efficient local refinement and coarsening routines. A (strong) local refinement close to the (evolving) interface in general enhances efficiency significantly. Furthermore, due to the nested hierarchy the use of efficient multigrid solvers is relatively easy.
- In our applications the surface tension is an important driving force. An accurate discretization of this force is essential for reliable simulation results. We use a *Laplace-Beltrami approach* in which the second derivative (curvature) in the surface tension functional can be avoided by partial integration. This technique is based on the representation of the curvature as a Laplace-Beltrami operator applied to the identity. We introduce and analyze an accurate variant of this method. In this method we use a hybrid version of the level set method in the sense that the level set equation is used to describe the evolution of the (implicitly given) interface but for the evaluation of the discrete surface tension functional we need an explicit reconstruction of the interface.
- In the class of two-phase flow problems that we consider there are several quantities which in general are discontinuous across the interface, namely viscosity, density, pressure and, if mass transport is considered, the concentration of a dissolved species. The (approximate) interface is not aligned

with the triangulation and thus we have unknowns (pressure, concentration) that are discontinuous within certain elements. For an accurate approximation of these unknowns we use the *extended finite element method* (XFEM), which has been used in the literature for other applications (e.g. crack propagation in continuum mechanics).

- If mass transport is considered then, due to the so-called Henry interface condition, the ratio of the unknown concentrations on the two sides of the interface has to be equal to a given constant. In general this (Henry) constant is not equal to one, which implies a discontinuity. To satisfy this interface condition we combine the XFEM approach with a *technique due to Nitsche*, in which the bilinear form that represents the partial differential equation is modified such that the jump relation is automatically satisfied in a certain weak sense.
- For one-phase flows and two-phase flows with a stationary interface we use the method of lines ("first space, then time") or the Rothe approach ("first time, then space") to obtain a fully discrete problem. For two-phase flow problems with a non-stationary interface the method of lines approach is not appropriate. The Rothe method is still useful but also *space-time finite element methods* are very suitable. We use the latter method class for the mass transport and for the surfactant transport equation.
- For the spatial discretization of the surfactant transport equation on the interface we introduce and analyze a *new interface finite element method.* The main idea of this method is the use of the trace of a standard outer finite element space (used for discretization of the flow variables) for discretization on the reconstructed approximate interface.
- In the time discretization we use implicit schemes in which the flow variables and the level set function are fully coupled. In each time step a nonlinear problem for these unknowns has to be solved. Due to the surface tension term there is a strongly nonlinear coupling between the flow variables and level set unknowns. We treat efficient *iterative decoupling strategies.*
- After discretization, decoupling and linearization one obtains large sparse linear systems with a saddle point structure. We use block preconditioners for the efficient solution of these linear systems. Special *Schur complement preconditioners* are presented.

Readership

We intended to make a monograph that is of interest for MSc and PhD students with a specialization in Numerical Analysis or Computational Engineering who want to get acquainted with numerical methods for two-phase incompressible flows. Basic knowledge of the numerical treatment of one-phase flow problems is assumed. Some of the topics presented may also be of interest for researchers already working in the field of numerical simulation of two-phase flows.

Further material

Most of the methods treated in this monograph have been implemented in a software package called DROPS, which has been developed at the Chair for Numerical Mathematics at RWTH Aachen. All numerical experiments, the results of which are given in this book, were performed with this package. More background information on DROPS and on publications from our research group is available on the website

www.igpm.rwth-aachen.de/DROPS/

The DROPS package is open source software under the GNU Lesser General Public License.

Acknowledgments

Both authors started their research on numerical methods for two-phase incompressible flow problems as members of the Collaborative Research Center 540 "Model-based experimental analysis of kinetic phenomena in fluid multiphase reactive systems" at RWTH Aachen. This interdisciplinary Research Center was initiated and coordinated by Wolfgang Marquardt. We are grateful to him and to other members of the Research Center for fruitful and inspiring collaborations. A further impulse for research on the topic of this monograph came from the initiative to establish the DFG Priority Programme "Transport Processes at Fluidic Interfaces", which started recently and is coordinated by Dieter Bothe and one of the authors. We thank our colleagues who contributed to this initiative. We also acknowledge financial support of the German Research Foundation (DFG) through funding of projects in the Collaborative Research Center 540, in which research related to this monograph has been done. Part of the results presented in this monograph, in particular those in the Chaps. 9 and 13, are based on joint work with Maxim Olshanskii. We acknowledge the pleasant collaboration. We thank Helmut Abels for proofreading Chap. 6. Many master and PhD students from the Chair of Numerical Mathematics at RWTH Aachen have contributed to the project of writing this monograph. These contributions range from joint work on the analysis of numerical methods, implementation of methods, the application to realistic two-phase flow models, code parallelization, to performing numerical experiments, the results of which are presented in this book, and proof-reading. In this respect we acknowledge the support of Patrick Esser, Oliver Fortmeier, Jörg Grande, Martin Horsky, Maxim Larin, Christoph Lehrenfeld, Eva Loch, Trung Hieu Nguyen, Volker Reichelt, Marcus Soemers and Yuanjun Zhang. In particular we thank Jörg Grande for the contribution to Chap. 13.

Aachen and Bonn
December 2010

Arnold Reusken
Sven Groß

Contents

1

Introduction

In this introductory chapter we describe the *models of one- and two-phase flow problems* that we consider, namely:

1) Navier-Stokes equations for one-phase flow (NS1),
2) Navier-Stokes equations for two-phase flow (NS2),
3) NS2 combined with transport of a dissolved species (NS2+T),
4) NS2 combined with transport of a surfactant *on* the interface (NS2+S).

These models are presented in Sect. 1.1 and consist of systems of coupled partial differential equations. To obtain a well-posed problem one has to add appropriate initial- and boundary conditions. This topic is briefly addressed in Sect. 1.2. An illustration of the type of two-phase flows that we are interested in is given in Sect. 1.3, where we present results of some numerical simulations. In Sect. 1.4 we give a schematic overview of the numerical methods that will be treated.

1.1 One- and two-phase flow models in strong formulation

In this section we give the partial differential equations corresponding to the models 1)-4). For ease of presentation these partial differential equations are given in the *strong* formulation. The numerical methods, in particular the finite element methods for spatial discretization, are based on the *weak* formulation of these partial differential equations. These weak formulations are given further on. In Sect. 1.2 we address the issue of initial and boundary conditions used in our models.

We always assume that the physical domain $\Omega \subset \mathbb{R}^3$ is an open bounded domain. This domain will also be the computational domain. We consider the flow problems for a fixed time interval denoted by $[0, T]$.

1.1.1 Navier-Stokes equations for one-phase flow

We derive the Navier-Stokes equations for modeling a laminar fluid flow. We assume the fluid to be incompressible, viscous, Newtonian and pure (i. e., no mixture of different components). Moreover we assume isothermal conditions and therefore neglect variations of density and dynamic viscosity due to temperature changes. Hence, dynamic viscosity and, due to incompressibility, also the density are constant (and positive).

The Eulerian coordinates of a point in Ω are denoted by $x = (x_1, x_2, x_3)$. We take a fixed $t_0 \in (0, T)$ and consider a time interval $(t_0 - \delta, t_0 + \delta)$, with $\delta > 0$ sufficiently small such that for $t \in (t_0 - \delta, t_0 + \delta)$ the quantities introduced below are well-defined. Let $\mathbf{X}$ denote a particle (also called "material point") in Ω at $t = t_0$, with Eulerian coordinates $\xi \in \mathbb{R}^3$. Let $X_\xi(t)$ denote the Eulerian coordinates of the particle $\mathbf{X}$ at time t. The mapping

$$t \to X_\xi(t), \quad t \in (t_0 - \delta, t_0 + \delta),$$

describes the trajectory of the particle $\mathbf{X}$. The particles are transported by a velocity field, which is denoted by $\mathbf{u} = \mathbf{u}(x,t) = \big(u_1(x,t), u_2(x,t), u_3(x,t)\big) \in \mathbb{R}^3$. Hence

$$\frac{d}{dt} X_\xi(t) = \mathbf{u}(X_\xi(t), t). \tag{1.1}$$

For the given $\mathbf{X}$, the solution of the system of ordinary differential equations

$$\frac{d}{dt} X_\xi(t) = \mathbf{u}\big(X_\xi(t), t\big), \quad t \in (t_0 - \delta, t_0 + \delta), \quad X_\xi(t_0) = \xi,$$

yields the trajectory of the particle $\mathbf{X}$.

Physical processes can be modeled in different coordinate systems. For flow problems, the two most important ones are (x, t) ("Eulerian") and (ξ, t) ("Lagrangian"):

- Euler coordinates (x, t): one takes an arbitrary fixed point x in space and considers the velocity $\mathbf{u}(x, t)$ at x. If time evolves *different* particles pass through x.
- Lagrange (or "material") coordinates (ξ, t): one takes an arbitrary fixed particle (material point) and considers its motion. If time evolves one thus follows the trajectory of a *fixed* particle.

Related to the Lagrangian coordinates we define the so-called *material derivative* of a (sufficiently smooth) function $f(x, t)$ on the trajectory of $\mathbf{X}$:

$$\dot{f}(X_\xi(t), t) := \frac{d}{dt} f(X_\xi(t), t).$$

If f is defined in a neighborhood of the trajectory we obtain from the chain rule and (1.1):

$$\dot{f} = \frac{\partial f}{\partial t} + \mathbf{u} \cdot \nabla f. \tag{1.2}$$

The derivation of partial differential equations that model the flow problem is based on conservation laws applied on a (small) subdomain, called a material volume, $W_0 \subset \Omega$. We derive these partial differential equations in Eulerian coordinates. Given W_0, define

$$W(t) := \{ X_\xi(t) : \xi \in W_0 \}.$$

$W(t)$ describes the position of the particles at time t, which were located in W_0 at time $t = t_0$. We need the following fundamental identity, which holds for a scalar sufficiently smooth function $f = f(x,t)$:

Reynolds' transport theorem:

$$\begin{aligned} \frac{d}{dt}\int_{W(t)} f(x,t)\,dx &= \int_{W(t)} \dot{f}(x,t) + f\,\mathrm{div}\,\mathbf{u}(x,t)\,dx \\ &= \int_{W(t)} \frac{\partial f}{\partial t}(x,t) + \mathrm{div}(f\mathbf{u})(x,t)\,dx, \end{aligned} \tag{1.3}$$

with $\dot{f} := \frac{\partial f}{\partial t} + \mathbf{u}\cdot\nabla f$ the material derivative.

First we consider the *conservation of mass* principle. Let $\rho(x,t)$ be the *density* of the fluid. If we take $f = \rho$ in (1.3) this yields

$$0 = \frac{d}{dt}\int_{W(t)} \rho\,dx = \int_{W(t)} \frac{\partial \rho}{\partial t} + \mathrm{div}(\rho\mathbf{u})\,dx,$$

which holds in particular for $t = t_0$ and for an arbitrary material volume $W(t_0) = W_0$ in Ω. Since also $t_0 \in (0,T)$ is arbitrary, we obtain the partial differential equation

$$\frac{\partial \rho}{\partial t} + \mathrm{div}(\rho\mathbf{u}) = 0 \quad \text{in} \quad \Omega \times (0,T).$$

Due to the assumption $\rho = const$ this simplifies to

$$\mathrm{div}\,\mathbf{u} = 0 \quad \text{in} \quad \Omega \times (0,T), \tag{1.4}$$

which is often called *mass conservation equation* or *continuity equation.*

We now consider *conservation of momentum.* The momentum of mass contained in $W(t)$ is given by

$$M(t) = \int_{W(t)} \rho\mathbf{u}\,dx.$$

Due to Newton's law the *change* of momentum $M(t)$ is equal to the force $F(t)$ acting on $W(t)$. This force is decomposed in a *volume* force $F_1(t)$ and a *boundary* force $F_2(t)$. We restrict ourselves to the case where the only volume force acting on the volume $W(t)$ is gravity:

$$F_1(t) = \int_{W(t)} \rho \mathbf{g}\, dx,$$

where $\mathbf{g} \in \mathbb{R}^3$ is the vector of gravitational acceleration. The boundary force $F_2(t)$ is used to describe internal forces, i.e., forces that a fluid exerts on itself. These include pressure and the viscous drag that a fluid element $W(t)$ gets from the adjacent fluid. These internal forces are *contact* forces: they act on the boundary $\partial W(t)$ of the fluid element $W(t)$. Let $\vec{t}$ denote this internal force vector, also called traction vector. Then we have

$$F_2(t) = \int_{\partial W(t)} \vec{t}\, ds.$$

Cauchy derived fundamental principles of continuum mechanics and in particular he derived the following law (often called *Cauchy's theorem*):

$$\vec{t} \text{ is a linear function of } \mathbf{n},$$

where $\mathbf{n} = \mathbf{n}(x,t) \in \mathbb{R}^3$ is the outer unit normal on $\partial W(t)$. For more explanation on this we refer to introductions to continuum mechanics, for example [130]. Thus it follows that there is a matrix $\boldsymbol{\sigma} = \boldsymbol{\sigma}(x,t) \in \mathbb{R}^{3\times 3}$, called the *stress tensor*, such that the boundary force can be represented as

$$F_2(t) = \int_{\partial W(t)} \boldsymbol{\sigma} \mathbf{n}\, ds. \tag{1.5}$$

Using these force representations in Newton's law and applying Stokes' theorem for $F_2(t)$ we get

$$\begin{aligned} \frac{d}{dt} M(t) &= F_1(t) + F_2(t) \\ &= \int_{W(t)} \rho \mathbf{g} + \operatorname{div} \boldsymbol{\sigma}\, dx. \end{aligned} \tag{1.6}$$

For a matrix $\mathbf{A}(x) \in \mathbb{R}^3$, $x \in \mathbb{R}^3$, its divergence is defined by

$$\operatorname{div} \mathbf{A}(x) = \begin{pmatrix} \operatorname{div}(a_{11}\ a_{12}\ a_{13}) \\ \operatorname{div}(a_{21}\ a_{22}\ a_{23}) \\ \operatorname{div}(a_{31}\ a_{32}\ a_{33}) \end{pmatrix} \in \mathbb{R}^3.$$

Using the transport theorem (1.3) in the left-hand side of (1.6) with $f = \rho\, u_i$, $i = 1, 2, 3$, we obtain

$$\int_{W(t)} \frac{\partial \rho\, u_i}{\partial t} + \operatorname{div}(\rho\, u_i\, \mathbf{u})\, dx = \int_{W(t)} \rho\, g_i + \operatorname{div} \boldsymbol{\sigma}_i\, dx, \qquad i = 1, 2, 3,$$

with $\boldsymbol{\sigma}_i$ the i-th row of $\boldsymbol{\sigma}$ and g_i the i-th component of $\mathbf{g}$. In vector notation, with $\mathbf{u} \otimes \mathbf{u} = (u_i u_j)_{1 \le i,j \le 3}$,

$$\int_{W(t)} \frac{\partial \rho \mathbf{u}}{\partial t} + \operatorname{div}(\rho \mathbf{u} \otimes \mathbf{u}) \, dx = \int_{W(t)} \rho \mathbf{g} + \operatorname{div} \boldsymbol{\sigma} \, dx, \tag{1.7}$$

which holds in particular for $t = t_0$ and for an arbitrary material volume $W(t_0) = W_0$ in Ω. Since $t_0 \in (0,T)$ is arbitrary, we obtain the partial differential equations

$$\frac{\partial \rho \mathbf{u}}{\partial t} + \operatorname{div}(\rho \mathbf{u} \otimes \mathbf{u}) = \rho \mathbf{g} + \operatorname{div} \boldsymbol{\sigma} \quad \text{in} \quad \Omega \times (0,T).$$

Note that $\operatorname{div}(\rho \mathbf{u} \otimes \mathbf{u}) = \rho(\mathbf{u} \cdot \nabla)\mathbf{u} + \rho \mathbf{u} \operatorname{div} \mathbf{u}$ and due to the continuity equation (1.4), the last summand vanishes, yielding the so-called *momentum equation*

$$\rho \frac{\partial \mathbf{u}}{\partial t} + \rho (\mathbf{u} \cdot \nabla)\mathbf{u} = \rho \mathbf{g} + \operatorname{div} \boldsymbol{\sigma}. \tag{1.8}$$

For viscous *Newtonian fluids* one assumes that the stress tensor $\boldsymbol{\sigma}$ is of the form

$$\boldsymbol{\sigma} = -p\mathbf{I} + L(\mathbf{D}), \tag{1.9}$$

where p is the pressure,

$$\mathbf{D}(\mathbf{u}) = \nabla \mathbf{u} + (\nabla \mathbf{u})^T$$

is the deformation tensor, $\nabla \mathbf{u} := \big(\nabla u_1 \; \nabla u_2 \; \nabla u_3 \big)$, and L is assumed to be a *linear* mapping. Based on this structural model for the stress tensor and using the additional assumptions that the medium is isotropic (i.e. its properties are the same in all space directions) and the action of the stress tensor is independent of the specific frame of reference ("invariance under a change in observer") it can be shown ([130, 107]) that the stress tensor must have the form

$$\boldsymbol{\sigma} = -p\mathbf{I} + \lambda \operatorname{div} \mathbf{u} \, \mathbf{I} + \mu \mathbf{D}(\mathbf{u}). \tag{1.10}$$

Further physical considerations lead to relations for the parameters μ, λ, e.g., $\mu > 0$ (for a viscous fluid), $\lambda \ge -\frac{2}{3}\mu$ or even $\lambda = -\frac{2}{3}\mu$. For the case of an incompressible fluid, i.e., $\operatorname{div} \mathbf{u} = 0$, the relation for the stress tensor simplifies to

$$\boldsymbol{\sigma} = -p\mathbf{I} + \mu \mathbf{D}(u), \tag{1.11}$$

with $\mu > 0$ the *dynamic viscosity*. Hence, we obtain the fundamental Navier-Stokes equations for incompressible flow:

$$\begin{aligned} \rho\Big(\frac{\partial \mathbf{u}}{\partial t} + (\mathbf{u} \cdot \nabla)\mathbf{u}\Big) &= -\nabla p + \operatorname{div}(\mu \mathbf{D}(\mathbf{u})) + \rho \mathbf{g} \text{ in } \Omega \\ \operatorname{div} \mathbf{u} &= 0 \text{ in } \Omega. \end{aligned} \tag{1.12}$$

These equations are considered for $t \in [0, T]$. Initial and boundary conditions corresponding to these Navier-Stokes equations are discussed in Sect. 1.2.

Remark 1.1.1 Using the assumption that μ is a strictly positive *constant* and the relation $\operatorname{div} \mathbf{u} = 0$ we get

$$\operatorname{div}(\mu \mathbf{D}(\mathbf{u})) = \mu \Delta \mathbf{u} = \mu \begin{pmatrix} \Delta u_1 \\ \Delta u_2 \\ \Delta u_3 \end{pmatrix}.$$

1.1.2 Navier-Stokes equations for two-phase flow

We now consider two-phase flows, i. e., Ω contains two different *immiscible* incompressible phases (liquid-liquid or liquid-gas) which may move in time and have different material properties ρ_i and μ_i, $i = 1, 2$. For each point in time, $t \in [0, T]$, Ω is partitioned into two open subdomains $\Omega_1(t)$ and $\Omega_2(t)$, $\overline{\Omega} = \overline{\Omega}_1(t) \cup \overline{\Omega}_2(t)$, $\Omega_1(t) \cap \Omega_2(t) = \emptyset$, each of them containing one of the phases, respectively. These phases are separated from each other by the interface $\Gamma(t) = \overline{\Omega}_1(t) \cap \overline{\Omega}_2(t)$, cf. Fig. 1.1. As mentioned before, we assume isothermal conditions and both phases to be pure substances. Furthermore, we do *not* consider reaction, mass transfer or phase transition.

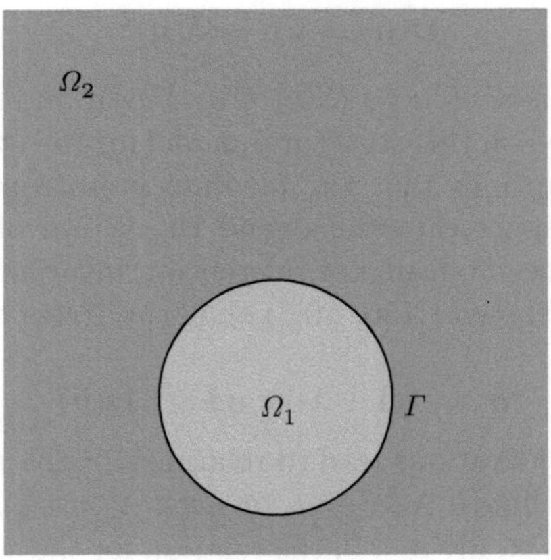

Fig. 1.1. 2D illustration of a domain Ω consisting of two phases Ω_1 and Ω_2 and interface Γ.

In each of the phases conservation of mass and momentum has to hold, yielding separate Navier-Stokes equations in the two domains $\Omega_i, i = 1, 2$:

$$\begin{cases} \rho_i(\frac{\partial \mathbf{u}}{\partial t} + (\mathbf{u} \cdot \nabla)\mathbf{u}) = \operatorname{div} \boldsymbol{\sigma}_i + \rho_i \mathbf{g} \\ \operatorname{div} \mathbf{u} = 0 \end{cases} \quad \text{in } \Omega_i, \ i = 1, 2, \tag{1.13}$$

with $\boldsymbol{\sigma}_i = -p\mathbf{I} + \mu_i(\nabla\mathbf{u} + (\nabla\mathbf{u})^T)$. We now derive *coupling conditions* at the interface. As the phases are viscous and no phase transition takes place, the velocity can be assumed to be continuous at the interface:

$$[\mathbf{u}] = 0 \quad \text{on } \Gamma. \tag{1.14}$$

Here for $x \in \Gamma$ and a function f defined in a neighborhood of Γ we define the jump across Γ by

$$[f](x) = [f]_\Gamma(x) := \lim_{h\downarrow 0} \left(f\left(x - h\,\mathbf{n}_\Gamma(x)\right) - f\left(x + h\,\mathbf{n}_\Gamma(x)\right)\right), \tag{1.15}$$

where $\mathbf{n}_\Gamma(x)$ denotes the unit normal on Γ at x, pointing from Ω_1 to Ω_2.

Remark 1.1.2 In the definition of the jump across the interface in (1.15) the normal is pointing from Ω_1 into Ω_2 and the jump is defined as the value close to the interface in Ω_1 minus the value close to the interface in Ω_2. In the literature sometimes the other sign convention (value in Ω_2 minus value in Ω_1) is used, leading to another sign in the interface condition (1.19) derived below. We choose this sign convention, since it is consistent with the standard form of the classical Laplace-Young pressure jump relation $[p]_\Gamma = \tau\kappa\mathbf{n}$, discussed in Remark 1.1.5.

Consider a fluid volume $W = W_1 \cup W_2$ as illustrated in Fig. 1.2 which contains a part γ of the interface Γ.

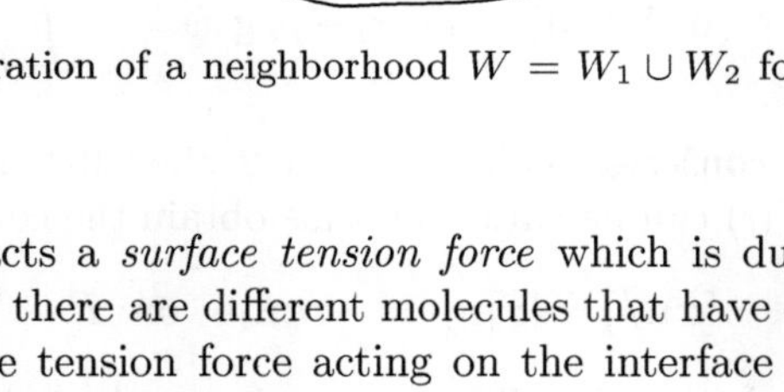

Fig. 1.2. 2D illustration of a neighborhood $W = W_1 \cup W_2$ for an interface point $x \in \Gamma$.

At the interface acts a *surface tension force* which is due to the fact that on both sides of Γ there are different molecules that have different attractive forces. The surface tension force acting on the interface segment γ can be modeled by (cf. [225, 47, 219] and Sect. 1.1.5)

$$F_3(t) = -\tau \int_{\gamma(t)} \kappa\mathbf{n}_\Gamma \, ds. \tag{1.16}$$

The parameter τ is the *surface tension coefficient*, which is a material property of the two-phase system. To simplify the presentation we assume τ to be constant. For many two-phase systems this is a reasonable assumption. The case of a variable surface tension coefficient τ is discussed in Remark 1.1.3. The scalar function $\kappa(x)$, $x \in \Gamma$, is the mean curvature, cf. Chap. 14, for which

$$\kappa(x) = \operatorname{div} \mathbf{n}_\Gamma(x), \qquad x \in \Gamma,$$

holds. Note that if at $x \in \Gamma$ the subdomain Ω_1 is locally convex, then κ is positive. This additional force term $F_3(t)$ has to be taken into account if we consider conservation of momentum, cf. Sect. 1.1.1. For a fluid volume $W = W_1 \cup W_2$ as in Fig. 1.2, instead of (1.5), (1.6) we now have

$$\begin{aligned}\frac{d}{dt}M(t) &= F_1(t) + F_2(t) + F_3(t) \\ &= \int_{W(t)} \rho \mathbf{g}\, dx + \int_{\partial W(t)} \boldsymbol{\sigma}\mathbf{n}\, ds - \tau \int_{\gamma(t)} \kappa \mathbf{n}_\Gamma\, ds.\end{aligned} \tag{1.17}$$

Since the stress tensor $\boldsymbol{\sigma}$ is not necessarily smooth across Γ we split ∂W into ∂W_1 and ∂W_2

$$\int_{\partial W(t)} \boldsymbol{\sigma}\mathbf{n}\, ds = \int_{\partial W_1(t)} \boldsymbol{\sigma}_1 \mathbf{n}_1\, ds + \int_{\partial W_2(t)} \boldsymbol{\sigma}_2 \mathbf{n}_2\, ds - \int_{\gamma(t)} [\boldsymbol{\sigma}]\mathbf{n}_\Gamma\, ds,$$

and apply the Stokes' theorem on W_1 and W_2 separately. Note that $\mathbf{n}_i$ is the outward normal on ∂W_i and $\mathbf{n}_\Gamma$ the normal at Γ, pointing from Ω_1 in Ω_2. Thus we obtain, cf. (1.7),

$$\begin{aligned}\int_{W(t)} \rho \left(\frac{\partial \mathbf{u}}{\partial t} + (\mathbf{u}\cdot\nabla)\cdot\mathbf{u}\right) dx &= \int_{W_1(t)} \operatorname{div} \boldsymbol{\sigma}_1\, dx + \int_{W_2(t)} \operatorname{div} \boldsymbol{\sigma}_2\, dx \\ &\quad + \int_{W(t)} \rho\, \mathbf{g}\, dx - \int_{\gamma(t)} [\boldsymbol{\sigma}]\mathbf{n}_\Gamma\, ds - \tau \int_{\gamma(t)} \kappa \mathbf{n}_\Gamma\, ds.\end{aligned}$$

This yields,

$$\sum_{i=1,2} \int_{W_i(t)} \rho_i \left(\frac{\partial \mathbf{u}}{\partial t} + (\mathbf{u}\cdot\nabla)\cdot\mathbf{u}\right) - \operatorname{div} \boldsymbol{\sigma}_i - \rho_i\, \mathbf{g}\, dx = -\int_{\gamma(t)} \tau\kappa\mathbf{n}_\Gamma + [\boldsymbol{\sigma}]\mathbf{n}_\Gamma\, ds.$$

Due to momentum conservation in W_i, $i = 1, 2$, the left-hand side equals zero, cf. (1.13). Since $W(t)$ can be varied we thus obtain the coupling condition:

$$[\boldsymbol{\sigma}\mathbf{n}_\Gamma] = [\boldsymbol{\sigma}]\mathbf{n}_\Gamma = -\tau\kappa\mathbf{n}_\Gamma \quad \text{on } \Gamma. \tag{1.18}$$

Finally, in view of the immiscibility assumption we introduce the normal velocity $V_\Gamma = V_\Gamma(x,t) \in \mathbb{R}$, which denotes the *size* of the velocity of the interface Γ at $x \in \Gamma(t)$ in normal direction, i.e., the movement of Γ in normal direction is given by $V_\Gamma \mathbf{n}$. The *immiscibility assumption* is modeled by the condition that the normal velocity of the interface should equal the normal component of the flow field at the interface, i.e. $V_\Gamma = \mathbf{u}\cdot\mathbf{n}_\Gamma$ at the interface. Summarizing, the latter condition, the equations in (1.13) and the coupling conditions in (1.14) and (1.18) lead to the following standard model for two-phase flows:

$$\begin{cases} \rho_i(\frac{\partial \mathbf{u}}{\partial t} + (\mathbf{u}\cdot\nabla)\mathbf{u}) = \operatorname{div}\boldsymbol{\sigma}_i + \rho_i\mathbf{g} \\ \operatorname{div}\mathbf{u} = 0 \end{cases} \quad \text{in } \Omega_i, \quad i = 1,2, \tag{1.19}$$

$$[\boldsymbol{\sigma}\mathbf{n}_\Gamma] = -\tau\kappa\mathbf{n}_\Gamma, \ [\mathbf{u}] = 0 \quad \text{on } \Gamma, \tag{1.20}$$

$$V_\Gamma = \mathbf{u}\cdot\mathbf{n}_\Gamma \quad \text{on } \Gamma. \tag{1.21}$$

We recall the Newtonian stress tensor model: $\boldsymbol{\sigma}_i = -p\mathbf{I} + \mu_i(\nabla\mathbf{u} + (\nabla\mathbf{u})^T)$. The density and viscosity, ρ_i and $\mu_i, i = 1,2$, are assumed to be constant in each phase.

Remark 1.1.3 In certain cases, for example in systems with significant surfactants ("surface active agents") one has to take a *variable* surface tension coefficient τ into account, cf. Sect. 1.1.5. In that case the surface tension force in (1.16) has to be replaced by its generalization

$$F_3(t) = -\int_\gamma \tau\kappa\mathbf{n}_\Gamma - \nabla_\Gamma\tau\, ds, \tag{1.22}$$

where $\nabla_\Gamma = P\nabla$, with $P = I - \mathbf{n}_\Gamma\mathbf{n}_\Gamma^T$, is the *tangential derivative.* The interface condition in (1.18) then generalizes to

$$[\boldsymbol{\sigma}\mathbf{n}_\Gamma] = -\tau\kappa\mathbf{n}_\Gamma + \nabla_\Gamma\tau. \tag{1.23}$$

In the remainder we often write $\mathbf{n}$ *instead of* $\mathbf{n}_\Gamma$.

1.1.3 Two-phase flow with transport of a dissolved species

We consider a two-phase flow problem as described above. We assume that one or both phases contain a dissolved species that is transported due to convection and molecular diffusion and does not adhere to the interface. The concentration of this species is denoted by $c(x,t)$. This flow problem can be modeled by the equations (1.19)-(1.21) for the flow variables and a convection-diffusion equation for the concentration c. At the interface we need interface conditions for c. The first interface condition comes from mass conservation, which implies flux continuity. The second condition results from a constitutive equation known as Henry's law, which states that the solubility of a gas in a liquid at a particular temperature is proportional to the pressure of that gas above the liquid. In mathematical terms this relation (at constant temperature) can be formulated as $p = k_{\mathrm{H}}\, c$ where p is the partial pressure of the solute in the gas above the solution, c is the concentration of the solute and k_H is known as the Henry's law constant and depends on the solute, the solvent and the temperature. The same solute in different solvents (at the same temperature) corresponds to different Henry constants, reflecting the different solubility properties of the two solvents. From this it can be deduced, that in a two-phase system with a solute, assuming instantaneous local equilibrium at the interface, there is a constant ratio between the concentrations of the solute on the two sides of the interface. Thus one obtains the following standard model:

Two-phase flow model (1.19) – (1.21) combined with:

$$\begin{aligned} \frac{\partial c}{\partial t} + \mathbf{u} \cdot \nabla c &= \operatorname{div}(D_i \nabla c) \quad \text{in} \quad \Omega_i, \quad i = 1, 2, \\ [D_i \nabla c \cdot \mathbf{n}]_\Gamma &= 0 \quad \text{on} \quad \Gamma, \\ c_1 &= C_H c_2 \quad \text{on} \quad \Gamma. \end{aligned} \tag{1.24}$$

The diffusion coefficient D_i is piecewise constant. In the interface condition we use the notation c_i for $c_{|\Omega_i}$ restricted to the interface. The constant $C_H > 0$ is given (Henry's constant). The Henry interface condition can also be written as $[\hat{C}c] = 0$, with $\hat{C} = 1$ in Ω_1, $\hat{C} = C_H$ in Ω_2. The model has to be combined with suitable initial and boundary conditions, cf. Sect. 1.2. In the formulation in (1.24) there is a coupling between fluid dynamics and mass transport only *in one direction*, in the sense that the velocity is used in the mass transport equation, but the concentration c does *not* influence the fluid dynamics. In certain systems it may be appropriate to consider a dependence of the surface tension coefficient on c, i.e. $\tau = \tau(c)$. In that case there is a coupling in two directions between fluid dynamics and mass transport.

1.1.4 Two-phase flow with transport of a surfactant on the interface

We consider a two-phase flow problem as described above in Sect. 1.1.2. We assume that there is a species (called tenside or surfactant) which adheres to the interface and is transported at the interface due to convection (movement of the interface) and due to diffusion (molecular diffusion on the interface). For simplicity we assume that there are no adsorption and desorption effects (i.e. no sources or sinks). The concentration of this surfactant is denoted by $S(x, t)$, $x \in \Gamma(t)$. A partial differential equation for this quantity can be derived from the conservation of mass principle (on subsets $\gamma(t)$ of the moving interface $\Gamma(t)$). For $t_0 \in (0, T)$, let γ_0 be a connected bounded subset of $\Gamma(t_0)$ and $\gamma(t) = \{ X_\xi(t) : \xi \in \gamma_0 \} \subset \Gamma(t)$, $t \in (t_0 - \delta, t_0 + \delta)$, with $\delta > 0$ sufficiently small. The conservation of mass property yields

$$\frac{d}{dt} \int_{\gamma(t)} S \, ds = - \int_{\partial\gamma(t)} q \cdot n \, d\tilde{s},$$

with n the unit normal to $\partial\gamma(t)$ lying in a tangent plane and pointing out of $\gamma(t)$.

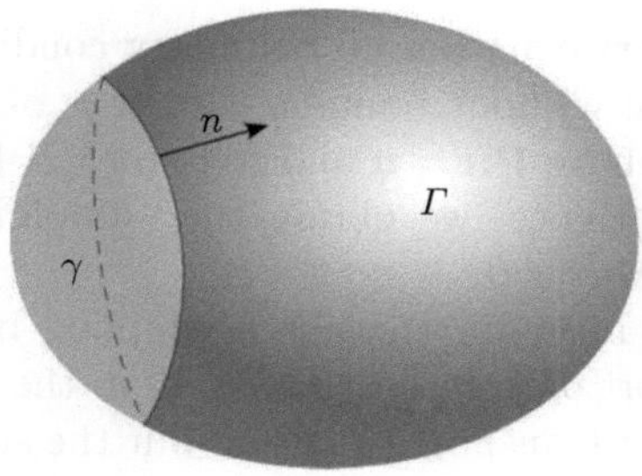

Fig. 1.3. Interface Γ and subset $\gamma \subset \Gamma$, with vector n which is normal to $\partial\gamma$ and tangential to Γ.

We restrict to the case of a *diffusive flux* $q := -D_\Gamma \nabla_\Gamma S$. Recall that the tangential derivative is defined by $\nabla_\Gamma := \mathbf{P}\nabla$ with $\mathbf{P} = \mathbf{I} - \mathbf{n}\mathbf{n}^T$. Note that the normal $\mathbf{n} = \mathbf{n}_\Gamma$ differs from the normal n. Using integration by parts on the manifold $\gamma(t)$ we obtain

$$\int_{\partial\gamma(t)} q \cdot n \, d\tilde{s} = \int_{\gamma(t)} \mathrm{div}_\Gamma q \, ds.$$

A variant of the transport theorem in (1.3), cf. (14.21b) and Remark 14.2.3, yields

$$\frac{d}{dt}\int_{\gamma(t)} S \, ds = \int_{\gamma(t)} \dot{S} + S \, \mathrm{div}_\Gamma \mathbf{u} \, ds$$

and thus we obtain

$$\int_{\gamma(t)} \dot{S} + S \, \mathrm{div}_\Gamma \mathbf{u} + \mathrm{div}_\Gamma q \, ds = 0,$$

which holds in particular for $\gamma(t_0) = \gamma_0$ arbitrary. Hence we obtain the following model for transport of surfactants:

Two-phase flow model (1.19) – (1.21) combined with:

$$\dot{S} + S \, \mathrm{div}_\Gamma \mathbf{u} = \mathrm{div}_\Gamma (D_\Gamma \nabla_\Gamma S) \quad \text{on } \Gamma. \tag{1.25}$$

If the diffusion coefficient D_Γ is constant on Γ we can reformulate the diffusion part as $\mathrm{div}_\Gamma(D_\Gamma \nabla_\Gamma S) = D_\Gamma \Delta_\Gamma S$. Using the definition of the material derivative the convection-diffusion equation in (1.25) can be written as

$$\frac{\partial S}{\partial t} + \mathbf{u} \cdot \nabla S + S \, \mathrm{div}_\Gamma \mathbf{u} = D_\Gamma \Delta_\Gamma S \quad \text{on } \Gamma.$$

In this formulation, for the partial derivatives $\frac{\partial}{\partial t}$ and $\mathbf{u} \cdot \nabla$ to be well-defined, one assumes that S is smoothly extended in a small neighborhood of Γ. For this surfactant transport equation no boundary conditions are needed if the interface Γ is a surface without boundary. In case of a stationary interface, i.e. $\mathbf{u} \cdot \mathbf{n} = 0$ on Γ, we have $\mathbf{P}\mathbf{u} = \mathbf{u}$ and thus $\mathbf{u} \cdot \nabla S + S\,\mathrm{div}_\Gamma \mathbf{u} = \mathbf{u} \cdot \nabla_\Gamma S + S\,\mathrm{div}_\Gamma \mathbf{u} = \mathrm{div}_\Gamma(\mathbf{u}S)$. Hence, we obtain the (simplified) diffusion equation $\frac{\partial S}{\partial t} + \mathrm{div}_\Gamma(\mathbf{u}S) - D_\Gamma \Delta_\Gamma S = 0$.

In the formulation in (1.25) there is a coupling between fluid dynamics and surfactant transport only *in one direction*, in the sense that the velocity is used in the surfactant transport equation, but the surfactant concentration S does *not* influence the fluid dynamics. In many systems with surfactants, there is a dependence of the surface tension coefficient on S, i.e. $\tau = \tau(S)$. In that case there is a coupling in two directions between fluid dynamics and surfactant transport.

1.1.5 Modeling of interfacial phenomena

The notion of interfacial transport phenomena usually refers to mass, momentum and energy transfer within a neighborhood of an interface, including the thermodynamics of the interface. Physico-chemical interfacial phenomena play a crucial role in high-tech applications like, for example, lab-on-a-chip systems, multiphase reactors in chemical engineering and micro process engineering. We refer to the (chemical) engineering literature for a treatment of these topics, e.g. [40], in which particle-stabilized foams and emulsions and new materials derived from such systems are studied. The understanding of most of these interfacial phenomena is still very poor. In particular there is a strong lack of (validated) mathematical models that describe interfacial processes appropriately. Research on modeling of interfacial transport phenomena is a very active and rapidly growing field. We do not treat modeling aspects in this monograph. In this section we only give a very brief introduction into basics related to the modeling of interfacial processes in two-phase incompressible immiscible flows. An extensive treatment of this topic and many references are given in [225].

Dividing surface and clean interface

There are continuum models in which an interface is represented as a *three*-dimensional region of very small thickness. One of the first models of this type was introduced by Korteweg [158]. The so-called phase field (or diffusive interface) models, treated in Sect. 6.2.4, belong to this class. More often models are used in which the interface is modeled by a (non-physical) *two*-dimensional *dividing surface*. This approach was first proposed by Gibbs. In such a *sharp interface model* for incompressible flows it is assumed that the dividing surface separates two homogeneous phases which both have a constant density. The effect of the interfacial region is taken into account by introducing so-called *excess* quantities (e.g., mass and energy) which are assigned to the dividing surface. We explain this by considering the excess mass density, denoted by ρ^Γ. For a given time t let $W(t)$ be a material volume, illustrated in Fig. 1.4, which contains two phases and an interfacial region of finite thickness. This interfacial region, denoted by R^I, is bounded by the surfaces Γ_1 and Γ_2 and is such that outside of R^I we have homogeneous phases, i.e., in the two subvolumes $W(t) \setminus R^I$, denoted by R_i, we have constant densities ρ_i, $i = 1, 2$. We choose a dividing surface Γ and assume the three surfaces to be parallel. The dividing surface is assumed to be transported with the flow velocity field $\mathbf{u}(x,t)$, $x \in \Gamma = \Gamma(t)$. In the interfacial region R^I we have a *mixture* of the two phases. The density of this mixture (total mass per volume) is denoted by ρ^I. Note that in general this density is *not* constant in R^I. This density function can be naturally extended outside R^I by $\rho^I = \rho_i$ in R_i, $i = 1, 2$.

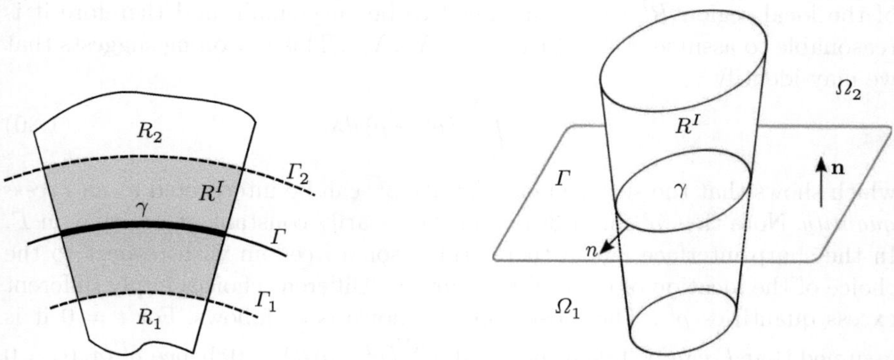

Fig. 1.4. Illustration of cylinder R^I in 2D (left) and 3D (right).

The mass conservation property is modeled by

$$\frac{d}{dt} \int_{W(t)} \rho^I \, dx = 0. \tag{1.26}$$

Let ρ be the piecewise constant function with constant values ρ_1, ρ_2 in the two subvolumes of $W(t)$ separated by the dividing surface Γ, which are denoted by $W_i(t)$, $i = 1, 2$. From (1.26) we obtain

$$\frac{d}{dt}\Big(\int_{W(t)} \rho \, dx + \int_{R^I} (\rho^I - \rho) \, dx \Big) = 0. \tag{1.27}$$

For a sharp interface model, with a dividing surface Γ and a piecewise constant density ρ, to be a good approximation of the model with an interfacial region and density ρ^I, we introduce a *surface* mass density ρ^Γ. Mass conservation in the sharp interface model then takes the form

$$\frac{d}{dt}\Big(\int_{W(t)} \rho \, dx + \int_{\gamma(t)} \rho^\Gamma \, ds \Big) = 0, \tag{1.28}$$

with $\gamma(t) := \Gamma \cap W(t)$ the part of the dividing surface contained in $W(t)$. Comparing (1.27) with (1.28) we see that we obtain the following relation between ρ^Γ and $\rho^I - \rho$:

$$\int_{\gamma(t)} \rho^\Gamma \, ds = \int_{R^I} (\rho^I - \rho) \, dx. \tag{1.29}$$

To elaborate this further we assume that the shape of the material volume is such that the line segment $\partial R^I \cap \partial W(t)$ in Fig. 1.4 (a manifold in the 3D case) is orthogonal to Γ. Then we have (for the 3D case)

$$\int_{R^I} (\rho^I - \rho) \, dx = \int_{\gamma(t)} \int_{\lambda_-}^{\lambda_+} (\rho^I - \rho)(1 - \kappa_1 \lambda)(1 - \kappa_2 \lambda) \, d\lambda \, ds,$$

where λ is the signed distance to the dividing surface Γ. The thickness $\lambda_+ - \lambda_-$ of the local region R^I can be assumed to be very small, and therefore it is reasonable to assume $|\kappa_i \lambda| \ll 1$ for $\lambda \in [\lambda_-, \lambda_+]$. This reasoning suggests that we may identify

$$\rho^\Gamma = \int_{\lambda_-}^{\lambda_+} (\rho^I - \rho) \, d\lambda, \tag{1.30}$$

which shows that the surface mass density ρ^Γ can be interpreted as an *excess quantity*. Note that ρ^Γ in (1.30) is not necessarily constant or positive on Γ. In the sharp interface model there still is some freedom with respect to the choice of the location of the dividing surface. Different choices imply different excess quantities ρ^Γ. The most popular choice is as follows. For $t = 0$ it is assumed that Γ can be taken such that $\int_{\lambda_-}^{\lambda^+} (\rho^I - \rho) \, d\lambda = 0$, hence $\rho^\Gamma(x, 0) = 0$ for $x \in \Gamma(0)$. For $t > 0$ the dividing surface is transported by the velocity field $\mathbf{u}$. From (1.28), Reynolds' transport theorem, the interface transport formula (14.21b) and $\dot{\rho} = 0$ in $W_i(t)$ we obtain

$$\sum_{i=1}^{2} \int_{W_i(t)} \rho \operatorname{div} \mathbf{u} \, dx + \int_{\gamma(t)} \dot{\rho}^\Gamma + \rho^\Gamma \operatorname{div}_\Gamma \mathbf{u} \, ds = 0.$$

The first term vanishes due to the assumption of incompressibility, i.e. $\operatorname{div} \mathbf{u} = 0$, in $W_i(t)$. Note that in general $\operatorname{div} \mathbf{u} = 0$ in $W_i(t)$ does not imply $\operatorname{div}_\Gamma \mathbf{u} = 0$. For the excess mass density ρ^Γ we thus obtain the equation

$$\dot{\rho}^\Gamma + \rho^\Gamma \operatorname{div}_\Gamma \mathbf{u} = 0 \quad \text{on} \quad \Gamma = \Gamma(t).$$

This equation and the initial condition $\rho^\Gamma(x, 0) = 0$ are fulfilled if we take $\rho^\Gamma \equiv 0$. This derivation motivates the so-called *clean interface assumption*: in the sharp interface model the excess mass density corresponding to the dividing surface is equal to zero. Then the mass conservation equation (1.28) can be simplified to $\frac{d}{dt} \int_{W(t)} \rho \, dx = 0$, which is consistent with the continuity equations $\operatorname{div} \mathbf{u} = 0$ in Ω_i, which are used in the sharp interface model (1.19). This clean interface assumption is a standard one in sharp interface models without tensides.

Surface tension as a contact force

In (1.16) we introduced surface tension as a force acting in a direction orthogonal to the (sharp) interface Γ. This orthogonality property holds only if we assume that the surface tension coefficient τ is *constant*, cf. Remark 1.1.3. We summarize some basic facts related to surface tension and from that derive the forces given in (1.16) and (1.22).

Surface tension is an excess force resulting from the fact that on both sides of Γ these are different phases with different molecular forces. Consider a two-phase system in equilibrium, with an interface Γ. Let γ be a (small) connected subset of Γ with boundary $\partial\gamma$ and a normal, denoted by n, which is orthogonal to $\partial\gamma$ and tangential to Γ, cf. Fig. 1.3. Surface tension is defined as a *force per unit of length on* $\partial\gamma$ *in the direction* n.

This surface tension force is given by $F_s = \tau\, n$, with τ the *surface tension coefficient.* Note that τ is the magnitude of the surface tension force. The SI unit of τ is Newton per meter. From this definition of F_s it follows that the surface tension force is a *contact force* within the interface Γ. Note, however, that this force is not an intrinsic property of Γ but induced by the two phases on both sides of Γ.

Remark 1.1.4 An equivalent definition of surface tension can be given in terms of energy. Considering the different molecular forces in the two phases it follows that the creation of more interface area is energetically costly and thus the two-phase system will try to (locally) minimize interface area. The amount of work needed to (locally) increase an interface area by an amount δA is given by $\tau\,\delta A$, with the same surface tension coefficient τ as in the definition used above. In this characterization the surface tension coefficient measures energy per unit of area and the SI unit is joule per square meter.

Let W be a fluid volume which is intersected by the interface Γ and define the interface segment $\gamma = W \cap \Gamma$. Surface tension exerts a *contact force* F_s on $\partial\gamma$. Using the partial integration rule (14.18) the total contact force F_s on $\partial\gamma$ can be rewritten as a force on Γ:

$$\int_{\partial\gamma} \tau n \, d\tilde{s} = -\int_{\gamma} \tau\kappa\mathbf{n}\, ds + \int_{\gamma} \nabla_\Gamma \tau \, ds. \tag{1.31}$$

Thus for the case of a constant surface tension coefficient τ we obtain the force as in (1.16) and for the general case the one in (1.22).

Remark 1.1.5 Using the surface tension force representation $F_s = \tau n$ introduced above we derive the classical *Laplace-Young law,* which for a two-phase system in equilibrium and with a spherical interface relates the pressure difference to the mean curvature. We consider a ball with center at the origin and radius R, the boundary of which is the interface of a two-phase system at equilibrium. The constant pressures within and outside the ball are given by p_1 and p_2, respectively. Note that $[p] = p_1 - p_2 > 0$. We use spherical coordinates:

$$(x_1, x_2, x_3) = r\big(\sin\theta\cos\phi,\ \sin\theta\sin\phi,\ \cos\theta\big), \quad r \geq 0,\ \theta \in [0, \pi],\ \phi \in [0, 2\pi).$$

We consider the upper half of the interface, i.e. the hemisphere given by $S = \{\,(r, \theta, \phi) : r = R,\ \theta \in [0, \frac{1}{2}\pi],\ \phi \in [0, 2\pi)\,\}$. There are (only) two forces exerted on S, namely a pressure difference force acting at each point of S and a surface tension force on ∂S. The former is in normal direction and has size $[p]$, the latter has direction $(0, 0, -1)^T$ and size τ, cf. Fig. 1.5. The x_3-component of the pressure force is given by $\cos\theta[p]$. The resulting total force in x_3 direction must be equal to zero, i.e.

$$2\pi R\tau = [p]\int_S \cos\theta\, ds = [p]\int_0^{2\pi}\int_0^{\frac{1}{2}\pi} \cos\theta\sin\theta R^2\, d\theta d\phi = [p]\pi R^2$$

must hold. From this we obtain the Laplace-Young law $[p] = \frac{2\tau}{R} = \tau\kappa$, with $\kappa = \frac{2}{R}$ the mean curvature of the sphere with radius R.

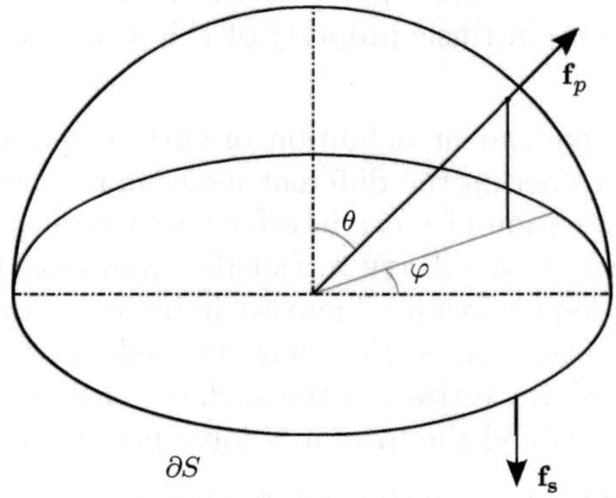

Fig. 1.5. Pressure force $\mathbf{f}_p$ and surface tension foce $\mathbf{f}_s$.

Variable surface tension coefficient: Langmuir model

It is generally accepted and experimentally verified that in many two-phase systems a surfactant changes the properties of the interface and through this can have a significant impact on the fluid dynamics of the system. One very important effect is that a surfactant can cause a change of the surface tension forces. In a system with a clean interface the surface tension coefficient τ is usually assumed to be constant, whereas in a system with surfactants the surface tension coefficient is often considered to be dependent on the local concentration of the surfactant, i.e, $\tau = \tau(S)$. A relatively simple and very popular model for $\tau(S)$ is due to Langmuir (also called Langmuir-Szyszkowski model in the literature). We briefly address the main ideas underlying this model. Consider a system consisting of one bulk phase and its surface Γ. In the bulk phase a component is dissolved which is adsorbed at the surface. We assume that there is no fluid dynamics and that at a given surface point $x \in \Gamma$ the surface surfactant concentration is locally constant. This surface concentration is denoted by $S(t) = S(x,t)$. Also the bulk concentration of the dissolved component is assumed to be locally constant, and is denoted by $S_b(t) = S_b(x,t)$. There is a maximal surface coverage denoted by S_∞. A very simple model for describing the ad- and desorption is given by

$$\frac{dS}{dt} = k_{\mathrm{ad}} S_b(t)\big(1 - \frac{S(t)}{S_\infty}\big) - k_{\mathrm{des}} \frac{S(t)}{S_\infty},$$

with positive constants k_{ad}, k_{des}. In equilibrium we have $\frac{dS}{dt} = 0$. Let $S_e = \lim_{t\to\infty} S_b(t)$ be the equilibrium local bulk concentration. For the equilibrium local surface concentration, denoted by S, we obtain $S = \frac{k_{\mathrm{ad}}}{k_{\mathrm{des}}} S_e(S_\infty - S)$ and thus

$$S = S_\infty \frac{S_e}{k_q + S_e}, \quad \text{with} \quad k_q := \frac{k_{\mathrm{des}}}{k_{\mathrm{ad}}}. \tag{1.32}$$

Hence we have a simple relation between the equilibrium states S (on Γ) and S_e (in the bulk phase). A relation between S, S_e and the surface tension energy τ is obtained from the Gibbs adsorption equation (or Gibbs isotherm), which is often used to relate the changes in concentration of a component in contact with a surface with changes in the surface tension. For $S \gg S_e$, which is the case for most surfactants, and assuming a constant temperature T, this Gibbs adsorption equation is given by

$$\frac{d\tau}{d \ln S_e} = -RTS,$$

with R the gas constant. Using $d \ln S_e = S_e^{-1} dS_e$ and the result in (1.32) we obtain

$$\frac{d\tau}{dS_e} = \frac{-RTS_\infty}{k_q + S_e},$$

and thus

$$\tau = \tau_0 - RTS_\infty \ln(1 + S_e/k_q).$$

From (1.32) we get $1 + S_e/k_q = (1 - S/S_\infty)^{-1}$ and thus we get the following relation between the surface tension coefficient τ and the surfactant concentration S:

$$\tau = \tau_0 + RTS_\infty \ln(1 - S/S_\infty), \tag{1.33}$$

which is the Langmuir model. Here τ_0 is the constant surface tension coefficient for the system with a clean interface. Note that $\tau = \tau(S)$ is a decreasing function of S, i.e., surface tension decreases if the concentration of the surfactant increases. In a realistic two-phase flow system with surfactant the concentration S will *not* be constant on the interface and based on this model one obtains a varying surface tension coefficient with a relatively small value in those parts of the interface where the surfactant concentration is relatively high.

Other models for $\tau = \tau(S)$ are derived in the literature, e.g. the Frumkin isotherm $\tau(S) = \tau_0 + RTS_\infty\Big(\ln(1-S/S_\infty) - K\big(\frac{S}{S_\infty}\big)^2\Big)$, where K is a measure for the interactions among the adsorbed surfactant particles with $K < 0$ ($K > 0$) if there are significant cohesive (repulsive) forces.

Surface viscosity: Boussinesq-Scriven model

If surfactants are present which cause a variable surface tension coefficient $\tau = \tau(S)$ this results in an effective elasticity of the interface. For certain two-phase flow systems, e.g. with suspended (nano)particles that reside on the interface, it is known that significant other effects also occur. Due to new high-tech applications (e.g., particle-stabilized emulsion, new materials) such systems with colloidal particles at liquid interfaces have attracted a strongly growing interest in the past decade. For modeling the rheological properties of such *particle-laden interfaces* one often introduces an *effective surface viscosity*, [168, 170]. The standard mathematical description of this is by means of the so-called *Boussinesq-Scriven model* which we now introduce, cf. also [225, 45, 219]. First we recall that for the bulk fluid, based on the Cauchy stress principle and assuming the Newtonian stress tensor form $\boldsymbol{\sigma} = -p\mathbf{I} + L(\mathbf{D}(\mathbf{u}))$, with a linear operator L, one can derive the stress tensor representation as in (1.10), i.e.,

$$\boldsymbol{\sigma} = -p\mathbf{I} + \lambda \operatorname{div} \mathbf{u}\, \mathbf{I} + \mu \mathbf{D}(\mathbf{u}). \tag{1.34}$$

The Boussinesq-Scriven model starts from the (rheological) assumption that the interface behaves like a two-dimensional Newtonian fluid. Recall that surface tension can be characterized as a contact force of the form $\int_{\partial\gamma} \tau n \, d\tilde{s}$, cf. (1.31). In analogy with the approach for a Newtonian fluid in the bulk phase, we start from the structural assumption that on each (small) connected surface segment $\gamma \subset \Gamma$, cf. Fig. 1.3, there is a contact force on $\partial\gamma$ of the form

$$\boldsymbol{\sigma}_\Gamma n, \quad \text{with } \boldsymbol{\sigma}_\Gamma = \tau\mathbf{P} + L(\mathbf{D}_\Gamma(\mathbf{u})), \quad \mathbf{D}_\Gamma(\mathbf{u}) := \mathbf{P}\big(\nabla_\Gamma \mathbf{u} + (\nabla_\Gamma \mathbf{u})^T\big)\mathbf{P},$$

with L a linear operator. Recall that $\mathbf{P} = \mathbf{I} - \mathbf{n}\mathbf{n}^T$ is the orthogonal projection onto Γ. This projection is used, since $\boldsymbol{\sigma}_\Gamma n = \boldsymbol{\sigma}_\Gamma \mathbf{P} n$ should represent only contact forces that are tangential to the surface. Note that for $L = 0$ this contact force reduces to the surface tension contact force $\boldsymbol{\sigma}_\Gamma n = \tau \mathbf{P} n = \tau n$. Using the same principles (isotropy, independence of the frame of reference) as in the derivation of (1.34) it can be shown, cf. [225, 15], that the interface stress tensor $\boldsymbol{\sigma}_\Gamma$ must have the following form:

$$\boldsymbol{\sigma}_\Gamma = \tau\mathbf{P} + \tilde{\lambda}_\Gamma \operatorname{div}_\Gamma \mathbf{u}\, \mathbf{P} + \mu_\Gamma \mathbf{D}_\Gamma(\mathbf{u}), \tag{1.35}$$

with parameters $\tilde{\lambda}_\Gamma$, μ_Γ. This is the interface analogon of the bulk stress tensor representation in (1.34). Note that in general $\operatorname{div}_\Gamma \mathbf{u} \neq 0$, even if $\operatorname{div} \mathbf{u} = 0$

holds. In case of viscous behavior of the interface one takes $\mu_\Gamma > 0$. For certain cases one can derive conditions on the parameter $\tilde{\lambda}_\Gamma$, for example $\tilde{\lambda}_\Gamma > -\mu_\Gamma$ ([225] Sect. 4.9.5). Therefore the interface stress tensor is also written in the form

$$\boldsymbol{\sigma}_\Gamma = \tau\mathbf{P} + (\lambda_\Gamma - \mu_\Gamma)\,\mathrm{div}_\Gamma\,\mathbf{u}\,\mathbf{P} + \mu_\Gamma\mathbf{D}_\Gamma(\mathbf{u}), \tag{1.36}$$

and one often assumes $\lambda_\Gamma = \tilde{\lambda}_\Gamma + \mu_\Gamma > 0$. The parameters μ_Γ and λ_Γ, which we assume to be constants, are referred to as the *interface shear viscosity* and *interface dilatational viscosity*, respectively. In the momentum balance we need the interface force as a force on the interface segment γ. Using the formula $\int_\gamma \mathrm{div}_\Gamma\,\mathbf{G}\mathbf{P}\,ds = \int_{\partial\gamma} \mathbf{G}n\,d\tilde{s}$, cf. (14.19), we obtain from (1.36) the interfacial force

$$\begin{aligned} F_3 &= \int_{\partial\gamma} \boldsymbol{\sigma}_\Gamma n\,d\tilde{s} \\ &= \int_\gamma \mathrm{div}_\Gamma\,(\tau\mathbf{P}) + (\lambda_\Gamma - \mu_\Gamma)\,\mathrm{div}_\Gamma\big(\mathrm{div}_\Gamma\,\mathbf{u}\,\mathbf{P}\big) + \mu_\Gamma\,\mathrm{div}_\Gamma\,\big(\mathbf{D}_\Gamma(\mathbf{u})\big)\,ds. \end{aligned}$$

Using this interface force F_3 and following the derivation in Sect. 1.1.2 we obtain a generalization of the interface condition in (1.20):

$$[\boldsymbol{\sigma}\mathbf{n}_\Gamma] = \mathrm{div}_\Gamma\,(\tau\mathbf{P}) + (\lambda_\Gamma - \mu_\Gamma)\,\mathrm{div}_\Gamma\big(\mathrm{div}_\Gamma\,\mathbf{u}\,\mathbf{P}\big) + \mu_\Gamma\,\mathrm{div}_\Gamma\,\big(\mathbf{D}_\Gamma(\mathbf{u})\big) \tag{1.37}$$

on Γ. This is the Boussinesq-Scriven model. For $\lambda_\Gamma = \mu_\Gamma = 0$ this model reduces to the one in (1.23) (or (1.20), if τ is constant) since

$$\mathrm{div}_\Gamma(\tau\mathbf{P}) = \tau\,\mathrm{div}_\Gamma\,\mathbf{P} + \nabla_\Gamma\tau = -\tau\kappa\mathbf{n} + \nabla_\Gamma\tau,$$

cf. (14.10). The generalized formulation (1.37) is used to model viscous effects in the interface.

1.2 Initial and boundary conditions

In this section we describe initial and boundary conditions that can be used in the models 1)-4) to make the problem well-posed.

For the NS1 model one needs suitable initial and boundary conditions only for the velocity $\mathbf{u}$. The initial condition is $\mathbf{u}(x,0) = \mathbf{u}_0(x)$ with a given function $\mathbf{u}_0$, which usually comes from the underlying physical problem. For the boundary conditions we distinguish between *essential* and *natural* boundary conditions. Let $\partial\Omega$ be subdivided into two parts $\partial\Omega = \partial\Omega_D \cup \partial\Omega_N$ with $\partial\Omega_D \cap \partial\Omega_N = \emptyset$. We use essential boundary conditions on $\partial\Omega_D$ that are of Dirichlet type. In applications these describe inflow conditions or conditions at walls (e.g., no-slip). Such Dirichlet conditions are of the form $\mathbf{u}(x,t) = \mathbf{u}_D(x,t)$ for $x \in \partial\Omega_D$, with a given function $\mathbf{u}_D$. If, for example, $\partial\Omega_D$ corresponds to a fixed wall, then a no-slip boundary condition is given by $\mathbf{u}(x,t) = 0$ for $x \in \partial\Omega_D$. On $\partial\Omega_N$ we prescribe natural boundary conditions, which are often used to describe outflow conditions. These natural boundary conditions are of the form

$$\boldsymbol{\sigma}\mathbf{n}_\Omega = -p_{ext}\mathbf{n}_\Omega, \quad \text{on } \partial\Omega_N, \tag{1.38}$$

with $\mathbf{n}_\Omega$ the outward pointing normal on $\partial\Omega_N$ and p_{ext} a given function (external pressure). For the case $p_{ext} = 0$ we thus obtain a homogeneous natural boundary condition.

Similar initial and boundary conditions can be used for the two-phase flow model NS2. In addition we then need the initial configuration, i.e., $\Gamma(0)$ must be given.

In the model NS2+T in (1.24) one needs in addition initial and boundary conditions for the concentration c. The initial condition is $c(x,0) = c_0(x)$ with a given initial concentration c_0. For the boundary conditions the standard ones, namely a Dirichlet (i.e., c given on part of $\partial\Omega$) and a Neumann ($\frac{\partial c}{\partial \mathbf{n}_\Omega}$ given on part of the boundary) condition can be used.

In model NS2+S in (1.25) one has to prescribe an initial concentration $S(x,0) = S_0(x)$, $x \in \Gamma$, for the surfactant. If Γ is a surface without boundary (droplet) no boundary conditions for S are needed.

1.3 Examples of two-phase flow simulations

In this section we give some simulation results for a two-phase system with a single droplet rising due to buoyancy forces where at the same time transport of some surface active agent (surfactant) on the interface is taking place. This application example is meant to give the reader a first impression of some features of two-phase flow systems and the challenges one is facing when treating such flows numerically.

Before giving details on this numerical simulation, we briefly address the importance of two-phase systems in chemical engineering. One example of such a system is a falling film which is used for cooling by heat transfer from a thin liquid layer to the gaseous phase (liquid-gas system). Another example is an extraction column where mass transport takes place between liquid bubbles and a surrounding liquid (liquid-liquid system). For the design of such bubble column reactors it is desirable to have a model that gives a detailed description of the transport phenomena between the bubbles and the surrounding fluid. Rather than considering the whole column reactor with swarms of bubbles, in a first step only a single droplet is investigated. Even for this simplified case the transport mechanisms are not well understood up to now. One interesting and important phenomenon is the formation of a so-called *stagnant cap* in the downstream part of the droplet. In this stagnant cap region the velocity is much smaller than in the region where the vortices occur, cf. Fig. 1.6. The formation of such stagnant caps has been observed in experiments. In Fig. 1.7

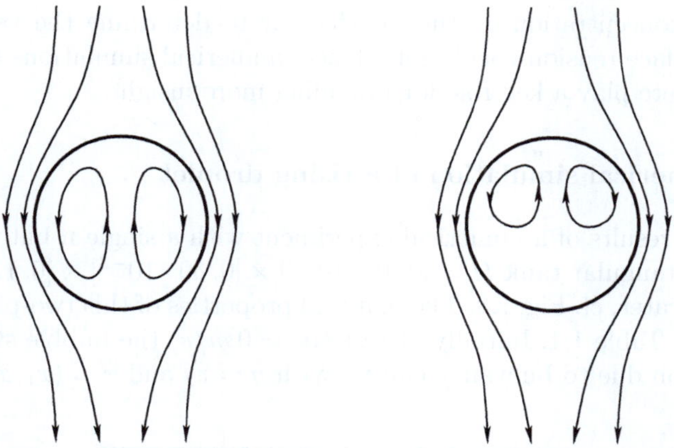

Fig. 1.6. Sketch of flow pattern inside and outside single droplet without (left) and with (right) stagnant cap.

a velocity distribution in a cross-section of a levitated toluene droplet is shown, measured by a fast nuclear magnetic resonance (NMR) technique (from [11]).

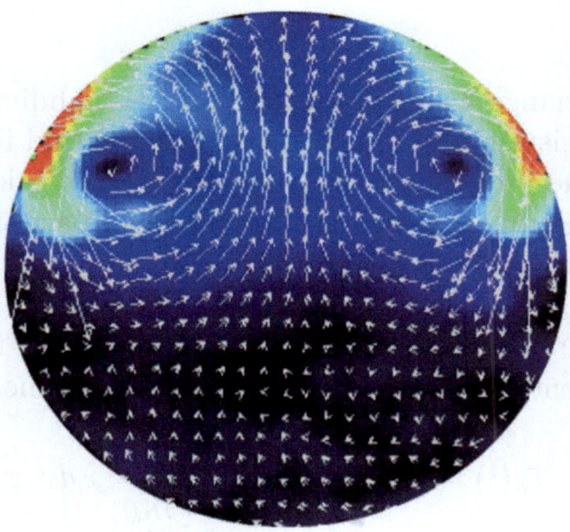

Fig. 1.7. NMR image of measured velocity field in toluene droplet.

There is (experimental) evidence that such regions with very low velocity are caused by surface active substances (surfactants) which adhere to the interface and due to the surrounding flow pattern are transported to the downstream part of the droplet. An interesting (modeling) question in this context is how this surfactant concentration affects the surface tension coefficient, i. e., to find an adequate model for $\tau = \tau(S)$. Furthermore one would like to understand how the variable surface tension coefficient $\tau(S)$ influences the velocity inside the droplet, in particular whether it induces a stagnant cap. It is very hard (for most systems even impossible) to measure in an experiment the surfactant concentration on the interface or to determine the values of the variable surface tension coefficient. Hence, numerical simulations like the one presented here play a key role for providing more insight.

1.3.1 Numerical simulation of a rising droplet

We present results of a numerical experiment with a single n-butanol droplet inside a rectangular tank $\Omega = [0, 12 \cdot 10^{-3}] \times [0, 30 \cdot 10^{-3}] \times [0, 12 \cdot 10^{-3}]\, m^3$ filled with water, cf. Fig. 1.8. The material properties of this two-phase system are given in Table 1.1. Initially at rest ($\mathbf{u}_0 = 0\, m/s$) the bubble starts to rise in y-direction due to buoyancy effects, with $y = x_2$ and $x = (x_1, x_2, x_3)$.

quantity (unit)	n-butanol	water
ρ (kg/m^3)	845.4	986.5
μ $(kg/m\,s)$	$3.281 \cdot 10^{-3}$	$1.388 \cdot 10^{-3}$
τ (N/m)	$1.63 \cdot 10^{-3}$	

Table 1.1. Material properties of the system n-butanol/water.

For the initial triangulation $\mathcal{T}_0$ the domain Ω is subdivided into $4 \times 10 \times 4$ sub-cubes each consisting of 6 tetrahedra. Then the grid is refined four times in the vicinity of the interface Γ. As time evolves the grid is adapted to the moving interface. Figure 1.9 shows the droplet and a part of the adaptive mesh for two different time steps. A movie of this numerical simulation is given on the website [90].

For a butanol droplet with radius $1\, mm$, in Fig. 1.10 the y-coordinate of the droplet's barycenter $\overline{x}_d$ is shown as a function of time, where

$$\overline{x}_d(t) = \mathrm{meas}_3(\Omega_1(t))^{-1} \int_{\Omega_1(t)} x\, dx.$$

The average velocity $\overline{\mathbf{u}}_d(t)$ of the drop is given by

$$\overline{\mathbf{u}}_d(t) = \mathrm{meas}_3(\Omega_1(t))^{-1} \int_{\Omega_1(t)} \mathbf{u}(x,t)\, dx.$$

Note that $\overline{x}_d'(t) = \overline{\mathbf{u}}_d(t)$ and, due to incompressibility and immiscibility, $\mathrm{meas}_3(\Omega_1(t)) = \mathrm{meas}_3(\Omega_1(0))$. For a butanol droplet with radius $1\, mm$ Fig. 1.11 shows the rise velocity, which is the second coordinate of the average velocity $\overline{\mathbf{u}}_d(t)$. After a certain time the rise velocity becomes almost constant and the bubble reaches a terminal rise velocity denoted by u_r. For the radius $r_d = 1\, mm$ we obtain $u_r = 53\, mm/s$. For technical applications the

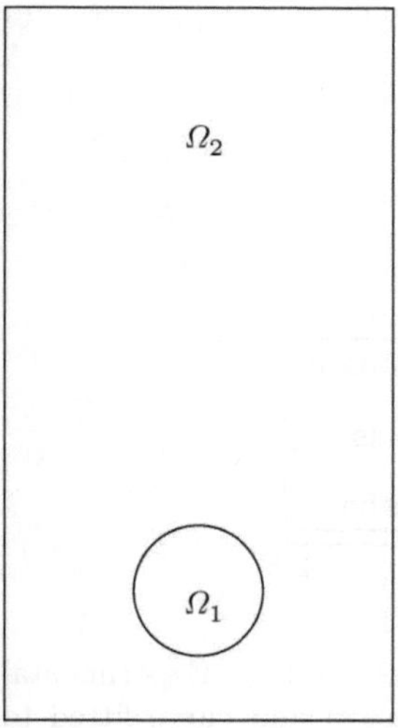

Fig. 1.8. 2D sketch of the rising bubble example.

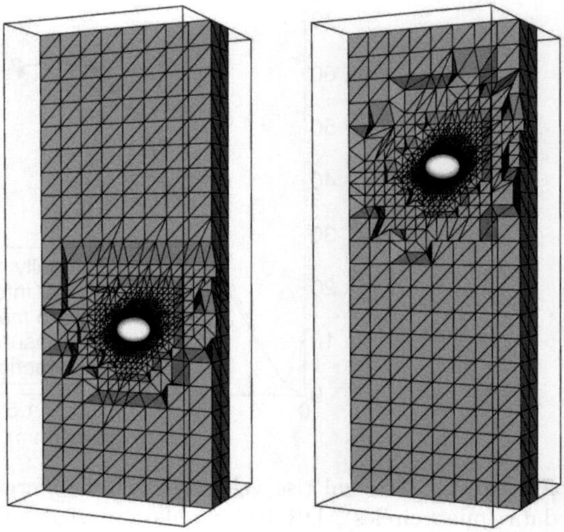

Fig. 1.9. Interface and part of the grid for a rising bubble with radius $r_d = 1\,mm$ at times $t = 0.2\,s$ (left) and $t = 0.4\,s$ (right).

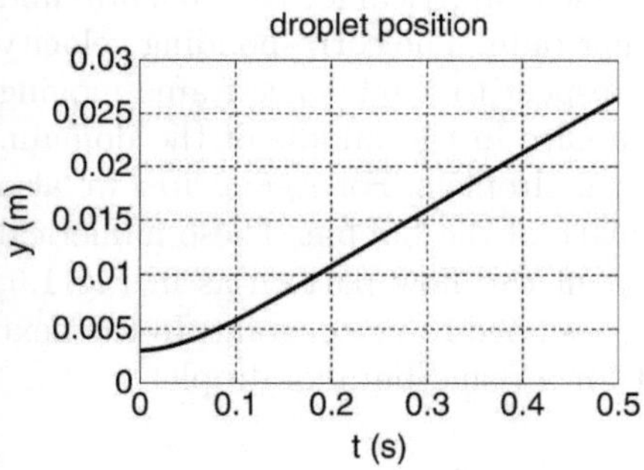

Fig. 1.10. y-coordinate of barycenter of a rising butanol droplet with radius $1\,mm$ as a function of time t.

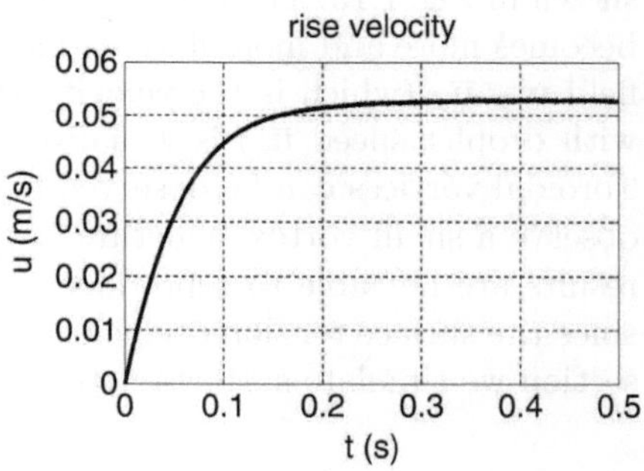

Fig. 1.11. Rise velocity of a butanol droplet with radius $1\,mm$ as a function of time t.

value of the terminal rise velocity is an important quantity, e. g., to predict the duration of a bubble's residence time inside a column reactor.

We computed the terminal rise velocities u_r of rising butanol droplets for different drop radii r_d. For larger droplets with $r_d \geq 1.5\,mm$ a coarser mesh was used (3 times local refinement instead of 4 times as for the smaller droplets) because of memory limitations. A validation of the simulation results by means of comparison with experimental data is given in [35]. In Fig. 1.12, which is taken from [35], the terminal rise velocity u_r is plotted versus the bubble radius r_d and a comparison of experimental and simulation results is shown. For a discussion of these results we refer to [35].

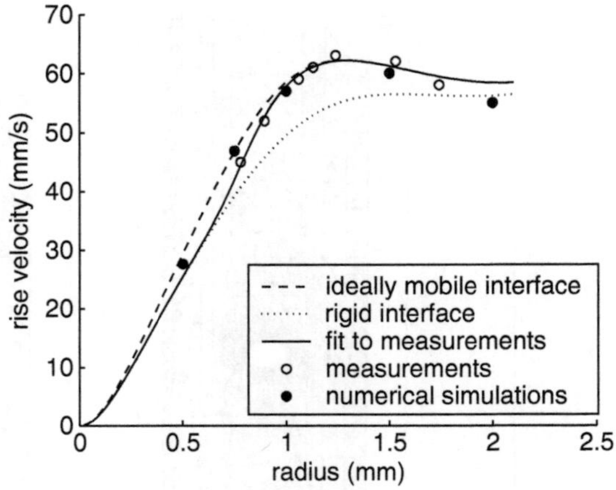

Fig. 1.12. Terminal rise velocities u_r for different droplet radii r_d. Experimental data (open circles), DROPS simulation results (filled circles) and curve fitted to experimental data (solid line).

The droplet shapes of rising butanol droplets for different radii r_d are shown in Fig. 1.13. The droplet shape is almost spherical for $r_d = 0.5\,mm$ and becomes more and more flattened for larger radii. The corresponding velocity field $\mathbf{u} - \overline{\mathbf{u}}_d$ (which is the velocity with respect to a reference frame moving with droplet speed $\overline{\mathbf{u}}_d$) is visualized on a slice in the middle of the domain. Toroidal vortices can be observed inside the droplets. For $r_d = 2\,mm$ we also observe a small vortex structure in the wake of the bubble. These numerical results are not able to reproduce a stagnant cap flow pattern as in Fig. 1.6, since the surface tension coefficient τ was assumed to be constant. In the next section we simulate surfactant transport for a rising butanol droplet.

1.3.2 Numerical simulation of a droplet with surfactant transport

We again consider the problem of a rising butanol droplet from the previous section, but now include surfactant transport on Γ. The model NS2+S consists of the two-phase flow problem (1.19)–(1.21) combined with the surfactant transport equation (1.25). The experimental setup and the numerical parameters are chosen as described in Sect. 1.3.1. We take a droplet radius $r_d = 1\,mm$. The initial constant surfactant concentration is chosen as $S_0 = 1$ and the surfactant diffusion coefficient is set to $D_\Gamma = 10^{-5}$.

As time evolves, the droplet starts to rise and changes its shape. The flow field $\mathbf{u}$ at the interface induces a surfactant transport from the top to the bottom of the droplet. Figure 1.14 shows the droplet's shape and surfactant concentration for $t = 0,\ 0.1,\ 0.2,\ 0.4\,s$, respectively. The surfactant is collected at the lower part of the droplet while the surfactant concentration at the upper part becomes relatively small. Figure 1.15 shows the surfactant concentration as a function of the vertical coordinate y, with $y = x_2$ and $x = (x_1, x_2, x_3)$,

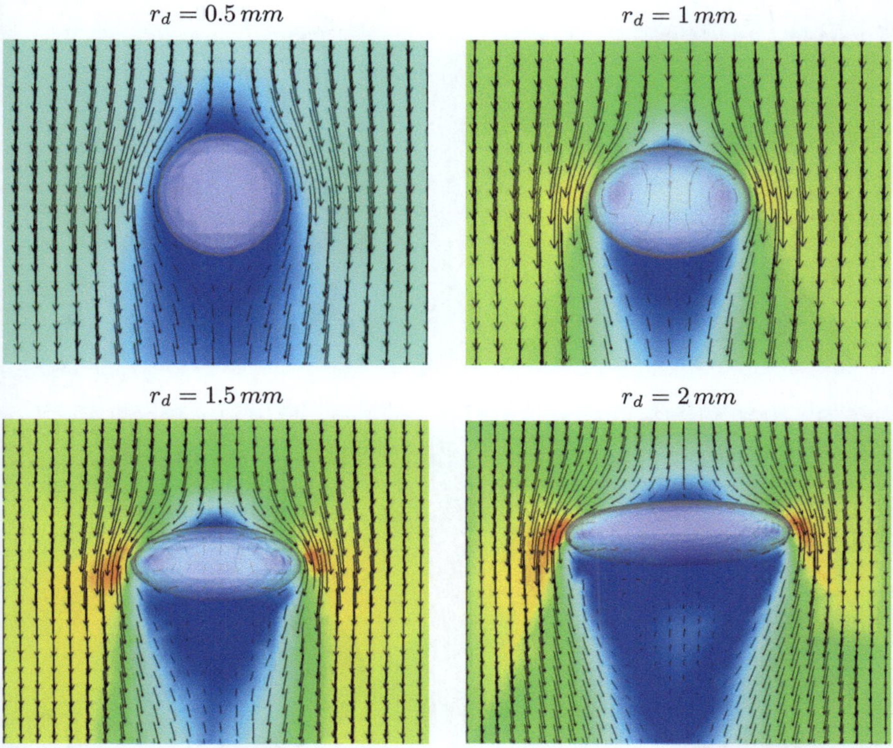

Fig. 1.13. Shape of n-butanol droplets for different radii r_d and velocity field $\mathbf{u} - \overline{\mathbf{u}}_d$ visualized on slice.

for each of the respective times. Hence, each snapshot of the rising droplet in Fig. 1.14 corresponds to one of the graphs in 1.14. For example, the constant surfactant concentration of the initial droplet ($t = 0\,s$) is represented in Fig. 1.15 as a straight vertical line of height $2\,mm$, which corresponds to the initial droplet diameter. The droplet's shape as well as the surfactant profile relative to the droplet becomes almost stationary for $t \geq 0.2\,s$, with $S \approx 3.2$ at the bottom and $S \approx 0.015$ at the top of the droplet, i. e., only 1.5% of the initial surfactant concentration.

The next step would be to consider a variable surface tension coefficient τ depending on the surfactant concentration S. At the top of the droplet this would be $\tau \approx 1.63\,mN/m$ as for the pure n-butanol/water system, while at the bottom the surface tension coefficient would be decreased as an effect of the high surfactant concentration. We do not consider this issue here. In Sect. 11.5.3 we give results of a numerical experiment in which the surface tension coefficient depends on the concentration of a dissolved species close to the interface, i.e. $\tau = \tau(c)$. In that experiment, the variable surface tension induces a stagnant cap as in the right picture in Fig. 1.6.

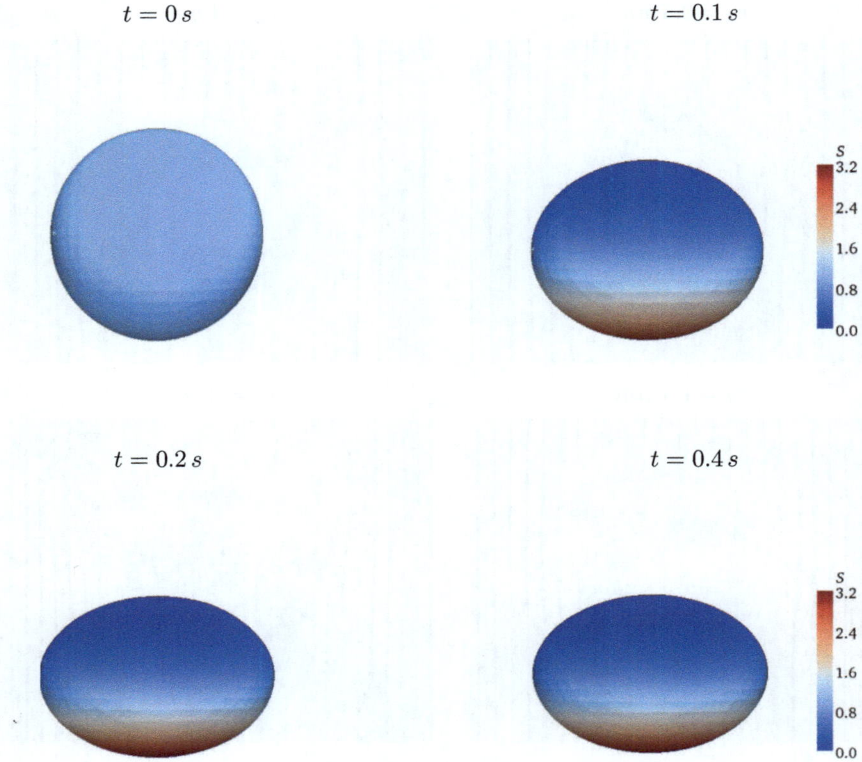

Fig. 1.14. Shape and surfactant concentration S (color-coded) of n-butanol droplet for different time steps $t = 0,\ 0.1,\ 0.2,\ 0.4\,s$.

1.4 Overview of numerical methods

In the following chapters many numerical methods for the simulation of the models introduced in Sect. 1.1 are treated. In this section we give an overview of important methods. for spatial discretization *only finite element methods* will be considered. Besides different finite element methods we also discuss algorithms for the construction of nested multilevel tetrahedral triangulations, implicit time discretization methods and iterative solvers for the resulting discrete problems. We list the main numerical methods:

- A level set method for interface capturing is used. We treat discretization methods for the linear hyperbolic level set equation. Also a fast-marching re-initialization algorithm is discussed.
- The construction of a multilevel hierarchy of nested tetrahedral triangulations is treated. Local refinement and coarsening routines are explained.

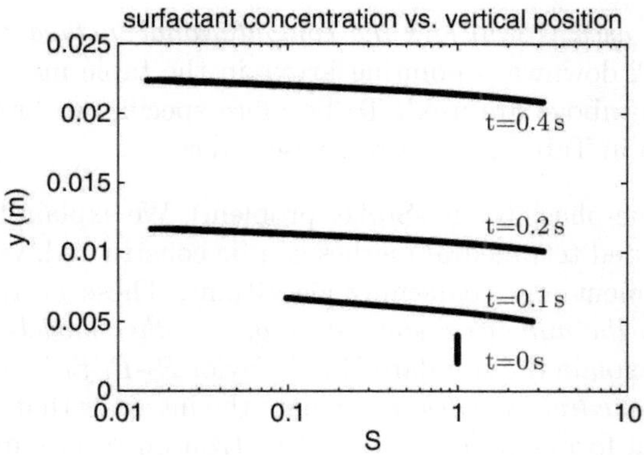

Fig. 1.15. Surfactant concentration S as function of vertical coordinate y for $t = 0,\ 0.1,\ 0.2,\ 0.4\,s$, respectively.

- Starting from suitable variational formulations of the models, for spatial discretization we apply finite element techniques based on conforming spaces. Special finite element spaces suitable for functions that are discontinuous across the interface are introduced. We use the XFEM ("*ex*tended *f*inite *e*lement *m*ethod") approach.
- For discretization of the surface tension force a special Laplace-Beltrami method is analyzed.
- For the fluid dynamics problem we derive several implicit time integration methods in which flow variables, surface tension forces and the level set function are strongly coupled.
- After space and time discretization of the fluid dynamics problem one obtains, in each time step, a nonlinear discrete problem in which the flow and level set unknowns are strongly coupled. We analyze an iterative decoupling strategy.
- For the solution of large sparse linear systems preconditioned Krylov subspace methods are discussed. Also inexact Uzawa type solvers for saddle point problems are analyzed. Several Schur complement preconditioners are considered. Multigrid solvers/preconditioners are explained.
- For the discretization of the mass transport equation we consider a method in which the XFEM technique is combined with a so-called Nitsche approach in order to satisfy the Henry interface condition.
- For the discretization of the surfactant convection-diffusion equation on the interface a special Eulerian finite element technique is introduced.
- For the space and time discretization of the mass transport and surfactant transport equations a space-time finite element approach appears to be very natural. We treat such space-time finite element methods.

In Table 1.2 we give a compact overview of all important methods considered, in the form of a matrix of methods. As can be seen from this table, we arrange the different methods according to two criteria, namely *the models for which they are used* (rows) and *the computational method class they belong to* (columns). A downward pointing arrow in the table means that methods from the row(s) above are used. To be more specific we briefly address the methods shown in Table 1.2 in a row wise order.

Model NS1 (one-phase Navier-Stokes problem). We explain how a multilevel hierarchy of nested tetrahedral meshes can be constructed, which allows simple local refinement and coarsening algorithms. These *grid related methods are also used in the numerical simulation of the other models.* For spatial discretization we explain the standard Hood-Taylor P_2-P_1 *finite element pair.* We briefly address *quadrature rules* to evaluate the integrals that occur in the discrete variational formulation. Spatial discretization results in a large system of nonlinear ordinary differential equations coupled with algebraic constraints (due to $\operatorname{div} \mathbf{u} = 0$), i.e., a DAE system (Differential Algebraic Equation). For this system we discuss several *numerical time integration rules.* A one-step θ-scheme and a fractional-step θ-scheme are treated. Per time step such a time integration rule results in a large nonlinear system of algebraic equations, in which velocity $\mathbf{u}$ and pressure p are coupled. For *linearization* of the term $(\mathbf{u} \cdot \nabla)\mathbf{u}$ a standard Picard iteration (with steplength optimization) is applied. After linearization we have a large sparse linear system of algebraic equations that is of saddle point type. Several efficient *iterative solvers*, like for example preconditioned minimal residual (MINRES), inexact Uzawa and multigrid methods are discussed.

Model NS2 (two-phase Navier-Stokes problem). For the interface representation ("interface capturing") we use a *level set approach.* In this method one uses a scalar level set function ϕ (which has no physical meaning) whose zero level coincides (approximately, due to discretization errors) with the interface. In the model (1.19)-(1.21) the immiscibility condition in (1.21) is then replaced by a linear hyperbolic partial differential equation for ϕ. An important issue is the *discretization of the level set equation.* For this we use piecewise quadratic finite elements combined with streamline diffusion stabilization (SDFEM). Another topic is the *approximation of the zero level* of this discretization ϕ_h of ϕ ($\Gamma \rightsquigarrow \Gamma_h$). Related to the level set function we also need a *re-initialization method.* We will reformulate the model NS2 such that the interface conditions (1.20) are eliminated and replaced by a *localized force term* (at the interface) in the momentum equation. A main issue is the *discretization of this localized surface tension force.* For this we introduce and analyze a Laplace-Beltrami technique. In this type of problems, due to surface tension, the pressure is discontinuous across the interface. For an appropriate treatment of this discontinuity we introduce a *special extended finite element space* (XFEM). Due to the pressure discontinuity and discontinuities in density and viscosity we need *special quadrature rules.* Application of a time integration rule results in a large nonlinear system of algebraic equations (per time step) in which $\mathbf{u}, p$ and ϕ are coupled. We explain an iterative *decoupling strategy* to split the coupled problem for $\mathbf{u}, p, \phi$ into two subproblems for $(\mathbf{u}, p)$ and ϕ, respectively. If in the flow problem there are very large jumps in density and viscosity across the interface (as, for example, in a liquid-gas system) then in order to obtain efficient iterative solvers we propose *special preconditioners* that are robust with respect to variation in the size of these jumps.

Model NS2+T (two-phase flow with mass transport). Due to the Henry interface condition $c_1 = C_H c_2$ (with $C_H \neq 1$) in (1.24), the concentration is discontinuous across the interface. For the spatial discretization of the convection-diffusion equation for the concentration c we use the XFEM technique. In order to satisfy the Henry jump condition at the interface a *technique due to Nitsche* is explained. For the *time integration* we distinguish two cases. If the interface is stationary, a standard method of lines approach can be applied in which the spatial finite element discretization is combined with a θ-scheme for time discretization. In case of a non-stationary interface this approach is not very satisfactory and we treat as alternatives a *Rothe method* (first time, then space) and a *space-time finite element* technique. A simple method for the *decoupling* of $(\mathbf{u}, p, \phi)$ and c in each time step is discussed.

Model NS2+S (two-phase flow with surfactant transport). In this model we have a convection-diffusion equation on the (evolving) interface Γ, cf. (1.25). For the spatial discretization we use a *special finite element space* that is obtained from suitable restriction of a standard finite element space used for discretization of the flow variables on the tetrahedral triangulation. Again for the time discretization we distinguish between a stationary and a non-stationary interface. For the latter case a *space-time finite element* method is discussed.

The methods addressed above are treated in this monograph and implemented in the DROPS package [90]. We mention a few other research groups in Numerical Analysis and Computational Engineering in which the numerical simulation of two-phase incompressible flow problems is an important research topic: the groups around Bothe [9, 10], Griebel [73], Herrmann [138, 139], Kuipers [80, 88], Lowengrub and Voigt [233, 234], Marchandise [173, 172], Tobiska [115, 118], Tryggvason [243, 182], Weigand [216, 215].

	Grids	**spatial discretization**	**time integration**	**couplings/linearization**	**iterative solvers**
NS1	multilevel tetrahedral hierarchy; local refinement and coarsening;	Hood-Taylor FE; quadrature;	θ-scheme; fract.-step scheme;	$(\mathbf{u}, p)$ fully coupled; Picard iteration for linearization;	inexact Uzawa; GMRES,GCR, MINRES; Schur compl. precond.; multigrid method;
NS2	↓	↓ XFEM for p; P_2+ SDFEM for ϕ; mass conservation; re-initialization of ϕ; $\Gamma \rightsquigarrow \Gamma_h$; discretization of f_Γ; special quadrature;	↓ generalized θ-scheme	↓ fixed point for decoupling of $(\mathbf{u}, p) \leftrightarrow \phi$;	↓ special preconditioners (large jumps);
NS2+T	↓	↓ Nitsche approach; XFEM for c;	↓ Rothe method; space-time FE;	↓ decoupling of $(\mathbf{u}, p, \phi) \leftrightarrow c$;	↓
NS2+S	↓	↓ Eulerian surface FE method;	↓ space-time FE;	↓ decoupling of $(\mathbf{u}, p, \phi) \leftrightarrow S$;	↓

Table 1.2. Overview of numerical methods.

Part I

One-phase incompressible flows

2

Mathematical models

2.1 Introduction

We recall the non-stationary Navier-Stokes equations for modeling a *one-phase* incompressible flow problem:

$$\begin{aligned} \rho(\frac{\partial \mathbf{u}}{\partial t} + (\mathbf{u} \cdot \nabla)\mathbf{u}) + \nabla p - \mu \Delta \mathbf{u} &= \rho \mathbf{g} \text{ in } \Omega \\ \operatorname{div} \mathbf{u} &= 0 \text{ in } \Omega, \end{aligned} \tag{2.1}$$

with given constants $\rho > 0$, $\mu > 0$. For simplicity we only consider homogeneous Dirichlet boundary conditions for the velocity (no-slip condition). Thus the boundary and initial conditions are given by

$$\mathbf{u} = 0 \quad \text{on } \partial\Omega, \quad \mathbf{u}(x, 0) = \mathbf{u}_0(x) \quad \text{for } x \in \Omega, \tag{2.2}$$

with a given initial condition $\mathbf{u}_0(x)$. In the discussion and analysis of numerical methods for this problem we will also use the following two simpler systems of partial differential equations. Firstly, the *non-stationary Stokes equations*

$$\begin{aligned} \rho\frac{\partial \mathbf{u}}{\partial t} + \nabla p - \mu \Delta \mathbf{u} &= \rho \mathbf{g} \text{ in } \Omega \\ \operatorname{div} \mathbf{u} &= 0 \text{ in } \Omega, \end{aligned} \tag{2.3}$$

with the same initial and boundary conditions as in (2.2). Note that opposite to the Navier-Stokes equations, the Stokes system is *linear* in the unknowns $\mathbf{u}$, p. Secondly, we consider the following type of *stationary* problem:

$$\begin{aligned} \xi \mathbf{u} + (\mathbf{w} \cdot \nabla)\mathbf{u} + \nabla p - \mu \Delta \mathbf{u} &= \rho \mathbf{g} \text{ in } \Omega \\ \operatorname{div} \mathbf{u} &= 0 \text{ in } \Omega, \end{aligned} \tag{2.4}$$

with a given constant $\xi \geq 0$ and a given vector function $\mathbf{w}(x) \in \mathbb{R}^3$. For the boundary condition we take the homogeneous Dirichlet condition $\mathbf{u} = 0$. This linear system of partial differential equations is called an *Oseen problem.* This type of problem occurs if in the non-stationary Navier-Stokes equation an implicit time discretization method is used and the nonlinearity is linearized via a fixed point strategy in which in the nonlinear term $(\mathbf{u}\cdot\nabla)\mathbf{u}$ is linearized by replacing the first $\mathbf{u}$ argument by an already computed approximation $\mathbf{u}^{\mathrm{old}} =: \mathbf{w}$. For the case $\mathbf{w}=0$ this problem reduces to the so-called *generalized Stokes equations* and for $\mathbf{w}=0$ and $\xi=0$ we obtain the *stationary Stokes problem.*

Formulation in dimensionless variables

For the derivation and analysis of numerical methods for the models presented above it is convenient to consider these models in a non-dimensionalized form. For this we introduce

$$\begin{aligned} L &: \text{ typical length scale (related to size of } \Omega), \\ U &: \text{ typical velocity size,} \end{aligned}$$

and dimensionless variables

$$\bar{x} = \frac{1}{L}x, \quad \bar{t} = \frac{U}{L}t, \quad \bar{\mathbf{u}}(\bar{x},\bar{t}) = \frac{\mathbf{u}(x,t)}{U}, \quad \bar{p}(\bar{x},\bar{t}) = \frac{p(x,t)}{\rho U^2}.$$

Furthermore, $\bar{\Omega} := \frac{1}{L}\Omega := \{\, \bar{x} \in \mathbb{R}^3 : L\bar{x} \in \Omega \,\}$, and a non-dimensional source term is defined as $\bar{\mathbf{g}} = \frac{L}{U^2}\mathbf{g}$. The partial differential equations in (2.1) can be written in these dimensionless quantities as follows, where differential operators w.r.t. $\bar{x}_i$ and $\bar{t}$ are denoted with a $^-$ (for example: $\bar{\nabla}$):

$$\begin{aligned} \frac{\partial \bar{\mathbf{u}}}{\partial \bar{t}} + (\bar{\mathbf{u}}\cdot\bar{\nabla})\bar{\mathbf{u}} + \bar{\nabla}\bar{p} - \frac{1}{Re}\bar{\Delta}\bar{\mathbf{u}} &= \bar{\mathbf{g}} \text{ in } \bar{\Omega} \\ \overline{\operatorname{div}}\bar{\mathbf{u}} &= 0 \text{ in } \bar{\Omega}, \end{aligned}$$

with the dimensionless *Reynolds number*

$$Re = \frac{\rho L U}{\mu}.$$

For notational simplicity, in the remainder we drop the bar notation, and thus obtain the following Navier-Stokes system in dimensionless variables:

Navier-Stokes

$$\begin{aligned} \frac{\partial \mathbf{u}}{\partial t} - \frac{1}{Re}\Delta\mathbf{u} + (\mathbf{u}\cdot\nabla)\mathbf{u} + \nabla p &= \mathbf{g} \text{ in } \Omega \\ \operatorname{div}\mathbf{u} &= 0 \text{ in } \Omega. \end{aligned} \tag{2.5}$$

We now list the above-mentioned related simpler models used in fluid dynamics. In this monograph we use these simpler models in the analysis of numerical methods. The non-stationary Stokes equations (2.3) in dimensionless formulation are as follows:

Stokes

$$\begin{aligned} \frac{\partial \mathbf{u}}{\partial t} - \frac{1}{Re}\Delta \mathbf{u} + \nabla p &= \mathbf{g} \text{ in } \Omega \\ \operatorname{div} \mathbf{u} &= 0 \text{ in } \Omega. \end{aligned} \tag{2.6}$$

Hence, the Stokes equations can be seen as a limit case of the Navier-Stokes equations if $Re \to 0$. In the remainder, if we refer to the (Navier-)Stokes problem we always mean the *non-stationary* (Navier-)Stokes equations. We now present three time *in*dependent models. The Oseen system (2.4) in dimensionless form is as follows:

Oseen

$$\begin{aligned} \xi \mathbf{u} - \frac{1}{Re}\Delta \mathbf{u} + (\mathbf{w} \cdot \nabla)\mathbf{u} + \nabla p &= \mathbf{g} \text{ in } \Omega \\ \operatorname{div} \mathbf{u} &= 0 \text{ in } \Omega. \end{aligned} \tag{2.7}$$

where now $\xi \geq 0$ is a dimensionless constant. *Special cases* of the Oseen problem are the generalized Stokes and the stationary Stokes problems:

Generalized Stokes

$$\begin{aligned} \xi \mathbf{u} - \frac{1}{Re}\Delta \mathbf{u} + \nabla p &= \mathbf{g} \text{ in } \Omega \\ \operatorname{div} \mathbf{u} &= 0 \text{ in } \Omega, \end{aligned} \tag{2.8}$$

Stationary Stokes

$$\begin{aligned} -\frac{1}{Re}\Delta \mathbf{u} + \nabla p &= \mathbf{g} \text{ in } \Omega \\ \operatorname{div} \mathbf{u} &= 0 \text{ in } \Omega. \end{aligned} \tag{2.9}$$

2.2 Weak formulation

We will use finite element methods for the discretization of the (Navier-)Stokes equations. These methods are based on the weak (or variational) formulation of the partial differential equations. In this section we discuss this weak formulation.

2.2.1 Function spaces

We only recall some basic facts from the theory on Sobolev spaces. For a detailed treatment of this subject we refer to the literature, e.g. [8].

One main motivation for using Sobolev spaces is that these are Banach spaces. Some of these are Hilbert spaces. In our treatment of elliptic boundary value problems we only need these Hilbert spaces and thus we restrict ourselves to the presentation of this subset of Sobolev Hilbert spaces.

First we introduce the concept of weak derivatives. Let $\Omega \subset \mathbb{R}^d$ be an open, bounded and connected domain, and take $u \in C^1(\Omega)$, $\phi \in C_0^\infty(\Omega)$, where $C_0^\infty(\Omega)$ consists of all functions in $C^\infty(\Omega)$ that have a compact support in Ω (and thus, since Ω is open, such functions are identically zero close to the boundary). Since ϕ vanishes identically outside some compact subset of Ω, one obtains by partial integration in the variable x_j:

$$\int_\Omega \frac{\partial u(x)}{\partial x_j}\phi(x)\,dx = -\int_\Omega u(x)\frac{\partial \phi(x)}{\partial x_j}\,dx$$

and thus

$$\int_\Omega D^\alpha u(x)\phi(x)\,dx = -\int_\Omega u(x)D^\alpha\phi(x)\,dx, \qquad |\alpha| = 1,$$

holds. Here $D^\alpha u$ with $\alpha = (\alpha_1, \ldots, \alpha_d)$, $|\alpha| = \alpha_1 + \ldots + \alpha_d$ denotes

$$D^\alpha u = \frac{\partial^{|\alpha|}}{\partial x_1^{\alpha_1}\ldots\partial x_d^{\alpha_d}}.$$

Repeated application of this result yields the fundamental *Green's formula*

$$\int_\Omega D^\alpha u(x)\phi(x)\,dx = (-1)^{|\alpha|}\int_\Omega u(x)D^\alpha\phi(x)\,dx, \tag{2.10}$$
$$\text{for all}\quad \phi \in C_0^\infty(\Omega),\ u \in C^k(\Omega),\ k = 1, 2, \ldots \quad\text{and}\quad |\alpha| \le k.$$

Based on this formula we introduce the notion of a weak derivative:

Definition 2.2.1 Consider $u \in L^2(\Omega)$ and $|\alpha| > 0$. If there exists $v \in L^2(\Omega)$ such that

$$\int_\Omega v(x)\phi(x)\,dx = (-1)^{|\alpha|}\int_\Omega u(x)D^\alpha\phi(x)\,dx \quad \text{for all}\quad \phi \in C_0^\infty(\Omega), \tag{2.11}$$

then v is called the αth *weak derivative* of u and is denoted by $D^\alpha u := v$.

Such weak derivatives are often introduced in the more general setting of so-called *distributions*. For our purposes, however, the definition above suffices. If for $u \in L^2(\Omega)$ the αth weak derivative exists then it is unique (in the usual Lebesgue sense). If $u \in C^k(\overline{\Omega})$ then for $0 < |\alpha| \leq k$ the αth weak derivative and the classical αth derivative coincide. The Sobolev space $H^m(\Omega)$, $m = 1, 2, \ldots$, consists of all functions in $L^2(\Omega)$ for which all αth weak derivatives with $|\alpha| \leq m$ exist:

$$H^m(\Omega) := \left\{ u \in L^2(\Omega) : D^\alpha u \text{ exists for all } 0 < |\alpha| \leq m \right\}. \tag{2.12}$$

For $m = 0$ we define $H^0(\Omega) := L^2(\Omega)$. In $H^m(\Omega)$ a natural inner product and corresponding norm are defined by

$$(u, v)_m := \sum_{|\alpha| \leq m} (D^\alpha u, D^\alpha v)_{L^2}, \qquad \|u\|_m := (u, u)_m^{\frac{1}{2}}, \qquad u, v \in H^m(\Omega). \tag{2.13}$$

It is easy to verify, that $(\cdot, \cdot)_m$ defines an inner product on $H^m(\Omega)$. We now formulate a main result:

Theorem 2.2.2 *The space* $\left(H^m(\Omega), (\cdot, \cdot)_m\right)$ *is a Hilbert space.*

Similar constructions can be applied if we replace the Hilbert space $L^2(\Omega)$ by the Banach space $L^p(\Omega)$, $1 \leq p < \infty$ of measurable functions for which $\|u\|_p := (\int_\Omega |u(x)|^p \, dx)^{1/p}$ is bounded. This results in Sobolev spaces which are usually denoted by $H_p^m(\Omega)$. For notational simplicity we deleted the index $p = 2$ in our presentation. For $p \neq 2$ the Sobolev space $H_p^m(\Omega)$ is a Banach space but *not* a Hilbert space.

One can also define these Sobolev spaces using a different technique, namely based on the concept of completion. Consider the function space

$$Z_m := \{ u \in C^\infty(\Omega) : \|u\|_m < \infty \}.$$

The completion of this space with respect to $\|\cdot\|_m$ yields the Sobolev space $H^m(\Omega)$:

$$H^m(\Omega) = \text{ completion of } (Z_m, (\cdot, \cdot)_m).$$

A compact notation is $H^m(\Omega) = \overline{Z_m}^{\|\cdot\|_m}$. Note that $C^\infty(\overline{\Omega}) \subset Z_m$. Under very mild assumptions on the domain Ω we even have

$$H^m(\Omega) = \overline{C^\infty(\overline{\Omega})}^{\|\cdot\|_m}.$$

Another space that plays an important role in the weak formulation of the partial differential equations is the following subspace of $H^1(\Omega)$:

$$H_0^1(\Omega) := \overline{C_0^\infty(\Omega)}^{\|\cdot\|_1}. \tag{2.14}$$

An important issue is the smoothness in the classical sense of functions from a Sobolev space. In this respect the following Sobolev embedding result is relevant:

$$H^m(\Omega) \hookrightarrow C^k(\overline{\Omega}) \quad \text{if} \;\; m - \frac{d}{2} > k \quad (d \text{ such that } \Omega \subset \mathbb{R}^d). \tag{2.15}$$

The symbol $\hookrightarrow$ is used to denote that the embedding between the two spaces is continuous, i.e. if m and k satisfy the condition in (2.15) then there exists a constant c such that

$$\|u\|_{C^k(\overline{\Omega})} \leq c\|u\|_m \quad \text{for all} \;\; u \in H^m(\Omega).$$

A basic result is the so-called Poincaré-Friedrichs inequality:

$$\|u\|_{L^2} \leq C \sqrt{\sum_{|\alpha|=1} \|D^\alpha u\|_{L^2}^2} \quad \text{for all} \;\; u \in H_0^1(\Omega). \tag{2.16}$$

Based on this we have the following norm equivalence:

$$|u|_1 \leq \|u\|_1 \leq C|u|_1 \quad \text{for all} \;\; u \in H_0^1(\Omega), \;\; |u|_1^2 := \sum_{|\alpha|=1} \|D^\alpha u\|_{L^2}^2, \tag{2.17}$$

i.e., $|\cdot|_1$ and $\|\cdot\|_1$ are *equivalent norms* on $H_0^1(\Omega)$.

In the weak formulation of elliptic boundary value problems one has to treat boundary conditions. For this the next result will be needed.

There exists a unique bounded linear operator

$$\gamma : H^1(\Omega) \to L^2(\partial\Omega), \quad \|\gamma(u)\|_{L^2(\partial\Omega)} \leq c\|u\|_1, \tag{2.18}$$

with the property that for all $u \in C^1(\overline{\Omega})$ the equality $\gamma(u) = u_{|\partial\Omega}$ holds. The operator γ is called the *trace operator*. For $u \in H^1(\Omega)$ the function $\gamma(u) \in L^2(\partial\Omega)$ represents the boundary "values" of u and is called the trace of u. For $\gamma(u)$ one often uses the notation $u_{|\partial\Omega}$. For example, for $u \in H^1(\Omega)$, the identity $u_{|\partial\Omega} = 0$ means that $\gamma(u) = 0$ in the $L^2(\partial\Omega)$ sense. Using the trace operator one can give another natural characterization of the space $H_0^1(\Omega)$:

$$H_0^1(\Omega) = \left\{ u \in H^1(\Omega) : u_{|\partial\Omega} = 0 \right\}.$$

The *dual space* of $H_0^1(\Omega)$, i.e. the space of all bounded linear functionals $H_0^1(\Omega) \to \mathbb{R}$ is denoted by

$$H^{-1}(\Omega) := H_0^1(\Omega)',$$

with norm

$$\|f\|_{-1} := \sup_{v \in H_0^1(\Omega)} \frac{f(v)}{\|v\|_1}, \quad f \in H^{-1}(\Omega).$$

Finally, we collect a few results on *Green's formulas* that hold in Sobolev spaces. For notational simplicity the function arguments x are deleted in the integrals, and in boundary integrals like, for example, $\int_{\partial\Omega} \gamma(u)\,\gamma(v)\,ds$ we delete the trace operator γ. The following identities hold, with $\mathbf{n} = (n_1, \ldots, n_d)$ the outward unit normal on $\partial\Omega$ and $H^m := H^m(\Omega)$:

$$\int_\Omega u \frac{\partial v}{\partial x_i}\,dx = -\int_\Omega \frac{\partial u}{\partial x_i} v\,dx + \int_{\partial\Omega} uvn_i\,ds, \quad u, v \in H^1, 1 \le i \le d$$

$$\int_\Omega \Delta u\, v\,dx = -\int_\Omega \nabla u \cdot \nabla v\,dx + \int_{\partial\Omega} \nabla u \cdot \mathbf{n} v\,ds, \quad u \in H^2, v \in H^1$$

$$\int_\Omega u \operatorname{div} \mathbf{v}\,dx = -\int_\Omega \nabla u \cdot \mathbf{v}\,dx + \int_{\partial\Omega} u\mathbf{v} \cdot \mathbf{n}\,ds, \quad u \in H^1, \mathbf{v} \in (H^1)^d.$$

The Sobolev spaces turn out to be the appropriate ones for the weak formulation of many partial differential equations. For the weak formulation of *time* dependent partial differential equations one needs in addition the concept of V-valued functions, where V is a given Banach space (for example $L^2(\Omega)$, or a Sobolev space). We now introduce this concept. For proofs and more information we refer to the literature, e.g. [256, 261].

Let V be a Banach space with norm denoted by $\|\cdot\|_V$. The space $L^2(0, T; V)$ consists of all functions $u : (0, T) \to V$ for which

$$\|u\|_{L^2(0,T;V)} := \Big(\int_0^T \|u(t)\|_V^2\,dt \Big)^{\frac{1}{2}} < \infty$$

holds. The space $L^2(0, T; V)$ is a Banach space. It is a Hilbert space if V is a Hilbert space. This definition applied to the dual space V' results in the space $L^2(0, T; V')$ of functional valued functions $u : (0, T) \to V'$ with

$$\|u\|_{L^2(0,T;V')} := \Big(\int_0^T \|u(t)\|_{V'}^2\,dt \Big)^{\frac{1}{2}} < \infty.$$

Recall that $\|u(t)\|_{V'} := \sup_{v \in V} \frac{|u(t)(v)|}{\|v\|_V}$. There is a linear bijective isometric mapping $j : L^2(0, T; V)' \to L^2(0, T; V')$, hence these spaces can be identified with each other. One often writes

$$L^2(0, T; V)' = L^2(0, T; V').$$

2.2.2 Oseen problem in weak formulation

In this section we treat the weak formulation of the Oseen problem in dimensionless form (2.7). For notational simplicity we only treat the three-dimensional case, i.e. $\Omega \subset \mathbb{R}^3$. We consider this problem with homogeneous Dirichlet boundary conditions $\mathbf{u} = 0$ on $\partial\Omega$. The Reynolds number $Re > 0$ and the problem parameter $\xi \ge 0$ are given constants. Furthermore we assume that the velocity field $\mathbf{w}$ satisfies $\mathbf{w} \in H^1(\Omega)^3$ and $\|\mathbf{w}\|_{L^\infty(\Omega)} < \infty$.

We introduce the spaces

$$\mathbf{V} := H_0^1(\Omega)^3, \quad Q := L_0^2(\Omega) = \left\{ q \in L^2(\Omega) : \int_\Omega q\, dx = 0 \right\}, \tag{2.19}$$

and the bilinear forms

$$m(\mathbf{u},\mathbf{v}) = (\mathbf{u},\mathbf{v})_{L^2} = \int_\Omega \mathbf{u}\cdot\mathbf{v}\, dx \tag{2.20a}$$

$$a(\mathbf{u},\mathbf{v}) = \frac{1}{Re}\int_\Omega \nabla\mathbf{u}\cdot\nabla\mathbf{v}\, dx = \frac{1}{Re}\sum_{i=1}^{3}\int_\Omega \nabla u_i\cdot\nabla v_i\, dx \tag{2.20b}$$

$$c(\mathbf{u},\mathbf{v}) = c(\mathbf{w};\mathbf{u},\mathbf{v}) = \int_\Omega (\mathbf{w}\cdot\nabla\mathbf{u})\cdot\mathbf{v}\, dx = \sum_{1\le i,j\le 3}\int_\Omega w_i\frac{\partial u_j}{\partial x_i}v_j\, dx \tag{2.20c}$$

$$b(\mathbf{v},q) = -\int_\Omega q\,\mathrm{div}\,\mathbf{v}\, dx. \tag{2.20d}$$

The weak formulation of (2.7) is as follows:

Find $\mathbf{u}\in\mathbf{V}$ and $p\in Q$ such that

$$\begin{aligned} \xi m(\mathbf{u},\mathbf{v}) + a(\mathbf{u},\mathbf{v}) + c(\mathbf{u},\mathbf{v}) + b(\mathbf{v},p) &= (\mathbf{g},\mathbf{v})_{L^2} \quad \text{for all} \;\; \mathbf{v}\in\mathbf{V},\\ b(\mathbf{u},q) &= 0 \quad \text{for all} \;\; q\in Q. \end{aligned} \tag{2.21}$$

We now turn to the analysis of this weak formulation. First we show that if the problem in strong formulation (2.7) has a sufficiently smooth solution $(\mathbf{u},p)$, then this pair is also a solution of (2.21). We start with the observation that the bilinear forms $m(\cdot,\cdot)$, $a(\cdot,\cdot)$, $c(\cdot,\cdot)$ are *continuous* on $\mathbf{V}\times\mathbf{V}$ and that $b(\cdot,\cdot)$ is *continuous* on $\mathbf{V}\times Q$.

Lemma 2.2.3 *Assume* $\mathbf{u}\in C^2(\overline{\Omega})^3$ *with* $\mathbf{u}=0$ *on* $\partial\Omega$ *and* $p\in C^1(\overline{\Omega})$ *with* $\int_\Omega p\, dx = 0$ *is a solution pair of* (2.7). *Then this pair also solves* (2.21).

Proof. From the assumptions on $\mathbf{u}$ and p it follows that $\mathbf{u}\in\mathbf{V}$, $p\in Q$. If we multiply the first equation in (2.7) by $\boldsymbol{\phi}\in C_0^\infty(\Omega)^3$, integrate over Ω and apply partial integration (for each of the three components) we obtain

$$\xi\int_\Omega \mathbf{u}\boldsymbol{\phi}\, dx + \frac{1}{Re}\int_\Omega \nabla\mathbf{u}\cdot\nabla\boldsymbol{\phi}\, dx + \int_\Omega (\mathbf{w}\cdot\nabla\mathbf{u})\cdot\boldsymbol{\phi}\, dx - \int_\Omega p\,\mathrm{div}\,\boldsymbol{\phi}\, dx = \int_\Omega \mathbf{g}\cdot\boldsymbol{\phi}\, dx.$$

Hence,

$$\xi m(\mathbf{u},\boldsymbol{\phi}) + a(\mathbf{u},\boldsymbol{\phi}) + c(\mathbf{u},\boldsymbol{\phi}) + b(\boldsymbol{\phi},p) = (\mathbf{g},\boldsymbol{\phi})_{L^2} \tag{2.22}$$

for all $\boldsymbol{\phi}\in C_0^\infty(\Omega)^3$. Using the continuity of the bilinear forms and of $\mathbf{v}\to(\mathbf{g},\mathbf{v})_{L^2}$ on $\mathbf{V}$, and the density of $C_0^\infty(\Omega)^3$ in $\mathbf{V}$ it follows that the identity in (2.22) even holds for all $\boldsymbol{\phi}\in\mathbf{V}$. Thus the first variational equation in (2.21) holds. Multiplication of the second equation in (2.7) by an arbitrary $q\in Q$ and integrating over Ω results in the second variational identity in (2.21). □

One very important property of the weak formulation (2.21) is that, opposite to the strong formulation in (2.21), under very mild assumptions it has a unique solution. The mathematical analysis of variational problems like the one in (2.21) is based on an abstract theory for saddle point problems as presented in the Appendix, Sect. 15.3. Theorem 15.3.1 can be applied to prove the well-posedness of the weak formulation of the Oseen problem (2.21). In the application of Theorem 15.3.1 we use the spaces $V = \mathbf{V}$, $M = Q$ and the bilinear forms

$$\hat{a}(\mathbf{u}, \mathbf{v}) = \xi m(\mathbf{u}, \mathbf{v}) + a(\mathbf{u}, \mathbf{v}) + c(\mathbf{u}, \mathbf{v}) \quad \text{on} \quad \mathbf{V} \times \mathbf{V}, \tag{2.23a}$$

$$\hat{b}(\mathbf{u}, q) = b(\mathbf{u}, q) \quad \text{on} \quad \mathbf{V} \times Q. \tag{2.23b}$$

A rather deep result from the theory on Sobolev spaces is the following:

$$\exists\, \beta > 0 : \quad \sup_{\mathbf{v} \in H_0^1(\Omega)^d} \frac{\int_\Omega q \operatorname{div} \mathbf{v} \, dx}{\|\mathbf{v}\|_1} \geq \beta \|q\|_{L^2} \quad \forall\, q \in L_0^2(\Omega). \tag{2.24}$$

A proof of this is given in [183, 91]. From this result we immediately obtain that for the bilinear form $b(\cdot, \cdot)$ on $\mathbf{V} \times Q$ the inf-sup condition in (15.17a) is satisfied. Using this the following theorem can be proved:

Theorem 2.2.4 *Consider the weak formulation of the Oseen problem in* (2.21). *Assume that ξ and $\mathbf{w}$ are such that $\xi - \frac{1}{2} \operatorname{div} \mathbf{w} \geq 0$ on Ω. Then this problem is well-posed.*

Proof. We apply Theorem 15.3.1 with the spaces $V = \mathbf{V}$, $M = Q$, the bilinear forms defined in (2.23) and the functionals $f_1(\mathbf{v}) := (\mathbf{g}, \mathbf{v})$, $f_2 = 0$. These bilinear forms and functionals are continuous. The inf-sup condition (15.17a) is satisfied due to property (2.24). We finally check the ellipticity condition (15.17b). For $\mathbf{u}, \mathbf{v} \in \mathbf{V}$ we have $\mathbf{u}_{|\partial\Omega} = \mathbf{v}_{|\partial\Omega} = 0$ and thus using partial integration we obtain

$$\begin{aligned} \int_\Omega (\mathbf{w} \cdot \nabla \mathbf{u}) \cdot \mathbf{v} \, dx &= \sum_{1 \leq i,j \leq 3} \int_\Omega w_i \frac{\partial u_j}{\partial x_i} v_j \, dx \\ &= -\int_\Omega \operatorname{div} \mathbf{w} (\mathbf{u} \cdot \mathbf{v}) \, dx - \int_\Omega (\mathbf{w} \cdot \nabla \mathbf{v}) \cdot \mathbf{u} \, dx. \end{aligned}$$

Hence, for $\mathbf{u} = \mathbf{v}$ we have

$$\int_\Omega (\mathbf{w} \cdot \nabla \mathbf{u}) \cdot \mathbf{u} \, dx = -\frac{1}{2} \int_\Omega \operatorname{div} \mathbf{w} (\mathbf{u} \cdot \mathbf{u}) \, dx.$$

This yields, using the assumption $\xi - \frac{1}{2} \operatorname{div} \mathbf{w} \geq 0$,

$$\begin{aligned}\hat{a}(\mathbf{u},\mathbf{u}) &= \xi \int_\Omega \mathbf{u}\cdot\mathbf{u}\,dx + \frac{1}{Re}\int_\Omega \nabla\mathbf{u}\cdot\nabla\mathbf{u}\,dx + \int_\Omega (\mathbf{w}\cdot\nabla\mathbf{u})\cdot\mathbf{u}\,dx \\ &= \int_\Omega (\xi - \frac{1}{2}\operatorname{div}\mathbf{w})\mathbf{u}\cdot\mathbf{u}\,dx + \frac{1}{Re}\int_\Omega \nabla\mathbf{u}\cdot\nabla\mathbf{u}\,dx \\ &\geq \frac{1}{Re}\int_\Omega \nabla\mathbf{u}\cdot\nabla\mathbf{u}\,dx \geq c\,\|\mathbf{u}\|_1^2,\end{aligned}$$

with a constant $c > 0$. In the last inequality we used the norm equivalence (2.17). □

We comment on the assumption $\xi - \frac{1}{2}\operatorname{div}\mathbf{w} \geq 0$ on Ω in Theorem 2.2.4. If we consider a stationary Stokes problem then we have $\xi = 0$, $\mathbf{w} = 0$ and thus this assumption is satisfied. For a generalized Stokes equation, which results from implicit time discretization of an non-stationary Stokes problem we have $\mathbf{w} = 0$, $\xi > 0$ and thus the assumption holds. Hence we conclude:

> *The stationary (generalized) Stokes equations in weak formulation, i.e.* (2.21) *with* $\xi \geq 0$, $\mathbf{w} = 0$, *have a unique solution pair* $(\mathbf{u}, p) \in \mathbf{V} \times Q$.

In the general case of an Oseen equation, which results after implicit time integration and linearization of a Navier-Stokes problem, it is reasonable to expect that $|\operatorname{div}\mathbf{w}|$, in which $\mathbf{w}$ is an approximation of the solution $\mathbf{u}$ (that satisfies $\operatorname{div}\mathbf{u} = 0$), is small compared to $\xi \sim \frac{1}{\Delta t}$ (Δt: time step in time discretization). Hence it is plausible that the condition $\xi - \frac{1}{2}\operatorname{div}\mathbf{w} \geq 0$ is satisfied.

2.2.3 Time dependent (Navier-)Stokes equations in weak formulation

In this section we treat the weak formulation of the non-stationary Stokes problem (2.6) and of the non-stationary Navier-Stokes problem (2.5). For both problems we restrict ourselves to the case with homogeneous Dirichlet boundary conditions, i.e., $\mathbf{u} = 0$ on $\partial\Omega$.

Compared to the weak formulation of the Oseen problem in Sect. 2.2.2 we now in addition have to address the issue of an appropriate treatment of the time derivative $\frac{\partial \mathbf{u}}{\partial t}$. For the weak formulation of many time dependent partial differential equations the Hilbert space $L^2(0,T;V)$ with a suitable Sobolev space V, cf. Sect. 2.2.1, turns out to be appropriate. For $u : (0,T) \to V$ one then needs a suitable weak derivative $u' = \frac{du}{dt}$. This can be defined by means of the very general and powerful concept of distributional derivatives, in which derivatives of linear mappings $L : C_0^\infty(0,T) \to V$ are defined, cf. [256, 261]. We will not use this (rather abstract) approach but introduce u' by means of weak derivatives as already presented in Definition 2.2.1. This (compared to the distributional concept) less general definition is sufficient for our purposes. We first present a weak variational formulation of a time dependent problem in an abstract setting and then apply this to derive weak formulations of the non-stationary Stokes- and Navier-Stokes equations.

An abstract variational formulation of a time dependent problem

Let V, H be Hilbert spaces such that $V \hookrightarrow H \hookrightarrow V'$ forms a Gelfand triple, which means that H is identified with its dual, $H \equiv H'$, the embedding $V \hookrightarrow H$ is continuous and V is dense in H. The scalar products in V and H are denoted by $(\cdot,\cdot)_V$ and $(\cdot,\cdot)_H$, respectively. In our applications we use, for example, the Gelfand triple $H_0^1(\Omega) \hookrightarrow L^2(\Omega) \hookrightarrow H^{-1}(\Omega)$.

We recall the definition of a weak derivative for a function $g \in L^2(0,T)$ as given in Definition 2.2.1: if for such a function g there exists $h \in L^2(0,T)$ such that

$$\int_0^T h(t)\phi(t)\,dt = -\int_0^T g(t)\phi'(t)\,dt \quad \text{for all} \quad \phi \in C_0^\infty(0,T), \tag{2.25}$$

then $h =: g'$ is the weak derivative of g.

We now introduce a weak time derivative for $u \in L^2(0,T;V)$. It is possible to define a weak derivative of u in the same space $L^2(0,T;V)$. However, for the time dependent problems that we consider it turns out to be more appropriate to define a weak derivative in the space $L^2(0,T;V')$. Using the Gelfand property $V \hookrightarrow H \equiv H' \hookrightarrow V'$ this can be done as follows. Take $t \in [0,T]$. The element $u(t) \in V$ can be identified with an element of V', through $v \to (u(t),v)_H$, $v \in V$. This identification of $u(t)$ with an element of V' leads to the following natural definition of a weak time derivative of u in $L^2(0,T;V')$.

Definition 2.2.5 Consider $u \in L^2(0,T;V)$. If there exists a $w \in L^2(0,T;V')$ such that

$$\int_0^T w(t)(v)\,\phi(t)\,dt = -\int_0^T (u(t),v)_H\phi'(t)\,dt, \tag{2.26}$$

for all $v \in V$ and all $\phi \in C_0^\infty(0,T)$, then w is called the weak (time) derivative of u and we write $u' = w$.

One can show, that if such a weak derivative exists, then it is unique. Furthermore, assume that $u : [0,T] \to H$ is smooth enough such that the classical (Fréchet) derivative exists in H. Denote this Fréchet derivative by $u'(t)$. Then

$$\int_0^T (u'(t),v)_H\,\phi(t)\,dt = -\int_0^T (u(t),v)_H\phi'(t)\,dt$$

holds for all $v \in V$ and all $\phi \in C_0^\infty(0,T)$, and thus $u'(t)$ identified with the functional $v \to (u'(t),v)_H$ is the weak derivative in the sense of Definition 2.2.5.

Lemma 2.2.6 *Assume that $u \in L^2(0,T;V)$ has a weak derivative $u' \in L^2(0,T;V')$. For arbitrary $v \in V$ define the function $g_v : t \to (u(t),v)_H$. Then $g_v \in L^2(0,T)$ and g_v has a weak derivative $g_v'(t) = \frac{d}{dt}(u(t),v)_H$ in the sense of* (2.25). *Furthermore,*

$$\frac{d}{dt}(u(t), v)_H = u'(t)(v) \quad \textit{for all} \quad v \in V \tag{2.27}$$

holds for almost all $t \in (0, T)$.

Proof. Take $v \in V$. Note that due to the Gelfand property $\|w\|_H \leq c\|w\|_V$ for all $w \in V$ holds. From

$$\int_0^T g_v(t)^2\, dt \leq \int_0^T \|u(t)\|_H^2 \|v\|_H^2\, dt \leq c^4 \int_0^T \|u(t)\|_V^2\, dt\, \|v\|_V^2$$

and $u \in L^2(0, T; V)$ it follows that $g_v \in L^2(0, T)$. For $h_v(t) := u'(t)(v)$ we have

$$\int_0^T h_v(t)^2\, dt \leq \int_0^T \|u'(t)\|_{V'}^2\, dt \|v\|_V^2 < \infty$$

and thus $h_v \in L^2(0, T)$. Using the property (2.26) we get

$$\begin{aligned}\int_0^T h_v(t)\phi(t)\, dt &= \int_0^T u'(t)(v)\phi(t)\, dt \\ &= -\int_0^T (u(t), v)_H \phi'(t)\, dt = -\int_0^T g_v(t)\phi'(t)\, dt\end{aligned}$$

for all $\phi \in C_0^\infty(0, T)$. Thus $h_v = g_v'$, i.e., $u'(t)(v) = \frac{d}{dt}(u(t), v)_H$ holds. □

Using the above notion of a weak derivative for functions $u \in L^2(0, T; V)$ we introduce the following space

$$W^1(0, T; V) := \left\{ v \in L^2(0, T; V) : \; v' \in L^2(0, T; V') \;\; \text{exists} \right\}.$$

We mention two important properties of this space. For proofs and further properties we refer to the literature, e.g. [256, 261]. Firstly, the space $W^1(0, T; V)$ is a Hilbert space w.r.t. the norm (and corresponding scalar product)

$$\|u\|_{W^1(0,T;V)} := \left(\|u\|_{L^2(0,T;V)}^2 + \|u'\|_{L^2(0,T;V')}^2\right)^{\frac{1}{2}}.$$

Secondly, there is a continuous embedding

$$W^1(0, T; V) \hookrightarrow C([0, T]; H), \tag{2.28}$$

where $C([0, T]; H)$ is the Banach space of all continuous functions $u : [0, T] \to H$ with norm $\|u\|_{C([0,T];H)} := \max_{0 \leq t \leq T} \|u(t)\|_H$. An important corollary of this embedding property is that for $u \in W^1(0, T; V)$ the values $u(t)$, $t \in [0, T]$, are well-defined in H.

Based on these preparations, we can introduce an abstract variational time dependent problem. Let $\hat{a} : V \times V \to \mathbb{R}$ be a bilinear form.

For $b(t) \in V'$, $t \in (0,T)$, $u_0 \in H$ we define the problem:

Find $u \in W^1(0,T;V)$ such that

$$\begin{aligned} &\frac{d}{dt}(u(t),v)_H + \hat{a}(u(t),v) = b(t)(v) \quad \text{for all } v \in V,\ t \in (0,T), \\ &u(0) = u_0. \end{aligned} \tag{2.29}$$

Due to Lemma 2.2.6 the term $\frac{d}{dt}(u(t),v)_H$ in (2.29) is well-defined. A main theorem is the following, cf. [256, 261] for a proof.

Theorem 2.2.7 *Take $u_0 \in H$, $b \in L^2(0,T;V')$ and assume that the bilinear form $\hat{a}(\cdot,\cdot)$ is continuous and elliptic on $V \times V$. Then the variational problem (2.29) is well-posed, i.e. it has a unique solution u and the linear mapping $(b,u_0) \to u$ is continuous from $L^2(0,T;V') \times H$ into $W^1(0,T;V)$.*

We summarize the essential ingredients for this abstract time dependent weak formulation to be well-posed:

- One uses a Gelfand triple of spaces $V \hookrightarrow H \hookrightarrow V'$ and uses the corresponding space $W^1(0,T;V)$. The weak derivative $u' \in L^2(0,T;V')$ is as in Definition 2.2.5.
- The data are from appropriate spaces: $b \in L^2(0,T;V')$, $u_0 \in H$.
- The bilinear form $\hat{a}(\cdot,\cdot)$ is continuous and elliptic on $V \times V$.

Below we use these abstract results for the derivation of appropriate weak formulations of the time dependent (Navier-)Stokes equations.

Remark 2.2.8 A proof of Theorem 2.2.7 is given in e.g. Theorem 26.1 in [256], Theorem 23.A in [261]. There it is also shown that the result still holds if the ellipticity condition $\hat{a}(v,v) \geq \gamma_V \|v\|_V^2$ for all $v \in V$ ($\gamma_V > 0$) is replaced by the weaker so-called Garding inequality:

$$\hat{a}(v,v) \geq \gamma_V \|v\|_V^2 - \gamma_H \|v\|_H^2 \quad \text{for all } v \in V,$$

with constants $\gamma_V > 0$ and γ_H independent of v.

Application to non-stationary Stokes equations

First we introduce suitable function spaces. Let

$$N(\Omega) := \{\, \mathbf{v} \in C_0^\infty(\Omega)^3 \mid \operatorname{div} \mathbf{v} = 0 \,\}$$

and

$$\mathbf{H}_{\text{div}} := \overline{N(\Omega)}^{\|\cdot\|_{L^2}} \quad (\text{closure of } N(\Omega) \text{ in } L^2(\Omega)^3\,) \tag{2.30}$$

$$\mathbf{V}_{\text{div}} := \overline{N(\Omega)}^{\|\cdot\|_1} \quad (\text{closure of } N(\Omega) \text{ in } H_0^1(\Omega)^3\,). \tag{2.31}$$

The spaces $(\mathbf{H}_{\rm div}, \|\cdot\|_{L^2})$, $(\mathbf{V}_{\rm div}, \|\cdot\|_1)$ are Hilbert spaces. From $\|\mathbf{v}\|_{L^2} \leq c\|\mathbf{v}\|_1$ for all $\mathbf{v} \in N(\Omega)$ and a density argument it follows that there is a continuous embedding $\mathbf{V}_{\rm div} \hookrightarrow \mathbf{H}_{\rm div}$. Using $N(\Omega) \subset \mathbf{V}_{\rm div} \subset \mathbf{H}_{\rm div}$ and the fact that $\mathbf{H}_{\rm div}$ is the closure of $N(\Omega)$ w.r.t. $\|\cdot\|_{L^2}$ it follows that $\mathbf{V}_{\rm div}$ is dense in $\mathbf{H}_{\rm div}$. Thus we have a Gelfand triple

$$\mathbf{V}_{\rm div} \hookrightarrow \mathbf{H}_{\rm div} \equiv \mathbf{H}'_{\rm div} \hookrightarrow \mathbf{V}'_{\rm div}. \tag{2.32}$$

The space $\mathbf{V}_{\rm div}$ can also be characterized as, cf. [235],

$$\mathbf{V}_{\rm div} = \left\{ \mathbf{v} \in H_0^1(\Omega)^3 \, : \, \operatorname{div} \mathbf{v} = 0 \right\}.$$

We take the bilinear form $a(\cdot,\cdot)$ as in (2.20b). It is continuous and elliptic on $H_0^1(\Omega)^3$, thus also on the subspace $\mathbf{V}_{\rm div}$ of $H_0^1(\Omega)^3$.

We introduce the following weak formulation of the non-stationary Stokes equations (2.6):

Determine $\mathbf{u} \in W^1(0,T;\mathbf{V}_{\rm div})$ such that

$$\begin{aligned} \frac{d}{dt}(\mathbf{u}(t),\mathbf{v})_{L^2} + a(\mathbf{u}(t),\mathbf{v}) &= (\mathbf{g},\mathbf{v})_{L^2} \quad \text{for all } \mathbf{v} \in \mathbf{V}_{\rm div}, \\ \mathbf{u}(0) &= \mathbf{u}_0 \end{aligned} \tag{2.33}$$

We first show that a smooth solution $(\mathbf{u},p)$ of (2.6) is also a solution of this weak formulation:

Lemma 2.2.9 *Let $(\mathbf{u},p)$ be a solution of (2.6). Define $\mathbf{u}(t) := \mathbf{u}(\cdot,t)$ and assume that $\mathbf{u} \in C^1\big([0,T];C^2(\overline{\Omega})^3\big)$, and $\mathbf{u}(0) = \mathbf{u}_0$. Then $\mathbf{u}$ satisfies (2.33).*

Proof. From $\mathbf{u} \in C^1\big([0,T];C^2(\overline{\Omega})^3\big)$ and $\operatorname{div}\mathbf{u} = 0$ it follows that $\mathbf{u} \in W^1(0,T;\mathbf{V}_{\rm div})$. We multiply the first equation in (2.6) by $\mathbf{v} \in \mathbf{V}_{\rm div}$ and integrate over Ω. Note that $\int_\Omega \nabla p \cdot \mathbf{v}\,dx = -\int_\Omega p \operatorname{div}\mathbf{v}\,dx = 0$, due to $\mathbf{v} \in \mathbf{V}_{\rm div}$. Using partial integration for the term $\frac{1}{Re}\int_\Omega \Delta\mathbf{u}\cdot\mathbf{v}\,dx$ we then obtain

$$\Big(\frac{d\mathbf{u}(t)}{dt},\mathbf{v}\Big)_{L^2} + a(\mathbf{u}(t),\mathbf{v}) = (\mathbf{g},\mathbf{v})_{L^2} \quad \text{for all } \mathbf{v} \in \mathbf{V}_{\rm div},$$

and thus (2.33) holds. □

As in Sect. 2.2.2, opposite to the strong formulation, the weak formulation has the nice property that well-posedness can be shown to hold under (very) mild assumptions on the data. To be more precise, the following theorem holds:

Theorem 2.2.10 *Assume $\mathbf{g} \in L^2(0,T;\mathbf{V}'_{\rm div})$ and $\mathbf{u}_0 \in \mathbf{H}_{\rm div}$. Then the weak formulation (2.33) is well-posed.*

Proof. We have a Gelfand triple $\mathbf{V}_{\mathrm{div}} \hookrightarrow \mathbf{H}_{\mathrm{div}} \equiv \mathbf{H}'_{\mathrm{div}} \hookrightarrow \mathbf{V}'_{\mathrm{div}}$ and the bilinear form $a(\cdot,\cdot)$ is continuous and elliptic on $\mathbf{V}_{\mathrm{div}}$. Application of Theorem 2.2.7 yields the desired result. □

The weak formulation in (2.33) has *no pressure variable*. This is due to the fact that the space $\mathbf{V}_{\mathrm{div}}$ contains only functions that satisfy the incompressibility condition $\operatorname{div}\mathbf{u} = 0$. For numerical purposes this weak formulation is less convenient due to the fact that in general it is hard to construct appropriate finite element subspaces of $\mathbf{V}_{\mathrm{div}}$. It turns out to be more convenient to have a weak formulation using $\mathbf{V} = H_0^1(\Omega)^3$ (instead of $\mathbf{V}_{\mathrm{div}}$). This can be achieved by introducing a suitable Lagrange-multiplier variable. In an abstract Hilbert space setting this is explained in Sect. 15.3, cf. Theorem 15.3.4. As in the strong formulation in (2.6), one can enforce the incompressibility condition by using a pressure variable $p \in Q = L_0^2(\Omega)$. The weak formulation in (2.33) can be used to derive a suitable weak formulation for $(\mathbf{u}, p) \in \mathbf{V} \times Q$, which is as follows:

Determine $\mathbf{u} \in W^1(0,T;\mathbf{V})$, $p \in L^2(0,T;Q)$ such that

$$\begin{aligned} \frac{d}{dt}(\mathbf{u}(t),\mathbf{v})_{L^2} + a(\mathbf{u}(t),\mathbf{v}) + b(\mathbf{v},p(t)) &= (\mathbf{g},\mathbf{v})_{L^2} \quad \text{for all } \mathbf{v} \in \mathbf{V}, \\ b(\mathbf{u}(t),q) &= 0 \quad \text{for all } q \in Q, \\ \mathbf{u}(0) &= \mathbf{u}_0. \end{aligned} \tag{2.34}$$

Theorem 2.2.11 *For* $\mathbf{g} \in L^2(0,T;L^2(\Omega)^3)$ *and* $\mathbf{u}_0 \in \mathbf{V}_{\mathrm{div}}$ *the weak formulation* (2.34) *has a unique solution* $(\mathbf{u}, p)$. *The solution* $\mathbf{u}$ *also solves the problem* (2.33).

Proof. A complete proof is given in [106] Proposition 6.38. We outline the main ideas of that proof. Assume that (2.34) has a solution $(\mathbf{u}, p) \in W^1(0,T;\mathbf{V}) \times L^2(0,T;Q)$. From the second equation in (2.34) we obtain $\operatorname{div}\mathbf{u} = 0$ in $L^2(\Omega)$. From this and from $\mathbf{u} \in L^2(0,T;\mathbf{V})$ it follows that $\mathbf{u} \in L^2(0,T;\mathbf{V}_{\mathrm{div}})$ holds. Note that $\mathbf{V}' \subset \mathbf{V}'_{\mathrm{div}}$ and thus from $\mathbf{u}' \in L^2(0,T;\mathbf{V}')$ it follows that $\mathbf{u}' \in L^2(0,T;\mathbf{V}'_{\mathrm{div}})$. Hence we have $\mathbf{u} \in W^1(0,T;\mathbf{V}_{\mathrm{div}})$. From the first equation in (2.34) with arbitrary $\mathbf{v} \in \mathbf{V}_{\mathrm{div}}$ (thus $b(\mathbf{v},p(t)) = 0$) it follows that $\frac{d}{dt}(\mathbf{u}(t),\mathbf{v})_{L^2} + a(\mathbf{u}(t),\mathbf{v}) = (\mathbf{g},\mathbf{v})_{L^2}$ for all $\mathbf{v} \in \mathbf{V}_{\mathrm{div}}$. We conclude that $\mathbf{u}$ is a solution of (2.33). Furthermore, due to Theorem 2.2.10, we have *uniqueness* of the solution $(\mathbf{u}, p)$. We now address existence of a solution $(\mathbf{u}, p)$ of (2.34). Let $\mathbf{u}$ be the unique solution of (2.33), which exists due to Theorem 2.2.10. From $\operatorname{div}\mathbf{u} = 0$ is follows that $\mathbf{u}$ satisfies the second equation in (2.34). For the solution $\mathbf{u}$ we have $\mathbf{u}' \in L^2(0,T;\mathbf{V}'_{\mathrm{div}})$. One can show (using the assumptions on the data, cf. [106]), that $\mathbf{u}' \in L^2(0,T;\mathbf{V}')$ holds. Hence, $\mathbf{u} \in W^1(0,T;\mathbf{V})$ and moreover,

$$\ell(t)(\mathbf{v}) := \frac{d}{dt}(\mathbf{u}(t),\mathbf{v})_{L^2} + a(\mathbf{u}(t),\mathbf{v}) - (\mathbf{g},\mathbf{v})_{L^2}$$

satisfies

$$\ell \in L^2(0,T;\mathbf{V}'), \quad \ell(t)(\mathbf{v}) = 0 \quad \text{for all} \quad t \in [0,T], \ \mathbf{v} \in \mathbf{V}_{\mathrm{div}}.$$

From a rather deep result (originally due to De Rham [77]), cf. Theorem 2.3 in [121], it follows that there exists $p(t) \in L_0^2(\Omega) = Q$ such that

$$(p(t), \operatorname{div}\mathbf{v})_{L^2} = \ell(t)(\mathbf{v}) \quad \text{for all} \quad \mathbf{v} \in \mathbf{V}.$$

And thus for $t \in [0,T]$ we have

$$\frac{d}{dt}(\mathbf{u}(t),\mathbf{v})_{L^2} + a(\mathbf{u}(t),\mathbf{v}) + b(\mathbf{v},p(t)) = (\mathbf{g},\mathbf{v})_{L^2} \quad \text{for all} \quad \mathbf{v} \in \mathbf{V},$$

i.e., the first equation in (2.34) holds, too. Using the inf-sup property (2.24) we obtain

$$\|p(t)\|_{L^2} \le c \sup_{\mathbf{v}\in\mathbf{V}} \frac{(p(t),\operatorname{div}\mathbf{v})_{L^2}}{\|\mathbf{v}\|_1} = c \sup_{\mathbf{v}\in\mathbf{V}} \frac{\ell(t)(\mathbf{v})}{\|\mathbf{v}\|_1} = c\,\|\ell(t)\|_{\mathbf{V}'}.$$

Using $\ell \in L^2(0,T,\mathbf{V}')$ it follows that $p \in L^2(0,T;Q)$ holds. Thus $(\mathbf{u},p)$ is a solution of (2.34). □

Note that in Theorem 2.2.11 the assumptions on the data $\mathbf{g}$ and $\mathbf{u}_0$ are somewhat stronger than in Theorem 2.2.10.

Application to non-stationary Navier-Stokes equations

We now turn to the weak formulation of the non-stationary Navier-Stokes equations (in dimensionless formulation)

$$\begin{aligned} \frac{\partial \mathbf{u}}{\partial t} - \frac{1}{Re}\Delta\mathbf{u} + (\mathbf{u}\cdot\nabla)\mathbf{u} + \nabla p &= \mathbf{f} \text{ in } \Omega \\ \operatorname{div}\mathbf{u} &= 0 \text{ in } \Omega, \end{aligned} \tag{2.35}$$

with a homogeneous Dirichlet boundary condition, $\mathbf{u} = 0$ on $\partial\Omega$, and an initial condition $\mathbf{u}(x,0) = \mathbf{u}_0(x)$ on Ω. The Reynolds number Re is a given strictly positive constant. The weak formulation of this problem takes a form very similar to that of the Stokes equations discussed above. The analysis of well-posedness of this weak formulation, however, is much more complicated as for the Stokes case. Due to the *non*linearity of the Navier-Stokes equations the weak formulation cannot be analyzed in the abstract framework of (2.29) and Theorem 2.2.7. Below we will only present a few main results. For proofs we refer to the literature, e.g. Chap. III in [235].

First we introduce a slight generalization of the weak derivative as defined in Definition 2.2.5 (still not using the concept of distributions). For a function $g \in L^2(0,T)$ a weak derivative $h = g' \in L^1(0,T)$ (instead of $\in L^2(0,T)$!) is still well-defined by the condition in (2.25). This is due to the fact that for $\phi \in C_0^\infty(0,T)$, $h \in L^1(0,T)$ we have

$$\left| \int_0^T h(t)\phi(t)\,dt \right| \leq \|\phi\|_{\infty,[0,T]} \int_0^T |h(t)|\,dt = \|\phi\|_{\infty,[0,T]}\|h\|_{L^1} < \infty.$$

Similarly, for $u \in L^2(0,T;V)$ (V a Hilbert space) we have a well-defined weak derivative $u' \in L^1(0,T;V')$ if in Definition 2.2.5 we replace "$w \in L^2(0,T;V')$" by "$w \in L^1(0,T;V')$". For this weak derivative the identity $\frac{d}{dt}(u(t),v)_H = u'(t)(v)$ as in (2.27) still holds. Instead of the space $W^1(0,T;V) = \{ v \in L^2(0,T;V) : v' \in L^2(0,T;V') \text{ exists} \}$, with V a Hilbert space, we need the larger space

$$W^1_*(0,T;V) := \{ v \in L^2(0,T;V) : v' \in L^1(0,T;V') \text{ exists} \},$$

which is a Banach space for $\|u\|_{W^1_*(0,T;V)} := \left(\|u\|^2_{L^2(0,T;V)} + \|u'\|^2_{L^1(0,T;V')}\right)^{\frac{1}{2}}$. For the weak formulation of the Navier-Stokes equations we use the same Gelfand triple (2.32) as for the Stokes problem with spaces $\mathbf{H}_{\text{div}}$, $\mathbf{V}_{\text{div}}$ as in (2.30)-(2.31). We use the same bilinear forms $a(\cdot,\cdot)$, $b(\cdot,\cdot)$ as for the Stokes problem and the trilinear form $c(\mathbf{w};\mathbf{u},\mathbf{v})$ as defined in (2.20c). The following weak formulation of (2.35) is similar to the weak Stokes problem (2.33) :

Determine $\mathbf{u} \in W^1_*(0,T;\mathbf{V}_{\text{div}})$ such that

$$\frac{d}{dt}(\mathbf{u}(t),\mathbf{v})_{L^2} + a(\mathbf{u}(t),\mathbf{v}) + c(\mathbf{u}(t);\mathbf{u}(t),\mathbf{v}) = (\mathbf{g},\mathbf{v})_{L^2} \;\forall \mathbf{v} \in \mathbf{V}_{\text{div}}, \qquad (2.36)$$
$$\mathbf{u}(0) = \mathbf{u}_0.$$

We comment on some properties of this weak formulation:

- For $\mathbf{g} \in L^2(0,T;\mathbf{V}_{\text{div}})$, $\mathbf{u}_0 \in \mathbf{H}_{\text{div}}$, there *exists* a solution $\mathbf{u}$ of the weak formulation (2.36). This solution has sufficient regularity such that the initial condition $\mathbf{u}(0) = \mathbf{u}_0$ is well-defined.
- *Uniqueness* of this solution is an open problem. Uniqueness of $\mathbf{u}$ can be shown to hold in special cases, for example:
  - If the boundary $\partial\Omega$ is sufficiently smooth, $\mathbf{u}_0 \in \mathbf{V}_{\text{div}}$, $\mathbf{g} \in L^\infty(0,T;\mathbf{H}_{\text{div}})$ then $\mathbf{u}(t)$ is unique on a time interval $[0,T^*]$ with T^* *sufficiently small.*
  - If the Reynolds number *Re* is sufficiently small, $\mathbf{u}_0$ and $\mathbf{g}$ are sufficiently smooth and these data are sufficiently small (in appropriate norms) then $\mathbf{u}(t)$ is unique for all $t \in [0,T]$.
  - *If* the weak solution $\mathbf{u}$ is sufficiently smooth ($\mathbf{u} \in L^\infty(0,T;\mathbf{H}_{\text{div}}) \cap L^8(0,T;L^4(\Omega))$) then $\mathbf{u}$ is unique. It is not known, however, whether in general this smoothness property holds.

There is an extensive literature on this topic of uniqueness of a weak solution in the three-dimensional case (i.e. $\Omega \subset \mathbb{R}^3$). Many other special cases are known in the literature. The general case, however, is still unsolved.

- If one considers the weak formulation (2.36) in the *two*-dimensional case (i.e. $\Omega \subset \mathbb{R}^2$) then *existence and uniqueness* have been proved.

Along the same lines as in Lemma 2.2.9 one can show that if the strong formulation (2.35) has a solution $(\mathbf{u}, p)$ that is sufficiently smooth ($\mathbf{u} \in C^1([0,T]; C^2(\overline{\Omega})^3)$) then $\mathbf{u}$ is also a solution of (2.36).

In the weak formulation (2.36) the incompressibility condition is fulfilled since it holds for all functions in the space $W_*^1(0,T;\mathbf{V}_{\mathrm{div}})$. As for the Stokes problem, a weak formulation in the larger velocity space $W_*^1(0,T;\mathbf{V})$ can be derived in which the incompressibility condition is enforced by using a pressure variable. This weak formulation is as follows:

Determine $\mathbf{u} \in W_*^1(0,T;\mathbf{V})$, $p \in L^2(0,T;Q)$ such that for all $\mathbf{v} \in \mathbf{V}$, $q \in Q$:

$$\begin{aligned} \frac{d}{dt}(\mathbf{u}(t),\mathbf{v})_{L^2} + a(\mathbf{u}(t),\mathbf{v}) + c(\mathbf{u}(t);\mathbf{u}(t),\mathbf{v}) + b(\mathbf{v},p(t)) &= (\mathbf{g},\mathbf{v})_{L^2}, \\ b(\mathbf{u}(t),q) &= 0, \\ \mathbf{u}(0) &= \mathbf{u}_0. \end{aligned} \tag{2.37}$$

As in the proof of Theorem 2.2.11 one can show that if $(\mathbf{u}, p)$ is a solution of (2.37) then $\mathbf{u}$ is a solution of (2.36). It can be shown that if the solution $\mathbf{u}$ of (2.37) is assumed to be sufficiently smooth then a pressure $p \in L^2(0,T;Q)$ exists such that the pair $(\mathbf{u}, p)$ is a solution of (2.37). For a treatment of this topic and a discussion of other similar weak formulations of the Navier-Stokes equations we refer to the literature, e.g. [235, 165].

3

Finite element discretization

In this chapter we treat finite element methods for the discretization of the variational Oseen problem (2.21) and for the *spatial* discretization of the variational formulation of the non-stationary Stokes- and Navier-Stokes equations. We restrict ourselves to the class of Hood-Taylor finite elements on tetrahedral grids. In order to perform local grid refinement/coarsening in an efficient way, which is very important in two-phase flow applications, and to be able to use fast multigrid iterative solution methods we will apply such finite element methods not on one grid but on a *hierarchy of nested triangulations*. The construction of such a multilevel grid hierarchy is discussed in Sect. 3.1. In Sect. 3.2 the Hood-Taylor finite element spaces are treated. We present a numerical example in Sect. 3.3, where the approximation order of such a Hood-Taylor finite element space is investigated.

3.1 Multilevel tetrahedral grid hierarchy

3.1.1 Multilevel triangulation

We first introduce some notions.

Definition 3.1.1 (Triangulation) A finite collection $\mathcal{T}$ of tetrahedra $T \subset \overline{\Omega}$ is called a *triangulation* of Ω (or $\overline{\Omega}$) if the following holds:

1. $\bigcup_{T \in \mathcal{T}} T = \overline{\Omega}$,
2. $\operatorname{int}(S) \cap \operatorname{int}(T) = \emptyset$ for all $S, T \in \mathcal{T}$ with $S \neq T$.

Here $\operatorname{int}(U)$ denotes the interior of the set $U \subset \overline{\Omega}$.

Definition 3.1.2 (Consistency) A triangulation $\mathcal{T}$ is called *consistent* if the intersection of any two tetrahedra in $\mathcal{T}$ is either empty, a common face, a common edge or a common vertex.

Definition 3.1.3 (Stability) A sequence of triangulations $(\mathcal{T}_0, \mathcal{T}_1, \mathcal{T}_2, \ldots)$ is called *stable* if all angles of all tetrahedra in this sequence are uniformly bounded away from zero.

This notion of stability is important in view of the approximation quality of finite element spaces which are based on these triangulations. It is known that a deterioration of the approximation quality may occur if the underlying triangulations contain (many) tetrahedra with very small angles. This undesirable effect is avoided if the triangulations are stable.

Definition 3.1.4 (Refinement) For a given tetrahedron T a triangulation $\mathcal{K}(T)$ of T is called a *refinement* of T if $|\mathcal{K}(T)| \geq 2$ and any vertex of any tetrahedron $T' \in \mathcal{K}(T)$ is either a vertex or an edge midpoint of T. In this case T' is called a *child* of T and T is called the *parent* of T'.

A refinement $\mathcal{K}(T)$ of T is called *regular* if $|\mathcal{K}(T)| = 8$, otherwise it is called *irregular.*

A triangulation $\mathcal{T}_{k+1}$ is called *refinement* of a triangulation $\mathcal{T}_k \neq \mathcal{T}_{k+1}$ if for every $T \in \mathcal{T}_k$ either $T \in \mathcal{T}_{k+1}$ or $\mathcal{K}(T) \subset \mathcal{T}_{k+1}$ for some refinement $\mathcal{K}(T)$ of T.

Definition 3.1.5 (Multilevel triangulation) A sequence of consistent triangulations $\mathcal{M} = (\mathcal{T}_0, \ldots, \mathcal{T}_J)$ is called a *multilevel triangulation* of Ω if the following holds:

1. For $0 \leq k < J$: $\mathcal{T}_{k+1}$ is a refinement of $\mathcal{T}_k$.
2. For $0 \leq k < J$: if $T \in \mathcal{T}_k \cap \mathcal{T}_{k+1}$, then $T \in \mathcal{T}_J$.

The tetrahedra $T \in \mathcal{T}_J$ are called the *leaves* of $\mathcal{M}$. Note that T is a leaf iff T has no children in $\mathcal{M}$.

A tetrahedron $T \in \mathcal{M}$ is called *regular* if $T \in \mathcal{T}_0$ or T resulted from a regular refinement of its parent. Otherwise T is called *irregular.*

A multilevel triangulation $\mathcal{M}$ is called *regular* if all irregular $T \in \mathcal{M}$ are leaves (i. e., have no children in $\mathcal{M}$).

$\mathcal{T}_0$ is called the *coarsest* or *initial triangulation*, $\mathcal{T}_J$ is called the *finest triangulation.*

Remark 3.1.6 Let $\mathcal{M}$ be a multilevel triangulation and V_k $(0 \leq k \leq J)$ be the corresponding finite element spaces of continuous functions $p \in C(\overline{\Omega})$ such that $p_{|T} \in \mathcal{P}_q$ for all $T \in \mathcal{T}_k$ $(q \geq 1)$. The refinement property 1 in Definition 3.1.5 implies *nestedness* of these finite element spaces: $V_k \subset V_{k+1}$.

Definition 3.1.7 (Hierarchical decomposition of $\mathcal{M}$) Consider a multilevel triangulation $\mathcal{M} = (\mathcal{T}_0, \ldots, \mathcal{T}_J)$ of Ω. For every tetrahedron $T \in \mathcal{M}$ a unique level number $\ell(T)$ is defined by

$$\ell(T) := \min \{ k : T \in \mathcal{T}_k \}.$$

The set $\mathcal{G}_k \subset \mathcal{T}_k$,

$$\mathcal{G}_k := \{ T \in \mathcal{T}_k : \ell(T) = k \}$$

is called the *hierarchical surplus* on level k, $k = 0, 1, \dots, J$. Note that

$$\mathcal{G}_0 = \mathcal{T}_0, \qquad \mathcal{G}_k = \mathcal{T}_k \setminus \mathcal{T}_{k-1} \quad \text{for } k = 1, \dots, J.$$

The sequence $\mathcal{H} = (\mathcal{G}_0, \dots, \mathcal{G}_J)$ is called the *hierarchical decomposition* of $\mathcal{M}$. Note that the multilevel triangulation $\mathcal{M}$ can be uniquely reconstructed from its hierarchical decomposition due to refinement property 2 in Definition 3.1.5.

Remark 3.1.8 The hierarchical decomposition induces simple data structures in a canonical way. The tetrahedra of each hierarchical surplus $\mathcal{G}_k$ are stored in a separate list. Thus every tetrahedron $T \in \mathcal{M}$ is stored exactly once since T has a unique level number $\ell(T)$. By introducing unique level numbers also for vertices, edges and faces, these sub-simplices can be stored in the same manner: For a sub-simplex S the level number $\ell(S)$ is defined as the level of its first appearance. In the implementation some of these objects are linked to others by pointers, for example, a tetrahedron is linked to its vertices, edges, faces, children and parent.

3.1.2 A multilevel refinement method

In this section we describe a *refinement and coarsening* algorithm which is essentially the method presented in [37, 38]. This method is based on similar ideas as the refinement algorithms in [21, 26]. We restrict ourselves to tetrahedral meshes. However, the method can easily be modified such that it is applicable to other element types such as, for example, hexahedra and pyramids. Although this method is usually called a *refinement* method, it is also applicable for *coarsening* triangulations. The *hierarchical* structure of $\mathcal{M}$ is essential for the coarsening strategy. In general, if only one (i.e. a one-level) triangulation is given it is relatively easy to apply a local refinement method to it, but a local coarsening algorithm is much harder to develop. The hierarchical structure, however, makes it possible to treat local coarsening in a similar manner as local refinement.

The refinement strategy is based on a set of regular and irregular refinement rules (also called *red* and *green* rules, due to [21]), which are described in the following two sections. The regular and irregular rules are local in the sense that they are applied to a single tetrahedron. These rules are applied in a (global) refinement algorithm that describes how the local rules can be combined to ensure consistency and stability, cf. Definitions 3.1.2 and 3.1.3.

The regular refinement rule

Let T be a given tetrahedron. For the construction of a regular refinement of T it is natural to connect midpoints of the edges of T by subdividing each of the faces into four congruent triangles. This yields four sub-tetrahedra at the corners of T (all similar to T) and an octahedron in the center. This octahedron is further subdivided into four sub-tetrahedra with equal volume (cf. Fig. 3.1). For this subdivision there are three different possibilities (each corresponding to one of three possible interior diagonals). This octahedron subdivision has to be chosen carefully in order to satisfy the stability condition. In [262] it is shown that always (i.e. in a repeated refinement procedure) selecting the longest diagonal will in general lead to non-stable triangulations. A *stable* tetrahedral regular refinement strategy, based on an idea from [112], is presented in [37, 39]. We recall this method.

Fig. 3.1. Regular refinement.

Let $T = [x^{(1)}, x^{(2)}, x^{(3)}, x^{(4)}]$ be a tetrahedron with *ordered* vertices $x^{(1)}$, $x^{(2)}$, $x^{(3)}$, $x^{(4)}$ and

$$x^{(ij)} := \frac{1}{2}(x^{(i)} + x^{(j)}), \qquad 1 \leq i < j \leq 4,$$

the midpoint of the edge between $x^{(i)}$ and $x^{(j)}$. The regular refinement $\mathcal{K}(T) := \{T_1, \ldots, T_8\}$ of T is constructed by the (red) rule

$$\begin{aligned}
T_1 &:= [x^{(1)}, x^{(12)}, x^{(13)}, x^{(14)}], & T_5 &:= [x^{(12)}, x^{(13)}, x^{(14)}, x^{(24)}],\\
T_2 &:= [x^{(12)}, x^{(2)}, x^{(23)}, x^{(24)}], & T_6 &:= [x^{(12)}, x^{(13)}, x^{(23)}, x^{(24)}],\\
T_3 &:= [x^{(13)}, x^{(23)}, x^{(3)}, x^{(34)}], & T_7 &:= [x^{(13)}, x^{(14)}, x^{(24)}, x^{(34)}],\\
T_4 &:= [x^{(14)}, x^{(24)}, x^{(34)}, x^{(4)}], & T_8 &:= [x^{(13)}, x^{(23)}, x^{(24)}, x^{(34)}].
\end{aligned} \tag{3.1}$$

$T_1, \ldots, T_4$ are the sub-tetrahedra at the corners of T, and $T_5, \ldots, T_8$ form the octahedron in the middle of T. In [39] it is shown that for any T the repeated application of this rule produces a sequence of consistent triangulations of T which is stable. For a given T all tetrahedra that are generated in such a recursive refinement process are elements from at most three different similarity classes. This guarantees stability of the resulting triangulations.

Irregular refinement rules

Let $\mathcal{T}$ be a given consistent triangulation. We select a subset $\mathcal{S}$ of tetrahedra from $\mathcal{T}$ and assume that the regular refinement rule is applied to each of the tetrahedra from $\mathcal{S}$. In general the resulting triangulation $\mathcal{T}'$ will not be consistent. The irregular (or green) rules are used to make this new triangulation consistent. For this we introduce the notion of an edge refinement pattern. Let $E_1, \ldots, E_6$ be the ordered edges of $T \in \mathcal{T}$. We define the 6-tuple

$$R(T) = (r_1, \ldots, r_6) \in \{0,1\}^6$$

by:

- $r_i = 1$ if E_i is an edge of a tetrahedron $S \in \mathcal{S}$ (i. e., edge E_i is refined and has two sub-edges in $\mathcal{T}'$) and
- $r_i = 0$ otherwise (i. e., edge E_i is not refined).

For $T \in \mathcal{S}$ we have $R(T) = (1, \ldots, 1)$. For $T \in \mathcal{T} \backslash \mathcal{S}$ the case $R(T) = (0, \ldots, 0)$ corresponds to the situation that the tetrahedron T does not contain any vertices from $\mathcal{T}'$ at the midpoints of its edges. For each of the $2^6 - 1$ possible patterns $R \neq (0, \ldots, 0)$ there exists a corresponding refinement $\mathcal{K}(T)$ of T (i.e. a rule of the form as in (3.1)) for which the vertices of the children coincide with vertices of T or with the vertices at the midpoints on the edges E_i with $r_i = 1$. This refinement, however, is not always unique. This is illustrated in Fig. 3.2.

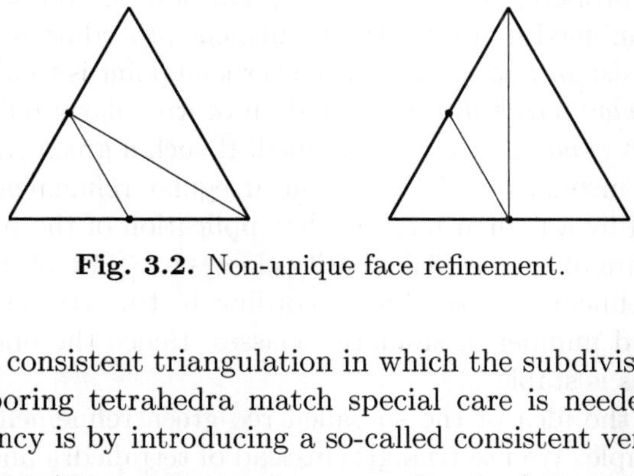

Fig. 3.2. Non-unique face refinement.

To obtain a consistent triangulation in which the subdivisions of adjacent faces of neighboring tetrahedra match special care is needed. One way to ensure consistency is by introducing a so-called consistent vertex numbering:

Definition 3.1.9 (Consistent vertex numbering) Let T_1 and T_2 be two adjacent tetrahedra with a common face $F = T_1 \cap T_2$ and local vertex ordering

$$T_l = [x_l^{(1)}, x_l^{(2)}, x_l^{(3)}, x_l^{(4)}], \qquad l = 1, 2.$$

The pair (T_1, T_2) has a *consistent vertex numbering*, if the ordering of the vertices of F induced by the vertex ordering of T_1 coincides with the one induced by the vertex ordering of T_2. A consistent triangulation $\mathcal{T}$ has a *consistent vertex numbering* if every two neighboring tetrahedra have this property.

Remark 3.1.10 A consistent vertex numbering can be constructed in a simple way. Consider an (initial) triangulation $\tilde{\mathcal{T}}$ with an arbitrary numbering of its vertices. This global numbering induces a canonical local vertex ordering which is a consistent vertex numbering of $\tilde{\mathcal{T}}$. Furthermore, each refinement rule can be defined such that the consistent vertex numbering property of the parent is inherited by its children by prescribing suitable local vertex orderings of the children. (3.1) is an example of such a rule. Using such a strategy a consistent triangulation $\tilde{\mathcal{T}}'$ that is obtained by refinement of $\tilde{\mathcal{T}}$ according to these rules also has a consistent vertex numbering.

Assume that the given triangulation $\mathcal{T}$ has a consistent vertex numbering. For a face with a pattern as in Fig. 3.2 one can then define a *unique* face refinement by connecting the vertex with the smallest number with the midpoint of the opposite edge. *For each edge refinement pattern* $R \in \{0,1\}^6$ *we then have a unique rule.* We emphasize that if for a given tetrahedron T the edge refinement pattern $R(T)$ is known, then *for the application of the regular or irregular rules to this tetrahedron no information from neighboring tetrahedra is needed.* Clearly, for parallelization this is a very nice property.

Multilevel refinement algorithm

Above we discussed how *consistency* of a triangulation can be achieved by the choice of suitable irregular refinement rules based on the consistent vertex numbering property. We will now explain how the regular and irregular rules can be combined in a repeated refinement procedure to obtain a *stable sequence of consistent triangulations.* The crucial point is to *allow only the refinement of regular tetrahedra*, i. e., children of irregularly refined tetrahedra, also called *green children*, are never refined. If such a green child T is marked for refinement, instead of refining T the irregular refinement of the parent will be replaced by a regular one. As the application of the regular rule (3.1) creates tetrahedra of at most 3 similarity classes (cf. [112, 39]), the tetrahedra created by a refinement procedure according to this strategy belong to an a-priori bounded number of similarity classes. Hence the obtained sequence of triangulations is stable.

We explain the idea of the so called red-green refinement strategy by a simple 2D example. We use triangles instead of tetrahedra and first illustrate the action of a *one-level* refinement method. Consider the triangulation $\mathcal{T}_1$ as depicted in Fig. 3.3.

In $\mathcal{T}_1$ two triangles are marked (by shading) for refinement. A one-level refinement algorithm (like the one described in [21]) only uses the finest triangulation $\mathcal{T}_1$ as input. It first applies the regular refinement rule (the so called "red refinement") to marked regular triangles and to the parents of green children, which are either marked or neighbors of marked triangles — green children are never refined because of stability reasons. This red refinement yields an inconsistent triangulation (cf. Fig. 3.4 in the middle). Thus in the

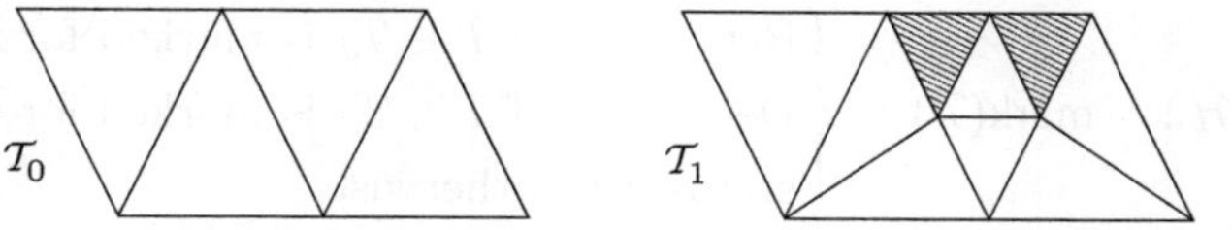

Fig. 3.3. Initial multilevel triangulation with some leaf tetrahedra marked for refinement (indicated by shading).

next step appropriate irregular refinement rules are applied to avoid hanging nodes ("green closure"). The output of the one-level refinement algorithm is the new triangulation $\mathcal{T}_2$ (cf. Fig. 3.4 on the right).

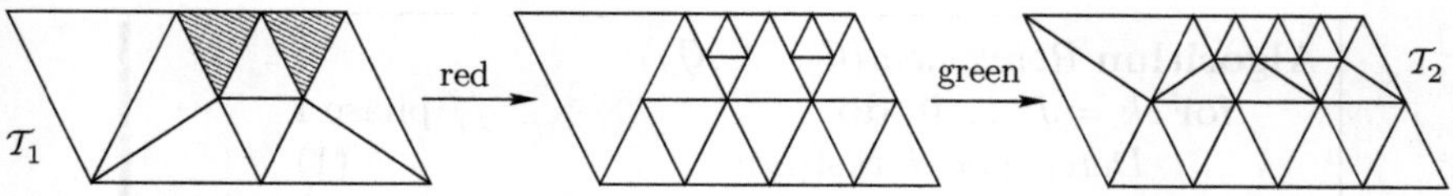

Fig. 3.4. One-level red/green refinement.

The new triangulation $\mathcal{T}_2$ is consistent, but *not* a refinement of $\mathcal{T}_1$ in the sense of Definition 3.1.4. Due to this, the corresponding FE spaces are not nested. Another disadvantage of this one-level approach is the fact that it is not obvious how to treat (local) coarsening of the triangulation.

We now turn to a *multilevel* refinement strategy. In such methods both input and output are multilevel triangulations (cf. Definition 3.1.5). In general such an algorithm not only affects the finest triangulation like in the case of the one-level method, but the whole multilevel triangulation. This is illustrated in the example below.

For the description of the multilevel refinement algorithm we introduce the notions of status and mark of a tetrahedron. Let $\mathcal{M} = (\mathcal{T}_0, \ldots, \mathcal{T}_J)$ be a multilevel triangulation that has been constructed by applying the regular and irregular refinement rules and let $\mathcal{H} = (\mathcal{G}_0, \ldots, \mathcal{G}_J)$ be the corresponding hierarchical decomposition. Every tetrahedron $T \in \mathcal{H}$ is either a leaf of $\mathcal{M}$ (i.e., $T \in \mathcal{T}_J$) or it has been refined. The label status is used to describe this property of T:

$$\text{For} \quad T \in \mathcal{H}: \quad \mathsf{status}(T) = \begin{cases} \mathsf{NoRef} & \text{if } T \text{ is a leaf of } \mathcal{M}, \\ \mathsf{RegRef} & \text{if } T \text{ is regularly refined in } \mathcal{M}, \\ \mathsf{IrregRef} & \text{if } T \text{ is irregularly refined in } \mathcal{M}. \end{cases}$$

The label IrregRef also contains the number of the irregular refinement rule (one out of 63) that has been used to refine T, i.e., the binary representation of status(T) coincides with the edge refinement pattern $R(T)$ of T.

In adaptive refinement an error estimator (or indicator) is used to mark certain elements of $\mathcal{T}_J$ for further refinement or for deletion. For this the label mark is used:

$$\text{For} \quad T \in \mathcal{H}: \quad \mathsf{mark}(T) = \begin{cases} \mathsf{Ref} & \text{if } T \in \mathcal{T}_J \text{ is marked for refinement,} \\ \mathsf{Del} & \text{if } T \in \mathcal{T}_J \text{ is marked for deletion,} \\ \mathsf{status}(T) & \text{otherwise.} \end{cases}$$

Hence, $\mathsf{mark}(T) \in \{\mathsf{Ref}, \mathsf{Del}, \mathsf{NoRef}, \mathsf{RegRef}, \mathsf{IrregRef}\}$. We describe a multilevel refinement algorithm known in the literature. The basic form of this method was introduced by BASTIAN [25] and developed further in the UG-group [26, 27, 245]. We use the presentation as in [37, 38], which is shown in Algorithm 3.1.11.

Algorithm 3.1.11 (Multilevel refinement)

```
Algorithm Refinement(G_0, ..., G_J)
    for k = J, ..., 0 do                                  // phase I
        DetermineMarks(G_k);                                  (1)
        MarksForClosure(G_k);                                 (2)
    for k = 0, ..., J do if G_k ≠ ∅ then                 // phase II
        if k > 0 then MarksForClosure(G_k);                   (3)
        if k < J then Unrefine(G_k);                          (4)
        Refine(G_k);                                          (5)
    if G_J = ∅ then J := J − 1;                               (6)
    else if G_{J+1} ≠ ∅ then J := J + 1;                      (7)
```

The input of the algorithm *Refinement* consists of a hierarchical decomposition

$$\mathcal{H} = (\mathcal{G}_0, \ldots, \mathcal{G}_J)$$

in which all refined tetrahedra T are labeled by $\mathsf{mark}(T) = \mathsf{status}(T)$ according to their status and the unrefined $T \in \mathcal{T}_J$ (i.e. the leaves) have $\mathsf{mark}(T) \in \{\mathsf{NoRef}, \mathsf{Ref}, \mathsf{Del}\}$. The output is again a hierarchical decomposition, where *all tetrahedra are marked according to their status.*

The main idea underlying the algorithm *Refinement* is illustrated using the multilevel triangulation $\mathcal{M} = (\mathcal{T}_0, \mathcal{T}_1)$ shown in Fig. 3.3. The hierarchical decomposition $\mathcal{H}$ and the corresponding marks are shown in Fig. 3.5.

Note that for the two shaded triangles in $\mathcal{G}_1$ we have $\mathsf{status}(T) \neq \mathsf{mark}(T)$. For all other triangles $\mathsf{status}(T) = \mathsf{mark}(T)$ holds. In phase I of the algorithm (top–down: (1),(2)) only marks are changed. In *DetermineMarks* some tetrahedra are labeled with new marks, which are of the type RegRef (for red refinement) or NoRef (for coarsening). The green closure marks are set in *MarksForClosure*, where appropriate irregular refinement marks are determined from the edge refinement patterns to avoid hanging nodes.

Once phase I has been completed the marks have been changed such that $\mathsf{mark}(T) \in \{\mathsf{NoRef}, \mathsf{RegRef}, \mathsf{IrregRef}\}$ holds for all $T \in \mathcal{H}$, cf. Fig. 3.6. All green children in $\tilde{\mathcal{G}}_1$ have $\mathsf{mark}(T) = \mathsf{NoRef}$, as they are not refined because of

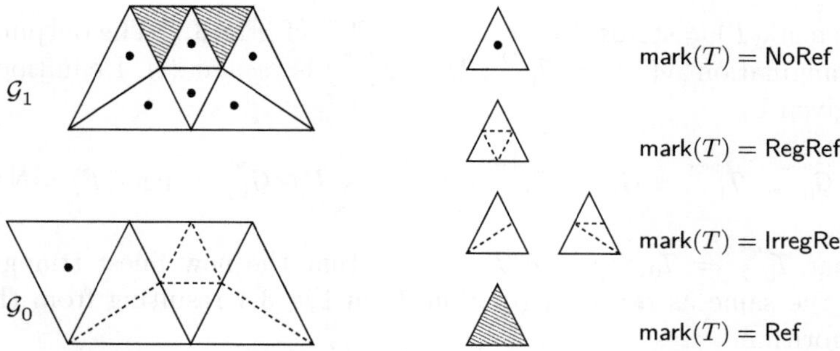

Fig. 3.5. Input hierarchical decomposition.

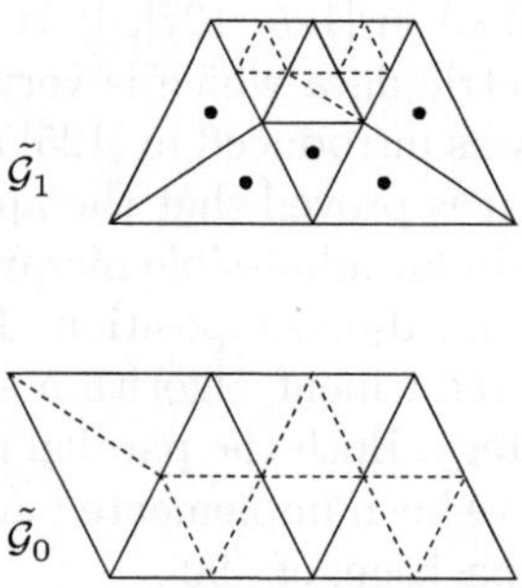

Fig. 3.6. Phase I: new marks, resulting in $\tilde{\mathcal{G}}_1$ and $\tilde{\mathcal{G}}_0$.

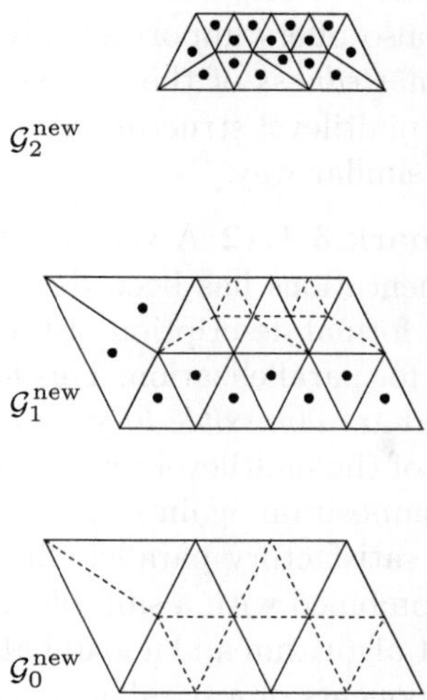

Fig. 3.7. Phase II: Output hierarchical decomposition $(\mathcal{G}_0^{\text{new}}, \mathcal{G}_1^{\text{new}}, \mathcal{G}_2^{\text{new}})$.

stability reasons. Instead the corresponding irregular refined parents in $\tilde{\mathcal{G}}_0$ are labeled by mark(T) = RegRef.

In the second phase (bottom–up: (3)-(5)) the actual refinement (coarsening is not needed in our example) is constructed: A call of *Unrefine*($\mathcal{G}_k$) deletes all tetrahedra, faces, edges and vertices on level $k+1$, which are not needed anymore due to changed marks. In the subroutine *Refine*($\mathcal{G}_k$) all $T \in \mathcal{G}_k$ with mark(T) $\neq$ status(T) are refined according to mark(T) and new objects (tetrahedra, faces, edges, vertices) on level $k+1$ are created. A subsequent call to *MarksForClosure* in (3) computes the appropriate refinement marks for the new created tetrahedra in the next sweep of the **for**-loop.

In the output hierarchical decomposition

$$\mathcal{H}^{\text{new}} = (\mathcal{G}_0^{\text{new}}, \mathcal{G}_1^{\text{new}}, \mathcal{G}_2^{\text{new}})$$

we have $\mathsf{mark}(T) = \mathsf{status}(T)$ for all $T \in \mathcal{H}^{\text{new}}$, cf. Fig. 3.7. The output multilevel triangulation $\mathcal{M}^{\text{new}} = (\mathcal{T}_0^{\text{new}}, \mathcal{T}_1^{\text{new}}, \mathcal{T}_2^{\text{new}})$ is *regular* (cf. Definition 3.1.5) and is given by

$$\mathcal{T}_0^{\text{new}} = \mathcal{G}_0^{\text{new}}, \quad \mathcal{T}_1^{\text{new}} = \mathcal{G}_1^{\text{new}}, \quad \mathcal{T}_2^{\text{new}} = \mathcal{G}_2^{\text{new}} \cup \left\{ T \in \mathcal{G}_1^{\text{new}} \ : \ \mathsf{mark}(T) = \mathsf{NoRef} \right\}.$$

Note that $\mathcal{T}_0^{\text{new}} = \mathcal{T}_0$, $\mathcal{T}_1^{\text{new}} \neq \mathcal{T}_1$ (!) and that the new finest triangulation $\mathcal{T}_2^{\text{new}}$ is the same as the triangulation $\mathcal{T}_2$ in Fig. 3.4 resulting from the one-level algorithm.

A more detailed discussion of the subroutines in algorithm *Refinement* is given in [37, 38, 125].

The multilevel method is more complicated than the one-level algorithm, but also offers important advantages: Property 1 of Definition 3.1.5 assures the *nestedness* of the corresponding finite element spaces, cf. Remark 3.1.6. The multilevel structure also allows to treat local refinement and *coarsening* in a similar way.

Remark 3.1.12 A version of the algorithm that is suitable for parallel implementations has been developed and is described in [125, 127]. It is based on a formal description of the distributed geometric data which is very suitable for parallelization. This formal description was introduced in [125] and is called an *admissible hierarchical decomposition*. It is proved that the application of the multilevel refinement algorithm to an input admissible hierarchical decomposition again yields an admissible hierarchical decomposition. To obtain satisfactory parallel efficiency this parallel refinement algorithm should be combined with a suitable load balancing strategy. Both the parallel refinement algorithm and a load balancing strategy have been implemented and are components of a parallel version of the DROPS package, cf. [90].

3.2 Hood-Taylor finite element spaces

In this section we treat the main topics related to the popular class of so-called Hood-Taylor finite elements for the (spatial) discretization of one phase incompressible flow problems. We introduce these spaces for the d-dimensional case, $d \leq 3$. In our applications we are particularly interested in $d = 3$.

3.2.1 Simplicial finite element spaces

Let Ω be a domain in $\mathbb{R}^d$ and $\mathcal{T}_h = \{T\}$ a subdivision (or triangulation) of Ω in a finite number of simplices T. This triangulation is called *consistent* if the following holds:

1. $\cup_{T \in \mathcal{T}_h} T = \overline{\Omega}$,
2. $\operatorname{int} T_1 \cap \operatorname{int} T_2 = \emptyset \quad$ for all $\; T_1, T_2 \in \mathcal{T}_h, \; T_1 \neq T_2$,

3. any $(d-1)$–dimensional subsimplex of any $T_1 \in \mathcal{T}_h$ is either a subset of $\partial\Omega$ or a subsimplex of another $T_2 \in \mathcal{T}_h$.

This definition generalizes the one given in Definition 3.1.2 for the case $d = 3$.

Let $\{\mathcal{T}_h\}$ be a family of triangulations and $h_T := \operatorname{diam}(T)$ for $T \in \mathcal{T}_h$. The index parameter h of $\mathcal{T}_h$ is taken such that

$$h = \max\{\, h_T : T \in \mathcal{T}_h \,\}.$$

Furthermore, for $T \in \mathcal{T}_h$ we define

$$\rho_T := \sup\{\, \operatorname{diam}(B) : B \text{ is a ball contained in } T \,\}.$$

A family of consistent triangulations $\{\mathcal{T}_h\}$ is called *regular* if

1. The parameter h approaches zero: $\inf\{\, h : \mathcal{T}_h \in \{\mathcal{T}_h\} \,\} = 0$,
2. $\exists\, \sigma$: $\frac{h_T}{\rho_T} \le \sigma$ for all $T \in \mathcal{T}_h$ and all $\mathcal{T}_h \in \{\mathcal{T}_h\}$.

For $d = 3$ a sequence of consistent tetrahedral triangulations satisfies condition 2 iff the sequence is stable (Definition 3.1.3).

The space of polynomials in $\mathbb{R}^d$ of degree less than or equal $k \ge 0$ is denoted by $\mathcal{P}_k$, i.e., $p \in \mathcal{P}_k$ is of the form

$$p(x) = \sum_{|\alpha| \le k} \gamma_\alpha x_1^{\alpha_1} x_2^{\alpha_2} \dots x_d^{\alpha_d}, \qquad \gamma_\alpha \in \mathbb{R}.$$

The dimension of $\mathcal{P}_k$ is

$$\dim \mathcal{P}_k = \binom{d+k}{k}. \tag{3.2}$$

The spaces of *simplicial finite elements* are given by

$$\mathbb{X}_h^0 := \{\, v \in L^2(\Omega) : v_{|T} \in \mathcal{P}_0 \ \text{ for all } \ T \in \mathcal{T}_h \,\}, \tag{3.3a}$$

$$\mathbb{X}_h^k := \{\, v \in C(\overline{\Omega}) : v_{|T} \in \mathcal{P}_k \ \text{ for all } \ T \in \mathcal{T}_h \,\}, \quad k \ge 1. \tag{3.3b}$$

These spaces consist of *piecewise polynomials* which, for $k \ge 1$, are continuous on Ω.

Remark 3.2.1 One can show that $\mathbb{X}_h^k \subset H^1(\Omega)$ holds for all $k \ge 1$.

We will also need simplicial finite element spaces with functions that are zero on $\partial\Omega$:

$$\mathbb{X}_{h,0}^k := \mathbb{X}_h^k \cap H_0^1(\Omega), \quad k \ge 1. \tag{3.4}$$

The values of $v_{|T} \in \mathcal{P}_k$ can be determined by using suitable interpolation points in the simplex T. For this it is convenient to use barycentric coordinates:

Definition 3.2.2 Let T be a non-degenerate d-simplex and $a_j \in \mathbb{R}^d$, $j = 1, \dots, d+1$ its vertices. Then T can be described by

$$T = \left\{ \sum_{j=1}^{d+1} \lambda_j a_j \, : \, 0 \leq \lambda_j \leq 1 \ \ \forall \, j, \ \sum_{j=1}^{d+1} \lambda_j = 1 \right\}. \tag{3.5}$$

To every $x \in T$ there corresponds a unique $(n+1)$-tuple $(\lambda_1, \dots, \lambda_{n+1})$ as in (3.5). These λ_j, $1 \leq j \leq d+1$, are called the *barycentric coordinates* of $x \in T$. The mapping $x \to (\lambda_1, \dots, \lambda_{d+1})$ is affine.

Using these barycentric coordinates we define the set

$$L_k(T) := \left\{ \sum_{j=1}^{d+1} \lambda_j a_j \, : \, \lambda_j \in \{0, \frac{1}{k}, \dots, \frac{k-1}{k}, 1\} \ \ \forall \, j, \ \sum_{j=1}^{d+1} \lambda_j = 1 \right\},$$

which is called the principal lattice of order k (in T). Examples for $d = 3$ and $k = 1, 2$ are given in Fig. 3.8.

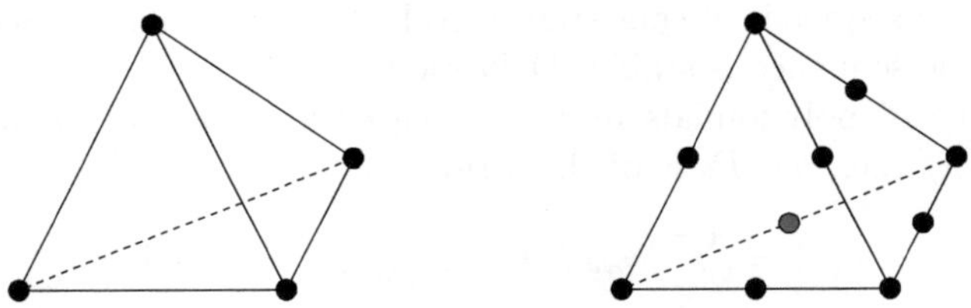

Fig. 3.8. Principal lattice, $d = 3$, $k = 1$ (left) and $k = 2$ (right).

This principal lattice consists of $\binom{d+k}{k}$ points (cf. (3.2)) and these can be used to determine a unique polynomial $p \in \mathcal{P}_k$:

Lemma 3.2.3 *Let T be a non-degenerated d-simplex. Then any polynomial $p \in \mathcal{P}_k$ is uniquely determined by its values on the principal lattice $L_k(T)$.*

Using this lattice, for $u \in C(\overline{\Omega})$ we define a corresponding function $I_{\mathbb{X}}^k u \in L^2(\Omega)$ by piecewise polynomial interpolation on each simplex $T \in \mathcal{T}_h$:

$$\forall \, T \in \mathcal{T}_h : \ (I_{\mathbb{X}}^k u)_{|T} \in \mathcal{P}_k \ \text{ such that } \ (I_{\mathbb{X}}^k u)(x_j) = u(x_j) \ \ \forall \, x_j \in L_k(T). \tag{3.6}$$

The piecewise polynomial function $I_{\mathbb{X}}^k u$ is continuous on Ω:

Lemma 3.2.4 *For $k \geq 1$ and $u \in C(\overline{\Omega})$ we have $I_{\mathbb{X}}^k u \in \mathbb{X}_h^k$.*

Proof. We consider the case $d = 3$, $k = 2$, cf. Fig. 3.8 (right). By definition we have $I_{\mathbb{X}}^2 u \in \mathcal{P}_2$, thus we only have to show that $I_{\mathbb{X}}^2 u$ is continuous across triangular faces between adjacent tetrahedra T_1, T_2. Define $p_i := (I_{\mathbb{X}}^2 u)_{|T_i}$, $i = 1, 2$. At the six interpolation points x_j, $j = 1, \dots, 6$, on the face $T_1 \cap T_2$ we have $p_1(x_j) = p_2(x_j) = u(x_j)$. The functions $(p_i)_{|T_1 \cap T_2}$ are two-dimensional polynomials of degree 2, which are uniquely determined by the six values

$p_i(x_j)$. We conclude that $p_1 = p_2$ on $T_1 \cap T_2$ and thus $I_{\mathbb{X}}^2 u$ is continuous across $T_1 \cap T_2$. Similar arguments can be applied to prove the result for the general case. □

Using the embedding result $H^m(\Omega) \hookrightarrow C(\overline{\Omega})$ for $m > \frac{d}{2}$, cf. (2.15), we obtain the following corollary.

Corollary 3.2.5 For $k \geq 1, m \geq 2$ we have:

$$I_{\mathbb{X}}^k u \in \mathbb{X}_h^k \quad \text{for all} \quad u \in H^m(\Omega),$$
$$I_{\mathbb{X}}^k u \in \mathbb{X}_{h,0}^k \quad \text{for all} \quad u \in H^m(\Omega) \cap H_0^1(\Omega).$$

Using interpolation error bounds one can derive the following main result.

Theorem 3.2.6 *Let* $\{\mathcal{T}_h\}$ *be a regular family of triangulations of* Ω *consisting of d-simplices and let* $\mathbb{X}_h^k$ *be the corresponding finite element space as in* (3.3b). *For* $2 \leq m \leq k+1$ *and* $t \in \{0,1\}$ *the following holds:*

$$\|u - I_{\mathbb{X}}^k u\|_t \leq C h^{m-t} |u|_m \quad \textit{for all} \quad u \in H^m(\Omega). \tag{3.7}$$

Recall that $\|\cdot\|_t$ and $|\cdot|_m$ are the norm on $H^t(\Omega)$ and the semi-norm on $H^m(\Omega)$, respectively, and that $\|\cdot\|_0 = \|\cdot\|_{L^2}$. A proof of this result can be found in many textbooks on finite element methods, e.g. [70, 48, 53, 106]. As a direct consequence of this interpolation error bound we obtain the following approximation error bound.

Theorem 3.2.7 *Let* $\{\mathcal{T}_h\}$ *be a regular family of triangulations of* Ω *consisting of d-simplices and let* $\mathbb{X}_h^k$, $\mathbb{X}_{h,0}^k$ *be the corresponding finite element space as in* (3.3b), (3.4). *For* $2 \leq m \leq k+1$ *and* $t \in \{0,1\}$ *the following holds:*

$$\inf_{v_h \in \mathbb{X}_h^k} \|u - v_h\|_t \leq C h^{m-t} |u|_m \ \textit{for all} \ \ u \in H^m(\Omega), \tag{3.8a}$$
$$\inf_{v_h \in \mathbb{X}_{h,0}^k} \|u - v_h\|_t \leq C h^{m-t} |u|_m \ \textit{for all} \ \ u \in H^m(\Omega) \cap H_0^1(\Omega). \tag{3.8b}$$

A function $u \in H^1(\Omega)$ is not necessarily continuous and therefore it may be that the nodal interpolation $I_{\mathbb{X}}^k u$ is not well-defined. Other quasi-interpolation operators (e.g. so-called Clement interpolation) have been developed that are well-defined for $u \in H^1(\Omega)$. Using these one can prove the approximation result

$$\inf_{v_h \in \mathbb{X}_h^k} \|u - v_h\|_{L^2} \leq C h |u|_1 \text{ for all } \ u \in H^1(\Omega), \quad k \geq 0. \tag{3.9}$$

3.2.2 Hood-Taylor finite element discretization of the Oseen problem

In this section we use the simplicial finite element spaces $\mathbb{X}_h^k$ introduced above for the discretization of the variational Oseen problem in (2.21). We apply the abstract results on the Galerkin discretization of saddle point problems given in the Appendix, Sect. 15.3, to the variational Oseen problem (2.21). Let $\mathbf{V}_h$ and Q_h be finite dimensional subspaces of $\mathbf{V} = H_0^1(\Omega)^3$ and $Q = L_0^2(\Omega)$, respectively. Then the *Galerkin discretization of the Oseen problem* (2.21) is as follows:

Find $\mathbf{u}_h \in \mathbf{V}_h$ and $p_h \in Q_h$, such that

$$\xi m(\mathbf{u}_h, \mathbf{v}_h) + a(\mathbf{u}_h, \mathbf{v}_h) + c(\mathbf{u}_h, \mathbf{v}_h) + b(\mathbf{v}_h, p_h) = (\mathbf{g}, \mathbf{v}_h)_{L^2} \quad \forall\, \mathbf{v}_h \in \mathbf{V}_h$$
$$b(\mathbf{u}_h, q_h) = 0 \quad \forall\, q_h \in Q_h. \tag{3.10}$$

This is of the form (15.20) with bilinear forms

$$\begin{aligned} \hat{a}(\mathbf{u}, \mathbf{v}) &= \xi m(\mathbf{u}, \mathbf{v}) + a(\mathbf{u}, \mathbf{v}) + c(\mathbf{u}, \mathbf{v}) \quad \text{on} \quad \mathbf{V} \times \mathbf{V}, \\ \hat{b}(\mathbf{u}, q) &= b(\mathbf{u}, q) \quad \text{on} \quad \mathbf{V} \times Q. \end{aligned} \tag{3.11}$$

The right-hand sides are given by $f_1(\mathbf{v}) = (\mathbf{g}, \mathbf{v})_{L^2}$, $f_2 = 0$. We recall the definitions of the bilinear forms:

$$m(\mathbf{u}, \mathbf{v}) = \int_\Omega \mathbf{u} \cdot \mathbf{v}\, dx, \quad a(\mathbf{u}, \mathbf{v}) = \frac{1}{Re} \int_\Omega \nabla \mathbf{u} \cdot \nabla \mathbf{v}\, dx,$$
$$c(\mathbf{u}, \mathbf{v}) = c(\mathbf{w}; \mathbf{u}, \mathbf{v}) = \int_\Omega (\mathbf{w} \cdot \nabla \mathbf{u}) \cdot \mathbf{v}\, dx, \quad b(\mathbf{u}, q) = -\int_\Omega q \operatorname{div} \mathbf{u}\, dx.$$

The bilinear form $\hat{b}(\cdot,\cdot)$ satisfies the inf-sup condition (15.21a) in Theorem 15.3.5, cf. (2.24) in Sect. 2.2.2. We make the assumption $\xi - \frac{1}{2} \operatorname{div} \mathbf{w} \geq 0$ on Ω, also used in Theorem 2.2.4. Then the bilinear form $\hat{a}(\cdot,\cdot)$ satisfies the ellipticity condition (15.21b) in Theorem 15.3.5. In view of the *discrete inf-sup condition* (15.21c) we introduce the following definition.

Definition 3.2.8 The pair $(\mathbf{V}_h, Q_h)$ is called *stable* if there exists a constant $\hat{\beta} > 0$ *independent* of h such that

$$\sup_{\mathbf{v}_h \in \mathbf{V}_h} \frac{\hat{b}(\mathbf{v}_h, q_h)}{\|\mathbf{v}_h\|_1} \geq \hat{\beta}\, \|q_h\|_{L^2} \quad \text{for all} \quad q_h \in Q_h. \tag{3.12}$$

□

In the literature this is also often called the *LBB stability* condition of the finite element pair $(\mathbf{V}_h, Q_h)$ (due to Ladyzenskaya, Babuska, Brezzi).

LBB-stable pairs of finite element spaces

We now turn to the question which pairs of finite element spaces can be used for the Galerkin discretization of the Oseen problem. In view of the simplicial finite element spaces introduced in Sect. 3.2.1 we consider the following so-called *Hood-Taylor* pair:

$$\left((\mathbb{X}_{h,0}^k)^d\,,\,\mathbb{X}_h^{k-1}\cap L_0^2(\Omega)\right),\quad k\geq 1. \tag{3.13}$$

The following remark shows that the issue of LBB-stability needs a careful analysis.

Remark 3.2.9 Take $d=2$, $\Omega=(0,1)^2$ and a uniform triangulation $\mathcal{T}_h$ of Ω that is defined as follows. For $N\in\mathbb{N}$ and $h:=\frac{1}{N+1}$ the domain Ω is subdivided in squares with sides of length h and vertices in the set $\{\,(ih,jh): 0\leq i,j\leq N+1\,\}$. The triangulation $\mathcal{T}_h$ is obtained by subdividing every square in two triangles by inserting a diagonal from (ih,jh) to $((i+1)h,(j+1)h)$. For the pair $(\mathbf{V}_h,Q_h)$ we take

$$(\mathbf{V}_h,Q_h):=\left((\mathbb{X}_{h,0}^1)^2\,,\,\mathbb{X}_h^0\cap L_0^2(\Omega)\right).$$

The space $\mathbf{V}_h$ has dimension $2N^2$ and $\dim(Q_h)=2(N+1)^2-1$. From $\dim(\mathbf{V}_h)<\dim(Q_h)$ and Remark 15.3.6 it follows that the condition (3.12) does not hold.

The same argument applies to the three dimensional case with a uniform triangulation of $(0,1)^3$ consisting of tetrahedra (every cube is subdivided in 6 tetrahedra). In this case we have $\dim(\mathbf{V}_h)=3N^3$ and $\dim(Q_h)=6(N+1)^3-1$.

This remark implies that in general for $k=1$ the Hood-Taylor pair in (3.13) is *not* LBB stable. However, for $k\geq 2$ this pair is stable:

Theorem 3.2.10 *Let $\{\mathcal{T}_h\}$ be a regular family of triangulations consisting of simplices. We assume that every $T\in\mathcal{T}_h$ has at least one vertex in the interior of Ω. Then the Hood-Taylor pair of finite element spaces with $k\geq 2$ is LBB stable.*

For a proof of this important result we refer to the literature, [41, 42, 54]. Using this stability property of the Hood-Taylor finite element spaces the following discretization error bound can be derived.

Theorem 3.2.11 *Let $\{\mathcal{T}_h\}$ be a regular family of triangulations as in Theorem* 3.2.10. *Consider the discrete Oseen problem* (3.10) *with Hood-Taylor finite element spaces as in* (3.13), $k\geq 2$. *Suppose that $\xi-\frac{1}{2}\operatorname{div}\mathbf{w}\geq 0$ holds and that the continuous solution $(\mathbf{u},p)$ lies in $H^m(\Omega)^3\times H^{m-1}(\Omega)$ with $m\geq 2$. For $2\leq m\leq k+1$ the following holds:*

$$\|\mathbf{u}-\mathbf{u}_h\|_1+\|p-p_h\|_{L^2}\leq C\,h^{m-1}\big(|\mathbf{u}|_m+|p|_{m-1}\big),$$

with a constant C independent of h and of $(\mathbf{u},p)$.

Proof. We consider the bilinear forms $\hat{a}(\cdot,\cdot)$ and $\hat{b}(\cdot,\cdot)$ as in (3.11) and apply Theorem 15.3.5. Due to the inf-sup property (2.24) and the assumption $\xi - \frac{1}{2}\operatorname{div}\mathbf{w} \geq 0$ the conditions (15.21a) and (15.21b) are satisfied (cf. Theorem 2.2.4). Due to the LBB stability of the Hood-Taylor pair (3.13) with $k \geq 2$ the discrete inf-sup condition (15.21c) is fulfilled with a constant β_h *in*dependent of h. For the approximation error corresponding to the Hood-Taylor finite element spaces the results in Theorem 3.2.7 and in (3.9) can be used:

$$\inf_{v_h \in \mathbb{X}^k_{h,0}} \|\mathbf{u} - \mathbf{v}_h\|_1 \leq Ch^{m-1}|\mathbf{u}|_m,$$

$$\inf_{q_h \in \mathbb{X}^{k-1}_h} \|p - q_h\|_{L^2} \leq Ch^{m-1}|p|_{m-1}.$$

This completes the proof. □

Note that in this theorem sufficient *regularity* of the Oseen problem is required, namely that the solution $(\mathbf{u}, p)$ lies in the space $H^m(\Omega)^3 \times H^{m-1}(\Omega)$ with $m \geq 2$. For a discussion of this regularity issue we refer to the literature. If this regularity assumption holds both for the Oseen problem (2.21) and for the corresponding adjoint problem, i.e. a problem as in (2.21) with $\mathbf{w}$ replaced by $-\mathbf{w}$, then a discretization error bound

$$\|\mathbf{u} - \mathbf{u}_h\|_{L^2} \leq C\, h^m \big(|\mathbf{u}|_m + |p|_{m-1}\big) \tag{3.14}$$

can be shown to hold, with C independent of h and of $(\mathbf{u}, p)$.

Remark 3.2.12 For the (generalized) Stokes equations we have $\mathbf{w} = 0$ and thus the problem is symmetric, i.e. the adjoint problem equals the original one. It is known ([160, 63, 75]) that for this case the regularity property $(\mathbf{u}, p) \in H^m(\Omega)^3 \times H^{m-1}(\Omega)$, with $m \geq 2$, is satisfied for $m = 2$ if Ω is convex and for the general case $m \geq 2$ if $\partial\Omega$ is sufficiently smooth.

3.2.3 Matrix-vector representation of the discrete problem

Consider the variational discrete Oseen problem (3.10) with Hood-Taylor finite element spaces, $(\mathbf{V}_h, Q_h) = \big((\mathbb{X}^k_{h,0})^3, \mathbb{X}^{k-1}_h \cap L^2_0(\Omega)\big)$, $k \geq 2$. For computing the unique discrete solution $(\mathbf{u}_h, p_h)$ we introduce the *nodal* basis functions in the simplicial finite element space $\mathbb{X}^k_h$, which are defined as follows. The union of all lattice points in $L_k(T)$, $T \in \mathcal{T}_h$, form the set of grid points $(\mathbf{x}_i)_{1\leq i\leq K}$. Note that $\dim(\mathbb{X}^k_h) = K$. To each of these grid points there corresponds a nodal finite element function $\phi_i \in \mathbb{X}^k_h$ with the property $\phi_i(\mathbf{x}_i) = 1$, $\phi_i(\mathbf{x}_j) = 0$ for all $j \neq i$. The set of functions $(\phi_i)_{1\leq i\leq K}$ forms the nodal basis of $\mathbb{X}^k_h$. In case of $\mathbb{X}^k_{h,0}$ only the nodal functions corresponding to grid points in the interior of Ω are used. In case of vector functions. i.e. $(\mathbb{X}^k_{h,0})^d$ with $d > 1$, to each grid point there correspond d of such nodal functions, namely one for each of the d components in the vector function. Let $\{\boldsymbol{\xi}_i\}_{1\leq i\leq N}$

and $\{\psi_i\}_{1\leq i\leq K}$ be such nodal bases of the finite element spaces $\mathbf{V}_h = (\mathbb{X}^k_{h,0})^3$ and $\mathbb{X}^{k-1}_h$, respectively. Hence, $\dim(\mathbb{X}^k_{h,0})^3 = N$, $\dim(\mathbb{X}^{k-1}_h) = K$. Consider the representations

$$\mathbf{u}_h = \sum_{j=1}^{N} u_j \boldsymbol{\xi}_j, \quad \vec{\mathbf{u}} := (u_1, \dots, u_N) \tag{3.15}$$

$$p_h = \sum_{j=1}^{K} p_j \psi_j, \quad \vec{\mathbf{p}} := (p_1, \dots, p_K). \tag{3.16}$$

Using this the discrete Oseen problem (3.10) can be reformulated as follows:

> Determine $\vec{\mathbf{u}} \in \mathbb{R}^N$, $\vec{\mathbf{p}} \in \mathbb{R}^K$ with $(p_h, 1)_{L^2} = 0$ such that
>
> $$\begin{pmatrix} \xi\mathbf{M} + \mathbf{A} + \mathbf{C} & \mathbf{B}^T \\ \mathbf{B} & 0 \end{pmatrix} \begin{pmatrix} \vec{\mathbf{u}} \\ \vec{\mathbf{p}} \end{pmatrix} = \begin{pmatrix} \vec{\mathbf{f}} \\ 0 \end{pmatrix}, \tag{3.17}$$

where

$$\mathbf{M} \in \mathbb{R}^{N\times N}, \quad \mathbf{M}_{ij} = \int_\Omega \boldsymbol{\xi}_i \cdot \boldsymbol{\xi}_j \, dx \tag{3.18a}$$

$$\mathbf{A} \in \mathbb{R}^{N\times N}, \quad \mathbf{A}_{ij} = \frac{1}{Re}\int_\Omega \nabla\boldsymbol{\xi}_i \cdot \nabla\boldsymbol{\xi}_j \, dx \tag{3.18b}$$

$$\mathbf{C} = \mathbf{C}(\mathbf{w}) \in \mathbb{R}^{N\times N}, \quad \mathbf{C}_{ij} = \int_\Omega (\mathbf{w}\cdot\nabla\boldsymbol{\xi}_j)\cdot\boldsymbol{\xi}_i \, dx \tag{3.18c}$$

$$\mathbf{B} \in \mathbb{R}^{K\times N}, \quad \mathbf{B}_{ij} = -\int_\Omega \psi_i \operatorname{div}\boldsymbol{\xi}_j \, dx \tag{3.18d}$$

$$\vec{\mathbf{f}} \in \mathbb{R}^N, \quad \vec{\mathbf{f}}_i = \int_\Omega \mathbf{g}\cdot\boldsymbol{\xi}_i \, dx. \tag{3.18e}$$

$\mathbf{M}, \mathbf{A}, \mathbf{C}$ are called mass matrix, diffusion matrix and convection matrix, respectively. Matrices with a block structure as in (3.17) are called saddle point matrices. Iterative solution methods for linear systems with such matrices will be treated in Chap. 5.

3.2.4 Hood-Taylor semi-discretization of the non-stationary (Navier-)Stokes problem

We recall the weak formulation of the non-stationary Stokes equations given (2.34): Find $\mathbf{u} \in W^1(0,T;\mathbf{V})$ and $p \in L^2(0,T;Q)$, such that $\mathbf{u}(0) = \mathbf{u}_0$ and

$$\begin{aligned} \frac{d}{dt} m(\mathbf{u}(t),\mathbf{v}) + a(\mathbf{u}(t),\mathbf{v}) + b(\mathbf{v},p(t)) &= (\mathbf{g},\mathbf{v})_{L^2} \quad \forall\, \mathbf{v}\in\mathbf{V}, \\ b(\mathbf{u}(t),q) &= 0 \quad \forall\, q\in Q, \end{aligned} \tag{3.19}$$

for almost all $t \in [0,T]$. For the *spatial* discretization of this problem we use the Galerkin approach with a stable Hood-Taylor pair from (3.13):

$$(\mathbf{V}_h, Q_h) = \big((\mathbb{X}_{h,0}^k)^3\,, \mathbb{X}_h^{k-1} \cap L_0^2(\Omega)\big), \quad k \geq 2.$$

Let $\mathbf{u}_{0,h} \in \mathbf{V}_h$ be an approximation of the initial condition $\mathbf{u}_0$. The Galerkin *semi*-discretization reads: Find $\mathbf{u}_h(t) \in \mathbf{V}_h$, with $\mathbf{u}_h(0) = \mathbf{u}_{0,h}$, and $p_h(t) \in Q_h$ such that:

$$\begin{aligned} \frac{d}{dt} m(\mathbf{u}_h(t), \mathbf{v}_h) + a(\mathbf{u}_h(t), \mathbf{v}_h) + b(\mathbf{v}_h, p_h(t)) &= (\mathbf{g}, \mathbf{v}_h)_{L^2} \ \forall\ \mathbf{v}_h \in \mathbf{V}_h, \\ b(\mathbf{u}_h(t), q_h) &= 0 \ \ \forall\ q_h \in Q_h, \end{aligned} \tag{3.20}$$

for all $t \in [0,T]$. Let $\{\boldsymbol{\xi}_i\}_{1\leq i\leq N}$ and $\{\psi_i\}_{1\leq i\leq K}$ be the standard nodal bases of the finite element spaces $\mathbf{V}_h$, $\mathbb{X}_h^{k-1}$ and consider the representations

$$\mathbf{u}_h(t) = \sum_{j=1}^{N} u_j(t)\boldsymbol{\xi}_j, \quad \vec{\mathbf{u}}(t) := (u_1(t), \ldots, u_N(t)) \tag{3.21a}$$

$$\mathbf{u}_{0,h} = \sum_{j=1}^{N} u_{0,j}\boldsymbol{\xi}_j, \quad \vec{\mathbf{u}}_0 := (u_{0,1}, \ldots, u_{0,N}) \tag{3.21b}$$

$$p_h(t) = \sum_{j=1}^{K} p_j(t)\psi_j, \quad \vec{\mathbf{p}}(t) := (p_1(t), \ldots, p_K(t)). \tag{3.21c}$$

Using this the Galerkin discretization can be rewritten as

Determine $\vec{\mathbf{u}}(t) \in \mathbb{R}^N$, $\vec{\mathbf{p}}(t) \in \mathbb{R}^K$ with $\vec{\mathbf{u}}(0) = \vec{\mathbf{u}}_0$ and $(p_h(t), 1)_{L^2} = 0$ such that

$$\begin{aligned} \mathbf{M}\frac{d\vec{\mathbf{u}}}{dt}(t) + \mathbf{A}\vec{\mathbf{u}}(t) + \mathbf{B}^T\vec{\mathbf{p}}(t) &= \vec{\mathbf{f}} \\ \mathbf{B}\vec{\mathbf{u}}(t) &= 0, \end{aligned} \tag{3.22}$$

for all $t \in [0,T]$,

with $\mathbf{M}, \mathbf{A}, \mathbf{B}$ and $\vec{\mathbf{f}}$ as in (3.18).

Thus we obtain a system of *differential algebraic equations* (DAEs) for the unknown vector functions $\vec{\mathbf{u}}(t)$, $\vec{\mathbf{p}}(t)$. Time discretization methods for this system are discussed in Chap. 4.

Lemma 3.2.13 *The problem in* (3.20), *or equivalently* (3.22), *has a unique solution.*

Proof. A proof is given in Sect. 4.2, Remark 4.2.1. □

We derive a bound for the discretization error of the semi-discrete Stokes problem in (3.20).

Theorem 3.2.14 *Let* $(\mathbf{u}, p)$ *and* $(\mathbf{u}_h, p_h)$ *be the solution of* (3.19) *and* (3.20), *respectively. Assume that* $(\mathbf{u}, p)$ *is sufficiently smooth:* $\mathbf{u} \in C^1([0,T]; H^m(\Omega)^3)$, $p \in C^1([0,T]; H^{m-1}(\Omega))$ *and that* $2 \leq m \leq k+1$, *with* $k \geq 2$. *Furthermore, we assume that for the discretization of the stationary Stokes problem with Hood-Taylor finite elements the error bound* (3.14) *holds (cf. Remark 3.2.12). Then the following holds for* $t \in [0,T]$*:*

$$\|\mathbf{u}(t) - \mathbf{u}_h(t)\|_{L^2} \leq e^{-c_0 t}\|\mathbf{u}_0 - \mathbf{u}_{0,h}\|_{L^2} + ch^m E_t(\mathbf{u}, p), \tag{3.23}$$

$$|\mathbf{u}(t) - \mathbf{u}_h(t)|_1 \leq |\mathbf{u}_0 - \mathbf{u}_{0,h}|_1 + c(1 + h\sqrt{t})h^{m-1}E_t(\mathbf{u}, p), \tag{3.24}$$

$$\Big(\int_0^t \|p(\tau) - p_h(\tau)\|_{L^2}^2 \, d\tau\Big)^{\frac{1}{2}} \leq c|\mathbf{u}_0 - \mathbf{u}_{0,h}|_1 + c\sqrt{t}\, h^{m-1}E_t(\mathbf{u}, p), \tag{3.25}$$

with constants $c_0 > 0$, c *independent of* h *and* t *and*

$$E_t(\mathbf{u}, p) := \sum_{\ell=0}^{1} \max_{0 \leq \tau \leq t} \Big(\big|\mathbf{u}^{(\ell)}(\tau)\big|_m + \big|p^{(\ell)}(\tau)\big|_{m-1}\Big).$$

Proof. Take $\mathbf{w} \in \mathbf{V}_{\text{div}} := \{\mathbf{v} \in \mathbf{V} : \operatorname{div} \mathbf{v} = 0\}$, $r \in Q$ and define $\ell(\mathbf{v}) := a(\mathbf{w}, \mathbf{v}) + b(\mathbf{v}, r)$, $\mathbf{v} \in \mathbf{V}$. Then $\ell \in \mathbf{V}'$ and $(\mathbf{w}, r)$ is the unique solution of the stationary Stokes problem

$$a(\mathbf{w}, \mathbf{v}) + b(\mathbf{v}, r) = \ell(\mathbf{v}) \quad \text{for all } \mathbf{v} \in \mathbf{V},$$
$$b(\mathbf{w}, q) = 0 \quad \text{for all } q \in Q.$$

Let $(\mathbf{w}_h, r_h) \in \mathbf{V}_h \times Q_h$ be the unique solution of the Galerkin discretization of this problem. The mapping $S : (\mathbf{w}, r) \to (\mathbf{w}_h, r_h)$ is linear on $\mathbf{V}_{\text{div}} \times Q$ and for $\mathbf{w}$ and r sufficiently smooth we have

$$\|\mathbf{w} - \mathbf{w}_h\|_1 + \|r - r_h\|_{L^2} \leq ch^{m-1}\big(|\mathbf{w}|_m + |r|_{m-1}\big), \tag{3.26}$$

$$\|\mathbf{w} - \mathbf{w}_h\|_{L^2} \leq ch^m\big(|\mathbf{w}|_m + |r|_{m-1}\big), \tag{3.27}$$

with constants c independent of h and of $(\mathbf{w}, r)$. Let $(\mathbf{u}, p)$ and $(\mathbf{u}_h, p_h)$ be as defined in the theorem. Define, for $t \in [0,T]$, $(\mathbf{w}_h(t), r_h(t)) := S(\mathbf{u}(t), p(t))$ and

$$\mathbf{e}_h(t) := \mathbf{u}(t) - \mathbf{u}_h(t) = \big(\mathbf{u}(t) - \mathbf{w}_h(t)\big) + \big(\mathbf{w}_h(t) - \mathbf{u}_h(t)\big) =: \rho_h(t) + \theta_h(t),$$
$$p(t) - p_h(t) = \big(p(t) - r_h(t)\big) + \big(r_h(t) - p_h(t)\big) =: \xi_h(t) + \eta_h(t).$$

Note that $\theta_h(t) \in \mathbf{V}_h$, $\eta_h(t) \in Q_h$ and $(\mathbf{w}_h'(t), r_h'(t)) = \frac{d}{dt}S(\mathbf{u}(t), p(t)) = S(\mathbf{u}'(t), p'(t))$. Due to (3.26), (3.27) we have

$$\|\rho_h(t)\|_1 \leq ch^{m-1}E_t(\mathbf{u}, p), \tag{3.28}$$

$$\|\rho_h(t)\|_{L^2} \leq ch^m E_t(\mathbf{u}, p), \tag{3.29}$$

$$\|\rho_h'(t)\|_{L^2} \leq ch^m E_t(\mathbf{u}, p), \tag{3.30}$$

$$\|\xi_h(t)\|_{L^2} \leq ch^{m-1}E_t(\mathbf{u}, p). \tag{3.31}$$

Subtraction of the variational equations for $(\mathbf{u}, p)$ and $(\mathbf{u}_h, p_h)$ results in

$$(\mathbf{e}'_h(t), \mathbf{v}_h)_{L^2} + a(\mathbf{e}_h(t), \mathbf{v}_h) + b(\mathbf{v}_h, p(t) - p_h(t)) = 0 \quad \text{for all } \ \mathbf{v}_h \in \mathbf{V}_h.$$

Using the definitions we obtain

$$a(\rho_h(t), \mathbf{v}_h) + b(\mathbf{v}_h, p(t) - p_h(t)) = b(\mathbf{v}_h, \eta_h(t)) \quad \text{for all } \ \mathbf{v}_h \in \mathbf{V}_h.$$

Using this and the splitting $\mathbf{e}_h(t) = \rho_h(t) + \theta_h(t)$ we get

$$(\theta'_h(t), \mathbf{v}_h)_{L^2} + a(\theta_h(t), \mathbf{v}_h) + b(\mathbf{v}_h, \eta_h(t)) = -(\rho'_h(t), \mathbf{v}_h)_{L^2} \tag{3.32}$$

for all $\mathbf{v}_h \in \mathbf{V}_h$. Based on this fundamental relation the following bounds can be derived:

$$\|\theta_h(t)\|_{L^2} \le e^{-c_0 t}\|\theta_h(0)\|_{L^2} + ch^m E_t(\mathbf{u}, p), \tag{3.33}$$

$$|\theta_h(t)|_1 \le |\theta_h(0)|_1 + c\sqrt{t}\, h^m E_t(\mathbf{u}, p), \tag{3.34}$$

$$\Big(\int_0^t |\theta_h(\tau)|_1^2 \, d\tau\Big)^{\frac{1}{2}} \le c\|\theta_h(0)\|_{L^2} + c\sqrt{t}\, h^m E_t(\mathbf{u}, p), \tag{3.35}$$

$$\Big(\int_0^t \|\theta'_h(\tau)\|_{L^2}^2 \, d\tau\Big)^{\frac{1}{2}} \le c|\theta_h(0)|_1 + c\sqrt{t}\, h^m E_t(\mathbf{u}, p), \tag{3.36}$$

with constants $c_0 > 0$ and c independent of h and t. We now prove these inequalities. In (3.32) we take $\mathbf{v}_h = \theta_h(t)$ and use that $b(\theta_h(t), \eta_h(t)) = 0$, resulting in

$$\frac{1}{2}\frac{d}{dt}\|\theta_h(t)\|_{L^2}^2 + \frac{1}{Re}|\theta_h(t)|_1^2 \le \|\rho'_h(t)\|_{L^2}\|\theta_h(t)\|_{L^2}. \tag{3.37}$$

Using the Poincaré-Friedrichs inequality $\|\theta_h(t)\|_{L^2} \le c|\theta_h(t)|_1$ we obtain from this

$$\frac{1}{2}\frac{d}{dt}\|\theta_h(t)\|_{L^2}^2 + \frac{1}{2}\frac{1}{Re}|\theta_h(t)|_1^2 \le c\|\rho'_h(t)\|_{L^2}^2.$$

Integration over $[0, t]$ results in

$$\begin{aligned}\int_0^t |\theta_h(\tau)|_1^2 \, d\tau &\le c\|\theta_h(0)\|_{L^2}^2 + c\int_0^t \|\rho'_h(\tau)\|_{L^2}^2 \, d\tau \\ &\le c\|\theta_h(0)\|_{L^2}^2 + c\,t h^{2m} E_t(\mathbf{u}, p)^2,\end{aligned}$$

which proves (3.35). From (3.37) we also obtain

$$\|\theta_h(t)\|_{L^2}\frac{d}{dt}\|\theta_h(t)\|_{L^2} + c_0\|\theta_h(t)\|_{L^2}^2 \le \|\rho'_h(t)\|_{L^2}\|\theta_h(t)\|_{L^2},$$

with $c_0 > 0$, and hence

$$\frac{d}{dt}\|\theta_h(t)\|_{L^2} + c_0\|\theta_h(t)\|_{L^2} \le \|\rho'_h(t)\|_{L^2}.$$

Multiplication by $e^{c_0 t}$ and integration over $[0,t]$ yields

$$\begin{aligned}\|\theta_h(t)\|_{L^2} &\leq e^{-c_0 t}\|\theta_h(0)\|_{L^2} + \int_0^t e^{-c_0(t-\tau)}\|\rho_h'(\tau)\|_{L^2}\,d\tau \\ &\leq e^{-c_0 t}\|\theta_h(0)\|_{L^2} + ch^m E_t(\mathbf{u},p)\int_0^t e^{-c_0(t-\tau)}\,d\tau \\ &\leq e^{-c_0 t}\|\theta_h(0)\|_{L^2} + ch^m E_t(\mathbf{u},p),\end{aligned}$$

with $c_0 > 0$ and c independent of h and t. Thus (3.33) holds. In (3.32) we now substitute $\mathbf{v}_h = \theta_h'(t)$. Using $b(\theta_h'(t), \eta_h(t)) = 0$ we get

$$\begin{aligned}\|\theta_h'(t)\|_{L^2}^2 + \frac{1}{2}\frac{1}{Re}\frac{d}{dt}|\theta_h(t)|_1^2 &\leq \|\rho_h'(t)\|_{L^2}\|\theta_h'(t)\|_{L^2} \\ &\leq \frac{1}{2}\|\rho_h'(t)\|_{L^2}^2 + \frac{1}{2}\|\theta_h'(t)\|_{L^2}^2,\end{aligned}$$

and thus

$$\|\theta_h'(t)\|_{L^2}^2 + \frac{1}{Re}\frac{d}{dt}|\theta_h(t)|_1^2 \leq \|\rho_h'(t)\|_{L^2}^2.$$

Integrating this results in

$$\int_0^t \|\theta_h'(\tau)\|_{L^2}^2\,d\tau + \frac{1}{Re}|\theta_h(t)|_1^2 \leq \frac{1}{Re}|\theta_h(0)|_1^2 + c\,th^{2m}E_t(\mathbf{u},p)^2,$$

which proves the results in (3.34) and (3.36).

Using (3.33)-(3.36) we derive the bounds stated in the theorem. Note that $\|\theta_h(0)\|_{L^2} \leq \|\mathbf{e}_h(0)\|_{L^2} + \|\rho_h(0)\|_{L^2} \leq \|\mathbf{e}_h(0)\|_{L^2} + ch^m E_t(\mathbf{u},p)$ holds, and thus, using (3.29) and (3.33) we get

$$\|\mathbf{e}_h(t)\|_{L^2} \leq \|\theta_h(t)\|_{L^2} + \|\rho_h(t)\|_{L^2} \leq e^{-c_0 t}\|\mathbf{e}_h(0)\|_{L^2} + ch^m E_t(\mathbf{u},p),$$

hence, the result in (3.23) holds. Similarly, from $|\theta_h(0)|_1 \leq |\mathbf{e}_h(0)|_1 + ch^{m-1}E_t(\mathbf{u},p)$ and (3.28), (3.34) we obtain

$$\begin{aligned}|\mathbf{e}_h(t)|_1 &\leq |\theta_h(t)|_1 + |\rho_h(t)|_1 \leq |\mathbf{e}_h(0)|_1 + c\sqrt{t}h^m E_t(\mathbf{u},p) + ch^{m-1}E_t(\mathbf{u},p) \\ &= |\mathbf{e}_h(0)|_1 + ch^{m-1}\big(1+\sqrt{t}h\big)E_t(\mathbf{u},p),\end{aligned}$$

which proves the result in (3.24). For the pressure error bound we use the LBB stability property of the Hood-Taylor pair, which implies

$$\|\eta_h(t)\|_{L^2} \leq c \sup_{\mathbf{v}_h \in \mathbf{V}_h} \frac{b(\mathbf{v}_h, \eta_h(t))}{\|\mathbf{v}_h\|_1}.$$

Using the fundamental relation (3.32) this implies

$$\begin{aligned}\|\eta_h(t)\|_{L^2} &\leq c \sup_{\mathbf{v}_h \in \mathbf{V}_h} \frac{(\theta_h'(t), \mathbf{v}_h)_{L^2} + a(\theta_h(t), \mathbf{v}_h) + (\rho_h'(t), \mathbf{v}_h)_{L^2}}{\|\mathbf{v}_h\|_1} \\ &\leq c\big(\|\theta_h'(t)\|_{L^2} + |\theta_h(t)|_1 + \|\rho_h'(t)\|_{L^2}\big),\end{aligned}$$

and thus using (3.35), (3.36), (3.30) we get

$$\Big(\int_0^t \|\eta_h(\tau)\|_{L^2}^2\,d\tau\Big)^{\frac12} \le c|\theta_h(0)|_1 + c\sqrt{t}h^m E_t(\mathbf{u},p).$$

Finally, using (3.31) yields

$$\begin{aligned}\Big(\int_0^t \|p(\tau)-p_h(\tau)\|_{L^2}^2\,d\tau\Big)^{\frac12} &\le \Big(\int_0^t \|\eta_h(\tau)\|_{L^2}^2\,d\tau\Big)^{\frac12} + \sqrt{t}\max_{0\le\tau\le t}\|\xi_h(\tau)\|_{L^2}\\ &\le c|\mathbf{e}_h(0)|_1 + c\sqrt{t}h^{m-1}E_t(\mathbf{u},p),\end{aligned}$$

and thus the pressure error bound (3.25) holds. □

Remark 3.2.15 The error bounds in this theorem show an optimal behavior w.r.t. the rate of convergence for $h \downarrow 0$. If we use the Hood-Taylor pair with index $k \ge 2$ (i.e., degree k polynomials for velocity and degree $k-1$ for pressure) and assume that the solution pair $(\mathbf{u},p)$ is sufficiently smooth ($m = k+1$) then Theorem 3.2.14 yields an h^{k+1} bound for the velocity L^2-error and an h^k bound both for the velocity H^1-error and the pressure L^2-error. From the result in (3.23) we see that in the L^2-norm the discretization error in the initial condition is exponentially damped for increasing t. The term $E_t(\mathbf{u},p)$ quantifies the smoothness of the solution pair $(\mathbf{u},p)$ on $[0,t]$. Furthermore note that in (3.24), (3.25) apart from a constant c (independent of t) and the smoothness measure $E_t(\mathbf{u},p)$ there are additional terms $h\sqrt{t}$ and $\sqrt{t}$, respectively, which grow as functions of t. Such a t-dependent factor does not occur in (3.23).

The same semi-discretization approach can be applied to the weak formulation of the *Navier*-Stokes problem (2.37). The semi-discrete Navier-Stokes problem is as follows: Find $\mathbf{u}_h(t) \in \mathbf{V}_h$, $p_h(t) \in Q_h$ such that $\mathbf{u}(0) = \mathbf{u}_{0,h}$ and for all $\mathbf{v}_h \in \mathbf{V}_h$ and all $q_h \in Q_h$:

$$\begin{aligned}\frac{d}{dt}m(\mathbf{u}_h(t),\mathbf{v}_h) + a(\mathbf{u}_h(t),\mathbf{v}_h) + c(\mathbf{u}_h(t);\mathbf{u}_h(t),\mathbf{v}_h) + b(\mathbf{v}_h,p_h(t)) &= (\mathbf{g},\mathbf{v}_h)_{L^2}\\ b(\mathbf{u}_h(t),q_h) &= 0,\end{aligned}$$

for all $t \in [0,T]$. Using the representations as in (3.21) this Galerkin discretization can be rewritten as

Determine $\vec{\mathbf{u}}(t) \in \mathbb{R}^N$, $\vec{\mathbf{p}}(t) \in \mathbb{R}^K$ with with $\vec{\mathbf{u}}(0) = \vec{\mathbf{u}}_0$ and $(p_h(t),1)_{L^2} = 0$ such that

$$\begin{aligned}\mathbf{M}\frac{d\vec{\mathbf{u}}}{dt}(t) + \mathbf{A}\vec{\mathbf{u}}(t) + \mathbf{N}(\vec{\mathbf{u}}(t))\vec{\mathbf{u}}(t) + \mathbf{B}^T\vec{\mathbf{p}}(t) &= \vec{\mathbf{f}}\\ \mathbf{B}\vec{\mathbf{u}}(t) &= 0,\end{aligned} \tag{3.38}$$

for all $t \in [0,T]$.

The nonlinear operator $\mathbf{N}$ is given by

$$\mathbf{N}(\vec{\mathbf{u}}) = \mathbf{N}(\mathbf{u}_h) \in \mathbb{R}^{N\times N}, \quad \mathbf{N}(\vec{\mathbf{u}})_{ij} = \int_\Omega (\mathbf{u}_h \cdot \nabla \boldsymbol{\xi}_j) \cdot \boldsymbol{\xi}_i \, dx.$$

We obtain a nonlinear system of differential algebraic equations (DAEs) for the unknown vector functions $\vec{\mathbf{u}}(t)$, $\vec{\mathbf{p}}(t)$. For a discretization error analysis of this problem we refer to the literature [141, 142].

3.3 Numerical experiments

After the theoretical analysis in Sect. 3.2 we now present a numerical study of the convergence behavior of the Hood-Taylor pair for $k = 2$, cf. Theorem 3.2.11. We consider two numerical experiments, namely the flow in a rectangular tube (Sect. 3.3.1) and the flow in a curved channel (Sect. 3.3.2).

3.3.1 Flow in a rectangular tube

Let $\Omega = (0, L) \times (0, 1)^2$ be a rectangular tube of length $L > 0$. Consider (3.10) with $\xi = 0, \mathbf{w} = 0$ (i. e., the stationary Stokes case) where we prescribe the boundary conditions $(\mathbf{u}, p)|_{x_1=0} = (\mathbf{u}, p)|_{x_1=L}$ (periodic boundary conditions in x_1-direction) and $\mathbf{u} = 0$ on the remaining boundaries. The right-hand side is set to $\mathbf{g} = (1, 0, 0)$, which can be interpreted as gravity force in x_1-direction. Then the analytic solution is given by $\mathbf{u}(x) = (Re\, s(x_2, x_3), 0, 0)$ and $p = 0$, where s is the solution of the 2D Poisson problem

$$-\Delta s = 1 \qquad \text{on } (0, 1)^2,$$

with homogeneous Dirichlet boundary conditions. By Fourier analysis, s can be expressed in terms of the Fourier series

$$s(x_2, x_3) = \frac{16}{\pi^4} \sum_{i,j=1}^{\infty} \alpha_{2i-1,2j-1} s_{2i-1,2j-1}(x_2, x_3), \quad (x_2, x_3) \in (0, 1)^2, \quad (3.39)$$

with $\alpha_{i,j} = \frac{1}{ij(i^2+j^2)}$ and $s_{i,j}(x_2, x_3) = \sin(i\pi x_2)\sin(j\pi x_3)$. In Fig. 3.9 we give a plot of s.

The initial triangulation $\mathcal{T}_0$ is constructed by subdividing Ω into $4 \times 1 \times 1$ sub-cubes each consisting of 6 tetrahedra. Then $\mathcal{T}_0$ is successively uniformly refined 5 times by applying the regular refinement rule yielding $\mathcal{T}_1, \ldots, \mathcal{T}_5$. In our experiments we used $L = 4$ and $Re = 1$. In Fig. 3.10 the discrete velocity $\mathbf{u}_h$ is illustrated for the triangulation $\mathcal{T}_3$.

The discrete pressure p_h is equal to zero (up to machine accuracy).Table 3.1 shows the dimension of the finite element spaces $\mathbf{V}_h$ and Q_h and the convergence of $\mathbf{u}_h$ to $\mathbf{u}$ w.r.t. the L^2 and H^1 norm for different refinement levels. We

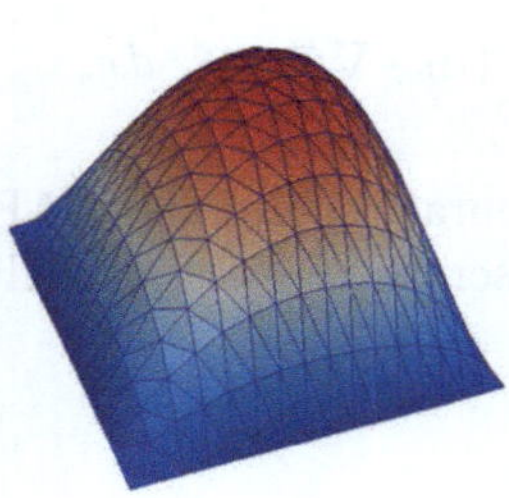

Fig. 3.9. Solution s of the Poisson equation on the unit square.

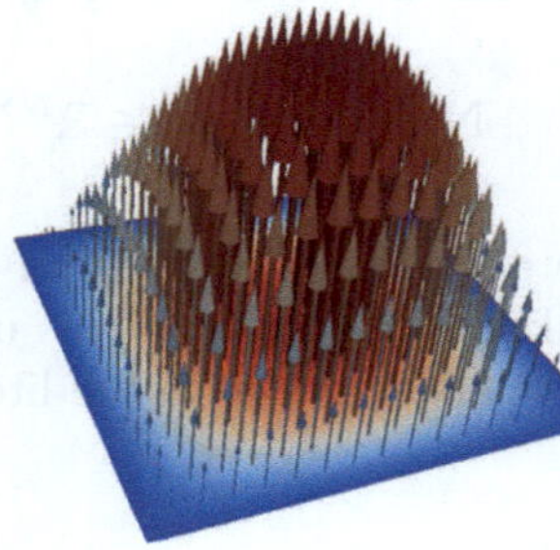

Fig. 3.10. Velocity $\mathbf{u}_h$ visualized on slice $x_1 = L/2$.

# ref.	dim $\mathbf{V}_h$	dim Q_h	$\|\mathbf{u}-\mathbf{u}_h\|_{L^2}$	order	$\|\mathbf{u}-\mathbf{u}_h\|_1$	order
0	24	16	9.63 E-3	—	7.17 E-2	—
1	432	72	2.06 E-3	2.22	1.89 E-2	1.92
2	4 704	400	2.18 E-4	3.24	4.01 E-3	2.24
3	43 200	2 592	2.49 E-5	3.13	9.40 E-4	2.09
4	369 024	18 496	3.02 E-6	3.04	2.30 E-4	2.03
5	3 048 192	139 392	3.01 E-7	3.33	5.75 E-5	2.00

Table 3.1. Dimension of the finite element spaces and convergence behavior w.r.t. L^2 and H^1 norm for different refinement levels.

observe third order convergence w.r.t. the L^2 norm and second order convergence w.r.t. the H^1 norm. These are the optimal rates which can be expected in view of Theorem 3.2.11 and the L^2 bound (3.14), as $\mathbf{u}$ and p are sufficiently smooth.

Note the (very) high dimension of the velocity space on level 5, due to the fact that the number of tetrahedra grows with a factor of 8 in each refinement. Furthermore it is clear that the dimension of the pressure space is much smaller than that of the velocity space.

3.3.2 Flow in a curved channel

Let $\Omega = \{\, x \in \mathbb{R}^3 : 0 < x_1 < L, -a(x_1) < x_2 < a(x_1), 0 < x_3 < 1 \,\}$ be a channel of length $L > 0$ with $a : [0, L] \to (0, \infty)$ defining the shape in x_2-direction, cf. Fig. 3.11. In our experiment we set $L = 4$ and $a(x_1) = e^{-\alpha x_1}$, $\alpha = 1/4$. Consider (3.10) with $\xi = 0, \mathbf{w} = 0$ (i. e., the stationary Stokes case). We take the pair $(\mathbf{u}, p)$ given by

$$u_1(x) = \frac{1 - \left(\frac{x_2}{a(x_1)}\right)^2}{a(x_1)}, \quad u_2(x) = \frac{u_1(x)\, x_2\, a'(x_1)}{a(x_1)}, \quad u_3(x) = 0,$$

$$p = \frac{1}{2} x_2^2 \alpha e^{\alpha x_1}(-\alpha^2 + 6e^{2\alpha x_1}) + \alpha e^{\alpha x_1} - \frac{2}{3\alpha} e^{3\alpha x_1}.$$

The velocity field $\mathbf{u}$ is divergence free. We take the right-hand side $\mathbf{f} := (-\frac{1}{2}x_2^2\alpha^2 e^{\alpha x_1}(\alpha^2 - 36e^{2\alpha x_1}), -9x_2^3\alpha^3 e^{3\alpha x_1}, 0)$ such that the pair $(\mathbf{u}, p)$ is a solution of (3.10). In our experiment we use boundary conditions $(\mathbf{u}, p)|_{x_3=0} = (\mathbf{u}, p)|_{x_3=1}$ (periodic boundary conditions in x_3-direction) and Dirichlet boundary conditions for $\mathbf{u}$ on the remaining part of the boundary.

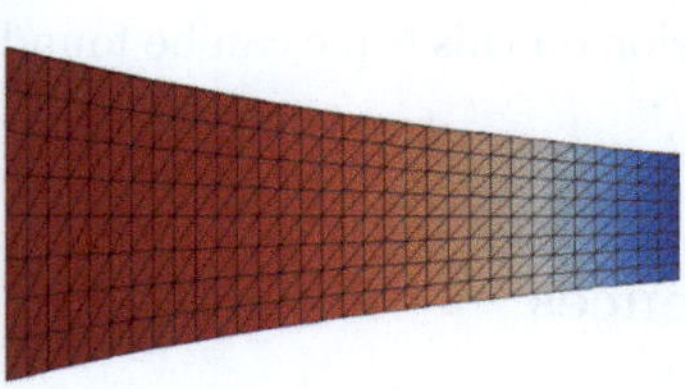

Fig. 3.11. Grid and pressure p_h visualized on slice $x_3 = 0.5$.

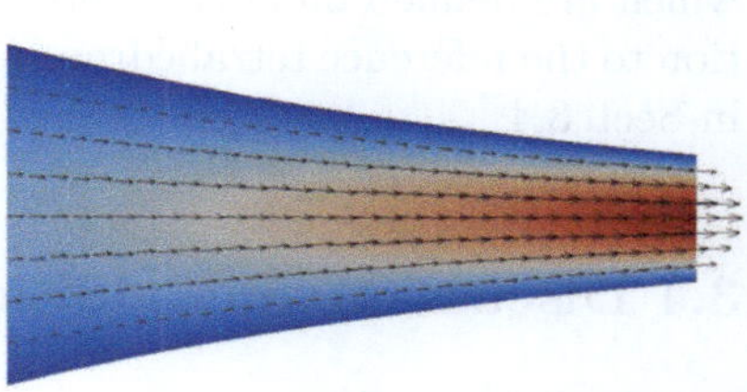

Fig. 3.12. Velocity $\mathbf{u}_h$ visualized on slice $x_3 = 0.5$.

For the discretization of Ω we first introduce the auxiliary domain $\hat{\Omega} = (0, L) \times (-1, 1) \times (0, 1)$, which is discretized by $4 \times 1 \times 1$ subcubes each subdivided into 6 tetrahedra. The resulting initial triangulation $\hat{\mathcal{T}}_0$ is then successively uniformly refined 5 times applying the regular refinement rule yielding $\hat{\mathcal{T}}_1, \ldots, \hat{\mathcal{T}}_5$. Using the mapping $F : \mathbb{R}^3 \to \mathbb{R}^3$, $F(x) = (x_1, a(x_1)x_2, x_3)$ the vertices of $\hat{\mathcal{T}}_\ell$ are mapped to the physical domain Ω. For each level $\ell = 0, \ldots, 5$, this induces corresponding triangulations $\mathcal{T}_\ell$ of polygonal domains Ω_ℓ *approximating* Ω. The mapping is such that all vertices on the boundary of the triangulation lie on the respective boundaries of Ω. At the $x_2 = \pm a(x_1)$ part of the boundary, however, the boundary faces of the triangulation do *not* coincide with the curved boundary. A 2D slice of the triangulation $\mathcal{T}_3$ and the corresponding pressure solution p_h is shown in Fig. 3.11. The velocity field $\mathbf{u}_h$ is depicted in Fig. 3.12.

Table 3.2 shows the dimension of the finite element spaces $\mathbf{V}_h$ and Q_h and the convergence of $\mathbf{u}_h$ to $\mathbf{u}$ w.r.t. L^2 and H^1 norm for different refinement levels. For small grid sizes the convergence order of the velocity error w.r.t. the H^1 norm tends to 1.5. This suboptimal behavior is due to the fact that a part of $\partial\Omega$ is curved and is approximated by a piecewise polygonal approximation. It is known, cf. [228], that due to this boundary approximation, with quadratic finite elements in general one has only a suboptimal $\mathcal{O}(h^{1.5})$ error behavior (w.r.t. the H^1 norm). The optimal order of 2 (cf. experiment in the previous section) can be achieved by using so-called *isoparametric* elements

# ref.	dim $\mathbf{V}_h$	dim Q_h	$\|\mathbf{u}-\mathbf{u}_h\|_{L^2}$	order	$\|\mathbf{u}-\mathbf{u}_h\|_1$	order	$\|p-p_h\|_{L^2}$	order
0	42	10	1.45 E-1	—	8.39 E-1	—	3.09 E+0	—
1	540	54	9.76 E-3	3.89	9.02 E-2	3.22	6.18 E-1	2.74
2	5 208	340	9.50 E-4	3.36	1.43 E-2	2.66	1.48 E-1	2.05
3	45 360	2 376	2.19 E-4	2.12	3.43 E-3	2.06	3.65 E-2	2.02
4	377 952	17 680	5.96 E-5	1.88	1.15 E-3	1.58	9.07 E-3	2.01
5	3 084 480	136 224	1.57 E-5	1.92	4.11 E-4	1.48	2.27 E-3	2.00

Table 3.2. Dimension of the finite element spaces and convergence behavior w.r.t. L^2 and H^1 norm for different refinement levels.

which are defined on curved tetrahedra by applying a non-affine transformation to the reference tetrahedron. A short discussion on this topic can be found in Sect. 3.4.

3.4 Discussion and additional references

In this monograph we restrict to *finite element* discretization approaches for the incompressible Navier-Stokes equations. Other important discretization methods, which are particularly popular among engineers, are based on *finite volume* techniques. For an introduction to these methods we refer to [252, 108, 248].

We treated only the (very popular) Hood-Taylor finite element spaces. These spaces are members of the family of *conforming* finite element spaces, which means that the spaces $(\mathbf{V}_h, Q_h)$ used for discretization are subspaces of the spaces in which the weak formulation is well-posed. In our setting we have $\mathbf{V}_h \subset H_0^1(\Omega)^3$, $Q_h \subset L_0^2(\Omega)$. Below we briefly address a few issues related to other finite element techniques for (Navier-)Stokes equations.

Other LBB stable pairs of conforming spaces. There are other conforming finite element spaces that are used for the discretization of (Navier-)Stokes equations. We mention two well-known techniques. For this we need the barycentric coordinates defined in Definition 3.2.2. Let $b_T(x) := \Pi_{i=1}^4 \lambda_i(x)$ be the product of the barycentric coordinates for $x \in T$. This "bubble function" is a polynomial of degree 4 in T that is zero on ∂T. It is extended by zero values outside T. Define

$$B_4 := \left\{ v \in C(\overline{\Omega}) : v_{|T} \in \operatorname{span}(b_T) \quad \text{for all} \quad T \in \mathcal{T}_h \right\}.$$

The *"mini-element"* is defined as the pair of spaces $(\mathbf{V}_h, Q_h)$ with

$$\mathbf{V}_h = (\mathbb{X}_{h,0}^1 \oplus B_4)^3, \quad Q_h = \mathbb{X}_h^1 \cap L_0^2(\Omega),$$

i.e., for the velocity we use the space of continuous piecewise linears extended by the space of bubble functions, and for the pressure we use the space of continuous piecewise linears. An advantage of this pair compared to the $P_2 - P_1$

Hood-Taylor pair is that the mini-element allows a simpler data structure: for the bubble functions one has one unknown per velocity component in each tetrahedron and all other unknowns for velocity and pressure are located at the vertices of the tetrahedra. The mini-element is one order less accurate than the $P_2 - P_1$ Hood-Taylor pair.

In the *Crouzeix-Raviart* pair of spaces one uses *discontinuous pressure* approximations. For the velocity one uses continuous piecewise polynomials, enriched (in order to guarantee stability) with bubble functions. More precisely:

$$\begin{aligned}\mathbf{V}_h &= \{\, \mathbf{v} \in C(\overline{\Omega})^3 : \mathbf{v}_{|T} \in (\mathbb{X}_{h,0}^2 \oplus B_4)^3 \quad \text{for all } \ T \in \mathcal{T}_h \,\} \\ Q_h &= \{\, q \in L_0^2(\Omega) : q_{|T} \in \mathcal{P}_1 \quad \text{for all } \ T \in \mathcal{T}_h \,\}.\end{aligned}$$

Similar Crouzeix-Raviart pairs are defined using higher order polynomials. Both the mini-element and the Crouzeix-Raviart pair are LBB-stable and conforming spaces. The enrichment of the velocity space using the bubble functions is important for the LBB stability property to hold. An extensive treatment of these spaces is given in e.g. [121, 206].

LBB stable nonconforming spaces. In the nonconforming case one uses a finite element pair $(\mathbf{V}_h, Q_h)$ with $\mathbf{V}_h \not\subset \mathbf{V}$ or $Q_h \not\subset Q$. An example from this class of finite element methods is the lowest order *nonconforming Crouzeix-Raviart* pair. We explain this pair. For a tetrahedron T the set of its 4 faces is denoted $\mathcal{F} = \{F\}$. The barycenter (center of gravity) of a triangle F is denoted by C_F. We introduce the space of piecewise linear functions:

$$\begin{aligned}\mathbf{V}_h^{CR} := \{\, \mathbf{v}_h \in L^2(\Omega)^3 \mid \forall\, T \in \mathcal{T}_h : \ (\mathbf{v}_h)_{|T} \in \mathcal{P}_1, \ \ [\mathbf{v}_h]_F(C_F) = 0 \quad \forall F \in \mathcal{F}, \\ \mathbf{v}_h(C_F) = 0 \quad \forall F \subset \partial\Omega \,\}.\end{aligned}$$

Here $[\mathbf{v}_h]_F$ denotes the jump of $\mathbf{v}_h$ across the face F. Due to the fact that functions from $\mathbf{V}_h^{CR}$ are not necessarily continuous across the faces of a tetrahedron, this space is nonconforming: $\mathbf{V}_h^{CR} \not\subset \mathbf{V}$. If for the pressure one uses the (conforming) space $\mathbb{X}_h^0$ of piecewise constants, then the pair $(\mathbf{V}_h^{CR}, \mathbb{X}_h^0)$ is LBB stable. A detailed treatment of this and other nonconforming pairs is given in [55, 74].

Unstable pairs: stabilization. Instead of using an LBB stable pair of finite element spaces for discretization of a saddle point problem, one can also use an unstable pair and apply the technique of stabilization. We outline a popular technique introduced in [146]. Let $(\mathbf{V}_h, Q_h)$ be a pair of conforming finite element spaces, not necessarily LBB stable. To simplify the presentation we assume that the pressure space Q_h contains only continuous pressure functions. We consider the stationary Stokes problem in weak formulation: determine $(\mathbf{u}, p) \in \mathbf{V} \times Q = H_0^1(\Omega)^3 \times L_0^2(\Omega)$ such that

$$a(\mathbf{u}, \mathbf{v}) + b(\mathbf{v}, p) - b(\mathbf{u}, q) = (\mathbf{g}, \mathbf{v})_{L^2} \quad \text{for all } \ (\mathbf{v}, q) \in \mathbf{V} \times Q. \tag{3.40}$$

The stabilized discretization reads as follows: determine $(\mathbf{u}_h, p_h) \in \mathbf{V}_h \times Q_h$ such that

$$a(\mathbf{u}_h, \mathbf{v}_h) + b(\mathbf{v}_h, p_h) - b(\mathbf{u}_h, q_h) + \delta s_h(\mathbf{u}_h, p_h; \mathbf{v}_h, q_h) = (\mathbf{g}, \mathbf{v}_h)_{L^2} + \delta g_h(\mathbf{v}_h, q_h)$$

for all $(\mathbf{v}_h, q_h) \in \mathbf{V}_h \times Q_h$, with the stabilization terms

$$\begin{aligned} & s_h(\mathbf{u}_h, p_h; \mathbf{v}_h, q_h) \\ & := \sum_{T\in\mathcal{T}_h} h_T^3 \int_T \big(-\frac{1}{Re}\Delta \mathbf{u}_h + \nabla p_h \big) \cdot \big(-\frac{1}{Re}\Delta \mathbf{v}_h + \nabla q_h \big) + \operatorname{div} \mathbf{u}_h \operatorname{div} \mathbf{v}_h \, dx, \\ & g_h(\mathbf{v}_h, q_h) := \sum_{T\in\mathcal{T}_h} h_T^3 \int_T \mathbf{g} \cdot \big(-\frac{1}{Re}\Delta \mathbf{v}_h + \nabla q_h \big) \, dx, \end{aligned}$$

and a stabilization parameter $\delta > 0$. Due to the stabilization term it can be shown that the bilinear form

$$(\mathbf{u}_h, p_h; \mathbf{v}_h, q_h) \to a(\mathbf{u}_h, \mathbf{v}_h) + b(\mathbf{v}_h, p_h) - b(\mathbf{u}_h, q_h) + \delta s_h(\mathbf{u}_h, p_h; \mathbf{v}_h, q_h)$$

satisfies a discrete inf-sup condition (in suitable norms) on $\mathbf{V}_h \times Q_h$ with a strictly positive inf-sup constant independent of h. This inf-sup property then leads to (optimal) discretization error bounds. Note that in such a method one has to choose an "appropriate" value for the stabilization parameter δ. In the literature this stabilization technique is known as a *Galerkin/Least-Squares method* (GaLS). To explain this name, we consider the Stokes problem in the formal operator form

$$\begin{pmatrix} A & B^T \\ B & 0 \end{pmatrix} \begin{pmatrix} \mathbf{u} \\ p \end{pmatrix} = \begin{pmatrix} \mathbf{g} \\ 0 \end{pmatrix}.$$

In the stabilization we add a term of the *least-squares* form

$$\Big\langle \begin{pmatrix} A & B^T \\ B & 0 \end{pmatrix} \begin{pmatrix} \mathbf{u}_h \\ p_h \end{pmatrix} - \begin{pmatrix} \mathbf{g} \\ 0 \end{pmatrix}, \begin{pmatrix} A & B^T \\ B & 0 \end{pmatrix} \begin{pmatrix} \mathbf{v}_h \\ q_h \end{pmatrix} \Big\rangle.$$

The discretization method is consistent in the sense that if in the stabilization term $(\mathbf{u}_h, p_h)$ is replaced by the solution $(\mathbf{u}, p)$ of the continuous problem (3.40) then the stabilization term is equal to zero. Due to this one still has the important Galerkin orthogonality property. We will use a variant of this method for the discretization of the level set equation in Sect. 7.2. For more explanation and other stabilization techniques we refer to the literature, e.g. [106, 110, 206, 239].

Spectral and hp-finite element methods In this monograph we restrict ourselves to the class of h-finite element methods. This means that in the finite element spaces we use polynomials of a fixed low degree (e.g. the P_2-P_1 Hood-Taylor pair) and a desired accuracy of the discretization is obtained by taking a mesh size that is sufficiently small ("h-refinement"). An alternative is to use a fixed mesh size h and then use polynomials of (very) high degree to obtain an accurate discretization ("p-refinement"). This leads to the class of so-called *spectral methods*, cf. [206]. If one uses a hybrid approach in the sense that in the discretization both the mesh size and the polynomial degree (per element T in the triangulation) are varied this leads to a hp-finite element method, cf. [217].

Discontinuous Galerkin techniques. In the past few years the discontinuous Galerkin method (DG) has received much attention. In this approach one uses finite element spaces consisting of piecewise polynomials without inter-element continuity requirement, i.e. instead of the space $\mathbb{X}_h^k$, $k \geq 1$, in (3.3b) one uses

$$\mathbb{X}_{h,DG}^k := \{\, v : \Omega \to \mathbb{R} : v_{|T} \in \mathcal{P}_k \quad \text{for all} \quad T \in \mathcal{T}_h \,\}.$$

In order to enforce smoothness across the faces of the elements T, the bilinear form is modified by adding suitable jump terms of the discrete test and trial functions across the element faces. The DG approach can be combined with stabilization techniques and due to the nice locality property (no continuity requirement across the faces in the finite element space) its use in an hp-finite element setting is very natural. Discontinuous Galerkin methods turn out to be particularly suitable for hyperbolic and convection-dominated problems. For further information we refer to the literature, e.g. [108, 16, 140].

Isoparametric finite elements. A further issue that is relevant in the context of finite element methods is how to treat *curved boundaries*. The effect of an inaccurate boundary approximation is seen in the numerical experiment in Sect. 3.3.2. For an accurate boundary approximation so-called isoparamtric finite elements are very useful. For a treatment of this standard finite element technique we refer to the literature, e.g. [70, 53, 48].

Mass conservation property. In an incompressible flow problem in weak formulation the mass conservation property is described by

$$\int_\Omega q \operatorname{div} \mathbf{u} \, dx = 0 \quad \text{for all} \quad q \in L_0^2(\Omega).$$

Hence, $\operatorname{div} \mathbf{u} = 0$ holds on Ω, in L^2-sense. In the discrete problem, with finite element spaces $\mathbf{V}_h$ and Q_h for velocity and pressure, respectively, one obtains the variational equation

$$\int_\Omega q_h \operatorname{div} \mathbf{u}_h \, dx = 0 \quad \text{for all} \quad q_h \in Q_h, \tag{3.41}$$

for the discrete velocity solution $\mathbf{u}_h \in \mathbf{V}_h$. Such a function $\mathbf{u}_h$ is also called *discretely divergence-free*. In general this does *not* imply $\operatorname{div} \mathbf{u}_h = 0$ in Ω (in L^2-sense) due to $Q_h \neq L_0^2(\Omega)$. This leads to the issue of how well mass is conserved in the finite element discretization. We briefly address two approaches that apply to the Hood-Taylor finite element method and result in discretizations with good mass conservation properties.

The first method is from [237] and is based on an extension of the pressure finite element space by piecewise constants. Let $\mathbb{X}_h^k$ and $\mathbb{X}_{h,0}^k$ be the polynomial finite element spaces introduced in Sect. 3.2. For the velocity discretization we use the same space as in the Hood-Taylor method , i.e., $\mathbf{V}_h = (\mathbb{X}_{h,0}^k)^d$, $k \geq 2$. For the pressure we extend the Hood-Taylor pressure space by the space of piecewise constants, i.e., we take $Q_h = (\mathbb{X}_h^{k-1} \cup \mathbb{X}_h^0) \cap L_0^2(\Omega)$. In [237] it is proved that for $k = 2$, $d = 2$, the pair $(\mathbf{V}_h, Q_h)$ is LBB-stable. In the discrete variational equation (3.41) corresponding to this finite element pair one can take, for $T \in \mathcal{T}_h$, the function $q_h(x) = 1 + c$ if $x \in T$, $q_h(x) = c$ otherwise, with a constant c such that $\int_\Omega q_h \, dx = 0$ holds. This results in

$$\begin{aligned} 0 &= \int_T \operatorname{div} \mathbf{u}_h \, dx + c \int_\Omega \operatorname{div} \mathbf{u}_h \, dx \\ &= \int_T \operatorname{div} \mathbf{u}_h \, dx + c \int_{\partial\Omega} \mathbf{u}_h \cdot \mathbf{n} \, ds = \int_T \operatorname{div} \mathbf{u}_h \, dx, \end{aligned}$$

where in the last identity we used that $\mathbf{u}_h = 0$ on $\partial\Omega$. Hence we obtain a *local mass-conservation property* in the sense that $\int_T \operatorname{div} \mathbf{u}_h \, dx = 0$ holds for all $T \in \mathcal{T}_h$.

The second method is based on the so-called Scott-Vogelius finite element pair [218]. In this pair, for the velocity one uses the same space as in the Hood-Taylor method, i.e., $\mathbf{V}_h = (\mathbb{X}_{h,0}^k)^d$, $k \geq 2$, and for the pressure one uses piecewise polynomials of degree $k - 1$ that are not necessarily continuous, i.e.

$$Q_h = \left\{ v \in L_0^2(\Omega) : v_{|T} \in \mathcal{P}_{k-1} \quad \text{for all} \quad T \in \mathcal{T}_h \right\}.$$

In [218] it is proved that for $d = 2$ the pair $(\mathbf{V}_h, Q_h)$ with $k \geq 4$ is LBB-stable provided the mesh does not contain any so-called nearly-singular vertices. Further analyses of this pair can be found in [263, 264, 265]. We outline two stability results for $d = 3$. For this we need special tetrahedral grids. Starting from a regular tetrahedral triangulation $\mathcal{T}_h$ a so-called Hsieh-Clough-Tocher triangulation is obtained by subdiving each tetrahedron $T \in \mathcal{T}_h$ into 4 subtetrahedra by connecting the barycenter of T to the four vertices of T. If each tetrahedron of this Hsieh-Clough-Tocher triangulation is further refined into 3 subtetrahedra by connecting the barycenter of T to each of the four barycenters of the four tetrahedra adjacent to T one obtains a so-called Powell-Sabin triangulation. In [263] it is proved that on Hsieh-Clough-Tocher triangulations the Scott-Vogelius pair $(\mathbf{V}_h, Q_h)$ is LBB-stable for $k \geq 3$. In [265] it is proved that on Powell-Sabin triangulations the pair $(\mathbf{V}_h, Q_h)$ is LBB-stable for $k = 2$. For the Scott-Vogelius pair $(\mathbf{V}_h, Q_h)$ one can take $q_h = \operatorname{div} \mathbf{u}_h + c$ in (3.41), with a constant c such that $\int_\Omega q_h \, dx = 0$ holds. Then one obtains

$$\int_\Omega (\operatorname{div} \mathbf{u}_h)^2 \, dx = 0,$$

and thus the discrete velocity is divergence-free, i.e., a discrete mass conservation property $\operatorname{div} \mathbf{u}_h = 0$ in Ω (in L^2-sense) holds. A relation between the Scott-Vogelius pair and the Hood-Taylor pair is derived in [62]. There it is shown that if the Hood-Taylor discretization is applied to the Navier-Stokes equation with grad-div stabilization, i. e., one adds a (consistent) term $\gamma(\operatorname{div} \mathbf{u}_h, \operatorname{div} \mathbf{v}_h)_{L^2}$ to the disrete momentum equation, then the resulting discrete velocity $\mathbf{u}_h$ tends to $\mathbf{u}_h^{SV}$ if $\gamma \to \infty$. Here $\mathbf{u}_h^{SV}$ denotes the divergence-free discrete velocity solution obtained by using the Scott-Vogelius pair Hence, for the Hood-Taylor pair the mass conservation property can be improved (significantly) if grad-div stabilization is used.

Type of triangulations. In this monograph we restrict ourselves to finite elements on *tetrahedral* triangulations. Finite element techniques, however, can also be applied using a subdivision of the three-dimensional domain into, for example, hexahedra or prisms. One can even use subdivisions consisting of combinations of tetrahedra, hexahedra and prisms. An important class are the tensor product finite elements (in the 2D case, these are also called quadrilateral finite elements). For a treatment of these we refer to standard finite element literature.

4

Time integration

4.1 Introduction

Let $I := [0, t_e]$, $f : I \to \mathbb{R}^N$, $F : \mathbb{R}^N \to \mathbb{R}^N$ and $u_0 \in \mathbb{R}^N$. Consider an initial value problem: determine $u(t) \in \mathbb{R}^N$ such that

$$\frac{du}{dt} + F(u) = f(t) \quad \text{for} \;\; t \in I, \quad u(0) = u_0. \tag{4.1}$$

As we will see further on, the Stokes- and Navier-Stokes systems of DAEs in (3.22) and (3.38) take this form if one eliminates the pressure variable by restricting to the subspace of (discrete) divergence free velocities. Related to existence and uniqueness of a solution of (4.1) we give a standard result from the literature (Picard-Lindelöf theorem). For $b > 0$ define $G_b := \{ v \in \mathbb{R}^N : \|v - u_0\| \leq b \}$ with $\| \cdot \|$ any given norm on $\mathbb{R}^N$. Assume that for $a > 0$ the function $f : [0, a] \to \mathbb{R}^N$ is continuous and that F satisfies the Lipschitz condition:

$$\|F(v) - F(w)\| \leq L\|v - w\| \quad \text{for all} \;\; v, w \in G_b. \tag{4.2}$$

Then the initial value problem in (4.1) *has a unique solution* $u(t)$ for $t \in [0, \alpha]$ with $\alpha := \min\{a, bL^{-1}\}$. In the remainder we assume that f and F satisfy these conditions (for suitable a, b, L) and that $t_e \leq \alpha$, i.e., (4.1) has a unique solution.

We discuss a few discretization methods for the general problem (4.1). In our applications the systems are very stiff and thus we need implicit methods. A classical and still very popular method is the θ-scheme:

$$\frac{u^{n+1} - u^n}{\Delta t} + \theta F(u^{n+1}) + (1 - \theta) F(u^n) = \theta f(t_{n+1}) + (1 - \theta) f(t_n), \tag{4.3}$$

with $\theta \in [0,1]$. For $\theta = 1$ this is the *implicit Euler scheme* and for $\theta = \frac{1}{2}$ this method is known as the *Crank-Nicolson method.* Another popular method is the *BDF2 scheme*:

$$\frac{3}{2}u^{n+1} - 2u^n + \frac{1}{2}u^{n-1} + \Delta t F(u^{n+1}) = \Delta t f(t_{n+1}). \tag{4.4}$$

Note that the θ-scheme is a *one*-step method, whereas the BDF2 method is a linear *two*-step scheme. Another method that is used in our applications is the following *fractional-step θ-scheme.* For a given $\theta \in (0, \frac{1}{2})$, the fractional-step θ-scheme is based on a subdivision of each time interval $[n\Delta t, (n+1)\Delta t]$ in three subintervals with endpoints $(n+\theta)\Delta t$, $(n+1-\theta)\Delta t$, $(n+1)\Delta t$. For given u^n the approximations $u^{n+\theta}$, $u^{n+1-\theta}$, u^{n+1} at these endpoints are defined by

$$\frac{u^{n+\theta} - u^n}{\theta \Delta t} + \alpha F(u^{n+\theta}) + (1-\alpha)F(u^n) = f(t_n) \tag{4.5a}$$

$$\frac{u^{n+1-\theta} - u^{n+\theta}}{(1-2\theta)\Delta t} + (1-\alpha)F(u^{n+1-\theta}) + \alpha F(u^{n+\theta}) = f(t_{n+1-\theta}) \tag{4.5b}$$

$$\frac{u^{n+1} - u^{n+1-\theta}}{\theta \Delta t} + \alpha F(u^{n+1}) + (1-\alpha)F(u^{n+1-\theta}) = f(t_{n+1-\theta}). \tag{4.5c}$$

Standard measures for the quality of discretization methods for (stiff) initial value problems are consistency, stability, a smoothing property and the amount of dissipativity. Below we treat these quality measures for the methods that we consider.

Consistency

The implicit Euler method has a consistency order of 1. The Crank-Nicolson, BDF2 and fractional-step θ-scheme, with $\theta = 1 \pm \frac{1}{2}\sqrt{2}$, all have consistency order 2. We derive this consistency result for the fractional-step θ-scheme.

Lemma 4.1.1 *Assume an arbitrary $f \in C^2([0,t_e])$ and $\lambda \in \mathbb{R}$. Let $u(t)$ be the solution of $\frac{du}{dt} - \lambda u = f$, $u(0) = u_0$. Let u^{n+1} be the result of the fractional-step θ-scheme* (4.5) *applied to this problem with $u^n := u(t_n)$. Then for $\theta = 1 \pm \frac{1}{2}\sqrt{2}$ we have*

$$|u(t_{n+1}) - u^{n+1}| \leq c\,(\Delta t)^3, \tag{4.6}$$

with a constant c independent of Δt and n.

Proof. We take $\theta = 1 \pm \frac{1}{2}\sqrt{2}$. For the solution $u(t)$ we have

$$u(t) = e^{\lambda(t-t_n)}u(t_n) + \int_{t_n}^{t} e^{\lambda(t-\tau)} f(\tau)\, d\tau, \quad t \geq t_n.$$

Hence, with $z := \lambda \Delta t$,

$$u(t_{n+1}) = e^z u(t_n) + \int_{t_n}^{t_{n+1}} e^{\lambda(t_{n+1}-\tau)} f(\tau)\, d\tau \,.$$

A straightforward calculation, in which we use that $2\theta^2 - 4\theta + 1 = 0$ holds, results in

$$\begin{aligned} u^{n+1} &= g(z)u^n + \Delta t\big[\theta(1+z) - (1-\alpha)\theta^2 z + \mathcal{O}(z^2)\big] f(t_n) \\ &+ \Delta t\big[(1-\theta)(1+\theta z) + (1-\alpha)(3\theta^2 - 4\theta + 1)z + \mathcal{O}(z^2)\big] f(t_{n+1-\theta}) \\ &= g(z)u^n + \Delta t\big(\theta(1+z) f(t_n) + (1-\theta)(1+\theta z) f(t_{n+1-\theta})\big) + \mathcal{O}(\Delta t^3), \end{aligned} \tag{4.7}$$

with

$$g(z) := \frac{\big(1 + (1-\alpha)\theta z\big)^2 \big(1 + \alpha(1-2\theta)z\big)}{\big(1 - \alpha\theta z\big)^2 \big(1 - (1-\alpha)(1-2\theta)z\big)}. \tag{4.8}$$

Taylor expansion results in

$$g(z) = 1 + z + \frac{1}{2} z^2 \big[1 + (1-2\alpha)(2\theta^2 - 4\theta + 1)\big] + \mathcal{O}(z^3) \quad (z \to 0).$$

For $\theta = 1 \pm \frac{1}{2}\sqrt{2}$ we have $2\theta^2 - 4\theta + 1 = 0$ and thus

$$g(z) = e^z + \mathcal{O}(z^3) \tag{4.9}$$

holds. The quadrature rule $\int_0^1 v(t)\, dt \approx \xi v(0) + (1-\xi) v(1-\xi)$ is exact for all linear functions v iff $\xi = 1 \pm \frac{1}{2}\sqrt{2}$. Thus for $\theta = 1 \pm \frac{1}{2}\sqrt{2}$ we have

$$\begin{aligned} \int_{t_n}^{t_{n+1}} e^{\lambda(t_{n+1}-\tau)} f(\tau)\, d\tau &= \Delta t\big(\theta e^z f(t_n) + (1-\theta) e^{\theta z} f(t_{n+1-\theta})\big) + \mathcal{O}(\Delta t^3) \\ &= \Delta t\big(\theta(1+z) f(t_n) + (1-\theta)(1+\theta z) f(t_{n+1-\theta})\big) + \mathcal{O}(\Delta t^3). \end{aligned}$$

Using this in combination with (4.7), (4.9) we get

$$\begin{aligned} u^{n+1} &= e^z u(t_n) + \Delta t\big(\theta(1+z) f(t_n) + (1-\theta)(1+\theta z) f(t_{n+1-\theta})\big) + \mathcal{O}(\Delta t^3) \\ &= e^z u(t_n) + \int_{t_n}^{t_{n+1}} e^{\lambda(t_{n+1}-\tau)} f(\tau)\, d\tau + \mathcal{O}(\Delta t^3) \\ &= u(t_{n+1}) + \mathcal{O}(\Delta t^3), \end{aligned}$$

and thus the result is proved. □

A similar bound as in (4.6) can be derived for the case that F is a nonlinear function which satisfies the Lipschitz condition in (4.2). Thus for $\theta = 1 \pm \frac{1}{2}\sqrt{2}$ the fractional-step θ-scheme has consistency order 2.

Remark 4.1.2 For the fractional-step θ-scheme to have consistency order 2 it is *necessary* to take the value $\theta = 1 \pm \frac{1}{2}\sqrt{2}$ in the following sense. Consider the special case $\lambda = 0$, $u_0 = 0$, $f(t) = t$, $n = 0$. Then we have $u(t_1) = u(\Delta t) = \frac{1}{2}(\Delta t)^2$ and a simple computation yields $u^1 = (1-\theta)^2(\Delta t)^2$. Thus for (4.6) to hold *with a* θ*-value independent of* Δt we need $(1-\theta)^2 = \frac{1}{2}$, i.e., $\theta = 1 \pm \frac{1}{2}\sqrt{2}$.

Remark 4.1.3 Consider the following variant of the fractional-step θ-scheme, with $G(u,t) := F(u) - f(t)$:

$$\frac{u^{n+\theta} - u^n}{\theta \Delta t} + \alpha G(u^{n+\theta}, t_{n+\theta}) + (1-\alpha)G(u^n, t_n) = 0$$

$$\frac{u^{n+1-\theta} - u^{n+\theta}}{(1-2\theta)\Delta t} + (1-\alpha)G(u^{n+1-\theta}, t_{n+1-\theta}) + \alpha G(u^{n+\theta}, t_{n+\theta}) = 0$$

$$\frac{u^{n+1} - u^{n+1-\theta}}{\theta \Delta t} + \alpha G(u^{n+1}, t_{n+1}) + (1-\alpha)G(u^{n+1-\theta}, t_{n+1-\theta}) = 0.$$

This scheme is equal to three steps of the θ-scheme (4.3), where α or $1-\alpha$ takes the role of θ in (4.3), and for the three substeps we use time steps $\theta\Delta t$, $(1-2\theta)\Delta t$ and $\theta\Delta t$, respectively. For an accuracy analysis we consider the same test problem as in Lemma 4.1.1 and take $u^n := u(t_n)$, $\theta = 1 \pm \frac{1}{2}\sqrt{2}$. Along the same lines as in the proof of Lemma 4.1.1 one can derive the following, with $z := \lambda\Delta t$:

$$\begin{aligned} u^{n+1} &= g(z)u^n + (1-\alpha)\Delta t\big[\theta(1+z)f(t_n) + (1-\theta)(1+\theta z)f(t_{n+1-\theta})\big] \\ &\quad + \alpha\Delta t\big[(1-\theta)\big(1+(1-\theta)z\big)f(t_{n+\theta}) + \theta f(t_{n+1})\big] + \mathcal{O}(\Delta t^3). \end{aligned}$$

For $\int_0^1 v(t)\,dt$ the quadrature rules $\theta v(0) + (1-\theta)v(1-\theta)$ and $(1-\theta)v(\theta) + \theta v(1)$ are exact for all linear functions. Hence, using the Taylor expansion $e^z = 1 + z + \mathcal{O}(z^2)$ we get

$$\begin{aligned} &\int_{t_n}^{t_{n+1}} e^{\lambda(t_{n+1}-\tau)} f(\tau)\,d\tau \\ &= (1-\alpha)\Delta t\big[\theta(1+z)f(t_n) + (1-\theta)(1+\theta z)f(t_{n+1-\theta})\big] \\ &+ \alpha\Delta t\big[(1-\theta)\big(1+(1-\theta)z\big)f(t_{n+\theta}) + \theta f(t_{n+1})\big] + \mathcal{O}(\Delta t^3). \end{aligned}$$

Thus as in the proof of Lemma 4.1.1 we obtain

$$u^{n+1} = e^z u(t_n) + \int_{t_n}^{t_{n+1}} e^{\lambda(t_{n+1}-\tau)} f(\tau)\,d\tau + \mathcal{O}(\Delta t^3) = u(t_{n+1}) + \mathcal{O}(\Delta t^3).$$

Hence, this variant has consistency order 2, too.

Stability

For an error analysis of time discretization methods for stiff problems *stability* properties have to be considered. For a stability analysis, these methods are applied to the test problem

$$\frac{du}{dt} = \lambda u, \qquad \lambda \in \mathbb{C},\ \mathrm{Re}(\lambda) \le 0. \tag{4.10}$$

A solution of this test problem satisfies the growth relation

$$|u(t_{n+1})| = |e^{\lambda \Delta t}||u(t_n)| = e^{\mathrm{Re}(\lambda)\Delta t}|u(t_n)|. \tag{4.11}$$

Due to $\mathrm{Re}(\lambda) \le 0$ the growth factor satisfies $0 \le e^{\mathrm{Re}(\lambda)\Delta t} \le 1$. For one-step methods applied to this test problem one obtains $|u^{n+1}| = g(\lambda \Delta t)|u^n|$, with a so-called *stability function* $g(z)$ which is an approximation of the growth factor $|e^z|$ in (4.11). For the implicit Euler, Crank-Nicolson and fractional-step θ-scheme (cf. (4.8)) the stability function is given by:

$$\begin{aligned} g_{EB}(z) &:= \left|\frac{1}{1-z}\right| \\ g_{CN}(z) &:= \left|\frac{1+\frac{1}{2}z}{1-\frac{1}{2}z}\right| \\ g_{FS}(z) &:= \left|\frac{\big(1+(1-\alpha)\theta z\big)^2\big(1+\alpha(1-2\theta)z\big)}{\big(1-\alpha\theta z\big)^2\big(1-(1-\alpha)(1-2\theta)z\big)}\right|. \end{aligned}$$

The variant of the fractional-step θ-scheme discussed in Remark 4.1.3 also has the stability function g_{FS}. For the BDF2 method one obtains $u^n = c_0\left(\frac{2+\sqrt{1+2z}}{3-2z}\right)^n + c_1\left(\frac{2-\sqrt{1+2z}}{3-2z}\right)^n$, with $z := \lambda\Delta t$ and constants c_0, c_1 that depend on the starting values u^0, u^1. The stability function of the BDF2 method is given by

$$g_{BDF}(z) := \max\left\{\left|\frac{2+\sqrt{1+2z}}{3-2z}\right|, \left|\frac{2-\sqrt{1+2z}}{3-2z}\right|\right\}.$$

For a given method with stability function g the so-called *stability region* is defined by

$$S := \{\, z \in \mathbb{C} :\ g(z) \le 1 \,\}.$$

The method is said to be A-stable if

$$\mathbb{C}^- := \{\, z \in \mathbb{C} :\ \mathrm{Re}(z) \le 0 \,\} \subset S$$

holds. From standard literature on time discretization methods for (stiff) initial value problems, cf. [134], it is known that the backward -Euler, Crank-Nicolson method and BDF2 method are A-stable.

We consider the fractional-step θ-scheme with $\theta = 1 \pm \frac{1}{2}\sqrt{2}$. Due to the structure of the fractional-step θ-scheme it is natural to restrict to $\alpha \in [0,1]$. First the case $\theta = 1 - \frac{1}{2}\sqrt{2}$ is treated.

Lemma 4.1.4 *Take $\alpha \in [0,1]$, $\theta = 1 - \frac{1}{2}\sqrt{2}$. The fractional-step θ-scheme is A-stable iff $\alpha \in [\frac{1}{2}, 1]$.*

Proof. For $\alpha > 0$ we have

$$\lim_{z \to -\infty} g_{FS}(z) = \left| \frac{1-\alpha}{\alpha} \right| = \frac{1-\alpha}{\alpha}.$$

Since $\frac{1-\alpha}{\alpha} > 1$ for $\alpha < \frac{1}{2}$ the method is not A-stable for $\alpha < \frac{1}{2}$. We consider $\alpha \geq \frac{1}{2}$. The denominator in the function g_{FS} has no zero in $\mathbb{C}^-$ and thus g_{FS} is the norm of a function that is analytic on $\mathbb{C}^-$. From the maximum principle for analytic functions it follows that

$$\max_{z \in \mathbb{C}^-} g_{FS}(z) = \max_{y \in \mathbb{R}} g_{FS}(iy).$$

Due to $g_{FS}(iy) = g_{FS}(-iy)$ we can restrict to $y \in [0, \infty)$. Note that $g_{FS}(0) = 1$ and $\lim_{y \to \infty} g_{FS}(iy) = \frac{1-\alpha}{\alpha} \leq 1$. A straightforward computation yields that on $[0, \infty)$ the derivative of the function $y \to g_{FS}(iy)$ is less than or equal to 0. Hence $\max_{y \in \mathbb{R}} g_{FS}(iy) \leq g_{FS}(0) = 1$. □

We now consider $\theta = 1 + \frac{1}{2}\sqrt{2}$, $\alpha \in [0, 1]$. For $\alpha < 1$ the denominator has a zero at $z_0 = (1-\alpha)^{-1}(1-2\theta)^{-1} < 0$. For the value $z = z_0$ the nominator is not equal to zero. Hence $\lim_{z \to z_0} g_{FS}(z) = \infty$ and thus the method is not A-stable. For $\alpha = 1$ it can be shown with the same arguments as in the proof of Lemma 4.1.4 that the method is A-stable. *Below, for the fractional-step θ-scheme we restrict to* $\theta = 1 - \frac{1}{2}\sqrt{2}$, $\alpha \in [\frac{1}{2}, 1]$, or $\theta = 1 + \frac{1}{2}\sqrt{2}$, $\alpha = 1$. For these parameter values the method has consistency order 2 and is A-stable.

Smoothing property

A further criterion which is relevant for comparing these methods is the notion of smoothing, which quantifies the amount of damping of the numerical solutions of (4.10) with λ such that $\mathrm{Re}(\lambda) \to -\infty$ (i.e. of "high" frequencies). For $\mathrm{Re}(\lambda) \to -\infty$ the growth factor $e^{\mathrm{Re}(\lambda)\Delta t}$, cf. (4.11), tends to zero. The smoothing property measures how well this strong damping behavior for $\mathrm{Re}(\lambda) \to -\infty$ is reflected in the numerical scheme. The method has a *smoothing property* if there exists a constant $\delta < 1$ such that for the corresponding stability function g we have

$$\lim_{\mathrm{Re}(z) \to -\infty} g(z) \leq \delta. \tag{4.12}$$

The size of δ is a measure for the strength of the smoothing: a small δ value corresponds to a strong smoothing. A strong smoothing is a desirable property of a numerical scheme. One easily verifies that for the backward Euler and the BDF2 methods we have a maximal smoothing effect, namely with $\delta = 0$. For the Crank-Nicolson method there is no smoothing at all: $\delta = 1$. For the fractional-step θ-method we have a smoothing effect with $\delta = \frac{1}{\alpha} - 1$, and thus the smoothing effect increases for larger α.

Dissipativity

The last property that we consider is the *amount of dissipativity* of a method. This is a measure for the quality of the numerical method when applied to (4.10) with a *periodic* solution of the form $u(t) = e^{ixt}$, $x \in \mathbb{R}$, i.e. with $\lambda = ix$. Hence, in (4.11) we then have a growth factor $e^{\mathrm{Re}(\lambda)\Delta t} = 1$. In this case we have to consider the corresponding stability functions $g(z)$ with $z = ix$, $x \in \mathbb{R}$. The amount of dissipativity is measured by the *deviation* of

$$d(x) := g(ix), \quad x \in \mathbb{R}, \tag{4.13}$$

from the optimal value 1. For the implicit Euler method we have

$$d_{EB}(x) = \frac{1}{\sqrt{1+x^2}},$$

and thus an increasing amount of dissipativity for larger x values. For the Crank-Nicolson we have

$$d_{CN}(x) = 1,$$

and thus *no dissipativity*. For the fractional-step θ-scheme the following holds. For $\theta = 1 + \frac{1}{2}\sqrt{2}$, $\alpha = 1$ we have $d_{FS}(x) = g_{FS}(ix) = \left(1 + (1-2\theta)x^2\right)^{\frac{1}{2}}\left(1 + \theta^2 x^2\right)^{-1}$, which is monotonically decreasing with value 0 for $x \to \infty$, thus in this case there is a large amount of dissipativity for large x values. For the case $\theta = 1 - \frac{1}{2}\sqrt{2}$, $\alpha \in [\frac{1}{2}, 1]$ the dissipativity function depends on α: $d_{FS,\alpha}(x) := g_{FS}(ix)$. We have $\lim_{x\to\infty} d_{FS,\alpha}(x) = \frac{1}{\alpha} - 1$. Inspection of the function $d_{FS,\alpha}$ yields that it is constant for $\alpha = \frac{1}{2}$ and strictly decreasing for $\alpha \in (\frac{1}{2}, 1]$. Furthermore, we have $d_{FS,\alpha'}(x) < d_{FS,\alpha}(x)$ if $\frac{1}{2} \le \alpha < \alpha' \le 1$ and $x > 0$. Thus we have more dissipativity for larger values of α. For a few cases the dissipativity function $d_{FS,\alpha}$ is illustrated in Fig. 4.1. Due to $d_{FS}(x) = d_{FS}(-x)$ it suffices to show results for $x \ge 0$.

Due to the fact that the stability function g_{FS} is the norm of a rational function we have

$$\lim_{\mathrm{Re}(z)\to-\infty} g_{FS}(z) = \lim_{x\to\infty} g_{FS}(ix).$$

This property also holds for the stability functions of the other three methods. Thus there is a conflict between good smoothing ($\lim_{\mathrm{Re}(z)\to-\infty} g_{FS}(z)$ close to zero, cf. (4.12)) and low dissipativity ($g_{FS}(ix) \approx 1$ for a large range of x values). From the analysis above and Fig. 4.1 we see that for $\theta = 1 - \frac{1}{2}\sqrt{2}$ and $\alpha = \frac{1}{2}$ the fractional-step θ-scheme has the same properties as the Crank-Nicolson method, namely no smoothing ($\delta = 1$) and no dissipativity. For $\theta = 1 + \frac{1}{2}\sqrt{2}$, $\alpha = 1$, the fractional-step θ-scheme has properties similar to those of the implicit Euler method: optimal smoothing ($\delta = 0$) and strong dissipativity. A good compromise is found by taking $\theta = 1 - \frac{1}{2}\sqrt{2}$ and $\alpha \in (0, \frac{1}{2})$. A popular parameter choice, cf. [207, 244], is

$$\theta := 1 - \frac{1}{2}\sqrt{2}, \quad \alpha := \frac{1-2\theta}{1-\theta} = 2 - \sqrt{2}. \tag{4.14}$$

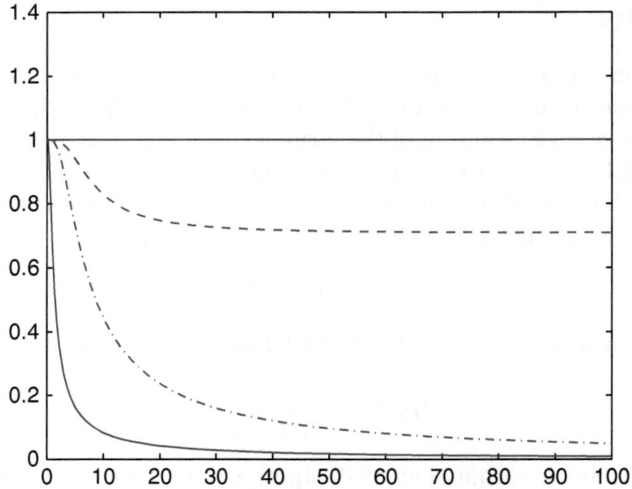

Fig. 4.1. Dissipativity functions $d_{FS,\alpha}$ for $\theta = 1 - \frac{1}{2}\sqrt{2}$, $\alpha \in \{\frac{1}{2}, 2 - \sqrt{2}, 1\}$, and $\theta = 1 + \frac{1}{2}\sqrt{2}$, $\alpha = 1$ (top to bottom).

For these values (cf. Fig. 4.1) the method has "modest" dissipativity and it has a "reasonable" smoothing property with $\delta = \frac{1}{\alpha} - 1 = \frac{1}{2}\sqrt{2}$. Furthermore, due to $\theta\alpha = (1 - 2\theta)(1 - \alpha)$ the systems in (4.5) for the unknowns $u^{n+\theta}$, $u^{n+1-\theta}$, u^{n+1}, respectively, have the same form. *In the remainder we only consider the fractional-step θ-scheme with the parameter values as in* (4.14).

The dissipativity functions $d(x)$ for the implicit Euler, Crank-Nicolson, BDF2 and fractional-step θ-scheme ($\theta = 1 - \frac{1}{2}\sqrt{2}$, $\alpha = 2 - \sqrt{2}$) are illustrated in Fig. 4.2. In all four cases we have $d(-x) = d(x)$ and therefore we show the functions only for $x \geq 0$.

In practice often the Crank-Nicolson method is used. A disadvantage of this method, however, is that it has no smoothing property. The fractional-step θ-schema is a method which has both a good smoothing property and modest dissipativity.

In our applications we will use the implicit Euler method (a simple method with a strong smoothing property), the Crank-Nicolson method and the fractional-step θ-scheme. Note that the implicit Euler and the Crank-Nicolson method are special cases of the θ-scheme.

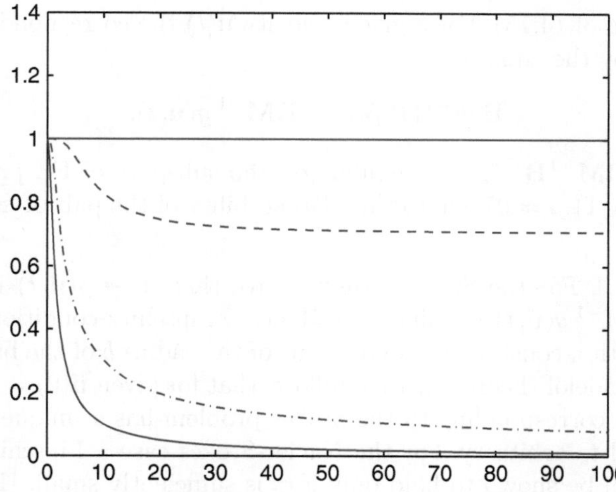

Fig. 4.2. Dissipativity functions d_{CN}, d_{FS}, d_{BDF} and d_{EB} (top to bottom).

4.2 The θ-scheme for the Navier-Stokes problem

The DAE system (3.38) is rewritten in the form

$$\begin{aligned}\frac{d\vec{\mathbf{u}}}{dt}(t) + \mathbf{M}^{-1}\mathbf{B}^T\vec{\mathbf{p}}(t) &= \mathbf{M}^{-1}\, g(\vec{\mathbf{u}}, t), \\ \mathbf{B}\,\vec{\mathbf{u}}(t) &= 0,\end{aligned} \tag{4.15}$$

where

$$g(\vec{\mathbf{u}}, t) := \vec{\mathbf{f}} - \mathbf{A}\vec{\mathbf{u}}(t) - \mathbf{N}(\vec{\mathbf{u}}(t))\vec{\mathbf{u}}(t).$$

The Stokes DAE system (3.22) has a similar form, with $g(\vec{\mathbf{u}}, t) := \vec{\mathbf{f}} - \mathbf{A}\vec{\mathbf{u}}(t)$. We eliminate the incompressibility constraint $\mathbf{B}\vec{\mathbf{u}}(t) = 0$ and the corresponding Lagrange multiplier $\vec{\mathbf{p}}(t)$ to replace the DAE system by an equivalent ODE system. This can be achieved by applying the $\mathbf{M}$-orthogonal projection $\mathbf{P}$ on $\ker \mathbf{B}$:

$$\mathbf{P} = \mathbf{I} - \mathbf{M}^{-1}\mathbf{B}^T(\mathbf{B}\mathbf{M}^{-1}\mathbf{B}^T)^{-1}\mathbf{B}.$$

The projection $\mathbf{P}$ is orthogonal w.r.t. the scalar product $\langle \cdot, \cdot \rangle_{\mathbf{M}} := \langle \mathbf{M}\,\cdot, \cdot \rangle$, and $\mathbf{P}\,\vec{\mathbf{v}} = \vec{\mathbf{v}}$ for all $\vec{\mathbf{v}} \in \ker \mathbf{B}$, furthermore $\mathbf{P}\mathbf{M}^{-1}\mathbf{B}^T = 0$. Hence, instead of a DAE system we obtain a system of ordinary differential equations:

A solution $\vec{\mathbf{u}}(t)$ of (4.15) satisfies

$$\frac{d\vec{\mathbf{u}}}{dt}(t) = \mathbf{P}\mathbf{M}^{-1}\, g(\vec{\mathbf{u}}, t). \tag{4.16}$$

If for a given initial condition $\vec{\mathbf{u}}(0) = \vec{\mathbf{u}}_0$, with $\mathbf{B}\mathbf{u}_0 = 0$, and $t \in [0, t_e]$ (with t_e sufficiently small) the problem in (4.16) has a unique solution, then this $\vec{\mathbf{u}}$ is also a solution of (4.15). For a given velocity $\vec{\mathbf{u}}(t)$ the corresponding pressure $\vec{\mathbf{p}}$ is defined by the equation

$$\mathbf{B}\mathbf{M}^{-1}\mathbf{B}^T\vec{\mathbf{p}}(t) = \mathbf{B}\mathbf{M}^{-1}\, g(\vec{\mathbf{u}}, t). \tag{4.17}$$

The matrix $\mathbf{B}\mathbf{M}^{-1}\mathbf{B}^T$ is nonsingular (on the subspace of FE pressure functions with $(p_h, 1)_{L^2} = 0$) due to the LBB stability of the pair of finite element spaces used.

Remark 4.2.1 For the Stokes case we have that $\vec{\mathbf{u}} \to g(\vec{\mathbf{u}}, t)$ is affine and thus $\vec{\mathbf{u}} \to \mathbf{B}\mathbf{M}^{-1}\, g(\vec{\mathbf{u}}, t)$ is affine, too. Hence a Lipschitz condition as in (4.2) is satisfied with a constant L independent of the radius b of the ball G_b. From the Picard-Lindelöf theorem it then follows that for given $\vec{\mathbf{u}}(0) = \vec{\mathbf{u}}_0$ the ODE system (4.16) corresponding to the Stokes problem has a unique solution for $t \in [0, t_e]$ and t_e arbitrary. For the Navier-Stokes case a Lipschitz condition as in (4.2) can be shown to hold only if t_e is sufficiently small. Hence in that case existence and uniqueness of a solution is guaranteed only for a sufficiently short time interval.

The θ-scheme (4.3) can be applied to the ODE system (4.16), which results in

$$\frac{\vec{\mathbf{u}}^{n+1} - \vec{\mathbf{u}}^n}{\Delta t} = \theta\, \mathbf{P}\mathbf{M}^{-1}\, g(\vec{\mathbf{u}}^{n+1}, t_{n+1}) + (1-\theta)\mathbf{P}\mathbf{M}^{-1}\, g(\vec{\mathbf{u}}^n, t_n). \tag{4.18}$$

We assume that, for a given $\vec{\mathbf{u}}^0$, this recursion has a unique solution (which holds for Δt sufficiently small). In addition we assume that $\mathbf{B}\vec{\mathbf{u}}^0 = 0$ holds. From (4.18) and $\mathbf{B}\mathbf{P} = 0$ it then follows that $\mathbf{B}\vec{\mathbf{u}}^n = 0$ holds for all n. Based on (4.17) we introduce a pressure variable $\vec{\mathbf{p}}^k$ such that the corresponding finite element pressure function p_h satisfies $(p_h, 1)_{L^2} = 0$ and such that $\mathbf{B}\mathbf{M}^{-1}\mathbf{B}^T\vec{\mathbf{p}}^k = \mathbf{B}\mathbf{M}^{-1}\, g(\vec{\mathbf{u}}^k, t_k)$ holds. Using the definition of the projection $\mathbf{P}$ the recurrence relation in (4.18) can be rewritten as

$$\begin{aligned}
&\frac{\vec{\mathbf{u}}^{n+1} - \vec{\mathbf{u}}^n}{\Delta t} + \mathbf{M}^{-1}\mathbf{B}^T\big(\theta\vec{\mathbf{p}}^{n+1} + (1-\theta)\vec{\mathbf{p}}^n\big) \\
&\qquad = \theta\mathbf{M}^{-1}\, g(\vec{\mathbf{u}}^{n+1}, t_{n+1}) + (1-\theta)\mathbf{M}^{-1}\, g(\vec{\mathbf{u}}^n, t_n).
\end{aligned}$$

Thus for given $\vec{\mathbf{u}}^n$ the pair $\vec{\mathbf{u}}^{n+1}$, $\vec{\tilde{\mathbf{p}}} := \theta\vec{\mathbf{p}}^{n+1} + (1-\theta)\vec{\mathbf{p}}^n$ is a solution of

$$\frac{\vec{\mathbf{u}}^{n+1} - \vec{\mathbf{u}}^n}{\Delta t} + \mathbf{M}^{-1}\mathbf{B}^T\vec{\tilde{\mathbf{p}}} = \theta\mathbf{M}^{-1}\, g(\vec{\mathbf{u}}^{n+1}, t_{n+1}) + (1-\theta)\mathbf{M}^{-1}\, g(\vec{\mathbf{u}}^n, t_n), \tag{4.19}$$

$$\mathbf{B}\,\vec{\mathbf{u}}^{n+1} = 0. \tag{4.20}$$

For given $\vec{\mathbf{u}}^n$ this saddle point problem has (for Δt sufficiently small) a unique solution pair $(\vec{\mathbf{u}}^{n+1}, \vec{\tilde{\mathbf{p}}})$ (on the subspace of pressure functions that satisfy $(p_h, 1)_{L^2} = 0$). Thus instead of (4.18) for computing $\vec{\mathbf{u}}^{n+1}$ we can use the equivalent formulation in (4.19)-(4.20) for computing $\vec{\mathbf{u}}^{n+1}$, $\vec{\tilde{\mathbf{p}}}$. An important advantage of the latter formulation is that the projection $\mathbf{P}$ has been eliminated. In the derivation it is essential that the mass matrix $\mathbf{M}$ does not depend on t. Summarizing, the θ-method for the Navier-Stokes DAE system takes the form

$$\begin{aligned}\mathbf{M}\frac{\vec{\mathbf{u}}^{n+1}-\vec{\mathbf{u}}^n}{\Delta t} &+ \theta[\mathbf{A}\vec{\mathbf{u}}^{n+1}+\mathbf{N}(\vec{\mathbf{u}}^{n+1})\vec{\mathbf{u}}^{n+1}]+\mathbf{B}^T\vec{\mathbf{p}} \\ &= \theta\vec{\mathbf{f}}^{n+1}-(1-\theta)[\mathbf{A}\vec{\mathbf{u}}^n+\mathbf{N}(\vec{\mathbf{u}}^n)\vec{\mathbf{u}}^n-\vec{\mathbf{f}}^n] \\ \mathbf{B}\vec{\mathbf{u}}^{n+1} &= 0.\end{aligned} \tag{4.21}$$

The θ-schema applied to the Stokes problem results in a system as in (4.21), with the two terms $\mathbf{N}(\cdot)$ replaced by 0. In each time step a system of equations for the unknowns $\vec{\mathbf{u}}^{n+1}, \vec{\mathbf{p}}$ has to be solved. For the (Navier-)Stokes problem this saddle point system is (non)linear. Iterative methods for solving this system are treated in Chap. 5.

Remark 4.2.2 In the derivation above, we applied the *method of lines* approach, in which we first discretize the space variable and then the time variable. In view of the time discretization for two-phase flow problems, treated in Chap. 8, we comment on an alternative approach, often called *Rothe's method*, in which first the time variable and then the space variable is discretized. We explain this for the Stokes case. Starting point is the time dependent Stokes problem in which the pressure has been eliminated, i.e. a formulation as in (2.33). This is a variational formulation of an ODE in the function space $\mathbf{V}_{\text{div}}$. To this problem one can apply the θ-scheme for discretization of the time variable, resulting in the following problem: given $\mathbf{u}^0 \in \mathbf{V}_{\text{div}}$, for $n \geq 0$, determine $\mathbf{u}^{n+1} \in \mathbf{V}_{\text{div}}$ such that

$$\begin{aligned}&\frac{1}{\Delta t}(\mathbf{u}^{n+1}-\mathbf{u}^n,\mathbf{v})_{L^2}+\theta a(\mathbf{u}^{n+1},\mathbf{v}) \\ &= \theta\mathbf{g}^{n+1}-(1-\theta)\big[a(\mathbf{u}^n,\mathbf{v})-\mathbf{g}^n\big] \quad \text{for all } \mathbf{v}\in\mathbf{V}_{\text{div}}.\end{aligned} \tag{4.22}$$

This is a "projected" (due to $\mathbf{V}_{\text{div}}$) *stationary* Stokes problem for the unknown function $\mathbf{u}^{n+1}$. Since finite element subspaces of $\mathbf{V}_{\text{div}}$ are in general difficult to construct, we reformulate this problem as a saddle point problem in $\mathbf{V}\times Q = H_0^1(\Omega)^3 \times L_0^2(\Omega)$. Define the bilinear form

$$\hat{a}(\mathbf{u},\mathbf{v}) = \frac{1}{\Delta t}(\mathbf{u},\mathbf{v})_{L^2}+\theta a(\mathbf{u},\mathbf{v}), \quad \mathbf{u},\mathbf{v}\in\mathbf{V},\ \theta\in(0,1].$$

This bilinear form is elliptic and continuous on $\mathbf{V}$. We can apply the abstract theory in Sect. 15.3, Theorems 15.3.1 and 15.3.4, from which it follows that the problem (4.22) has a unique solution $\mathbf{u}^{n+1}$ which can also be characterized by the following Oseen problem: determine $\mathbf{u}^{n+1}\in\mathbf{V}$ and $p\in Q$ such that

$$\begin{aligned}&\frac{1}{\Delta t}(\mathbf{u}^{n+1}-\mathbf{u}^n,\mathbf{v})_{L^2}+\theta a(\mathbf{u}^{n+1},\mathbf{v})+b(\mathbf{v},p) \\ &= \theta\mathbf{g}^{n+1}-(1-\theta)\big[a(\mathbf{u}^n,\mathbf{v})-\mathbf{g}^n\big] \quad \text{for all } \mathbf{v}\in\mathbf{V}, \\ &b(\mathbf{u},q)=0 \quad \text{for all } q\in Q.\end{aligned} \tag{4.23}$$

To this problem we can apply a Galerkin discretization with spaces $\mathbf{V}_h \subset \mathbf{V}$, $Q_h \subset Q$. Using standard nodal bases, we then obtain a fully discrete problem as in (4.21), with $N(\cdot) = 0$. The mass and stiffness matrices $\mathbf{M}$ and $\mathbf{A}$ and the right-hand sides $\vec{\mathbf{f}}^n$ are the same in the two approaches. Hence, the two methods yield the same results.

Although these two approaches turn out to be equivalent in case of an non-stationary Stokes problem with Hood-Taylor finite element spaces for spatial discretization and the θ-scheme for time discretization we comment on a subtle difference between the methods that will become important if we treat two-phase flow problems. For the method of lines the approach is as follows: we start with a saddle point problem for $(\mathbf{u}, p)$, apply spatial Galerkin discretization, eliminate p_h, apply time discretization, introduce p_h again. For Rothe's method: start with a saddle point problem for $(\mathbf{u}, p)$, eliminate p, apply time discretization, introduce p again, apply spatial Galerkin discretization. We see that in the former method we eliminate and re-introduce the spatially discrete pressure variable p_h, whereas in the latter this is done for the spatially continuous variable p. If in a time step $t_n \to t_{n+1}$ one wants to use *different* pressure finite element spaces, then this pressure elimination and re-introduction can be problematic for the method of lines approach, whereas this is not the case for Rothe's method.

4.3 Fractional-step θ-scheme for the Navier-Stokes problem

Applying the fractional-step θ-scheme to the Navier-Stokes problem in ODE form (4.16) and transforming it back to its original DAE form along the same lines as in Sect. 4.2 results in

$$\begin{cases} \mathbf{M}\frac{\vec{\mathbf{u}}^{n+\theta}-\vec{\mathbf{u}}^n}{\theta\Delta t} + \alpha[\mathbf{A}\vec{\mathbf{u}}^{n+\theta} + \mathbf{N}(\vec{\mathbf{u}}^{n+\theta})\vec{\mathbf{u}}^{n+\theta}] + \mathbf{B}^T\vec{\mathbf{p}}^1 \\ \qquad = \vec{\mathbf{f}}^n - (1-\alpha)[\mathbf{A}\vec{\mathbf{u}}^n + \mathbf{N}(\vec{\mathbf{u}}^n)\vec{\mathbf{u}}^n] \\ \mathbf{B}\vec{\mathbf{u}}^{n+\theta} = 0 \end{cases} \tag{4.24}$$

$$\begin{cases} \mathbf{M}\frac{\vec{\mathbf{u}}^{n+1-\theta}-\vec{\mathbf{u}}^{n+\theta}}{(1-2\theta)\Delta t} + (1-\alpha)[\mathbf{A}\vec{\mathbf{u}}^{n+1-\theta} + \mathbf{N}(\vec{\mathbf{u}}^{n+1-\theta})\vec{\mathbf{u}}^{n+1-\theta}] + \mathbf{B}^T\vec{\mathbf{p}}^2 \\ \qquad = \vec{\mathbf{f}}^{n+1-\theta} - \alpha[\mathbf{A}\vec{\mathbf{u}}^{n+\theta} + \mathbf{N}(\vec{\mathbf{u}}^{n+\theta})\vec{\mathbf{u}}^{n+\theta}] \\ \mathbf{B}\vec{\mathbf{u}}^{n+1-\theta} = 0 \end{cases} \tag{4.25}$$

$$\begin{cases} \mathbf{M}\frac{\vec{\mathbf{u}}^{n+1}-\vec{\mathbf{u}}^{n+1-\theta}}{\theta\Delta t} + \alpha[\mathbf{A}\vec{\mathbf{u}}^{n+1} + \mathbf{N}(\vec{\mathbf{u}}^{n+1})\vec{\mathbf{u}}^{n+1}] + \mathbf{B}^T\vec{\mathbf{p}}^3 \\ \qquad = \vec{\mathbf{f}}^{n+1-\theta} - (1-\alpha)[\mathbf{A}\vec{\mathbf{u}}^{n+1-\theta} + \mathbf{N}(\vec{\mathbf{u}}^{n+1-\theta})\vec{\mathbf{u}}^{n+1-\theta}] \\ \mathbf{B}\vec{\mathbf{u}}^{n+1} = 0. \end{cases} \tag{4.26}$$

If we take parameter values as in (4.14) then the *non*linear problems for the pairs $(\vec{\mathbf{u}}^{n+\theta}, \vec{\mathbf{p}}^1)$, $(\vec{\mathbf{u}}^{n+1-\theta}, \vec{\mathbf{p}}^2)$, $(\vec{\mathbf{u}}^{n+1}, \vec{\mathbf{p}}^3)$ in these three substeps have a similar form. We obtain the fractional-step θ-scheme for the Stokes by replacing all terms $\mathbf{N}(\cdot)$ by 0.

Remark 4.3.1 If one uses the variant of the fractional-step θ-scheme as described in Remark 4.1.3 then in each time interval $[n\Delta t, (n+1)\Delta t]$ three successive substeps of the θ-scheme (4.21) (with different values for θ) are applied.

4.4 Numerical experiments

To analyze the time discretization error for different time integration schemes, we reconsider the test case of a rectangular tube described in Sect. 3.3.1. Instead of a stationary Stokes problem we now consider the non-stationary Stokes problem (2.34) on $\Omega \times [0, T]$ with $T = 2$ for different time step sizes Δt. To obtain a time-dependent velocity and pressure field, we prescribe an oscillating boundary condition $\mathbf{u}(0, x_2, x_3) = s(x_2, x_3)\big(1 + 0.25\sin(2\pi t)\big)$ at the inflow boundary $x_1 = 0$, with s defined in (3.39), an outflow boundary condition $\boldsymbol{\sigma}\mathbf{n} = 0$ for $x_1 = L$ and $\mathbf{u} = 0$ on the remaining boundaries.

For spatial discretization we use the Hood-Taylor finite element pair for $k = 2$ on a triangulation $\mathcal{T}$ which is constructed by subdividing Ω into $16\times4\times4$ sub-cubes each consisting of 6 tetrahedra. For this fixed spatial discretization different time integration schemes are analyzed for different time step sizes $\Delta t = T/n_t$ where $n_t = 25, 50, 100, 200, 400, 800$ denotes the number of time steps applied to obtain the approximations $\vec{\mathbf{u}}^{n_t}$, $\vec{p}^{\,n_t}$ to $\vec{\mathbf{u}}(T)$, $\vec{p}(T)$, respectively. As the exact solutions $\vec{\mathbf{u}}(T), \vec{p}(T)$ of the DAE system (3.22) are not available we instead use reference solutions $\vec{\mathbf{u}}^{\mathrm{ref}}$, $\vec{p}^{\,\mathrm{ref}}$ obtained by applying 2000 steps of the fractional-step θ-scheme with step size $\Delta t = 10^{-3}$.

n_t	$\|\vec{\mathbf{u}}^{\mathrm{ref}} - \vec{\mathbf{u}}^{n_t}\|_{L^2}$	order	$\|\vec{\mathbf{u}}^{\mathrm{ref}} - \vec{\mathbf{u}}^{n_t}\|_1$	order	$\|\vec{p}^{\,\mathrm{ref}} - \vec{p}^{\,n_t}\|_{L^2}$	order
25	3.69 E-5	—	3.54 E-4	—	7.71 E-3	—
50	1.38 E-5	1.42	1.32 E-4	1.43	2.17 E-3	1.83
100	5.69 E-6	1.29	5.40 E-5	1.29	6.51 E-4	1.73
200	2.53 E-6	1.17	2.40 E-5	1.17	2.17 E-4	1.59
400	1.19 E-6	1.09	1.12 E-5	1.09	8.13 E-5	1.42
800	5.75 E-7	1.05	5.43 E-6	1.05	3.38 E-5	1.26

Table 4.1. Convergence behavior of the implicit Euler scheme w.r.t. time step size.

For a fixed spatial coordinate $x = (2, 0.5, 0.5)$ in the center of the domain Ω, the first velocity component $u_1(x,t)$ and pressure $p(x,t)$ are shown as a function of time $t \in [0, 2]$ in Fig. 4.3. Also given are the results for the implicit Euler scheme ($\theta = 1$), the Crank-Nicolson scheme ($\theta = 0.5$) and the

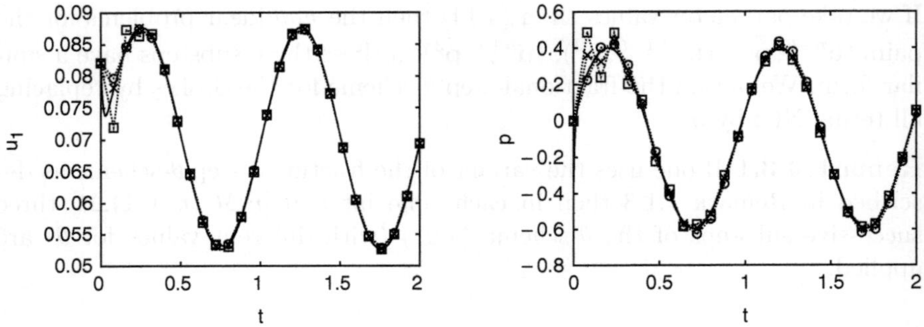

Fig. 4.3. Velocity u_1 (left) and pressure p (right) at point $x = (2, 0.5, 0.5) \in \Omega$ as a function of time. Shown are the reference solution (solid line) and implicit Euler (circles), Crank-Nicolson (squares) and fractional-step (crosses) solutions for 25 time steps, respectively.

n_t	$\|\vec{\mathbf{u}}^{\text{ref}} - \vec{\mathbf{u}}^{n_t}\|_{L^2}$	order	$\|\vec{\mathbf{u}}^{\text{ref}} - \vec{\mathbf{u}}^{n_t}\|_1$	order	$\|\vec{p}^{\text{ ref}} - \vec{p}^{\text{ } n_t}\|_{L^2}$	order
25	2.75 E-5	—	6.87 E-4	—	9.16 E-3	—
50	3.84 E-6	2.84	1.38 E-4	2.31	9.65 E-5	6.57
100	6.56 E-7	2.55	6.66 E-6	4.38	4.21 E-5	1.20
200	1.63 E-7	2.00	1.60 E-6	2.06	1.04 E-5	2.02
400	4.07 E-8	2.01	3.98 E-7	2.01	2.57 E-6	2.01
800	9.99 E-9	2.03	9.78 E-8	2.03	6.56 E-7	1.97

Table 4.2. Convergence behavior of the Crank-Nicolson scheme w.r.t. time step size.

n_t	$\|\vec{\mathbf{u}}^{\text{ref}} - \vec{\mathbf{u}}^{n_t}\|_{L^2}$	order	$\|\vec{\mathbf{u}}^{\text{ref}} - \vec{\mathbf{u}}^{n_t}\|_1$	order	$\|\vec{p}^{\text{ ref}} - \vec{p}^{\text{ } n_t}\|_{L^2}$	order
25	5.85 E-7	—	7.93 E-6	—	2.67 E-4	—
50	3.40 E-7	0.78	3.31 E-6	1.26	5.50 E-5	2.28
100	9.49 E-8	1.84	9.08 E-7	1.87	1.14 E-5	2.27
200	2.37 E-8	2.00	2.30 E-7	1.98	2.56 E-6	2.15
400	5.70 E-9	2.06	5.58 E-8	2.05	6.02 E-7	2.09
800	1.24 E-9	2.20	1.21 E-9	2.20	1.31 E-7	2.20

Table 4.3. Convergence behavior of the fractional-step θ-scheme w.r.t. time step size.

fractional-step θ-scheme applying 25 steps with time step size $\Delta t = 0.08$. We notice an oscillatory behavior of the Crank-Nicolson scheme in the first time steps which is probably due to the fact that this method does not have a good smoothing property, as explained in Sect. 4.1.

Tables 4.1–4.3 show the convergence w.r.t. time step size for the different time discretization schemes. The numerical experiments confirm the first order convergence of the implicit Euler scheme and second order convergence of the Crank-Nicolson and the fractional-step θ-scheme.

Comparing the second order schemes we observe that the errors for the fractional-step scheme are smaller than those of the Crank-Nicolson scheme by about a factor of 10. Note, however, that in the fractional-step scheme three macro-steps are performed per time step, and thus for a fair comparison it should be compared to a Crank-Nicolson scheme with a time step size divided by 3. This would lead to Crank-Nicolson errors which are roughly $3^2 = 9$ times smaller than those in Table 4.2 and thus are of the same order of magnitude as the errors in the fractional-step scheme given in Table 4.3.

4.5 Discussion and additional references

In this chapter we restricted ourselves to basic, but still very popular, time discretization methods for non-stationary (Navier-)Stokes equations. We briefly discuss a few related aspects.

Error analyses of a fully (space and time) discrete problem as in (4.21) are presented in [141, 142, 143]. In the literature there are only very few studies on *adaptive* time stepping for solving non-stationary Navier-Stokes equations; a recent paper is [154]. There are several variants of the fractional-step θ-scheme for the Navier-Stokes equations based on different operator splittings. Some of these are discussed in [122]. A popular variant is based on a semi-implict treatment of the nonlinear term in the Navier-Stokes equations. In such a method one replaces the term $\mathbf{N}(\vec{\mathbf{u}}^{n+\theta})\vec{\mathbf{u}}^{n+\theta}$ in (4.24) by $\mathbf{N}(\vec{\mathbf{u}}^{n})\vec{\mathbf{u}}^{n+\theta}$, and similarly for the nonlinear terms in (4.25), (4.26), cf. [206]. Alternatively, instead of replacing $\vec{\mathbf{u}}^{n+\theta}$ by $\vec{\mathbf{u}}^{n}$ one can also replace it by a more accurate extrapolation of $\vec{\mathbf{u}}^{n}$ and $\vec{\mathbf{u}}^{n-1}$. Other semi-implicit methods are explained in [206].

A class of methods that is particularly popular in the engineering literature are the so-called projection methods. These methods have a predictor-corrector structure, in which in the predictor step, which does not involve pressure, a new velocity field is determined and in the corrector step, which involves a pressure variable, this new velocity field is "projected" onto the subspace of divergence free functions. To explain the main idea we consider a basic variant of this method in semi-discrete form only, i.e. we discretize in time but not in space, and formulate it in strong formulation. Let $\mathbf{u}^n \in \mathbf{V} := H_0^1(\Omega)^3$ be a given approximation of $\mathbf{u}(\cdot, t_n)$. We define $\tilde{\mathbf{u}}^{n+1} \in \mathbf{V}$ as the solution of

$$\frac{1}{\Delta t}\left(\tilde{\mathbf{u}}^{n+1} - \mathbf{u}^n\right) - \frac{1}{Re}\Delta\tilde{\mathbf{u}}^{n+1} + (\mathbf{u}^n \cdot \nabla)\tilde{\mathbf{u}}^{n+1} = \mathbf{g}(t_{n+1}).$$

The approximation $\tilde{\mathbf{u}}^{n+1} \approx \mathbf{u}(\cdot, t_{n+1})$ is projected onto the space of divergence free functions by solving a saddle point problem: determine $\mathbf{u}^{n+1} \in \mathbf{V}$ and $q \in L_0^2(\Omega)$ such that

$$\begin{aligned} \frac{1}{\Delta t}(\mathbf{u}^{n+1} - \tilde{\mathbf{u}}^{n+1}) + \nabla q &= 0 \quad \text{in } \Omega \\ \operatorname{div} \mathbf{u}^{n+1} &= 0 \quad \text{in } \Omega \\ \mathbf{u}^{n+1} \cdot \mathbf{n} &= 0 \quad \text{on } \partial\Omega. \end{aligned}$$

An detailed study of this projection method and variants of it can be found in [176].

5

Iterative solvers

In this chapter we address the issue of iterative solvers for the large sparse (nonlinear) systems of equations that arise after space and time discretization (using an implicit time integration method) of the non-stationary Stokes and Navier-Stokes equations.

5.1 Linearization method

Using an implicit time-stepping scheme for the Navier-Stokes problem we obtain a nonlinear system of algebraic equations in each time step. As an example, we consider the first step (4.24) in the fractional-step θ-scheme (note that the other two steps have a similar form). The nonlinear system is as follows:

$$\left\{\begin{aligned} &\left(\tfrac{1}{\theta\Delta t}\mathbf{M} + \alpha\mathbf{A}\right)\vec{\mathbf{u}}^{n+\theta} + \alpha\mathbf{N}(\vec{\mathbf{u}}^{n+\theta})\vec{\mathbf{u}}^{n+\theta} + \mathbf{B}^T\vec{\mathbf{p}}^1 \\ &\qquad = \vec{\mathbf{f}}^n + \left(\tfrac{1}{\theta\Delta t}\mathbf{M} - (1-\alpha)(\mathbf{A} + \mathbf{N}(\vec{\mathbf{u}}^n))\right)\vec{\mathbf{u}}^n \\ &\mathbf{B}\vec{\mathbf{u}}^{n+\theta} = 0. \end{aligned}\right. \tag{5.1}$$

This is a nonlinear system of the form

$$\mathbf{K}(\mathbf{x})\begin{pmatrix}\mathbf{x}\\ \mathbf{y}\end{pmatrix} = \begin{pmatrix}\mathbf{b}\\ \mathbf{c}\end{pmatrix}, \quad \text{with } \mathbf{K}(\mathbf{x}) = \begin{pmatrix}\alpha\mathbf{A} + \frac{1}{\theta\Delta t}\mathbf{M} + \alpha\mathbf{N}(\mathbf{x}) & \mathbf{B}^T \\ \mathbf{B} & 0\end{pmatrix}. \tag{5.2}$$

Systems arising in other time discretization methods have a very similar structure. If instead of a Navier-Stokes problem we consider a Stokes equation, the resulting discrete problem has a structure as in (5.2) but without the term $\mathbf{N}(\mathbf{x})$.

As a first step in solving the nonlinear system of equations in (5.2) a linearization approach has to be applied. One possibility is to use a Newton linearization. Here we consider a simpler method that is often used in practice and in which the computation of Jacobians is avoided. This Richardson type of method is as follows, cf. [244]:

$$\begin{pmatrix} \mathbf{x}^{k+1} \\ \mathbf{y}^{k+1} \end{pmatrix} = \begin{pmatrix} \mathbf{x}^{k} \\ \mathbf{y}^{k} \end{pmatrix} - \omega_{k+1} \mathbf{K}(\mathbf{x}^k)^{-1} \left[\mathbf{K}(\mathbf{x}^k) \begin{pmatrix} \mathbf{x}^{k} \\ \mathbf{y}^{k} \end{pmatrix} - \begin{pmatrix} \mathbf{b} \\ \mathbf{c} \end{pmatrix} \right]. \tag{5.3}$$

For a compact notation the correction is denoted by

$$\begin{pmatrix} \Delta\mathbf{x}^{k} \\ \Delta\mathbf{y}^{k} \end{pmatrix} := \mathbf{K}(\mathbf{x}^k)^{-1} \left[\mathbf{K}(\mathbf{x}^k) \begin{pmatrix} \mathbf{x}^{k} \\ \mathbf{y}^{k} \end{pmatrix} - \begin{pmatrix} \mathbf{b} \\ \mathbf{c} \end{pmatrix} \right].$$

An optimality criterion for fixing the step length ω_{k+1} is given by the one dimensional optimization problem

$$\omega_{opt} = \arg\min_{\omega} \left\| \mathbf{K}(\mathbf{x}^k - \omega\Delta\mathbf{x}^k) \begin{pmatrix} \mathbf{x}^{k} - \omega\Delta\mathbf{x}^k \\ \mathbf{y}^{k} - \omega\Delta\mathbf{y}^k \end{pmatrix} - \begin{pmatrix} \mathbf{b} \\ \mathbf{c} \end{pmatrix} \right\|. \tag{5.4}$$

Here $\|\cdot\|$ denotes the Euclidean norm. This strategy is known as *line search.* This approach, however, is not feasible in our applications as the optimization problem in (5.4) is nonlinear in ω and computing the optimal ω-value would require too much computational work. Therefore we modify the criterion in (5.4) to obtain a computationally more feasible one:

$$\tilde{\omega}_{opt} = \arg\min_{\omega} \left\| \tilde{\mathbf{K}} \begin{pmatrix} \mathbf{x}^{k} - \omega\Delta\mathbf{x}^k \\ \mathbf{y}^{k} - \omega\Delta\mathbf{y}^k \end{pmatrix} - \begin{pmatrix} \mathbf{b} \\ \mathbf{c} \end{pmatrix} \right\| \tag{5.5}$$

with $\tilde{\mathbf{K}} := \mathbf{K}(\mathbf{x}^k - \omega_k\Delta\mathbf{x}^k)$, using the step length ω_k from the previous iteration (and $\omega_0 := 1$ at the beginning). This modified optimization problem is linear in ω and its solution is given by

$$\tilde{\omega}_{opt} = \frac{\left\langle \tilde{\mathbf{K}} \begin{pmatrix} \Delta\mathbf{x}^{k} \\ \Delta\mathbf{y}^{k} \end{pmatrix}, \tilde{\mathbf{K}} \begin{pmatrix} \mathbf{x}^{k} \\ \mathbf{y}^{k} \end{pmatrix} - \begin{pmatrix} \mathbf{b} \\ \mathbf{c} \end{pmatrix} \right\rangle}{\left\| \tilde{\mathbf{K}} \begin{pmatrix} \Delta\mathbf{x}^{k} \\ \Delta\mathbf{y}^{k} \end{pmatrix} \right\|^2}, \tag{5.6}$$

where $\langle\cdot,\cdot\rangle$ denotes the Euclidean scalar product. Note that the evaluation of (5.6) only requires the construction of $\tilde{\mathbf{K}} = \mathbf{K}(\mathbf{x}^k - \omega_k\Delta\mathbf{x}^k)$ and the computation of some matrix-vector multiplications and scalar products. This is negligible effort compared to the most time consuming part in the iteration (5.3), namely the solution of a linear problem with matrix $\mathbf{K}(\mathbf{x}^k)$. The method in (5.3) with the choice $\omega_{k+1} = \tilde{\omega}_{opt}$ from (5.6) is called *adaptive defect correction method.*

In our applications we experienced that the step length control given by (5.6) is more robust than using a fixed step length $\omega_{k+1} = \omega$. The method is implemented in the following form.

Algorithm 5.1.1 (adaptive defect correction)
Set $\omega_0 = 1$. *Repeat until desired accuracy:*

1. *Calculate the defect*

$$\begin{pmatrix} \mathbf{d}_1 \\ \mathbf{d}_2 \end{pmatrix} = \begin{pmatrix} (\alpha\mathbf{A} + \frac{1}{\theta\Delta t}\mathbf{M} + \alpha\mathbf{N}(\mathbf{x}^k))\mathbf{x}^k + \mathbf{B}^T\mathbf{y}^k - \mathbf{b} \\ \mathbf{B}\mathbf{x}^k - \mathbf{c} \end{pmatrix}.$$

2. *Solve the discrete Oseen system for the corrections* $\Delta \mathbf{x}^k$ *and* $\Delta \mathbf{y}^k$*:*

$$[\alpha \mathbf{A} + \frac{1}{\theta \Delta t}\mathbf{M} + \alpha \mathbf{N}(\mathbf{x}^k)]\Delta \mathbf{x}^k + \mathbf{B}^T \Delta \mathbf{y}^k = \mathbf{d}_1$$
$$\mathbf{B}\Delta \mathbf{x}^k = \mathbf{d}_2,$$

with accuracy tol^k.

3. *Step size control: Calculate the step length parameter*

$$\omega_{k+1} := \frac{\left\langle \tilde{\mathbf{K}}\begin{pmatrix}\Delta \mathbf{x}^k \\ \Delta \mathbf{y}^k\end{pmatrix}, \tilde{\mathbf{K}}\begin{pmatrix}\mathbf{x}^k \\ \mathbf{y}^k\end{pmatrix} - \begin{pmatrix}\mathbf{b} \\ \mathbf{c}\end{pmatrix}\right\rangle}{\left\| \tilde{\mathbf{K}}\begin{pmatrix}\Delta \mathbf{x}^k \\ \Delta \mathbf{y}^k\end{pmatrix}\right\|^2} \tag{5.7}$$

with

$$\tilde{\mathbf{K}} := \begin{pmatrix} \alpha \mathbf{A} + \frac{1}{\theta \Delta t}\mathbf{M} + \alpha \mathbf{N}(\mathbf{x}^k - \omega_k \Delta \mathbf{x}^k) & \mathbf{B}^T \\ \mathbf{B} & 0 \end{pmatrix}.$$

4. *Update* $\mathbf{x}^k$, $\mathbf{y}^k$

$$\begin{pmatrix}\mathbf{x}^{k+1} \\ \mathbf{y}^{k+1}\end{pmatrix} = \begin{pmatrix}\mathbf{x}^k \\ \mathbf{y}^k\end{pmatrix} - \omega_{k+1}\begin{pmatrix}\Delta \mathbf{x}^k \\ \Delta \mathbf{y}^k\end{pmatrix}.$$

The computational costs of this linearization method are dominated by the arithmetic work needed for solving the linear system in step 2. The discrete Oseen system in step 2 can be solved using iterative solvers that are treated in Sect. 5.3.

Remark 5.1.2 If we take $\omega_k = 1$ for all k then the linearization takes the simpler form

$$\begin{pmatrix} \alpha \mathbf{A} + \frac{1}{\theta \Delta t}\mathbf{M} + \alpha \mathbf{N}(\mathbf{x}^k) & \mathbf{B}^T \\ \mathbf{B} & 0 \end{pmatrix}\begin{pmatrix}\mathbf{x}^{k+1} \\ \mathbf{y}^{k+1}\end{pmatrix} = \begin{pmatrix}\mathbf{b} \\ \mathbf{c}\end{pmatrix}.$$

This fixed point method is also often used for linearization of the discrete Navier-Stokes equations.

Note that in case of a Stokes problem a linearization method is not needed. In the following sections we treat iterative solvers that can be used for solving the large sparse *linear* systems that arise after discretization and linearization of (Navier-)Stokes equations.

5.2 Iterative solvers for symmetric saddle point problems

The discrete (generalized) Stokes problem has a matrix-vector representation of the form

$$\mathbf{K}\begin{pmatrix}\mathbf{x}\\ \mathbf{y}\end{pmatrix} := \begin{pmatrix}\tilde{\mathbf{A}} & \mathbf{B}^T\\ \mathbf{B} & 0\end{pmatrix}\begin{pmatrix}\mathbf{x}\\ \mathbf{y}\end{pmatrix} = \begin{pmatrix}\mathbf{f}_1\\ \mathbf{f}_2\end{pmatrix}, \qquad (5.8)$$
$$\tilde{\mathbf{A}} := \alpha\mathbf{A} + \beta\mathbf{M} \in \mathbb{R}^{m\times m}, \quad \mathbf{B} \in \mathbb{R}^{n\times m},$$

where either the parameters are $\alpha = 1$, $\beta = 0$ (stationary Stokes) or $\alpha, \beta > 0$ are determined by the time integration scheme used. In that case the parameter α is of order 1 and β is proportional to $1/\Delta t \in (0,\infty)$. From a rescaling argument it follows that for the analysis it is no restriction to assume $\alpha = 1$.

The matrix $\tilde{\mathbf{A}}$ is *symmetric positive definite.* An important role is played by the so-called *Schur complement* matrix

$$\mathbf{S} = \mathbf{B}\tilde{\mathbf{A}}^{-1}\mathbf{B}^T.$$

This matrix is symmetric positive semi-definite. To avoid technical details in the analysis we assume that

$$\operatorname{rank}(\mathbf{B}) = n, \qquad (5.9)$$

i.e., the matrix $\mathbf{B}$ has full rank. Then the Schur complement $\mathbf{S}$ is *symmetric positive definite.*

Remark 5.2.1 In our applications the assumption $\operatorname{rank}(\mathbf{B}) = n$ usually does *not* hold. Consider a Hood-Taylor finite element pair and a matrix $\mathbf{B}$ as in (3.18d). Let $\mathbf{e} := (1,1,\ldots,1)^T \in \mathbb{R}^n$ be the vector representation of the constant finite element pressure function $1 \in \mathbb{X}_h^{k-1}$. Using the LBB stability of the finite element spaces it follows that $\mathbf{B}^T\mathbf{y} = 0$ iff $\mathbf{y} \in \operatorname{span}(\mathbf{e})$. Hence $\operatorname{rank}(\mathbf{B}) = n-1$ holds. Let $\mathbf{M}_p$ be the mass matrix in the pressure finite element space $\mathbb{X}_h^{k-1}$. The pressure finite element subspace $\{\, p_h \in \mathbb{X}_h^{k-1} : (p_h, 1)_{L^2} = 0 \,\}$ corresponds to the subspace $(\mathbf{M}_p\mathbf{e})^\perp := \{\, \mathbf{y} \in \mathbb{R}^n : \langle \mathbf{y}, \mathbf{M}_p\mathbf{e}\rangle = 0 \,\}$. The Schur complement $\mathbf{S}$ is an isomorphism $(\mathbf{M}_p\mathbf{e})^\perp \to \mathbf{e}^\perp$. The results (and methods) discussed below for the case $\operatorname{rank}(\mathbf{B}) = n$ also apply (with minor modifications) to the case discussed in this remark with $\operatorname{rank}(\mathbf{B}) = n-1$, provided instead of $\mathbf{S} : \mathbb{R}^n \to \mathbb{R}^n$ one considers $\mathbf{S} : (\mathbf{M}_p\mathbf{e})^\perp \to \mathbf{e}^\perp$.

The matrix $\mathbf{K}$ is symmetric and strongly indefinite and has a *saddle point structure.*

Lemma 5.2.2 *The matrix* $\mathbf{K}$ *has* m *strictly positive and* n *strictly negative eigenvalues.*

Proof. From the factorization

$$\mathbf{K} = \begin{pmatrix}\tilde{\mathbf{A}} & 0\\ \mathbf{B} & \mathbf{I}\end{pmatrix}\begin{pmatrix}\tilde{\mathbf{A}}^{-1} & 0\\ 0 & -\mathbf{S}\end{pmatrix}\begin{pmatrix}\tilde{\mathbf{A}} & \mathbf{B}^T\\ 0 & \mathbf{I}\end{pmatrix}$$

it follows that $\mathbf{K}$ is congruent to the matrix $\operatorname{blockdiag}(\tilde{\mathbf{A}}^{-1}, -\mathbf{S})$ which has m strictly positive and n strictly negative eigenvalues. Now apply Sylvester's inertia theorem. □

Remark 5.2.3 Consider the linear system

$$\mathbf{K}\begin{pmatrix}\mathbf{x}\\ \mathbf{y}\end{pmatrix} = \begin{pmatrix}\mathbf{f}_1\\ \mathbf{f}_2\end{pmatrix}. \tag{5.10}$$

Define the functional $\mathcal{L} : \mathbb{R}^m \times \mathbb{R}^n \to \mathbb{R}$ by $\mathcal{L}(\mathbf{x},\mathbf{y}) = \frac{1}{2}\langle \tilde{\mathbf{A}}\mathbf{x},\mathbf{x}\rangle + \langle \mathbf{B}\mathbf{x},\mathbf{y}\rangle - \langle \mathbf{f}_1,\mathbf{x}\rangle - \langle \mathbf{f}_2,\mathbf{y}\rangle$. It can be shown that $(\mathbf{x}^*,\mathbf{y}^*)$ is a solution of the problem (5.10) iff

$$\mathcal{L}(\mathbf{x}^*,\mathbf{y}) \le \mathcal{L}(\mathbf{x}^*,\mathbf{y}^*) \le \mathcal{L}(\mathbf{x},\mathbf{y}^*) \quad \text{for all} \;\; \mathbf{x}\in\mathbb{R}^m, \;\; \mathbf{y}\in\mathbb{R}^n.$$

Due to this property the linear system is called a saddle-point problem. This saddle point property is discussed in a general Hilbert space setting in Theorem 15.3.3.

In Sect. 5.2.1 we treat the preconditioned MINRES method for solving a linear system with a symmetric indefinite matrix. This method can be applied to the system in (5.8). In Sect. 5.2.2 we discuss an alternative, namely the inexact Uzawa method. In Sect. 5.4 we address the issue of preconditioning.

5.2.1 Preconditioned MINRES

In this section we treat the preconditioned MINRES method that can be used for solving the saddle point system in (5.8).

First we recall some basic ideas underlying this method. For this we consider a linear system

$$\mathbf{A}\mathbf{x} = \mathbf{b}, \tag{5.11}$$

where $\mathbf{A}$ is a general regular *symmetric* $n \times n$-matrix. Note the abuse of notation: in this section $\mathbf{A}$ is not (necessarily) the same as the matrix $\mathbf{A}$ in (5.8). We emphasize that $\mathbf{A}$ in (5.11) is allowed to be *in*definite. The MINRES method, introduced in [199], is based on the following residual minimization problem:

$$\begin{cases} \text{Given } \mathbf{x}^0 \in \mathbb{R}^n, \text{ determine } \mathbf{x}^k \in \mathbf{x}^0 + \mathcal{K}^k(\mathbf{A};\mathbf{r}^0) \text{ such that} \\ \|\mathbf{A}\mathbf{x}^k - \mathbf{b}\| = \min\{\, \|\mathbf{A}\mathbf{x} - \mathbf{b}\| \;\mid\; \mathbf{x} \in \mathbf{x}^0 + \mathcal{K}^k(\mathbf{A};\mathbf{r}^0)\,\}, \end{cases} \tag{5.12}$$

where $\mathbf{r}^0 := \mathbf{b} - \mathbf{A}\mathbf{x}^0$ and $\mathcal{K}^k(\mathbf{A};\mathbf{r}^0) := \mathrm{span}\{\mathbf{r}^0, \mathbf{A}\mathbf{r}^0, \ldots, \mathbf{A}^{k-1}\mathbf{r}^0\}$ is the *Krylov subspace.* Note that the Euclidean norm is used and that for *any* regular $\mathbf{A}$ this minimization problem has a unique solution $\mathbf{x}^k$, which is illustrated in Fig. 5.1.

We have a *projection*: $\mathbf{r}^0 - \mathbf{r}^k$ is the orthogonal projection (with respect to $\langle\cdot,\cdot\rangle$) of $\mathbf{r}^0$ on $\mathbf{A}(\mathcal{K}^k(\mathbf{A};\mathbf{r}^0)) = \mathrm{span}\{\mathbf{A}\mathbf{r}^0, \mathbf{A}^2\mathbf{r}^0, \ldots, \mathbf{A}^k\mathbf{r}^0\}$. For the MINRES algorithm for computing this $\mathbf{x}^k$ we refer to the literature, e.g. [199, 124]. *In the derivation of an efficient* MINRES *algorithm it is essential that* $\mathbf{A}$ *is symmetric.* The resulting method has low arithmetic costs per iteration, which

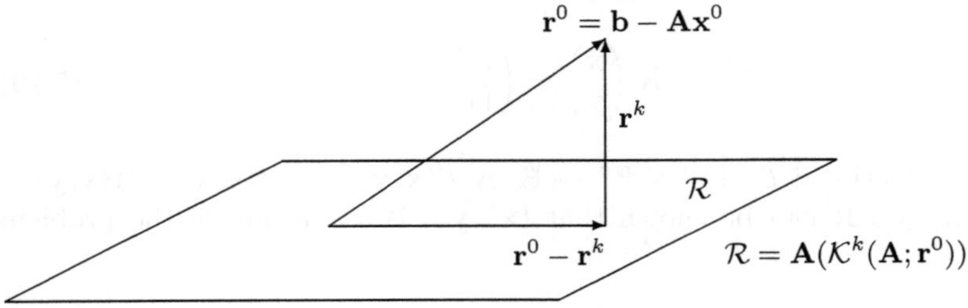

Fig. 5.1. Residual minimization.

are of the same order of magnitude as those in a CG iteration. Per iteration one has to perform only one matrix-vector multiplication $\mathbf{Ay}$, and one has to compute a few inner products and a few vector additions.

This method can be combined with *preconditioning*. For this we assume a *symmetric positive definite* preconditioner $\mathbf{Q}$. Let $\mathbf{Q}^{\frac{1}{2}}$ be the symmetric positive definite matrix such that $\mathbf{Q} = \mathbf{Q}^{\frac{1}{2}}\mathbf{Q}^{\frac{1}{2}}$. For the derivation of the preconditioned MINRES method we consider the preconditioned system

$$\mathbf{Q}^{-\frac{1}{2}}\mathbf{A}\mathbf{Q}^{-\frac{1}{2}}\mathbf{z} = \mathbf{Q}^{-\frac{1}{2}}\mathbf{b}, \qquad \mathbf{z} = \mathbf{Q}^{\frac{1}{2}}\mathbf{x}.$$

Note that $\hat{\mathbf{A}} := \mathbf{Q}^{-\frac{1}{2}}\mathbf{A}\mathbf{Q}^{-\frac{1}{2}}$ is symmetric. Thus the MINRES algorithm can be applied to this, resulting in the *preconditioned* MINRES method, denoted by PMINRES. The residual minimization criterion in (5.12) applied to the transformed system can be reformulated as follows:

$$\begin{cases} \text{Given } \mathbf{x}^0 \in \mathbb{R}^n, \text{ compute } \mathbf{x}^k \in \mathbf{x}^0 + \mathcal{K}^k(\mathbf{Q}^{-1}\mathbf{A};\mathbf{Q}^{-1}\mathbf{r}^0) \text{ such that} \\ \|\mathbf{Q}^{-1}\mathbf{A}\mathbf{x}^k - \mathbf{Q}^{-1}\mathbf{b}\|_Q \\ = \min\{\, \|\mathbf{Q}^{-1}\mathbf{A}\mathbf{x} - \mathbf{Q}^{-1}\mathbf{b}\|_Q \mid \mathbf{x} \in \mathbf{x}^0 + \mathcal{K}^k(\mathbf{Q}^{-1}\mathbf{A};\mathbf{Q}^{-1}\mathbf{r}^0)\,\}, \end{cases}$$

with $\mathbf{r}^0 = \mathbf{b} - \mathbf{A}\mathbf{x}^0$ and $\langle\cdot,\cdot\rangle_Q = \langle \mathbf{Q}\cdot,\cdot\rangle$. From this we see that now the preconditioned residual $\mathbf{Q}^{-1}(\mathbf{b}-\mathbf{A}\mathbf{x})$ is minimized (in $\|\cdot\|_Q$) over a transformed Krylov subspace. An efficient implementation of the preconditioned MINRES method can be derived in which one needs per iteration one evaluation of the preconditioner, i.e. the computation of $\mathbf{Q}^{-1}\mathbf{y}$ for a given $\mathbf{y}$, and one matrix-vector product with $\mathbf{A}$. Note that $\mathbf{w} = \mathbf{Q}^{-1}\mathbf{y} \Leftrightarrow \mathbf{Q}\mathbf{w} = \mathbf{y}$ holds, i.e. the evaluation of the preconditioner may be realized by solving a linear system with the matrix $\mathbf{Q}$. The matrix $\mathbf{Q}^{\frac{1}{2}}$ is *not* used in the algorithm. For $\mathbf{Q} = \mathbf{I}$ the PMINRES reduces to MINRES. A convergence result is given in the following theorem.

Theorem 5.2.4 *Let* $\mathbf{A} \in \mathbb{R}^{n\times n}$ *be symmetric and* $\mathbf{Q} \in \mathbb{R}^{n\times n}$ *symmetric positive definite. For* $\mathbf{x}^k$, $k \geq 0$, *computed in the preconditioned MINRES algorithm we define* $\hat{\mathbf{r}}^k = \mathbf{Q}^{-1}(\mathbf{b} - \mathbf{A}\mathbf{x}^k)$. *The following holds:*

$$\begin{aligned} \|\hat{\mathbf{r}}^k\|_Q &= \min_{p_k \in \mathcal{P}_k; p_k(0)=1} \|p_k(\mathbf{Q}^{-1}\mathbf{A})\hat{\mathbf{r}}^0\|_Q \\ &\leq \min_{p_k \in \mathcal{P}_k; p_k(0)=1} \max_{\lambda \in \sigma(\mathbf{Q}^{-1}\mathbf{A})} |p_k(\lambda)| \, \|\hat{\mathbf{r}}^0\|_Q. \end{aligned} \tag{5.13}$$

Proof. Using the fact that the transformed residual $\mathbf{Q}^{-1}(\mathbf{b} - \mathbf{A}\mathbf{x})$ is minimized with $\mathbf{x} \in \mathbf{x}^0 + \mathcal{K}^k(\mathbf{Q}^{-1}\mathbf{A}; \mathbf{Q}^{-1}\mathbf{r}^0)$ we get

$$\begin{aligned} \|\hat{\mathbf{r}}^k\|_Q &= \min_{p_{k-1} \in \mathcal{P}_{k-1}} \|\mathbf{Q}^{-1}\mathbf{b} - \mathbf{Q}^{-1}\mathbf{A}\big(\mathbf{x}^0 + p_{k-1}(\mathbf{Q}^{-1}\mathbf{A})\hat{\mathbf{r}}^0\big)\|_Q \\ &= \min_{p_{k-1} \in \mathcal{P}_{k-1}} \|\hat{\mathbf{r}}^0 - \mathbf{Q}^{-1}\mathbf{A}p_{k-1}(\mathbf{Q}^{-1}\mathbf{A})\hat{\mathbf{r}}^0\|_Q \\ &= \min_{p_k \in \mathcal{P}_k; p_k(0)=1} \|p_k(\mathbf{Q}^{-1}\mathbf{A})\hat{\mathbf{r}}^0\|_Q. \end{aligned}$$

Note that $\mathbf{Q}^{-1}\mathbf{A}$ is symmetric with respect to $\langle\cdot,\cdot\rangle_Q$. And thus

$$\|p_k(\mathbf{Q}^{-1}\mathbf{A})\hat{\mathbf{r}}^0\|_Q \leq \|p_k(\mathbf{Q}^{-1}\mathbf{A})\|_Q \|\hat{\mathbf{r}}^0\|_Q = \max_{\lambda \in \sigma(\mathbf{Q}^{-1}\mathbf{A})} |p_k(\lambda)| \, \|\hat{\mathbf{r}}^0\|_Q$$

holds. □

For constructing a good preconditioner $\mathbf{Q}$ there are the following two main criteria:

- for arbitrary $\mathbf{z} \in \mathbb{R}^n$ one can determine $\mathbf{Q}^{-1}\mathbf{z}$ with "low" computational costs, and
- $\mathbf{Q}$ is a good approximation of $\mathbf{A}$ in the sense that

$$\hat{m}(k) := \min_{p_k \in \mathcal{P}_k; p_k(0)=1} \max_{\lambda \in \sigma(\mathbf{Q}^{-1}\mathbf{A})} |p_k(\lambda)|$$

  has a "fast decrease" as a function of k.

By using Chebyshev polynomials and assumptions on the spectrum $\sigma(\mathbf{Q}^{-1}\mathbf{A})$ one can derive bounds for $\hat{m}(k)$. We present two results that are well-known in the literature. Proofs are given in, for example, [124].

Theorem 5.2.5 *Let* $\mathbf{A}, \mathbf{Q}$ *and* $\hat{\mathbf{r}}^k$ *be as in Theorem 5.2.4. Assume that all eigenvalues of* $\mathbf{Q}^{-1}\mathbf{A}$ *are positive. Then*

$$\frac{\|\hat{\mathbf{r}}^k\|_Q}{\|\hat{\mathbf{r}}^0\|_Q} \leq \hat{m}(k) \leq 2\left(1 - \frac{2}{\sqrt{\kappa(\mathbf{Q}^{-1}\mathbf{A})} + 1}\right)^k, \qquad k = 0, 1, \ldots$$

holds.

We note that in this bound the dependence on the condition number $\kappa(\mathbf{Q}^{-1}\mathbf{A})$ is the same as in well-known bounds for the preconditioned CG method. In particular the bound decreases (i.e., the rate of convergence is expected to be higher) for a decreasing condition number $\kappa(\mathbf{Q}^{-1}\mathbf{A})$.

Theorem 5.2.6 *Let* $\mathbf{A}, \mathbf{Q}$ *and* $\hat{\mathbf{r}}^k$ *be as in Theorem 5.2.4. Assume that* $\sigma(\mathbf{Q}^{-1}\mathbf{A}) \subset [a,b] \cup [c,d]$ *with* $a < b < 0 < c < d$ *and* $b - a = d - c$. *Then*

$$\frac{\|\hat{\mathbf{r}}^k\|_Q}{\|\hat{\mathbf{r}}^0\|_Q} \leq \hat{m}(k) \leq 2\left(1 - \frac{2}{\sqrt{\frac{ad}{bc}} + 1}\right)^{[k/2]}, \qquad k = 0, 1, \ldots \tag{5.14}$$

holds ($[k/2]$ *denotes the largest* $n \in \mathbb{N}$ *such that* $n \leq k/2$).

In the special case $a = -d$, $b = -c$ the reduction factor in (5.14) takes the form $1 - 2/(\kappa(\mathbf{Q}^{-1}\mathbf{A}) + 1)$. Note that here the dependence on $\kappa(\mathbf{Q}^{-1}\mathbf{A})$ is different from the positive definite case in Theorem 5.2.5.

Preconditioned MINRES for saddle point systems

The PMINRES method can be applied to the symmetric system (5.8) which has a special block structure. Based on this block structure we introduce a preconditioner with a block diagonal structure

$$\begin{aligned} &\mathbf{Q} := \begin{pmatrix} \mathbf{Q}_A & 0 \\ 0 & \mathbf{Q}_S \end{pmatrix} \\ &\mathbf{Q}_A \in \mathbb{R}^{m \times m}, \ \mathbf{Q}_A = \mathbf{Q}_A^T > 0, \quad \mathbf{Q}_S \in \mathbb{R}^{n \times n}, \ \mathbf{Q}_S = \mathbf{Q}_S^T > 0. \end{aligned} \tag{5.15}$$

Here and in the remainder the notation $\mathbf{A} > \mathbf{C}$ ($\mathbf{A} \geq \mathbf{C}$), with symmetric matrices $\mathbf{A}$ and $\mathbf{C}$, means that $\mathbf{A} - \mathbf{C}$ is symmetric positive (semi-)definite, i.e., all eigenvalues of $\mathbf{A} - \mathbf{C}$ are > 0 (≥ 0). Such matrix inequalities are called *spectral inequalities.*

The preconditioned matrix is given by

$$\begin{aligned} &\hat{\mathbf{K}} = \mathbf{Q}^{-\frac{1}{2}}\mathbf{K}\mathbf{Q}^{-\frac{1}{2}} = \begin{pmatrix} \hat{\mathbf{A}} & \hat{\mathbf{B}}^T \\ \hat{\mathbf{B}} & 0 \end{pmatrix}, \\ &\hat{\mathbf{A}} := \mathbf{Q}_A^{-\frac{1}{2}}\tilde{\mathbf{A}}\mathbf{Q}_A^{-\frac{1}{2}}, \qquad \hat{\mathbf{B}} := \mathbf{Q}_S^{-\frac{1}{2}}\mathbf{B}\mathbf{Q}_A^{-\frac{1}{2}}. \end{aligned} \tag{5.16}$$

Theorem 5.2.4 yields that the rate of convergence of the preconditioned MINRES method depends on $\sigma(\mathbf{Q}^{-1}\mathbf{K}) = \sigma(\hat{\mathbf{K}})$. We now derive bounds for this spectrum. First we consider a very special preconditioner, which in a certain sense is optimal:

Lemma 5.2.7 *For* $\mathbf{Q}_A = \tilde{\mathbf{A}}$ *and* $\mathbf{Q}_S = \mathbf{S}$ *we have*

$$\sigma(\hat{\mathbf{K}}) \subset \{\, \tfrac{1}{2}(1 - \sqrt{5})\, ,\, 1\, ,\, \tfrac{1}{2}(1 + \sqrt{5})\, \}.$$

Proof. Note that in this case

$$\hat{\mathbf{K}} = \begin{pmatrix} \mathbf{I} & \hat{\mathbf{B}}^T \\ \hat{\mathbf{B}} & 0 \end{pmatrix}, \qquad \hat{\mathbf{B}} = \mathbf{S}^{-\frac{1}{2}}\mathbf{B}\tilde{\mathbf{A}}^{-\frac{1}{2}}.$$

Take $\mu \in \sigma(\hat{\mathbf{K}})$ with corresponding eigenvector $\begin{pmatrix} \mathbf{v} \\ \mathbf{w} \end{pmatrix} \neq \begin{pmatrix} 0 \\ 0 \end{pmatrix}$, i.e.,

$$\begin{pmatrix} \mathbf{I} & \hat{\mathbf{B}}^T \\ \hat{\mathbf{B}} & 0 \end{pmatrix} \begin{pmatrix} \mathbf{v} \\ \mathbf{w} \end{pmatrix} = \mu \begin{pmatrix} \mathbf{v} \\ \mathbf{w} \end{pmatrix}. \tag{5.17}$$

If $\mathbf{w} = 0$ holds, then $\mathbf{v} \neq 0$, and from $\mathbf{v} = \mu\mathbf{v}$ it follows that $\mu = 1$. Now assume $\mathbf{w} \neq 0$. From (5.17) we get $\hat{\mathbf{B}}\hat{\mathbf{B}}^T\mathbf{w} = \mu(\mu - 1)\mathbf{w}$. Note that $\hat{\mathbf{B}}\hat{\mathbf{B}}^T = \mathbf{I}$ and thus we conclude $\mu(\mu - 1) = 1$, hence $\mu = \frac{1}{2}(1 \pm \sqrt{5})$. □

Note that from the result in (5.13) it follows that the preconditioned MINRES method with the preconditioner as in Lemma 5.2.7 yields (in exact arithmetic) the exact solution in at most three iterations. In the evaluation of the preconditioner, however, we have to solve linear systems with the matrix $\tilde{\mathbf{A}}$ and with the matrix $\mathbf{S}$ exactly (up to rounding errors). In most applications (e.g., the Stokes problem) this is extremely costly. Hence this preconditioner *is not feasible.*

Remark 5.2.8 In the applications that we consider the Schur complement matrix $\mathbf{S} = \mathbf{B}\tilde{\mathbf{A}}^{-1}\mathbf{B}^T$ is *never computed explicitly.* This is due to the fact that the matrix $\tilde{\mathbf{A}}^{-1}$ is *not sparse* anymore: it contains much more nonzero entries than the matrix $\tilde{\mathbf{A}}$. Depending on the particular problem, the triangulation and the finite element spaces used, the matrix $\tilde{\mathbf{A}}^{-1}$ has between $\mathcal{O}(m^{1.5})$ and $\mathcal{O}(m^2)$ nonzero entries, whereas the sparse matrix $\tilde{\mathbf{A}}$ has only $\mathcal{O}(m)$ nonzero entries. For a given vector $\mathbf{y}$ the matrix-vector product $\mathbf{Sy}$ can be computed by using $\mathbf{Sy} = \mathbf{B}(\tilde{\mathbf{A}}^{-1}(\mathbf{B}^T\mathbf{y}))$. The exact computation of this (up to machine accuracy), however, is still very costly, because it requires to solve a linear system of the form $\tilde{\mathbf{A}}\mathbf{z} = \mathbf{B}^T\mathbf{y}$ up to machine accuracy. These observations explain, why in our applications we often use *approximate* Schur complements.

Instead of the preconditioning approach in Lemma 5.2.7 we will use approximations $\mathbf{Q}_A$ of $\tilde{\mathbf{A}}$ and $\mathbf{Q}_S$ of $\mathbf{S}$. The quality of these approximations is measured by using spectral inequalities. For given preconditioners $\mathbf{Q}_A$ and $\mathbf{Q}_S$ let Γ_A, Γ_S, γ_A, $\gamma_S > 0$ be such that:

$$\begin{aligned} \gamma_A\mathbf{Q}_A &\leq \tilde{\mathbf{A}} \leq \Gamma_A\mathbf{Q}_A, \\ \gamma_S\mathbf{Q}_S &\leq \mathbf{S} \leq \Gamma_S\mathbf{Q}_S. \end{aligned} \tag{5.18}$$

Using an analysis as in [212, 224] we obtain a result for the eigenvalues of the preconditioned matrix:

Theorem 5.2.9 *For the matrix* $\hat{\mathbf{K}}$ *as in* (5.16) *with preconditioners that satisfy* (5.18) *we have:*

$$\sigma(\hat{\mathbf{K}}) \subset \left[\frac{1}{2}(\gamma_A - \sqrt{\gamma_A^2 + 4\Gamma_S\Gamma_A})\,,\,\frac{1}{2}(\gamma_A - \sqrt{\gamma_A^2 + 4\gamma_S\gamma_A})\right] \\ \cup\left[\gamma_A\,,\,\frac{1}{2}(\Gamma_A + \sqrt{\Gamma_A^2 + 4\Gamma_S\Gamma_A})\right].$$

Proof. We use the following inequalities

$$\gamma_A\mathbf{I} \le \hat{\mathbf{A}} \le \Gamma_A\mathbf{I} \tag{5.19a}$$
$$\gamma_A\tilde{\mathbf{A}}^{-1} \le \mathbf{Q}_A^{-1} \le \Gamma_A\tilde{\mathbf{A}}^{-1} \tag{5.19b}$$
$$\gamma_S\mathbf{I} \le \mathbf{Q}_S^{-\frac{1}{2}}\mathbf{S}\mathbf{Q}_S^{-\frac{1}{2}} \le \Gamma_S\mathbf{I}. \tag{5.19c}$$

Note that $\hat{\mathbf{B}}\hat{\mathbf{B}}^T = \mathbf{Q}_S^{-\frac{1}{2}}\mathbf{B}\mathbf{Q}_A^{-1}\mathbf{B}^T\mathbf{Q}_S^{-\frac{1}{2}}$. Using (5.19b) and (5.19c) we get

$$\gamma_A\gamma_S\mathbf{I} \le \hat{\mathbf{B}}\hat{\mathbf{B}}^T \le \Gamma_A\Gamma_S\mathbf{I}. \tag{5.20}$$

Take $\mu \in \sigma(\hat{\mathbf{K}})$. Then $\mu \neq 0$ and there exists $(\mathbf{v},\mathbf{w}) \neq (0,0)$ such that

$$\begin{aligned}\hat{\mathbf{A}}\mathbf{v} + \hat{\mathbf{B}}^T\mathbf{w} &= \mu\mathbf{v} \\ \hat{\mathbf{B}}\mathbf{v} &= \mu\mathbf{w}.\end{aligned} \tag{5.21}$$

From $\mathbf{v} = 0$ it follows that $\mathbf{w} = 0$, hence, $\mathbf{v} \neq 0$ must hold. From (5.21) we obtain $(\hat{\mathbf{A}} + \frac{1}{\mu}\hat{\mathbf{B}}^T\hat{\mathbf{B}})\mathbf{v} = \mu\mathbf{v}$ and thus $\mu \in \sigma(\hat{\mathbf{A}} + \frac{1}{\mu}\hat{\mathbf{B}}^T\hat{\mathbf{B}})$. Note that all nonzero eigenvalues of $\hat{\mathbf{B}}^T\hat{\mathbf{B}}$ are also eigenvalues of $\hat{\mathbf{B}}\hat{\mathbf{B}}^T$. We first consider the case $\mu > 0$. Using (5.20) and (5.19a) we get

$$\gamma_A\mathbf{I} \le \hat{\mathbf{A}} + \frac{1}{\mu}\hat{\mathbf{B}}^T\hat{\mathbf{B}} \le (\Gamma_A + \frac{1}{\mu}\Gamma_S\Gamma_A)\mathbf{I},$$

and thus $\gamma_A \le \mu \le \Gamma_A + \frac{1}{\mu}\Gamma_S\Gamma_A$ holds. This yields

$$\mu \in \left[\gamma_A\,,\,\frac{1}{2}(\Gamma_A + \sqrt{\Gamma_A^2 + 4\Gamma_S\Gamma_A})\right].$$

We now consider the case $\mu < 0$. From (5.20) and (5.19a) it follows that

$$\hat{\mathbf{A}} + \frac{1}{\mu}\hat{\mathbf{B}}^T\hat{\mathbf{B}} \ge (\gamma_A + \frac{1}{\mu}\Gamma_S\Gamma_A)\mathbf{I},$$

and thus $\mu \ge \gamma_A + \frac{1}{\mu}\Gamma_S\Gamma_A$. This yields $\mu \ge \frac{1}{2}(\gamma_A - \sqrt{\gamma_A^2 + 4\Gamma_S\Gamma_A})$. Finally we derive an upper bound for $\mu < 0$. We introduce $\nu := -\mu > 0$. From (5.21) it follows that for $\mu < 0$, $\mathbf{w} \neq 0$ must hold. Furthermore, we have

$$\hat{\mathbf{B}}(\hat{\mathbf{A}} + \nu\mathbf{I})^{-1}\hat{\mathbf{B}}^T\mathbf{w} = \nu\mathbf{w},$$

and thus $\nu \in \sigma(\hat{\mathbf{B}}(\hat{\mathbf{A}} + \nu\mathbf{I})^{-1}\hat{\mathbf{B}}^T)$. From $\mathbf{I} + \nu\hat{\mathbf{A}}^{-1} \leq (1 + \frac{\nu}{\gamma_A})\mathbf{I}$ and (5.19c) we obtain

$$\begin{aligned}\hat{\mathbf{B}}(\hat{\mathbf{A}} + \nu\mathbf{I})^{-1}\hat{\mathbf{B}}^T &= \hat{\mathbf{B}}\hat{\mathbf{A}}^{-\frac{1}{2}}(\mathbf{I} + \nu\hat{\mathbf{A}}^{-1})^{-1}\hat{\mathbf{A}}^{-\frac{1}{2}}\hat{\mathbf{B}}^T \geq (1 + \frac{\nu}{\gamma_A})^{-1}\hat{\mathbf{B}}\hat{\mathbf{A}}^{-1}\hat{\mathbf{B}}^T \\ &= (1 + \frac{\nu}{\gamma_A})^{-1}\mathbf{Q}_S^{-\frac{1}{2}}\mathbf{S}\mathbf{Q}_S^{-\frac{1}{2}} \geq (1 + \frac{\nu}{\gamma_A})^{-1}\gamma_S\mathbf{I}.\end{aligned}$$

We conclude that $\nu \geq (1 + \frac{\nu}{\gamma_A})^{-1}\gamma_S$ holds. Hence, for $\mu = -\nu$ we get $\mu \leq \frac{1}{2}(\gamma_A - \sqrt{\gamma_A^2 + 4\gamma_S\gamma_A})$. □

Remark 5.2.10 Note that if $\gamma_A = \Gamma_A = \gamma_S = \Gamma_S = 1$, i.e., $\mathbf{Q}_A = \tilde{\mathbf{A}}$ and $\mathbf{Q}_S = \mathbf{S}$, we obtain $\sigma(\hat{\mathbf{K}}) = \{\frac{1}{2}(1 - \sqrt{5})\} \cup [\,1\,,\,\frac{1}{2}(1 + \sqrt{5})\,]$, which is sharp (cf. Lemma 5.2.7).

From the results in Theorem 5.2.9 and Theorem 5.2.4 it follows that one can expect *fast convergence of the preconditioned MINRES method if the spectral constants* γ_A, Γ_A, γ_S *and* Γ_S *are close to* 1, in other words, if we have *good preconditioners* $\mathbf{Q}_A$ and $\mathbf{Q}_S$ of $\tilde{\mathbf{A}}$ and $\mathbf{S}$, respectively. In particular, we have robustness with respect to variation of parameters (for example, mesh size h or parameters α, β in (5.8)) if these spectral constants are independent of the parameters.

The (P)MINRES method treated above is a popular method from the class of so-called Krylov subspace methods. The basic idea of the MINRES method (and other Krylov subspace methods) is that one determines a certain "best" approximation of the solution in the Krylov subspace, cf. (5.12). The error in this approximation decreases due to the fact that the dimension of the Krylov subspace increases if the iteration index k increases. The special saddle point block structure of the matrix $\mathbf{K}$ is *not* used in the MINRES method. The block structure *is* used in the block diagonal preconditioner in (5.15). In the next section we discuss an iterative method that is based on a comletely different idea, namely an approximate block LU-factorization of the matrix $\mathbf{K}$.

5.2.2 Inexact Uzawa method

The matrix $\mathbf{K}$ has a block factorization

$$\mathbf{K} = \begin{pmatrix} \tilde{\mathbf{A}} & 0 \\ \mathbf{B} & -\mathbf{I} \end{pmatrix} \begin{pmatrix} \mathbf{I} & \tilde{\mathbf{A}}^{-1}\mathbf{B}^T \\ 0 & \mathbf{S} \end{pmatrix}, \quad \mathbf{S} = \mathbf{B}\tilde{\mathbf{A}}^{-1}\mathbf{B}^T.$$

Solving the problem $\mathbf{K}\begin{pmatrix} \mathbf{x} \\ \mathbf{y} \end{pmatrix} = \begin{pmatrix} \mathbf{f}_1 \\ \mathbf{f}_2 \end{pmatrix}$ by block forward-backward substitution yields the equivalent problem:

$$1. \text{ Solve } \tilde{\mathbf{A}}\mathbf{z} = \mathbf{f}_1. \tag{5.22}$$

$$2. \text{ Solve } \mathbf{S}\mathbf{y} = \mathbf{B}\mathbf{z} - \mathbf{f}_2. \tag{5.23}$$

$$3. \text{ Solve } \tilde{\mathbf{A}}\mathbf{x} = \mathbf{f}_1 - \mathbf{B}^T\mathbf{y}. \tag{5.24}$$

In the Uzawa method one applies an iterative solver (e.g., CG) to the Schur complement system in step 2. Note that the matrix $\mathbf{S}$ in this system is symmetric positive definite. The $\tilde{\mathbf{A}}$-systems that occur in each iteration of this method and in the steps 1 and 3 are solved sufficiently accurate using some fast Poisson solver.

We consider a simple variant of this method in which the solutions of the $\tilde{\mathbf{A}}$-systems are replaced by approximate ones. Let $\mathbf{Q}_A$ be a preconditioner of $\tilde{\mathbf{A}}$. We use this preconditioner in the steps 1 and 3 and also for the approximation of the Schur complement in step 2. For this we introduce the notation

$$\hat{\mathbf{S}} := \mathbf{B}\mathbf{Q}_A^{-1}\mathbf{B}^T \tag{5.25}$$

for the *approximate* Schur complement. For solving linear systems with the matrix $\hat{\mathbf{S}}$ we use an (possibly nonlinear) iterative method denoted by Ψ: for each $\mathbf{w}$, $\Psi(\mathbf{w})$ is an approximation to the solution $\mathbf{z}^*$ of $\hat{\mathbf{S}}\mathbf{z} = \mathbf{w}$. We assume that

$$\|\Psi(\mathbf{w}) - \mathbf{z}^*\|_{\hat{S}} \leq \delta \|\mathbf{z}^*\|_{\hat{S}} \quad \text{for all } \mathbf{w} \tag{5.26}$$

holds with a given tolerance parameter $0 < \delta < 1$. The most important example for Ψ is a (preconditioned) CG method. Let $(\mathbf{x}^k, \mathbf{y}^k)$ be a given approximation to the solution $(\mathbf{x}, \mathbf{y})$. Note that using the block factorization of $\mathbf{K}$ we get

$$\begin{pmatrix} \mathbf{x} \\ \mathbf{y} \end{pmatrix} = \begin{pmatrix} \mathbf{x}^k \\ \mathbf{y}^k \end{pmatrix} + \begin{pmatrix} \mathbf{I} & -\tilde{\mathbf{A}}^{-1}\mathbf{B}^T\mathbf{S}^{-1} \\ 0 & \mathbf{S}^{-1} \end{pmatrix} \begin{pmatrix} \tilde{\mathbf{A}}^{-1} & 0 \\ \mathbf{B}\tilde{\mathbf{A}}^{-1} & -\mathbf{I} \end{pmatrix} \left[\begin{pmatrix} \mathbf{f}_1 \\ \mathbf{f}_2 \end{pmatrix} - \mathbf{K} \begin{pmatrix} \mathbf{x}^k \\ \mathbf{y}^k \end{pmatrix} \right]. \tag{5.27}$$

Using the approximations $\tilde{\mathbf{A}}^{-1} \approx \mathbf{Q}_A^{-1}$, $\mathbf{S}^{-1}\mathbf{w} \approx \hat{\mathbf{S}}^{-1}\mathbf{w} \approx \Psi(\mathbf{w})$ and the notation $\mathbf{r}_1^k := \mathbf{f}_1 - \tilde{\mathbf{A}}\mathbf{x}^k - \mathbf{B}^T\mathbf{y}^k$ we obtain the (nonlinear) iterative method

$$\begin{aligned} \mathbf{x}^{k+1} &= \mathbf{x}^k + \mathbf{Q}_A^{-1}\mathbf{r}_1^k - \mathbf{Q}_A^{-1}\mathbf{B}^T\Psi\big(\mathbf{B}(\mathbf{Q}_A^{-1}\mathbf{r}_1^k + \mathbf{x}^k) - \mathbf{f}_2\big) \\ \mathbf{y}^{k+1} &= \mathbf{y}^k + \Psi\big(\mathbf{B}(\mathbf{Q}_A^{-1}\mathbf{r}_1^k + \mathbf{x}^k) - \mathbf{f}_2\big). \end{aligned} \tag{5.28}$$

Thus we obtain an *inexact Uzawa method* with the following algorithmic structure:

$$\begin{cases} \begin{pmatrix} \mathbf{x}^0 \\ \mathbf{y}^0 \end{pmatrix} : \text{starting vector; } \mathbf{r}_1^0 := \mathbf{f}_1 - \tilde{\mathbf{A}}\mathbf{x}^0 - \mathbf{B}^T\mathbf{y}^0 \\ \text{for } k \geq 0 \ : \\ \qquad \mathbf{w} := \mathbf{x}^k + \mathbf{Q}_A^{-1}\mathbf{r}_1^k \\ \qquad \mathbf{z} := \Psi(\mathbf{B}\mathbf{w} - \mathbf{f}_2) \\ \qquad \mathbf{x}^{k+1} := \mathbf{w} - \mathbf{Q}_A^{-1}\mathbf{B}^T\mathbf{z} \\ \qquad \mathbf{y}^{k+1} := \mathbf{y}^k + \mathbf{z} \\ \qquad \mathbf{r}_1^{k+1} := \mathbf{r}_1^k - \tilde{\mathbf{A}}(\mathbf{x}^{k+1} - \mathbf{x}^k) - \mathbf{B}^T\mathbf{z}. \end{cases} \tag{5.29}$$

We take $\Psi(\cdot)$ is as follows:

$$\Psi(\mathbf{B}\mathbf{w} - \mathbf{f}_2) = \begin{cases} \text{Result of } \ell \text{ PCG iter. applied to } \hat{\mathbf{S}}\mathbf{v} = \mathbf{B}\mathbf{w} - \mathbf{f}_2 \\ \text{with initialization 0 and preconditioner } \mathbf{Q}_S \ . \end{cases} \tag{5.30}$$

This algorithm consists of an inner-outer iteration. In the algorithm we use preconditioners $\mathbf{Q}_A$ of $\tilde{\mathbf{A}}$ and $\mathbf{Q}_S$ of $\hat{\mathbf{S}} \approx \mathbf{S}$. An analysis of this method is given in [202] where it is shown that in the inner iteration (5.30) one should use very small ℓ-values ($\ell = 1, 2$), cf. Remark 5.2.16 below, and that this inexact Uzawa method is an efficient solver for the saddle point problem (5.8) *provided*

$$\textit{we have good preconditioners } \mathbf{Q}_A \textit{ of } \tilde{\mathbf{A}} \textit{ and } \mathbf{Q}_S \textit{ of } \mathbf{S}. \tag{5.31}$$

Remark 5.2.11 In [202] it is shown that the algorithm (5.29) can be implemented in such a way that per outer iteration one needs $\ell + 1$ evaluations of the preconditioner $\mathbf{Q}_A^{-1}$, ℓ evaluations of $\mathbf{Q}_S^{-1}$, $\ell+1$ matrix-vector multiplications with $\mathbf{B}$, ℓ matrix-vector multiplications with $\mathbf{B}^T$ and one matrix-vector multiplication with $\tilde{\mathbf{A}}$.

We discuss the convergence analysis from [202]. We assume that for the preconditioner $\mathbf{Q}_A$ we have

$$\gamma_A \mathbf{Q}_A \leq \tilde{\mathbf{A}} \leq \mathbf{Q}_A. \tag{5.32}$$

We make the scaling assumption $\tilde{\mathbf{A}} \leq \mathbf{Q}_A$ because it simplifies the analysis and it is often satisfied in practice, for example it is fulfilled for a multigrid preconditioner, cf. Sect. 5.4.2. This scaling assumption, however, is not essential neither for the algorithm nor for the convergence analysis.
In the analysis we use the following natural norms:

$$\|\mathbf{x}\|_Q := \|\mathbf{Q}_A^{\frac{1}{2}}\mathbf{x}\| = \langle \mathbf{Q}_A\mathbf{x}, \mathbf{x}\rangle^{\frac{1}{2}} \ \text{ for } \ \mathbf{x} \in \mathbb{R}^m,$$
$$\|\mathbf{y}\|_{\hat{S}} := \|\hat{\mathbf{S}}^{\frac{1}{2}}\mathbf{y}\| = \langle \hat{\mathbf{S}}\mathbf{y}, \mathbf{y}\rangle^{\frac{1}{2}} \ \text{ for } \ \mathbf{y} \in \mathbb{R}^n.$$

For the error in algorithm (5.29) we use the notation

$$\mathbf{e}^k = \begin{pmatrix} \mathbf{x} \\ \mathbf{y} \end{pmatrix} - \begin{pmatrix} \mathbf{x}^k \\ \mathbf{y}^k \end{pmatrix} =: \begin{pmatrix} \mathbf{e}_1^k \\ \mathbf{e}_2^k \end{pmatrix}.$$

Lemma 5.2.12 *For* $\mathbf{w}$ *as in* (5.29) *and* $\mathbf{d}_2 := \mathbf{B}\mathbf{w} - \mathbf{f}_2$ *we have the bound*

$$\langle \hat{\mathbf{S}}^{-1}\mathbf{d}_2, \mathbf{d}_2 \rangle^{\frac{1}{2}} \leq (1 - \gamma_A)\|\mathbf{e}_1^k\|_Q + \|\mathbf{e}_2^k\|_{\hat{S}}.$$

Proof. Define $\|\mathbf{y}\|_{\hat{S}^{-1}} = \langle \hat{\mathbf{S}}^{-1}\mathbf{y}, \mathbf{y} \rangle^{\frac{1}{2}}$. Note that for the exact discrete solution $\mathbf{x}$ we have $\mathbf{B}\mathbf{x} = \mathbf{f}_2$. Using this and the definition of $\mathbf{w}$ we get

$$\mathbf{B}\mathbf{w} - \mathbf{f}_2 = \mathbf{B}\mathbf{x}^k - \mathbf{B}\mathbf{x} + \mathbf{B}\mathbf{Q}_A^{-1}(\tilde{\mathbf{A}}\mathbf{e}_1^k + \mathbf{B}^T\mathbf{e}_2^k) = -\mathbf{B}(\mathbf{I} - \mathbf{Q}_A^{-1}\tilde{\mathbf{A}})\mathbf{e}_1^k + \hat{\mathbf{S}}\mathbf{e}_2^k.$$

Hence,

$$\begin{aligned} \langle \hat{\mathbf{S}}^{-1}\mathbf{d}_2, \mathbf{d}_2 \rangle^{\frac{1}{2}} &= \|\mathbf{d}_2\|_{\hat{S}^{-1}} \leq \|\mathbf{B}(\mathbf{I} - \mathbf{Q}_A^{-1}\tilde{\mathbf{A}})\mathbf{e}_1^k\|_{\hat{S}^{-1}} + \|\hat{\mathbf{S}}\mathbf{e}_2^k\|_{\hat{S}^{-1}} \\ &\leq \|\hat{\mathbf{S}}^{-\frac{1}{2}}\mathbf{B}(\mathbf{I} - \mathbf{Q}_A^{-1}\tilde{\mathbf{A}})\mathbf{Q}_A^{-\frac{1}{2}}\|\|\mathbf{Q}_A^{\frac{1}{2}}\mathbf{e}_1^k\| + \|\hat{\mathbf{S}}^{\frac{1}{2}}\mathbf{e}_2^k\|. \end{aligned}$$

Now note that

$$\|\hat{\mathbf{S}}^{-\frac{1}{2}}\mathbf{B}(\mathbf{I} - \mathbf{Q}_A^{-1}\tilde{\mathbf{A}})\mathbf{Q}_A^{-\frac{1}{2}}\| \leq \|\hat{\mathbf{S}}^{-\frac{1}{2}}\mathbf{B}\mathbf{Q}_A^{-\frac{1}{2}}\|\|\mathbf{Q}_A^{\frac{1}{2}}(\mathbf{I} - \mathbf{Q}_A^{-1}\tilde{\mathbf{A}})\mathbf{Q}_A^{-\frac{1}{2}}\|,$$

and

$$\begin{aligned} \|\hat{\mathbf{S}}^{-\frac{1}{2}}\mathbf{B}\mathbf{Q}_A^{-\frac{1}{2}}\|^2 &= \rho(\hat{\mathbf{S}}^{-\frac{1}{2}}\mathbf{B}\mathbf{Q}_A^{-1}\mathbf{B}^T\hat{\mathbf{S}}^{-\frac{1}{2}}) = \rho(\mathbf{I}) = 1, \\ \|\mathbf{Q}_A^{\frac{1}{2}}(\mathbf{I} - \mathbf{Q}_A^{-1}\tilde{\mathbf{A}})\mathbf{Q}_A^{-\frac{1}{2}}\| &= \rho(\mathbf{I} - \mathbf{Q}_A^{-\frac{1}{2}}\tilde{\mathbf{A}}\mathbf{Q}_A^{-\frac{1}{2}}) \leq 1 - \gamma_A. \end{aligned}$$

This completes the proof. □

We now derive a main result on the error reduction in the inexact Uzawa method.

Theorem 5.2.13 *Consider the inexact Uzawa method* (5.29) *with* Ψ *such that* (5.26) *holds. For the error* $\mathbf{e}^k = (\mathbf{e}_1^k, \mathbf{e}_2^k)$ *we have the bounds*

$$\|\mathbf{e}_1^{k+1}\|_Q \leq (1 - \gamma_A)\|\mathbf{e}_1^k\|_Q + \|\mathbf{e}_2^{k+1}\|_{\hat{S}}, \tag{5.33}$$

$$\|\mathbf{e}_2^{k+1}\|_{\hat{S}} \leq (1 - \gamma_A)(1 + \delta)\|\mathbf{e}_1^k\|_Q + \delta\|\mathbf{e}_2^k\|_{\hat{S}}, \tag{5.34}$$

with γ_A *from* (5.32) *and* δ *from* (5.26).

Proof. For the error component $\mathbf{e}_1^{k+1}$ we have the relations

$$\begin{aligned} \mathbf{e}_1^{k+1} &= \mathbf{x} - \mathbf{x}^{k+1} = \mathbf{x} - \mathbf{w} + \mathbf{Q}_A^{-1}\mathbf{B}^T\mathbf{z} \\ &= \mathbf{x} - \mathbf{x}^k - \mathbf{Q}_A^{-1}(\tilde{\mathbf{A}}\mathbf{e}_1^k + \mathbf{B}^T\mathbf{e}_2^k - \mathbf{B}^T\mathbf{z}) \\ &= (\mathbf{I} - \mathbf{Q}_A^{-1}\tilde{\mathbf{A}})\mathbf{e}_1^k - \mathbf{Q}_A^{-1}\mathbf{B}^T(\mathbf{e}_2^k - \mathbf{z}) \\ &= (\mathbf{I} - \mathbf{Q}_A^{-1}\tilde{\mathbf{A}})\mathbf{e}_1^k - \mathbf{Q}_A^{-1}\mathbf{B}^T\mathbf{e}_2^{k+1}. \end{aligned}$$

Hence,

$$\|\mathbf{e}_1^{k+1}\|_Q \le \|\mathbf{I} - \mathbf{Q}_A^{-\frac{1}{2}} \tilde{\mathbf{A}} \mathbf{Q}_A^{-\frac{1}{2}}\| \|\mathbf{e}_1^k\|_Q + \|\mathbf{Q}_A^{-\frac{1}{2}} \mathbf{B}^T \hat{\mathbf{S}}^{-\frac{1}{2}}\| \|\mathbf{e}_2^{k+1}\|_{\hat{S}}.$$

In combination with

$$\|\mathbf{I} - \mathbf{Q}_A^{-\frac{1}{2}} \tilde{\mathbf{A}} \mathbf{Q}_A^{-\frac{1}{2}}\| \le 1 - \gamma_A, \qquad \|\mathbf{Q}_A^{-\frac{1}{2}} \mathbf{B}^T \hat{\mathbf{S}}^{-\frac{1}{2}}\| = 1,$$

this proves the inequality in (5.33). For the error component $\mathbf{e}_2^{k+1}$ we obtain, with $\mathbf{d}_2 := \mathbf{B}\mathbf{w} - \mathbf{f}_2$,

$$\mathbf{e}_2^{k+1} = \mathbf{y} - \mathbf{y}^{k+1} = \mathbf{e}_2^k - \mathbf{z} = (\mathbf{e}_2^k - \hat{\mathbf{S}}^{-1}\mathbf{d}_2) + (\hat{\mathbf{S}}^{-1}\mathbf{d}_2 - \mathbf{z}). \tag{5.35}$$

For the first term we get

$$\begin{aligned}
\|\mathbf{e}_2^k - \hat{\mathbf{S}}^{-1}\mathbf{d}_2\|_{\hat{S}} &= \|\mathbf{e}_2^k - \hat{\mathbf{S}}^{-1}\mathbf{B}(\mathbf{x}^k + \mathbf{Q}_A^{-1}\tilde{\mathbf{A}}\mathbf{e}_1^k + \mathbf{Q}_A^{-1}\mathbf{B}^T\mathbf{e}_2^k)\|_{\hat{S}} \\
&= \|\mathbf{e}_2^k + \hat{\mathbf{S}}^{-1}\mathbf{B}(\mathbf{I} - \mathbf{Q}_A^{-1}\tilde{\mathbf{A}})\mathbf{e}_1^k - \hat{\mathbf{S}}^{-1}\mathbf{B}\mathbf{Q}_A^{-1}\mathbf{B}^T\mathbf{e}_2^k\|_{\hat{S}} \\
&= \|\hat{\mathbf{S}}^{-1}\mathbf{B}(\mathbf{I} - \mathbf{Q}_A^{-1}\tilde{\mathbf{A}})\mathbf{e}_1^k\|_{\hat{S}} \\
&\le \|\hat{\mathbf{S}}^{-\frac{1}{2}}\mathbf{B}\mathbf{Q}_A^{-\frac{1}{2}}\| \|\mathbf{I} - \mathbf{Q}_A^{-\frac{1}{2}}\tilde{\mathbf{A}}\mathbf{Q}_A^{-\frac{1}{2}}\| \|\mathbf{e}_1^k\|_Q \\
&\le (1 - \gamma_A)\|\mathbf{e}_1^k\|_Q.
\end{aligned} \tag{5.36}$$

Furthermore, using (5.26) and Lemma 5.2.12 we also have

$$\begin{aligned}
\|\hat{\mathbf{S}}^{-1}\mathbf{d}_2 - \mathbf{z}\|_{\hat{S}} &= \|\hat{\mathbf{S}}^{-1}\mathbf{d}_2 - \Psi(\mathbf{d}_2)\|_{\hat{S}} \le \delta\, \|\hat{\mathbf{S}}^{-1}\mathbf{d}_2\|_{\hat{S}} \\
&= \delta\, \langle \hat{\mathbf{S}}^{-1}\mathbf{d}_2, \mathbf{d}_2 \rangle^{\frac{1}{2}} \le \delta\big((1 - \gamma_A)\|\mathbf{e}_1^k\|_Q + \|\mathbf{e}_2^k\|_{\hat{S}}\big).
\end{aligned} \tag{5.37}$$

Combination of the results in (5.35), (5.36), (5.37) proves the inequality in (5.34). □

As a simple consequence of this theorem we obtain the following convergence result:

Theorem 5.2.14 *Define*

$$\mu_A := 1 - \gamma_A, \quad g(\mu_A, \delta) := 2\mu_A + \delta(1 + \mu_A). \tag{5.38}$$

Consider the inexact Uzawa method (5.29) *with* Ψ *such that* (5.26) *holds. For the error* $\mathbf{e}^k = (\mathbf{e}_1^k, \mathbf{e}_2^k)$ *we have the bounds*

$$\max\big\{\|\mathbf{e}_1^{k+1}\|_Q, \|\mathbf{e}_2^{k+1}\|_{\hat{S}}\big\} \le g(\mu_A, \delta) \max\big\{\|\mathbf{e}_1^k\|_Q, \|\mathbf{e}_2^k\|_{\hat{S}}\big\}, \tag{5.39}$$

$$\|\mathbf{e}_1^k\|_Q + \|\mathbf{e}_2^k\|_{\hat{S}} \le 3\frac{1}{2}\Big(\frac{g(\mu_A, \delta) + \sqrt{g(\mu_A, \delta)^2 - 4\mu_A\delta}}{2}\Big)^k (\|\mathbf{e}_1^0\|_Q + \|\mathbf{e}_2^0\|_{\hat{S}}). \tag{5.40}$$

Proof. Define the matrix

$$\mathrm{C} = \begin{pmatrix} \mu_A(2+\delta) & \delta \\ \mu_A(1+\delta) & \delta \end{pmatrix}.$$

Due to Theorem 5.2.13 we obtain

$$\begin{pmatrix} \|\mathbf{e}_1^{k+1}\|_Q \\ \|\mathbf{e}_2^{k+1}\|_{\hat{S}} \end{pmatrix} \leq \mathrm{C} \begin{pmatrix} \|\mathbf{e}_1^{k}\|_Q \\ \|\mathbf{e}_2^{k}\|_{\hat{S}} \end{pmatrix},$$

where "$\leq$" is meant entrywise. From $\|\mathrm{C}\|_\infty = g(\mu_A, \delta)$ we obtain the result in (5.39). The eigenvalues of the matrix $\mathbf{C}$ are given by $\lambda_{1,2} = \frac{1}{2}g(\mu_A,\delta) \pm \frac{1}{2}\sqrt{g(\mu_A,\delta)^2 - 4\mu_A\delta}$. Thus we have

$$\rho(\mathrm{C}) = \frac{1}{2}\Big(g(\mu_A,\delta) + \sqrt{g(\mu_A,\delta)^2 - 4\mu_A\delta}\Big).$$

We have $\lambda_1 = \lambda_2$ iff $\mu = \delta = 0$. Hence there exists an eigenvector decomposition $\mathrm{C} = \mathrm{V}\,\mathrm{diag}(\lambda_1,\lambda_2)\mathrm{V}^{-1}$ and

$$\|\mathbf{e}_1^k\|_Q + \|\mathbf{e}_2^k\|_{\hat{S}} \leq \rho(C)^k \|\mathrm{V}\|_1 \|\mathrm{V}^{-1}\|_1 \big(\|\mathbf{e}_1^0\|_Q + \|\mathbf{e}_2^0\|_{\hat{S}}\big)$$

holds. A MAPLE computation yields that for the condition number (in the 1-norm) of the eigenvector matrix we have the uniform bound

$$\max_{0 \leq \mu_A, \delta \leq 1} \|\mathrm{V}\|_1 \|\mathrm{V}^{-1}\|_1 \leq 3\frac{1}{2}.$$

Hence the result in (5.40) holds. □

Corollary 5.2.15 Clearly, the bound for the contraction factor in (5.39) and the bound for the asymptotic convergence factor in (5.40) depend only on μ_A and δ and the bounds are monotonic functions of these parameters. Note that for $\mu_A \to 0$ we obtain the contraction factor of the exact Uzawa method: $g(0,\delta) = \delta$. We also have $g(\mu_A,\delta) \geq \frac{1}{2}\big(g(\mu_A,\delta) + \sqrt{g(\mu_A,\delta)^2 - 4\mu_A\delta}\big)$ and

$$g(\mu_A,\delta) < 1, \quad \text{iff } 0 \leq \delta < \frac{1-2\mu_A}{1+\mu_A}, \tag{5.41}$$

$$\frac{1}{2}\big(g(\mu_A,\delta) + \sqrt{g(\mu_A,\delta)^2 - 4\mu_A\delta}\big) < 1, \quad \text{iff } 0 \leq \delta < 1 - 2\mu_A. \tag{5.42}$$

Hence, for $\mu_A < \frac{1}{2}$ and δ sufficiently small (as quantified in (5.41), (5.42)) we have a convergent method.

Remark 5.2.16 We comment on the important special case where we take for $\mathbf{Q}_A^{-1}$ a symmetric multigrid V-cycle, cf. Sect. 5.4.2. Then μ_A is the contraction number (w.r.t. the norm $\|\cdot\|_A$) of this multigrid method. It is known from numerical experiments (for our problem class) that in many cases $0.1 \leq \mu_A \leq 0.2$ holds. The result in (5.42) shows that for $\mu_A \approx 0.15$ we have a convergent method if for the accuracy of the inner method we take $\delta \lesssim 0.7$. Clearly it is not efficient to use a very small value for δ. We comment on the choice of this parameter $\delta \in (0, 0.7)$. As an example, we consider the case $\mu_A = 0.1$. For the contraction number in (5.39) we then have $g(0.1, \delta) = 0.2 + 1.1\delta$. For a given accuracy eps $\ll 1$ that is required in the inexact Uzawa method let k be such that

$$g(0.1, \delta)^k \approx \text{eps}.$$

Then

$$k \approx \frac{\ln(\text{eps})}{\ln(0.2 + 1.1\delta)} \tag{5.43}$$

holds. We use the ansatz (cf. (5.30)) $\delta = \beta^\ell$ with some $\beta < 1$, which implies $\ell = \ln \delta (\ln \beta)^{-1}$. Note that ℓ (# of inner iterations) is a decreasing function of δ and k (# of outer iterations) as in (5.43) is an increasing function of δ. It is reasonable to assume, cf. Remark 5.2.11, that the arithmetic costs in k iterations of the inexact Uzawa method are dominated by $k(\ell+1)$ evaluations of $\mathbf{Q}_A^{-1}$. Hence, one wants to minimize

$$k(\ell + 1) \approx \frac{\ln(\text{eps})}{\ln \beta} \frac{\ln \delta + \ln \beta}{\ln(0.2 + 1.1\delta)} =: k_{\text{Uzawa}} K(\beta, \delta),$$

where $k_{\text{Uzawa}} := \ln(\text{eps}) / \ln \beta$ is the number of iterations of the *exact* Uzawa method ($\mu_A = 0$). For different β values the function $\delta \to K(\beta, \delta)$ is given in Fig. 5.2. We see that a close to optimal value for δ is obtained in a broad

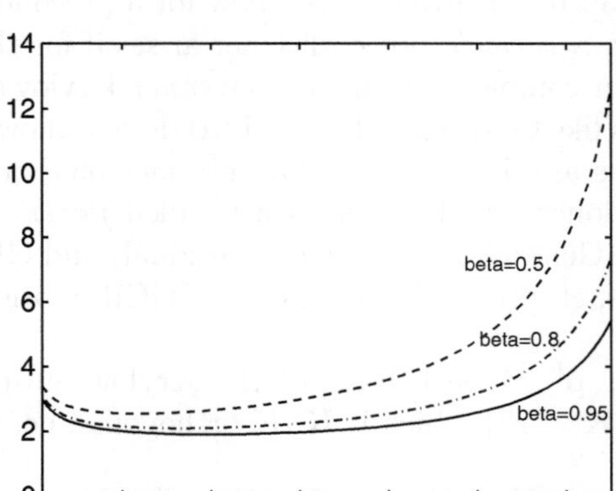

Fig. 5.2. Function $\delta \to K(\beta, \delta)$ for $\beta = 0.5, 0.8, 0.95$.

range, say (in this example) $\delta \in [0.1, 0.5]$. We conclude that in order to maximize the efficiency of the inexact Uzawa method one should take a low accuracy in the inner solver Ψ (for example, $\delta = 0.4$, cf. (5.26)). Moreover, the efficiency is not very sensitive with respect to the precise choice of this inner accuracy parameter δ. For other μ_A values (< 0.2) one observes similar effects.

5.3 Iterative Solvers for Oseen problems

The discrete Oseen problem has a matrix-vector representation of the form

$$\mathbf{K}\begin{pmatrix}\mathbf{x}\\ \mathbf{y}\end{pmatrix} := \begin{pmatrix}\tilde{\mathbf{A}} & \mathbf{B}^T\\ \mathbf{B} & 0\end{pmatrix}\begin{pmatrix}\mathbf{x}\\ \mathbf{y}\end{pmatrix} = \begin{pmatrix}\mathbf{f}_1\\ \mathbf{f}_2\end{pmatrix}, \qquad \tilde{\mathbf{A}} := \alpha\left[\mathbf{A} + \mathbf{N}(\mathbf{x}^{\text{old}})\right] + \beta\mathbf{M}, \tag{5.44}$$

where the parameter β is proportional to $1/\Delta t$, cf. (5.1). This linear system has a saddle point structure but opposite to the discrete generalized Stokes equation in (5.8) the system matrix $\mathbf{K}$ in this case is *non-symmetric*. For this type of linear systems, methods similar to the ones used for the Stokes problem in Sect. 5.2 have been developed. We briefly address two classes of methods, namely preconditioned Krylov subspace solvers and inexact Uzawa methods.

Preconditioned Krylov subspace methods

One can apply a Krylov subspace method directly to the saddle point system (5.44) and combine this with a block preconditioner, i.e. a preconditioner that uses the block structure of the saddle point matrix. For standard Krylov subspace methods applicable to systems with a non-symmetric matrix like GMRES or BiCGSTAB, and their preconditioned variants, we refer to the literature, e.g. [213]. It is convenient to allow for a preconditioner that varies per iteration. Such a *variable* preconditioner arises if for the preconditioning of $\tilde{\mathbf{A}}$ or of the Schur complement one uses an *inner* Krylov subspace iteration. Standard methods like GMRES or BiCGSTAB do not allow such variable preconditioners. There are, however, Krylov subspace methods that can handle variable preconditioners. Such methods are called *flexible* iterations. Examples are the GCR (Generalized Conjugate Residual) and GMRESR (recursive GMRES) methods, cf. [213]. The main idea of GCR is briefly outlined in the following.

Let $(\mathbf{p}^0, \mathbf{p}^1, \ldots, \mathbf{p}^{k-1})$ be a basis of the Krylov subspace $\mathcal{K}^k(\mathbf{K}; \mathbf{r}^0) := \text{span}\{\mathbf{r}^0, \mathbf{K}\mathbf{r}^0, \ldots, \mathbf{K}^{k-1}\mathbf{r}^0\}$ which is $\mathbf{K}^T\mathbf{K}$-orthogonal, i.e. such that

$$\langle \mathbf{K}\mathbf{p}_j, \mathbf{K}\mathbf{p}_i\rangle = 0 \qquad \text{for all } i \neq j.$$

Due to this orthogonality property it is easy to compute the solution of a *residual minimization problem* as in (5.12). The unique

$$\begin{pmatrix}\mathbf{x}^k\\ \mathbf{y}^k\end{pmatrix} =: \mathbf{z}^k \in \mathbf{z}^0 + \mathcal{K}^k(\mathbf{K}; \mathbf{r}^0) \quad \text{such that } \left\|\mathbf{K}\mathbf{z}^k - \mathbf{f}\right\| \text{ is minimal} \tag{5.45}$$

can be computed recursively by

$$\mathbf{z}^k = \mathbf{z}^{k-1} + \frac{\langle \mathbf{r}^{k-1}, \mathbf{K}\mathbf{p}^{k-1} \rangle}{\langle \mathbf{K}\mathbf{p}^{k-1}, \mathbf{K}\mathbf{p}^{k-1} \rangle}\, \mathbf{p}^{k-1},$$

with $\mathbf{r}^{k-1} := \mathbf{f} - \mathbf{K}\mathbf{z}^{k-1}$. Computing the $\mathbf{K}^T\mathbf{K}$-orthogonal basis vectors turns out to be computationally expensive. The next $\mathbf{K}^T\mathbf{K}$-orthogonal basis vector $\mathbf{p}^k$ can be computed as a linear combination of the residual $\mathbf{r}^k$ and *all* previous basis vectors $\mathbf{p}^0, \ldots, \mathbf{p}^{k-1}$. The resulting GCR method has the following algorithmic structure:

$$\begin{cases} \mathbf{z}^0 := \begin{pmatrix} \mathbf{x}^0 \\ \mathbf{y}^0 \end{pmatrix} : \text{starting vector; } \mathbf{r}^0 := \begin{pmatrix} \mathbf{f}_1 \\ \mathbf{f}_2 \end{pmatrix} - \mathbf{K}\mathbf{z}^0;\ \mathbf{p}^0 = \mathbf{r}^0 \\ \text{for } k \geq 0 : \\ \quad \alpha_k := \dfrac{\langle \mathbf{r}^k, \mathbf{K}\mathbf{p}^k \rangle}{\langle \mathbf{K}\mathbf{p}^k, \mathbf{K}\mathbf{p}^k \rangle} \\ \quad \mathbf{z}^{k+1} := \mathbf{z}^k + \alpha_k \mathbf{p}^k \\ \quad \mathbf{r}^{k+1} := \mathbf{r}^k - \alpha_k \mathbf{K}\mathbf{p}^k \\ \quad \text{Compute } \beta_{i,k} := -\dfrac{\langle \mathbf{K}\mathbf{r}^{k+1}, \mathbf{K}\mathbf{p}^i \rangle}{\langle \mathbf{K}\mathbf{p}^i, \mathbf{K}\mathbf{p}^i \rangle} \quad \text{for } i = 0, 1, \ldots, k \\ \quad \mathbf{p}^{k+1} := \mathbf{r}^{k+1} + \displaystyle\sum_{i=0}^{k} \beta_{i,k} \mathbf{p}^i. \end{cases} \tag{5.46}$$

Opposite to the MINRES method, in the GCR algorithm *there is a significant growth in the arithmetic costs per iteration* if the iteration index k increases. This is essentially due to the fact that in (5.45) a residual minimization criterion is used. It is known, that for a *non*symmetric matrix in general this leads to an "expensive" algorithm. Recall that the (cheap) MINRES method is applicable to *symmetric* matrices only. As in the case of the methods discussed in the previous section, the GCR method can be combined with a preconditioner. Furthermore, different mathematically equivalent algorithms of (preconditioned) GCR exist. Below we give a preconditioned GCR algorithm that is often used in practice. We use the notation $\mathbf{Q}_k$ (instead of $\mathbf{Q}$) for the preconditioner to indicate that the preconditioner may change per iteration.

$$
\begin{cases}
\mathbf{z}^0 := \begin{pmatrix} \mathbf{x}^0 \\ \mathbf{y}^0 \end{pmatrix} : \text{starting vector; } \mathbf{r}^0 := \begin{pmatrix} \mathbf{f}_1 \\ \mathbf{f}_2 \end{pmatrix} - \mathbf{K}\mathbf{z}^0; \\
\text{for } k \geq 0 : \\
\qquad \mathbf{s}^{k+1} := \mathbf{Q}_k^{-1}\mathbf{r}^k \\
\qquad \tilde{\mathbf{v}}^{k+1} := \mathbf{K}\mathbf{s}^{k+1} \\
\qquad \mathbf{s}^{k+1} := \mathbf{s}^{k+1} - \sum_{i=1}^{k} \langle \tilde{\mathbf{v}}^{k+1}, \mathbf{v}^i \rangle \, \mathbf{s}^i \\
\qquad \hat{\mathbf{v}}^{k+1} := \tilde{\mathbf{v}}^{k+1} - \sum_{i=1}^{k} \langle \tilde{\mathbf{v}}^{k+1}, \mathbf{v}^i \rangle \, \mathbf{v}^i \\
\qquad \mathbf{s}^{k+1} := \mathbf{s}^{k+1} / \|\hat{\mathbf{v}}^{k+1}\|_2 \\
\qquad \mathbf{v}^{k+1} := \hat{\mathbf{v}}^{k+1} / \|\hat{\mathbf{v}}^{k+1}\|_2 \\
\qquad \mathbf{z}^{k+1} := \mathbf{z}^k + \langle \mathbf{r}^k, \mathbf{v}^{k+1} \rangle \, \mathbf{s}^{k+1} \\
\qquad \mathbf{r}^{k+1} := \mathbf{r}^k - \langle \mathbf{r}^k, \mathbf{v}^{k+1} \rangle \, \mathbf{v}^{k+1}.
\end{cases}
\tag{5.47}
$$

For the case $\mathbf{Q}_k = \mathbf{I}$ (no preconditioning) one can verify that the algorithms in (5.47) and (5.46) produce the same sequence of approximations $\mathbf{z}^k$ (in exact arithmetic). Relations between these two variants, for the case $\mathbf{Q}_k = \mathbf{I}$, are $\mathbf{v}^{k+1} = \frac{\mathbf{K}\mathbf{p}^k}{\|\mathbf{K}\mathbf{p}^k\|}$ and $\mathbf{s}^{k+1} = \frac{\mathbf{p}^k}{\|\mathbf{K}\mathbf{p}^k\|}$.

As for the generalized Stokes problem, the issue of *preconditioning*, i.e., the choice of $\mathbf{Q}_k^{-1}$ in (5.47), is of great importance for the efficiency of such a Krylov subspace method. *The use of an appropriate preconditioner is often much more important than the choice of the particular Krylov subspace method* (e.g. GMRESR or GCR).

Concerning suitable block preconditioners for the Oseen problem we start with the block factorizations

$$
\begin{aligned}
&\begin{pmatrix} \tilde{\mathbf{A}} & \mathbf{B}^T \\ \mathbf{B} & 0 \end{pmatrix} \begin{pmatrix} \tilde{\mathbf{A}} & \mathbf{B}^T \\ 0 & -\mathbf{S} \end{pmatrix}^{-1} = \begin{pmatrix} \mathbf{I} & 0 \\ \mathbf{B}\tilde{\mathbf{A}}^{-1} & \mathbf{I} \end{pmatrix} =: \hat{\mathbf{K}}_r, \\
&\begin{pmatrix} \tilde{\mathbf{A}} & 0 \\ \mathbf{B} & -\mathbf{S} \end{pmatrix}^{-1} \begin{pmatrix} \tilde{\mathbf{A}} & \mathbf{B}^T \\ \mathbf{B} & 0 \end{pmatrix} = \begin{pmatrix} \mathbf{I} & \tilde{\mathbf{A}}^{-1}\mathbf{B}^T \\ 0 & \mathbf{I} \end{pmatrix} =: \hat{\mathbf{K}}_l, \quad \mathbf{S} := \mathbf{B}\tilde{\mathbf{A}}^{-1}\mathbf{B}^T,
\end{aligned}
\tag{5.48}
$$

where we assumed that $\text{rank}(\mathbf{B}) = n$, and thus $\mathbf{S}$ is regular, cf. Remark 5.2.1. The preconditioned matrices $\hat{\mathbf{K}}_r$, $\hat{\mathbf{K}}_l$ result from "right" and "left" preconditioning" of the matrix $\mathbf{K}$ with preconditioners

$$
\mathbf{Q}_r := \begin{pmatrix} \tilde{\mathbf{A}} & \mathbf{B}^T \\ 0 & -\mathbf{S} \end{pmatrix}, \quad \mathbf{Q}_l := \begin{pmatrix} \tilde{\mathbf{A}} & 0 \\ \mathbf{B} & -\mathbf{S} \end{pmatrix}.
$$

Note that $(\hat{\mathbf{K}}_p - \mathbf{I})^2 = 0$, $p \in \{r, l\}$ and that both preconditioned matrices have all eigenvalues equal to 1. If the GMRES method is applied to the left preconditioned system $\mathbf{Q}_l^{-1}\mathbf{K}\mathbf{x} = \mathbf{Q}_l^{-1}\mathbf{f}$ then the preconditioned residual $\hat{\mathbf{r}}^k = \mathbf{Q}_l^{-1}(\mathbf{f} - \mathbf{K}\mathbf{x}^k)$ satisfies the optimality property

$$\|\hat{\mathbf{r}}^k\| = \min_{p_k \in \mathcal{P}_k; p_k(0)=1} \|p_k(\mathbf{Q}_l^{-1}\mathbf{K})\mathbf{r}^0\| = \min_{p_k \in \mathcal{P}_k; p_k(0)=1} \|p_k(\hat{\mathbf{K}}_l)\mathbf{r}^0\|.$$

For $p_2(z) = (z-1)^2$ we have $p_2(\hat{\mathbf{K}}_l) = 0$ and thus one obtains (in exact arithmetic) the solution of a linear system $\mathbf{K}\mathbf{x} = \mathbf{b}$ in (at most) two iterations. The same holds for the GMRES method applied to the right preconditioned system with matrix $\hat{\mathbf{K}}_r$. Clearly the preconditioners $\mathbf{Q}_r$ and $\mathbf{Q}_l$ are *not feasible.* In practice one often uses preconditioners of the form

$$\mathbf{Q} := \begin{pmatrix} \mathbf{Q}_A & \mathbf{B}^T \\ 0 & -\mathbf{Q}_S \end{pmatrix}, \quad \text{or} \quad \mathbf{Q} := \begin{pmatrix} \mathbf{Q}_A & 0 \\ \mathbf{B} & -\mathbf{Q}_S \end{pmatrix}, \tag{5.49}$$

with preconditioners $\mathbf{Q}_A$ of $\tilde{\mathbf{A}}$ and $\mathbf{Q}_S$ of the Schur complement $\mathbf{S}$. For $\mathbf{Q}_A$ one can use a multigrid preconditioner (resulting from a multigrid method for each of the scalar diffusion-convection-reaction type of problems in the block-diagonal matrix $\tilde{\mathbf{A}}$) or some other preconditioned Krylov subspace method (BiCGSTAB, for example), cf. Sect. 5.4. In the latter case the preconditioner is nonlinear and changes per iteration, i.e. $\mathbf{Q} = \mathbf{Q}_k$ in (5.49). In such a situation one needs a flexible Krylov subspace method, like GCR. The issue of Schur complement preconditioning in the non-symmetric case is much more difficult as in the symmetric case. This topic is briefly addressed in Sect. 5.4.3.

For *symmetric* saddle point problems there is extensive theory with analyses of the rate of convergence of iterative solvers and of the effect of preconditioning, cf. [31]. An example of such an analysis is given in Sect. 5.2. For *non*symmetric saddle point problems like the Oseen problem the state of the art concerning theoretical analyses is still quite unsatisfactory. Although many algorithms (iterative methods and preconditioners) have been developed there are only very few significant theoretical results. Moreover, these results typically hold for rather special cases. We do not present examples of such results here but refer to the literature, for example the overview paper [31].

Inexact Uzawa method

For the discrete Oseen problem in (5.44) an inexact Uzawa method as in (5.29) can be used. For the Oseen problem the Schur complement is *non*symmetric and therefore one should not apply a preconditioned CG method in (5.30) but for Ψ one can use, for example, a preconditioned GCR method. For the preconditioner $\mathbf{Q}_A$ one can apply a multigrid method or a (preconditioned) Krylov subspace iteration. More delicate is the choice of $\mathbf{Q}_S$, which is briefly addressed in Sect. 5.4.3.

5.4 Preconditioners

5.4.1 Introduction

For efficient iterative solvers for the (generalized) Stokes and Oseen problems it is essential to have good preconditioners $\mathbf{Q}_A$ and $\mathbf{Q}_S$ of $\tilde{\mathbf{A}}$ and $\mathbf{S}$ (or $\hat{\mathbf{S}}$), respectively.

Consider a *general* nonsingular $n \times n$-matrix $\mathbf{A}$ (not necessarily the same as the matrix $\mathbf{A}$ in (5.8) or (5.44)). Assume a nonsingular matrix $\mathbf{Q} \approx \mathbf{A}$, which is called a *preconditioner* of $\mathbf{A}$. For a preconditioner to be useful for the acceleration of iterative methods one often requires that for all $\mathbf{y} \in \mathbb{R}^n$

$$\mathbf{x} = \mathbf{Q}^{-1}\mathbf{y} \text{ can be determined with "low" costs } (cn \text{ flops}), \tag{5.50a}$$
$$\text{and} \quad \kappa(\mathbf{Q}^{-1}\mathbf{A}) < \kappa(\mathbf{A}). \tag{5.50b}$$

The concept of preconditioning is very general and it can be found at many places in the literature. An overview of several preconditioning techniques for discretized scalar elliptic partial differential equations is given in [57, 124, 213]. For example, a very popular approach is to use $\mathbf{Q} = \mathbf{LU} \approx \mathbf{A}$, where $\mathbf{L}$ is a sparse lower triangular matrix and $\mathbf{U}$ a sparse upper triangular matrix ("Incomplete LU factorization").

A well-known technique is to use linear or nonlinear iterative methods as preconditioners. We briefly explain this for the case of a linear iterative method:

$$\mathbf{x}^{k+1} = \mathbf{x}^k - \mathbf{W}^{-1}(\mathbf{A}\mathbf{x}^k - \mathbf{b}). \tag{5.51}$$

The iteration matrix (which determines the error propagation) of this method is given by $\mathbf{C} = \mathbf{I} - \mathbf{W}^{-1}\mathbf{A}$. If one uses this iterative method for *preconditioning* then $\mathbf{Q} := \mathbf{W}$ is taken as the preconditioner for $\mathbf{A}$. If the method (5.51) converges then $\mathbf{W}$ (and thus $\mathbf{Q}$) is a reasonable approximation for $\mathbf{A}$ in the sense that $\rho(\mathbf{I} - \mathbf{W}^{-1}\mathbf{A}) < 1$.

The iteration in (5.51) corresponds to an iterative method and thus $\mathbf{W}^{-1}\mathbf{y}$ ($\mathbf{y} \in \mathbb{R}^n$) can be computed with acceptable arithmetic costs. Hence the condition in (5.50a), with $\mathbf{Q} = \mathbf{W}$, is satisfied.

Related to the implementation of such a preconditioner we note the following. In an iterative method the matrix $\mathbf{W}$ is usually *not* used in its implementation (cf. symmetric Gauss-Seidel or the multigrid method in Sect. 5.4.2), i.e. the iteration (5.51) is implemented without explicitly computing $\mathbf{W}$. The vector $\mathbf{x} = \mathbf{W}^{-1}\mathbf{y}$ is the result of (5.51) with $k = 0$, $\mathbf{x}^0 = 0$, $\mathbf{b} = \mathbf{y}$. From this it follows that the computation of $\mathbf{x} = \mathbf{Q}^{-1}\mathbf{y}$ can be implemented by *performing one iteration of the iterative method applied to* $\mathbf{Az} = \mathbf{y}$ *with starting vector* 0.

A bound for $\kappa(\mathbf{W}^{-1}\mathbf{A})$ (cf. (5.50b)) is presented in the following lemma.

Lemma 5.4.1 *We assume that* $\mathbf{A}$ *and* $\mathbf{W}$ *are symmetric positive definite matrices and that the method in* (5.51) *is convergent, i.e.* $\rho(\mathbf{C}) < 1$ *with* $\mathbf{C} := \mathbf{I} - \mathbf{W}^{-1}\mathbf{A}$. *Then the following holds:*

$$\kappa(\mathbf{W}^{-1}\mathbf{A}) \leq \frac{1+\rho(\mathbf{C})}{1-\rho(\mathbf{C})}. \tag{5.52}$$

Proof. Because $\mathbf{A}$ and $\mathbf{W}$ are symmetric positive definite it follows that

$$\sigma(\mathbf{W}^{-1}\mathbf{A}) = \sigma(\mathbf{W}^{-\frac{1}{2}}\mathbf{A}\mathbf{W}^{-\frac{1}{2}}) \subset (0,\infty) .$$

Using $\rho(\mathbf{I} - \mathbf{W}^{-1}\mathbf{A}) < 1$ we obtain that $\sigma(\mathbf{W}^{-1}\mathbf{A}) \subset (0,2)$. The eigenvalues of $\mathbf{W}^{-1}\mathbf{A}$ are denoted by μ_i:

$$0 < \mu_1 \leq \mu_2 \leq \ldots \leq \mu_n < 2 .$$

Hence $\rho(\mathbf{C}) = \max\{|1-\mu_1|, |1-\mu_n|\}$ holds and

$$\kappa(\mathbf{W}^{-1}\mathbf{A}) = \frac{\mu_n}{\mu_1} = \frac{1+(\mu_n-1)}{1-(1-\mu_1)} \leq \frac{1+|1-\mu_n|}{1-|1-\mu_1|} .$$

Thus

$$\kappa(\mathbf{W}^{-1}\mathbf{A}) \leq \frac{1+\rho(\mathbf{C})}{1-\rho(\mathbf{C})}$$

holds. □

With respect to the bound in (5.52) we note that the function $x \to \frac{1+x}{1-x}$ increases monotonically on $[0,1)$ and thus we have a bound on $\kappa(\mathbf{W}^{-1}\mathbf{A})$ that decreases if $\rho(\mathbf{C})$ decreases. This indicates that *the higher the convergence rate of the iterative method in* (5.51), *the better the quality of* $\mathbf{W}$ *as a preconditioner for* $\mathbf{A}$. In Sect. 5.4.2 we will introduce multigrid methods, which are known to have a very high rate of convergence when applied to linear systems with a matrix of the form $\tilde{\mathbf{A}}$ as in (5.8) or (5.44). Such multigrid methods are linear iterative methods as in (5.51), with a matrix $\mathbf{W} = \mathbf{W}_{MG}$ that is given only implicitly. Using a multigrid method one then obtains a (very) good preconditioner $\mathbf{Q}_A = \mathbf{W}_{MG}$ of $\tilde{\mathbf{A}}$.

The above preconditioning approach can be generalized to *non*linear iterative methods, for example Krylov subspace methods. For a given $\mathbf{y} \in \mathbb{R}^n$ the vector $\mathbf{x} = \Psi(\mathbf{y}) \approx \mathbf{A}^{-1}\mathbf{y}$ is then determined by applying one or a few iterations of the nonlinear method to the system $\mathbf{A}\mathbf{z} = \mathbf{y}$ with starting vector 0.

5.4.2 Multigrid preconditioner

In this section we give an introduction to multigrid methods. A detailed treatment can be found in, for example [51, 133, 132, 210, 251]. The remaining part of this section is organized as follows. First, in Subsection I, for a very simple one-dimensional diffusion problem we introduce the main idea underlying multigrid solvers. Then this idea is generalized to d-dimensional ($d = 2, 3$) elliptic boundary value problems (Subsection II). After that, in Subsection III, we illustrate the efficiency of multigrid methods by presenting results of a numerical experiment for the three-dimensional Poisson equation. Finally, in Subsection IV we briefly address some theoretical convergence results.

The application of multigrid as a preconditioner $\mathbf{Q}_A$ in the generalized Stokes and Oseen saddle point problems is discussed in Remark 5.4.26 below.

I. Multigrid for a one-dimensional model problem

We consider a very simple model problem to show the basic principle behind the multigrid approach, namely the two-point boundary value model problem

$$\begin{cases} -u''(x) = f(x), & x \in \Omega := (0,1). \\ u(0) = u(1) = 0\,. \end{cases} \tag{5.53}$$

We will use a finite element method for the discretization of this problem. This, however, is *not* essential: other discretization methods (finite differences, finite volumes) result in discrete problems that are very similar. The corresponding multigrid methods have properties very similar to those in the case of a finite element discretization.

For the finite element discretization we need the variational formulation of this boundary value problem: find $u \in H_0^1(\Omega)$ such that

$$\int_0^1 u'v'\,dx = \int_0^1 fv\,dx \quad \text{for all} \quad v \in H_0^1(\Omega).$$

For the discretization we introduce a *sequence* of nested uniform grids. For $\ell = 0, 1, 2, \ldots$, we define

$$h_\ell = 2^{-\ell-1} \qquad (\text{"mesh size"})\,, \tag{5.54a}$$
$$n_\ell = h_\ell^{-1} - 1 \qquad (\text{"number of interior grid points"})\,, \tag{5.54b}$$
$$\xi_{\ell,i} = ih_\ell\,, \quad i = 0, 1, ..., n_\ell + 1 \qquad (\text{"grid points"})\,, \tag{5.54c}$$
$$\Omega_\ell^{\text{int}} = \{\, \xi_{\ell,i} : 1 \le i \le n_\ell \,\} \qquad (\text{"interior grid"})\,, \tag{5.54d}$$
$$\mathcal{T}_{h_\ell} = \cup \{\, [\xi_{\ell,i}, \xi_{\ell,i+1}] : 0 \le i \le n_\ell \,\} \qquad (\text{"triangulation"})\,. \tag{5.54e}$$

The space of linear finite elements corresponding to the triangulation $\mathcal{T}_{h_\ell}$ is given by

$$V_\ell = \mathbb{X}_{h_\ell,0}^1 = \left\{ v \in C(\Omega) : v_{|[\xi_{\ell,i},\xi_{\ell,i+1}]} \in \mathcal{P}_1\,,\, i = 0, \ldots, n_\ell,\ v(0) = v(1) = 0 \right\}.$$

The standard nodal basis in this space is denoted by $(\phi_i)_{1 \le i \le n_\ell}$. These functions satisfy $\phi_i(\xi_{\ell,i}) = 1$, $\phi_i(\xi_{\ell,j}) = 0$ for all $j \ne i$. This basis induces an isomorphism

$$P_\ell : \mathbb{R}^{n_\ell} \to V_\ell \ , \quad P_\ell \mathbf{x} = \sum_{i=1}^{n_\ell} x_i \phi_i. \tag{5.55}$$

The Galerkin discretization in the space V_ℓ is as follows: determine $u_\ell \in V_\ell$ such that

$$\int_0^1 u'_\ell v'_\ell \, dx = \int_0^1 f v_\ell \, dx \quad \text{for all} \quad v_\ell \in V_\ell.$$

Using the representation $u_\ell = \sum_{j=1}^{n_\ell} x_j \phi_j$ this yields a linear system

$$\mathbf{A}_\ell \mathbf{x}_\ell = \mathbf{b}_\ell \ , \quad (\mathbf{A}_\ell)_{ij} = \int_0^1 \phi'_i \phi'_j \, dx, \quad (\mathbf{b}_\ell)_i = \int_0^1 f \phi_i \, dx. \tag{5.56}$$

The solution of this discrete problem is denoted by $\mathbf{x}^*_\ell$. The solution of the Galerkin discretization in the function space V_ℓ is given by $u_\ell = P_\ell \mathbf{x}^*_\ell$. A simple computation shows that

$$\mathbf{A}_\ell = h_\ell^{-1} \text{tridiag}(-1, 2, -1) \in \mathbb{R}^{n_\ell \times n_\ell}.$$

Note that, apart from a scaling factor, the same matrix results from a standard discretization with finite differences of the problem (5.53).
Clearly, in practice one should not solve the problem in (5.56) using an iterative method (a Cholesky factorization $\mathbf{A} = \mathbf{L}\mathbf{L}^T$ is stable and efficient). However, we do apply a basic iterative method here, to illustrate a certain "smoothing" property which plays an important role in multigrid methods. We consider the damped Jacobi method

$$\mathbf{x}_\ell^{k+1} = \mathbf{x}_\ell^k - \frac{1}{2}\omega h_\ell (\mathbf{A}_\ell \mathbf{x}_\ell^k - \mathbf{b}_\ell) \quad \text{with} \quad \omega \in (0, 1] \ . \tag{5.57}$$

The iteration matrix of this method, which describes the error propagation $\mathbf{e}_\ell^{k+1} = \mathbf{C}_\ell \mathbf{e}_\ell^k$, $\mathbf{e}_\ell^k := \mathbf{x}_\ell^* - \mathbf{x}_\ell^k$, is given by

$$\mathbf{C}_\ell = \mathbf{C}_\ell(\omega) = \mathbf{I} - \frac{1}{2}\omega h_\ell \mathbf{A}_\ell \ .$$

In this simple model problem an orthogonal eigenvector basis of $\mathbf{A}_\ell$, and thus of $\mathbf{C}_\ell$, too, is known. This basis is closely related to the "Fourier modes":

$$w^\nu(x) = \sin(\nu \pi x), \quad x \in [0, 1], \quad \nu = 1, 2, \ldots .$$

Note that w^ν satisfies the boundary conditions in (5.53) and that $-(w^\nu)''(x) = (\nu\pi)^2 w^\nu(x)$ holds, and thus w^ν is an eigenfunction of the problem in (5.53). We introduce vectors $\mathbf{z}_\ell^\nu \in \mathbb{R}^{n_\ell}$, $1 \le \nu \le n_\ell$, which correspond to the Fourier modes w^ν restricted to the interior grid Ω_ℓ^{int}:

$$\mathbf{z}_\ell^\nu := \left(w^\nu(\xi_{\ell,1}), w^\nu(\xi_{\ell,2}), \ldots, w^\nu(\xi_{\ell,n_\ell}) \right)^T .$$

These vectors form an orthogonal basis of $\mathbb{R}^{n_\ell}$. For $\ell = 2$ we give an illustration in Fig. 5.3.

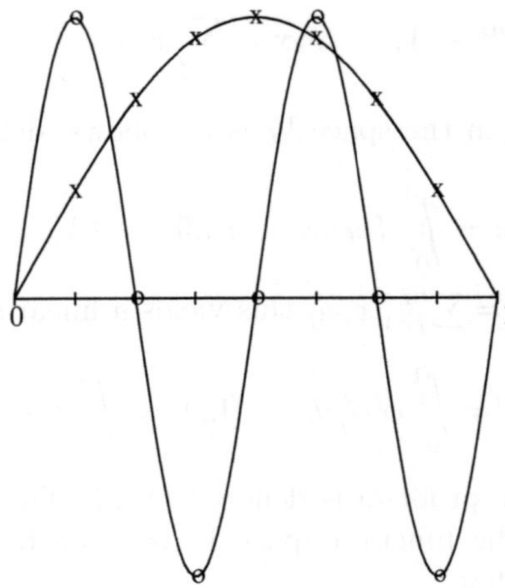

Fig. 5.3. Two discrete Fourier modes.

To a vector $\mathbf{z}_\ell^\nu$ there corresponds a frequency ν. For $\nu < \frac{1}{2}n_\ell$ the vector $\mathbf{z}_\ell^\nu$, or the corresponding finite element function $P_\ell \mathbf{z}_\ell^\nu$, is called a "*low frequency mode*", and for $\nu \geq \frac{1}{2}n_\ell$ this vector (or the corresponding finite element function, respectively) is called a "*high frequency mode*". The vectors $\mathbf{z}_\ell^\nu$ are eigenvectors of the matrix $\mathbf{A}_\ell$:

$$\mathbf{A}_\ell \mathbf{z}_\ell^\nu = \frac{4}{h_\ell}\sin^2(\nu\frac{\pi}{2}h_\ell)\mathbf{z}_\ell^\nu,$$

and thus we have

$$\mathbf{C}_\ell \mathbf{z}_\ell^\nu = \big(1 - 2\omega\sin^2(\nu\frac{\pi}{2}h_\ell)\big)\mathbf{z}_\ell^\nu \ . \tag{5.58}$$

From this we obtain

$$\begin{aligned}\|\mathbf{C}_\ell\|_2 &= \max\nolimits_{1\leq\nu\leq n_\ell}\ |1 - 2\omega\sin^2(\nu\tfrac{\pi}{2}h_\ell)| \\ &= 1 - 2\omega\sin^2(\tfrac{\pi}{2}h_\ell) = 1 - \tfrac{1}{2}\omega\pi^2h_\ell^2 + \mathcal{O}(h_\ell^4)\ .\end{aligned} \tag{5.59}$$

From this we see that the damped Jacobi method is convergent ($\|\mathbf{C}_\ell\|_2 < 1$), but that the rate of convergence will be very low for h_ℓ small.

Note that the eigenvalues and the eigenvectors of $\mathbf{C}_\ell$ are functions of the parameter $\nu h_\ell \in [0,1]$:

$$\lambda_{\ell,\nu} := 1 - 2\omega\sin^2(\nu\frac{\pi}{2}h_\ell) =: g_\omega(\nu h_\ell)\ ,\ \text{with} \tag{5.60a}$$

$$g_\omega(y) = 1 - 2\omega\sin^2(\frac{\pi}{2}y), \qquad y \in [0,1]. \tag{5.60b}$$

Hence, the size of the eigenvalues $\lambda_{\ell,\nu}$ can directly be obtained from the graph of the function g_ω. In Fig. 5.4 we show the graph of the function g_ω for a few values of ω.

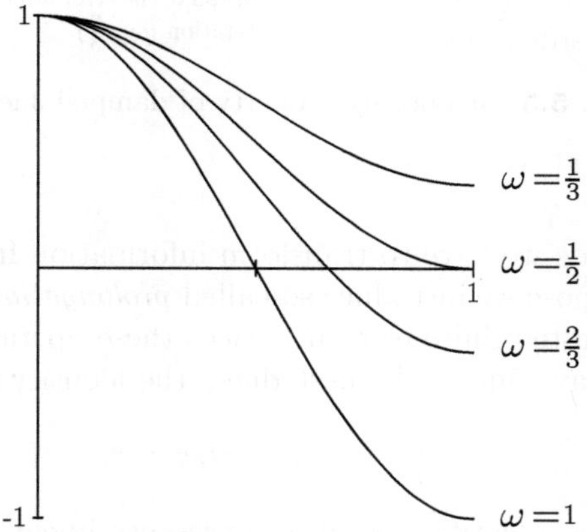

Fig. 5.4. Graph of g_ω.

From the graphs in this figure we conclude that for a suitable choice of ω we have $|g_\omega(y)| \ll 1$ if $y \in [\frac{1}{2}, 1]$. We choose $\omega = \frac{2}{3}$ (then $|g_\omega(\frac{1}{2})| = |g_\omega(1)|$ holds). Then we have $|g_{\frac{2}{3}}(y)| \leq \frac{1}{3}$ for $y \in [\frac{1}{2}, 1]$. Using this and the result in (5.60a) we obtain

$$|\lambda_{\ell,\nu}| \leq \frac{1}{3} \quad \text{for } \nu \geq \frac{1}{2} n_\ell \ .$$

Hence:

the high frequency modes are strongly damped by the iteration matrix $\mathbf{C}_\ell$.

From Fig. 5.4 it is also clear that the low rate of convergence of the damped Jacobi method is caused by the low frequency modes ($\nu h_\ell \ll 1$).

Summarizing, we draw the conclusion that in this example the damped Jacobi method will "smooth" the error. This elementary observation is of great importance for the two-grid method introduced below. In the setting of multigrid methods the damped Jacobi method is called a "smoother". The smoothing property of damped Jacobi is illustrated in Fig. 5.5. It is important to note that the discussion above concerning smoothing is related to the iteration matrix $\mathbf{C}_\ell$, which means that the *error* will be made smoother by the damped Jacobi method, but not (necessarily) the new iterand $\mathbf{x}^{k+1}$.

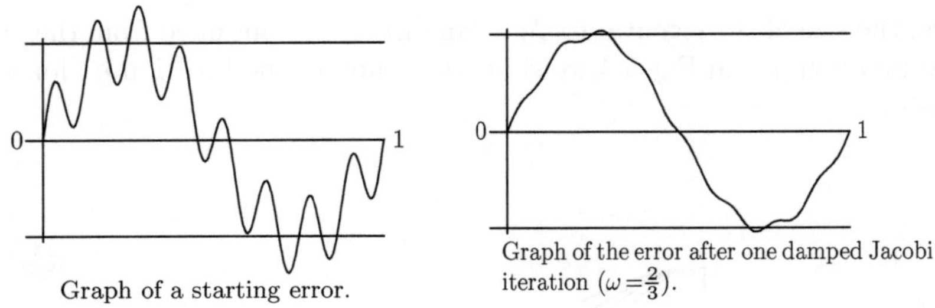

Fig. 5.5. Smoothing property of damped Jacobi.

In multigrid methods we have to transform information from one grid to another. For that purpose we introduce so-called *prolongations* and *restrictions*. In a setting with nested finite element spaces these operators can be defined in a very natural way. Due to the nestedness the identity operator

$$I_\ell : V_{\ell-1} \to V_\ell, \quad I_\ell v = v,$$

is well-defined. This identity operator represents linear interpolation as is illustrated for $\ell = 2$ in Fig. 5.6. The matrix representation of this interpolation operator is given by

$$\mathbf{p}_\ell : \mathbb{R}^{n_{\ell-1}} \to \mathbb{R}^{n_\ell}, \quad \mathbf{p}_\ell := P_\ell^{-1} P_{\ell-1}. \tag{5.61}$$

A simple computation yields

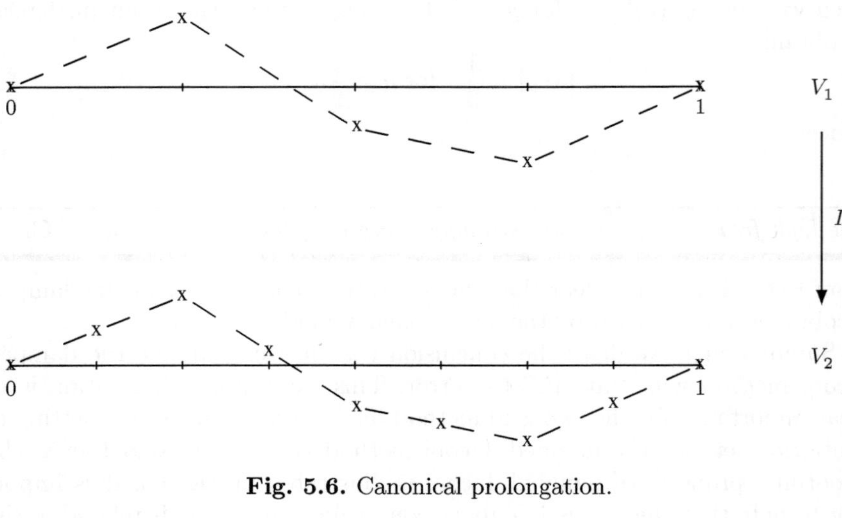

Fig. 5.6. Canonical prolongation.

$$\mathbf{p}_\ell = \begin{bmatrix} \frac{1}{2} & & & \emptyset \\ 1 & & & \\ \frac{1}{2} & \frac{1}{2} & & \\ & 1 & & \\ & \frac{1}{2} & & \\ & & \ddots & \\ & & & \frac{1}{2} \\ & & & 1 \\ \emptyset & & & \frac{1}{2} \end{bmatrix}_{n_\ell \times n_{\ell-1}} . \tag{5.62}$$

We can also restrict a given grid function v_ℓ on Ω_ℓ^{int} to a grid function on $\Omega_{\ell-1}^{\text{int}}$. An obvious approach is to use a restriction r based on simple injection:

$$(r_{\text{inj}} v_\ell)(\xi) = v_\ell(\xi) \quad \text{if } \xi \in \Omega_{\ell-1}^{\text{int}} .$$

When used in a multigrid method then often this restriction based on injection is not satisfactory (cf. [133], Sect. 3.5). A better method is obtained if a natural Galerkin property is satisfied. It can easily be verified (cf. also Lemma 5.4.3) that with $\mathbf{A}_\ell$, $\mathbf{A}_{\ell-1}$ and $\mathbf{p}_\ell$ as defined in (5.56), (5.61) we have

$$\mathbf{r}_\ell \mathbf{A}_\ell \mathbf{p}_\ell = \mathbf{A}_{\ell-1} \quad \text{iff} \quad \mathbf{r}_\ell = \mathbf{p}_\ell^T . \tag{5.63}$$

Thus the natural Galerkin condition $\mathbf{r}_\ell \mathbf{A}_\ell \mathbf{p}_\ell = \mathbf{A}_{\ell-1}$ implies the choice

$$\mathbf{r}_\ell = \mathbf{p}_\ell^T \tag{5.64}$$

for the restriction operator. In the remainder we use this restriction.

The *two-grid* method is based on the idea that a *smooth* error, which results from the application of one or a few damped Jacobi iterations, can be approximated fairly well on a *coarser* grid. We now introduce this two-grid method.

Consider $\mathbf{A}_\ell \mathbf{x}_\ell^* = \mathbf{b}_\ell$ and let $\overline{\mathbf{x}}_\ell$ be the result of one or a few damped Jacobi iterations applied to a given starting vector $\mathbf{x}_\ell^0$. For the error $\mathbf{e}_\ell := \mathbf{x}_\ell^* - \overline{\mathbf{x}}_\ell$ we have

$$\mathbf{A}_\ell \mathbf{e}_\ell = \mathbf{b}_\ell - \mathbf{A}_\ell \overline{\mathbf{x}}_\ell =: \mathbf{d}_\ell \quad (\text{"residual" or "defect"}). \tag{5.65}$$

Based on the assumption that $\mathbf{e}_\ell$ is smooth it seems reasonable to use an approximation $\mathbf{e}_\ell \approx \mathbf{p}_\ell \tilde{\mathbf{e}}_{\ell-1}$ with an appropriate vector (grid function) $\tilde{\mathbf{e}}_{\ell-1} \in \mathbb{R}^{n_{\ell-1}}$. To determine the vector $\tilde{\mathbf{e}}_{\ell-1}$ we use the equation (5.65) and the Galerkin property (5.63). This results in the equation

$$\mathbf{A}_{\ell-1} \tilde{\mathbf{e}}_{\ell-1} = \mathbf{r}_\ell \mathbf{d}_\ell$$

for the vector $\tilde{\mathbf{e}}_{\ell-1}$. Note that $\mathbf{x}^* = \overline{\mathbf{x}}_\ell + \mathbf{e}_\ell \approx \overline{\mathbf{x}}_\ell + \mathbf{p}_\ell \tilde{\mathbf{e}}_{\ell-1}$. Thus for the new iterand we take $\mathbf{x}_\ell := \overline{\mathbf{x}}_\ell + \mathbf{p}_\ell \tilde{\mathbf{e}}_{\ell-1}$. In a more compact formulation this two-grid method is as follows:

$$
\left\{
\begin{array}{l}
\textit{procedure}\ \mathrm{TGM}_\ell(\mathbf{x}_\ell, \mathbf{b}_\ell) \\
\quad \text{if } \ell = 0 \text{ then } \mathbf{x}_0 := \mathbf{A}_0^{-1}\mathbf{b}_0 \text{ else} \\
\quad \text{begin} \\
\qquad \mathbf{x}_\ell := J_\ell^\nu(\mathbf{x}_\ell, \mathbf{b}_\ell) \quad (*\ \nu \text{ smoothing it., e.g. damped Jacobi } *) \\
\qquad \mathbf{d}_{\ell-1} := \mathbf{r}_\ell(\mathbf{b}_\ell - \mathbf{A}_\ell\mathbf{x}_\ell) \quad (*\ \text{restriction of defect } *) \\
\qquad \tilde{\mathbf{e}}_{\ell-1} := \mathbf{A}_{\ell-1}^{-1}\mathbf{d}_{\ell-1} \quad (*\ \text{solve coarse grid problem } *) \\
\qquad \mathbf{x}_\ell := \mathbf{x}_\ell + \mathbf{p}_\ell\tilde{\mathbf{e}}_{\ell-1} \quad (*\ \text{add correction } *) \\
\qquad \mathrm{TGM}_\ell := \mathbf{x}_\ell \\
\quad \text{end;}
\end{array}
\right. \tag{5.66}
$$

Often, after the coarse grid correction $\mathbf{x}_\ell := \mathbf{x}_\ell + \mathbf{p}_\ell\tilde{\mathbf{e}}_{\ell-1}$, one or a few smoothing iterations are applied. Smoothing before/after the coarse grid correction is called pre/post-smoothing. Besides the smoothing property a second property which is of great importance for a multigrid method is the following:

The coarse grid system $\mathbf{A}_{\ell-1}\tilde{\mathbf{e}}_{\ell-1} = \mathbf{d}_{\ell-1}$ *is of the same form as* $\mathbf{A}_\ell\mathbf{x}_\ell = \mathbf{b}_\ell$.

Thus for solving the problem $\mathbf{A}_{\ell-1}\tilde{\mathbf{e}}_{\ell-1} = \mathbf{d}_{\ell-1}$ *approximately* we can apply the two-grid algorithm in (5.66) recursively. This results in the following *multigrid method* for solving $\mathbf{A}_\ell\mathbf{x}_\ell^* = \mathbf{b}_\ell$:

$$
\left\{
\begin{array}{l}
\textit{procedure}\ \mathrm{MGM}_\ell(\mathbf{x}_\ell, \mathbf{b}_\ell) \\
\quad \text{if } \ell = 0 \text{ then } \mathbf{x}_0 := \mathbf{A}_0^{-1}\mathbf{b}_0 \text{ else} \\
\quad \text{begin} \\
\qquad \mathbf{x}_\ell := J_\ell^{\nu_1}(\mathbf{x}_\ell, \mathbf{b}_\ell) \quad (*\ \text{presmoothing } *) \\
\qquad \mathbf{d}_{\ell-1} := \mathbf{r}_\ell(\mathbf{b}_\ell - \mathbf{A}_\ell\mathbf{x}_\ell) \\
\qquad \mathbf{e}_{\ell-1}^0 := \mathbf{0};\ \text{ for } i = 1 \text{ to } \tau \text{ do } \mathbf{e}_{\ell-1}^i := \mathrm{MGM}_{\ell-1}(\mathbf{e}_{\ell-1}^{i-1}, \mathbf{d}_{\ell-1}); \\
\qquad \mathbf{x}_\ell := \mathbf{x}_\ell + \mathbf{p}_\ell\mathbf{e}_{\ell-1}^\tau \\
\qquad \mathbf{x}_\ell := J_\ell^{\nu_2}(\mathbf{x}_\ell, \mathbf{b}_\ell) \quad (*\ \text{postsmoothing } *) \\
\qquad \mathrm{MGM}_\ell := \mathbf{x}_\ell \\
\quad \text{end;}
\end{array}
\right. \tag{5.67}
$$

If one wants to solve the system on a given finest grid, say with level number $\overline{\ell}$, i.e. $\mathbf{A}_{\overline{\ell}}\mathbf{x}_{\overline{\ell}}^* = \mathbf{b}_{\overline{\ell}}$, then we apply some iterations of $\mathrm{MGM}_{\overline{\ell}}(\mathbf{x}_{\overline{\ell}}, \mathbf{b}_{\overline{\ell}})$.

Based on efficiency considerations (explained below) we usually take $\tau = 1$ ("V-cycle") or $\tau = 2$ ("W-cycle") in the recursive call in (5.67). For the case $\ell = 3$ the structure of one multigrid iteration with $\tau \in \{1, 2\}$ is illustrated in Fig. 5.7.

II. Multigrid for scalar elliptic problems

We introduce multigrid methods which can be used for solving discretized scalar elliptic boundary value problems. A model example from this problem class is the Poisson equation

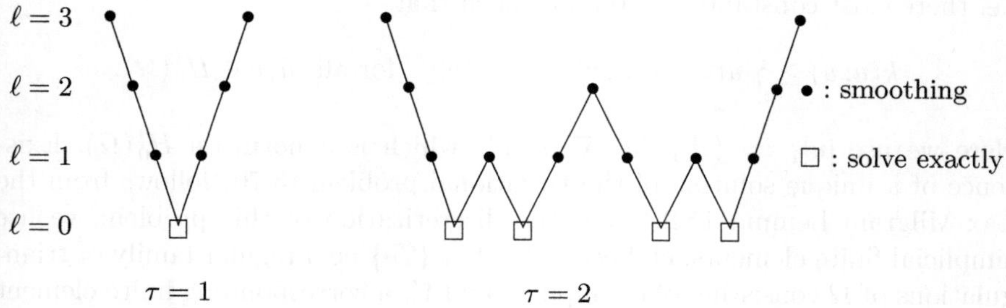

Fig. 5.7. Structure of one multigrid iteration

$$
\begin{aligned}
-\Delta u &= f \quad \text{in} \quad \Omega \subset \mathbb{R}^d, \\
u &= 0 \quad \text{on} \quad \partial\Omega.
\end{aligned}
$$

We emphasize, however, that opposite to the CG method, the *applicability of multigrid methods is not restricted to symmetric problems.* Multigrid methods can also be used for solving problems which are non-symmetric (e.g., with a convection term in the equation). If the problem is strongly non-symmetric (e.g., the underlying equation has a dominating convection term) one usually has to modify the standard multigrid approach in the sense that special smoothers and/or special prolongations and restrictions should be used. We do not treat this topic here.

We will introduce the two-grid and multigrid method by generalizing the approach presented for the 1D diffusion problem above to the higher (i.e., two and three) dimensional case. We consider a scalar elliptic boundary value problem of the form

$$
\begin{aligned}
-\operatorname{div}(a\nabla u) + \mathbf{b}\cdot\nabla u + cu = f \quad &\text{in} \quad \Omega, \\
u = 0 \quad &\text{on} \quad \partial\Omega.
\end{aligned}
\tag{5.68}
$$

This problem is considered in a domain $\Omega \subset \mathbb{R}^d$, $d = 2$ or 3. We assume that the functions a, c and the vector function $\mathbf{b}$ are sufficiently smooth on Ω and

$$
a(x) \geq a_0 > 0, \quad c(x) - \frac{1}{2}\operatorname{div}\mathbf{b}(x) \geq 0 \quad \text{for all} \quad x \in \overline{\Omega}. \tag{5.69}
$$

These assumptions guarantee that the problem is elliptic and well-posed. In view of the finite element discretization we introduce the variational formulation of this problem: Find $u \in H_0^1(\Omega)$ such that

$$
k(u, v) = f(v) \quad \text{for all} \quad v \in H_0^1(\Omega), \tag{5.70}
$$

with a bilinear form and right-hand side

$$
k(u, v) = \int_\Omega a\nabla u \cdot \nabla v + \mathbf{b}\cdot\nabla u v + cuv\, dx\ , \quad f(v) = \int_\Omega f v\, dx.
$$

If (5.69) holds then this bilinear form is *continuous and elliptic* on $H_0^1(\Omega)$, i.e. there exist constants $\gamma > 0$ and c such that

$$k(u,u) \geq \gamma |u|_1^2, \quad k(u,v) \leq c|u|_1|v|_1 \quad \text{for all} \quad u,v \in H_0^1(\Omega).$$

Here we use $|u|_1 := \left(\int_\Omega \nabla u \cdot \nabla u \, dx\right)^{\frac{1}{2}}$, which is a norm on $H_0^1(\Omega)$. Existence of a unique solution of the variational problem (5.70) follows from the Lax-Milgram Lemma 15.2.1. For the discretization of this problem we use simplicial finite elements, cf. Sect. 3.2.1. Let $\{\mathcal{T}_h\}$ be a regular family of triangulations of Ω consisting of d-simplices and V_h a corresponding finite element space. For simplicity we only consider *linear* finite elements:

$$V_h = \mathbb{X}_{h,0}^1 = \left\{ v \in C(\Omega) : v_{|T} \in \mathcal{P}_1 \text{ for all } T \in \mathcal{T}_h \quad v_{|\partial\Omega} = 0 \right\}.$$

The presentation and implementation of the multigrid method is greatly simplified if we assume a given sequence of *nested* finite element spaces.

Assumption 5.4.2 In the remainder of this section we assume that we have a sequence V_ℓ, $\ell = 0, 1, \ldots$, of simplicial finite element spaces which are nested:

$$V_\ell \subset V_{\ell+1} \quad \text{for all} \quad \ell. \tag{5.71}$$

This assumption holds for a multilevel tetrahedral grid hierarchy as explained in Sect. 3.1, cf. Remark 3.1.6. This assumption is not necessary for a successful application of multigrid methods. For a treatment of multigrid methods in case of non-nestedness we refer to [242].

In V_ℓ we use the standard nodal basis $(\phi_i)_{1 \leq i \leq n_\ell}$. This basis induces an isomorphism

$$P_\ell : \mathbb{R}^{n_\ell} \to V_\ell \,, \quad P_\ell \mathbf{x} = \sum_{i=1}^{n_\ell} x_i \phi_i.$$

The Galerkin discretization: Find $u_\ell \in V_\ell$ such that

$$k(u_\ell, v_\ell) = f(v_\ell) \quad \text{for all} \quad v_\ell \in V_\ell, \tag{5.72}$$

can be represented as a linear system

$$\mathbf{A}_\ell \mathbf{x}_\ell = \mathbf{b}_\ell \,, \text{ with } (\mathbf{A}_\ell)_{ij} = k(\phi_j, \phi_i), \; (\mathbf{b}_\ell)_i = f(\phi_i), \; 1 \leq i,j \leq n_\ell. \tag{5.73}$$

The solution $\mathbf{x}_\ell^*$ of this linear system yields the Galerkin finite element solution $u_\ell = P_\ell \mathbf{x}_\ell^*$. Along the same lines as in the one-dimensional case we introduce a multigrid method for solving this system of equations on an arbitrary level $\ell \geq 0$.

For the *smoother* we use a basic iterative method such as, for example, a *Richardson method*

$$\mathbf{x}^{k+1} = \mathbf{x}^k - \omega_\ell(\mathbf{A}_\ell \mathbf{x}^k - \mathbf{b}),$$

a *damped Jacobi method*

$$\mathbf{x}^{k+1} = \mathbf{x}^k - \omega \mathbf{D}_\ell^{-1}(\mathbf{A}_\ell \mathbf{x}^k - \mathbf{b}), \tag{5.74}$$

or a *Gauss-Seidel method*

$$\mathbf{x}^{k+1} = \mathbf{x}^k - (\mathbf{D}_\ell - \mathbf{L}_\ell)^{-1}(\mathbf{A}_\ell \mathbf{x}^k - \mathbf{b}), \tag{5.75}$$

where $\mathbf{D}_\ell - \mathbf{L}_\ell$ is the lower triangular part of the matrix $\mathbf{A}_\ell$. For such a method we use the general notation

$$\mathbf{x}^{k+1} = \mathcal{S}_\ell(\mathbf{x}^k, \mathbf{b}_\ell) = \mathbf{x}^k - \mathbf{W}_\ell^{-1}(\mathbf{A}_\ell \mathbf{x}^k - \mathbf{b}), \quad k = 0, 1, \ldots$$

The corresponding iteration matrix is denoted by

$$\mathbf{S}_\ell = \mathbf{I} - \mathbf{W}_\ell^{-1}\mathbf{A}_\ell.$$

For the *prolongation* we use the matrix representation of the identity $I_\ell : V_{\ell-1} \to V_\ell$, i.e.,

$$\mathbf{p}_\ell := P_\ell^{-1} P_{\ell-1}. \tag{5.76}$$

The choice of the restriction is based on the following elementary lemma:

Lemma 5.4.3 *Let* $\mathbf{A}_\ell$, $\ell \geq 0$, *be the stiffness matrix defined in* (5.73) *and* $\mathbf{p}_\ell$ *as in* (5.76). *Then for* $\mathbf{r}_\ell : \mathbb{R}^{n_\ell} \to \mathbb{R}^{n_{\ell-1}}$ *we have:*

$$\mathbf{r}_\ell \mathbf{A}_\ell \mathbf{p}_\ell = \mathbf{A}_{\ell-1} \quad \textit{if and only if} \quad \mathbf{r}_\ell = \mathbf{p}_\ell^T.$$

Proof. For the stiffness matrix matrix the identity

$$\langle \mathbf{A}_\ell \mathbf{x}, \mathbf{y} \rangle = k(P_\ell \mathbf{x}, P_\ell \mathbf{y}) \quad \text{for all} \quad \mathbf{x}, \mathbf{y} \in \mathbb{R}^{n_\ell}$$

holds. From this we get

$$\begin{aligned} & \mathbf{r}_\ell \mathbf{A}_\ell \mathbf{p}_\ell = \mathbf{A}_{\ell-1} \\ \Leftrightarrow \quad & \langle \mathbf{A}_\ell \mathbf{p}_\ell \mathbf{x}, \mathbf{r}_\ell^T \mathbf{y} \rangle = \langle \mathbf{A}_{\ell-1} \mathbf{x}, \mathbf{y} \rangle \quad \text{for all} \quad \mathbf{x}, \mathbf{y} \in \mathbb{R}^{n_{\ell-1}} \\ \Leftrightarrow \quad & k(P_{\ell-1}\mathbf{x}, P_\ell \mathbf{r}_\ell^T \mathbf{y}) = k(P_{\ell-1}\mathbf{x}, P_{\ell-1}\mathbf{y}) \quad \text{for all} \quad \mathbf{x}, \mathbf{y} \in \mathbb{R}^{n_{\ell-1}}. \end{aligned}$$

Using the ellipticity of $k(\cdot, \cdot)$ it now follows that

$$\begin{aligned} & \mathbf{r}_\ell \mathbf{A}_\ell \mathbf{p}_\ell = \mathbf{A}_{\ell-1} \\ \Leftrightarrow \quad & P_\ell \mathbf{r}_\ell^T \mathbf{y} = P_{\ell-1}\mathbf{y} \quad \text{for all} \quad \mathbf{y} \in \mathbb{R}^{n_{\ell-1}} \\ \Leftrightarrow \quad & \mathbf{r}_\ell^T \mathbf{y} = P_\ell^{-1} P_{\ell-1}\mathbf{y} = \mathbf{p}_\ell \mathbf{y} \quad \text{for all} \quad \mathbf{y} \in \mathbb{R}^{n_{\ell-1}} \\ \Leftrightarrow \quad & \mathbf{r}_\ell^T = \mathbf{p}_\ell. \end{aligned}$$

Thus the claim is proved. □

This motivates that for the *restriction* we take:

$$\mathbf{r}_\ell := \mathbf{p}_\ell^T . \tag{5.77}$$

Using these components we can define a multigrid method with exactly the same structure as in (5.67):

$$\begin{array}{l}
\textit{procedure}\ \mathrm{MGM}_\ell(\mathbf{x}_\ell, \mathbf{b}_\ell) \\
\quad \text{if}\ \ \ell = 0\ \ \text{then}\ \ \mathbf{x}_0 := \mathbf{A}_0^{-1}\mathbf{b}_0\ \ \text{else} \\
\quad \text{begin} \\
\qquad \mathbf{x}_\ell := \mathcal{S}_\ell^{\nu_1}(\mathbf{x}_\ell, \mathbf{b}_\ell) \quad (*\ \text{presmoothing}\ *) \\
\qquad \mathbf{d}_{\ell-1} := \mathbf{r}_\ell(\mathbf{b}_\ell - \mathbf{A}_\ell \mathbf{x}_\ell) \\
\qquad \mathbf{e}_{\ell-1}^0 := \mathbf{0};\ \ \text{for}\ i = 1\ \text{to}\ \tau\ \text{do}\ \ \mathbf{e}_{\ell-1}^i := \mathrm{MGM}_{\ell-1}(\mathbf{e}_{\ell-1}^{i-1}, \mathbf{d}_{\ell-1}); \\
\qquad \mathbf{x}_\ell := \mathbf{x}_\ell + \mathbf{p}_\ell \mathbf{e}_{\ell-1}^\tau \\
\qquad \mathbf{x}_\ell := \mathcal{S}_\ell^{\nu_2}(\mathbf{x}_\ell, \mathbf{b}_\ell) \quad (*\ \text{postsmoothing}\ *) \\
\qquad \mathrm{MGM}_\ell := \mathbf{x}_\ell \\
\quad \text{end;}
\end{array} \tag{5.78}$$

We briefly comment on some important issues related to this multigrid method.

Smoothers

For many problems basic iterative methods provide good smoothers. In particular the Gauss-Seidel method is often a very effective smoother. Other smoothers used in practice are the damped Jacobi method and the ILU method, cf. [255].

Prolongation and restriction

If instead of a discretization with nested finite element spaces one uses a finite difference or a finite volume method then one can not use the approach in (5.76) to define a prolongation. However, for these cases other canonical constructions for the prolongation operator exist. We refer to [133], [242] or [251] for a treatment of this topic. A general technique for the construction of a prolongation operator in case of non-nested finite element spaces is given in [49].

Arithmetic costs per iteration

We discuss the arithmetic costs of one MGM_ℓ iteration as defined in (5.78). For this we introduce a unit of arithmetic work on level ℓ:

$$WU_\ell := \#\ \text{flops needed for}\ \ \mathbf{A}_\ell \mathbf{x}_\ell - \mathbf{b}_\ell\ \ \text{computation.} \tag{5.79}$$

We assume:

$$WU_{\ell-1} \lesssim g\, WU_\ell \quad \text{with} \quad g < 1 \quad \text{independent of } \ell. \tag{5.80}$$

Note that if $\mathcal{T}_\ell$ is constructed through a uniform global grid refinement of $\mathcal{T}_{\ell-1}$ (for $d = 3$: subdivision of each tetrahedron $T \in \mathcal{T}_{\ell-1}$ into 8 smaller tetrahedra by regular refinement) then (5.80) holds with $g = (\frac{1}{2})^d$. Furthermore we make the following assumptions concerning the arithmetic costs of each of the substeps in the procedure MGM$_\ell$:

$$\mathbf{x}_\ell := \mathcal{S}_\ell(\mathbf{x}_\ell, \mathbf{b}_\ell) \;:\; \text{ costs } \lesssim WU_\ell$$

$$\left.\begin{array}{r}\mathbf{d}_{\ell-1} := \mathbf{r}_\ell(\mathbf{b}_\ell - \mathbf{A}_\ell \mathbf{x}_\ell) \\ \mathbf{x}_\ell := \mathbf{x}_\ell + \mathbf{p}_\ell \mathbf{e}_{\ell-1}^{\tau}\end{array}\right\} \text{total costs } \lesssim 2\, WU_\ell .$$

For the amount of work in one multigrid V-cycle ($\tau = 1$) on level ℓ, which is denoted by VMG_ℓ, we get, using $\nu := \nu_1 + \nu_2$:

$$\begin{aligned} VMG_\ell &\lesssim \nu WU_\ell + 2WU_\ell + VMG_{\ell-1} = (\nu+2)WU_\ell + VMG_{\ell-1} \\ &\lesssim (\nu+2)\big(WU_\ell + WU_{\ell-1} + \ldots + WU_1\big) + VMG_0 \\ &\lesssim (\nu+2)\big(1 + g + \ldots + g^{\ell-1}\big)WU_\ell + VMG_0 \\ &\lesssim \frac{\nu+2}{1-g}\, WU_\ell . \end{aligned} \tag{5.81}$$

In the last inequality we assumed that the costs for computing $\mathbf{x}_0 = \mathbf{A}_0^{-1}\mathbf{b}_0$ (i.e., VMG_0) are negligible compared to WU_ℓ. The result in (5.81) shows that the arithmetic costs for one V-cycle are proportional (also if $\ell \to \infty$) to the costs of a residual computation. For example, for $g = \frac{1}{8}$ (uniform refinement in 3D) the arithmetic costs of a V-cycle with $\nu_1 = \nu_2 = 1$ on level ℓ are comparable to $4\frac{1}{2}$ times the costs of a residual computation on level ℓ.
For the W-cycle ($\tau = 2$) the arithmetic costs on level ℓ are denoted by WMG_ℓ. We have:

$$\begin{aligned} WMG_\ell &\lesssim \nu WU_\ell + 2WU_\ell + 2WMG_{\ell-1} = (\nu+2)WU_\ell + 2WMG_{\ell-1} \\ &\lesssim (\nu+2)\big(WU_\ell + 2WU_{\ell-1} + 2^2 WU_{\ell-2} + \ldots + 2^{\ell-1}WU_1\big) + WMG_0 \\ &\lesssim (\nu+2)\big(1 + 2g + (2g)^2 + \ldots + (2g)^{\ell-1}\big)WU_\ell + WMG_0 . \end{aligned}$$

From this we see that to obtain a bound proportional to WU_ℓ we have to assume

$$g < \frac{1}{2}.$$

Under this assumption we get for the W-cycle

$$WMG_\ell \lesssim \frac{\nu+2}{1-2g}\, WU_\ell$$

(again we neglected WMG_0). Similar bounds can be obtained for $\tau \geq 3$, provided $\tau g < 1$ holds.

III. Numerical experiment: Multigrid applied to a Poisson equation

In this section we present results of a standard multigrid solver applied to the model problem of the Poisson equation:

$$\begin{aligned} -\Delta u &= f \quad \text{in} \quad \Omega := (0,1)^3, \\ u &= 0 \quad \text{on} \quad \partial\Omega. \end{aligned}$$

We take $f(x_1, x_2, x_3) = x_1^2 + e^{x_2}x_1 + x_3^2 x_2$. For the discretization we start with a uniform subdivision of Ω into cubes with edges of length $h_0 := \frac{1}{4}$. Each cube is subdivided into six tetrahedra. This yields the starting triangulation $\mathcal{T}_0$ of Ω. The triangulation $\mathcal{T}_1$ with mesh size $h_1 = \frac{1}{8}$ is constructed by regular subdivision of each tetrahedron in $\mathcal{T}_0$ into 8 child tetrahedra. This uniform refinement strategy is repeated, resulting in a family of triangulations $(\mathcal{T}_\ell)_{\ell \geq 0}$ with corresponding mesh size $h_\ell = 2^{-\ell-2}$. For discretization of this problem we use the space of linear finite elements on these triangulations. The resulting linear system is denoted by $\mathbf{A}_\ell \mathbf{x}_\ell = \mathbf{b}_\ell$. We consider the problem of solving this linear system on a fixed finest level $\ell = \bar{\ell}$. Below we consider $\bar{\ell} = 1, \ldots, 5$. For $\bar{\ell} = 5$ the multilevel triangulation contains 14.380.416 tetrahedra and in the linear system we have 2.048.383 unknowns.

We briefly discuss the components used in the multigrid method for solving this linear system. For the prolongation and restriction we use the canonical ones as in (5.76), (5.77). For the smoother we use two different methods, namely a damped Jacobi method and a symmetric Gauss-Seidel method (SGS). The damped Jacobi method is as in (5.74) with $\omega := 0.7$. The symmetric Gauss-Seidel method consists of two substeps. In the first step we use a Gauss-Seidel iteration as in (5.75). In the second step we apply this method with a reversed ordering of the equations and the unknowns. The arithmetic costs per iteration for such a symmetric Gauss-Seidel smoother are roughly twice as high as for a damped Jacobi method. In the experiment we use the same number of pre- and post-smoothing iterations, i.e. $\nu_1 = \nu_2$. The total number of smoothing iterations per multigrid iteration is $\nu := \nu_1 + \nu_2$. We use a multigrid V-cycle. i.e., $\tau = 1$ in the recursive call in (5.78). The coarsest grid used in the multigrid method is $\mathcal{T}_0$, i.e. with a mesh size $h_0 = \frac{1}{4}$. In all experiments we use a starting vector $\mathbf{x}^0 := 0$. The rate of convergence is measured by looking at relative residuals:

$$r_k := \frac{\|\mathbf{A}_{\bar{\ell}} \mathbf{x}^k - \mathbf{b}_{\bar{\ell}}\|_2}{\|\mathbf{b}_{\bar{\ell}}\|_2}.$$

In Fig. 5.8 (left) we show results for SGS with $\nu = 4$. For $\bar{\ell} = 1, \ldots, 5$ we plotted the relative residuals r_k for $k = 1, \ldots, 8$. In Fig. 5.8 (right) we show results for the SGS method with varying number of smoothing iterations, namely $\nu = 2, 4, 6$. For $\bar{\ell} = 1, \ldots, 5$ we give the average residual reduction per iteration $r := (r_8)^{\frac{1}{8}}$.

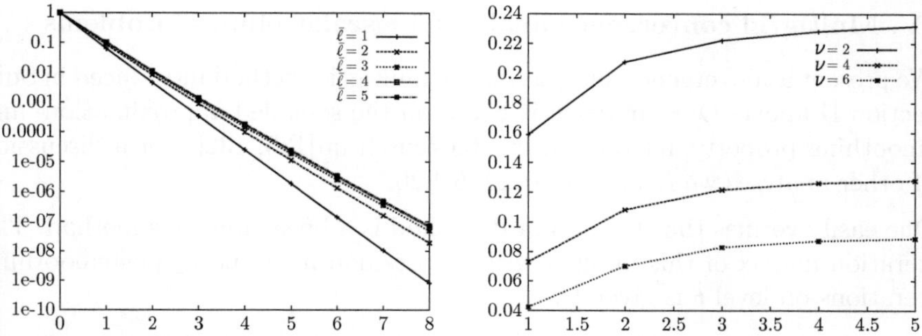

Fig. 5.8. Convergence of multigrid V-cycle with SGS smoother. Left: r_k, for $k = 0, \ldots, 8$ and $\bar{\ell} = 1, \ldots, 5$. Right: $(r_8)^{\frac{1}{8}}$ for $\bar{\ell} = 1, \ldots, 5$ and $\nu = 2, 4, 6$.

These results show the very fast and essentially level independent rate of convergence of this multigrid method. For a larger number of smoothing iterations the convergence is faster. On the other hand, also the costs per iteration then increase, cf. (5.81) (with $g = \frac{1}{8}$). Usually, in practice the number of smoothings per iteration is not taken very large. Typical values are $\nu = 2$ or $\nu = 4$. In the Fig. 5.9 we show similar results but now for the damped Jacobi smoother (damping with $\omega = 0.7$) instead of the SGS method.

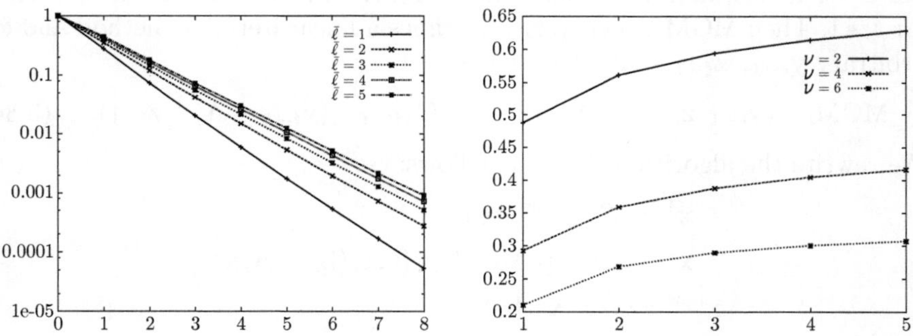

Fig. 5.9. Convergence of multigrid V-cycle with damped Jacobi smoother. Left: r_k, for $k = 0, \ldots, 8$ and $\bar{\ell} = 1, \ldots, 5$. Right: $(r_8)^{\frac{1}{8}}$ for $\bar{\ell} = 1, \ldots, 5$ and $\nu = 2, 4, 6$.

For the method with damped Jacobi smoothing we also observe an essentially level independent rate of convergence. Furthermore there is an increase in the rate of convergence when the number ν of smoothing step gets larger. Comparing the results of the multigrid method with Jacobi smoothing to those with SGS smoothing we see that the latter method has a significantly faster convergence. Note, however, that the arithmetic costs per iteration for the latter method are higher (the ratio lies between 1.5 and 2).

IV. Multigrid convergence analysis for scalar elliptic problems

We present a convergence analysis for the multigrid method introduced in Subsection II above. Our approach is based on the so-called approximation- and smoothing property, introduced by Hackbusch in [133, 132]. For a discussion of other analyses we refer to Remark 5.4.20.

One easily verifies that the two-grid method is a linear iterative method. The iteration matrix of this method with ν_1 presmoothing and ν_2 postsmoothing iterations on level ℓ is given by

$$\mathbf{C}_{TG,\ell} = \mathbf{C}_{TG,\ell}(\nu_2, \nu_1) = \mathbf{S}_\ell^{\nu_2}(\mathbf{I} - \mathbf{p}_\ell \mathbf{A}_{\ell-1}^{-1} \mathbf{r}_\ell \mathbf{A}_\ell)\mathbf{S}_\ell^{\nu_1}, \tag{5.82}$$

with $\mathbf{S}_\ell = \mathbf{I} - \mathbf{W}_\ell^{-1}\mathbf{A}_\ell$ the iteration matrix of the smoother.

Theorem 5.4.4 *The multigrid method* (5.78) *is a linear iterative method with iteration matrix $\mathbf{C}_{MG,\ell}$ given by*

$$\mathbf{C}_{MG,0} = 0 \tag{5.83a}$$

$$\mathbf{C}_{MG,\ell} = \mathbf{S}_\ell^{\nu_2}\big(\mathbf{I} - \mathbf{p}_\ell(\mathbf{I} - \mathbf{C}_{MG,\ell-1}^{\tau})\mathbf{A}_{\ell-1}^{-1}\mathbf{r}_\ell \mathbf{A}_\ell\big)\mathbf{S}_\ell^{\nu_1} \tag{5.83b}$$

$$= \mathbf{C}_{TG,\ell} + \mathbf{S}_\ell^{\nu_2}\mathbf{p}_\ell \mathbf{C}_{MG,\ell-1}^{\tau}\mathbf{A}_{\ell-1}^{-1}\mathbf{r}_\ell \mathbf{A}_\ell \mathbf{S}_\ell^{\nu_1}, \quad \ell = 1, 2, \ldots \tag{5.83c}$$

Proof. The result in (5.83a) is trivial. The result in (5.83c) follows from (5.83b) and the definition of $\mathbf{C}_{TG,\ell}$. We now prove the result in (5.83b) by induction. For $\ell = 1$ it follows from (5.83a) and (5.82). Assume that the result is correct for $\ell - 1$. Then $\mathrm{MGM}_{\ell-1}(\mathbf{y}_{\ell-1}, \mathbf{z}_{\ell-1})$ defines a linear iterative method and for arbitrary $\mathbf{y}_{\ell-1}$, $\mathbf{z}_{\ell-1} \in \mathbb{R}^{n_{\ell-1}}$ we have

$$\mathrm{MGM}_{\ell-1}(\mathbf{y}_{\ell-1}, \mathbf{z}_{\ell-1}) - \mathbf{A}_{\ell-1}^{-1}\mathbf{z}_{\ell-1} = \mathbf{C}_{MG,\ell-1}(\mathbf{y}_{\ell-1} - \mathbf{A}_{\ell-1}^{-1}\mathbf{z}_{\ell-1}). \tag{5.84}$$

We rewrite the algorithm (5.78) as follows:

$$\begin{aligned}
\mathbf{x}^1 &:= \mathcal{S}_\ell^{\nu_1}(\mathbf{x}_\ell^{\mathrm{old}}, \mathbf{b}_\ell) \\
\mathbf{x}^2 &:= \mathbf{x}^1 + \mathbf{p}_\ell \mathrm{MGM}_{\ell-1}^{\tau}\big(0, \mathbf{r}_\ell(\mathbf{b}_\ell - \mathbf{A}_\ell \mathbf{x}^1)\big) \\
\mathbf{x}_\ell^{\mathrm{new}} &:= \mathcal{S}_\ell^{\nu_2}(\mathbf{x}^2, \mathbf{b}_\ell).
\end{aligned}$$

From this we get

$$\begin{aligned}
\mathbf{x}_\ell^{\mathrm{new}} - \mathbf{x}_\ell^* &= \mathbf{x}_\ell^{\mathrm{new}} - \mathbf{A}_\ell^{-1}\mathbf{b}_\ell = \mathbf{S}_\ell^{\nu_2}(\mathbf{x}^2 - \mathbf{x}_\ell^*) \\
&= \mathbf{S}_\ell^{\nu_2}\big(\mathbf{x}^1 - \mathbf{x}_\ell^* + \mathbf{p}_\ell \mathrm{MGM}_{\ell-1}^{\tau}\big(0, \mathbf{r}_\ell(\mathbf{b}_\ell - \mathbf{A}_\ell \mathbf{x}^1)\big)\big).
\end{aligned}$$

Now we use the result (5.84) with $\mathbf{y}_{\ell-1} = 0$, $\mathbf{z}_{\ell-1} := \mathbf{r}_\ell(\mathbf{b}_\ell - \mathbf{A}_\ell \mathbf{x}^1)$. This yields

$$\begin{aligned}
\mathbf{x}_\ell^{\mathrm{new}} - \mathbf{x}_\ell^* &= \mathbf{S}_\ell^{\nu_2}\big(\mathbf{x}^1 - \mathbf{x}_\ell^* + \mathbf{p}_\ell(\mathbf{A}_{\ell-1}^{-1}\mathbf{z}_{\ell-1} - \mathbf{C}_{MG,\ell-1}^{\tau}\mathbf{A}_{\ell-1}^{-1}\mathbf{z}_{\ell-1})\big) \\
&= \mathbf{S}_\ell^{\nu_2}\big(\mathbf{I} - \mathbf{p}_\ell(\mathbf{I} - \mathbf{C}_{MG,\ell-1}^{\tau})\mathbf{A}_{\ell-1}^{-1}\mathbf{r}_\ell \mathbf{A}_\ell\big)(\mathbf{x}^1 - \mathbf{x}_\ell^*) \\
&= \mathbf{S}_\ell^{\nu_2}\big(\mathbf{I} - \mathbf{p}_\ell(\mathbf{I} - \mathbf{C}_{MG,\ell-1}^{\tau})\mathbf{A}_{\ell-1}^{-1}\mathbf{r}_\ell \mathbf{A}_\ell\big)\mathbf{S}_\ell^{\nu_1}(\mathbf{x}^{\mathrm{old}} - \mathbf{x}_\ell^*).
\end{aligned}$$

This completes the proof. □

The convergence analysis is based on the following splitting of the two-grid iteration matrix, with $\nu_2 = 0$, i.e. no postsmoothing:

$$\begin{aligned}\|\mathbf{C}_{TG,\ell}(0,\nu_1)\|_2 &= \|(\mathbf{I}-\mathbf{p}_\ell\mathbf{A}_{\ell-1}^{-1}\mathbf{r}_\ell\mathbf{A}_\ell)\mathbf{S}_\ell^{\nu_1}\|_2 \\ &\leq \|\mathbf{A}_\ell^{-1}-\mathbf{p}_\ell\mathbf{A}_{\ell-1}^{-1}\mathbf{r}_\ell\|_2\,\|\mathbf{A}_\ell\mathbf{S}_\ell^{\nu_1}\|_2.\end{aligned} \tag{5.85}$$

We will discuss suitable bounds for $\|\mathbf{A}_\ell^{-1}-\mathbf{p}_\ell\mathbf{A}_{\ell-1}^{-1}\mathbf{r}_\ell\|_2$ (called *approximation property*) and for $\|\mathbf{A}_\ell\mathbf{S}_\ell^{\nu_1}\|_2$ (called *smoothing property*). In the convergence analysis we need the following:

Assumption 5.4.5 In the remainder of this section we assume that the family of triangulations $\{\mathcal{T}_{h_\ell}\}$ corresponding to the finite element spaces V_ℓ, $\ell = 0,1,\ldots$, is *quasi-uniform* and that $h_{\ell-1} \leq c h_\ell$ with a constant c independent of ℓ.

We give some results that will be used in the analysis further on. First we recall an *inverse inequality* that is known from the analysis of finite element methods:

$$|v_\ell|_1 \leq c\, h_\ell^{-1}\|v_\ell\|_{L^2} \quad \text{for all } \; v_\ell \in V_\ell,$$

with a constant c independent of ℓ. For this result to hold we need Assumption 5.4.5.

We now show that, apart from a scaling factor, the isomorphism $P_\ell : (\mathbb{R}^{n_\ell}, \langle\cdot,\cdot\rangle) \to (V_\ell, (\cdot,\cdot)_{L^2})$ and its inverse are uniformly (w.r.t. ℓ) bounded:

Lemma 5.4.6 *There exist constants $c_1 > 0$ and c_2 independent of ℓ such that*

$$c_1\|P_\ell\mathbf{x}\|_{L^2} \leq h_\ell^{\frac{1}{2}d}\|\mathbf{x}\|_2 \leq c_2\|P_\ell\mathbf{x}\|_{L^2} \quad \textit{for all } \; \mathbf{x}\in\mathbb{R}^{n_\ell}. \tag{5.86}$$

Proof. The definition of P_ℓ yields $P_\ell\mathbf{x} = \sum_{i=1}^{n_\ell} x_i\phi_i =: v_\ell \in V_\ell$ and $v_\ell(\xi_i) = x_i$, where ξ_i is the vertex in the triangulation which corresponds to the nodal basis function ϕ_i. Note that

$$\|P_\ell\mathbf{x}\|_{L^2}^2 = \|v_\ell\|_{L^2}^2 = \sum_{T\in\mathcal{T}_\ell}\|v_\ell\|_{L^2(T)}^2.$$

Since v_ℓ is linear on each simplex T in the triangulation $\mathcal{T}_\ell$ there are constants $\tilde{c}_1 > 0$ and $\tilde{c}_2$ independent of h_ℓ such that

$$\tilde{c}_1\|v_\ell\|_{L^2(T)}^2 \leq |T| \sum_{\xi_j\in V(T)} v_\ell(\xi_j)^2 \leq \tilde{c}_2\|v_\ell\|_{L^2(T)}^2,$$

where $V(T)$ denotes the set of vertices of the simplex T. Summation over all $T\in\mathcal{T}_\ell$, using $v_\ell(\xi_j) = x_j$ and $|T| \sim h_\ell^d$ we obtain

$$\hat{c}_1\|v_\ell\|_{L^2}^2 \leq h_\ell^d\sum_{i=1}^{n_\ell} x_i^2 \leq \hat{c}_2\|v_\ell\|_{L^2}^2,$$

with constants $\hat{c}_1 > 0$ and $\hat{c}_2$ independent of h_ℓ and thus we get the result in (5.86). □

The third preliminary result concerns the scaling of the stiffness matrix:

Lemma 5.4.7 *Let* $\mathbf{A}_\ell$ *be the stiffness matrix as in* (5.73)*. Assume that the bilinear form is such that the usual conditions* (5.69) *are satisfied. Then there exist constants* $c_1 > 0$ *and* c_2 *independent of* ℓ *such that*

$$c_1 h_\ell^{d-2} \leq \|\mathbf{A}_\ell\|_2 \leq c_2 h_\ell^{d-2}.$$

Proof. First note that

$$\|\mathbf{A}_\ell\|_2 = \max_{\mathbf{x},\mathbf{y}\in\mathbb{R}^{n_\ell}} \frac{\langle \mathbf{A}_\ell \mathbf{x}, \mathbf{y}\rangle}{\|\mathbf{x}\|_2 \|\mathbf{y}\|_2}.$$

Using the result in Lemma 5.4.6, the continuity of the bilinear form and the inverse inequality we get

$$\begin{aligned}
\max_{\mathbf{x},\mathbf{y}\in\mathbb{R}^{n_\ell}} \frac{\langle \mathbf{A}_\ell \mathbf{x}, \mathbf{y}\rangle}{\|\mathbf{x}\|_2 \|\mathbf{y}\|_2} &\leq c h_\ell^d \max_{v_\ell, w_\ell \in V_\ell} \frac{k(v_\ell, w_\ell)}{\|v_\ell\|_{L^2} \|w_\ell\|_{L^2}} \\
&\leq c h_\ell^d \max_{v_\ell, w_\ell \in V_\ell} \frac{|v_\ell|_1 |w_\ell|_1}{\|v_\ell\|_{L^2} \|w_\ell\|_{L^2}} \leq c\, h_\ell^{d-2},
\end{aligned}$$

and thus the upper bound is proved. The lower bound follows from

$$\max_{\mathbf{x},\mathbf{y}\in\mathbb{R}^{n_\ell}} \frac{\langle \mathbf{A}_\ell \mathbf{x}, \mathbf{y}\rangle}{\|\mathbf{x}\|_2 \|\mathbf{y}\|_2} \geq \max_{1\leq i\leq n_\ell} \langle \mathbf{A}_\ell \mathbf{e}_i, \mathbf{e}_i\rangle = k(\phi_i, \phi_i) \geq c|\phi_i|_1^2 \geq c h_\ell^{d-2}.$$

The last inequality can be shown by using for $T \subset \operatorname{supp}(\phi_i)$ the affine transformation from the unit simplex to T. □

Approximation property

We derive a bound for the first factor in the splitting (5.85). We start with two important assumptions that are crucial for the analysis. The first one concerns *regularity of the continuous problem*, the second one is a *discretization error bound.*

Assumption 5.4.8 We assume that the continuous problem in (5.70) is H^2-regular, i.e. for $f \in L^2(\Omega)$ the corresponding solution u satisfies

$$|u|_2 \leq c\, \|f\|_{L^2},$$

with a constant c independent of f. Furthermore we assume a finite element discretization error bound for the Galerkin discretization (5.72):

$$\|u - u_\ell\|_{L^2} \leq c h_\ell^2 |u|_2,$$

with c independent of u and of ℓ. Here $|\cdot|_2$ denotes the semi-norm (second derivatives only) on $H^2(\Omega)$.

In the analysis we will use the adjoint operator $P_\ell^* : V_\ell \to \mathbb{R}^{n_\ell}$ which satisfies $(P_\ell \mathbf{x}, v_\ell)_{L^2} = \langle \mathbf{x}, P_\ell^* v_\ell \rangle$ for all $\mathbf{x} \in \mathbb{R}^{n_\ell}$, $v_\ell \in V_\ell$. As a direct consequence of Lemma 5.4.6 we obtain

$$c_1 \|P_\ell^* v_\ell\|_2 \leq h_\ell^{\frac{1}{2}d} \|v_\ell\|_{L^2} \leq c_2 \|P_\ell^* v_\ell\|_2 \quad \text{for all} \quad v_\ell \in V_\ell, \tag{5.87}$$

with constants $c_1 > 0$ and c_2 independent of ℓ. We now formulate a main result for the convergence analysis of multigrid methods:

Theorem 5.4.9 (Approximation property.) *Consider* $\mathbf{A}_\ell$, $\mathbf{p}_\ell$, $\mathbf{r}_\ell$ *as defined in* (5.73), (5.76),(5.77). *Assume that the variational problem* (5.70) *is such that the usual conditions* (5.69) *are satisfied. Moreover, Assumption 5.4.8 holds. Then there exists a constant* C_A *independent of* ℓ *such that*

$$\|\mathbf{A}_\ell^{-1} - \mathbf{p}_\ell \mathbf{A}_{\ell-1}^{-1} \mathbf{r}_\ell\|_2 \leq C_A \|\mathbf{A}_\ell\|_2^{-1} \quad \textit{for} \quad \ell = 1, 2, \ldots \tag{5.88}$$

Proof. Let $\mathbf{b}_\ell \in \mathbb{R}^{n_\ell}$ be given. The constants in the proof are independent of $\mathbf{b}_\ell$ and of ℓ. Consider the variational problems:

$$\begin{aligned} u \in H_0^1(\Omega): &\quad k(u, v) = ((P_\ell^*)^{-1} \mathbf{b}_\ell, v)_{L^2} \quad \text{for all} \quad v \in H_0^1(\Omega), \\ u_\ell \in V_\ell: &\quad k(u_\ell, v_\ell) = ((P_\ell^*)^{-1} \mathbf{b}_\ell, v_\ell)_{L^2} \quad \text{for all} \quad v_\ell \in V_\ell, \\ u_{\ell-1} \in V_{\ell-1}: &\quad k(u_{\ell-1}, v_{\ell-1}) = ((P_\ell^*)^{-1} \mathbf{b}_\ell, v_{\ell-1})_{L^2} \quad \text{for all} \quad v_{\ell-1} \in V_{\ell-1}. \end{aligned}$$

Then

$$\mathbf{A}_\ell^{-1} \mathbf{b}_\ell = P_\ell^{-1} u_\ell \quad \text{and} \quad \mathbf{A}_{\ell-1}^{-1} \mathbf{r}_\ell \mathbf{b}_\ell = P_{\ell-1}^{-1} u_{\ell-1}$$

hold. Hence we obtain, using Lemma 5.4.6,

$$\|(\mathbf{A}_\ell^{-1} - \mathbf{p}_\ell \mathbf{A}_{\ell-1}^{-1} \mathbf{r}_\ell) \mathbf{b}_\ell\|_2 = \|P_\ell^{-1}(u_\ell - u_{\ell-1})\|_2 \leq c\, h_\ell^{-\frac{1}{2}d} \|u_\ell - u_{\ell-1}\|_{L^2}. \tag{5.89}$$

Now we use the assumptions on the discretization error bound and on the H^2-regularity of the problem. This yields

$$\begin{aligned} \|u_\ell - u_{\ell-1}\|_{L^2} &\leq \|u_\ell - u\|_{L^2} + \|u_{\ell-1} - u\|_{L^2} \\ &\leq c h_\ell^2 |u|_2 + + c h_{\ell-1}^2 |u|_2 \leq c h_\ell^2 \|(P_\ell^*)^{-1} \mathbf{b}_\ell\|_{L^2}. \end{aligned} \tag{5.90}$$

We combine (5.89) with (5.90) and use (5.87), and get

$$\|(\mathbf{A}_\ell^{-1} - \mathbf{p}_\ell \mathbf{A}_{\ell-1}^{-1} \mathbf{r}_\ell) \mathbf{b}_\ell\|_2 \leq c\, h_\ell^{2-d} \|\mathbf{b}_\ell\|_2$$

and thus $\|\mathbf{A}_\ell^{-1} - \mathbf{p}_\ell \mathbf{A}_{\ell-1}^{-1} \mathbf{r}_\ell\|_2 \leq c\, h_\ell^{2-d}$. The proof is completed if we use Lemma 5.4.7. □

Note that in the proof of the approximation property we use the underlying continuous problem.

Smoothing property

We derive inequalities of the form

$$\|\mathbf{A}_\ell \mathbf{S}_\ell^\nu\|_2 \leq g(\nu)\|\mathbf{A}_\ell\|_2$$

where $g(\nu)$ is a monotonically decreasing function with $\lim_{\nu\to\infty} g(\nu) = 0$. This function g does *not* depend on ℓ. We restrict ourselves to the case that $\mathbf{A}_\ell$ is symmetric positive definite.

We start with an elementary lemma:

Lemma 5.4.10 *Let* $\mathbf{B} \in \mathbb{R}^{m\times m}$ *be a symmetric positive definite matrix with* $\sigma(\mathbf{B}) \subset (0,1]$*. Then we have*

$$\|\mathbf{B}(\mathbf{I}-\mathbf{B})^\nu\|_2 \leq \frac{1}{2(\nu+1)} \quad \textit{for} \quad \nu = 1,2,\ldots$$

Proof. Note that

$$\|\mathbf{B}(\mathbf{I}-\mathbf{B})^\nu\|_2 = \max_{x\in(0,1]} x(1-x)^\nu = \frac{1}{\nu+1}\Big(\frac{\nu}{\nu+1}\Big)^\nu.$$

A simple computation shows that $\nu \to \big(\frac{\nu}{\nu+1}\big)^\nu$ is decreasing on $[1,\infty)$. □

Below for a few basic iterative methods we derive the smoothing property for the symmetric case, i.e., $\mathbf{b} = 0$ in the bilinear form $k(\cdot,\cdot)$. We first consider the Richardson method:

Theorem 5.4.11 *Assume that in the bilinear form we have* $\mathbf{b} = 0$ *and that the usual conditions* (5.69) *are satisfied. Let* $\mathbf{A}_\ell$ *be the stiffness matrix in* (5.73)*. For all* $c_0 \in (0,1]$ *the smoothing property*

$$\|\mathbf{A}_\ell(\mathbf{I} - \frac{c_0}{\rho(\mathbf{A}_\ell)}\mathbf{A}_\ell)^\nu\|_2 \leq \frac{1}{2c_0(\nu+1)}\|\mathbf{A}_\ell\|_2\,, \quad \nu = 1,2,\ldots$$

holds.

Proof. Note that $\mathbf{A}_\ell$ is symmetric positive definite. Apply Lemma 5.4.10 with $\mathbf{B} := \omega_\ell \mathbf{A}_\ell$, $\omega_\ell := c_0\,\rho(\mathbf{A}_\ell)^{-1}$. This yields

$$\|\mathbf{A}_\ell(\mathbf{I}-\omega_\ell\mathbf{A}_\ell)^\nu\|_2 \leq \omega_\ell^{-1}\frac{1}{2(\nu+1)} \leq \frac{1}{2c_0(\nu+1)}\rho(\mathbf{A}_\ell) = \frac{1}{2c_0(\nu+1)}\|\mathbf{A}_\ell\|_2,$$

and thus the result is proved. □

A similar result can be shown for the damped Jacobi method:

Theorem 5.4.12 *Assume that in the bilinear form we have* $\mathbf{b} = 0$ *and that the usual conditions* (5.69) *are satisfied. Let* $\mathbf{A}_\ell$ *be the stiffness matrix in* (5.73) *and* $\mathbf{D}_\ell := \operatorname{diag}(\mathbf{A}_\ell)$. *There exists a constant* c_0, *independent of* ℓ, *with* $0 < c_0 \leq \rho(\mathbf{D}_\ell^{-1}\mathbf{A}_\ell)^{-1}$, *such that for all* $\omega \in (0, c_0]$ *the smoothing property*

$$\|\mathbf{A}_\ell(\mathbf{I} - \omega\mathbf{D}_\ell^{-1}\mathbf{A}_\ell)^\nu\|_2 \leq \frac{1}{2\omega(\nu+1)}\|\mathbf{A}_\ell\|_2 \ , \quad \nu = 1, 2, \ldots$$

holds.

Proof. Define the symmetric positive definite matrix $\tilde{\mathbf{B}} := \mathbf{D}_\ell^{-\frac{1}{2}}\mathbf{A}_\ell\mathbf{D}_\ell^{-\frac{1}{2}}$. Note that

$$(\mathbf{D}_\ell)_{ii} = (\mathbf{A}_\ell)_{ii} = k(\phi_i, \phi_i) \geq c\,|\phi_i|_1^2 \geq c\,h_\ell^{d-2} \ , \tag{5.91}$$

with $c > 0$ independent of ℓ and i. Using this in combination with Lemma 5.4.7 we get

$$\|\tilde{\mathbf{B}}\|_2 \leq \frac{\|\mathbf{A}_\ell\|_2}{\lambda_{\min}(\mathbf{D}_\ell)} \leq \hat{c} \ , \quad \hat{c} \text{ independent of } \ell.$$

Take $c_0 := \hat{c}^{-1}$ and note that $\hat{c} \geq \rho(\mathbf{D}_\ell^{-1}\mathbf{A}_\ell)$ holds. For all $\omega \in (0, c_0]$ we have $\sigma(\omega\tilde{\mathbf{B}}) \subset (0, 1]$. Application of Lemma 5.4.10, with $\mathbf{B} = \omega\tilde{\mathbf{B}}$, yields

$$\begin{aligned}\|\mathbf{A}_\ell(\mathbf{I} - \omega\mathbf{D}_\ell^{-1}\mathbf{A}_\ell)^\nu\|_2 &\leq \omega^{-1}\|\mathbf{D}_\ell^{\frac{1}{2}}\|_2\|\omega\tilde{\mathbf{B}}(\mathbf{I} - \omega\tilde{\mathbf{B}})^\nu\|_2\|\mathbf{D}_\ell^{\frac{1}{2}}\|_2 \\ &\leq \frac{\|\mathbf{D}_\ell\|_2}{2\omega(\nu+1)} \leq \frac{1}{2\omega(\nu+1)}\|\mathbf{A}_\ell\|_2,\end{aligned}$$

and thus the result is proved. □

Remark 5.4.13 The value of the parameter ω used in Theorem 5.4.12 is such that $\omega\rho(\mathbf{D}_\ell^{-1}\mathbf{A}_\ell) = \omega\rho(\mathbf{D}_\ell^{-\frac{1}{2}}\mathbf{A}_\ell\mathbf{D}_\ell^{-\frac{1}{2}}) \leq 1$ holds. Note that

$$\rho(\mathbf{D}_\ell^{-\frac{1}{2}}\mathbf{A}_\ell\mathbf{D}_\ell^{-\frac{1}{2}}) = \max_{\mathbf{x}\in\mathbb{R}^{n_\ell}} \frac{\langle\mathbf{A}_\ell\mathbf{x}, \mathbf{x}\rangle}{\langle\mathbf{D}_\ell\mathbf{x}, \mathbf{x}\rangle} \geq \max_{1\leq i\leq n_\ell} \frac{\langle\mathbf{A}_\ell\mathbf{e}_i, \mathbf{e}_i\rangle}{\langle\mathbf{D}_\ell\mathbf{e}_i\mathbf{e}_i\rangle} = 1$$

and thus we have $\omega \leq 1$. This explains why in multigrid methods one usually uses a *damped* Jacobi method as a smoother.

We finally consider the symmetric Gauss-Seidel method. If $\mathbf{A}_\ell = \mathbf{A}_\ell^T$ this method has an iteration matrix

$$\mathbf{S}_\ell = \mathbf{I} - \mathbf{W}_\ell^{-1}\mathbf{A}_\ell, \quad \mathbf{W}_\ell = (\mathbf{D}_\ell - \mathbf{L}_\ell)\mathbf{D}_\ell^{-1}(\mathbf{D}_\ell - \mathbf{L}_\ell^T) \ , \tag{5.92}$$

where we use the decomposition $\mathbf{A}_\ell = \mathbf{D}_\ell - \mathbf{L}_\ell - \mathbf{L}_\ell^T$ with $\mathbf{D}_\ell$ a diagonal matrix and $\mathbf{L}_\ell$ a strictly lower triangular matrix.

Theorem 5.4.14 *Assume that in the bilinear form we have* $\mathbf{b} = 0$ *and that the usual conditions* (5.69) *are satisfied. Let* $\mathbf{A}_\ell$ *be the stiffness matrix in* (5.73) *and* $\mathbf{W}_\ell$ *as in* (5.92). *The smoothing property*

$$\|\mathbf{A}_\ell(\mathbf{I} - \mathbf{W}_\ell^{-1}\mathbf{A}_\ell)^\nu\|_2 \leq \frac{c}{\nu+1}\|\mathbf{A}_\ell\|_2 \ , \quad \nu = 1, 2, \ldots$$

holds with a constant c *independent of* ν *and* ℓ.

Proof. Note that $\mathbf{W}_\ell = \mathbf{A}_\ell + \mathbf{L}_\ell\mathbf{D}_\ell^{-1}\mathbf{L}_\ell^T$ and thus $\mathbf{W}_\ell$ is symmetric positive definite. Define the symmetric positive definite matrix $\mathbf{B} := \mathbf{W}_\ell^{-\frac{1}{2}}\mathbf{A}_\ell\mathbf{W}_\ell^{-\frac{1}{2}}$. From

$$\begin{aligned}
0 < \max_{\mathbf{x}\in\mathbb{R}^{n_\ell}} \frac{\langle \mathbf{Bx}, \mathbf{x}\rangle}{\langle \mathbf{x}, \mathbf{x}\rangle} &= \max_{\mathbf{x}\in\mathbb{R}^{n_\ell}} \frac{\langle \mathbf{A}_\ell\mathbf{x}, \mathbf{x}\rangle}{\langle \mathbf{W}_\ell\mathbf{x}, \mathbf{x}\rangle} \\
&= \max_{\mathbf{x}\in\mathbb{R}^{n_\ell}} \frac{\langle \mathbf{A}_\ell\mathbf{x}, \mathbf{x}\rangle}{\langle \mathbf{A}_\ell\mathbf{x}, \mathbf{x}\rangle + \langle \mathbf{D}_\ell^{-1}\mathbf{L}_\ell^T\mathbf{x}, \mathbf{L}_\ell^T\mathbf{x}\rangle} \leq 1
\end{aligned}$$

it follows that $\sigma(\mathbf{B}) \subset (0, 1]$. Application of Lemma 5.4.10 yields

$$\|\mathbf{A}_\ell(\mathbf{I} - \mathbf{W}_\ell^{-1}\mathbf{A}_\ell)^\nu\|_2 \leq \|\mathbf{W}_\ell^{\frac{1}{2}}\|_2^2\,\|\mathbf{B}(\mathbf{I} - \mathbf{B})^\nu\|_2 \leq \|\mathbf{W}_\ell\|_2 \frac{1}{2(\nu+1)}.$$

From (5.91) we have $\|\mathbf{D}_\ell^{-1}\|_2 \leq c\, h_\ell^{2-d}$. Using the sparsity of $\mathbf{A}_\ell$ we obtain

$$\|\mathbf{L}_\ell\|_2\|\mathbf{L}_\ell^T\|_2 \leq \|\mathbf{L}_\ell\|_\infty\|\mathbf{L}_\ell\|_1 \leq c(\max_{i,j}|(\mathbf{A}_\ell)_{ij}|)^2 \leq c\|\mathbf{A}_\ell\|_2^2.$$

In combination with Lemma 5.4.7 we then get

$$\begin{aligned}
\|\mathbf{W}_\ell\|_2 &\leq \|\mathbf{A}_\ell\|_2 + \|\mathbf{D}_\ell^{-1}\|_2\|\mathbf{L}_\ell\|_2\|\mathbf{L}_\ell^T\|_2 \\
&\leq \|\mathbf{A}_\ell\|_2 + c\, h_\ell^{2-d}\|\mathbf{A}_\ell\|_2^2 \leq c\|\mathbf{A}_\ell\|_2,
\end{aligned} \tag{5.93}$$

and this completes the proof. □

For the symmetric positive definite case smoothing properties have also been proved for other iterative methods. For example, in [255, 253] a smoothing property is proved for a variant of the ILU method and in [56] it is shown that the SPAI (sparse approximate inverse) preconditioner satisfies a smoothing property.

Smoothing properties of the Richardson, damped Jacobi and Gauss-Seidel methods can also be proved for the *non*symmetric case, i.e, $\mathbf{b} \neq 0$ in the bilinear form $k(\cdot,\cdot)$. We do not treat this here, but refer to the literature, e.g. [133, 132, 210].

Multigrid contraction number

Based on the approximation and smoothing property we prove a bound for the contraction number in the Euclidean norm of the multigrid algorithm (5.78) with $\tau \geq 2$. We follow the analysis introduced in [133, 132].
Apart from the approximation and smoothing property that have been proved above we also need the following stability bound for the iteration matrix of the smoother:

$$\exists\, C_S \,:\, \|\mathbf{S}_\ell^\nu\|_2 \leq C_S \quad \text{for all } \ell \text{ and } \nu. \tag{5.94}$$

Lemma 5.4.15 *For the Richardson method as in Theorem 5.4.11 the inequality* (5.94) *holds with* $C_S = 1$.

Proof. Follows from

$$\|\mathbf{S}_\ell\|_2 = \|\mathbf{I} - \frac{c_0}{\rho(\mathbf{A}_\ell)}\mathbf{A}_\ell\|_2 = \max_{\lambda \in \sigma(\mathbf{A}_\ell)} \left|1 - c_0 \frac{\lambda}{\rho(\mathbf{A}_\ell)}\right| \leq 1.$$

□

Lemma 5.4.16 *For the damped Jacobi method as in Theorem 5.4.12 the inequality* (5.94) *holds.*

Proof. Due to the choice of ω we have

$$\|\mathbf{S}_\ell\|_D = \|\mathbf{D}_\ell^{\frac{1}{2}}(\mathbf{I} - \omega \mathbf{D}_\ell^{-1}\mathbf{A}_\ell)\mathbf{D}_\ell^{-\frac{1}{2}}\|_2 \leq 1,$$

and thus

$$\|\mathbf{S}_\ell^\nu\|_2 \leq \|\mathbf{D}_\ell^{-\frac{1}{2}}(\mathbf{D}_\ell^{\frac{1}{2}}\mathbf{S}_\ell\mathbf{D}_\ell^{-\frac{1}{2}})^\nu \mathbf{D}_\ell^{\frac{1}{2}}\|_2 \leq \kappa(\mathbf{D}_\ell^{\frac{1}{2}})\, \|\mathbf{S}_\ell\|_D^\nu \leq \kappa(\mathbf{D}_\ell^{\frac{1}{2}}).$$

Now note that $\mathbf{D}_\ell$ is uniformly (w.r.t. ℓ) well-conditioned. □

Similar results hold for the *non*symmetric case.
Using Lemma 5.4.6 it follows that for $\mathbf{p}_\ell = P_\ell^{-1}P_{\ell-1}$ we have

$$C_{p,1}\|\mathbf{x}\|_2 \leq \|\mathbf{p}_\ell \mathbf{x}\|_2 \leq C_{p,2}\|\mathbf{x}\|_2 \quad \text{for all } \mathbf{x} \in \mathbb{R}^{n_{\ell-1}}. \tag{5.95}$$

with constants $C_{p,1} > 0$ and $C_{p,2}$ independent of ℓ.
We now formulate a main convergence result for the multigrid method.

Theorem 5.4.17 *Consider the multigrid method with iteration matrix given in* (5.83) *and parameter values* $\nu_2 = 0$, $\nu_1 = \nu > 0$, $\tau \geq 2$. *Assume that there are constants* C_A, C_S *and a monotonically decreasing function* $g(\nu)$ *with* $g(\nu) \to 0$ *for* $\nu \to \infty$ *such that for all* ℓ:

$$\|\mathbf{A}_\ell^{-1} - \mathbf{p}_\ell \mathbf{A}_{\ell-1}^{-1} \mathbf{r}_\ell\|_2 \leq C_A \|\mathbf{A}_\ell\|_2^{-1} \tag{5.96a}$$

$$\|\mathbf{A}_\ell \mathbf{S}_\ell^\nu\|_2 \leq g(\nu)\, \|\mathbf{A}_\ell\|_2\,, \quad \nu \geq 1 \tag{5.96b}$$

$$\|\mathbf{S}_\ell^\nu\|_2 \leq C_S\,, \quad \nu \geq 1. \tag{5.96c}$$

For any $\xi^* \in (0,1)$ *there exists a* ν^* *such that for all* $\nu \geq \nu^*$

$$\|\mathbf{C}_{MG,\ell}\|_2 \leq \xi^*\,, \quad \ell = 0, 1, \ldots$$

holds.

Proof. For the two-grid iteration matrix we have

$$\|\mathbf{C}_{TG,\ell}\|_2 \leq \|\mathbf{A}_\ell^{-1} - \mathbf{p}_\ell \mathbf{A}_{\ell-1}^{-1} \mathbf{r}_\ell\|_2 \|\mathbf{A}_\ell \mathbf{S}_\ell^\nu\|_2 \leq C_A g(\nu).$$

Define $\xi_\ell = \|\mathbf{C}_{MG.\ell}\|_2$. From (5.83) we obtain $\xi_0 = 0$ and for $\ell \geq 1$:

$$\begin{aligned}
\xi_\ell &\leq C_A g(\nu) + \|\mathbf{p}_\ell\|_2 \xi_{\ell-1}^\tau \|\mathbf{A}_{\ell-1}^{-1} \mathbf{r}_\ell \mathbf{A}_\ell \mathbf{S}_\ell^\nu\|_2 \\
&\leq C_A g(\nu) + C_{p,2} C_{p,1}^{-1} \xi_{\ell-1}^\tau \|\mathbf{p}_\ell \mathbf{A}_{\ell-1}^{-1} \mathbf{r}_\ell \mathbf{A}_\ell \mathbf{S}_\ell^\nu\|_2 \\
&\leq C_A g(\nu) + C_{p,2} C_{p,1}^{-1} \xi_{\ell-1}^\tau \big(\|(\mathbf{I} - \mathbf{p}_\ell \mathbf{A}_{\ell-1}^{-1} \mathbf{r}_\ell \mathbf{A}_\ell) \mathbf{S}_\ell^\nu\|_2 + \|\mathbf{S}_\ell^\nu\|_2\big) \\
&\leq C_A g(\nu) + C_{p,2} C_{p,1}^{-1} \xi_{\ell-1}^\tau \big(C_A g(\nu) + C_S\big) \leq C_A g(\nu) + C^* \xi_{\ell-1}^\tau,
\end{aligned}$$

with $C^* := C_{p,2} C_{p,1}^{-1} (C_A g(1) + C_S)$. Elementary analysis shows that for $\tau \geq 2$ and any $\xi^* \in (0,1)$ the sequence $x_0 = 0$, $x_i = C_A g(\nu) + C^* x_{i-1}^\tau$, $i \geq 1$, is bounded by ξ^* for $g(\nu)$ sufficiently small. □

Remark 5.4.18 Consider $\mathbf{A}_\ell$, $\mathbf{p}_\ell$, $\mathbf{r}_\ell$ as defined in (5.73), (5.76),(5.77). Assume that the variational problem (5.70) is such that the usual conditions (5.69) are satisfied. Moreover, the problem (5.70) is such that Assumption 5.4.8 is satisfied. In the multigrid method we use the Richardson or the damped Jacobi method. *Then the assumptions* (5.96) *are fulfilled and thus for* $\nu_2 = 0$ *and* ν_1 *sufficiently large the multigrid W-cycle has a contraction number smaller than one independent of* ℓ.

Remark 5.4.19 Let $\mathbf{C}_{MG,\ell}(\nu_2, \nu_1)$ be the iteration matrix of the multigrid method with ν_1 pre- and ν_2 postsmoothing iterations. With $\nu := \nu_1 + \nu_2$ we have

$$\rho\big(\mathbf{C}_{MG,\ell}(\nu_2, \nu_1)\big) = \rho\big(\mathbf{C}_{MG,\ell}(0, \nu)\big) \leq \|\mathbf{C}_{MG,\ell}(0, \nu)\|_2.$$

Using Theorem 5.4.17 we thus get, for $\tau \geq 2$, a bound for the *spectral radius* of the iteration matrix $\mathbf{C}_{MG,\ell}(\nu_2, \nu_1)$.

Remark 5.4.20 The multigrid convergence analysis presented above assumes sufficient regularity (namely H^2-regularity) of the elliptic boundary value problem. There have been developed convergence analyses in which this regularity assumption is avoided and an h-independent convergence rate of multigrid is proved. These analyses are based on so-called subspace decomposition techniques. Two review papers on multigrid convergence proofs are [260] and [257].

Convergence analysis for symmetric positive definite problems

The multigrid convergence analysis presented above applies to the general, i.e. possibly non-symmetric, boundary value problem as in (5.68). The elliptic problem with $\mathbf{b} = 0$ (no convection) results in a stiffness matrix $\mathbf{A}_\ell$ that is symmetric positive definite. This property allows a refined analysis which proves that the contraction number of the multigrid method with $\tau \geq 1$ (the V-cycle is included!) and $\nu_1 = \nu_2 \geq 1$ pre- and postsmoothing iterations is bounded by a constant smaller than one independent of ℓ. The basic idea of this analysis is due to [50] and is further simplified by [133, 132].
We outline a main result of this analysis. For this we make the following

Assumption 5.4.21 In the bilinear form $k(\cdot,\cdot)$ in (5.70) we have $\mathbf{b} = 0$ and the conditions (5.69) are satisfied.

Due to this the stiffness matrix $\mathbf{A}_\ell$ is symmetric positive definite and we can define the energy scalar product and corresponding norm:

$$\langle \mathbf{x}, \mathbf{y} \rangle_A := \langle \mathbf{A}_\ell \mathbf{x}, \mathbf{y} \rangle \ , \quad \|\mathbf{x}\|_A := \langle \mathbf{x}, \mathbf{x} \rangle_A^{\frac{1}{2}} \quad \mathbf{x}, \mathbf{y} \in \mathbb{R}^{n_\ell}.$$

We only consider smoothers with an iteration matrix $\mathbf{S}_\ell = \mathbf{I} - \mathbf{W}_\ell^{-1}\mathbf{A}_\ell$ in which $\mathbf{W}_\ell$ is symmetric positive definite. Important examples are the Richardson, damped Jacobi and symmetric Gauss-Seidel smoothers:

$$\text{Richardson method}: \ \mathbf{W}_\ell = c_0^{-1}\rho(\mathbf{A}_\ell)\mathbf{I} \ , \quad c_0 \in (0,1], \tag{5.97a}$$

$$\text{Damped Jacobi}: \ \mathbf{W}_\ell = \omega^{-1}\mathbf{D}_\ell, \ \omega \ \text{ as in Theorem 5.4.12}, \tag{5.97b}$$

$$\text{Symm. Gauss-Seidel}: \ \mathbf{W}_\ell = (\mathbf{D}_\ell - \mathbf{L}_\ell)\mathbf{D}_\ell^{-1}(\mathbf{D}_\ell - \mathbf{L}_\ell^T). \tag{5.97c}$$

For symmetric matrices $\mathbf{B}, \mathbf{C} \in \mathbb{R}^{m \times m}$ we use the notation $\mathbf{B} \leq \mathbf{C}$ iff $\langle \mathbf{B}\mathbf{x}, \mathbf{x} \rangle \leq \langle \mathbf{C}\mathbf{x}, \mathbf{x} \rangle$ for all $\mathbf{x} \in \mathbb{R}^m$, i.e., $\mathbf{C} - \mathbf{B}$ is symmetric positive semi-definite.

Lemma 5.4.22 *For $\mathbf{W}_\ell$ as in (5.97) the following properties hold:*

$$\mathbf{A}_\ell \leq \ \mathbf{W}_\ell \quad \textit{for all } \ell \tag{5.98a}$$

$$\exists\, C_W : \quad \|\mathbf{W}_\ell\|_2 \leq C_W \|\mathbf{A}_\ell\|_2 \quad \textit{for all } \ell. \tag{5.98b}$$

Proof. For the Richardson method the result is trivial. For the damped Jacobi method we have $\omega \in (0, \rho(\mathbf{D}_\ell^{-1}\mathbf{A}_\ell)^{-1}]$ and thus $\omega\rho(\mathbf{D}_\ell^{-\frac{1}{2}}\mathbf{A}_\ell\mathbf{D}_\ell^{-\frac{1}{2}}) \leq 1$. This yields $\mathbf{A}_\ell \leq \omega^{-1}\mathbf{D}_\ell = \mathbf{W}_\ell$. The result in (5.98b) follows from $\|\mathbf{D}_\ell\|_2 \leq \|\mathbf{A}_\ell\|_2$. For the symmetric Gauss-Seidel method the results (5.98a) follows from $\mathbf{W}_\ell = \mathbf{A}_\ell + \mathbf{L}_\ell\mathbf{D}_\ell^{-1}\mathbf{L}_\ell^T$ and the result in (5.98b) is proved in (5.93). □

We introduce the following *modified approximation property*:

$$\exists \tilde{C}_A : \quad \|\mathbf{W}_\ell^{\frac{1}{2}}(\mathbf{A}_\ell^{-1} - \mathbf{p}_\ell\mathbf{A}_{\ell-1}^{-1}\mathbf{r}_\ell)\mathbf{W}_\ell^{\frac{1}{2}}\|_2 \leq \tilde{C}_A \quad \text{for } \ell = 1, 2, \ldots \tag{5.99}$$

We note that the standard approximation property (5.88) implies the result (5.99) if we consider the smoothers in (5.97):

Lemma 5.4.23 *Consider* $\mathbf{W}_\ell$ *as in* (5.97) *and assume that the approximation property* (5.88) *holds. Then* (5.99) *holds with* $\tilde{C}_A = C_W C_A$*, where* C_W *is as in* (5.98b).

Proof. Trivial. □

One easily verifies that for the smoothers in (5.97) the modified approximation property (5.99) implies the standard approximation property (5.88) if $\kappa(\mathbf{W}_\ell)$ is uniformly (w.r.t. ℓ) bounded. The latter property holds for the Richardson and the damped Jacobi method.

The following result is proved in [132].

Theorem 5.4.24 *We take* $\nu_1 = \nu_2 = \frac{1}{2}\nu$ *and consider the multigrid algorithm with iteration matrix* $\mathbf{C}_{MG,\ell} = \mathbf{C}_{MG,\ell}(\nu, \tau)$ *as in* (5.83)*. Assume that* (5.98a) *and* (5.99) *hold. For* $\nu \geq 2$ *and* $\tau \geq 1$ *the inequality*

$$\|\mathbf{C}_{MG,\ell}\|_A \leq \frac{\tilde{C}_A}{\tilde{C}_A + \nu}$$

holds. The matrix $\mathbf{A}_\ell^{\frac{1}{2}}\mathbf{C}_{MG,\ell}\mathbf{A}_\ell^{-\frac{1}{2}}$ *is symmetric positive definite.*

Corollary 5.4.25 Consider $\mathbf{A}_\ell$, $\mathbf{p}_\ell$, $\mathbf{r}_\ell$ as defined in (5.73), (5.76),(5.77). Assume that the variational problem (5.70) is such that $\mathbf{b} = 0$ and that the usual conditions (5.69) are satisfied. Moreover, the problem is such that Assumption 5.4.8 is satisfied. In the multigrid method we use one of the smoothers (5.97). Then the assumptions (5.98a) and (5.99) are satisfied and thus for $\nu_1 = \nu_2 \geq 1$ the multigrid V-cycle has a contraction number (w.r.t. $\|\cdot\|_A$) smaller than one independent of ℓ. □

For the symmetric case ($\mathbf{b} = 0$) it is shown in [190] that the multigrid rate of convergence is robust with respect to the size of the reaction term. The contraction number (w.r.t. $\|\cdot\|_A$) of the multigrid method can be bounded by a constant smaller than one independent of ℓ *and* of $c \in [0, \infty)$, where c

is the coefficient in (5.68) which determines the size of the reaction term.

We now discuss the use of a multigrid method as preconditioner for saddle point problems resulting from generalized Stokes and Oseen problems.

Remark 5.4.26 We consider the matrix $\tilde{\mathbf{A}}$ in the saddle point system (5.8) or (5.44). This matrix has a diagonal block-structure: $\tilde{\mathbf{A}} = \text{blockdiag}(\tilde{\mathbf{A}}_1, \ldots, \tilde{\mathbf{A}}_d)$ with identical blocks $\tilde{\mathbf{A}}_1 = \ldots = \tilde{\mathbf{A}}_d =: \tilde{\mathbf{A}}_{\text{block}}$. In a three-dimensional flow problem we have $d = 3$. The diagonal block $\tilde{\mathbf{A}}_{\text{block}}$ is the stiffness matrix corresponding to the Galerkin discretization, in the space $\mathbb{X}^k_{h,0}$ of simplicial finite elements, of a diffusion-convection-reaction problem of the form as in (5.68). In case of a (generalized) Stokes problem there is no convection term. The reaction term is scaled with the parameter $\beta \sim \frac{1}{\Delta t} \in [0, \infty)$. Without loss of generality we can assume that $\alpha = 1$. For β sufficiently large (i.e., the time step Δt sufficiently small) the matrix $\tilde{\mathbf{A}}_{\text{block}}$ is very similar to the mass matrix in the space $\mathbb{X}^k_{h,0}$, which is *well-conditioned.* In that case preconditioning is not really an issue. The choice $\mathbf{Q}_A = \beta\, \text{diag}(\mathbf{M})$ then results in a satisfactory preconditioner for $\tilde{\mathbf{A}}$. In practice, however, it is usually not clear whether "Δt sufficiently small" is satisfied. Therefore a preconditioner that is good uniformly in β is desirable. Such a preconditioner for the matrix $\tilde{\mathbf{A}}$ can be obtained using a multigrid method applied to the diagonal block $\tilde{\mathbf{A}}_{\text{block}}$. For simplicity we consider the symmetric case, i.e., the generalized Stokes problem. Let $\mathbf{C}_{MG}$ be the iteration matrix of a symmetric multigrid method applied to the symmetric positive definite matrix $\tilde{\mathbf{A}}_{\text{block}}$. The matrix $\mathbf{Q}_{MG}$ is defined by $\mathbf{C}_{MG} =: \mathbf{I} - \mathbf{Q}_{MG}^{-1}\tilde{\mathbf{A}}_{\text{block}}$, cf. Sect. 5.4.1. As explained in Sect. 5.4.1, the matrix $\mathbf{Q}_{MG}$, although *not* explicitly available, can be used as a preconditioner for $\tilde{\mathbf{A}}_{\text{block}}$: For given $\mathbf{y}$ the vector $\mathbf{Q}_{MG}^{-1}\mathbf{y}$ is the result of one multigrid iteration with starting vector equal to zero applied to the system $\mathbf{A}_{\text{block}}\mathbf{z} = \mathbf{y}$. From multigrid convergence analyses it follows that under certain reasonable assumptions, for example those in Theorem 5.4.24, the matrix $\mathbf{Q}_{MG}$ is symmetric positive definite and $\sigma(\mathbf{I} - \mathbf{Q}_{MG}^{-1}\tilde{\mathbf{A}}_{\text{block}}) \subset [0, \rho_{MG}]$ holds with the contraction number $\rho_{MG} < 1$ independent of the mesh size parameter h and independent of β. For the preconditioner $\mathbf{Q}_A$ of $\tilde{\mathbf{A}}$ we take

$$\mathbf{Q}_A := \text{blockdiag}(\mathbf{Q}_{MG}) \quad (d \text{ blocks}).$$

For this multigrid preconditioner we then have the following spectral inequalities

$$(1 - \rho_{MG})\mathbf{Q}_A \leq \tilde{\mathbf{A}} \leq \mathbf{Q}_A, \qquad \rho_{MG} < 1 \quad \text{independent of } h \text{ and } \beta, \tag{5.100}$$

i.e., we obtain spectral inequalities as in (5.18), (5.32) with $\Gamma_A = 1$ and $\gamma_A > 0$ independent of h, β. In this sense the multigrid method yields an optimal preconditioner.

5.4.3 Preconditioners for the Schur complement

In this section we discuss the choice of an appropriate Schur complement preconditioner $\mathbf{Q}_S$. In the first part we give a detailed treatment for the case of the stationary Stokes problem. After that we discuss a Schur complement preconditioner for the generalized Stokes problem. Finally we address the issue of Schur complement preconditioning for the Oseen problem.

Stationary Stokes problem

We consider the discrete problem resulting from the Galerkin discretization of the stationary Stokes problem with Hood-Taylor finite element spaces. As preliminaries for the analysis of the Schur complement preconditioner we recall a few definitions and notations. The Hood-Taylor pair is given by

$$(\mathbf{V}_h, Q_h) = \left((\mathbb{X}_{h,0}^k)^d, \mathbb{X}_h^{k-1} \cap L_0^2(\Omega)\right), \quad k \geq 2.$$

Here we use the notation d for the dimension of the velocity vector ($\Omega \subset \mathbb{R}^d$). For the bases in these spaces we use standard nodal basis functions. In the velocity space $\mathbf{V}_h = (\mathbb{X}_{h,0}^k)^d$ the set of basis functions is denoted by $(\boldsymbol{\xi}_i)_{1 \leq i \leq m}$. The basis in the pressure space $\mathbb{X}_h^{k-1}$ is denoted by $(\psi_i)_{1 \leq i \leq n}$. The corresponding isomorphisms are given by

$$P_{h,1} : \mathbb{R}^m \to \mathbf{V}_h, \qquad P_{h,1}\mathbf{x} = \sum_{i=1}^m x_i \boldsymbol{\xi}_i, \tag{5.101}$$

$$P_{h,2} : \mathbb{R}^n \to \mathbb{X}_h^{k-1}, \qquad P_{h,2}\mathbf{y} = \sum_{i=1}^n y_i \psi_i. \tag{5.102}$$

The stiffness matrix for the *Stokes* problem is given by

$$\mathbf{K} = \begin{pmatrix} \mathbf{A} & \mathbf{B}^T \\ \mathbf{B} & 0 \end{pmatrix} \in \mathbb{R}^{(m+n)\times(m+n)}, \text{ with}$$

$$\langle \mathbf{A}\mathbf{x}, \tilde{\mathbf{x}} \rangle = a(P_{h,1}\mathbf{x}, P_{h,1}\tilde{\mathbf{x}}) = \int_\Omega (\nabla P_{h,1}\mathbf{x}) \cdot (\nabla P_{h,1}\tilde{\mathbf{x}})\, dx \quad \forall\, \mathbf{x}, \tilde{\mathbf{x}} \in \mathbb{R}^m,$$

$$\langle \mathbf{B}\mathbf{x}, \mathbf{y} \rangle = b(P_{h,1}\mathbf{x}, P_{h,2}\mathbf{y}) = -\int_\Omega P_{h,2}\mathbf{y}\, \operatorname{div}(P_{h,1}\mathbf{x})\, dx \quad \forall\, \mathbf{x} \in \mathbb{R}^m,\ \mathbf{y} \in \mathbb{R}^n.$$

The matrix $\mathbf{A} = \text{blockdiag}(\mathbf{A}_1, \ldots, \mathbf{A}_d)$ is symmetric positive definite and $\mathbf{A}_1 = \ldots = \mathbf{A}_d =: \mathbf{A}_{\text{block}}$ is the stiffness matrix corresponding to the Galerkin discretization of the Poisson equation in the space $\mathbb{X}_{h,0}^k$ of simplicial finite elements.

For the preconditioner $\mathbf{Q}_S$ of the Schur complement $\mathbf{S}$ we use the *mass matrix in the pressure space*, which is defined by

$$\langle \mathbf{M}_p\mathbf{y}, \mathbf{z} \rangle = (P_{h,2}\mathbf{y}, P_{h,2}\mathbf{z})_{L^2} \quad \text{for all } \mathbf{y}, \mathbf{z} \in \mathbb{R}^n. \tag{5.103}$$

This mass matrix is symmetric positive definite and $\mathrm{diag}(\mathbf{M}_p)^{-1}\mathbf{M}_p$ is in general well-conditioned. Therefore $\mathrm{diag}(\mathbf{M}_p)$ is spectrally equivalent to $\mathbf{M}_p$ and solving a linear system with matrix $\mathbf{M}_p$ to reasonable accuracy can be realized efficiently by applying a preconditioned CG method with $\mathrm{diag}(\mathbf{M}_p)$ as preconditioner.

We recall the LBB stability property of the Hood-Taylor finite element pair $(\mathbf{V}_h, Q_h)$:

$$\exists\, \hat{\beta} > 0 \ : \quad \sup_{\mathbf{v}_h \in \mathbf{V}_h} \frac{b(\mathbf{v}_h, q_h)}{\|\mathbf{v}_h\|_1} \geq \hat{\beta} \|q_h\|_{L^2} \quad \text{for all} \ \ q_h \in Q_h, \tag{5.104}$$

with $\hat{\beta}$ independent of h. Using this stability property we get the following spectral inequalities for the preconditioner $\mathbf{M}_p$:

Theorem 5.4.27 *Let* $\mathbf{M}_p$ *be the pressure mass matrix defined in* (5.103). *Assume that the stability property* (5.104) *holds. Then*

$$\hat{\beta}^2 \, \mathbf{M}_p \leq \mathbf{S} \leq d\, \mathbf{M}_p \quad \mathit{on} \ \ \big(\mathbf{M}_p(1, \ldots, 1)\big)^{\perp} \tag{5.105}$$

holds.

Proof. For $\mathbf{y} \in \mathbb{R}^n$ we have:

$$\max_{\mathbf{x} \in \mathbb{R}^m} \frac{\langle \mathbf{B}\mathbf{x}, \mathbf{y}\rangle}{\langle \mathbf{A}\mathbf{x}, \mathbf{x}\rangle^{\frac{1}{2}}} = \max_{\mathbf{x} \in \mathbb{R}^m} \frac{\langle \mathbf{B}\mathbf{A}^{-\frac{1}{2}}\mathbf{x}, \mathbf{y}\rangle}{\|\mathbf{x}\|} = \max_{\mathbf{x} \in \mathbb{R}^m} \frac{\langle \mathbf{x}, \mathbf{A}^{-\frac{1}{2}}\mathbf{B}^T\mathbf{y}\rangle}{\|\mathbf{x}\|}$$
$$= \|\mathbf{A}^{-\frac{1}{2}}\mathbf{B}^T\mathbf{y}\| = \langle \mathbf{S}\mathbf{y}, \mathbf{y}\rangle^{\frac{1}{2}}.$$

Hence, for arbitrary $\mathbf{y} \in \mathbb{R}^n$:

$$\langle \mathbf{S}\mathbf{y}, \mathbf{y}\rangle^{\frac{1}{2}} = \max_{\mathbf{u}_h \in \mathbf{V}_h} \frac{b(\mathbf{u}_h, P_{h,2}\mathbf{y})}{|\mathbf{u}_h|_1}. \tag{5.106}$$

Note that $\mathbf{y} \in \big(\mathbf{M}_p(1, \ldots, 1)\big)^{\perp}$ iff $(P_{h,2}\mathbf{y}, 1)_{L^2} = 0$. Using this, (5.106) and the stability bound (5.104) we get

$$\langle \mathbf{S}\mathbf{y}, \mathbf{y}\rangle^{\frac{1}{2}} \geq \hat{\beta}\, \|P_{h,2}\mathbf{y}\|_{L^2} = \hat{\beta}\, \langle \mathbf{M}_p\mathbf{y}, \mathbf{y}\rangle^{\frac{1}{2}}$$

for all $\mathbf{y} \in \big(\mathbf{M}_p(1, \ldots, 1)\big)^{\perp}$ and thus the first inequality in (5.105) holds. Note that

$$\begin{aligned} |b(\mathbf{u}_h, P_{h,2}\mathbf{y})| &\leq \|\operatorname{div} \mathbf{u}_h\|_{L^2} \|P_{h,2}\mathbf{y}\|_{L^2} \\ &\leq \sqrt{d}\, |\mathbf{u}_h|_1 \|P_{h,2}\mathbf{y}\|_{L^2} = \sqrt{d}\, |\mathbf{u}_h|_1 \langle \mathbf{M}_p\mathbf{y}, \mathbf{y}\rangle^{\frac{1}{2}} \end{aligned}$$

holds. Combining this with (5.106) proves the second inequality in (5.105). □

The result in Theorem 5.4.27 is a special case of the general abstract result given in Remark 15.5.1.

Corollary 5.4.28 Suppose that for solving a discrete Stokes problem with stiffness matrix $\mathbf{K}$ we use a preconditioned MINRES method with a multigrid preconditioner $\mathbf{Q}_A$ (for $\mathbf{A}$) as in Remark 5.4.26 and a Schur complement preconditioner $\mathbf{Q}_S = \mathbf{M}_p$ (for $\mathbf{S}$) as defined above. Then the inequalities (5.18) hold with constants $\gamma_A, \Gamma_A, \gamma_S, \Gamma_S$ that are independent of h. From Theorem 5.2.9 it follows that the spectrum of the preconditioned matrix $\tilde{\mathbf{K}}$ is contained in a set $[a,b] \cup [c,d]$ with $a < b < 0 < c < d$, all independent of h. From Theorem 5.2.6 we then conclude that the *residual reduction factor can be bounded by a constant smaller than one independent of* h. □

Generalized Stokes problem

We briefly address the issue of Schur complement preconditioning for the *generalized* Stokes case, i.e., with $\tilde{\mathbf{A}} = \alpha\mathbf{A} + \beta\mathbf{M}$, $\alpha, \beta > 0$, with $\alpha = \mathcal{O}(1)$ and $\beta \sim \frac{1}{\Delta t}$. In Sect. 15.5 an abstract analysis related to Schur complement preconditioning of a parameter dependent saddle point problem is presented. In Sect. 15.5.3 this abstract analysis is applied to a *continuous* generalized Stokes problem. In the continuous setting, the Schur complement operator $S : L_0^2(\Omega) \to L_0^2(\Omega)$ is given by $S = -\operatorname{div}(-\Delta + \tau I)^{-1}\nabla$, cf. (15.55). As Schur complement preconditioner the operator $\tilde{S}^{-1} = I - \tau\Delta_N^{-1}$ is proposed, where Δ_N^{-1} is the solution operator of a Neumann problem in the pressure space $H^1(\Omega) \cap L_0^2(\Omega)$, cf. Theorem 15.5.14. The abstract analysis can also be applied to the finite element *discretization* of such a generalized Stokes problem, cf. [189]. This results in a discrete analogon of $\tilde{S}^{-1}$ (denoted by $\tilde{\mathbf{Q}}_S^{-1}$ below) that was introduced by Cahouet and Chabard [61]. We briefly explain this method. For more details and a convergence analysis we refer to the literature [189]. For $g \in L^2(\Omega)$ consider the Neumann problem: find $w \in H^1(\Omega) \cap L_0^2(\Omega)$ such that

$$(\nabla w, \nabla \phi)_{L^2} = (g, \phi)_{L^2} \quad \text{for all} \quad \phi \in H^1(\Omega) \cap L_0^2(\Omega). \tag{5.107}$$

Let $\mathbf{T}_h$ be the stiffness matrix of the Galerkin discretization of this problem in $\mathbb{X}_h^{k-1} \subset H^1(\Omega)$:

$$\langle \mathbf{T}_h \mathbf{x}, \mathbf{y} \rangle = (\nabla P_{h,2}\mathbf{x}, \nabla P_{h,2}\mathbf{y})_{L^2} \quad \text{for all} \quad \mathbf{x}, \mathbf{y} \in \mathbb{R}^n,$$

with $P_{h,2} : \mathbb{R}^n \to \mathbb{X}_h^{k-1}$ the finite element isomorphism as in (5.102). Note that $\ker(\mathbf{T}_h) = \operatorname{span}(\mathbf{e})$, with $\mathbf{e} := (1,1,\ldots,1)^T$, and $\mathbf{T}_h : (\mathbf{M}_p\mathbf{e})^\perp \to \mathbf{e}^\perp$ is bijective. We define $\tilde{\mathbf{Q}}_S^{-1} : \mathbf{e}^\perp \to \mathbb{R}^n$ by:

$$\tilde{\mathbf{Q}}_S^{-1} = \alpha\,\mathbf{M}_p^{-1} + \beta\,\mathbf{T}_h^{-1}. \tag{5.108}$$

Here $\mathbf{M}_p$ denotes the mass matrix in the pressure space as defined in (5.103). Note, that for $\alpha = 1$, $\beta = 0$ (which corresponds to the discrete *stationary* Stokes equation) we get $\tilde{\mathbf{Q}}_S = \mathbf{M}_p$. In [52, 189] it is shown that $\tilde{\mathbf{Q}}_S$ is uniformly in h and α, β spectrally equivalent to the Schur complement $\mathbf{S} = \mathbf{B}\tilde{\mathbf{A}}^{-1}\mathbf{B}^T$

(on the subspace $(\mathbf{M}_p\mathbf{e})^\perp$). For the *continuous* case the corresponding spectral equivalence result is stated in Theorem 15.5.14. For the evaluation of $\tilde{\mathbf{Q}}_S^{-1}\mathbf{y}$ one has to solve (approximately) a linear system with the mass matrix $\mathbf{M}_p$ and one with the stiffness matrix $\mathbf{T}_h$. For the mass matrix this is easy, cf. the discussion for the stationary Stokes case above. For the stiffness matrix $\mathbf{T}_h$ it suffices to apply one or a few iterations of a standard multigrid method.

Oseen problem

The issue of Schur complement preconditioning in the *non*symmetric case is much more difficult as in the symmetric case. In case of the Oseen problem an "optimal" Schur complement preconditioner is not known, yet. For "small" time steps Δt (i.e., β large) and "small" Reynolds numbers the matrix $\tilde{\mathbf{A}} = \alpha\left[\mathbf{A} + \mathbf{N}(\mathbf{x}^{\mathrm{old}})\right] + \beta\mathbf{M}$ in the Oseen problem is "close to" the matrix $\tilde{\mathbf{A}} = \alpha\mathbf{A} + \beta\mathbf{M}$ that arises in the generalized Stokes problem. In that case the Schur complement preconditioner in (5.108) can be satisfactory for the Oseen problem, too.

We briefly explain another Schur complement preconditioner that is efficient for a much larger range of time steps and of Reynolds numbers than the one in (5.108). This preconditioner is introduced in [99] and further developed in [102, 100]. Define $\mathbf{M}_1 := \mathrm{diag}(\mathbf{M})$, with $\mathbf{M}$ the mass matrix in the velocity space. The so-called *BFBt* preconditioner (or least-squares commutator preconditioner) is given by

$$\mathbf{Q}_S^{-1} = (\mathbf{B}\mathbf{M}_1^{-1}\mathbf{B}^T)^{-1}\,\mathbf{B}\,\mathbf{M}_1^{-1}\,\tilde{\mathbf{A}}\,\mathbf{M}_1^{-1}\,\mathbf{B}^T\,(\mathbf{B}\mathbf{M}_1^{-1}\mathbf{B}^T)^{-1}. \tag{5.109}$$

The expression for this preconditioner is rather complicated, but its implementation can be realized fairly efficiently. Since $\mathbf{M}_1$ is diagonal the matrix $\mathbf{B}\mathbf{M}_1^{-1}\mathbf{B}^T$ is still sparse. This matrix has properties similar to a discrete Laplace operator in the pressure finite element space. For solving linear systems with the symmetric positive definite matrix $\mathbf{B}\mathbf{M}_1^{-1}\mathbf{B}^T$ (approximately) one can use a preconditioned CG method or a multigrid method. Note that this matrix has dimension $n \times n$ and typically $n \ll (n+m)$ holds.

Remark 5.4.29 For the *BFBt* preconditioner there is no rigorous theory that shows that this is a good preconditioner for the Schur complement. A motivation for the structure of this preconditioner and results of numerical experiments that indicate the efficiency of this Schur complement preconditioner are given in [100]. A simple heuristic explanation of this preconditioner as follows. The Schur complement can be represented as

$$\begin{aligned}\mathbf{S} = \mathbf{B}\tilde{\mathbf{A}}^{-1}\mathbf{B}^T &= (\mathbf{B}\mathbf{M}_1^{-\frac{1}{2}})(\mathbf{M}_1^{\frac{1}{2}}\tilde{\mathbf{A}}^{-1}\mathbf{M}_1^{\frac{1}{2}})(\mathbf{B}\mathbf{M}_1^{-\frac{1}{2}})^T \\ &=: \hat{\mathbf{B}}\mathbf{M}_1^{\frac{1}{2}}\tilde{\mathbf{A}}^{-1}\mathbf{M}_1^{\frac{1}{2}}\hat{\mathbf{B}}^T.\end{aligned} \tag{5.110}$$

The matrix $\mathbf{M}_1^{\frac{1}{2}}\tilde{\mathbf{A}}^{-1}\mathbf{M}_1^{\frac{1}{2}}$ is invertible. The matrix $\mathbf{B}$ (or $\hat{\mathbf{B}}$) is in general not invertible, but we assume that $\mathrm{rank}(\mathbf{B}) = n$, or equivalently, $\mathrm{rank}(\hat{\mathbf{B}}) = n$

holds. The *generalized inverse* (or Moore-Penrose inverse) $\hat{\mathbf{B}}^\dagger$ of $\hat{\mathbf{B}}$ is given by

$$\hat{\mathbf{B}}^\dagger = \hat{\mathbf{B}}^T(\hat{\mathbf{B}}\hat{\mathbf{B}}^T)^{-1} = \mathbf{M}_1^{-\frac{1}{2}}\mathbf{B}^T(\mathbf{B}\mathbf{M}_1^{-1}\mathbf{B}^T)^{-1}.$$

If on the right-hand side in (5.110) we formally apply inversion and use this generalized inverse we obtain

$$(\hat{\mathbf{B}}^\dagger)^T\mathbf{M}_1^{-\frac{1}{2}}\tilde{\mathbf{A}}\mathbf{M}_1^{-\frac{1}{2}}\hat{\mathbf{B}}^\dagger = (\mathbf{B}\mathbf{M}_1^{-1}\mathbf{B}^T)^{-1}\mathbf{B}\mathbf{M}_1^{-1}\tilde{\mathbf{A}}\mathbf{M}_1^{-1}\mathbf{B}^T(\mathbf{B}\mathbf{M}_1^{-1}\mathbf{B}^T)^{-1},$$

i.e., the preconditioner in (5.109).

5.5 Numerical experiments

We reconsider the example from Sect. 3.3.1, i. e., flow through a rectangular tube $\Omega = (0, L) \times (0, 1)^2$ with $L = 4$. The right-hand side is set to $\mathbf{g} = (g, 0, 0)$ with $g \in \mathbb{R}$. Note that the solution of the corresponding stationary Stokes problem is given by $\mathbf{u}_{\mathrm{St}}(x) := (Re \cdot g \cdot s(x_2, x_3), 0, 0)$, cf. (3.39). We consider the Navier-Stokes equation in dimensionless variables, with $Re = 1$. We prescribe the boundary conditions $\mathbf{u} = \mathbf{u}_{\mathrm{St}}$ for $x_1 = 0, L$ and $\mathbf{u} = 0$ on the remaining boundaries. After spatial discretization and applying the implicit Euler scheme for time discretization we obtain the following discrete problem,

$$\left\{\begin{array}{l} (\frac{1}{\Delta t}\mathbf{M} + \mathbf{A})\vec{\mathbf{u}} + \mathbf{C}(\vec{\mathbf{w}})\vec{\mathbf{u}} + \mathbf{B}^T\vec{\mathbf{p}} \\ \qquad = \vec{\mathbf{f}} + \frac{1}{\Delta t}\mathbf{M}\vec{\mathbf{u}}^{\mathrm{old}}, \\ \mathbf{B}\vec{\mathbf{u}} = 0. \end{array}\right. \tag{5.111}$$

In the numerical experiments we will study the iterative solution of the system (5.111) with $\mathbf{u}^{\mathrm{old}} = \mathbf{u}_{\mathrm{St}}$. In the following sections we consider three different cases: the Stokes case ($\vec{\mathbf{w}} = 0$), the Oseen case ($\vec{\mathbf{w}} = \mathbf{u}^{\mathrm{old}}$) and the Navier-Stokes case ($\vec{\mathbf{w}} = \vec{\mathbf{u}}$). The parameter g, which can be interpreted as some artificial gravitational constant and influences the size of the velocity $\mathbf{u}$ (i.e. the amount of convection), is varied in order to study the impact on the convergence rates of the iterative solver. The inexact Uzawa method is used in all of the three examples, using appropriate preconditioners $\mathbf{Q}_A$ and $\mathbf{Q}_S$. The starting vector is chosen as $(\vec{\mathbf{u}}_0, \vec{p}_0) = (\vec{\mathbf{u}}^{\mathrm{old}}, 0)$. The iteration is stopped after a reduction of the Euclidean norm of the starting residual by a factor of 10^6, i. e., $\|\mathbf{r}^k\| \leq 10^{-6}\|\mathbf{r}^0\|$. To measure the arithmetic costs, which are dominated by the application of the preconditioners, the number of evaluations of $\mathbf{Q}_A^{-1}$ is counted. Note that the number of $\mathbf{Q}_S^{-1}$ and $\mathbf{Q}_A^{-1}$ evaluations are almost the same, cf. Remark 5.2.11, so we will not report the numbers of $\mathbf{Q}_S^{-1}$ evaluations in the following.

5.5.1 Stokes case

Taking $\vec{\mathbf{w}} = 0$, the (discrete) convection term $\mathbf{C}(\vec{\mathbf{w}})\vec{\mathbf{u}}$ in (5.111) vanishes, yielding a generalized discrete Stokes problem. The preconditioners for the inexact Uzawa algorithm are chosen as follows. For $\mathbf{Q}_A$ one multigrid V-cycle iteration is applied to $\mathbf{A}$. The Schur complement is preconditioned by the Cahouet-Chabard preconditioner $\mathbf{Q}_S$ given in (5.108), but with $\mathbf{T}_h^{-1}$ replaced by the application of one multigrid V-cycle iteration to $\mathbf{T}_h$. Setting $g = 1$, computations are performed for time step sizes $\Delta t = 1/10, 1/20, 1/40$ and mesh sizes $h = 1/4, 1/8, 1/16$. The number of unknowns of the corresponding systems is given in Table 3.1 (cf. refinement levels 2, 3, 4, respectively). Table 5.1 gives the number of $\mathbf{Q}_A^{-1}$ evaluations for the different time step sizes and mesh sizes.

	Cahouet-Chabard			*BFBt*		
	$h = 1/4$	$h = 1/8$	$h = 1/16$	$h = 1/4$	$h = 1/8$	$h = 1/16$
$\Delta t = 1/10$	44	47	46	36	58	89
$\Delta t = 1/20$	42	49	45	32	56	101
$\Delta t = 1/40$	35	40	45	27	50	86

Table 5.1. Stokes case: number of $\mathbf{Q}_A^{-1}$ evaluations for different h, Δt.

These results indicate good robustness of the rate of convergence of the iterative solver w. r. t. variation in h and Δt. Even though the *BFBt* preconditioner (5.109) is originally designed for Oseen problems, it can also be applied to the Stokes case. We repeated the computations, but this time used the *BFBt* preconditioner as Schur complement preconditioner $\mathbf{Q}_S$, where for the solution of the arising systems a CG method is applied such that Euclidean norm of the residual is reduced by a factor of 10. We observe that the *BFBt* preconditioner is not robust w. r. t. variation of h, as the number of $\mathbf{Q}_A^{-1}$ evaluations increases significantly for smaller h, cf. Table 5.1.

	Cahouet-Chabard			*BFBt*		
	$h = 1/4$	$h = 1/8$	$h = 1/16$	$h = 1/4$	$h = 1/8$	$h = 1/16$
$\Delta t = 1/10$	86	100	98	36	58	108
$\Delta t = 1/20$	93	113	113	34	60	119
$\Delta t = 1/40$	88	120	112	28	54	89

Table 5.2. Stokes case, mixed boundary conditions: number of $\mathbf{Q}_A^{-1}$ evaluations for different h, Δt.

We repeat the computations for slightly different boundary conditions, namely homogeneous natural boundary conditions $\boldsymbol{\sigma}\mathbf{n} = 0$ for $x_1 = L$ (outflow boundary conditions) while on the remaining boundaries the boundary conditions are chosen as in the previous case. The results are given in Table 5.2. Note that for the Cahouet-Chabard preconditioner we still observe a robust behavior, although the number of $\mathbf{Q}_A^{-1}$ evaluations is almost doubled. The *BFBt* preconditioner shows almost the same behavior as before. We conclude that for the Cahouet-Chabard preconditioner there is a significant dependence of the preconditioning quality on the type of boundary conditions used.

5.5.2 Oseen case

Choosing $\vec{\mathbf{w}} = \vec{\mathbf{u}}^{\mathrm{old}} = (g \cdot s(x_2, x_3), 0, 0)$ in (5.111), a discrete Oseen problem is obtained, where the strength of the convective part can be controlled by the parameter g. We compare the cases $g = 1$ and $g = 10^3$ for two different choices for the Schur complement preconditioner $\mathbf{Q}_S$, namely the Cahouet-Chabard and the $BFBt$ preconditioner also used in the previous section. The matrix $\tilde{\mathbf{A}}$ and the Schur complement are both nonsymmetric. For the solution of the Schur complement system in (5.30), instead of a preconditioned CG solver a preconditioned GCR iteration is used. For $\mathbf{Q}_A$ a Jacobi-preconditioned BiCGStab iteration with initial vector equal to zero is applied, until the Euclidean norm of the residual is reduced by a factor of 100. To obtain such a residual reduction, the BiCGStab solver requires about 20 iterations for $g = 1$ and about 50 iterations for $g = 10^3$. Taking one multigrid V-cycle for $\mathbf{Q}_A$ (as in the Stokes case) would be sufficient for the problem with mild convection ($g = 1$) yielding similar results as in Table 5.1, but the method either fails to converge or has very slow convergence for the convection-dominated case ($g = 10^3$). The reason for this poor multigrid performance comes from the fact that we used standard multigrid components (smoother, prolongation) that are appropriate for diffusion dominated problems but not suitable for problems with strong convection. This poor behavior can be avoided by using modified components, cf. the discussion in Sect. 5.6.

	Cahouet-Chabard			$BFBt$		
	$h = 1/4$	$h = 1/8$	$h = 1/16$	$h = 1/4$	$h = 1/8$	$h = 1/16$
$\Delta t = 1/10$	43	40	38	25	43	81
$\Delta t = 1/20$	40	33	37	25	41	71
$\Delta t = 1/40$	39	36	42	23	36	63

Table 5.3. Oseen case, $g = 1$: number of $\mathbf{Q}_A^{-1}$ evaluations for different h, Δt.

	Cahouet-Chabard			$BFBt$		
	$h = 1/4$	$h = 1/8$	$h = 1/16$	$h = 1/4$	$h = 1/8$	$h = 1/16$
$\Delta t = 1/10$	85	99	107	51	101	171
$\Delta t = 1/20$	76	87	107	45	88	166
$\Delta t = 1/40$	58	76	97	45	86	146

Table 5.4. Oseen case, $g = 1000$: number of $\mathbf{Q}_A^{-1}$ evaluations for different h, Δt.

Tables 5.3 and 5.4 show the number of $\mathbf{Q}_A^{-1}$ evaluations for the two preconditioners and two different choices of g. As expected, the convection-dominated problem ($g = 1000$) results in higher computational costs. The numerical results indicate that again the $BFBt$ preconditioner deteriorates for smaller mesh sizes h. For the Cahouet-Chabard preconditioner and $g = 1$ we observe robustness w.r.t. h as in the Stokes case, whereas for $g = 1000$ there is a mild increase of the number of $\mathbf{Q}_A^{-1}$ evaluations for decreasing h. For the case $g = 1000$, $h = 1/8$, both Schur complement preconditioners lead to comparable overall computing times, whereas for a finer grid size $h = 1/16$ the Cahouet-Chabard preconditioner outperforms the $BFBt$ preconditioner.

5.5.3 Navier-Stokes case

We now turn to the Navier-Stokes case by setting $\vec{\mathbf{w}} = \vec{\mathbf{u}}$ in (5.111). For a fixed mesh size $h = 1/16$ and time step size $\Delta t = 1/40$, the parameter g is varied within a range of $1 \dots 10^3$. The adaptive defect correction method (cf. Algorithm 5.1.1) is used for linearization of the nonlinear problem. The nonlinear iteration is stopped if $\|\mathbf{r}^k\| \le 10^{-6}\|\mathbf{r}^0\|$ holds. The linearized problems are of Oseen type and are solved as outlined in the previous section, reducing the Euclidean norm of the residual by a factor of 10 in each linearization step. Table 5.5 shows the number of fixed point iterations and the accumulated number of $\mathbf{Q}_A^{-1}$ evaluations for different values of g and for the two Schur complement preconditioners.

g	1	10	100	1000
Cahouet-Chabard	4 (37)	4 (37)	6 (63)	10 (120)
$BFBt$	5 (63)	5 (75)	6 (84)	9 (174)

Table 5.5. Navier-Stokes case: number of fixed point iterations (and $\mathbf{Q}_A^{-1}$ evaluations) for $h = 1/8$, $\Delta t = 1/10$ and different values of g.

As expected, the number of fixed point iterations increases with increasing g due to the stronger nonlinearity of the problem. Comparing both Schur complement preconditioners, the Cahouet-Chabard preconditioner turns out to be more efficient than the $BFBt$ preconditioner for all choices of g. In recent literature there have been developed (improved) variants of the $BFBt$ preconditioner that might perform better, cf. [103].

5.6 Discussion and additional references

There is an extensive literature on efficient iterative solvers for large sparse (linear) systems. We only address some issues that are directly related to the material treated in this chapter.

In recent years there has been much progress concerning the analysis of iterative methods and preconditioners for *symmetric* saddle point problems, we refer to the overview paper [31], which contains many references on this subject. Another interesting survey paper is [175]. In Sect. 5.4.3 we treated the issue of Schur complement preconditioning. For the generalized Stokes problem (which is symmetric) the Cahouet-Chabard preconditioner is an efficient one. For the (non-symmetric) Oseen problem, however, an "optimal" Schur complement preconditioner, i.e. one with a good approximation quality uniformly w.r.t. the mesh size, the time step and the Reynolds number, is not known yet. Several Oseen Schur complement preconditioners, that perform well in certain parameter ranges, have been studied in the literature. One class of such methods consists of variants of the $BFBt$ preconditioner (cf. Sect. 5.4.3). This preconditioner belongs to the class of so-called *approximate commutator preconditioners*, which consists of two subclasses namely of the pressure convection-diffusion preconditioners (PCD) and of the least-squares commutator methods (LSC). The $BFBt$ preconditioner belongs to the latter subclass. An overview of the main ideas underlying these approximate commutator preconditioners can be found in [102]. Recent work on these preconditioners, resulting in improvements of the $BFBt$ method is presented in [101, 103].

Another approach for preconditioning the discrete Oseen problem is based on the *augmented Lagrangian technique* (AL). This technique is based on a rather general idea, and is used for other applications in [109]. We briefly outline the main idea of the AL applied to the Oseen problem. Consider the Oseen saddle point system

$$\begin{pmatrix} \tilde{\mathbf{A}} & \mathbf{B}^T \\ \mathbf{B} & 0 \end{pmatrix} \begin{pmatrix} \mathbf{x} \\ \mathbf{y} \end{pmatrix} = \begin{pmatrix} \mathbf{b} \\ 0 \end{pmatrix}, \tag{5.112}$$

and assume that $\mathbf{B}$ has full row rank n. Let $\mathbf{M}_p$ be the mass matrix in the pressure space. Note that $\mathbf{B}\mathbf{x} = 0$ and thus the solution of (5.112) is also the solution of the saddle point system

$$\begin{pmatrix} \tilde{\mathbf{A}} + \gamma \mathbf{B}^T \mathbf{M}_p^{-1} \mathbf{B} & \mathbf{B}^T \\ \mathbf{B} & 0 \end{pmatrix} \begin{pmatrix} \mathbf{x} \\ \mathbf{y} \end{pmatrix} = \begin{pmatrix} \mathbf{b} \\ 0 \end{pmatrix}, \tag{5.113}$$

with a given parameter $\gamma > 0$. The Schur complement of the matrix in (5.112) is given by $\mathbf{S} = \mathbf{B}\tilde{\mathbf{A}}^{-1}\mathbf{B}^T$, whereas the Schur complement of the matrix in (5.113) is given by $\hat{\mathbf{S}}_\gamma = \mathbf{B}(\tilde{\mathbf{A}} + \gamma \mathbf{B}^T \mathbf{M}_p^{-1} \mathbf{B})^{-1} \mathbf{B}^T$. For the inverse of this Schur complement we have the identity (cf. [32])

$$\hat{\mathbf{S}}_\gamma^{-1} = \mathbf{S}^{-1} + \gamma \mathbf{M}_p^{-1}.$$

Hence, for γ sufficiently large the problem of Schur complement preconditioning for the matrix in (5.113) is easy to solve: one can take $\mathbf{M}_p$ as a preconditioner for $\hat{\mathbf{S}}_\gamma$. There is, however, a price to pay, caused by the fact that for a block preconditioner of the matrix in (5.113) one needs, besides a preconditioner for $\hat{\mathbf{S}}_\gamma$, also a preconditioner for the block $\hat{\mathbf{A}}_\gamma := \tilde{\mathbf{A}} + \gamma \mathbf{B}^T \mathbf{M}_p^{-1} \mathbf{B}$. The matrix $\mathbf{B}^T \mathbf{M}_p^{-1} \mathbf{B}$ has a large null space, and thus for γ large the matrix $\hat{\mathbf{A}}_\gamma$ is very ill conditioned and the problem of finding a good preconditioner for $\hat{\mathbf{A}}_\gamma$ is (much) more difficult than that of finding a good preconditioner for $\tilde{\mathbf{A}}$. Such augmented Lagrangian based preconditioners for the Oseen problem are studied in [32, 33].

In Sect. 5.4.2 we treated the multigrid method as a preconditioner for $\tilde{\mathbf{A}}$. In that treatment we restricted ourselves to the diffusion-dominated case (small Reynolds number), which corresponds to a matrix $\tilde{\mathbf{A}}$ that is "almost symmetric". In case of strong convection (large Reynolds number) the standard multigrid technique as explained in Sect. 5.4.2 in general does not work properly. For convection-dominated problems, special efficient multigrid techniques have been developed in the literature. Key ingredients are the use of special (often called "robust") smoothers and/or problem adapted prolongations and restrictions. We refer to the literature for more information, e.g. [133, 191, 242]. Another multigrid approach that often is very efficient in case of strong convection is the so-called *algebraic multigrid method* (AMG). A treatment of this method can be found in, e.g., [159, 242]. In Sect. 5.4.2 we only considered multigrid for *scalar* elliptic problems. In Remark 5.4.26 it is explained how this method can be used for preconditioning the $\tilde{\mathbf{A}}$ block in the saddle point system. Multigrid methods can also be applied directly, either as solver or as preconditioner, to the whole saddle point system. An early contribution in this field of so-called *coupled multigrid* is [254]; further studies of this approach are [151, 150, 164, 251].

In this monograph we do not treat the topic of *parallelization of iterative solvers.* In particular for three-dimensional two-phase flow problems (coupled with mass or surfactant transport) in many cases it may be necessary to use a parallel code in order to obtain acceptable computing times.

An issue that we did not address at all, but is important for an efficient numerical simulation, is how to choose the tolerances in the iterative methods. In the type of solvers that we use in our one- or two-phase flow simulations there are many tolerance parameters that have to be set. Given the spatial mesh size resolution, the time step Δt has to be chosen. Then, in case of a two-phase flow problem, in each time step there is an iteration that decouples the movement of the interface from the fluid dynamics unknowns $(\mathbf{u}, p)$, for which one needs a stopping criterion. Then in the Navier-Stokes subproblem one has a linearization (e.g. adaptive defect correction) and in each linearization step a discrete Oseen problem is solved iteratively. If one uses in the preconditioner an iterative method, then yet another appropriate stopping criterion is needed. Thus there are many (say between 6 and 9) tolerance parameters that have to chosen. Clearly, the stopping criteria are not independent. If, for example, one wants to have a very high time discretization accuracy (very small time step) then one needs a more accurate solution of the nonlinear discrete problem per time step. If all tolerance parameters are set such that one has very high accuracy in all iterative methods, then this can easily result in a very inefficient overall solver. If on the other hand one (or more) of the accuracies are "too low" this can lead to inaccurate results or a breakdown of the simulation. We consider this problem of parameter tuning to be very relevant and largely unsolved. We are not aware of literature in which, for an interesting class of CFD problems, this topic is systematically studied.

Part II

Two-phase incompressible flows

6

Mathematical model

6.1 Introduction

We recall the Navier-Stokes model (1.19)-(1.21) for two-phase incompressible flows:

$$\begin{cases} \rho_i\left(\frac{\partial \mathbf{u}}{\partial t} + (\mathbf{u}\cdot\nabla)\mathbf{u}\right) = -\nabla p + \rho_i \mathbf{g} + \operatorname{div}(\mu_i \mathbf{D}(\mathbf{u})) \\ \operatorname{div}\mathbf{u} = 0 \end{cases} \quad \text{in } \Omega_i, \tag{6.1}$$

$$[\boldsymbol{\sigma}\mathbf{n}]_\Gamma = -\tau\kappa\mathbf{n}, \quad [\mathbf{u}] = 0 \quad \text{on } \Gamma, \tag{6.2}$$

$$V_\Gamma = \mathbf{u}\cdot\mathbf{n} \quad \text{on } \Gamma. \tag{6.3}$$

We recall the definition of the stress tensor $\boldsymbol{\sigma} := -p\mathbf{I} + \mu\mathbf{D}(\mathbf{u})$ and the deformation tensor $\mathbf{D}(\mathbf{u}) := \nabla\mathbf{u} + \nabla\mathbf{u}^T$. For the velocity we use Dirichlet boundary conditions $\mathbf{u} = \mathbf{u}_D$ on $\partial\Omega_D$ and natural boundary conditions on $\partial\Omega \setminus \partial\Omega_D$. The initial condition for the velocity is $\mathbf{u}(x,0) = \mathbf{u}_0(x),\ x \in (\Omega_1 \cup \Omega_2)(0)$, with a given function $\mathbf{u}_0 : \Omega \to \mathbb{R}^3$. Furthermore, we assume that the initial interface $\Gamma(0)$ is given. Note that the model (6.1)-(6.3) is *not* in dimensionless form.

Remark 6.1.1 We address the formulation of the two-phase flow model (6.1)-(6.3) in dimensionless variables. In this model we have two Navier-Stokes equations in the two subdomains Ω_i, $i = 1, 2$. Therefore it is an option to consider a subdomain dependent scaling. For the dimensionless variables we use the same notation as in the derivation of the one-phase dimensionless Navier-Stokes equation in Sect. 2.1: $\bar{x},\ \bar{t},\ \bar{\mathbf{u}},\ \bar{p}$. It does not make sense to use different spatial scales in the two subdomains. Hence we choose *one* typical length scale, denoted by L. We want to maintain the continuity property $[\mathbf{u}]_\Gamma = 0$ also in the transformed variables and thus we choose *one* typical velocity size U. In the pressure rescaling we allow a subdomain dependent rescaling with $\tilde{\rho}_i > 0$

a given constant in Ω_i (with unit kg/m^3). The corresponding piecewise constant function on Ω is denoted by $\tilde{\rho}$. Based on this, the dimensionless variables are given by

$$\bar{x} = \frac{1}{L}x, \quad \bar{t} = \frac{U}{L}t, \quad \bar{\mathbf{u}}(\bar{x}, \bar{t}) = \frac{\mathbf{u}(x,t)}{U}, \quad \bar{p}_i(\bar{x}, \bar{t}) = \frac{p(x,t)}{\tilde{\rho}_i U^2}, \ i = 1, 2.$$

Furthermore, $\bar{\Omega} := \frac{1}{L}\Omega := \{\, \bar{x} \in \mathbb{R}^3 : L\bar{x} \in \Omega \,\}$ and $\bar{\mathbf{g}} := \frac{L}{U^2}\mathbf{g}$. The partial differential equations in (6.1) can be written in these dimensionless quantities as follows, where differential operators w.r.t. $\bar{x}_i$ and $\bar{t}$ are denoted with a $^-$ (for example: $\bar{\nabla}$):

$$\begin{aligned} \frac{\rho_i}{\tilde{\rho}_i}\Big(\frac{\partial \bar{\mathbf{u}}}{\partial \bar{t}} + (\bar{\mathbf{u}} \cdot \bar{\nabla})\bar{\mathbf{u}}\Big) &= -\bar{\nabla}\bar{p} + \frac{\rho_i}{\tilde{\rho}_i}\bar{\mathbf{g}} + \overline{\mathrm{div}}\big(\frac{1}{Re_i}\bar{\mathbf{D}}(\bar{\mathbf{u}})\big) \\ &= \overline{\mathrm{div}}(\bar{\boldsymbol{\sigma}}) + \frac{\rho_i}{\tilde{\rho}_i}\bar{\mathbf{g}} \quad \text{in} \quad \bar{\Omega}_i, \\ \overline{\mathrm{div}}\, \bar{\mathbf{u}} &= 0 \quad \text{in} \quad \bar{\Omega}_i, \end{aligned} \tag{6.4}$$

with the dimensionless Reynolds numbers $Re_i = \frac{\tilde{\rho}_i LU}{\mu_i}$, $i = 1, 2$, and

$$\bar{\boldsymbol{\sigma}} = -\bar{p}\mathbf{I} + \frac{1}{Re_i}\big(\bar{\nabla}\bar{\mathbf{u}} + \bar{\nabla}\bar{\mathbf{u}}^T\big).$$

Considering this rescaled problem it is tempting to choose $\tilde{\rho}_i = \rho_i$, since this leads to a simplification. In particular one then has a constant 1 in front of the material derivative. However, it is also necessary to rescale the interface conditions in (6.2)-(6.3). The conditions $[\mathbf{u}] = 0$, $V_\Gamma = \mathbf{u} \cdot \mathbf{n}$ transform to

$$[\bar{\mathbf{u}}] = 0, \quad \bar{V}_\Gamma = \bar{\mathbf{u}} \cdot \bar{\mathbf{n}}, \tag{6.5}$$

with $\bar{\mathbf{n}}(\bar{x}) = \mathbf{n}(x)$. The momentum balance condition $[\boldsymbol{\sigma}\mathbf{n}] = -\tau\kappa\mathbf{n}$ takes the form

$$[\tilde{\rho}U^2\bar{\boldsymbol{\sigma}}\bar{\mathbf{n}}] = -\frac{\tau}{L}\bar{\kappa}\bar{\mathbf{n}},$$

with $\bar{\kappa} = \overline{\mathrm{div}}_\Gamma \bar{\mathbf{n}}$, the curvature in transformed variables. To be able to write this momentum balance condition in the usual form $[\bar{\boldsymbol{\sigma}}\bar{\mathbf{n}}] = \alpha\,\bar{\kappa}\bar{\mathbf{n}}$, $\alpha \in \mathbb{R}$, the scaling function $\tilde{\rho}$ has to be taken constant across Γ, and thus $\tilde{\rho}_1 = \tilde{\rho}_2 = \tilde{\rho}$. Therefore, in the transformation to dimensionless variables one normally takes a *constant* density scaling factor (e.g., $\tilde{\rho} = \frac{1}{2}(\rho_1 + \rho_2)$) and then the momentum interface condition is given by

$$[\bar{\boldsymbol{\sigma}}\bar{\mathbf{n}}] = -\frac{1}{We}\bar{\kappa}\bar{\mathbf{n}}, \quad We := \frac{\tilde{\rho}U^2 L}{\tau}. \tag{6.6}$$

The dimensionless so-called *Weber number* is a measure for the relative size of inertial and surface tension forces. The model in dimensionless variables is given by (6.4), (6.5), (6.6). Note that similar to (6.1), in (6.4) one has a piecewise constant density $\rho/\tilde{\rho}$ and a piecewise constant viscosity $1/Re_i$. This is an important difference compared to the dimensionless one-phase Navier-Stokes problem in (2.5).

We discuss a weak formulation of the Navier-Stokes equations and the interface conditions in (6.1)-(6.2). We consider the model in physical dimensions since there is no significant advantage if one instead uses the model in dimensionless variables (6.4), (6.5), (6.6).

We use the Sobolev spaces

$$\begin{aligned}
\mathbf{V} &:= H^1(\Omega)^3, \\
\mathbf{V}_0 &:= \{ \mathbf{v} \in \mathbf{V} : \mathbf{v} = 0 \text{ on } \partial\Omega_D \}, \\
\mathbf{V}_D &:= \{ \mathbf{v} \in \mathbf{V} : \mathbf{v} = \mathbf{u}_D \text{ on } \partial\Omega_D \}, \\
Q &:= L_0^2(\Omega) = \left\{ q \in L^2(\Omega) : \int_\Omega q \, dx = 0 \right\},
\end{aligned}$$

and define the bilinear forms

$$\begin{aligned}
m : \mathbf{V} \times \mathbf{V} \to \mathbb{R} : \quad & m(\mathbf{u}, \mathbf{v}) := \int_\Omega \rho \, \mathbf{u}\mathbf{v} \, dx, \\
a : \mathbf{V} \times \mathbf{V} \to \mathbb{R} : \quad & a(\mathbf{u}, \mathbf{v}) := \frac{1}{2} \int_\Omega \mu \, \mathrm{tr} \left(\mathbf{D}(\mathbf{u}) \mathbf{D}(\mathbf{v}) \right) dx, \qquad (6.7) \\
& = \frac{1}{2} \int_\Omega \mu \sum_{i,j=1}^{3} \left(\mathbf{D}(\mathbf{u}) \right)_{ij} \left(\mathbf{D}(\mathbf{v}) \right)_{ij} dx \\
b : \mathbf{V} \times Q \to \mathbb{R} : \quad & b(\mathbf{v}, q) := -\int_\Omega q \, \mathrm{div} \, \mathbf{v} \, dx,
\end{aligned}$$

and the trilinear form

$$c : \mathbf{V} \times \mathbf{V} \times \mathbf{V} \to \mathbb{R} : \quad c(\mathbf{u}; \mathbf{v}, \mathbf{w}) := \int_\Omega \rho (\mathbf{u} \cdot \nabla \mathbf{v}) \mathbf{w} \, dx.$$

For the weak formulation of the interface condition $[\boldsymbol{\sigma}\mathbf{n}]_\Gamma = -\tau\kappa\mathbf{n}$ in (6.2) we introduce the linear functional

$$f_\Gamma : \mathbf{V} \to \mathbb{R} : \quad f_\Gamma(\mathbf{v}) := -\int_\Gamma \tau\kappa\mathbf{n} \cdot \mathbf{v} \, ds. \qquad (6.8)$$

If the curvature κ is bounded on Γ we have

$$|f_\Gamma(\mathbf{v})| \leq c \, \|\kappa\|_{L^\infty(\Gamma)} \|\mathbf{v}\|_{L^2(\Gamma)} \leq \tilde{c} \|\mathbf{v}\|_1 \quad \text{for all } \mathbf{v} \in \mathbf{V},$$

where in the last inequality we used a trace theorem on Γ. Hence we get that f_Γ is a *bounded* linear functional on $\mathbf{V}$, i.e., $f_\Gamma \in \mathbf{V}'$.

Remark 6.1.2 We restrict ourselves to the model with a constant surface tension coefficient τ. More general models with an interface momentum balance of the form $[\boldsymbol{\sigma}\mathbf{n}] = \mathrm{div}_\Gamma(\boldsymbol{\sigma}_\Gamma)$ are discussed in Sect. 7.6.1. For such an interface condition the surface tension functional generalizes to $f_\Gamma(\mathbf{v}) = \int_\Gamma \mathrm{div}_\Gamma(\boldsymbol{\sigma}_\Gamma) \cdot \mathbf{v} \, ds$.

A weak formulation of the Navier-Stokes equations and the coupling conditions in (6.2) is as follows:

Find $\mathbf{u}(t) = \mathbf{u}(\cdot,t) \in \mathbf{V}_D$, $p(t) = p(\cdot,t) \in Q$ such that for almost all $t \in [0,T]$

$$m(\frac{\partial \mathbf{u}}{\partial t}, \mathbf{v}) + c(\mathbf{u};\mathbf{u},\mathbf{v}) \\ +a(\mathbf{u},\mathbf{v}) + b(\mathbf{v},p) = (\rho\mathbf{g},\mathbf{v})_{L^2} + f_\Gamma(\mathbf{v}) \quad \text{for all } \mathbf{v} \in \mathbf{V}_0, \tag{6.9}$$

$$b(\mathbf{u},q) = 0 \quad \text{for all } q \in Q, \tag{6.10}$$

and initial condition $\mathbf{u}(0) = \mathbf{u}_0$ in Ω.

The time derivative has to be taken in a suitable weak sense, cf. below. Note that in this model we have the weak formulation of *one* Navier-Stokes equation in *the whole domain* Ω. The *localized force term* f_Γ originates from the first interface coupling condition in (6.2). Also note that in general the bilinear forms $m(\cdot,\cdot)$, $a(\cdot,\cdot)$ and the trilinear form $c(\cdot\,;\cdot,\cdot)$ depend on t, due to the fact that we have $\Omega_i = \Omega_i(t)$ and thus the density and viscosity coefficients (which are piecewise constant in Ω_i) are time dependent.

The surface tension functional f_Γ will play an important role in the remainder of this monograph, both in the analysis of the models considered and in the numerical methods that will be treated. This functional has other useful representations, for example those given in Lemma 14.1.2.

Remark 6.1.3 The idea to replace the first interface coupling condition in (6.2) by a localized force term in the momentum equation was introduced in [47]. In the (engineering) literature this is known as the CSF ("Continuum Surface Force") approach. In [47] and in most other papers in which such a localized surface tension force is used, this force at the interface is *approximated* by some *volume* force (hence, *continuum* surface force). We briefly explain the main idea, for details we refer to [47, 64]. Take $x \in \Gamma$ and let $U \subset \Omega$ be a (small) neighborhood of x. Define $\gamma := \Gamma \cap U$. Let $\mathbf{g} : \gamma \to \mathbb{R}^3$ be a smooth vector function ("force at the interface"), for example $\mathbf{g}(x) = \tau\kappa(x)\mathbf{n}_\Gamma(x)$, and $\tilde{\mathbf{g}} : U \to \mathbb{R}^3$ a suitable smooth extension of $\mathbf{g}$. Furthermore, let d_Γ be the signed distance function: $d_\Gamma(x) = \text{dist}(\Gamma, x)$ for $x \in U \cap \Omega_2$, $d_\Gamma(x) = -\text{dist}(\Gamma, x)$ for $x \in U \cap \Omega_1$. For the "force acting on γ" we have:

$$\int_\gamma \mathbf{g}(s)\,ds = \lim_{\epsilon\downarrow 0} \int_U \delta_\epsilon(d_\Gamma(x))\tilde{\mathbf{g}}(x)\,dx,$$

with a one-dimensional smoothed Dirac delta function δ_ϵ, i.e. for $\xi > 0$ we have $\lim_{\epsilon\downarrow 0} \int_{-\xi}^{\xi} \delta_\epsilon(s)h(s)\,ds = h(0)$ for smooth functions h. Then in the spirit of the derivation of the Navier-Stokes equations in the *strong* formulation (as in (6.1)), based on conservation laws and forces on "arbitrary" neighborhoods

U, the volume force at $x \in \gamma$ is taken as $\delta_\epsilon(d_\Gamma(x))\tilde{\mathbf{g}}(x)$. In this approach one has freedom in choosing the extension $\tilde{\mathbf{g}}$ of $\mathbf{g}$ and in choosing the regularization of the Dirac delta function. It is shown in [241, 104] that the latter issue is nontrivial: seemingly natural regularizations, which work well in 1D, may lead to large errors in higher dimensions. In [64] an extension $\tilde{\mathbf{g}}$ of $\mathbf{g}$ based on the level set method is introduced, cf. Remark 6.2.4 below. All Dirac delta function regularizations lead to functions δ_ϵ with unbounded derivatives for $\epsilon \downarrow 0$, and thus such a regularization requires a high mesh resolution close to Γ.

In the *weak* formulation (6.9) this regularization (and extension) issue does not occur. The localized surface tension force is represented as a well-defined functional $f_\Gamma \in \mathbf{V}'$ (provided the curvature is bounded). In this sense the weak formulation is better suited for describing surface tension forces than the strong formulation.

The following lemma indicates that (6.9)-(6.10) is a correct weak formulation for the Navier-Stokes problems in the two subdomains with coupling conditions as in (6.2).

Lemma 6.1.4 *Assume that* (6.1)-(6.2) *has a solution* $(\mathbf{u}, p)$ *with* $\mathbf{u}_{|\partial\Omega_D} = \mathbf{u}_D$, $\int_\Omega p\,dx = 0$, $\Gamma(t)$ *is sufficiently smooth and* $\mathbf{u}$, p *are sufficiently smooth:*

$$\mathbf{u} \in C^1\big(0,T; C^2(\overline{\Omega}_i)^3\big), \quad p \in C\big(0,T; C^1(\overline{\Omega}_i)\big), \quad i = 1,2.$$

Then $(\mathbf{u}, p)$ *solves* (6.9)-(6.10).

Proof. Due to the smoothness assumption on $\mathbf{u}$ in the subdomains Ω_i and $[\mathbf{u}]_\Gamma = 0$ we have $\mathbf{u} \in \mathbf{V}_D$. Furthermore, $p(\cdot, t) \in Q$ holds. From $\operatorname{div} \mathbf{u} = 0$ it follows that (6.10) holds. We now consider the variational equation in (6.9):

$$\begin{aligned} &\int_\Omega \rho \frac{\partial \mathbf{u}}{\partial t} \mathbf{v}\,dx + \int_\Omega \rho(\mathbf{u} \cdot \nabla \mathbf{u})\mathbf{v}\,dx + \frac{1}{2}\int_\Omega \mu \operatorname{tr}\big(\mathbf{D}(\mathbf{u})\mathbf{D}(\mathbf{v})\big)\,dx \\ &- \int_\Omega p \operatorname{div} \mathbf{v}\,dx = \int_\Omega \rho \mathbf{g} \cdot \mathbf{v}\,dx - \int_\Gamma \tau\kappa \mathbf{n} \cdot \mathbf{v}\,ds. \end{aligned} \tag{6.11}$$

We need the following partial integration rules, which hold for functions $q : U \to \mathbb{R}$ and $\mathbf{w}, \mathbf{v} : U \to \mathbb{R}^3$ that are sufficiently smooth on $U \subset \Omega$:

$$- \int_U q \operatorname{div} \mathbf{w}\,dx = \int_U \nabla q \cdot \mathbf{w}\,dx - \int_{\partial U} q\,\mathbf{w} \cdot \mathbf{n}\,ds,$$

$$\frac{1}{2}\int_U \operatorname{tr}\big(\mathbf{D}(\mathbf{w})\mathbf{D}(\mathbf{v})\big)\,dx = -\int_U \big(\operatorname{div} \mathbf{D}(\mathbf{w})\big) \cdot \mathbf{v}\,dx + \int_{\partial U} \big(\mathbf{D}(\mathbf{w})\,\mathbf{n}\big) \cdot \mathbf{v}\,ds.$$

In the equation in (6.11) we take a test function $\mathbf{v} \in C_0^\infty(\Omega)^3$, split the integrals over Ω into integrals over Ω_i, $i = 1,2$, and use the partial integration rules (with $U = \Omega_i$). Thus (6.11) can be rewritten as

$$\sum_{i=1}^{2}\int_{\Omega_i}\Big(\rho_i\big(\frac{\partial \mathbf{u}}{\partial t}+\mathbf{u}\cdot\nabla\mathbf{u}\big)-\operatorname{div}\big(\mu_i\mathbf{D}(\mathbf{u})\big)+\nabla p\Big)\cdot\mathbf{v}\,dx$$
$$=\sum_{i=1}^{2}\int_{\Omega_i}\rho_i\mathbf{g}\cdot\mathbf{v}\,dx-\int_{\Gamma}[\mu\mathbf{D}(\mathbf{u})\mathbf{n}-p\mathbf{n}]_{\Gamma}\cdot\mathbf{v}\,ds-\int_{\Gamma}\tau\kappa\mathbf{n}\cdot\mathbf{v}\,ds. \tag{6.12}$$

Due to the interface condition $[\boldsymbol{\sigma}\mathbf{n}]_\Gamma=-\tau\kappa\mathbf{n}$ the last two terms on the right-hand side cancel. From (6.1) it then follows that (6.12) and thus (6.11) holds. Since $C_0^\infty(\Omega)^3$ is dense in $\mathbf{V}_0$ we conclude that (6.9) is satisfied. □

Remark 6.1.5 From the proof above one can infer why we use the bilinear form $a(\cdot,\cdot)$ as in (6.7) and not the simpler one $\hat{a}(\mathbf{u},\mathbf{v})=\int_\Omega\mu\nabla\mathbf{u}\cdot\nabla\mathbf{v}\,dx$, which is used in the weak formulation of the one-phase Navier-Stokes equations. Partial integration on the subdomains applied to this bilinear form results in

$$\sum_{i=1}^{2}\int_{\Omega_i}\mu_i\nabla\mathbf{u}\cdot\nabla\mathbf{v}\,dx=-\sum_{i=1}^{2}\int_{\Omega_i}\mu_i\Delta\mathbf{u}\cdot\mathbf{v}\,dx+\int_{\Gamma}[\mu(\nabla\mathbf{u})^T\mathbf{n}]_{\Gamma}\cdot\mathbf{v}\,ds,$$

and thus in (6.12) instead of the term $\int_\Gamma[\mu\mathbf{D}(\mathbf{u})\mathbf{n}-p\mathbf{n}]_\Gamma\cdot\mathbf{v}\,ds=\int_\Gamma[\boldsymbol{\sigma}\mathbf{n}]_\Gamma\cdot\mathbf{v}\,ds$ one would obtain $\int_\Gamma[\mu(\nabla\mathbf{u})^T\mathbf{n}-p\mathbf{n}]_\Gamma\cdot\mathbf{v}\,ds$, which is *not* consistent with the interface condition $[\boldsymbol{\sigma}\mathbf{n}]_\Gamma=-\tau\kappa\mathbf{n}$ in (6.2).

For the above variational Navier-Stokes problem with the surface tension functional f_Γ one can derive the following energy estimate.

Lemma 6.1.6 *Consider the variational problem (6.9)-(6.10), with ρ constant, say $\rho=1$, with $\mathbf{u}_D=0$ (homogeneous Dirichlet boundary condition) and $\mathbf{g}=0$ (no external forces). Assume that for $0\le t\le T$ the interface $\Gamma(t)$ is a sufficiently smooth compact manifold. Let $(\mathbf{u},p)$ be a solution of (6.9)-(6.10) with $\mathbf{u}\in L^2(0,T;\mathbf{V}_0)$. Then the following holds:*

$$\frac{1}{2}\|\mathbf{u}(T)\|_{L^2}^2+\int_0^T a(\mathbf{u}(t),\mathbf{u}(t))\,dt+\tau\,\mathrm{meas}_2\big(\Gamma(T)\big)$$
$$=\frac{1}{2}\|\mathbf{u}_0\|_{L^2}^2+\tau\,\mathrm{meas}_2\big(\Gamma(0)\big). \tag{6.13}$$

Proof. We take $\mathbf{v}=\mathbf{u}$ in (6.9). Using partial integration, $\rho=1$, $\mathbf{u}_{|\partial\Omega}=0$ we get $c(\mathbf{u};\mathbf{u},\mathbf{u})=0$. Furthermore, due to $\operatorname{div}\mathbf{u}=0$ we have $b(\mathbf{u},p)=0$. Thus we obtain

$$\int_\Omega\frac{\partial\mathbf{u}}{\partial t}\mathbf{u}\,dx+a(\mathbf{u},\mathbf{u})=-\tau\int_\Gamma\kappa\,\mathbf{n}\cdot\mathbf{u}\,ds.$$

Integration over $t\in[0,T]$ and applying partial integration in t results in

$$\frac{1}{2}\|\mathbf{u}(T)\|_{L^2}^2+\int_0^T a(\mathbf{u}(t),\mathbf{u}(t))\,dt=\frac{1}{2}\|\mathbf{u}_0\|_{L^2}^2-\tau\int_0^T\int_\Gamma\kappa\,\mathbf{n}\cdot\mathbf{u}\,ds\,dt. \tag{6.14}$$

From (14.15) and Lemma 14.2.2 we obtain

$$\int_{\Gamma(t)} \kappa \mathbf{n} \cdot \mathbf{u} \, ds = \int_{\Gamma(t)} \operatorname{div}_\Gamma \mathbf{u} \, ds = \frac{d}{dt} \int_{\Gamma(t)} 1 \, ds.$$

Using this yields

$$-\tau \int_0^T \int_\Gamma \kappa \mathbf{n} \cdot \mathbf{u} \, ds dt. = -\tau \big(\operatorname{meas}_2 \Gamma(T) - \operatorname{meas}_2 \Gamma(0) \big),$$

which, combined with (6.14), completes the proof. □

Remark 6.1.7 The result in this lemma has a physical interpretation: The kinetic energy difference $\frac{1}{2}\|\mathbf{u}(T)\|_{L^2}^2 - \frac{1}{2}\|\mathbf{u}(0)\|_{L^2}^2$ is balanced by the sum of kinetic energy dissipation $\int_0^T a(\mathbf{u}(t), \mathbf{u}(t)) \, dt$ and the change in surface tension energy $\tau\big(\operatorname{meas}_2 \Gamma(T) - \operatorname{meas}_2 \Gamma(0) \big)$.

The question of well-posedness of the variational problem (6.9)-(6.10) combined with the immiscibility condition $V_\Gamma = \mathbf{u} \cdot \mathbf{n}$ is a very difficult one. Below we briefly address some known results.

First we consider a strongly simplified case, namely a Stokes problem with a *stationary* interface Γ. In that case the term with the trilinear form $c(\cdot;\cdot,\cdot)$ vanishes and the bilinear forms $m(\cdot,\cdot)$, $a(\cdot,\cdot)$ do *not* depend on t. We assume $\partial\Omega_D = \partial\Omega$ and $\mathbf{u}_D = 0$, i.e., a problem with homogeneous Dirichlet boundary conditions on $\partial\Omega$. We introduce the weighted L^2-scalar product $(\mathbf{v}, \mathbf{w})_{L^2,\rho} := (\rho\mathbf{v}, \mathbf{w})_{L^2}$. In this simplified case the variational problem (6.9)-(6.10) reduces to: determine $\mathbf{u}(t) = \mathbf{u}(\cdot, t) \in \mathbf{V}_{\mathrm{div}} = \{ \mathbf{v} \in \mathbf{V}_0 : \operatorname{div} \mathbf{v} = 0 \}$ with $\mathbf{u}(0) = \mathbf{u}_0$ and

$$(\frac{\partial \mathbf{u}}{\partial t}, \mathbf{v})_{L^2,\rho} + a(\mathbf{u}(t), \mathbf{v}) = (\mathbf{g}, \mathbf{v})_{L^2,\rho} + f_\Gamma(\mathbf{v}) \quad \text{for all } \mathbf{v} \in \mathbf{V}_{\mathrm{div}}, \tag{6.15}$$

for almost all $t \in [0, T]$. This variational problem is very similar to the *one-phase* Stokes problem in (2.33). Compared to (2.33), in (6.15) we have a slightly different bilinear form $a(\cdot,\cdot)$ in which a weighting with the piecewise constant viscosity μ is used, a modified L^2-scalar product (namely $(\cdot,\cdot)_{L^2,\rho}$) and an additional functional f_Γ on the right-hand side. The analysis of well-posedness of the one-phase variational Stokes problem in (2.33), cf. Theorem 2.2.10, can also be applied to the two-phase variational Stokes problem in (6.15) (notation as in Sect. 2.2.3):

Theorem 6.1.8 *Assume* $\mathbf{g} \in L^2(0, T; \mathbf{V}_{\mathrm{div}}')$, $\|\kappa\|_{L^\infty(\Gamma)} < \infty$ *and* $\mathbf{u}_0 \in \mathbf{H}_{\mathrm{div}}$. *Then the variational problem* (6.15) *is well-posed.*

Proof. Use the same arguments as in the proof of Theorem 2.2.10. Note that the norms induced by the standard L^2-scalar product and by $(\cdot,\cdot)_{L^2,\rho}$ are equivalent. Furthermore, $\|\kappa\|_{L^\infty(\Gamma)} < \infty$ implies that $f_\Gamma \in \mathbf{V}' \subset \mathbf{V}_{\mathrm{div}}'$. □

For the above result to hold, the weak derivative $\mathbf{u}' = \frac{\partial \mathbf{u}}{\partial t}$ in (6.15) is defined as explained in Sect. 2.2.3. For the unique solution $\mathbf{u}$ we have

$$\mathbf{u} \in W^1(0,T;\mathbf{V}_{\text{div}}) = \left\{ \mathbf{v} \in L^2(0,T;\mathbf{V}_{\text{div}}) : \mathbf{v}' \in L^2(0,T;\mathbf{V}'_{\text{div}}) \right\}. \quad (6.16)$$

Well-posedness results for the general Navier-Stokes case in (6.9)-(6.10) are known in the literature, however, only for special cases. In [84] well-posedness of a Navier-Stokes problem as in (6.9)-(6.10) combined with the interface condition $V_\Gamma = \mathbf{u} \cdot \mathbf{n}$ is analyzed. Instead of a bounded domain Ω, the case $\Omega = \mathbb{R}^3$ is considered (with "boundary" condition $\lim_{|x|\to\infty} \mathbf{u}(x,t) = 0$). The initial interface $\Gamma(0)$ is assumed to be a closed manifold. The main result in [84] can be summarized as follows. If the data $\Gamma(0)$, $\mathbf{u}_0$ and $\mathbf{g}$ are sufficiently smooth then for $t \in [0,T]$, *with T sufficiently small*, the two-phase Navier-Stokes problem in a weak formulation similar to (6.9)-(6.10) and with the interface condition $V_\Gamma = \mathbf{u}\cdot\mathbf{n}$ has a unique solution. The analysis is in Sobolev spaces similar to the one in (6.16). The analysis is quite technical, the main underlying idea, however, is rather easy to explain. We outline this idea. For $\xi \in \Omega$ and a given velocity field $\mathbf{u}(x,t)$ we define the characteristic $X_\xi(\tau)$:

$$\begin{cases} \frac{d}{d\tau} X_\xi(\tau) = \mathbf{u}(X_\xi(\tau),\tau), & \tau \geq 0, \\ X_\xi(0) = \xi. \end{cases} \quad (6.17)$$

$X_\xi(\tau)$ can be interpreted as the path of an infinitely small particle with initial position ξ. For $\mathbf{u}(x,t)$ sufficiently smooth (Lipschitz with respect to x) this system of ODEs has a unique solution. The smoothness of X_ξ depends on the smoothness of $\mathbf{u}$. To each $(x,t) \in \Omega \times [0,T]$, with T sufficiently small, there corresponds a unique $\xi \in \Omega$ such that $x = X_\xi(t)$. Physically this means that starting from (x,t) one follows the flow field backwards in time resulting in $(\xi,0)$:

$$x = \xi + \int_0^t \mathbf{u}(X_\xi(\tau),\tau)\, d\tau. \quad (6.18)$$

This defines the coordinate transformation $(x,t) = (X_\xi(t),t) \to (\xi,t)$ from *Eulerian coordinates* (x,t) to *Lagrangian coordinates* (ξ,t). The problem (6.1)-(6.3) can be transformed in Lagrangian coordinates (ξ,t) resulting in an non-stationary *Stokes* type of problem with a *stationary* interface $\Gamma(0)$. For this transformed problem well-posedness (in suitable Sobolev spaces) is shown in [85]. The length T of the time interval should be such that the coordinate transformation $(x,t) \to (\xi,t)$ is well-defined and the Jacobian of this transformation is bounded (in a suitable norm). This depends on norms of the data $\mathbf{g}$, $\mathbf{u}_0$ and the curvature of $\Gamma(0)$.

In [232] a well-posedness result for the Navier-Stokes problem on *arbitrary* time intervals $[0,T]$ is proved, using the same Euler $\to$ Lagrange coordinate transformation. In that paper the case with a bounded domain Ω is treated. We summarize its main result. For arbitrary $T > 0$ the Navier-Stokes problem in a weak formulation similar to (6.9)-(6.10) and with the interface condition $V_\Gamma = \mathbf{u} \cdot \mathbf{n}$ has a unique solution (in suitable Sobolev spaces) if the data $\mathbf{g}$, $\mathbf{u}_0$ are sufficiently small and the initial interface $\Gamma(0)$ is sufficiently close to a sphere.

The analyses addressed above are applicable only in cases with *sufficiently smooth data* (initial and boundary data, source terms and initial interface). They do not cover situations in which the smoothness of the interface deteriorates, for example in a problem with colliding droplets. In such cases it may well happen that interface quantities like the curvature κ or the normal velocity V_Γ are not well-defined in the strong sense and suitable weak alternatives must be considered. Only few theoretical results that deal with well-posedness issues for such less regular problems are known in the literature. The analyses for less regular cases are based on *alternative characterizations of the interface.* These interface representations induce corresponding numerical techniques for the simulation of two-phase flow problems. In Sect. 6.2 we treat the most important approaches for interface representation. In the remainder of this monograph we then restrict ourselves to one of these, namely the level set representation. A suitable weak formulation of the level set interface representation combined with the weak formulation of the Navier-Stokes problem in (6.9)-(6.10) leads to the weak model that we consider for our numerical simulations. This model is presented in Sect. 6.3.

6.2 Interface representation

In this section we discuss the most important approaches for characterizing the interface. These techniques play a role both in the theoretical analysis of well-posedness and in numerical methods for simulating the two-phase flow problem.

6.2.1 Explicit interface representation: interface tracking

If the interface is sufficiently smooth then its curvature and other interface quantities like V_Γ, $\mathbf{n}_\Gamma$ are well-defined in the classical sense. For a velocity field $\mathbf{u} \in \mathbf{V}$ and a smooth interface $\Gamma(t)$ the trace $\mathbf{u}_{|\Gamma}$ and the immiscibility condition $V_\Gamma = \mathbf{u}\cdot\mathbf{n}$ in (6.3) are well-defined. The evolution of the interface can be described by using the Lagrangian coordinates. Take a (virtual) particle $\mathbf{X}$ on the interface at $t = t_0$ with Eulerian coordinates $\xi \in \Gamma(t_0)$. For $t \geq t_0$, let $X_\xi(t)$ be the Eulerian coordinates of this particle. The particles on the interface are transported by the flow field, hence for $X_\xi(t)$ we have the ODE system (6.17) and the interface $\Gamma(t)$ can be characterized as follows, cf. (6.18):

$$x \in \Gamma(t) \;\Leftrightarrow\; x = \xi + \int_{t_0}^{t} \mathbf{u}(X_\xi(\tau), \tau)\, d\tau, \quad \xi \in \Gamma(t_0), \;\; t \geq t_0. \qquad (6.19)$$

This Lagrangian point of view is essential for the analyses of well-posedness for two-phase flow problems with sufficient smoothness, as briefly addressed above in Sect. 6.1 (cf. [84, 232]). The interface representation in (6.19) also forms the basis for a class of numerical methods, known as *interface tracking.*

In these methods a collection of markers is put on a given interface $\Gamma(t_0)$ and then transported (numerically) by the flow field $\mathbf{u}$ to obtain the markers on the interface $\Gamma(t_0 + \Delta t)$. The collection of markers on $\Gamma(t_0)$ could be the set of vertices of a triangulation of $\Gamma(t_0)$. In such methods one usually has to redistribute the markers after a certain number of time steps. In general it is rather difficult to treat topology changes (e.g. collision of droplets) in a systematic and accurate way. Usually in interface tracking methods for two phase flows the Lagrangian approach is used only for the propagation of the interface markers. The Navier-Stokes equations are solved on a fixed grid (i.e., an Eulerian approach), cf. Fig. 6.1.

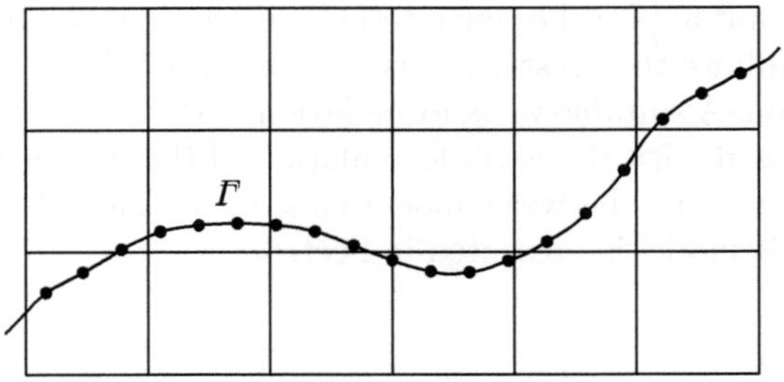

Fig. 6.1. Front tracking on an Eulerian grid for the flow problem. The interface Γ is represented by connected marker points.

Thus one needs operators for the transfer of information between the (moving) interface and the underlying fixed grid. Such front tracking methods have been successfully applied in the simulation of two-phase flows. An overview and detailed treatment of this technique can be found in [246, 243, 182]. Hybrid variants of this technique have been developed, for example so-called arbitrary Lagrangian-Eulerian (ALE) methods in which the interface (or surface) is resolved by a mesh and this mesh is moved with the flow velocity (Lagrangian interface tracking). In the interior flow domain a moving mesh is used with a mesh velocity that generally differs from the flow velocity and is taken such that strong mesh distortions are avoided. Such a mesh velocity can be obtained, for example, as the solution of a linear elasticity equation with a prescribed displacement on the boundary. Often the Navier-Stokes equations are then formulated using a *relative* velocity, which is the difference between the flow and the mesh velocity. Such ALE methods are very popular for the simulation of fluid structure interaction (FSI) problems, in which typically the movement of the boundary of the fluid domain is relatively small. ALE techniques have also been applied in the numerical simulation of one-phase flows with a free surface or of two-phase flows, e.g. [23, 28, 29, 117, 118, 185]. The Lagrangian interface tracking method can also be combined with a pure Lagrangian approach for the Navier-Stokes equations, based on an interior mesh movement that is based on the flow velocity field, cf. for example [152, 145].

6.2.2 Volume tracking based on the characteristic function

Let $\chi_1(\cdot,t) : \Omega \to \mathbb{R}$ be the characteristic function corresponding to the subdomain $\Omega_1(t)$, $t \geq 0$, i.e., $\chi_1(x,t) = 1$ for $x \in \Omega_1(t)$, $\chi_1(x,t) = 0$ otherwise. In this section we treat methods based on the simple observation $\Gamma(t) = \partial\Omega_1(t) = \partial\mathrm{supp}(\chi_1(\cdot,t))$. The function χ_1 characterizes the subdomain Ω_1 and we track this function (and thus the boundary of its support) to follow the evolution of the interface. We will derive a transport equation for χ_1 induced by the immiscibility condition $V_\Gamma = \mathbf{u}\cdot\mathbf{n}$. For this we need some additional notation. We introduce the space-time subdomain and interface:

$$\begin{aligned}
\Omega_T &:= \Omega \times [0,T] \subset \mathbb{R}^4, \\
\Omega_{i,T} &:= \left\{\, (x,t) \in \mathbb{R}^4 : x \in \Omega_i(t),\ 0 \leq t \leq T \,\right\},\ i = 1,2, \\
\Gamma_T &:= \left\{\, (x,t) \in \mathbb{R}^4 : x \in \Gamma(t),\ 0 \leq t \leq T \,\right\}.
\end{aligned}$$

The outward normal (not necessarily normalized) on $\partial\Omega_{1,T} \cap \Gamma_T$ is given by

$$\hat{\mathbf{n}} = \hat{\mathbf{n}}_\Gamma(x,t) = \begin{pmatrix} \mathbf{n}_\Gamma(x) \\ -V_\Gamma(x) \end{pmatrix} \in \mathbb{R}^4, \quad (x,t) \in \Gamma_T.$$

The immiscibility condition $V_\Gamma = \mathbf{u}\cdot\mathbf{n}$ on Γ can be written as:

$$\hat{\mathbf{n}} \cdot \begin{pmatrix} \mathbf{u} \\ 1 \end{pmatrix} = 0 \quad \text{on } \Gamma_T. \tag{6.20}$$

Lemma 6.2.1 *Let $\chi_1(\cdot,t)$ be the characteristic function corresponding to $\Omega_1(t)$ and $\mathbf{u} \in L^2(0,T;\mathbf{V})$ with $\mathrm{div}\,\mathbf{u} = 0$. The condition in (6.20) holds iff*

$$\int_{\Omega_T} \chi_1\Big(\frac{\partial\phi}{\partial t} + \mathbf{u}\cdot\nabla_x\phi\Big)\,dx\,dt = 0 \quad \text{for all } \phi \in C_0^\infty(\Omega_T), \tag{6.21}$$

i.e., in the sense of distributional derivatives,

$$\frac{\partial\chi_1}{\partial t} + \mathbf{u}\cdot\nabla_x\chi_1 = 0 \quad \text{in } D'(\Omega_T) := C_0^\infty(\Omega_T)'. \tag{6.22}$$

Proof. Using $\mathrm{div}\,\mathbf{u} = 0$ and the definition of distributional derivatives we have

$$\begin{aligned}
&\int_{\Omega_T} \chi_1\Big(\frac{\partial\phi}{\partial t} + \mathbf{u}\cdot\nabla\phi\Big)\,dx\,dt = 0 \quad \text{for all } \phi \in C_0^\infty(\Omega_T) \\
\text{iff}\quad & \frac{\partial\chi_1}{\partial t} + \mathrm{div}(\mathbf{u}\chi_1) = 0 \quad \text{in } D'(\Omega_T), \\
\text{iff}\quad & \frac{\partial\chi_1}{\partial t} + \mathbf{u}\cdot\nabla_x\chi_1 = 0 \quad \text{in } D'(\Omega_T),
\end{aligned}$$

thus the equivalence between (6.21) and (6.22) holds. For $\mathbf{u} \in L^2(0,T;\mathbf{V})$ its trace on Γ_T, denoted by $\mathbf{u}_{|\Gamma_T}$ is well-defined. The boundary of $\Omega_{1,T}$ can be partitioned as $\partial\Omega_{1,T} = (\partial\Omega_{1,T} \cap \partial\Omega_T) \cup (\partial\Omega_{1,T} \cap \Gamma_T)$. For $\phi \in C_0^\infty(\Omega_T)$ we have $\phi = 0$ on $\partial\Omega_{1,T} \cap \partial\Omega_T$. Using partial integration and $\chi_1 = 1$ on $\Omega_{1,T}$, $\chi_1 = 0$ on $\Omega_T \setminus \Omega_{1,T}$ we obtain

$$\int_{\Omega_T} \chi_1 (\frac{\partial\phi}{\partial t} + \mathbf{u} \cdot \nabla_x \phi)\, dx\, dt = \int_{\Omega_{1,T}} \frac{\partial\phi}{\partial t} + \mathbf{u} \cdot \nabla_x \phi \, dx\, dt$$
$$= \int_{\Omega_{1,T}} \begin{pmatrix} \mathbf{u} \\ 1 \end{pmatrix} \cdot \nabla_{x,t}\phi \, dx\, dt = \int_{\Gamma_T} \begin{pmatrix} \mathbf{u}_{|\Gamma_T} \\ 1 \end{pmatrix} \cdot \hat{\mathbf{n}}\, \phi \, dx\, dt.$$

In the last equality we used $\operatorname{div} \mathbf{u} = 0$. Since $\phi \in C_0^\infty(\Omega_T)$ is arbitrary it follows that (6.21) holds if and only if $\begin{pmatrix} \mathbf{u}_{|\Gamma_T} \\ 1 \end{pmatrix} \cdot \hat{\mathbf{n}} = 0$ holds (in $L^2(\Gamma_T)$ sense).

□

From this lemma it follows that the *immiscibility condition is satisfied if we solve a (weak) transport equation for the characteristic function* χ_1. The equivalence in this lemma holds in cases where the quantities that occur in the immiscibility condition (6.20) are well-defined. If this is not the case (as for example in a colliding droplet problem) then the result of this lemma offers a possibility to generalize the immiscibility condition by considering a suitable weak transport equation for the characteristic function of the subdomains. This idea is the basis of the analysis of well-posedness presented in [81] (for a two-phase Stokes problem) and [188] (for a two-phase Navier-Stokes problem).
If the velocity field $\mathbf{u}$ is *sufficiently smooth*, e.g. continuous in t and Lipschitz continuous w.r.t. x, then a strong formulation of the transport equation in the Lagrangian form

$$\dot{\chi}_1 = \frac{d}{dt}\chi_1(X_\xi(t), t) = 0,$$

with $\chi_1(X_\xi(0), 0) = 1$ if $\xi \in \Omega_1(0)$ and zero otherwise, is well-defined and has a unique solution. For general flow problems, however, one wants to relax the smoothness assumption on $\mathbf{u}$ and then for the transport equation weaker solution concepts are needed. One such a concept, namely of so-called *renormalized solutions of transport equations*, is introduced in the fundamental paper [89]. Using this, the following result can be proved (Proposition 3.3. from [188]).

Proposition 6.2.2 *Take* $\mathbf{u} \in L^2(0, \infty; \mathbf{V})$ *with* $\operatorname{div} \mathbf{u} = 0$ *and* $\mu_0 \in L^\infty(\Omega)$. *Then there is a unique weak solution* $\mu \in L^\infty(\Omega_T)$ *in the following sense:*

$$\int_0^\infty \int_\Omega \mu (\frac{\partial\phi}{\partial t} + \mathbf{u} \cdot \nabla_x \phi)\, dx\, dt = \int_\Omega \mu_0 \phi(x, 0)\, dx \quad \forall\, \phi \in C_0^\infty(\mathbb{R}^4). \quad (6.23)$$

Moreover, if μ_0 *is piecewise constant, i.e.,* $\mu_0 \in \{c_1, \dots c_M\}$ *a.e., with constants* c_i, *then* $\mu \in \{c_1, \dots c_M\}$ *a.e..*

Remark 6.2.3 The concept of renormalized solutions allows unique weak solutions of (6.23) even for velocity fields $\mathbf{u}$ with less regularity than $\mathbf{u} \in$

$L^2(0,\infty;\mathbf{V})$. Starting with the paper [89] there have appeared a lot of studies on well-posedness of the weak formulation (6.23). In [89] existence and uniqueness of a weak (renormalized) solution is proved for velocity fields $\mathbf{u} \in L^1(0,\infty;H^1_p(\mathbb{R}^d)^d)$. Results with even less smooth velocity fields have been derived. For example, in [12] well-posedness for velocity fields from the class of functions of bounded variation (BV) is proved. For an overview and further references we refer to [76].

A weak formulation of a transport equation as in Proposition 6.2.2 can be combined with a standard weak formulation of a two-phase Stokes problem as follows. We take an initial velocity field $\mathbf{u}_0 \in \mathbf{V}_0$ (i.e., homogeneous Dirichlet boundary conditions on $\partial\Omega$) with $\operatorname{div}\mathbf{u}_0 = 0$. Let $\mu_0 \in \{\mu_1,\mu_2\}$ be the piecewise constant viscosity in the two initial subdomains: $\mu_0(x) = \mu_i > 0$ for $x \in \Omega_i(0)$, $i = 1,2$. In [81] it is proved that there exists *at least one solution* $(\mathbf{u},p,\mu)$ with $\mathbf{u} \in L^\infty(0,\infty;\mathbf{V}_0)$, $\mathbf{u}(\cdot,0) = \mathbf{u}_0$, $p \in L^2(0,\infty;Q)$, $\mu \in L^\infty(Q_T)$, $\mu(\cdot,0) = \mu_0$ and $\mu \in \{\mu_1,\mu_2\}$ a.e. such that

$$
\begin{aligned}
&\frac{1}{2}\int_\Omega \mu \operatorname{tr}\big(\mathbf{D}(\mathbf{u})\mathbf{D}(\mathbf{v})\big)\,dx + b(\mathbf{v},p) = 0 \quad \text{for all } \mathbf{v} \in \mathbf{V}_0,\ t \ge 0,\\
&b(\mathbf{u},q) = 0 \quad \text{for all } q \in Q,\ t \ge 0,\\
&\frac{\partial \mu}{\partial t} + \mathbf{u}\cdot\nabla\mu = 0 \quad \text{in the weak sense as in (6.23).}
\end{aligned} \tag{6.24}
$$

Thus we have *existence of a weak solution of a two-phase flow problem.* We briefly address some issues related to this result. In [188] a similar weak formulation of a Navier-Stokes two-phase problem is considered and an existence result is proved. The transport equation for the viscosity μ "replaces" the immiscibility condition, cf. Lemma 6.2.1. The analysis only yields existence of a weak solution; uniqueness is still an open problem. This concept of weak solutions allows singularities of the interface (e.g. collision of droplets) and yields existence global in time for "general" initial data. If we define the sets $\Omega_i(t) := \{\, x \in \Omega : \mu(t) = \mu_i \,\}$, $i = 1,2$, then, due to $\mu(\cdot,t) \in \{\mu_1,\mu_2\}$ a.e., we have $\overline{\Omega_1(t)} \cup \overline{\Omega_2(t)} = \overline{\Omega}$. It is, however, in general not clear what the "interface" should be. If we take $\Gamma(t) := \partial\Omega_1(t)$, then $\Gamma(t)$ can have a strictly positive Lebesgue measure. This effect is called "interface flattening" (or interface thickening). The fact that the interface can be "rough" and/or "flat" may be related to the fact that in the weak formulation above we do *not* take surface tension into account (which has a smoothing effect). It is, however, not known whether the analysis can be extended to the case with surface tension. An extensive treatment of several topics related to weak (or "generalized") solutions of two-phase flows with incompressible immiscible fluids which allow singular interfaces is given in [1, 2]. In particular it is remarked in these papers that if surface tension is taken into account, the existence of weak solutions (in the sense as explained above) is still an open problem. In [2] an even weaker concept of so-called measure-valued varifold solutions is introduced. Within that framework existence of a solution can be shown to hold for a suitable weak formulation of a two-phase flow problem which allows for singularities of the interface and takes surface tension into account.

The analysis of well-posedness addressed above relies on a weak formulation of the transport equation for the viscosity μ, cf. (6.23). Since μ is piecewise constant one can equivalently consider a transport equation for the characteristic function χ_1 corresponding to the subdomain Ω_1. There is an important class of numerical methods in which the treatment of the interface is based on a weak formulation of the transport equation

$$\frac{\partial \chi_1}{\partial t} + \mathbf{u} \cdot \nabla \chi_1 = 0. \tag{6.25}$$

This is the class of *VOF-methods* (Volume of Fluid), which we now introduce. The original idea of this approach goes back to [187]. Note that the equation in (6.25) is not well-defined in the classical sense, since χ_1 is discontinuous across the interface Γ. Instead of using a weak formulation of (6.25) based on distributional derivatives one can also (formally) eliminate the gradient operator by integrating this transport equation. Take an arbitrary (small, connected) fluid volume $W \subset \Omega$. Integrating over W and formally applying partial integration results in

$$\frac{\partial}{\partial t} \int_W \chi_1 \, dx + \int_{\partial W} \chi_1 \mathbf{u} \cdot \mathbf{n} \, ds = 0. \tag{6.26}$$

Here $\mathbf{n}$ denotes the outward unit normal on ∂W. This equation can be seen as a weak formulation of (6.25) and has a clear physical interpretation: it describes *volume conservation*. The change of volume of fluid 1 (i.e. the one in Ω_1) contained in W equals the volume flux (induced by the velocity field $\mathbf{u}$) across the boundary ∂W. Note that for an incompressible fluid conservation of volume is equivalent to mass conservation. *In VOF-methods one constructs approximations of the characteristic function χ_1 based on discretization of the conservation law* (6.26). We explain the main idea for a simple 2D case, namely with $\Omega = (0,1)^2$. Assume that Ω is partitioned in square cells $W_{ij} := [ih, (i+1)h] \times [jh, (j+1)h]$, $0 \leq i, j \leq m-1$ with $mh = 1$. We introduce the color function (or area fraction in 2D, volume fraction in 3D):

$$C_{ij}(t) := |W_{ij}|^{-1} \int_{W_{ij}} \chi_1(x,t) \, dx = h^{-2} \int_{W_{ij}} \chi_1(x,t) \, dx.$$

We have $0 < C_{ij} < 1$ in cells W_{ij} cut by the interface and $C_{ij} = 0$ or 1 away from it, cf. Fig. 6.2. Assume that for time $t = t_n$ (an approximation of) the color function is known in all cells, i.e., we have known values $C_{ij}^n \approx C_{ij}(t_n)$, $0 \leq i, j \leq m-1$. The values for the next time level $t_{n+1} = t_n + \Delta t$ are obtained by discretization of (6.26) using a standard finite volume approach:

$$C_{ij}^{n+1} = C_{ij}^n + h^{-2} \int_{t_n}^{t_{n+1}} \int_{\partial W_{ij}} \tilde{\chi}_1 \mathbf{u} \cdot \mathbf{n} \, ds \, dt, \tag{6.27}$$

where $\tilde{\chi}_1 = \tilde{\chi}_1(x, t_n)$ is a known approximation of the characteristic function χ_1, cf. below.

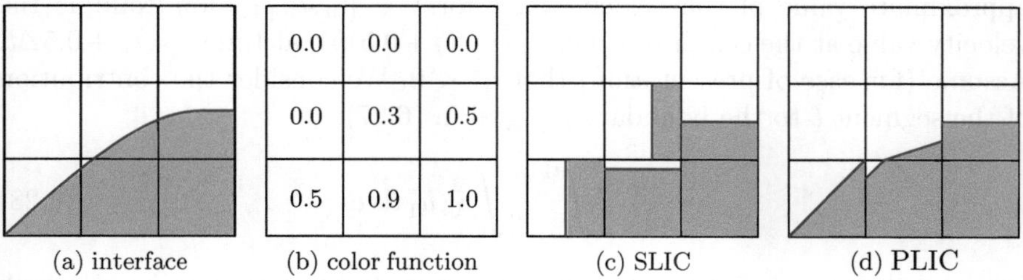

Fig. 6.2. Illustration of color function C and interface reconstruction: a) interface Γ and cells W_{ij}, b) values C_{ij} of color function, c) SLIC approximation, d) PLIC approximation.

In a VOF method one distinguishes the following two steps:

1. *Reconstruction of the interface*: given the values C_{ij}^n, $0 \le i, j \le m-1$, of the color function, an approximate interface is computed. This then determines the approximate characteristic function $\tilde{\chi}_1(x, t_n)$, used in (6.27).
2. *Color function advection step*: given the function $\tilde{\chi}_1(x, t_n)$ the boundary fluxes in (6.27) are approximated, resulting in the updated values C_{ij}^{n+1}, $0 \le i, j \le m-1$.

The reconstruction is such that the consistency property

$$C_{ij}^n = h^{-2} \int_{W_{ij}} \tilde{\chi}_1(x, t_n)\, dx$$

holds (volume conservation). Since the introduction of the VOF-method there have been many papers in which interface reconstruction techniques have been treated. The earliest algorithm, denoted by SLIC ("simple line interface calculation") was introduced in the paper [187]. In this approach the reconstructed interface consists of line segments that are parallel to one of the coordinate axes, cf. Fig. 6.2. This method is only first order accurate, i.e. $\mathcal{O}(h)$, in the accuracy of the reconstruction of the interface. Modifications of this technique can be found in [144, 69, 163]. More accurate (namely second order) reconstruction methods use piecewise linear segments that are not necessarily aligned with the coordinate axes. This technique is known as PLIC ("piecewise linear interface construction). Methods of this type are studied in, for example, [204, 214, 17]. We do not treat such reconstruction methods here, but refer to the above-mentioned literature.

We briefly address the advection step. The methods known in the literature can be divided into two categories: unsplit schemes and operator split schemes.

We only discuss the latter, for the former we refer to the literature, e.g. [204]. Consider one side of the boundary of W_{ij}, say the line segment connecting (ih, jh) with $(ih, (j+1)h)$, which is denoted by ℓ. Let $\mathbf{u}^* = (u_1^*, u_2^*)$ be an approximate value of the velocity on ℓ for $t \in [t_n, t_{n+1}]$, for example the velocity value at the center of ℓ, i.e. at $(ih, (j+0.5)h)$, at time $t = t_n + 0.5\Delta t$. Assume (for ease of presentation) that $u_1^* < 0$. We consider the contribution of the segment ℓ to the boundary integral in (6.27):

$$-h^{-2}\int_{t_n}^{t_{n+1}}\int_\ell \tilde{\chi}_1 u_1^*\,ds\,dt \tag{6.28}$$

(we used that $\mathbf{n} = (-1, 0)$ on ℓ). After the reconstruction step we have in each cell W_{ij} an approximation $\tilde{\chi}_1(x, t_n)$ of the characteristic function $\chi_1(x, t_n)$. In (6.28) we need values for $\tilde{\chi}_1(x, t)$, $x \in \ell$, $t \in [t_n, t_{n+1}]$. For this we take the values of $\tilde{\chi}_1(\cdot, t_n)$ transported by the flow field $(u_1^*, 0)^T$ during the time $t - t_n$, i.e. we take

$$\tilde{\chi}_1(x,t) := \tilde{\chi}_1\big(x - (t - t_n)\begin{pmatrix} u_1^* \\ 0 \end{pmatrix}, t_n\big), \quad x \in \ell,\ t \in [t_n, t_{n+1}]. \tag{6.29}$$

For this to be well-defined one has to satisfy the CFL-condition

$$\Delta t\, |u_1^*| \le h. \tag{6.30}$$

Introduce $z := -(t - t_n)u_1^*$. Using (6.29) we obtain

$$\begin{aligned} -h^{-2}\int_{t_n}^{t_{n+1}}\int_\ell \tilde{\chi}_1(s,t)u_1^*\,ds\,dt &= h^{-2}\int_0^{\Delta t|u_1^*|}\int_\ell \tilde{\chi}_1(s + \begin{pmatrix} z \\ 0 \end{pmatrix}, t_n)\,ds\,dz \\ &= h^{-2}\big|\mathrm{supp}\,\{\, \tilde{\chi}_1(x,t_n) : x \in [ih, ih + \Delta t|u_1^*|] \times [jh, (j+1)h]\ \}\,\big|. \end{aligned}$$

Hence, for the area flux across the side ℓ we obtain h^{-2} times the area of the support of the reconstructed characteristic function $\tilde{\chi}_1(\cdot, t_n)$ in the cell W_{ij} between the vertical lines with x_1-coordinates ih and $ih + \Delta t|u_1^*|$. This area flux leads to new intermediate values for the color function values in the cells W_{ij} and $W_{i-1,j}$. The same is done for all other vertical cell sides in the grid. These area fluxes in the horizontal direction lead to intermediate values $C_{ij}^{n+1,*}$, $0 \le i, j \le m - 1$. Based on these new values of the color function the reconstruction step is repeated, resulting in a new characteristic function $\tilde{\chi}_1(\cdot, t_n)$, which is then used to compute area fluxes in the vertical direction (i.e. across horizontal cell sides). Thus we obtain the final new values C_{ij}^{n+1}, $0 \le i, j \le m - 1$. Due to this two-step procedure (three-step in 3D), first the fluxes in one direction and then those in the other direction, this approach is called an operator split scheme.

We give some comments on the VOF technique. The method is very popular for the simulation of two-phase flows, in particular in the engineering community. Most of these methods have very good mass-conservation properties. In principle topology changes (droplet collisions) can be handled easily.

Usually the method is applied on (logically) rectangular grids; it is difficult to apply an accurate VOF technique on unstructured triangular or tetrahedral grids. In the method a CFL condition as in (6.30) must be satisfied, which may lead to severe (undesirable) restrictions on the size of the time step. In general it is difficult to obtain accurate approximations of intrinsic geometric properties of the interface, such as curvature and normal direction.

6.2.3 Volume tracking based on the level set function

An important difference between the interface tracking approach in Sect. 6.2.1 and the volume tracking approach in Sect. 6.2.2 is that the former is based on a Lagrangian ODE technique, cf. (6.19), and the latter on an Eulerian PDE approach, cf. (6.25). The method presented in this section is also of Eulerian PDE type. The approach discussed in the previous section is based on (the weak formulation of) the transport equation (6.25) for the characteristic function χ_1. This function is discontinuous across the interface, which requires a special numerical treatment of the transport equation. Furthermore, the interface is *not* characterized by values of χ_1 but by the boundary of its support.

An alternative is to use instead of χ_1 another indicator function. In the level set approach a *smooth* initial function $\phi_0(x)$, $x \in \Omega$ is chosen such that

$$\phi_0(x) < 0 \Leftrightarrow x \in \Omega_1(0), \quad \phi_0(x) > 0 \Leftrightarrow x \in \Omega_2(0), \quad \phi_0(x) = 0 \Leftrightarrow x \in \Gamma(0).$$

A popular choice is to take ϕ_0 (approximately) equal to a signed distance function to the initial interface, cf. Fig. 6.3.

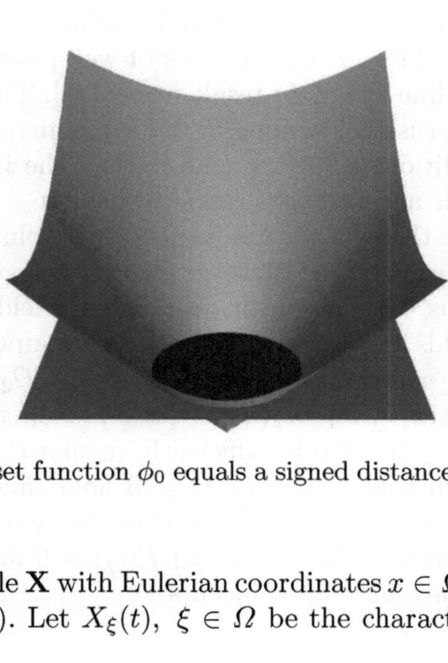

Fig. 6.3. Initial level set function ϕ_0 equals a signed distance function, 2D example.

A (virtual) particle $\mathbf{X}$ with Eulerian coordinates $x \in \Omega$ has a corresponding indicator value $\phi_0(x)$. Let $X_\xi(t)$, $\xi \in \Omega$ be the characteristics as defined in

(6.17), assuming that the velocity field $\mathbf{u}(x,t)$ is sufficiently smooth. For $t > 0$ the level set function values $\phi(x,t)$ are defined by keeping the values constant along characteristics, i.e.,

$$\phi(X_\xi(t),t) := \phi_0(\xi), \quad \xi \in \Omega, \ t \geq 0.$$

Differentiating this with respect to t results in the transport equation

$$\frac{\partial \phi}{\partial t} + \mathbf{u} \cdot \nabla \phi = 0 \quad \text{in} \ \ \Omega, \ \ t \geq 0. \tag{6.31}$$

This transport equation is of the same form as the one in (6.25). There are, however, important differences. Firstly, if the velocity field $\mathbf{u}(x,t)$ is sufficiently smooth (Lipschitz with respect to x) then the equation in (6.31) is well-defined in its *strong* formulation, due to the fact that the initial condition ϕ_0 is continuous. Due to the discontinuity in the characteristic function, the equation (6.25) is not well-defined in its strong form. Secondly, the interface $\Gamma(t)$ *can be characterized by values of the level set function* at time t:

$$\Gamma(t) = \{\, x \in \Omega : \ \phi(x,t) = 0 \,\}.$$

As already mentioned above, this is not the case for the characteristic function. For the linear hyperbolic partial differential equation in (6.31), besides the initial condition one needs suitable boundary conditions, for example, a Dirichlet boundary condition $\phi(x,t) = \phi_D(x)$ on the inflow boundary $\partial\Omega_{in} := \{\, x \in \partial\Omega : \ \mathbf{u} \cdot \mathbf{n}_\Omega < 0 \,\}$.

A suitable weak formulation of the level set equation, i.e., the transport equation (6.31) combined with continuous initial condition ϕ_0 as defined above, is used in the literature [119] for the analysis of well-posedness of a two-phase flow problem. We outline the main result from [119]. The domain Ω is taken as a d-dimensional torus (corresponding to a rectangular domain with periodic boundary conditions). For the transport of the interface the level set equation is used with a *continuous* velocity field $\mathbf{u} \in C(\overline{\Omega} \times [0,T])^d$. For general continuous $\mathbf{u}$ there is no uniqueness of a solution of the equation (6.31) in its strong formulation. However, the concept of *viscosity solutions* of transport equations with a continuous velocity field can be applied, cf. [72]. This theory yields unique so-called sub- and supersolutions of (6.31), which induce unique generalized evolutions $\Omega_1(t)$, $\Omega_2(t)$, $t \in [0,T]$, with $\Omega_1(0) = \{\, x \in \Omega : \ \phi_0(x) < 0 \,\}, \Omega_2(0) = \{\, x \in \Omega : \ \phi_0(x) > 0 \,\}$. One defines $\Gamma(t) := \Omega \setminus (\Omega_1(t) \cup \Omega_2(t))$. If $\mathbf{u}$ is sufficiently regular (Lipschitz with respect to x) it can be shown that $\text{meas}_d(\Gamma(t)) = 0$ holds and $\Gamma(t)$ describes the interface in the usual strong sense. If, however, the velocity field $\mathbf{u}$ is (only) continuous it is not known whether $\text{meas}_d(\Gamma(t)) = 0$ for $t > 0$ holds, i.e., it might be that *"interface flattening"* occurs.

For the two-phase flow problem a Stokes model of the form

$$\frac{\partial \mathbf{u}}{\partial t} - \operatorname{div}(\mu \nabla \mathbf{u}) + \nabla p = \mathbf{g} \quad \text{in } \Omega,$$
$$\operatorname{div} \mathbf{u} = 0 \quad \text{in } \Omega,$$

is considered. A weak formulation of this problem in the Sobolev space $H_p^1(\Omega)$ with $p > 2(d+1)$ is analyzed with a piecewise constant viscosity function $\mu(x,t) = \mu_i$ for $x \in \Omega_i(t)$, $i = 1,2$, and $\mu(x,t) = \frac{1}{2}(\mu_1 + \mu_2)$ for $x \in \Gamma(t)$. It is proved that for $|\mu_1 - \mu_2|$ sufficiently small there exists for almost all $t \in [0,T]$ a solution $\mathbf{u}$ with $\mathbf{u} \in C(\overline{\Omega} \times [0,T])^d$, $\mathbf{u}(\cdot,t) \in H_p^1(\Omega)^d$ of the Stokes problem coupled with an appropriate weak formulation of the transport problem for the evolution of the level set function ϕ. We refer to [119] for the precise results. Here we restrict ourselves to a few further comments related to this analysis. The continuity property for the velocity field $\mathbf{u}$ is needed to be able to apply the theory of viscosity solutions of transport equations as in (6.31). This theory also requires the initial condition ϕ_0 to be continuous. Note that this holds for the level set function ϕ_0 but not for the characteristic function χ_1 used in Sect. 6.2.2. For the solution $\mathbf{u}$ of the Stokes problem the regularity $\mathbf{u}(\cdot,t) \in H_p^1(\Omega)^d$ with $p > 2(d+1)$ (> 2) is proved, which due to a Sobolev embedding result implies continuity of $\mathbf{u}$. For the regularity property $\mathbf{u} \in H_p^1(\Omega)^d$ with $p > 2(d+1)$ to hold one needs that the jump in the viscosity $|\mu_1 - \mu_2|$ is sufficiently small. The analysis only applies to the case *without* surface tension and it only yields existence of a solution of the two-phase Stokes problem; uniqueness is an open problem. The existence is global in time ($t \in [0,T]$) and allows singularities of the interface (colliding droplets). However, "interface flattening" might occur, i.e., it is not clear whether the interface remains sharp.

The level set equation (6.31) is not only used in the analysis of well-posedness of a two-phase flow problem but also forms the basis of an important class of *numerical techniques* for representing the interface. These *level set methods* are used not only in two-phase flow simulations but also in many other applications with interfaces or free boundaries, cf. the overview paper [221] and the monographs [222, 198]. We outline the main ideas. The linear hyperbolic transport equation (6.31), or a weak variant of it, is considered with an initialization $\phi_0(x)$ that is continuous, close to a signed distance function and such that $\Gamma(0) = \{\, x \in \Omega : \phi_0(x) = 0 \,\}$ holds. The velocity $\mathbf{u}$ results from the Navier-Stokes flow problem. The transport equation is discretized in space and time using appropriate numerical methods. We will treat this issue in more detail in Sect. 7.2. The accurate discretization of the level set equation is (much) easier than that of the transport equation considered in the VOF method in Sect. 6.2.2 because in the latter one has to approximate the discontinuous characteristic function χ_1 whereas in the level set method the solution ϕ is smooth (close to the interface, for a sufficiently short time interval). During time evolution, in a neighborhood of the zero level it is monitored how much the discrete solution $\phi_h(x,t)$ deteriorates from a signed distance function. For this one can use, for example, the quantity $\|\nabla\phi_h(x,t)\|_2$ as an indicator. If the deterioration exceeds a given tolerance a *re-initialization* of the given level set function, say $\phi_h(x,t_n)$, is performed. In this re-initialization one determines a new level set function $\phi_h^{\text{new}}(x,t_n)$ such that $\|\nabla\phi_h^{\text{new}}(x,t_n)\|_2 \approx 1$ (in a neighborhood of the zero level) and $\{\, x \in \Omega : \phi_h(x,t_n) = 0 \,\} \approx \{\, x \in \Omega : \phi_h^{\text{new}}(x,t_n) = 0 \,\}$ holds, i.e. one determines a re-initialization ϕ_h^{new} with (approximately) the same zero level as the current level set function but which is much closer to a signed distance function. This topic of re-initialization is addressed in more detail in Sect. 7.4.1.

Remark 6.2.4 The level set method is often combined with the CSF approach explained in Remark 6.1.3. This idea was introduced in [64]. There it is shown that (under certain smoothness assumptions) the following holds, with notation as in Remark 6.1.3:

$$\int_\gamma \kappa \mathbf{n}_\Gamma \, ds = \lim_{\epsilon \downarrow 0} \int_U \kappa(\phi)\delta_\epsilon(\phi(x))\nabla\phi \, dx, \tag{6.32}$$

with the level set function ϕ and δ_ϵ a one-dimensional smoothed Dirac delta function. For the approximation of the curvature term $\kappa(\phi)$ one can use, cf. (14.7),

$$\kappa(x) = \operatorname{div} \mathbf{n}_\Gamma(x) = \operatorname{div}\big(\frac{\nabla\phi}{|\nabla\phi|}\big), \quad x \in \Gamma,$$

and extend this relation to $x \in U$ ($|\nabla\phi|^2 := \nabla\phi \cdot \nabla\phi$). This leads to a *volume* surface tension force term of the form

$$-\,\tau \operatorname{div}\big(\frac{\nabla\phi}{|\nabla\phi|}\big)\delta_\epsilon(\phi(x))\nabla\phi \tag{6.33}$$

in the *strong* formulation of the momentum equation, which acts in an ϵ-neighborhood of the interface Γ. Clearly this approach induces an error due to numerical regularization with the smoothed Dirac delta function.

6.2.4 Phase field representation

In the interface representations treated above the interface is either tracked explicitly or "captured" implicitly as the discontinuity of a characteristic function or the zero level of an approximate signed distance function. In all three cases one typically has a *sharp interface*. This sharp interface property may be lost due to numerical effects, for example if one combines the level set method with the CSF technique as described in Remark 6.2.4, in which the surface tension force is approximated by a volume force using a smoothed Dirac delta function. In that approach, although the interface is represented sharply as the zero level of the level set function there is an interface smearing effect due to the smoothing of the surface tension force. In the continuous model the sharp interface property may be lost if in cases with interface singularities an interface flattening effect occurs. In the approach discussed in this section the model is such that one *always has a non sharp or diffusive interface*. These so-called phase field models are based on the observation that even for two (macroscopically) immiscible fluids there is a very thin interfacial region in which partial mixing of the two fluids occurs, cf. Sect. 1.1.5. In this sense, the physical interface is not sharp but diffusive. The interfacial mixing region has nonzero thickness but is extremely thin (about 100 nm). Hence modeling it as a sharp interface (as is done in the methods discussed above) seems reasonable. There are, however, mechanisms, for example in droplet collision, that are relevant and act on length scales comparable to that of interface thickness. For an accurate modeling of these mechanisms a diffusive interface representation may be more appropriate. Quantities that in the sharp interface formulation are localized at the interface, such as surface tension or surfactant transport, are distributed in a narrow interfacial region in a phase field model. The idea of diffusive interface modeling is an old one and was already used in [208, 247]. An overview on diffusive interface methods is given in [13]. Below we describe one popular phase field model for two-phase incompressible flows, namely the Navier-Stokes equations combined with the *Cahn-Hilliard* equation for the representation of the interface.

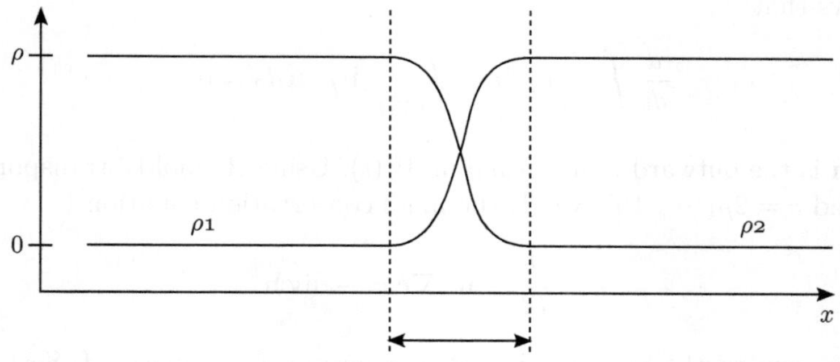

Fig. 6.4. Partial densities ρ_j, $j = 1, 2$, and diffusive interface (region between dashed lines) in the phase field representation.

Throughout this section, let $\rho_j = \rho_j(x)$, $x \in \Omega$, $j = 1, 2$, denote the *partial density* (or mass concentration) of the fluid j, i.e., for $W \subset \Omega$ the quantity $\int_W \rho_j(x)\,dx$ is the mass of the fluid j contained in W. Note that this notation differs from the one previously used, where ρ_j denoted the (constant) density of fluid j as a pure substance. The partial densities $\rho_1(x)$ and $\rho_2(x)$ are in general not constant, i.e., there is a mixing region representing the diffusive interface, cf. Fig. 6.4. The density of the *mixture* is denoted by $\rho(x)$, $x \in \Omega$, i.e., $\int_W \rho(x)\,dx$ is the total mass of the fluid contained in W. Clearly $\rho = \rho_1 + \rho_2$ holds. We restrict ourselves to the case of *matched densities*, i.e. we assume

$$\rho \text{ is constant in } \Omega. \tag{6.34}$$

It is no restriction to take $\rho = 1$. We introduce a so-called order parameter

$$c := \rho_1 - \rho_2 = 2\rho_1 - 1 \in [-1, 1].$$

This concentration (or density) difference has value -1 in regions filled by fluid 2, value 1 in regions filled by fluid 1 and values in between in the mixing region. It is assumed to be a smooth function of (x, t). Note that opposite to the characteristic function χ_1 and the level set function ϕ, which are used as indicator functions in the approaches treated above, *the order parameter c has a physical meaning.* A main issue is to derive an appropriate model for the evolution of c. We outline the derivation of the Navier-Stokes/Cahn-Hilliard model as given in [131]. It is based on a local dissipation inequality (corresponding to the second law of thermodynamics) and basic concepts from continuum mechanics, such as mass and momentum conservation, cf. Sect. 1.1.1. The mixture is considered as *one* incompressible Newtonian fluid. Its velocity field is denoted by $\mathbf{u}(x, t)$. Due to the incompressibility assumption we have $\operatorname{div} \mathbf{u} = 0$. Let $W(t)$ a material volume that is advected by the velocity field $\mathbf{u}$ and $\mathbf{h}_j$ the mass flux of fluid j, measured *relative* to the gross motion of the fluid. Due to incompressibility ($\rho = 1$) the relation $\mathbf{h}_1 = -\mathbf{h}_2$ holds. We define the (relative) mass flux quantity $\mathbf{h} := \mathbf{h}_1 - \mathbf{h}_2 = 2\mathbf{h}_1$. From mass conservation it follows that

$$\frac{d}{dt} \int_{W(t)} \rho_j \, dx + \int_{\partial W(t)} \mathbf{h}_j \cdot \mathbf{n} \, ds = 0,$$

where $\mathbf{n}$ is the outward unit normal on $W(t)$. Using Reynolds' transport theorem and $c = 2\rho_1 - 1$ this yields the mass conservation equation

$$\dot{c} = \frac{\partial c}{\partial t} + \mathbf{u} \cdot \nabla c = -\operatorname{div} \mathbf{h}. \tag{6.35}$$

We now turn to the conservation of momentum $\int_{W(t)} \rho \mathbf{u} \, dx$, cf. Sect. 1.1.1. For simplicity we assume that there are no external forces like gravity. Based on fundamental principles from continuum mechanics (Cauchy's theorem) one obtains from momentum conservation and using $\rho = 1$ the equation

$$\rho \dot{\mathbf{u}} = \frac{\partial \mathbf{u}}{\partial t} + (\mathbf{u} \cdot \nabla)\mathbf{u} = \operatorname{div} \boldsymbol{\sigma}, \tag{6.36}$$

with a symmetric stress tensor $\boldsymbol{\sigma}$ that is associated with the macroscopic motion of the fluid. For a one-phase incompressible Newtonian fluid one has $\boldsymbol{\sigma} = -p\mathbf{I} + \mu \mathbf{D}(\mathbf{u})$, cf. (1.11). In [131] a stress tensor for the case of a fluid mixture is derived using local energy inequalities that are based on the second law of thermodynamics. The resulting stress tensor is given in (6.42) below. We sketch the main idea, for more details we refer to [131, 3]. We consider a total energy $e_W(\mathbf{u}, c)$ in a volume W which is the sum of a kinetic energy and a *free energy*:

$$e_W(\mathbf{u}, c) = \int_W \rho \frac{1}{2} \mathbf{u} \cdot \mathbf{u} + \psi(c, \nabla c)\, dx.$$

The free energy $\int_W \psi(c, \nabla c)\, dx$ is used to describe energy changes due to mixing of the fluids. Since there is significant mixing only in a very thin interfacial region, this energy is also called surface energy. Cahn and Hilliard [60] proposed the following form for the mixing energy density

$$\psi(c, \nabla c) = \varepsilon \frac{1}{2} |\nabla c|^2 + \varepsilon^{-1} \psi_0(c), \tag{6.37}$$

with $|\nabla c|^2 = \nabla c \cdot \nabla c$, $\varepsilon > 0$ a small parameter and ψ_0 a double well potential. The latter means that ψ_0 should have exactly two global minima, namely at $c = \pm 1$ in our case. For this double well potential there are several possibilities used in the literature, e.g.,

$$\psi_0(c) = (1 - c^2)^2, \quad c \in \mathbb{R},$$

$$\psi_0(c) = \frac{\theta}{2}\big((1 + c)\ln(1 + c) + (1 - c)\ln(1 - c)\big) - \frac{\theta_c}{2} c^2, \quad 0 < \theta < \theta_c, \ |c| < 1,$$

$$\psi_0(c) = -\frac{\theta_c}{2} c^2 \quad \text{if } c \in [-1, 1], \ \infty \text{ otherwise},$$

cf. [3, 6] for a discussion of these and other double well potentials. Since both fluids are assumed to be present, it follows that $0 < |\Omega|^{-1} \int_\Omega \rho_1\, dx < 1$ and thus $-1 < |\Omega|^{-1} \int_\Omega c\, dx < 1$ holds. Hence, $c(x)$ can not have a constant value equal to -1 or 1, which corresponds to a minimum of ψ_0. A "diffusive interface" is represented by the region where $c(x)$ varies between $-1 + \xi$ and $1 - \xi$ (with $0 < \xi \ll 1$).

Remark 6.2.5 As an example, consider for given $a > 0$,

$$e_{\text{free}}(c) := \int_{-a}^{a} \varepsilon \frac{1}{2} c'(x)^2 + \varepsilon^{-1}(1 - c(x)^2)^2\, dx,$$

and minimization of this functional over the set V consisting of all piecewise linear $c = c_\delta$ with $c(x) = -1$ if $x \le -\delta$, $c(x) = 1$ if $x \ge \delta$, $c(x) = \frac{x}{\delta}$ if $-\delta \le x \le \delta$, and with $\delta \in (0, a)$ arbitrary. A straightforward computation yields

$$\min_{c_\delta \in V} e_{\text{free}}(c_\delta) = e_{\text{free}}(c_{\delta^*}) = \frac{8}{\sqrt{15}}, \quad \text{for } \delta^* = \frac{1}{4}\sqrt{15}\, \varepsilon.$$

Hence, in this example we have a transition region of width $2\delta^* = \frac{1}{2}\sqrt{15}\, \varepsilon$ between the extrema ± 1.

The macroscopic stresses in the mixture, modeled by contact forces $\boldsymbol{\sigma}\mathbf{n}$ that act on $\partial W(t)$, induce a corresponding energy exchange across $\partial W(t)$ (force times distance, per time unit), also called working, given by

$$\int_{\partial W(t)} \boldsymbol{\sigma}\mathbf{n} \cdot \mathbf{u}\, ds. \tag{6.38}$$

A second energy transport is due to the microscopic diffusion (i.e. diffusive effect within the mixture). This can be modeled as follows. Let μ_j^{chem} be the chemical potential of fluid j and $\mu^{\text{chem}} := \frac{1}{2}(\mu_1^{\text{chem}} - \mu_2^{\text{chem}})$. Related to the notation we remark that μ^{chem} should not be confused with $\mu = \mu(x,t)$, which we use to denote the viscosity of a fluid. Using $\mathbf{h}_1 + \mathbf{h}_2 = 0$ (as ρ is constant) and $\mathbf{h} = 2\mathbf{h}_1$ we obtain that

$$-\sum_{j=1}^{2} \int_{\partial W(t)} \mu_j^{\text{chem}} \mathbf{h}_j \cdot \mathbf{n}\, ds = -\int_{\partial W(t)} \mu^{\text{chem}} \mathbf{h} \cdot \mathbf{n}\, ds, \tag{6.39}$$

which models the energy transported into $W(t)$ due to microscopic diffusion. The third energy exchange is related to so-called microforces. In [131] these forces are introduced and it is assumed that the working of these forces accompanies changes in the concentration c. i.e., these forces cause the microscopic mixing. These forces are modeled as *contact forces* and denoted by $\boldsymbol{\xi}$. A corresponding scalar body force is given by

$$\pi := -\operatorname{div} \boldsymbol{\xi}, \tag{6.40}$$

i.e., we have a microforce balance $\int_{W(t)} \pi\, dx + \int_{\partial W(t)} \boldsymbol{\xi} \cdot \mathbf{n}\, ds = 0$. The energy exchange induced by these microforces is given by

$$\int_{\partial W(t)} \dot{c} \boldsymbol{\xi} \cdot \mathbf{n}\, ds. \tag{6.41}$$

The three energies in (6.38), (6.39) and (6.41) are related to the total energy $e_W(\mathbf{u}, c)$ of the system, and based on the second law of thermodynamics (increase of entropy) the following *energy dissipation inequality* is assumed to hold:

$$\frac{d}{dt} e_W(\mathbf{u}, c) \leq \int_{\partial W(t)} \boldsymbol{\sigma}\mathbf{n} \cdot \mathbf{u}\, ds + \int_{\partial W(t)} \dot{c} \boldsymbol{\xi} \cdot \mathbf{n}\, ds - \int_{\partial W(t)} \mu^{\text{chem}} \mathbf{h} \cdot \mathbf{n}\, ds.$$

Based on this, in [131] the following constitutive relations are derived:

$$\begin{aligned} \boldsymbol{\sigma} &= -p\mathbf{I} + \mu \mathbf{D}(\mathbf{u}) - \varepsilon \nabla c \nabla c^T, \\ \boldsymbol{\xi} &= \varepsilon \nabla c, \\ \mathbf{h} &= -m(c) \nabla \mu^{\text{chem}}, \\ \pi &= \mu^{\text{chem}} - \varepsilon^{-1} \psi_0'(c). \end{aligned} \tag{6.42}$$

The relations $\boldsymbol{\xi} = \varepsilon \nabla c$ and $\mathbf{h} = -m(c) \nabla \mu^{\text{chem}}$ can be seen as generalized Fick's laws. For simplicity we *assume the mobility constant* $m = m(c) > 0$ *to be a constant.* Using this and (6.40) we obtain

$$\mu^{\text{chem}} = \varepsilon^{-1}\psi_0'(c) - \varepsilon\Delta c.$$

Using the conservation laws (6.35), (6.36) and the generalized Fick's law for **h** in (6.42) we obtain the following *Navier-Stokes/Cahn-Hilliard phase field model.*

$$\frac{\partial \mathbf{u}}{\partial t} + (\mathbf{u}\cdot\nabla)\mathbf{u} = -\nabla p + \operatorname{div}\left(\mu(c)\mathbf{D}(\mathbf{u})\right) - \varepsilon\operatorname{div}(\nabla c\nabla c^T), \tag{6.43a}$$

$$\operatorname{div}\mathbf{u} = 0, \tag{6.43b}$$

$$\frac{\partial c}{\partial t} + \mathbf{u}\cdot\nabla c = m\Delta\mu^{\text{chem}}, \tag{6.43c}$$

$$\mu^{\text{chem}} = \varepsilon^{-1}\psi_0'(c) - \varepsilon\Delta c. \tag{6.43d}$$

For the general case in which $m = m(c)$ is *not* constant, the term on the right-hand side in (6.43c) has to be replaced by $\operatorname{div}(m(c)\nabla\mu^{\text{chem}})$. Suitable initial and boundary conditions for the functions c and μ^{chem} are needed, for example, homogeneous Neumann boundary conditions both for c and μ^{chem} ($\nabla c\cdot\mathbf{n} = \nabla\mu^{\text{chem}}\cdot\mathbf{n} = 0$ on $\partial\Omega$) and an initial condition $c(x,0) = c_0(x)$ for $x\in\Omega$. A simple model for $\mu(c)$ is to use a convex combination between the constant viscosities μ_1, μ_2 of the pure fluids:

$$\mu(c) = \frac{c+1}{2}\mu_1 + \frac{1-c}{2}\mu_2.$$

The model (6.43) is a fundamental phase field model that occurs at many places in the literature and forms the basis for many other phase field models. Below we briefly address the following issues: theoretical results, generalizations, numerical methods, relation to other approaches.

Theoretical results. First we consider an alternative form of (6.43a) and an energy conservation result. For general smooth scalar functions v the identity

$$\operatorname{div}(\nabla v\nabla v^T) = \frac{1}{2}\nabla|\nabla v|^2 + \Delta v\,\nabla v \tag{6.44}$$

holds. Using this (with $v = c$), the definition of the free energy density ψ in (6.37) and the expression in (6.43d) we obtain

$$\begin{aligned}\nabla\psi(c,\nabla c) &= \varepsilon\frac{1}{2}\nabla|\nabla c|^2 + \varepsilon^{-1}\psi_0'(c)\nabla c\\ &= \varepsilon\operatorname{div}(\nabla c\nabla c^T) - \varepsilon\Delta c\nabla c + \varepsilon^{-1}\psi_0'(c)\nabla c\\ &= \varepsilon\operatorname{div}(\nabla c\nabla c^T) + \mu^{\text{chem}}\nabla c.\end{aligned}$$

From this we see that the Navier-Stokes equation in (6.43a) can be replaced by

$$\frac{\partial \mathbf{u}}{\partial t} + (\mathbf{u} \cdot \nabla)\mathbf{u} = -\nabla \tilde{p} + \operatorname{div}\big(\mu(c)\mathbf{D}(\mathbf{u})\big) + \mu^{\text{chem}} \nabla c, \tag{6.45}$$

where we introduced a new pressure variable $\tilde{p} := p + \psi(c, \nabla c)$. Using this Navier-Stokes equation we derive the following lemma.

Lemma 6.2.6 *Assume that for* $t \in [0, T]$ *the Navier-Stokes/Cahn-Hilliard equations* (6.43a)-(6.43d) *have a sufficiently smooth solution* $(\mathbf{u}, c, \mu^{\text{chem}})$ *with boundary conditions* $\mathbf{u}_{|\partial\Omega} = 0$, $\nabla c \cdot \mathbf{n} = \nabla \mu^{\text{chem}} \cdot \mathbf{n} = 0$ *on* $\partial\Omega$*. Then for the total energy*

$$e_\Omega(\mathbf{u}, c) = \int_\Omega \frac{1}{2}\mathbf{u} \cdot \mathbf{u} + \psi(c, \nabla c)\, dx = \frac{1}{2}\|\mathbf{u}\|_{L^2}^2 + \int_\Omega \psi(c, \nabla c)\, dx$$

the following holds:

$$\frac{d}{dt} e_\Omega\big(\mathbf{u}(t), c(t)\big) = -a\big((\mathbf{u}(t), \mathbf{u}(t)\big) - m\|\nabla \mu^{\text{chem}}(t)\|_{L^2}^2, \tag{6.46}$$

$$\begin{aligned} &e_\Omega\big(\mathbf{u}(T), c(T)\big) + \int_0^T a\big((\mathbf{u}(t), \mathbf{u}(t)\big)\, dt + m \int_0^T \|\nabla \mu^{\text{chem}}(t)\|_{L^2}^2\, dt \\ &= e_\Omega\big(\mathbf{u}(0), c(0)\big). \end{aligned} \tag{6.47}$$

Proof. First we consider the time derivative of the free energy part in the total energy:

$$\begin{aligned} \frac{d}{dt} \int_\Omega \psi(c, \nabla c)\, dx &= \int_\Omega \varepsilon \frac{d}{dt} \frac{1}{2} |\nabla c|^2 + \varepsilon^{-1} \psi_0'(c) \frac{\partial c}{\partial t}\, dx \\ &= \int_\Omega -\varepsilon \Delta c \frac{\partial c}{\partial t} + \varepsilon^{-1} \psi_0'(c) \frac{\partial c}{\partial t}\, dx = \int_\Omega \mu^{\text{chem}} \frac{\partial c}{\partial t}\, dx. \end{aligned}$$

Thus

$$\frac{d}{dt} e_\Omega\big(\mathbf{u}(t), c(t)\big) = \int_\Omega \frac{\partial \mathbf{u}}{\partial t} \cdot \mathbf{u} + \mu^{\text{chem}} \frac{\partial c}{\partial t}\, dx \tag{6.48}$$

holds. We multiply the Navier-Stokes equation (6.45) by $\mathbf{u}$, integrate over Ω, use that $c(\mathbf{u}; \mathbf{u}, \mathbf{u}) = b(\tilde{p}, \mathbf{u}) = 0$ and obtain

$$\int_\Omega \frac{\partial \mathbf{u}}{\partial t} \cdot \mathbf{u}\, dx = -a\big((\mathbf{u}(t), \mathbf{u}(t)\big) + \int_\Omega \mu^{\text{chem}} \nabla c \cdot \mathbf{u}\, dx.$$

Using (6.43c) results in

$$\begin{aligned} \int_\Omega \frac{\partial \mathbf{u}}{\partial t} \cdot \mathbf{u}\, dx &= -a\big((\mathbf{u}(t), \mathbf{u}(t)\big) - \int_\Omega \mu^{\text{chem}} \frac{\partial c}{\partial t}\, dx + m \int_\Omega \mu^{\text{chem}} \Delta \mu^{\text{chem}}\, dx \\ &= -a\big((\mathbf{u}(t), \mathbf{u}(t)\big) - \int_\Omega \mu^{\text{chem}} \frac{\partial c}{\partial t}\, dx - m\|\nabla \mu^{\text{chem}}\|_{L^2}^2, \end{aligned}$$

and using this in (6.48) proves the result in (6.46). The result in (6.47) is a direct consequence of the one in (6.46). □

The result in this lemma can be compared with the one in Lemma 6.1.6. The result in (6.47) has a physical interpretation: the energy difference $e_\Omega(\mathbf{u}(T), c(T)) - e_\Omega(\mathbf{u}(0), c(0))$ is balanced by the sum of the kinetic energy dissipation $\int_0^T a((\mathbf{u}(t), \mathbf{u}(t))\, dt$ and energy dissipation $m \int_0^T \|\nabla \mu^{\mathrm{chem}}(t)\|_{L^2}^2\, dt$ that is related to the microscopic diffusion of the two phases close to the interface.

In the phase field model that we consider the free energy $\int_\Omega \psi(c, \nabla c)\, dx$ replaces the interfacial energy $\tau \int_\Gamma 1\, ds$ that occurs in a sharp interface model. For $\varepsilon \to 0$ this free energy tends to $\tau \int_\Gamma 1\, ds$, in some suitable weak sense, cf. the following remark.

Remark 6.2.7 We discuss a result from [179] on properties of the Cahn-Hilliard free energy functional. We introduce a scaled version of this functional and for simplicity we choose a specific form of ψ_0, namely $\psi_0(c) = (1-c^2)^2$. The corresponding *scaled* (by ε) free energy is given by

$$\tilde{e}_{\mathrm{free}}(c) = \varepsilon e_{\mathrm{free}}(c) := \int_\Omega \varepsilon^2 |\nabla c|^2 + (1-c^2)^2\, dx.$$

We consider minimization of this functional over the set

$$V_\alpha := \left\{ c \in L^1(\Omega) : -1 \le c(x) \le 1 \ \text{ a.e. }, \ \ |\Omega|^{-1} \int_\Omega c\, dx = \alpha \right\},$$

with $-1 < \alpha < 1$. For $\varepsilon = 0$ the problem $\min \{ \tilde{e}_{\mathrm{free}}(c) : c \in V_\alpha \}$ has infinitely many solutions, namely all "piecewise constant" functions c, with $c(x) = 1$ for $x \in \Omega_1$, and $\Omega_1 \subset \Omega$ an arbitrary measurable set with $|\Omega|^{-1}|\Omega_1| = \frac{1}{2}(\alpha+1)$, $c(x) = -1$ for $x \in \Omega_2 := \Omega \setminus \Omega_1$. For such a function c we have $|\Omega|^{-1} \int_\Omega c\, dx = |\Omega|^{-1}(|\Omega_1| - |\Omega_2|) = 2|\Omega|^{-1}|\Omega_1| - 1 = \alpha$, i.e., $c \in V_\alpha$ and $\tilde{e}_{\mathrm{free}}(c) = 0$ (if $\varepsilon = 0$). We define $\Gamma := \partial\Omega_1 \cap \Omega$. Even if we restrict to cases in which this boundary Γ is assumed to be sufficiently smooth it can have *arbitrary area*. We now treat $\varepsilon > 0$ with $\varepsilon \downarrow 0$ and show that then the situation is quite different. We outline a main result from [179]. Consider the minimization problem

$$\min \{ \tilde{e}_{\mathrm{free}}(c) : c \in V_\alpha \}. \tag{6.49}$$

Let $(\varepsilon_n)_{n \ge 0}$ be a sequence of strictly positive numbers with $\lim_{n\to\infty} \varepsilon_n = 0$ and (u_{ε_n}) a sequence of solutions of (6.49) with $\varepsilon = \varepsilon_n$. Then there exists a subsequence, which we also denote by (u_{ε_n}), which tends to a limit u_0 in $L^1(\Omega)$, i.e. $\lim_{n\to\infty} \int_\Omega (u_{\varepsilon_n} - u_0)\, dx = 0$. The limit function takes only the extremum values 1 or -1 on Ω: $u_0 = \pm 1$ a.e. on Ω. Define $\Omega_1 := \{ x \in \Omega : u_0(x) = 1 \}$, and Γ, Ω_2 as above. Thus $\frac{1}{2}(u_{\varepsilon_n} + 1)$ tends to the characteristic function corresponding to Ω_1. To simplify the presentation we assume that Γ is smooth, say Lipschitz continuous (cf. [179] for the general case). Note that $u_0 \in V_\alpha$ and from $|\Omega|^{-1} \int_\Omega u_0\, dx = \alpha$ it follows that $|\Omega|^{-1}(|\Omega_1| - (|\Omega| - |\Omega_1|)) = \alpha$, and thus $|\Omega|^{-1}|\Omega_1| = \frac{1}{2}(\alpha+1)$.

The following holds:

$$\lim_{n\to\infty} \varepsilon_n^{-1} \tilde{e}_{\text{free}}(u_{\varepsilon_n}) = 2c_0 \int_\Gamma 1\, ds, \quad c_0 := \int_{-1}^{1} \psi_0(c)^{\frac{1}{2}}\, dc = \frac{4}{3}, \tag{6.50}$$

$$\int_\Gamma 1\, ds = \min\left\{ \int_{F\cap\Omega} 1\, ds : F = \partial W,\ W \subset \Omega,\ \frac{|W|}{|\Omega|} = \frac{1}{2}(\alpha+1) \right\}. \tag{6.51}$$

We refer to [179] for more details. This means that the free energy $e_{\text{free}}(u_{\varepsilon_n}) = \int_\Omega \psi(u_{\varepsilon_n})\, dx$ converges to $2c_0 \int_\Gamma 1\, ds$ and that the interface Γ *has minimal area*, in the sense as in (6.51). For these results to hold, *it is essential that in the free energy functional* e_{free} *the "regularization term" with* ∇c *is included.*

In the Navier-Stokes/Cahn-Hilliard model the chemical potential μ^{chem} can be eliminated by substitution of (6.43d) into (6.43c). Furthermore, in the Navier-Stokes problem the pressure can be eliminated by restricting to the subspace of divergence-free velocity fields. This results in a strongly coupled highly nonlinear system of PDEs for the unknowns $\mathbf{u}$ and c. For the analysis it may be convenient not to eliminate μ^{chem}. For the Navier-Stokes/Cahn-Hilliard model (6.43a)-(6.43d), with suitable initial and boundary conditions, the state of the art concerning existence and uniqueness of solutions is much better than for the models considered in Sects. 6.2.2 and 6.2.3. To a large extent this is due to the following two (related) facts. Firstly, if we substitute (6.43d) into (6.43c) this results in a time-dependent *convection-diffusion* problem for the concentration c. The diffusion term $-\varepsilon m \Delta^2 c$ that occurs in this equation is not present in the pure convection equations (6.25) and (6.31) and has a regularizing effect. Secondly, in this diffusive interface model we have an energy estimate such that $\|\nabla \mu^{\text{chem}}\|_{L^2}$ can be controlled, cf. Lemma 6.2.6. Such a term is not present in the energy estimate for a sharp interface model as in Lemma 6.1.6. Recently, an extensive analysis of the model (6.43a)-(6.43d) has been given in [3]. Results on existence and uniqueness of weak solutions of this model are presented which are comparable to the results for the *one-phase* Navier-Stokes model for an incompressible Newtonian fluid. For $d = 2$ existence and uniqueness of a weak solution $(\mathbf{u}, c, \mu^{\text{chem}})$ has been proved, provided the initial data for $\mathbf{u}$ and c are sufficiently regular. For $d = 3$ existence is shown to hold, but (as for the one-phase Navier-Stokes equations) uniqueness only in special cases, for example for $t \in [0, T]$ with T sufficiently small. We refer to [3, 5] for precise statements and a discussion of further results.

Generalizations. Above we considered the case of *matched densities*, cf. (6.34). In many two-phase systems the assumption ρ = constant is not reasonable, since it implies that both pure fluids must have (approximately) the same density. There are generalizations of the Navier-Stokes/Cahn-Hilliard model (6.43a)-(6.43d) for the case that the two fluids have different (or "non-matched") densities, cf. e.g. [171, 3]. These models are much more complicated as in the case of matched densities.

One important difference is that instead of the equation $\operatorname{div} \mathbf{u} = 0$ one obtains $\frac{\partial \rho}{\partial t} + \operatorname{div}(\rho \mathbf{u}) = 0$, which implies that in general $\operatorname{div} \mathbf{u} = 0$ does not hold. Hence, although the two pure fluids are incompressible, the mixture does not have this property; it is called quasi-incompressible. This complicates the analysis because one can not eliminate the pressure unknown by restricting to the space of divergence free velocity fields. Another difference is that the pressure p enters the partial differential equation for c, which makes the coupling between the Navier-Stokes and the Cahn-Hilliard equations even less transparent. There are existence results for weak solutions of the diffuse interface model for the case with non-matched densities only for some special cases, cf. [4] for a discussion. A phase field model based on the Cahn-Hilliard free energy functional for a three-(or more) phase flow problem, in which the three fluids can have different densities, is derived in [156].

In the literature there are studies on phase field models with free energies that differ from the Cahn-Hilliard form.

Numerical methods. In order to have a proper modeling of relevant physical phenomena, the parameter ε in the Navier-Stokes/Cahn-Hilliard model (6.43a)-(6.43d) has to be taken extremely small. As a consequence the order parameter c has very large gradients that must be resolved numerically. The equation for c (after elimination of μ^{chem}) is of convection-diffusion type with a *fourth* order diffusion term $-\varepsilon m \Delta^2$. The numerical treatment of such biharmonic type of equations is known to be difficult. Furthermore there is a strong nonlinear coupling between the Navier-Stokes (for $(\mathbf{u}, p)$) and Cahn-Hilliard (for (c, μ^{chem})) equations. Hence, even for the case of matched densities the Navier-Stokes/Cahn-Hilliard model has a very high numerical complexity. For non-matched densities (which are physically much more relevant) there is a further significant increase in the numerical complexity. Some early numerical results for a Navier-Stokes/Cahn-Hilliard model (6.43a)-(6.43d) of a two-dimensional two-phase flow problem are presented in [149]. This model is also simulated, again for a two-dimensional problem, in [155]. In both cases uniform grids and finite difference or finite volume discretization methods are used. Numerical simulations of a spatially three-dimensional Navier-Stokes/Cahn-Hilliard model with matched densities are given in [20]. Recent work on numerical simulations of two-phase flows based on phase field models is found in [233, 234]. The numerical simulation of a spatially two-dimensional *three*-phase system with matched densities is treated in [156].

It appears that up to now numerical simulations of two-phase flows based on phase field interface representations are used and studied much less than those based on interface tracking or interface capturing (with VOF or level set) techniques.

Comparison to other approaches. One important difference between the diffusive and the sharp interface approach was already mentioned above: the energy estimates are different, cf. Lemma 6.1.6 and Lemma 6.2.6. This has strong implications for the theoretical analysis. Furthermore, in the volume tracking techniques (VOF or level set) discussed above we have a pure transport equation for an indicator function (characteristic function or level set function), whereas as in the phase field method we have a convection-diffusion type of equation for the concentration c. In the Navier-Stokes equation (6.43a) of the phase field model a "localized force" term

$$-\varepsilon \operatorname{div}(\nabla c \nabla c^T) \tag{6.52}$$

occurs. We comment on this term and its relation to the surface tension forces used in other approaches. First we consider the force term that occurs in the *strong* formulation if one uses a level set CSF approach, as explained in Remark 6.2.4. We assume a (highly) idealized situation in which the zero level of c describes the interface Γ and the scaled order function $\tilde{c} := \varepsilon c$ is a signed distance function to Γ in a neighborhood U_ε of Γ. Hence, $\varepsilon|\nabla c| = |\nabla \tilde{c}| = 1$ holds in U_ε. Using $\phi = \tilde{c}$ in (6.33) and $\delta_\epsilon(\phi(z)) \approx \epsilon^{-1}$ for z sufficiently close to Γ, we obtain

$$\begin{aligned} -\tau \operatorname{div}\Big(\frac{\nabla \phi}{|\nabla \phi|}\Big)\delta_\epsilon(\phi(z))\nabla \phi &\approx -\tau \epsilon^{-1} \operatorname{div}(\nabla \tilde{c})\nabla \tilde{c} \\ &= -\tau \epsilon^{-1}\varepsilon^2 \Delta c \nabla c = -\tau \epsilon^{-1}\varepsilon^2 \operatorname{div}(\nabla c \nabla c^T), \end{aligned}$$

in a sufficiently small neighborhood of Γ. In the last equality we used the relation (6.44) and $\nabla|\nabla c|^2 = 0$. This agrees with the localized force as in (6.52) if we take ϵ (in Dirac delta function) and ε (in Cahn-Hilliard) such that $\epsilon \sim \tau\varepsilon$.

In the *weak* formulation the force term in the Cahn-Hilliard model (6.43a) takes the form

$$\tilde{f}_\Omega(\mathbf{v}) := -\varepsilon \int_\Omega \operatorname{div}(\nabla c \nabla c^T)\mathbf{v}\, dx, \tag{6.53}$$

which we now compare with the functional $f_\Gamma(\mathbf{v}) = -\tau \int_\Gamma \kappa \mathbf{n} \cdot \mathbf{v}\, ds$ in (6.8) that is used for surface tension representation in the weak formulation of a sharp interface representation. For this we use another formula, cf. Lemma 14.1.2:

$$f_\Gamma(\mathbf{v}) = -\tau \int_\Gamma \operatorname{tr}(\mathbf{P}\nabla \mathbf{v})\, ds, \quad \mathbf{P} = \mathbf{I} - \mathbf{n}\mathbf{n}^T.$$

We consider only test functions with $\operatorname{div} \mathbf{v} = 0$ (which is reasonable, since by elimination of the pressure one can restrict to the subspace of divergence free velocities). Then we have $\operatorname{tr} \nabla \mathbf{v} = \operatorname{div} \mathbf{v} = 0$ and thus we get

$$f_\Gamma(\mathbf{v}) = \tau \int_\Gamma \operatorname{tr}(\mathbf{n}\mathbf{n}^T \nabla \mathbf{v})\, ds = \tau \int_\Gamma \mathbf{n}^T \nabla \mathbf{v}\, \mathbf{n}\, ds, \tag{6.54}$$

with $\mathbf{n} = \mathbf{n}_\Gamma$. Using partial integration the volume force in (6.53) can be reformulated as

$$\tilde{f}_\Omega(\mathbf{v}) = \varepsilon \int_\Omega \operatorname{tr}(\nabla c \nabla c^T \nabla \mathbf{v})\, dx = \varepsilon \int_\Omega \nabla c^T \nabla \mathbf{v} \nabla c\, dx.$$

As above we assume that $\tilde{c} = \varepsilon c$ is a signed distance function to Γ in U_ε. Note that then $\nabla\tilde{c}(x) = \mathbf{n}(x)$ for $x \in \Gamma$ and $\nabla\tilde{c}(y) = \mathbf{n}(x)$ for $y \in U_\varepsilon$, $x \in \Gamma$ and $y - x = \alpha\mathbf{n}(x)$ with $\alpha \in \mathbb{R}$. Assume that $\mathbf{v}$ is sufficiently smooth and that $|\nabla\tilde{c}(y)| \ll 1$ for $y \notin U_\varepsilon$, cf. Fig. 6.4. Using a suitable coordinate transformation one obtains

$$\begin{aligned}\tilde{f}_\Omega(\mathbf{v}) &= \varepsilon^{-1}\int_\Omega \nabla\tilde{c}^T\nabla\mathbf{v}\nabla\tilde{c}\,dx \approx \varepsilon^{-1}\int_{-\varepsilon}^{\varepsilon}\int_\Gamma \nabla\tilde{c}(s,0)^T\nabla\mathbf{v}(s,r)\nabla\tilde{c}(s,0)\,ds\,dr\\ &\approx \varepsilon^{-1}\int_{-\varepsilon}^{\varepsilon}\int_\Gamma \mathbf{n}(s,0)^T\nabla\mathbf{v}(s,0)\mathbf{n}(s,0)\,ds\,dr \approx 2\int_\Gamma \mathbf{n}^T\nabla\mathbf{v}\mathbf{n}\,ds,\end{aligned}$$

which is of the same form as the functional in (6.54).

6.3 Weak formulation

In the remainder of this monograph we restrict to the level set method for interface representation. Our choice is motivated by the following requirements. We want to develop a solver that can handle interface singularities (droplet collision), too. Therefore an interface tracking approach based on the interface condition $V_\Gamma = \mathbf{u}\cdot\mathbf{n}$ in (6.3) is less suitable. A representation of the interface as a surface in $\mathbb{R}^3$, which corresponds to a sharp interface model, is desirable since we want to use a model for mass transport between the phases, in which a Henry condition *at* the interface occurs, and a model for surfactant transport *on* the interface. It is not clear how (variants of) these models can be combined with a phase field approach. Therefore, we decided not to use a phase field method for interface representation. Comparing the VOF and level set interface capturing methods we decided to use the latter, since the numerical treatment of the transport equation for the level set function (which is smooth) is easier than for the characteristic function (which is discontinuous) and because the level set approach fits better in a finite element discretization framework than the VOF approach. The latter is more natural in a finite volume discretization context. Furthermore, the task of interface reconstruction is much easier using the discrete level set function instead of the discrete VOF color function. A disadvantage of the level set method compared to VOF is that it has a worse mass conservation property.

In this section we present a two-phase Navier-Stokes/level set model. We start with the strong formulation. Then a weaker variational model is formulated, which forms the basis of the finite element discretization method treated in the next chapter.

The jumps in the coefficients ρ and μ can be described using the level set function ϕ in combination with the Heaviside function $H : \mathbb{R} \to \mathbb{R}$:

$$H(\zeta) = 0 \quad \text{for} \quad \zeta < 0, \qquad H(\zeta) = 1 \quad \text{for} \quad \zeta > 0.$$

For ease one can set $H(0) = \frac{1}{2}$. We define

$$\begin{aligned} \rho(\phi) &:= \rho_1 + (\rho_2 - \rho_1)H(\phi), \\ \mu(\phi) &:= \mu_1 + (\mu_2 - \mu_1)H(\phi). \end{aligned} \tag{6.55}$$

We reconsider the strong formulation of the two-phase flow problem in (6.1)-(6.2). Instead of the Lagrangian interface propagation condition $V_\Gamma = \mathbf{u} \cdot \mathbf{n}$ in (6.3) we use the level set function for the representation of the interface and therefore add the level set equation (6.31) to the model. This results in the following model for the two-phase problem in $\Omega \times [0, T]$:

$$\begin{gathered} \begin{cases} \rho(\phi)\left(\frac{\partial \mathbf{u}}{\partial t} + (\mathbf{u} \cdot \nabla)\mathbf{u}\right) = -\nabla p + \operatorname{div}(\mu(\phi)\mathbf{D}(\mathbf{u})) + \rho(\phi)\mathbf{g} \\ \operatorname{div} \mathbf{u} = 0 \end{cases} \quad \text{in } \Omega_i,\ i = 1, 2, \\ [\boldsymbol{\sigma}\mathbf{n}]_\Gamma = -\tau\kappa\mathbf{n}, \quad [\mathbf{u}]_\Gamma = 0 \quad \text{on } \Gamma, \\ \frac{\partial \phi}{\partial t} + \mathbf{u} \cdot \nabla \phi = 0 \quad \text{in } \Omega, \end{gathered} \tag{6.56}$$

together with suitable initial and boundary conditions for $\mathbf{u}$ and ϕ, cf. Sect. 1.2. For the level set function ϕ, the initial condition is $\phi(x, 0) = \phi_0(x)$, in which ϕ_0 is given and should be such that $\{ x \in \mathbb{R}^3 : \phi_0(x) = 0 \} = \Gamma(0)$. Moreover, ϕ_0 should be an (approximate) signed distance function to $\Gamma(0)$. To make the problem with the linear hyperbolic level set equation well-posed one needs boundary conditions on the inflow boundary $\partial\Omega_{in} := \{ x \in \partial\Omega : \mathbf{u} \cdot \mathbf{n}_\Omega < 0 \}$. There are no natural (e.g., physics based) boundary conditions for ϕ at the inflow boundary. We are only interested in values of ϕ close to the interface (= zero level on ϕ) and ϕ is evolved according to $\frac{\partial \phi}{\partial t} + \mathbf{u} \cdot \nabla \phi = 0$ only for a short time interval. After this short time a re-initialization of ϕ is applied, cf. Sect. 7.4.1. Due to this the issue of the choice of the boundary condition for the level set function on $\partial\Omega_{in}$ is of minor importance.

Note that the model (6.56) is *not* in dimensionless form. A dimensionless formulation can be derived in a similar way as in Remark 6.1.1.

As discussed in Sect. 6.2.3 a general weak formulation of the model (6.56) for which well-posedness has been proved, is not known in the literature. As basis for the finite element discretization we will use the weak formulation of the Navier-Stokes problem given in (6.9)-(6.10) and combine it with a weak formulation of the level set equation.

We address this weak formulation of the level set equation. We do not apply the approach based on viscosity solutions of transport equations, briefly discussed in Sect. 6.2.3, since this requires the velocity field to be continuous: $\mathbf{u} \in C(\overline{\Omega} \times [0, T])^3$, which is not compatible with the usual weak formulations of the Navier-Stokes equation.

We also do not use a weak formulation as in Proposition 6.2.2 which is based on the concept of renormalized solutions of transport equations. A disadvantage of this formulation is that it leads to a space-time variational problem. In our setting we want to have a variational formulation only in space, cf. (6.9)-(6.10). We introduce a "space-only" variational formulation of the level set equation as in [106], Sect. 6.3. We consider a transport equation of the form

$$\frac{\partial \phi}{\partial t} + \mathbf{w} \cdot \nabla \phi = 0, \tag{6.57}$$

with $\mathbf{w} \in H^1(\Omega)^3$, $\operatorname{div} \mathbf{w} = 0$ and (for simplicity) $\mathbf{w}_{|\partial\Omega} = 0$. Then $\partial\Omega_{in} = \emptyset$ and thus we do not need boundary conditions for ϕ. The initial condition is given by $\phi(x, 0) = \phi_0(x)$. Note that opposite to $\mathbf{u}$ used in the level set equation in (6.56) the velocity field $\mathbf{w}$ is *independent of* t. We introduce a so-called anisotropic Sobolev space, in which only derivatives in a particular direction, namely the flow direction $\mathbf{w}$, are considered. On $C^\infty(\Omega)$ we introduce the norm (and corresponding scalar product) $\|u\|_{1,\mathbf{w}}^2 := \|u\|_{L^2}^2 + \|\mathbf{w} \cdot \nabla u\|_{L^2}^2$. Let $W_{\mathbf{w}}$ be the completion of $C^\infty(\Omega)$ with respect to this norm. Then $W_{\mathbf{w}}$ is a Hilbert space and this space can also be characterized as

$$W_{\mathbf{w}} = \{\, u \in L^2(\Omega) : \mathbf{w} \cdot \nabla u \in L^2(\Omega) \,\}$$

(where the derivative is defined in a distributional sense). This appears to be an appropriate space for a weak formulation of the transport equation (6.57):

Proposition 6.3.1 *Take $\phi_0 \in W_{\mathbf{w}}$. There exists a unique $\phi(t) = \phi(\cdot, t) \in C^1([0,T]; L^2(\Omega)) \cap C([0,T]; W_{\mathbf{w}})$ such that $\phi(0) = \phi_0$ and*

$$\left(\frac{d\phi}{dt}, v\right)_{L^2} + (\mathbf{w} \cdot \nabla\phi, v)_{L^2} = 0 \quad \text{for all } v \in L^2(\Omega), \quad t \in [0,T].$$

Proof. This result is given in Theorem 6.52 in [106]. Its proof is based on a fundamental result known as the Hille-Yosida theorem. We only present some key ingredients which indicate that $W_{\mathbf{w}}$ and $L^2(\Omega)$ are the right spaces for the variational formulation. For a complete proof we refer to [106] and the references therein.

We outline the Hille-Yosida theorem. Let H be a Hilbert space and $C : D(C) \subset H \to H$ a linear operator. This operator is called monotone if $(Cv, v)_H \geq 0$ for all $v \in D(C)$ holds, and maximal if $I + C : D(C) \to H$ is bijective. The operator is maximal monotone if both properties hold. The Hille-Yosida theorem essentially states that if C is maximal monotone then an initial value problem of the form

$$\frac{du}{dt} + Cu = f, \quad t \in [0,T], \quad u(0) = u_0,$$

with $f \in H$ and $u_0 \in D(C)$, has a unique solution $u \in C^1([0,T]; H) \cap C([0,T]; D(C))$. In our context we have $H = L^2(\Omega)$ and $C : W_{\mathbf{w}} = D(C) \to$

$L^2(\Omega)$ is defined by $(C\phi, v)_H = (\mathbf{w} \cdot \nabla\phi, v)_{L^2} =: c(\phi, v)$. The operator C is monotone, since:

$$(C\phi, \phi)_H = \int_\Omega \mathbf{w} \cdot \nabla\phi \, \phi \, dx = -\int_\Omega \phi(\phi \operatorname{div} \mathbf{w} + \mathbf{w} \cdot \nabla\phi) \, dx = -(C\phi, \phi)_H,$$

and thus $(C\phi, \phi)_H = 0$ for all $\phi \in W_\mathbf{w}$. In order to show that C is also maximal we consider the bilinear form $id + c : W_\mathbf{w} \times L^2(\Omega) \to \mathbb{R}$ given by $id(\phi, v) + c(\phi, v) = (\phi + \mathbf{w} \cdot \nabla\phi, v)_{L^2}$. This bilinear form is bounded on $W_\mathbf{w} \times L^2(\Omega)$. Now note that for $\phi \in W_\mathbf{w}$ we have

$$\begin{aligned}
\sup_{v \in L^2(\Omega)} \frac{id(\phi, v) + c(\phi, v)}{\|v\|_{L^2}} &= \sup_{v \in L^2(\Omega)} \frac{(\phi + \mathbf{w} \cdot \nabla\phi, v)_{L^2}}{\|v\|_{L^2}} = \|\phi + \mathbf{w} \cdot \nabla\phi\|_{L^2} \\
&= \left(\|\phi\|_{L^2}^2 + 2(\phi, \mathbf{w} \cdot \nabla\phi)_{L^2} + \|\mathbf{w} \cdot \nabla\phi\|_{L^2}^2\right)^{\frac{1}{2}} \\
&= \left(\|\phi\|_{L^2}^2 + \|\mathbf{w} \cdot \nabla\phi\|_{L^2}^2\right)^{\frac{1}{2}} = \|\phi\|_{1,\mathbf{w}},
\end{aligned}$$

and thus the inf-sup property

$$\inf_{\phi \in W_\mathbf{w}} \sup_{v \in L^2(\Omega)} \frac{id(\phi, v) + c(\phi, v)}{\|\phi\|_{1,\mathbf{w}} \|v\|_{L^2}} \geq 1$$

holds. Furthermore, it can be shown that $id(\phi, v) + c(\phi, v) = 0$ for all $\phi \in W_\mathbf{w}$ implies $v = 0$. From the boundedness of the bilinear form $id + c$, the inf-sup bound and the latter result it follows that $I + C : W_\mathbf{w} \to L^2(\Omega)$ is bijective, cf. Theorem 15.1.1. Hence, C is maximal monotone and the Hille-Yosida theorem yields existence and uniqueness of $\phi \in C^1([0,T]; L^2(\Omega)) \cap C([0,T]; W_\mathbf{w})$ such that $(\frac{d\phi}{dt}, v)_{L^2} + c(\phi, v)_{L^2} = 0$ for all $v \in L^2(\Omega)$. □

Motivated by this result we introduce a weak formulation of the level set equation in (6.56). We want to allow $\mathbf{u}_{\partial\Omega} \neq \emptyset$ and therefore use a subspace $W_{\mathbf{u},D} := \{ w \in W_\mathbf{u} : w_{|\partial\Omega_{in}} = \phi_D \}$ of $W_\mathbf{u}$. The weak formulation is as follows: find $\phi(\cdot, t) \in W_{\mathbf{u},D}$ such that $\phi(\cdot, 0) = \phi_0$ and

$$(\frac{\partial\phi}{\partial t}, v)_{L^2} + (\mathbf{u} \cdot \nabla\phi, v)_{L^2} = 0 \quad \text{for all } \; v \in L^2(\Omega), \quad t \in [0,T]. \tag{6.58}$$

Note that in this problem the velocity $\mathbf{u}$ *depends on* t and therefore Proposition 6.3.1 can not be applied. Related to this, the space $W_{\mathbf{u},D}$ depends on $\mathbf{u}$ and thus on t.

Summarizing, we obtain the following two-phase incompressible flow model:

Find $\mathbf{u}(t) = \mathbf{u}(\cdot, t) \in \mathbf{V}_D$, $p(t) = p(\cdot, t) \in Q$, $\phi(t) = \phi(\cdot, t) \in W_{\mathbf{u},D}$ such that for almost all $t \in [0, T]$

$$m(\frac{\partial \mathbf{u}}{\partial t}, \mathbf{v}) + c(\mathbf{u}; \mathbf{u}, \mathbf{v}) \\ + a(\mathbf{u}, \mathbf{v}) + b(\mathbf{v}, p) = (\rho \mathbf{g}, \mathbf{v})_{L^2} + f_\Gamma(\mathbf{v}) \quad \text{for all} \quad \mathbf{v} \in \mathbf{V}_0, \tag{6.59a}$$

$$b(\mathbf{u}, q) = 0 \quad \text{for all } q \in Q, \tag{6.59b}$$

$$(\frac{\partial \phi}{\partial t}, v)_{L^2} + (\mathbf{u} \cdot \nabla \phi, v)_{L^2} = 0 \quad \text{for all } v \in L^2(\Omega), \tag{6.59c}$$

$$\text{initial conditions} \quad \mathbf{u}(0) = \mathbf{u}_0, \quad \phi(0) = \phi_0 \quad \text{in } \Omega.$$

Thus we have a Navier-Stokes equation (in weak form) in the *whole domain* Ω coupled with a linear hyperbolic equation (in weak form) for the level set function ϕ. The spaces used for velocity $\mathbf{u}$ and pressure p are the same as those used in the weak formulation of a one-phase flow problem. In the two-phase Navier-Stokes equation we have *discontinuous* viscosity and density coefficients. Furthermore, we have a source term f_Γ which is (only) a functional and which requires integration over the (unknown) interface Γ. This is a *sharp interface model*: there is no regularization ("numerical diffusion") caused by a smoothed Dirac delta function, cf. Remark 6.2.4.

The issue of well-posedness of this weak formulation is largely unsolved. Only under strong (unrealistic) smoothness assumptions on the data (including the initial interface) well-posedness results for (6.59a)-(6.59b) are known in the literature, cf. the discussion in Sect. 6.1. If the velocity field $\mathbf{u}(x, t)$ is sufficiently smooth (Lipschitz w.r.t. x) then the strong formulation of the level set equation is well-posed and thus also the weaker formulation in (6.59c). This weak model, however, is supposed to be suitable for less regular problems, too. Theoretical analysis that shows correctness of this claim is lacking.

In the next chapter we treat finite element methods for the discretization of this model.

7

Finite element discretization of two-phase flow model

7.1 Introduction

In this chapter we treat finite element methods for the two-phase flow model in (6.59). We use a nested family of multilevel triangulations $\{\mathcal{T}_h\}$ as explained in Sect. 3.1. In our applications these grids will be locally refined in a (small) neighborhood of the interface. In Sect. 7.2 we discuss a finite element method for discretization of the level set equation. In Sect. 7.3 it is explained how for a resulting approximation ϕ_h of the level set function ϕ a corresponding approximation Γ_h (= approximate zero level of ϕ_h) of the interface Γ can be constructed. Other important issues related to the level set function, such as re-initialization, are treated in Sect. 7.4. Results of experiments with numerical methods applied to the level set equation are given in Sect. 7.5. In Sect. 7.6 a method for discretization of the surface tension force f_Γ is presented. An error analysis of this method is given in Sect. 7.7 and results of numerical experiments with this method are presented in Sect. 7.8. In Sect. 7.9 we treat a special finite element space for the discretization of the pressure variable. Results of numerical experiments with this space are given in Sect. 7.10. Finally, in Sect. 7.11 we apply the methods treated in this chapter for the discretization of the two-phase flow model (6.59).

7.2 Discretization of the level set equation

The level set equation is of linear hyperbolic type. It is well-known that standard conforming finite element discretization methods are in general not very suitable for such partial differential equations, since these methods can be unstable. There is extensive literature on finite element techniques for hyperbolic problems. We do not give an overview here, but refer to monographs in which this topic is treated, e.g., [108, 211, 206]. One popular strategy is to combine standard finite element spaces with a stabilization technique. A fundamental stabilization method, that is very often used in practice, is the *streamline-diffusion finite element method* (SDFEM). We will apply this method for the discretization of the level set equation. In Sect. 7.2.1 we explain the basic idea of the technique using a simple 1D problem. In Sect. 7.2.2 we apply this method for the discretization of the level set equation.

7.2.1 Introduction to stabilization

We consider the very simple one-dimensional (hyperbolic) problem

$$\begin{aligned} bu'(x) + u(x) &= f(x), \quad x \in I := (0,1), \quad b > 0 \text{ a given constant}, \\ u(0) &= 0. \end{aligned} \tag{7.1}$$

For the weak formulation we introduce the Hilbert spaces

$$H_1 = \left\{\, v \in H^1(I) : v(0) = 0 \,\right\}, \quad H_2 = L^2(I).$$

The norm on H_1 is $\|v\|_1^2 = \|v\|_{L^2}^2 + \|v'\|_{L^2}^2$. We define the bilinear form

$$k(u,v) = \int_0^1 bu'v + uv\,dx$$

on $H_1 \times H_2$.

Theorem 7.2.1 *Take $f \in L^2(I)$. There exists a unique $u \in H_1$ such that*

$$k(u,v) = (f,v)_{L^2} \quad \textit{for all} \quad v \in H_2. \tag{7.2}$$

Moreover, $\|u\|_1 \leq c\|f\|_{L^2}$ holds with c independent of f.

Proof. The proof is based on an application of Theorem 15.1.1. The bilinear form $k(\cdot,\cdot)$ is continuous on $H_1 \times H_2$:

$$|k(u,v)| \leq b\|u'\|_{L^2}\|v\|_{L^2} + \|u\|_{L^2}\|v\|_{L^2} \leq \sqrt{2}\max\{1,b\}\|u\|_1\|v\|_{L^2},$$

for $u \in H_1$, $v \in H_2$. For $u \in H_1$ we have

$$\begin{aligned} \sup_{v\in H_2} \frac{k(u,v)}{\|v\|_{L^2}} &= \sup_{v\in H_2} \frac{(bu'+u,v)_{L^2}}{\|v\|_{L^2}} = \|bu'+u\|_{L^2} \\ &= \left(b^2\|u'\|_{L^2}^2 + \|u\|_{L^2}^2 + 2b(u',u)_{L^2}\right)^{\frac{1}{2}}. \end{aligned}$$

Using $u(0) = 0$ we get $(u',u)_{L^2} = u(1)^2 - (u,u')_{L^2}$ and thus $(u',u)_{L^2} \geq 0$. Hence we get

$$\sup_{v\in H_2} \frac{k(u,v)}{\|v\|_{L^2}} \geq \min\{1,b\}\|u\|_1 \quad \text{for all} \quad u \in H_1,$$

i.e., the inf-sup condition for $k(\cdot,\cdot)$ is satisfied. We now prove that if $v \in H_2$ is such that $k(u,v) = 0$ for all $u \in H_1$, this implies $v = 0$. Take $v \in H_2$ with $k(u,v) = 0$ for all $u \in H_1$. This implies $b\int_0^1 u'v\,dx = -\int_0^1 uv\,dx$ for all $u \in C_0^\infty(I)$ and thus $v \in H^1(I)$ with $v' = \frac{1}{b}v$ (weak derivative). Using this we obtain

$$\begin{aligned} -\int_0^1 uv\,dx &= b\int_0^1 u'v\,dx = bu(1)v(1) - b\int_0^1 uv'\,dx \\ &= bu(1)v(1) - \int_0^1 uv\,dx \quad \text{for all } \ u \in H_1, \end{aligned}$$

and thus $u(1)v(1) = 0$ for all $u \in H_1$. This implies $v(1) = 0$. Using this and $bv' - v = 0$ yields

$$\begin{aligned} \|v\|_{L^2}^2 &= (v,v)_{L^2} + (bv' - v, v)_{L^2} \\ &= b(v',v)_{L^2} = \frac{b}{2}\big(v(1)^2 - v(0)^2\big) = -\frac{b}{2}v(0)^2 \le 0. \end{aligned}$$

This implies $v = 0$. Application of Theorem 15.1.1 yields existence and uniqueness of a solution $u \in H_1$ and $\|u\|_1 \le c\,\|f\|_{L^2}$, which completes the proof. □

Remark 7.2.2 The analysis in the proof above is essentially the same as that used in the proof of Proposition 6.3.1 to show that the operator $I + C : W_{\mathbf{w}} \to L^2(\Omega)$ is bijective. This operator corresponds to the bilinear form $(\phi, v) \to (\phi + \mathbf{w}\cdot\nabla\phi, v)_{L^2}$ on $W_{\mathbf{w}} \times L^2(\Omega)$, which is the higher dimensional generalization of the bilinear form $k(\cdot,\cdot)$ used in the proof above.

For the discretization of the well-posed variational problem (7.2) we use a Galerkin method with a standard finite element space. To simplify the notation we use a uniform grid and consider only linear finite elements. Let $h = \frac{1}{n}$, $x_i = ih$, $0 \le i \le n$, and

$$\mathbb{X}_h = \big\{\, v \in C(I) : v(0) = 0,\ v_{|[x_i,x_{i+1}]} \in \mathcal{P}_1 \ \text{ for } \ 0 \le i \le n-1 \,\big\}.$$

Note that $\mathbb{X}_h \subset H_1$ and $\mathbb{X}_h \subset H_2$. The discretization is as follows:

$$\text{determine } \ u_h \in \mathbb{X}_h \ \text{ such that } \ k(u_h, v_h) = (f, v_h)_{L^2} \quad \text{for all } \ v_h \in \mathbb{X}_h. \tag{7.3}$$

For the error analysis of this method we apply Céa's lemma 15.1.3. It remains to verify the discrete inf-sup condition:

$$\exists\, \varepsilon_h > 0 \ : \quad \sup_{v_h \in \mathbb{X}_h} \frac{k(u_h, v_h)}{\|v_h\|_{L^2}} \ge \varepsilon_h\,\|u_h\|_1 \quad \text{for all } \quad u_h \in \mathbb{X}_h. \tag{7.4}$$

Related to this we give the following lemma:

Lemma 7.2.3 *The inf-sup property* (7.4) *holds with* $\varepsilon_h = c\,h$, $c > 0$ *independent of* h.

Proof. For $u_h \in \mathbb{X}_h$ we have $(u_h', u_h)_{L^2} = \frac{1}{2}u_h(1)^2 \geq 0$ and thus

$$\sup_{v_h \in \mathbb{X}_h} \frac{k(u_h, v_h)}{\|v_h\|_{L^2}} \geq \frac{k(u_h, u_h)}{\|u_h\|_{L^2}} = \frac{b(u_h', u_h)_{L^2} + \|u_h\|_{L^2}^2}{\|u_h\|_{L^2}} \geq \|u_h\|_{L^2}.$$

Now apply an inverse inequality, $\|v_h'\|_{L^2} \leq ch^{-1}\|v_h\|_{L^2}$ for all $v_h \in \mathbb{X}_h$, resulting in $\|u_h\|_{L^2} \geq \frac{1}{2}\|u_h\|_{L^2} + ch\|u_h'\|_{L^2} \geq ch\,\|u_h\|_1$ with a constant $c > 0$ independent of h. □

It can be shown that the result in this lemma is sharp in the sense that the best (i.e. largest) inf-sup constant ε_h in (7.4) in general satisfies $\varepsilon_h \leq c\,h$. This indicates that the standard linear finite element method is *unstable* in the sense that the inf-sup constant deteriorates for $h \downarrow 0$. This instability can be observed in numerical experiments with the discretization (7.3) for this simple 1D problem.

We will show how a satisfactory discretization with the space $\mathbb{X}_h$ of linear finite elements can be obtained by using the concept of *stabilization.*

If $u \in H_1$ satisfies (7.2), then

$$\int_0^1 (bu' + u)bv'\,dx = (f, bv')_{L^2} \quad \text{for all } \ v \in H_1 \tag{7.5}$$

also holds. We add this equation δ-times, with δ a parameter in $[0, 1]$, to the one in (7.2). Thus the solution $u \in H_1$ of (7.2) also satisfies

$$k_\delta(u, v) = f_\delta(v) \quad \text{for all } \ v \in H_1, \text{ with} \tag{7.6a}$$

$$k_\delta(u, v) := (bu' + u, \delta bv' + v)_{L^2}, \quad f_\delta(v) := (f, \delta bv' + v)_{L^2}. \tag{7.6b}$$

Note that for $\delta = 0$ we have the original bilinear form and that $\delta = 1$ results in a problem with a *symmetric* bilinear form . For $\delta \neq 1$ the bilinear form $k_\delta(\cdot, \cdot)$ is not symmetric. For all $\delta \in [0, 1]$ we have $f_\delta \in H_1'$. The stabilizing effect for $\delta > 0$ is seen from the ellipticity estimate:

$$\begin{aligned} k_\delta(u, u) &= \delta b^2 \int_0^1 (u')^2\,dx + \int_0^1 u^2\,dx + b(\delta + 1)\int_0^1 u'u\,dx \\ &\geq \delta b^2 |u|_1^2 + \|u\|_{L^2}^2 \quad \text{for all } \ u \in H_1. \end{aligned} \tag{7.7}$$

Note that for $\delta > 0$ the norm $|u|_1$ occurs in this stability estimate. The discrete problem is as follows:

$$\text{determine } \ u_h \in \mathbb{X}_h \ \text{ such that } \ k_\delta(u_h, v_h) = f_\delta(v_h) \quad \text{for all } \ v_h \in \mathbb{X}_h. \tag{7.8}$$

The discrete solution u_h depends on δ. Using the stability estimate (7.7), approximation properties of the finite element space $\mathbb{X}_h$ and a variant of Céa's lemma the following (sharp) result on the discretization error can be proved:

Proposition 7.2.4 *Let $u \in H_1$ and $u_h \in \mathbb{X}_h$ be the solutions of (7.2) and (7.8), respectively, and assume that $u \in H^2(I)$. For all $\delta \in [0,1]$ the error bound*

$$b\sqrt{\delta}|u-u_h|_1 + \|u-u_h\|_{L^2} \leq Ch\big[h + b\sqrt{\delta} + b\min\{1, \frac{h}{b\sqrt{\delta}}\}\big]\|u''\|_{L^2} \quad (7.9)$$

holds with a constant C independent of h, δ, b and u.

The term between square brackets in (7.9) is minimal for $h \leq b$ if we take

$$\delta = \delta_{\mathrm{opt}} = \frac{h}{b}. \quad (7.10)$$

We consider three cases:
$\underline{\delta = 0}$ (no stabilization): Then we get $\|u-u_h\|_{L^2} \leq ch\|u''\|_{L^2}$. We can not control the discretization error in the stronger H^1-norm.
$\underline{\delta = 1}$ (full stabilization): Then we obtain

$$|u-u_h|_1 \leq ch\|u''\|_{L^2}, \quad \|u-u_h\|_{L^2} \leq ch\|u''\|_{L^2}.$$

$\underline{\delta = \delta_{\mathrm{opt}}}$ (optimal value): This results in

$$|u-u_h|_1 \leq ch\|u''\|_{L^2}, \quad \|u-u_h\|_{L^2} \leq ch^{\frac{3}{2}}\|u''\|_{L^2}. \quad (7.11)$$

Hence, in the latter case the bound for the norm $|\cdot|_1$ is the same as for $\delta = 1$, but we have an improvement in the L^2-error bound. The best stability property, in the sense of (7.7), is for the case $\delta = 1$. A somewhat weaker stability property but a better approximation property is obtained for $\delta = \delta_{\mathrm{opt}}$. For $\delta = \delta_{\mathrm{opt}}$ we have a *good compromise between sufficient stability and high approximation quality.*

The concept of stabilization as explained in this section is a very general one. It can be applied in higher dimensions, using finite elements of degree larger than one and also if instead of a hyperbolic equation one has to discretize a convection-diffusion problem in which convection is dominating. An extensive analysis of stabilization techniques is given in [211].

7.2.2 Discretization of the level set equation by the streamline diffusion finite element method

In this section we treat a stabilization approach, the so-called streamline diffusion stabilization method (SDFEM), for the discretization (in space) of the level set equation (6.59c). This method is based on the same approach as presented for a relatively simple one-dimensional hyperbolic problem in the previous section.

We introduce the finite element space of continuous piecewise polynomial functions. Let $\mathcal{V}(\partial\Omega_{in})$ be the set of vertices on the inflow boundary $\partial\Omega_{in} :=$

$\{x \in \partial\Omega : \mathbf{u}\cdot\mathbf{n}_\Omega < 0\}$. Given the Dirichlet boundary data ϕ_D on $\partial\Omega_{in}$ we define, for $k \geq 1$, the (affine) finite element space

$$V_h(\phi_D) := \left\{ v \in C(\Omega) : v_{|T} \in \mathcal{P}_k \ \forall\, T \in \mathcal{T}_h, \ \ v(x) = \phi_D(x) \ \forall\, x \in \mathcal{V}(\partial\Omega_{in}) \right\}.$$

The choice of the boundary data ϕ_D will be addressed in Remark 7.5.1. We use the notation $V_h = V_h(0)$. The latter space is independent of t, whereas $V_h(\phi_D)$ depends on t if the boundary data ϕ_D are time dependent. Note that $V_h(\phi_D) = V_h$ holds if $\phi_D = 0$ or $\partial\Omega_{in} = \emptyset$. As we will see later on, for the quality of the curvature approximation of the interface it is important to use finite elements of *degree at least two* for the approximation of the level set function. As explained in Sect. 7.2.1, cf. (7.6), for the spatial discretization of the level set equation (6.59c) we use test functions $\hat{v}_h \in L^2(\Omega)$ of the form

$$\hat{v}_h|_T := v_h + \delta_T \mathbf{u}\cdot\nabla v_h, \qquad T \in \mathcal{T}_h, \ \ v_h \in V_h. \tag{7.12}$$

The *streamline diffusion finite element* discretization of the level set equation is as follows:

Let $\phi_{0,h} \in V_h(\phi_D)$ be an approximation of the initial condition $\phi_0 = \phi(0)$. Determine $\phi_h(t) \in V_h(\phi_D)$ with $\phi_h(0) = \phi_{0,h}$ and such that

$$\sum_{T\in\mathcal{T}_h} \Big(\frac{\partial \phi_h}{\partial t} + \mathbf{u}\cdot\nabla\phi_h, \, v_h + \delta_T \mathbf{u}\cdot\nabla v_h\Big)_{L^2(T)} = 0 \quad \text{for all } v_h \in V_h, \tag{7.13}$$

and $t \in [0, T]$.

Note that compared to the standard Galerkin finite element discretization ($\delta_T = 0$ for all T) in (7.13) we have added a stabilizing term of the form $(\mathbf{u}\cdot\nabla\phi_h, \mathbf{u}\cdot\nabla v_h)_{L^2}$, which is the variational form of a diffusion acting only in the direction $\mathbf{u}$. This explains the name of this finite element method. Based on a theoretical error bound and numerical experiments for model problems the parameter δ_T is often taken as

$$\delta_T = c\,\frac{h_T}{\max\{\varepsilon_0, \|\mathbf{u}\|_{\infty,T}\}} \tag{7.14}$$

with a given small $\varepsilon_0 > 0$ and $c = \mathcal{O}(1)$. This streamline diffusion discretization is consistent in the following sense.

Lemma 7.2.5 *Let $\phi(t)$ be a solution of* (6.59c). *Then ϕ satisfies*

$$\sum_{T\in\mathcal{T}_h} \Big(\frac{\partial \phi}{\partial t} + \mathbf{u}\cdot\nabla\phi, \, v_h + \delta_T \mathbf{u}\cdot\nabla v_h\Big)_{L^2(T)} = 0 \quad \textit{for all } v_h \in V_h,$$

for all $t \in [0, T]$.

Proof. This immediately follows from the fact that for the test functions $\hat{v}_h$ as in (7.12) we have $\hat{v}_h \in L^2(\Omega)$. □

Matrix-vector representation

For the matrix-vector formulation of the semidiscrete problem (7.13) we use the standard nodal basis in V_h, which is denoted by $\{\xi_i\}_{i=1,\dots,N_{V_h}}$. Hence, for all i we have $\xi_i(x) = 0$ for all $x \in \mathcal{V}(\partial\Omega_{in})$. The vector representation of $v_h \in V_h(\phi_D)$ is given by

$$v_h = \sum_{i=1}^{N_{V_h}} v_i\, \xi_i + b_h, \quad v_i \in \mathbb{R}, \tag{7.15}$$

with $b_h = b_h(x,t) \in V_h(\phi_D)$ such that $b_h(x,t) = \phi_D(x,t)$ for all $x \in \mathcal{V}(\partial\Omega_{in})$ and $b_h(x_i,t) = 0$ at all other vertices x_i, $i = 1,\dots,N_{V_h}$. We define the matrices $\mathbf{E} = \mathbf{E}(\mathbf{u}) \in \mathbb{R}^{N_{V_h}\times N_{V_h}}$ and $\mathbf{H} = \mathbf{H}(\mathbf{u}) \in \mathbb{R}^{N_{V_h}\times N_{V_h}}$:

$$\mathbf{E}_{ij} := \sum_{T\in\mathcal{T}_h} \big(\xi_j,\, \xi_i + \delta_T \mathbf{u}\cdot\nabla\xi_i\big)_{L^2(T)} \qquad \text{(stabilized mass matrix)},$$

$$\mathbf{H}_{ij} := \sum_{T\in\mathcal{T}_h} \big(\mathbf{u}\cdot\nabla\xi_j,\, \xi_i + \delta_T \mathbf{u}\cdot\nabla\xi_i\big)_{L^2(T)} \quad \text{(stabilized convection)},$$

$$\mathbf{b}_i = \sum_{T\in\mathcal{T}_h} \big(\frac{\partial b_h}{\partial t} + \mathbf{u}\cdot\nabla b_h,\, \xi_i + \delta_T \mathbf{u}\cdot\nabla\xi_i\big)_{L^2(T)} \quad \text{(boundary data)},$$

for $1 \le i,j \le N_{V_h}$. Thus, using

$$\phi_h(t) = \sum_{i=1}^{N_{V_h}} \phi_i(t)\xi_i + b_h, \quad \vec{\phi}(t) := \big(\phi_1(t),\dots,\phi_{N_{V_h}}(t)\big),$$

and $\vec{\phi}_0$ the vector representation of the initial value $\phi_{0,h} - b_h(\cdot,0) \in V_h$ we can reformulate (7.13) in matrix-vector notation:

> Find $\vec{\phi}(t) \in \mathbb{R}^{N_{V_h}}$ with $\vec{\phi}(0) = \vec{\phi}_0$ and for all $t \in [0,T]$
>
> $$\mathbf{E}(\mathbf{u})\frac{d\phi}{dt}(t) + \mathbf{H}(\mathbf{u})\vec{\phi}(t) = -\mathbf{b}(t). \tag{7.16}$$

Note that in general the velocity field $\mathbf{u}$ depends on t and thus the matrices $\mathbf{E}(\mathbf{u})$ and $\mathbf{H}(\mathbf{u})$ are time dependent. In practice, the velocity field $\mathbf{u}$ will be replaced by a finite element approximation $\mathbf{u}_h$.

Time discretization

The discretization in (7.13), or in (7.16), can be combined with standard time discretization techniques (method of lines). If the velocity $\mathbf{u}$ depends on t then for the method of lines approach the formulation in (7.16) is more natural, since in (7.13) the test functions $\hat{v}_h$ then depend on t. The θ-schema applied to (7.16) results in

$$\frac{\vec{\phi}^{\,n+1} - \vec{\phi}^{\,n}}{\Delta t} = -\theta \mathbf{E}(\vec{\mathbf{u}}^{n+1})^{-1}\big(\mathbf{H}(\vec{\mathbf{u}}^{n+1})\vec{\phi}^{\,n+1} + \mathbf{b}(t_{n+1})\big) - (1-\theta)\mathbf{E}(\vec{\mathbf{u}}^{n})^{-1}\big(\mathbf{H}(\vec{\mathbf{u}}^{n})\vec{\phi}^{\,n} + \mathbf{b}(t_{n})\big).$$

This can be reformulated in a computationally more favorable form using a new variable

$$\vec{\mathbf{w}}^k = -\mathbf{E}(\vec{\mathbf{u}}^k)^{-1}\big(\mathbf{H}(\vec{\mathbf{u}}^k)\vec{\phi}^{\,k} + \mathbf{b}(t_k)\big),$$

which satisfies (for $\theta \neq 0$)

$$\theta \vec{\mathbf{w}}^{n+1} = \frac{\vec{\phi}^{\,n+1} - \vec{\phi}^{\,n}}{\Delta t} - (1-\theta)\vec{\mathbf{w}}^n,$$

resulting in

$$\mathbf{E}(\vec{\mathbf{u}}^{n+1})\frac{\vec{\phi}^{\,n+1} - \vec{\phi}^{\,n}}{\Delta t} = -\theta\big(\mathbf{H}(\vec{\mathbf{u}}^{n+1})\vec{\phi}^{\,n+1} + \mathbf{b}(t_{n+1})\big) + (1-\theta)\mathbf{E}(\vec{\mathbf{u}}^{n+1})\vec{\mathbf{w}}^n. \tag{7.17}$$

Discretization error bound

A discretization error analysis of the fully discrete problem (7.17) for the general case of a time-dependent velocity field $\mathbf{u}$ is not known, yet. For the case of a *stationary* and divergence free velocity field $\mathbf{u} = \mathbf{u}(x)$ an error analysis has recently been given in [59]. We outline the main result of that analysis. We assume that the Dirichlet data ϕ_D are also independent of t. Instead of a local stability parameter $\delta = \delta_T$ we assume quasi-uniformity of the triangulation and use one global parameter $\delta = h\|\mathbf{u}\|^{-1}_{L^\infty(\Omega)}$. For a stationary velocity field $\mathbf{u}$ the matrices $\mathbf{E}(\mathbf{u})$ and $\mathbf{H}(\mathbf{u})$ are independent of t and the scheme (7.17) is the matrix-vector representation of the following discrete problem, cf. (7.13): let $\phi_h^0 := \phi_{0,h} \in V_h(\phi_D)$ be an approximation of the initial condition $\phi_0 = \phi(0)$; for $n \geq 0$ determine $\phi_h^n \in V_h(\phi_D)$ such that

$$\Big(\frac{\phi_h^{n+1} - \phi_h^n}{\Delta t} + \mathbf{u}\cdot\nabla\big(\theta\phi_h^{n+1} + (1-\theta)\phi_h^n\big),\, v_h + \delta \mathbf{u}\cdot\nabla v_h\Big)_{L^2} = 0, \tag{7.18}$$

for all $v_h \in V_h$. In [59] an analysis for $\theta \in (0,1]$ is presented. Here we restrict to the Crank-Nicolson method, i.e., $\theta = \frac{1}{2}$. In the analysis it is assumed that the solution ϕ is sufficiently smooth, such that higher order derivatives are bounded. Let $N\Delta t = T$, i.e. ϕ_h^N is the numerical approximation of $\phi(\cdot, T)$. The following error bound can be shown to hold:

$$\|\phi_h^N - \phi(\cdot,T)\|_{L^2} + \sqrt{\delta}\big\|\mathbf{u}\cdot\nabla\big(\phi_h^N - \phi(\cdot,T)\big)\big\|_{L^2} \leq cT(h^{k+\frac{1}{2}} + \Delta t^2), \tag{7.19}$$

with a constant c that depends on the smoothness of the data ϕ_D and of the solution ϕ but not on T, h, Δt. In [59] this result is proved for the case of homogeneous inflow data $\phi_D = 0$, but the analysis can easily be extended to the inhomogeneous case $\phi_D \neq 0$, cf. [169]. As in the analysis presented for the simple hyperbolic problem in Sect. 7.2, the bound in (7.19) reflects that due to the stabilization not only the L^2-norm of the error but also its derivative in streamline direction can be controlled. The estimate (7.19) is similar to the one given in (7.11). The term Δt^2 in the error bound is of optimal order. The term $h^{k+\frac{1}{2}}$ is optimal for the error in the streamline derivative (recall: $\delta \sim h$) and suboptimal (by a factor $\sqrt{h}$) for the L^2-norm of the error.

7.3 Construction of an approximate interface Γ_h

As we will see further on, at several places (e.g., in the discretization of the surface tension force functional $f_\Gamma(\mathbf{v})$) we will need an approximation $\Gamma_h(t)$ of the interface $\Gamma(t)$. In this section we discuss a simple method for constructing such an approximation.

Assume that for a fixed $t \in [0, T]$ we have a finite element approximation $\phi_h(\cdot, t) \in V_h = \mathbb{X}_h^2$ of the level set function $\phi(x, t)$. To simplify the notation, in the remainder of this section we write $\phi(x, t) =: \phi(x)$, $\phi_h(x, t) =: \phi_h(x)$. In Sect. 7.4.1 we address the issue of re-initialization of the level set function, which is introduced to assure that ϕ (or ϕ_h) remains, in a neighborhood of the interface, close to a signed distance function. Thus ϕ and its approximation $\phi_h \in V_h$ can be assumed to be close to a signed distance function. Let $\tilde{\Gamma}_h$ be the *zero level of* ϕ_h and

$$\mathcal{T}_h^\Gamma := \left\{ T \in \mathcal{T}_h : \operatorname{meas}_2(T \cap \tilde{\Gamma}_h) > 0 \right\} \tag{7.20}$$

the collection of tetrahedra which contain the approximate interface $\tilde{\Gamma}_h$. Let $\mathcal{T}_{h'}^\Gamma$ be the collection of tetrahedra obtained by one further regular refinement of all $T \in \mathcal{T}_h^\Gamma$ (subdivision of each tetrahedron in 8 child tetrahedra, cf. Fig. 3.1). Furthermore, $I(\phi_h)$ is the continuous piecewise *linear* function on $\mathcal{T}_{h'}^\Gamma$ which interpolates ϕ_h at all vertices of all tetrahedra in $\mathcal{T}_{h'}^\Gamma$. Note that the degrees of freedom of the P_1 finite element functions on $\mathcal{T}_{h'}^\Gamma$ (located at the vertices) coincide with the degrees of freedom of the P_2 finite element functions on $\mathcal{T}_h^\Gamma$ (located at the vertices and midpoints of edges).

The approximation Γ_h of the interface Γ is defined by

$$\Gamma_h := \{ x \in \Omega : I(\phi_h)(x) = 0 \}. \tag{7.21}$$

Hence, Γ_h consists of piecewise planar segments $\Gamma_T \subset \Gamma_h$, where

$$\Gamma_T := T \cap \Gamma_h \tag{7.22}$$

for $T \in \mathcal{T}_{h'}^\Gamma$.

The *interface mesh size parameter* h_Γ is the maximal diameter of these segments. Thus h_Γ is approximately the maximal diameter of the tetrahedra in $\mathcal{T}_{h'}^\Gamma$, i.e., $2h_\Gamma$ is approximately the maximal diameter of the tetrahedra in $\mathcal{T}_h$ *that are close to the interface.* In our applications we use local refinement close to the interface, which implies $h_\Gamma \ll h$. In Fig. 7.1 we illustrate this construction for the two-dimensional case. An illustration for a 3D case is given in Fig. 7.5. Note that in general the segments of Γ_h *are not aligned with the faces of the tetrahedral triangulation* $\mathcal{T}_{h'}^\Gamma$.

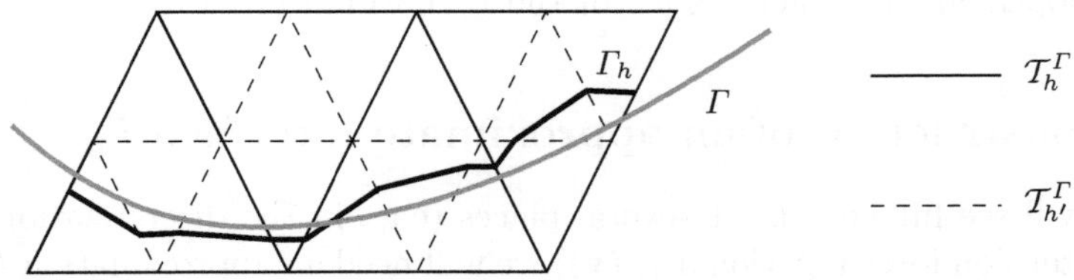

Fig. 7.1. Construction of approximate interface for 2D case.

Each of the planar segments Γ_T of Γ_h is either a triangle or a quadrilateral, depending on the sign pattern of ϕ_h on the corresponding $T \in \mathcal{T}_{h'}^\Gamma$, cf. Fig. 7.2. By construction the vertices of a planar segment Γ_T are located on those edges of T along which ϕ_h changes its sign. If there are two positive and two negative values of ϕ_h on the vertices of T, then the corresponding interface segment Γ_T is a quadrilateral. In all other cases Γ_T is a triangle. The quadrilaterals can (formally) be divided into two triangles. Thus Γ_h consists of a set of triangular faces, which is denoted by $\mathcal{F}_h$.

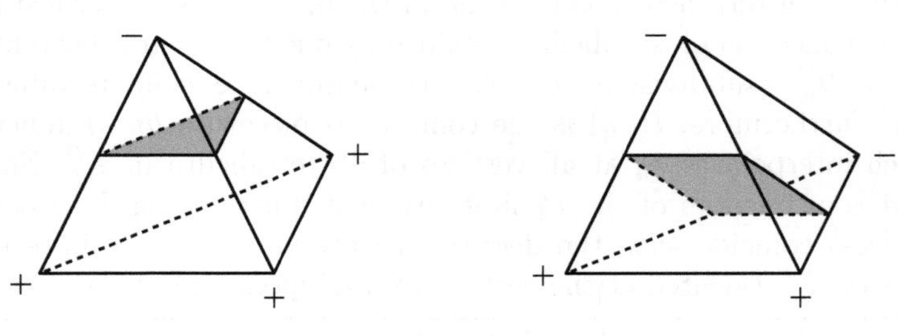

Fig. 7.2. Sign pattern of ϕ_h on $T \in \mathcal{T}_{h'}^\Gamma$ and corresponding interface segment $\Gamma_T = T \cap \Gamma$ (in gray): either a triangle or a quadrilateral.

Special cases may occur if some of the values of ϕ_h on the vertices of T are equal to zero (or below a given tolerance). Let $0 \leq n_0 \leq 4$ be the number of these (close to) zero values. In the following we discuss the shape of Γ_T in all the cases $n_0 = 0, 1, 2, 3, 4$.

- $n_0 = 0$ is not a special case, the situation is as depicted in Fig. 7.2 which was discussed in the foregoing paragraph.
- For $n_0 = 1, 2$ we distinguish two cases: If the other $4 - n_0$ non-zero values have the same sign, then Γ_T is a point ($n_0 = 1$) or a line segment ($n_0 = 2$) and can be ignored as $\text{meas}_2(\Gamma_T) = 0$. Otherwise the non-zero values are of different sign yielding $3 - n_0$ edges with a change of sign, as a simple case differentiation shows. Thus Γ_T has 3 vertices, hence Γ_T is a triangle.
- In the case $n_0 = 3$ the interface segment Γ_T is equal to a face of T. Then one has to take care that this face is not counted twice when computing a surface integral on Γ_h using an assembling strategy over $T \in \mathcal{T}_{h'}^{\Gamma}$.
- If $n_0 = 4$ then either the intersection of $T \in \mathcal{T}_{h'}^{\Gamma}$ with Γ_h is empty (cf. for example the left upper triangle in Fig. 7.1) or there is a degeneration, namely $\Gamma_T = T$, i.e. the interface segment is three-dimensional. Of course, the latter makes not much sense. Such a situation typically indicates that the grid is too coarse to represent the interface properly, cf. Fig. 7.3.

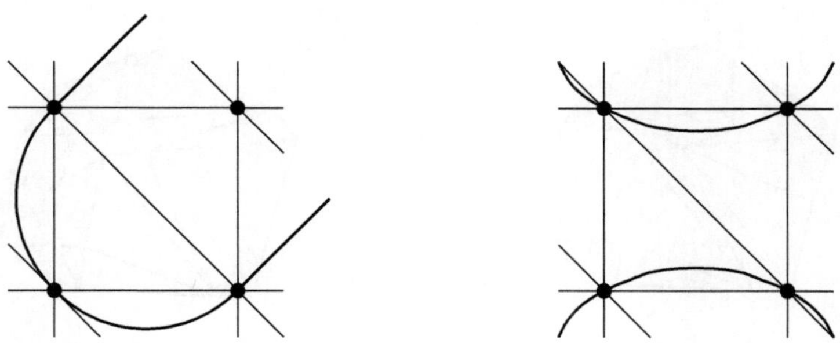

Fig. 7.3. 2D examples for interface degeneration such that the interface reconstruction fails. Left: curvature κ too large compared to grid resolution ($|\kappa| \geq \frac{2}{h}$). Right: distance d between interfaces too small compared to grid resolution ($d \leq h$).

If a situation as on the right in Fig. 7.3 occurs, then in the interface reconstruction special measures have to be taken to handle the (almost) topological singularity.

For an example in which Γ is a sphere, the resulting polygonal approximations Γ_h for $h = \frac{1}{5}$ and $h = \frac{1}{10}$ are shown in Fig. 7.4. A detail of such a polygonal interface approximation is shown in Fig. 7.5.

7.3.1 Error in approximation of Γ by Γ_h

We assume a fixed sufficiently smooth interface Γ, which is the zero level of ϕ, and a mesh size h_Γ that is sufficiently small such that degenerations as in Fig. 7.3 do not occur. We analyze the quality of Γ_h as an approximation

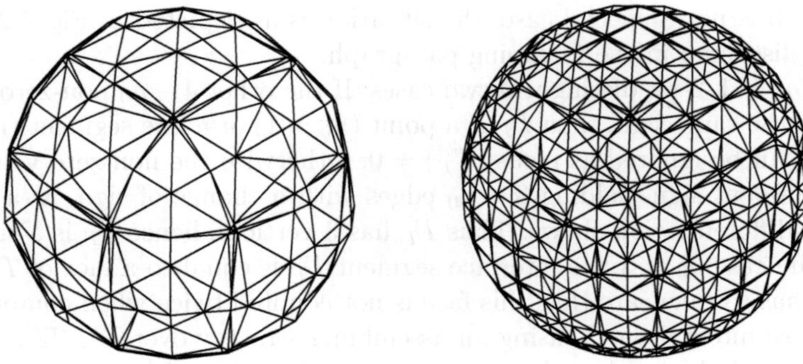

Fig. 7.4. Approximate interface Γ_h for an example with a sphere, on a coarse grid (left) and after one refinement (right).

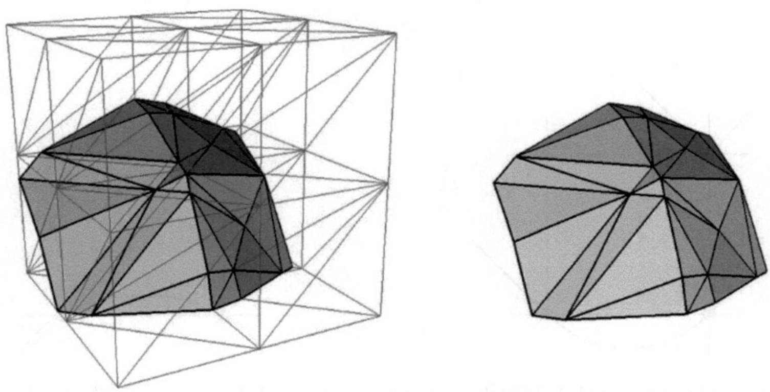

Fig. 7.5. Detail of the interface triangulation Γ_h. On the left, also the outer triangulation $\mathcal{T}_{h'}^{\Gamma}$ is shown.

of Γ. For this we first introduce some notation and further assumptions. Let $U := \{ x \in \mathbb{R}^3 : \operatorname{dist}(x, \Gamma) < c \}$ be a sufficiently small neighborhood of Γ. We define $\mathcal{T}_h^{\Gamma}$ as in (7.20), i.e., the collection of tetrahedra which intersect the zero level $\tilde{\Gamma}_h$ of ϕ_h, and assume that $\mathcal{T}_h^{\Gamma} \subset U$. Let d be the signed distance function

$$d : U \to \mathbb{R}, \qquad |d(x)| := \operatorname{dist}(x, \Gamma) \quad \text{for all } x \in U.$$

Thus Γ is the zero level set of d. Note that $\mathbf{n}_\Gamma = \nabla d$ on Γ. We define $\mathbf{n}(x) := \nabla d(x)$ for $x \in U$. Thus $\mathbf{n} = \mathbf{n}_\Gamma$ on Γ and $\|\mathbf{n}(x)\| = 1$ for all $x \in U$. Here and in the remainder of this section $\|\cdot\|$ denotes the Euclidean norm on $\mathbb{R}^3$. We introduce a local orthogonal coordinate system by using the projection $\mathbf{p} : U \to \Gamma$:

$$\mathbf{p}(x) = x - d(x)\mathbf{n}(x) \quad \text{for all } x \in U.$$

We assume that the decomposition $x = \mathbf{p}(x) + d(x)\mathbf{n}(x)$ is unique for all $x \in U$. Note that

$$\mathbf{n}(x) = \mathbf{n}\big(\mathbf{p}(x)\big) \quad \text{for all } x \in U.$$

The unit normal on Γ_h (pointing outward from Ω_1) is denoted by $\mathbf{n}_h$. Using these preliminaries we can derive the following approximation result.

Theorem 7.3.1 *Assume that $\phi \in H^2_\infty(U)$ and that for $c_1, c_0 > 0$*

$$c_0 \le \|\nabla\phi(x)\| \le c_1 \quad \text{for all } \; x \in U. \tag{7.23}$$

Furthermore, we assume that the approximation $\phi_h \in V_h = \mathbb{X}^2_h$ of ϕ satisfies

$$\|\phi_h - \phi\|_{L^\infty(U)} + h_\Gamma \|\phi_h - \phi\|_{H^1_\infty(U)} \le c h_\Gamma^m \|\phi\|_{H^m_\infty(U)}, \quad m = 1, 2. \tag{7.24}$$

Then the following holds:

$$|d(x)| \le c h_\Gamma^2 \quad \text{for all } x \in \Gamma_h, \tag{7.25a}$$

$$\|\mathbf{n}(x) - \mathbf{n}_h(x)\| \le c h_\Gamma \quad \text{for all } x \in \Gamma_h. \tag{7.25b}$$

Proof. Let I be the linear interpolation operator corresponding to $\mathcal{T}^\Gamma_{h'}$, used in (7.21), and define the piecewise linear function $\tilde{\phi}_h = I\phi_h$. Recall that $\Gamma_h = \left\{ x \in \mathbb{R}^3 : \tilde{\phi}_h(x) = 0 \right\}$. Using standard properties of I and the error bound in (7.24) one obtains

$$\begin{aligned} \|\tilde{\phi}_h - \phi\|_{L^\infty(\mathcal{T}^\Gamma_h)} &\le \|I(\phi_h - \phi)\|_{L^\infty(\mathcal{T}^\Gamma_h)} + \|I\phi - \phi\|_{L^\infty(\mathcal{T}^\Gamma_h)} \\ &\le \|\phi_h - \phi\|_{L^\infty(U)} + c h_\Gamma^2 \|\phi\|_{H^2_\infty(U)} \\ &\le c\, h_\Gamma^2 \|\phi\|_{H^2_\infty(U)}. \end{aligned}$$

Due to $\tilde{\phi}_h(x) = 0$ for $x \in \Gamma_h$ this yields

$$|\phi(x)| \le c h_\Gamma^2 \quad \text{for } \; x \in \Gamma_h. \tag{7.26}$$

Take $x \in \Gamma_h$ and introduce the notation $y = \mathbf{p}(x) = x - d(x)\mathbf{n}(x) = x - d(x)\mathbf{n}(y) \in \Gamma$. For suitable s with $|s| \le |d(x)|$ and $\tilde{y} = y + s\mathbf{n}(y)$ we get

$$\begin{aligned} \phi(x) &= \phi(x) - \phi(y) = \phi(y + d(x)\mathbf{n}(y)) - \phi(y) \\ &= d(x)\nabla\phi(y + s\mathbf{n}(y)) \cdot \mathbf{n}(y) = d(x)\nabla\phi(\tilde{y}) \cdot \mathbf{n}(y) \\ &= d(x)\big((\nabla\phi(\tilde{y}) - \nabla\phi(y)) \cdot \mathbf{n}(y) + \|\nabla\phi(y)\|\big). \end{aligned} \tag{7.27}$$

Due to (7.23) we have $\|\nabla\phi(y)\| \ge c_0$. We assume that U is sufficiently small such that $\|\nabla\phi(\tilde{y}) - \nabla\phi(y)\| \le \|\phi\|_{H^2_\infty(U)}|d(x)| \le \frac{1}{2}c_0$ holds. Hence we obtain from (7.27) that $|\phi(x)| \ge \frac{1}{2}c_0|d(x)|$ holds, and using (7.26) yields

$$|d(x)| \le c|\phi(x)| \le ch_\Gamma^2, \quad x \in \Gamma_h,$$

i.e., the result in (7.25a). We also have, using Assumption (7.24)

$$\begin{aligned}\|\tilde{\phi}_h - \phi\|_{H^1_\infty(\mathcal{T}_h^\Gamma)} &\le \|I(\phi_h - \phi)\|_{H^1_\infty(\mathcal{T}_h^\Gamma)} + \|I\phi - \phi\|_{H^1_\infty(\mathcal{T}_h^\Gamma)} \\ &\le c\,\|\phi_h - \phi\|_{H^1_\infty(U)} + ch_\Gamma\|\phi\|_{H^2_\infty(U)} \le ch_\Gamma\|\phi\|_{H^2_\infty(U)}.\end{aligned}$$

This implies

$$\|\nabla\tilde{\phi}_h(x)\| = \|\nabla\phi(x)\| + \mathcal{O}(h), \quad x \in \Gamma_h.$$

Using this, we obtain for $x \in \Gamma_h$ (not on an edge) and $y = \mathbf{p}(x) \in \Gamma$

$$\begin{aligned}\|\mathbf{n}(x) - \mathbf{n}_h(x)\| = \|\mathbf{n}(y) - \mathbf{n}_h(x)\| &= \left\| \frac{\nabla\phi(y)}{\|\nabla\phi(y)\|} - \frac{\nabla\tilde{\phi}_h(x)}{\|\nabla\tilde{\phi}_h(x)\|} \right\| \\ \le \left\| \frac{\nabla\phi(y)}{\|\nabla\phi(y)\|} - \frac{\nabla\phi(x)}{\|\nabla\phi(x)\|} \right\| &+ \left\| \frac{\nabla\phi(x)}{\|\nabla\phi(x)\|} - \frac{\nabla\tilde{\phi}_h(x)}{\|\nabla\tilde{\phi}_h(x)\|} \right\|.\end{aligned}$$

For the first term we obtain, from a Taylor expansion, Assumption (7.23) and $\|x - y\| \le ch_\Gamma^2$:

$$\left\| \frac{\nabla\phi(y)}{\|\nabla\phi(y)\|} - \frac{\nabla\phi(x)}{\|\nabla\phi(x)\|} \right\| \le ch_\Gamma^2.$$

For the second term we get

$$\begin{aligned}&\left\| \frac{\nabla\phi(x)}{\|\nabla\phi(x)\|} - \frac{\nabla\tilde{\phi}_h(x)}{\|\nabla\tilde{\phi}_h(x)\|} \right\| = \left\| \frac{\nabla\phi - \nabla\tilde{\phi}\frac{\|\nabla\phi\|}{\|\nabla\tilde{\phi}\|}}{\|\nabla\phi\|} \right\| \\ &\le c_0^{-1} \left| \frac{\|\nabla\phi(x)\|}{\|\nabla\tilde{\phi}_h(x)\|} - 1 \right| \cdot \|\nabla\tilde{\phi}_h(x)\| + c_0^{-1}\,\|\nabla\tilde{\phi}_h(x) - \nabla\phi(x)\| \le c\,h_\Gamma,\end{aligned}$$

which completes the proof. □

The results in (7.25a) and (7.25b) give (satisfactory) quantitative results on the approximation quality of Γ_h. The result in (7.25b) implies that the tangent planes are close in a certain sense, and that "zigzag" effects in the approximation Γ_h do not occur. These bounds will play a crucial role in the analysis of the surface tension force discretization in Sect. 7.7. We comment on the assumptions in (7.23) and (7.24). Due to re-initialization, in a neighborhood of the interface the level set function ϕ is close to a signed distance function and thus $\|\nabla\phi\| \approx 1$ can be expected to hold. We claim that the assumption on the discretization error bound in (7.24) is also reasonable. Due to the re-initialization it is reasonable to assume that ϕ is (very) smooth. Hence, using quadratic finite elements, an optimal order discretization method would have an error bound of the form (7.24) with $m = 3$. We do not know whether the SDFEM applied to the hyperbolic level set equation is of optimal order. In (7.24), however, we only assume an h_Γ^2 error bound (instead of the optimal h_Γ^3

bound) to be satisfied. The only rigorous discretization error bounds for the level set equation known to us are given in [59], cf. the discussion in Sect. 7.2.2, in particular the result in (7.19). For quadratic finite elements and a suitable time step Δt the latter result yields a bound $ch_\Gamma^{2\frac{1}{2}}$. The discretization error, however, is measured in a weaker norm as the one used in (7.24).

For a sufficiently smooth level set function ϕ and an approximation $\phi_h \in \mathbb{X}_h^2$ one could consider the case in which the approximation error bound in (7.24) even holds for an $m \in (2,3]$. Such a stronger assumption, however, would not improve the bounds in (7.25). This is due to the fact that *linear* interpolation is used in the construction of Γ_h. In the proof this is reflected by the terms $\|I\phi - \phi\|_{L^\infty(\mathcal{T}_h^\Gamma)}$ and $\|I\phi - \phi\|_{H^1_\infty(\mathcal{T}_h^\Gamma)}$ for which the optimal error bounds are of the form ch_Γ^2 and ch_Γ, respectively. Below we present a result which shows that by using ϕ_h, instead of $I\phi_h$, a better normal approximation than the one in (7.25b) can be obtained. This result will be crucial in the discretization of the surface tension force treated in Sect. 7.6.

Lemma 7.3.2 *Assume that $\phi \in H^3(U)$ and that (7.23) holds. Furthermore, we assume that $\phi_h \in \mathbb{X}_h^2$ satisfies*

$$\|\phi_h - \phi\|_{H^1_\infty(U)} \leq ch_\Gamma^p, \quad \text{for a} \;\; p \in (0,2]. \tag{7.28}$$

For $x \in U$ define $\tilde{\mathbf{n}}_h(x) := \frac{\nabla\phi_h(x)}{\|\nabla\phi_h(x)\|}$. The following holds:

$$\|\mathbf{n}(x) - \tilde{\mathbf{n}}_h(x)\| \leq ch_\Gamma^p \quad \text{for all } x \in \Gamma_h. \tag{7.29}$$

Proof. We use similar arguments as in the proof of Theorem 7.3.1. For $x \in \Gamma_h$ (not on an edge)

$$\|\mathbf{n}(x) - \tilde{\mathbf{n}}_h(x)\| = \|\mathbf{n}(y) - \tilde{\mathbf{n}}_h(x)\| = \left\| \frac{\nabla\phi(y)}{\|\nabla\phi(y)\|} - \frac{\nabla\phi_h(x)}{\|\nabla\phi_h(x)\|} \right\|$$
$$\leq \left\| \frac{\nabla\phi(y)}{\|\nabla\phi(y)\|} - \frac{\nabla\phi(x)}{\|\nabla\phi(x)\|} \right\| + \left\| \frac{\nabla\phi(x)}{\|\nabla\phi(x)\|} - \frac{\nabla\phi_h(x)}{\|\nabla\phi_h(x)\|} \right\|.$$

Using a Taylor expansion and $\|x - y\| \leq ch_\Gamma^2$ we obtain a bound ch_Γ^2 for the first term. Since $\|\nabla\phi_h(x) - \nabla\phi(x)\| \leq ch_\Gamma^p$, the second term can be bounded by ch_Γ^p using the same arguments as in the proof of Theorem 7.3.1. □

Let $\tilde{\Gamma}_h$ be the zero level set of $\phi_h \in \mathbb{X}_h^2$. This zero level is difficult to compute; therefore, in practice we use its piecewise planar approximation Γ_h. For $x \in \tilde{\Gamma}_h$, the quantity $\tilde{\mathbf{n}}_h(x)$ is the unit outward normal on $\tilde{\Gamma}_h$. Note that, for a given $\phi_h \in \mathbb{X}_h^2$ and $x \in U$ it is easy to compute the quantity $\tilde{\mathbf{n}}_h(x)$. In the sense as in Theorem 7.3.1 and Lemma 7.3.2, $\tilde{\mathbf{n}}_h(x)$ is a better approximation to the normal $\mathbf{n}(x)$ than $\mathbf{n}_h(x)$.

7.4 Corrections of the level set function

During the evolution of the level set function ϕ or of its finite element approximation ϕ_h, which is driven by the velocity field $\mathbf{u}$, the property of ϕ (ϕ_h) being close to a signed distance function is lost. This has undesirable effects, which can be avoided by using a re-initialization technique as explained in Sect. 7.4.1. In general the (spatial and temporal) discretizations of the level set equation are such that mass conservation is not guaranteed on the discrete level (only for $h,\ \Delta t \downarrow 0$). This issue of loss of mass is briefly addressed in Sect. 7.4.2.

7.4.1 Re-initialization

Assume that the initial data $\phi_0(x)$, $x \in \Omega$, for the level set equation are such that (locally, close to the initial interface) ϕ_0 is a signed distance function to $\Gamma(0)$. Then in general during the evolution of the level set function ϕ the property of ϕ being close to a (signed) distance function is lost, which has undesirable effects. For example, an accurate spatial discretization of ϕ becomes hard in regions where ϕ has a very strong variation, and the problem of finding the zero level set of ϕ becomes ill conditioned in regions where ϕ is very flat. Therefore, often level set methods are combined with a re-initialization (also called "reparametrization") technique.

Assume that for a *given* $t_0 \in [0,T]$ an approximation $\phi_h(\cdot)$ of the level set function $\phi(\cdot,t_0)$ is known. Given this ϕ_h a re-initialization method results in $\tilde{\phi}_h$ such that:

1. The zero level of $\tilde{\phi}_h$ is (approximately) equal to that of ϕ_h.
2. The function $\tilde{\phi}_h$ is close to a signed distance function: $\|\nabla\tilde{\phi}_h\| \approx 1$ (close to the interface).

The function $\tilde{\phi}_h$ is then used as re-initialization in the evolution of the level set function: $\tilde{\phi}_h$ is taken as "initial" data to solve the level set equation for $t \geq t_0$.

Different re-initialization techniques are known in the literature, cf. [221, 222, 148, 266, 205]. A popular method is based on solving the Eikonal equation

$$\|\nabla\psi\| = 1$$

by introducing a pseudo-time evolution as follows. Let ϕ_h be the given approximation of the level set function, and consider the first order partial differential equation for $\psi = \psi(x,\tau)$:

$$\begin{aligned} \frac{\partial\psi}{\partial\tau} &= S_\alpha(\phi_h)(1 - \|\nabla\psi\|), \qquad \tau \geq 0,\ x \in \Omega, \\ \psi(\cdot,0) &= \phi_h, \end{aligned} \tag{7.30}$$

with

$$S_\alpha(\zeta) = \frac{\zeta}{\sqrt{\zeta^2 + \alpha^2}}, \qquad \zeta \in \mathbb{R},$$

where α is a regularization parameter ($0 < \alpha \ll 1$). The function S_α is a smoothed sign function. Due to $S_\alpha(0) = 0$ the zero level set of ψ remains equal to that of ϕ_h. A stationary solution $\psi(x) = \lim_{\tau\to\infty} \psi(x,\tau)$ of (7.30) solves the Eikonal equation and thus ψ is a signed distance function. The equation (7.30) can be reformulated in the more convenient form

$$\frac{\partial \psi}{\partial \tau} + \mathbf{w}(\psi)\cdot\nabla\psi = S_\alpha(\phi_h) \quad \text{with } \mathbf{w}(\psi) := S_\alpha(\phi_h)\frac{\nabla\psi}{\|\nabla\psi\|}. \tag{7.31}$$

In practice, the equation (7.31) is discretized in space and time and for sufficiently large $\tau_f > 0$ one can use the computed discrete solution $\tilde{\phi}_h := \psi_h(\cdot, \tau_f)$ as a re-initialization of ϕ_h. For a further discussion of this re-initialization method we refer to the literature [231, 230, 240]. Using this technique one faces the following difficulties. Firstly, the method contains control parameters α and τ_f and there are no good practical criteria on how to select these. Secondly, the partial differential equation (7.31) is nonlinear and hyperbolic; accurate discretization of this type of partial differential equation is rather difficult. Finally, the invariance property of the zero level holds for the stationary solution $\psi(x,\tau)$ of the continuous problem in (7.30), but after discretization it is usually lost. It may well happen that the difference between the zero levels of ϕ_h and $\psi_h(\cdot,\tau_f)$ is "large".

Another technique for re-initialization is the Fast Marching Method (cf. [157, 220]). In [148] a survey and comparison of different re-initialization methods is given, where (for a certain class of problems) the Fast Marching Method turns out to be the most accurate and efficient one. In our level set method we use a variant of the Fast Marching Method that is explained in detail below.

Fast Marching Method (FMM)

In our level set method we have a piecewise quadratic function $\phi_h \in V_h = \mathbb{X}_h^2$ for which a re-initialization should be determined. We first describe the FMM applied to a piecewise *linear* function and then explain how this method is applied to the piecewise quadratic ϕ_h.

Let ψ_h be a piecewise linear function on the tetrahedral triangulation $\mathcal{T}_h$. The zero level of ψ_h is denoted by Γ_h. This zero level consists of planar segments Γ_T:

$$\Gamma_h = \bigcup_{T\in\mathcal{T}_h^\Gamma} \Gamma_T, \quad \text{with } \Gamma_T := \Gamma_h \cap T, \tag{7.32}$$

and $\mathcal{T}_h^\Gamma$ the collection of all tetrahedra that have a nonempty intersection with Γ_h. The planar segment Γ_T is either a triangle or a quadrilateral, cf. Fig. 7.2. We introduce some notation. For $T \in \mathcal{T}_h$, $\mathcal{V}(T)$ is the set of the four vertices of T. More general, for a collection of tetrahedra $\mathcal{S}$, the set of all vertices contained in $\mathcal{S}$ is denoted by $\mathcal{V}(\mathcal{S})$. We write $\mathcal{V} := \mathcal{V}(\mathcal{T}_h)$. Furthermore, for $v \in \mathcal{V}$, $\mathcal{T}(v)$ is the set of all tetrahedra which have v as a vertex:

$$\mathcal{T}(v) = \{ T \in \mathcal{T}_h : v \in \mathcal{V}(T) \} .$$

We also need the recursively defined larger neighborhoods:

$$\mathcal{T}^1(v) := \mathcal{T}(v), \quad \mathcal{T}^{k+1}(v) = \big\{ T \in \mathcal{T}_h : \mathcal{V}(T) \cap \mathcal{V}(\mathcal{T}^k(v)) \neq \emptyset \big\}, \quad k \geq 1.$$

Finally, for $v \in \mathcal{V}$, $\mathcal{N}(v)$ is the collection of all neighboring vertices of v (i. e., for each $w \in \mathcal{N}(v)$ there is an edge in $\mathcal{T}_h$ connecting v and w):

$$\mathcal{N}(v) := \Big(\bigcup_{T \in \mathcal{T}(v)} \mathcal{V}(T) \Big) \setminus \{v\}.$$

As input for the FMM we need $\mathcal{T}_h$, the zero level set Γ_h and $\mathrm{sign}(\psi_h(v))$, $v \in \mathcal{V}$. Thus only $\mathrm{sign}(\psi_h(v))$ is needed, and not $\psi_h(v)$ itself. The FMM consists of two phases: The *initialization phase*, where the values at vertices close to the interface are determined, and the *extension phase*, where the information is propagated from the interface to the vertices in the far field.

Initialization phase. We define the set of vertices corresponding to $\mathcal{T}_h^\Gamma$:

$$\mathcal{V}_\Gamma := \mathcal{V}(\mathcal{T}_h^\Gamma) = \big\{ v \in \mathcal{V}(T) : T \in \mathcal{T}_h^\Gamma \big\} . \tag{7.33}$$

The aim of the initialization phase is to define a discrete (approximate) distance function $\hat{d}(v)$ for each $v \in \mathcal{V}_\Gamma$. To this end, we present two possible strategies, a geometry-based approach and a weighted scaling approach.

We first consider the geometry-based approach. For $v \in \mathcal{V}_\Gamma$ and $T \in \mathcal{T}(v) \cap \mathcal{T}_h^\Gamma$ let Γ_T be the planar segment as in (7.32). This segment is either a triangle or a quadrilateral. In the latter case Γ_T can be subdivided into two triangles. If Δ is a triangle in $\mathbb{R}^3$ and $p \in \mathbb{R}^3$ then the distance between Δ and p

$$d(p, \Delta) := \min_{x \in \Delta} \|p - x\|, \tag{7.34}$$

can be computed using elementary geometry, for example as follows. If $\Delta = \mathrm{conv}\{v_1, v_2, v_3\}$ and $\mathbf{A} = (v_2 - v_1, v_3 - v_1)$ one first solves the 3×2 least squares problem $\|\mathbf{A}z - (p - v_1)\| \to \min$. This results in the orthogonal projection of p on the plane that contains Δ. If this orthogonal projection is contained in the triangle Δ then the residual of the least squares problem equals $d(p, \Delta)$. Otherwise, $d(p, \Delta) = \mathrm{dist}(p, \partial\Delta)$, and thus the distance of p to the three edges of Δ has to be determined.

Hence, for arbitrary $v \in \mathcal{V}_\Gamma$, $T \in \mathcal{T}_h^\Gamma$ we can compute

$$d_T(v) := d(v, \Gamma_T).$$

For a *given* $k \geq 1$ the (approximate) distance between v and Γ_h is defined by

$$\hat{d}(v) := \min_{T \in \mathcal{T}^k(v) \cap \mathcal{T}_h^\Gamma} d_T(v) = \min_{x \in \Gamma_h \cap \mathcal{T}^k(v)} \|v - x\| \quad \text{for } v \in \mathcal{V}_\Gamma. \tag{7.35}$$

In practice we typically use $k = 2$. Properties of the geometry based initialization are discussed in Remark 7.4.3.

Another approach is based on a scaling of the level set function at the vertices $v \in \mathcal{V}_\Gamma$. One motivation for this approach comes from the following observation. If one wants to guarantee that the approximated interface Γ_h is not moved by re-initialization, then the only choice is $\hat{d}(v) = \alpha^{-1}\phi_h(v)$ for $v \in \mathcal{V}_\Gamma$ with a suitable scalar $\alpha > 0$. Achieving the (distance) property $\|\nabla \hat{d}\| \approx 1$ by scaling with a *single* scalar, however, is not possible in general. Consider, for example, the case of a level set function ϕ_h which has a large gradient $\|\nabla\phi_h\| \approx 10^3$ in one part of $\mathcal{T}_h^\Gamma$ and a small gradient $\nabla\phi_h\| \approx 10^{-3}$ in another part.

This leads to the idea to use a vertex-dependent scalar α_v, i.e., to define

$$\hat{d}(v) = \alpha_v^{-1}\phi_h(v), \quad v \in \mathcal{V}_\Gamma. \tag{7.36}$$

We propose to use

$$\alpha_v := \frac{\sum_{T \in \mathcal{T}(v)} \int_T \|\nabla\phi_h\|\, dx}{\sum_{T \in \mathcal{T}(v)} \int_T 1\, dx}, \quad v \in \mathcal{V}_\Gamma, \tag{7.37}$$

i.e., α_v is an average of the gradients of ϕ_h on $\mathcal{T}(v)$. Compared to the geometry-based approach described above there is no need to reconstruct Γ_h from ϕ_h, allowing for a relatively simple implementation of the method. A comparison of both methods in a numerical example is given in Sect. 7.5.2. After completion of the initialization phase the values $\{ (v, \hat{d}(v)) \; : \; v \in \mathcal{V}_\Gamma \}$ determine an approximate distance grid function for the vertices $v \in \mathcal{V}_\Gamma$.

Extension phase. The second phase consists of a greedy algorithm in which the approximate distance function $\hat{d}$ is extended to neighbor vertices of $\mathcal{V}_\Gamma$ and then to neighbors of neighbors, etc. To explain this more precisely we introduce two sets of vertices.

The first set $\hat{\mathcal{V}} \subset \mathcal{V}$ contains the vertices where the values of the approximate distance function $\hat{d} : \mathcal{V} \to \mathbb{R}$ have already been computed. As initialization we take $\hat{\mathcal{V}} := \mathcal{V}_\Gamma$. We call $\hat{\mathcal{V}}$ the *finalized set.*

The second one is the set of so-called *active* vertices $\mathcal{A} \subset \mathcal{V} \setminus \hat{\mathcal{V}}$, which consists of vertices $v \notin \hat{\mathcal{V}}$ that have a neighboring vertex in $\hat{\mathcal{V}}$:

$$\mathcal{A} := \left\{ v \in \mathcal{V} \setminus \hat{\mathcal{V}} : \mathcal{N}(v) \cap \hat{\mathcal{V}} \neq \emptyset \right\}. \tag{7.38}$$

$\mathcal{A}$ is called the *active set.* After the initialization phase, the initial active set $\mathcal{A}_0$ is given by

$$\mathcal{A}_0 := \{ v \in \mathcal{V} \setminus \mathcal{V}_\Gamma : \mathcal{N}(v) \cap \mathcal{V}_\Gamma \neq \emptyset \}. \tag{7.39}$$

For $v \in \mathcal{A}$ we define an approximate distance function in a similar way as in the initialization phase. Since its values may change if the finalized and active set are updated, we denote it by $\tilde{d} : \mathcal{A} \to \mathbb{R}$. We emphasize that $\tilde{d}$ has tentative character in contrast to $\hat{d}$, which will be the final outcome of the algorithm. The construction of $\tilde{d}$ is described in the following.

Take $v \in \mathcal{A}$ and $T \in \mathcal{T}(v)$ with $\mathcal{V}(T) \cap \hat{\mathcal{V}} \neq \emptyset$. Note that such a T exists if $\mathcal{A}$ is nonempty. There are three possible cases, namely $|\mathcal{V}(T) \cap \hat{\mathcal{V}}| \in \{1, 2, 3\}$.

- If $|\mathcal{V}(T) \cap \hat{\mathcal{V}}| = 1$, say $\mathcal{V}(T) \cap \hat{\mathcal{V}} = \{w\}$, we define

$$\tilde{d}_T(v) := \hat{d}(w) + \|v - w\|.$$

- For the other two cases, i. e., $\mathcal{V}(T) \cap \hat{\mathcal{V}} = \{w_i\}_{1 \le i \le m}$ with $m = 2$ or $m = 3$, we use a distance function to the line segment $W = \text{conv}(w_1, w_2)$ (for $m = 2$) or the triangle $W = \text{conv}(w_1, w_2, w_3)$ (for $m = 3$), which is denoted by $d(v, W)$. In the case $m = 3$, this is the distance function as in (7.34). Let $P_W : \mathbb{R}^3 \to W$ be such that $d(v, W) = \|v - P_W v\|$. We define

$$\tilde{d}_T(v) := \hat{d}(P_W v) + \|v - P_W v\| = \hat{d}(P_W v) + d(v, W).$$

  The value $\hat{d}(P_W v)$ is determined by linear interpolation of the *known* values $\hat{d}(w_j)$, $1 \le j \le m$. This is well-defined since $w_j \in \hat{\mathcal{V}}$ for $1 \le j \le m$ and $\hat{d}$ is already defined on $\hat{\mathcal{V}}$.

The tentative approximate distance function $\tilde{d} : \mathcal{A} \to \mathbb{R}$ at active vertices $v \in \mathcal{A}$ is defined by

$$\tilde{d}(v) := \min \left\{ \tilde{d}_T(v) : T \in \mathcal{T}(v) \text{ with } \mathcal{V}(T) \cap \hat{\mathcal{V}} \neq \emptyset \right\}. \tag{7.40}$$

The complete re-initialization method is as follows:

Algorithm 7.4.1 (Fast Marching Method)

1. Initialization: $\mathcal{V}_\Gamma$ as in (7.33), compute $\hat{d}(\mathcal{V}_\Gamma)$ as in (7.35) (or (7.36)).
2. Initialize finalized set $\hat{\mathcal{V}} := \mathcal{V}_\Gamma$ and active set $\mathcal{A} := \mathcal{A}_0$, cf. (7.39).
3. For the initial active set $\mathcal{A}_0$, compute $\tilde{d}(\mathcal{A}_0)$ as in (7.39), (7.40).
4. While $\mathcal{A} \neq \emptyset$, repeat the following steps:
   a) Determine $v_{\min} \in \mathcal{A}$ such that $\tilde{d}(v_{\min}) = \min_{v \in \mathcal{A}} \tilde{d}(v)$.
   b) Update finalized set $\hat{\mathcal{V}} := \hat{\mathcal{V}} \cup \{v_{\min}\}$ and define $\hat{d}(v_{\min}) := \tilde{d}(v_{\min})$.
   c) Update active set $\mathcal{A} := (\mathcal{A} \cup \tilde{\mathcal{N}}) \setminus \{v_{\min}\}$ where $\tilde{\mathcal{N}} := \mathcal{N}(v_{\min}) \setminus \hat{\mathcal{V}}$.
   d) (Re)compute $\tilde{d}(v)$ for $v \in \tilde{\mathcal{N}}$.
5. For all $v \in \mathcal{V}$, set $\hat{d}(v) := \text{sign}(\psi_h(v)) \cdot \hat{d}(v)$.

After this re-initialization we have $\hat{\mathcal{V}} = \mathcal{V}$ and a grid function $\hat{d}(v)$, $v \in \mathcal{V}$, which uniquely determines a continuous piecewise linear approximate signed distance function. This function is defined to be the re-initialization of ψ_h, denoted by $\hat{\psi}_h$. The zero level set of $\hat{\psi}_h$ is denoted by $\hat{\Gamma}_h$. Below, in Remark 7.4.3 we discuss important approximation properties of the re-initialization $\hat{\psi}_h$ and its zero level.

Remark 7.4.2 (Complexity) The number of (arithmetic) operations for the initialization phase (steps 1–3 in Algorithm 7.4.1) is $\mathcal{O}(|\mathcal{V}_\Gamma| + |\mathcal{A}_0|)$. For the extension phase (steps 4–5 in Algorithm 7.4.1) the sorting and updating in the steps 4.a)-c) can be realized with $\mathcal{O}(\log|\mathcal{A}|)$ complexity using a heap data structure for $\mathcal{A}$. Step 4 is repeated $N_\mathcal{V} := |\mathcal{V} \setminus \mathcal{V}_\Gamma|$ times, and thus this FMM has an overall complexity of the order $\mathcal{O}(N_\mathcal{V} \log N_\mathcal{V})$.

Remark 7.4.3 (Approximation properties) A detailed analysis of the FMM in Algorithm 7.4.1 using the geometry-based initialization phase (7.35) is given in [123]. We outline some main results. By construction each $T \in \mathcal{T}_h^\Gamma$ contains a segment of the new zero level $\hat{\Gamma}_h$ and thus $\mathrm{dist}(\Gamma_h, \hat{\Gamma}_h) \leq h_\Gamma$ holds. In practice, however, one typically observes $\mathrm{dist}(\Gamma_h, \hat{\Gamma}_h) \ll h_\Gamma$, which can be explained by the results from [123]. We use notation and assumptions as in Sect. 7.3.1. We assume that Γ_h approximates a smooth interface Γ and define d to be the signed distance function to Γ. The approximation error is assumed to be sufficiently small in the following sense:

$$|d(x)| \leq ch_\Gamma^2 \quad \text{for all } x \in \Gamma_h, \tag{7.41a}$$

$$\|\mathbf{n}(x) - \mathbf{n}_h(x)\| \leq ch_\Gamma \quad \text{for all } x \in \Gamma_h. \tag{7.41b}$$

In our setting these are reasonable assumptions, cf. Theorem 7.3.1. The approximate interface Γ_h is the zero level of the given piecewise linear function ψ_h. Let d_h be the signed distance function to Γ_h. After the initialization phase, for $v \in \mathcal{V}_\Gamma$ the re-initialization $\hat{\psi}_h(v)$ is determined by the function $\hat{d}$ in (7.35): $\hat{\psi}_h(v) = \mathrm{sign}(\psi_h(v))\hat{d}(v)$ for $v \in \mathcal{V}_\Gamma$. Note that $\hat{d}$ in (7.35) depends on $k \geq 1$. It is obvious that for k sufficiently large we have

$$\hat{\psi}_h(v) = d_h(v) \quad \text{for all } \; v \in \mathcal{V}_\Gamma, \tag{7.42}$$

i.e. at the vertices in $\mathcal{V}_\Gamma$ we have determined the *exact* signed distance to Γ_h. For the theoretical analysis we assume that (7.42) holds. Based on experience, in computations we take $k = 2$. Note that a larger k value induces higher computational costs for the re-initialization.

Based on the assumptions in (7.41), (7.42) one can derive the following results:

$$|d(x)| \leq ch_\Gamma^2 \quad \text{for all } x \in \hat{\Gamma}_h, \tag{7.43a}$$

$$\|\nabla\hat{\psi}_h - \mathbf{n}\|_{L^\infty(\mathcal{T}_h^\Gamma)} \leq ch_\Gamma. \tag{7.43b}$$

The result in (7.43a) shows that in the re-initialization the accuracy of the zero level set as an approximation of Γ is maintained. Due to $\|\mathbf{n}\| = 1$, we conclude from (7.43b) that $\|\nabla\hat{\psi}_h(x)\| = 1 + \mathcal{O}(h_\Gamma)$ for $x \in \mathcal{T}_h^\Gamma$, i.e., $\hat{\psi}_h$ is, at least in a neighborhood of its zero level, close to a signed distance function. These approximation properties of the re-initialization are illustrated in a numerical experiment in Sect. 7.5.2.

Application of FMM to a piecewise quadratic function

Let $\phi_h \in \mathbb{X}_h^2$ be a piecewise quadratic function corresponding to the triangulation $\mathcal{T}_h$. The regular refinement of $\mathcal{T}_h$ is denoted by $\mathcal{T}_{h'} := \{T' \in \mathcal{K}(T) : T \in \mathcal{T}_h\}$ and $I(\phi_h)$ is the continuous piecewise linear function on $\mathcal{T}_{h'}$ that interpolates ϕ_h at all vertices of all tetrahedra in $\mathcal{T}_{h'}$. The approximate interface Γ_h is the zero level of $I(\phi_h)$. We can apply the FMM given above to the function $I(\phi_h)$, which results in the function $\widehat{I(\phi_h)}$ that is piecewise linear on $\mathcal{T}_{h'}$. The values at the vertices of this function uniquely define a piecewise *quadratic* function on $\mathcal{T}_h$, which is denoted by $\hat{\phi}_h =: FMM(\phi_h)$ and is defined to be the re-initialization of ϕ_h.

Remark 7.4.4 There is a need for re-initialization, only if the size of the gradient of ϕ_h is "too small" or "too large". One possibility to quantify this is the following. For $c > 1$ define the subset

$$V_c := \left\{ \phi_h \in \mathbb{X}_h^2 : \|\nabla\phi_h\|_{L^2(T)} < c|T|^{\frac{1}{2}} \quad \forall\ T \in \mathcal{T}_{h'} \right\}$$
$$\cap \left\{ \phi_h \in \mathbb{X}_h^2 : \frac{1}{c}|T|^{\frac{1}{2}} < \|\nabla\phi_h\|_{L^2(T)} \quad \forall\ T \in \mathcal{T}_{h'} \right\}.$$

In practice we take $c \in [5, 10]$ and apply the FMM only if $\phi_h \notin V_c$. This defines a re-initialization mapping $\mathrm{ReInit} : \mathbb{X}_h^2 \to \mathbb{X}_h^2$:

$$\mathrm{ReInit}(\phi_h) = \begin{cases} \phi_h & \text{if } \phi_h \in V_c, \\ \hat{\phi}_h = FMM(\phi_h) & \text{otherwise.} \end{cases} \tag{7.44}$$

The FMM is such that $\|\nabla\hat{\phi}_h\|$ is close to one, in particular we have (for c not too close to one) $\hat{\phi}_h \in V_c$. Hence one can expect $\mathrm{ReInit}(\mathrm{ReInit}(\phi_h)) = \mathrm{ReInit}(\phi_h)$ to hold for all $\phi_h \in \mathbb{X}_h^2$. Furthermore, one can check that the re-initialization mapping ReInit is *continuous* on V_c.

7.4.2 Mass conservation

Due to immiscibility the mass of the phase contained in $\Omega_i(t)$ is constant. Using the incompressibility of the phases, it follows that the volume $V_i(t) := \int_{\Omega_i(t)} 1\,dx$ is conserved, i.e. $\frac{d}{dt}V_i(t) = 0$ for $i = 1, 2$. Due to $\overline{\Omega_1(t)} \cup \overline{\Omega_2(t)} = \overline{\Omega}$ it suffices to consider $i = 1$ (or $i = 2$). For the level set function the quantity $\int_{\Omega_i(t)} \phi(x,t)\,dx$ is conserved:

$$\frac{d}{dt}\int_{\Omega_i(t)} \phi(x,t)\,dx = 0, \quad i = 1, 2, \tag{7.45}$$

which follows from the Reynolds' transport theorem and $\operatorname{div}\mathbf{u} = 0$. There is, however, no natural relation between this conservation property and mass conservation. The VOF method for interface capturing is based on a discretization of a transport equation for the characteristic function corresponding to Ω_1 (denoted by χ_1) and thus this method has a natural discrete mass conservation property. Recall the transport equation for χ_1:

$$\frac{\partial}{\partial t}\int_W \chi_1\,dx + \int_{\partial W} \chi_1 \mathbf{u}\cdot\mathbf{n}\,ds = 0, \quad W \subset \Omega,$$

cf. (6.26) ($\mathbf{n}$ is the outward unit normal on ∂W). Applying a conservative finite volume method to this problem results in a discretization $\tilde{\chi}_1(x, t_n)$ of $\chi_1(x, t_n)$, for which

$$\int_\Omega \tilde{\chi}_1(x, t_n) - \tilde{\chi}_1(x, t_{n-1})\,dx = 0, \quad n = 1, \ldots \frac{T}{\Delta t},$$

holds. Hence the volume conservation property holds on the discrete level.

In general the (temporal and spatial) discretization of the level set equation do *not* guarantee a volume conservation property. Also the FMM for re-initialization of the level set function is not volume-conserving. This is a disadvantage of the level set method compared to the VOF method. Since for $\Delta t \downarrow 0$, $h \downarrow 0$ the discrete level set function converges to the exact solution ϕ of the level set equation, one may expect that the amount of change in volume is reduced if the time step and mesh size are taken smaller. For the SDFEM method combined with Crank-Nicolson time discretization this is analyzed in [169]. The analysis is based on a discretization error bound

$$\|\phi_h^N - \phi(\cdot, T)\|_{L^2} \leq cT(h^{k+\frac{1}{2}} + \Delta t^2), \tag{7.46}$$

for the finite element approximation $\phi_h^N(x)$, cf. (7.19). Let $V_1(\phi_h^N)$ be the numerical approximation of the exact volume $V_1(T) = V_1(0) = V_1$, i.e., $V_1(\phi_h^N) = \int_{\Omega_{1,h}^N} 1\,dx$, with $\Omega_{1,h}^N := \{\, x \in \Omega : \phi_h^N(x) < 0 \,\}$. In [169] it is shown that from the discretization error bound (7.46) and with $\Delta t \sim h^{\frac{1}{2}k+\frac{1}{4}}$ one can derive the volume error estimate

$$|V_1(\phi_h^N) - V_1| \leq ch^k.$$

This estimate shows how the volume error can be controlled by reducing the mesh size.

In recent years there have appeared studies in which modifications of the level set method are presented that have better volume conservation properties. Often these modifications are based on combining the level set approach with a VOF technique. We do not treat this topic here, but refer to the literature, e.g. [229, 203, 87, 196].

In the literature also the following very simple (but less satisfactory) strategy, which guarantees volume conservation for the level set method, can be found. Let $V_1(0) = \int_{\Omega_1(0)} 1\,dx$ be the volume of Ω_1 at $t = 0$, that is assumed to be known. For a given $t > 0$, let $\phi_h(x) \approx \phi(x,t)$ be a computed approximation of the level set function and introduce

$$\Omega_{1,h}(t) := \{\, x \in \Omega : \phi_h(x,t) < 0 \,\}, \quad V_1(\phi_h;t) := \int_{\Omega_{1,h}(t)} 1\,dx.$$

We assume that the quantity $V_1(\phi_h;t)$ can easily be determined (sufficiently accurate). In our applications, where ϕ_h is piecewise quadratic on $\mathcal{T}_h$, we use the interpolation $I(\phi_h)$, which is piecewise linear on the refined triangulation $\mathcal{T}_{h'}$ and take

$$\Omega_{1,h}(t) := \big\{\, x \in \Omega : I\big(\phi_h(\cdot,t)\big)(x) < 0 \,\big\}.$$

Then $\partial\Omega_{1,h}(t) = \Gamma_h(t)$ and the integral $\int_{\Omega_{1,h}(t)} 1\,dx$ can be determined exactly (apart from rounding errors) using a simple quadrature rule on tetrahedra. In general one has no volume conservation, i.e. there may be a significant difference between $V_1(0)$ and $V_1(\phi_h;t)$. Due to the fact that ϕ_h is close to a signed distance function, a shift of the interface over a distance δ in outward normal direction can be realized (approximately) if one subtracts δ from the approximate level set function ϕ_h. For (exact!) volume conservation one has to find $\delta \in \mathbb{R}$ such that

$$V_1(\phi_h - \delta;t) - V_1(0) = 0$$

holds. In a method for computing a zero of this scalar function it is important to keep the number of evaluations of $V_1(\cdot;t)$ low. In our numerical simulations we use the Anderson-Björck method [14] to solve this equation. Let δ^* be a solution of this problem. We then set $\phi_h^{\text{new}} := \phi_h - \delta^*$ and discard ϕ_h.

Note that this strategy only works if Ω_1 consists of a single component. If there are multiple components, volume must be preserved for each of them. In this case the algorithm can be modified to shift ϕ_h only locally.

Clearly, using this simple strategy we have optimal volume conservation for the discrete level set function. Nevertheless, this approach is not very satisfactory since it introduces an additional discretization error source which is very hard to control.

7.5 Numerical experiments with the level set equation

In this section we present results of two numerical experiments to illustrate the performance of the discretization method for the level set equation and of the fast marching re-initialization technique.

7.5.1 Discretization using the SDFEM

We take $\Omega = [0,1]^3$ and the ball $\Omega_1 := \{ x \in \mathbb{R}^3 : \|x - x_M\| < 0.2 \}$ with center $x_M = (\frac{1}{2}, \frac{1}{4}, \frac{1}{2})$. We use the stationary velocity field

$$\hat{\mathbf{u}}(x) = c(y) \cdot (y_2, \; -y_1, \; 0)$$

where $y := x - (\frac{1}{2}, \frac{1}{2}, \frac{1}{2})$ and

$$c(y) = \begin{cases} 4\|y\|(0.5 - \|y\|) & \text{if } \|y\| \le \frac{1}{2}, \\ 0 & \text{otherwise.} \end{cases}$$

Hence, $\hat{\mathbf{u}}$ is a circular velocity field, cf. Fig. 7.6 for an illustration of $\hat{\mathbf{u}}$. Furthermore, $\hat{\mathbf{u}} = 0$ on $\partial\Omega$ holds. We consider the time interval $[0, T_{\text{end}}]$ and define the velocity field

$$\mathbf{u}(x,t) = \begin{cases} \hat{\mathbf{u}}(x) & t \le \frac{1}{2}T_{\text{end}}, \\ -\hat{\mathbf{u}}(x) & t > \frac{1}{2}T_{\text{end}}. \end{cases}$$

We consider the level set equation

$$\frac{\partial \phi}{\partial t} + \mathbf{u} \cdot \nabla \phi = 0 \quad \text{in} \quad \Omega, \quad t \in [0, T_{\text{end}}],$$

with initial condition $\phi(x,0) = d(x)$, where d is the signed distance function to $\Gamma(0) := \partial\Omega_1$. In Fig. 7.7, for $T_{\text{end}} = 20$, an illustration of the computed zero level of $\phi(x,t)$ at different times is shown. Clearly, $\phi(x, T_{\text{end}}) = d(x)$, $x \in \Omega$, holds. Thus the exact solution is known for $t = T_{\text{end}}$.

The initial tetrahedral triangulation is obtained by subdividing Ω into 8 subcubes, each of which is subdivided into six tetrahedra. Then repeated global regular refinement is applied to this initial triangulation. This results in nested triangulations $\mathcal{T}_{h_\ell}$ with mesh size parameter $h_\ell = (\frac{1}{2})^\ell$. On each triangulation we apply a method of lines discretization.

For the space discretization we use the SDFEM with quadratic finite elements. Due to $\mathbf{u} = 0$ on $\partial\Omega$ we do not need boundary conditions for ϕ. Therefore we can use the finite space $V_h := \mathbb{X}_h^2$ of piecewise quadratic finite elements, without any boundary conditions.

Remark 7.5.1 In other problems we usually have $\mathbf{u} \neq 0$ on $\partial\Omega$. In that case, in the weak formulation of the level set equation in (6.59c) we use a trial space with Dirichlet boundary conditions on the inflow boundary $\partial\Omega_{in}$ to make this hyperbolic problem well-posed. We discuss a possible choice of these (artificial) boundary conditions. Let $\phi_0 = \phi_0(x)$, $x \in \Omega$, be the initialization for the level set function. The Dirichlet data for the level set function can be taken as follows:

$$\phi_D(x,t) = \phi_0(x) - \mathbf{u}(x,0) \cdot \nabla \phi_0(x)\, t, \quad x \in \partial\Omega_{in}, \; t \in [0, t_0]. \tag{7.47}$$

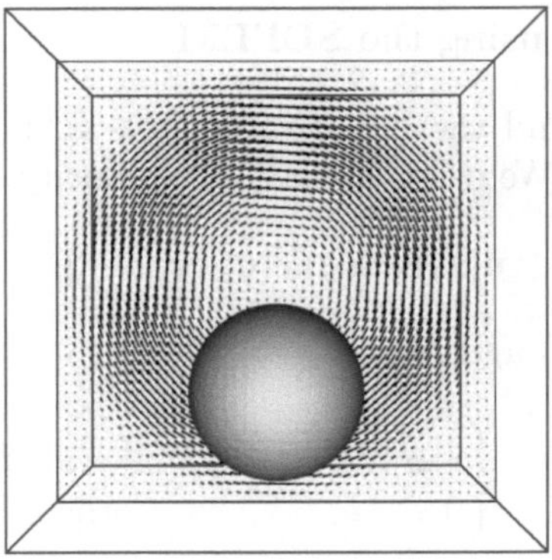

Fig. 7.6. Interface $\Gamma(0)$. Also shown is the velocity field $\hat{\mathbf{u}}$ on the slice $x_3 = 0$.

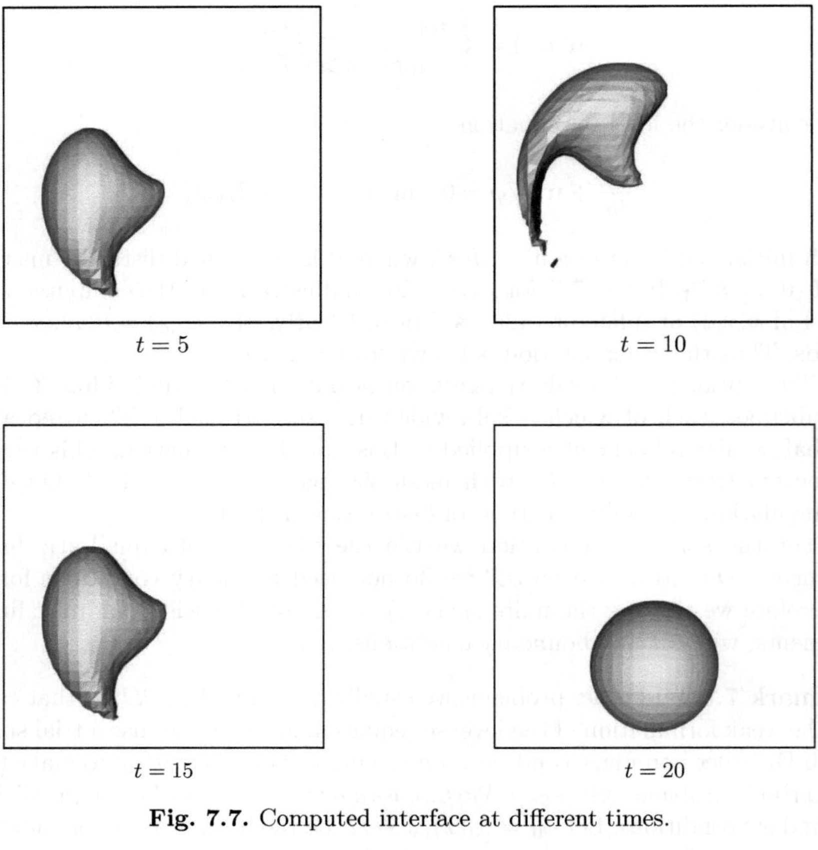

Fig. 7.7. Computed interface at different times.

These data are used until $t = t_0$, the point in time at which a re-initialization is applied. The re-initialization results in a modified level set function $\tilde{\phi}_h(x)$ and for $t \in (t_0, t_1]$ Dirichlet boundary data are defined as in (7.47) but with ϕ_0 replaced by $\tilde{\phi}_h$ and $\mathbf{u}(x, 0)$ replaced by $\mathbf{u}(x, t_0)$, etc. Note that this Dirichlet boundary condition is time dependent. We explain the heuristics leading to a boundary condition as in (7.47). For this we assume the inflow boundary $\partial\Omega_{in}$ to be planar and $\mathbf{u}(x, 0)$, $x \in \partial\Omega_{in}$, to be normal to the inflow boundary, i.e. $\mathbf{u}(x, 0) = -\|\mathbf{u}(x, 0)\|\mathbf{n}(x)$, with $\mathbf{n}$ the outward pointing unit normal on the boundary. The initial data ϕ_0 are extrapolated linearly by $\phi_0(x + \alpha\mathbf{n}(x)) := \phi_0(x) + \alpha\frac{\partial\phi_0(x)}{\partial n} = \phi_0(x) + \alpha\mathbf{n}(x) \cdot \nabla\phi_0(x)$, $\alpha \geq 0$. This defines initial data in the inflow region, outside the domain Ω. The velocity field $\mathbf{u}$ has a natural constant extension given by $\mathbf{u}(x + \alpha\mathbf{n}(x), 0) := \mathbf{u}(x, 0)$, $x \in \partial\Omega_{in}$, $\alpha \geq 0$. Solving the level set equation, which describes transport of the initial data by the velocity field $\mathbf{u}$, results in

$$\phi(x, t) = \phi_0(x - t\,\mathbf{u}(x, 0)) = \phi_0(x) - t\,\mathbf{u}(x, 0) \cdot \nabla\phi_0(x), \quad x \in \partial\Omega_{in}, \ t \geq 0,$$

which is the boundary condition proposed in (7.47).

The streamline diffusion finite element method is as explained in Sect. 7.2.2: Determine $\phi_h(t) \in V_h$ with $\phi_h(0) = d_h$ and such that

$$\sum_{T \in \mathcal{T}_h} \left(\frac{\partial\phi_h}{\partial t} + \mathbf{u} \cdot \nabla\phi_h,\, v_h + \delta_T \mathbf{u} \cdot \nabla v_h\right)_{L^2(T)} = 0 \quad \text{for all } v_h \in V_h,$$

and $t \in [0, T_{\text{end}}]$. Here d_h is the nodal interpolation of the initial condition d in the finite element space V_h. For the parameter δ_T in this method we take, cf. (7.14),

$$\delta_T = SD\frac{h_T}{\max\{10^{-3}, \|\mathbf{u}\|_{\infty,T}\}},$$

with a constant SD that is varied below. The resulting system of ordinary differential equations (7.16) is discretized using the implicit Euler or Crank-Nicolson method with a time step size Δt. Because in this experiment we want to study the accuracy of the discretization method, we do *not* use re-initialization of the level set function. In the Crank-Nicolson method we do *not* apply any volume correction procedure. In the implicit Euler method, however, the results without volume correction turn out to be very poor. Therefore we applied the simple method described in Sect. 7.4.2 at $t = 4, 8, \ldots, 20$. The space and time discretization results in an approximation $\phi_h^n \in V_h$ of $\phi(\cdot, t_n)$, $t_n := n\Delta t$. Let N be such that $N\Delta t = T_{\text{end}}$, i.e, ϕ_h^N is an approximation of $\phi(\cdot, T_{\text{end}}) = d$. The approximate zero level of ϕ_h^n is constructed as explained in Sect. 7.3 and is denoted by $\Gamma_h(t_n)$. This approximate interface consists of a set $\{\, \Gamma_T : T \in \mathcal{T}_{h'}^{\Gamma} \,\}$ of planar segments $\Gamma_T = T \cap \Gamma_h$, $T \in \mathcal{T}_{h'}^{\Gamma}$.

We now turn to a quantitative evaluation of the discretization method. Since $\Gamma(T_{\text{end}}) = \Gamma(0)$ and d is the signed distance function to the *exact* initial zero level $\Gamma(0)$, as a measure for the quality of $\Gamma_h(t_n) \approx \Gamma(0)$, $n = 0$, $n = N$, we introduce

$$\|d\|_{L^2(\Gamma_h(t_n))} := \sqrt{\sum_{\Gamma_T \subset \Gamma_h(t_n)} \int_{\Gamma_T} d(x)^2 \, dx}, \quad n = 0, \; n = N. \tag{7.48}$$

As a second error measure we use the global $L^2(\Omega)$ error,

$$\|d - \phi_h^n\|_{L^2(\Omega)} = \sqrt{\sum_{T \in \mathcal{T}_h} \int_T |d(x) - \phi_h^n(x)|^2 \, dx}, \quad n = 0, \; n = N, \tag{7.49}$$

which can be determined accurately using suitable quadrature. Note that for $n = 0$ we have $\phi_h^0 = d_h$ and in these error quantities only the interpolation error $d - d_h$ and the approximation of the zero level of the interpolant d_h plays a role. In Table 7.1 we give these quantities for $n = 0$ and different grid sizes h_ℓ.

ℓ	$\|d\|_{L^2(\Gamma_h(0))}$	order
1	1.77 E-2	-
2	3.90 E-3	2.18
3	9.37 E-4	2.06
4	2.31 E-4	2.02
5	5.82 E-5	1.99

ℓ	$\|d - \phi_h^0\|_{L^2(\Omega)}$	order
1	4.90 E-2	-
2	1.28 E-2	1.94
3	3.27 E-3	1.97
4	8.17 E-4	2.00
5	2.04 E-4	2.00

Table 7.1. Approximation errors for different mesh sizes $h = h_\ell$.

These results are consistent with the expected h_ℓ^2 convergence.

The quality of the space and time discretization is measured by these quantities for $n = N$. We only consider the implicit Euler method, since for the Crank-Nicolson method this test case is not representative, cf. Remark 7.5.2. Results for $T_{\text{end}} = 20$ and several mesh and time step sizes are given in Table 7.2 and Table 7.3. Note that for $SD = 0$ there is no stabilization in the spatial finite element discretization.

A first (surprising) observation is that the method without stabilization ($SD = 0$) produces very good results. One observes an $\mathcal{O}(\Delta t)$ error behavior if the time step is reduced. For $\Delta t = 2^{-6}$ one obtains close to optimal errors; for example, for $\ell = 3$ we have $\|d\|_{L^2(\Gamma_h(20))} = 1.07$ E-3 and $\|d - \phi_h^N\|_{L^2(\Omega)} = 4.21$ E-3, which have to be compared with the interpolation errors 9.37 E-4 and 3.27 E-3 of the initial data from Table 7.1. The performance of the method with stabilization is worse. In case of the stabilization with $SD = \frac{1}{2}$ we do not observe an $\mathcal{O}(\Delta t)$ error reduction behavior. Furthermore, for Δt "small" the error is dominated by the spatial discretization error and stagnates at a level higher than the interpolation error level. An explanation of this behavior is a topic for further research.

	$\ell = 3$		$\ell = 4$	
Δt	$SD = 0$	$SD = \frac{1}{2}$	$SD = 0$	$SD = \frac{1}{2}$
2^{-1}	1.34 E-2	1.42 E-2	1.40 E-2	1.32 E-2
2^{-2}	7.04 E-3	1.03 E-2	7.24 E-3	7.34 E-3
2^{-3}	3.87 E-3	8.84 E-3	3.87 E-3	4.55 E-3
2^{-4}	2.20 E-3	8.31 E-3	2.04 E-3	3.27 E-3
2^{-5}	1.39 E-3	8.11 E-3	1.05 E-3	2.80 E-3
2^{-6}	1.07 E-3	8.02 E-3	5.64 E-4	2.66 E-3

Table 7.2. Implicit Euler: error $\|d\|_{L^2(\Gamma_h(20))}$ for different $h = h_\ell$ and Δt

	$\ell = 3$		$\ell = 4$	
Δt	$SD = 0$	$SD = \frac{1}{2}$	$SD = 0$	$SD = \frac{1}{2}$
2^{-1}	3.77 E-2	4.15 E-2	3.61 E-2	3.72 E-2
2^{-2}	2.14 E-2	2.54 E-2	1.96 E-2	2.06 E-2
2^{-3}	1.22 E-2	1.66 E-2	1.03 E-2	1.13 E-2
2^{-4}	7.54 E-3	1.23 E-2	5.49 E-3	6.62 E-3
2^{-5}	5.27 E-3	1.02 E-2	3.12 E-3	4.37 E-3
2^{-6}	4.21 E-3	9.29 E-3	1.94 E-3	3.33 E-3

Table 7.3. Implicit Euler: error $\|d - \phi_h^N\|_{L^2(\Omega)}$ for different $h = h_\ell$ and Δt

Remark 7.5.2 Consider a system of ODEs of the form $y'(t) + F(t)y(t) = 0$, $y(0) = y_0$, $t \in [0, T_{\text{end}}]$ with $F(t) = A$ for $t \in [0, \frac{1}{2}T_{\text{end}}]$, $F(t) = -A$ for $t \in (\frac{1}{2}T_{\text{end}}, T_{\text{end}}]$ and $A \in \mathbb{R}^{n\times n}$ a given matrix. We apply the Crank-Nicolson method with a time step size $\Delta t = T_{\text{end}}/N$ and N even, resulting in approximations y^n of $y(t_n)$, $n = 1, 2, \ldots, N$. One easily checks that, due to the special symmetry in the problem and in the Crank-Nicolson method, we have $y^N = y(0)$, i.e., the initial condition is exactly reproduced. In our test example we have such a symmetry in the spatially discretized problem (for the case $SD = 0$). Therefore, if we repeat the numerical experiment described above using the Crank-Nicolson method instead of the implicit Euler method we obtain $\|d\|_{L^2(\Gamma_h(20))} = \|d - \phi_h^N\|_{L^2(\Omega)} = 0$ for the case $SD = 0$ and very small errors for the case $SD = \frac{1}{2}$.

In order to compare the Crank-Nicolson method to the implicit Euler method we performed an experiment in which only the time discretization error is measured. On a fixed triangulation with mesh size $h = \frac{1}{8}$ we applied the SDFEM with $SD = 0.1$. We take $T_{\text{end}} = 20$ and on the time interval $[0, 10]$ the resulting system of ODEs is solved with a "small" time step $\frac{1}{10}2^{-5}$ resulting in a reference solution at $t = 10$ denoted by $\tilde{\phi}_h \approx \phi(\cdot, 10)$. Note that for $t \in [0, 10]$ the droplet is transported with the velocity field $\hat{\mathbf{u}}$ and does not move back to the initial position. Hence the symmetry property addressed in Remark 7.5.2 does not hold. The system of ODEs is solved using both the Crank-Nicolson and the implicit Euler method for time step sizes $\Delta t = 2^{-k}$, $k = 0, \ldots, 5$. The computed result at $t = 10$ is denoted by $\phi_h^{\frac{1}{2}N}$. In the Tables 7.4 and 7.5 we give the errors $\|\tilde{\phi}_h - \phi_h^{\frac{1}{2}N}\|_{L^2(\Omega)}$. These results show the expected rate of convergence, namely $\mathcal{O}(\Delta t)$ for the Euler method and $\mathcal{O}(\Delta t^2)$ for the Crank-Nicolson method.

Δt	2^0	2^{-1}	2^{-2}	2^{-3}	2^{-4}	2^{-5}
$\ell = 3$	2.88 E-2	1.54 E-2	7.85 E-3	3.93 E-3	1.96 E-3	9.65 E-4

Table 7.4. Implicit Euler: discretization error $\|\tilde{\phi}_h - \phi_h^{\frac{1}{2}N}\|_{L^2(\Omega)}$.

Δt	2^0	2^{-1}	2^{-2}	2^{-3}	2^{-4}	2^{-5}
$\ell = 3$	1.42 E-3	3.87 E-4	9.82 E-5	2.82 E-5	6.14 E-6	1.53 E-6

Table 7.5. Crank-Nicolson: discretization error $\|\tilde{\phi}_h - \phi_h^{\frac{1}{2}N}\|_{L^2(\Omega)}$.

7.5.2 Re-initialization by the Fast Marching Method

In this section we present some quantitative results related to the quality of the Fast Marching re-initialization method. More results are presented [123]. We consider the cubic domain $\Omega = (-1, 1)^3$ and the quadratic level set function

$$\phi(x) = \|x\|^2 - r^2, \quad x \in \Omega, \quad r = 0.6.$$

The zero level of ϕ, denoted by Γ, is given by the sphere centered at the origin with radius r. On Γ we have $\|\nabla\phi\| = 2r = 1.2$. The signed distance function to Γ is denoted by $d(x)$. The domain Ω is subdivided into 8 subcubes, each subdivided into 6 tetrahedra. This defines the level $\ell = 0$ triangulation. The level $\ell \geq 1$ triangulation $\mathcal{T}_{h_\ell}$ is obtained by ℓ local refinements in the neighborhood $\{\, x \in \Omega : \, |d(x)| \leq 0.1 \,\}$. The level ℓ triangulation has local mesh size parameter $h_{\Gamma,\ell} = (\frac{1}{2})^\ell$. To the quadratic function ϕ on a given triangulation $\mathcal{T}_h = \mathcal{T}_{h_\ell}$ we apply the FMM as discussed at the end of Sect. 7.4.1. As output of the initialization phase in the FMM one obtains an approximate signed distance function, denoted by $\hat{\psi}_h$, which is piecewise linear on $\mathcal{T}_{h'}^{\Gamma}$ ($h' = \frac{1}{2}h$; we use notation as in Sect. 7.4.1). The zero level of $\hat{\psi}_h$, which is contained in $\mathcal{T}_{h'}^{\Gamma}$, defines the new approximate interface denoted by $\hat{\Gamma}_h$. Let $\hat{\mathcal{F}}$ denote the set of triangles forming $\hat{\Gamma}_h$, i.e., $\hat{\Gamma}_h = \bigcup_{F \in \hat{\mathcal{F}}} F$, and $\hat{\mathcal{P}} = \left\{ v \in \hat{\Gamma}_h : v \text{ is vertex of triangle } F \in \hat{\mathcal{F}} \right\}$ the set of their vertices.

To measure the quality of this re-initialization we computed the quantities

$$e_{h,\infty} := \max_{v \in \hat{\mathcal{P}}} |d(v) - \hat{\psi}_h(v)| = \max_{v \in \hat{\mathcal{P}}} |d(v)|, \tag{7.50a}$$

$$\nabla e_{h,\infty} := \max_{T \in \mathcal{T}_{h'}^{\Gamma}} \left\{ \|\nabla d(c) - \nabla \hat{\psi}_h(c)\| \, : \, c \text{ barycenter of } T \right\}. \tag{7.50b}$$

An important parameter in the geometry-based initialization phase is k, cf. (7.35). Below we consider $k = 1$ and $k = 2$. The results are compared to the scaling approach, cf. (7.36). In Tables 7.6 and 7.7 we present values of $e_{h,\infty}$ and $\nabla e_{h,\infty}$, respectively, for different levels ℓ and the different initialization strategies. Using the geometry-based initialization phase, for $k = 1$ we do

ℓ	geom, $k = 1$	order	geom, $k = 2$	order	scale	order
1	5.49 E-2	–	5.49 E-2	–	3.55 E-2	–
2	1.34 E-2	2.03	1.34 E-2	2.03	1.01 E-2	1.81
3	3.62 E-3	1.89	3.62 E-3	1.89	2.55 E-3	1.99
4	9.04 E-4	2.00	9.04 E-4	2.00	6.60 E-4	1.95
5	3.18 E-4	1.51	2.27 E-4	2.00	1.66 E-4	1.99
6	1.42 E-4	1.17	5.67 E-5	2.00	4.16 E-5	1.99

Table 7.6. Error measures $e_{h,\infty}$ for different initialization approaches and $h = h_\ell$.

not observe second order convergence for $e_{h,\infty}$. Moreover, we see a stagnation for $\nabla e_{h,\infty}$, meaning that the re-initialized interface does not get smoother on finer grids which renders the choice $k = 1$ unfeasible. Taking $k = 2$ instead, we do get quadratic convergence for $e_{h,\infty}$ and linear convergence for $\nabla e_{h,\infty}$, showing that the choice $k = 2$ should be preferred. The same convergence properties hold for the scaling approach which is less expensive in terms of computational effort. The results confirm the theoretical error bounds discussed in Remark 7.4.3. We briefly comment on this. Take $T \in \mathcal{T}_{h'}^{\Gamma}$ and let I_T be the linear interpolation operator on T that interpolates at the vertices of T. For $x \in T$ we have $(d - \hat{\psi}_h)(x) = I_T(d - \hat{\psi}_h)(x) + \mathcal{O}(h_T^2)$. The results in the fifth column in Table 7.6 indicate $|I_T(d - \hat{\psi}_h)(x)| \le c h_T^2$ uniformly in T and $x \in \Gamma_T = \Gamma \cap T$. Hence for $x \in \hat{\Gamma}_h$, i.e. $\hat{\psi}_h(x) = 0$, we get $|d(x)| \le c h_\Gamma^2$, which is the same as the bound in (7.43a). Since $\nabla d = \mathbf{n}$ the results in the fifth column in Table 7.7 are consistent with the error bound given in (7.43b).

7.6 Discretization of the surface tension functional

In this section we explain how the localized surface tension force term $f_\Gamma(\mathbf{v})$ in (6.59a) can be approximated. We use the approach presented in [23, 93, 126].

Let $\mathbf{V}_h$ be the finite element space that is used for the discretization of the velocity unknown. In our simulations we use for $\mathbf{V}_h$ the standard conforming space of continuous piecewise quadratic functions. Applying the Galerkin

ℓ	geom, $k=1$	order	geom, $k=2$	order	scale	order
1	3.07 E-1	–	2.09 E-1	–	1.87 E-1	–
2	2.70 E-1	0.19	1.34 E-1	0.64	1.03 E-1	0.87
3	1.73 E-1	0.64	7.12 E-2	0.91	5.62 E-2	0.87
4	3.78 E-1	-1.13	3.69 E-2	0.95	2.79 E-2	1.01
5	4.69 E-1	-0.31	1.80 E-2	1.04	1.42 E-2	0.98
6	5.11 E-1	-0.12	8.95 E-3	1.01	7.04 E-3	1.01

Table 7.7. Error measures $\nabla e_{h,\infty}$ for different initialization approaches and $h = h_\ell$.

discretization to the variational (momentum) equation (6.59a) results in a *surface tension functional* of the form

$$f_\Gamma(\mathbf{v}_h) = -\tau \int_\Gamma \kappa \mathbf{v}_h \cdot \mathbf{n}\, ds, \quad \mathbf{v}_h \in \mathbf{V}_h. \tag{7.51}$$

In many numerical simulations of two-phase flows, the discretization of the curvature κ is a very delicate problem. This is related to the fact that κ contains *second* derivatives. One way to express these second derivatives is by means of the Laplace-Beltrami characterization of the mean curvature, cf. (14.9):

$$-\Delta_\Gamma \,\mathrm{id}_\Gamma(x) = \kappa(x)\mathbf{n}(x), \quad x \in \Gamma. \tag{7.52}$$

In the variational formulation we have the possibility to lower the order of differentiation by shifting one of the derivatives to the test function, as is shown in Lemma 14.1.2. Using this, we see that (7.51) can be rewritten as follows:

$$f_\Gamma(\mathbf{v}_h) = -\tau \int_\Gamma \nabla_\Gamma \,\mathrm{id}_\Gamma \cdot \nabla_\Gamma \mathbf{v}_h \, ds, \quad \mathbf{v}_h \in \mathbf{V}_h. \tag{7.53}$$

In this variational setting it is natural to use the expression on the right-hand side in (7.53) as a starting point for the discretization. This idea is used in, for example, [93, 23, 116, 126, 147, 177]. In this discretization we use the approximation Γ_h of Γ. Given this approximate interface Γ_h,

the localized force term $f_\Gamma(\mathbf{v}_h)$ is approximated by

$$f_{\Gamma_h}(\mathbf{v}_h) := -\tau \int_{\Gamma_h} \nabla_{\Gamma_h} \mathrm{id}_{\Gamma_h} \cdot \nabla_{\Gamma_h} \mathbf{v}_h \, ds, \qquad \mathbf{v}_h \in \mathbf{V}_h. \tag{7.54}$$

In Sect. 7.7 we will derive a bound for the error quantity

$$\|f_\Gamma - f_{\Gamma_h}\|_{\mathbf{V}_h'} = \sup_{\mathbf{v}_h \in \mathbf{V}_h} \frac{f_\Gamma(\mathbf{v}_h) - f_{\Gamma_h}(\mathbf{v}_h)}{\|\mathbf{v}_h\|_1}, \tag{7.55}$$

with f_{Γ_h} as in (7.54). Note that this quantity is essential in the analysis of discretization errors in velocity and pressure, cf. Corollary 7.10.5 (Strang-lemma).

Remark 7.6.1 Assume that Γ is sufficiently smooth.Then

$$f_\Gamma(\mathbf{v}) = -\tau \int_\Gamma \kappa \mathbf{v} \cdot \mathbf{n} \, ds \tag{7.56}$$

is a bounded linear functional on $\mathbf{V}_0 = H_0^1(\Omega)^3$. From Lemma 14.1.2 it follows that for this functional we have the *equivalent* representation

$$f_\Gamma : \mathbf{v} \to -\tau \int_\Gamma \nabla_\Gamma \operatorname{id}_\Gamma \cdot \nabla_\Gamma \mathbf{v} \, ds. \tag{7.57}$$

Such an equivalence, however, does *not* hold for f_{Γ_h}. Because Γ_h is not sufficiently smooth, a partial integration result as in Lemma 14.1.2 does not hold. The linear functional

$$\mathbf{v} \to -\tau \int_{\Gamma_h} \nabla_{\Gamma_h} \operatorname{id}_{\Gamma_h} \cdot \nabla_{\Gamma_h} \mathbf{v} \, ds$$

is *not* necessarily bounded on $\mathbf{V}_0$. Due to this, in (7.54) and (7.55) we only consider $\mathbf{v}_h \in \mathbf{V}_h$.

We also introduce a modified (improved) variant of the functional f_{Γ_h}. Define the orthogonal projection

$$\mathbf{P}_h(x) := \mathbf{I} - \mathbf{n}_h(x)\mathbf{n}_h(x)^T \quad \text{for } x \in \Gamma_h, \ x \text{ not on an edge},$$

where $\mathbf{n}_h$ is the unit normal on Γ_h (pointing outward from Ω_1). The tangential derivative along Γ_h can be written as $\nabla_{\Gamma_h} g = \mathbf{P}_h \nabla g$. Note that

$$\nabla_{\Gamma_h} \operatorname{id}_{\Gamma_h} = \mathbf{P}_h \nabla \operatorname{id}_{\Gamma_h} = (\mathbf{P}_h e_1, \mathbf{P}_h e_2, \mathbf{P}_h e_3)^T,$$

with e_i the i-th standard basis vector in $\mathbb{R}^3$. Thus the functional f_{Γ_h} can be written as

$$\begin{aligned} f_{\Gamma_h}(\mathbf{v}_h) &= -\tau \int_{\Gamma_h} \mathbf{P}_h \nabla \operatorname{id}_{\Gamma_h} \cdot \nabla_{\Gamma_h} \mathbf{v}_h \, ds \\ &= -\tau \sum_{i=1}^3 \int_{\Gamma_h} \mathbf{P}_h e_i \cdot \nabla_{\Gamma_h} v_i \, ds, \quad v_i := (\mathbf{v}_h)_i. \end{aligned} \tag{7.58}$$

The discrete interface Γ_h is constructed as the zero level of $I\phi_h$, where ϕ_h is a piecewise *quadratic* function, cf. Sect. 7.3. This piecewise quadratic function contains better information about the curvature of Γ than its piecewise linear interpolation $I\phi_h$ that is used for the construction of Γ_h. An improved projection $\tilde{\mathbf{P}}_h$ based on ϕ_h can be defined as follows:

$$\tilde{\mathbf{n}}_h(x) := \frac{\nabla \phi_h(x)}{\|\nabla \phi_h(x)\|}, \qquad \tilde{\mathbf{P}}_h(x) := \mathbf{I} - \tilde{\mathbf{n}}_h(x)\tilde{\mathbf{n}}_h(x)^T, \quad x \in \Gamma_h. \tag{7.59}$$

Hence an obvious modification is given by

$$
\begin{aligned}
\tilde{f}_{\Gamma_h}(\mathbf{v}_h) &= -\tau \int_{\Gamma_h} \tilde{\mathbf{P}}_h \nabla \operatorname{id}_{\Gamma_h} \cdot \nabla_{\Gamma_h} \mathbf{v}_h \, ds \\
&= -\tau \sum_{i=1}^{3} \int_{\Gamma_h} \tilde{\mathbf{P}}_h e_i \cdot \nabla_{\Gamma_h} v_i \, ds, \qquad v_i := (\mathbf{v}_h)_i \qquad (7.60) \\
&= -\tau \int_{\Gamma_h} \operatorname{tr}\big(\tilde{\mathbf{P}}_h \nabla_{\Gamma_h} \mathbf{v}_h\big) \, ds.
\end{aligned}
$$

In Sect. 7.7 it is shown that this discretization of the surface tension force is (significantly) better than the one in (7.54), namely with an error bound $\mathcal{O}(h)$ instead of $\mathcal{O}(\sqrt{h})$. This is confirmed by numerical experiments in Sect. 7.8.

7.6.1 Treatment of general surface tension tensors

In the previous section we restricted ourselves to the case of a *constant* surface tension coefficient τ, i.e. with an interface condition of the form

$$[\boldsymbol{\sigma}\mathbf{n}_\Gamma] = -\tau\kappa\mathbf{n}_\Gamma = \operatorname{div}_\Gamma(\tau\mathbf{P}).$$

We now consider the more general case with an interface condition of the form

$$[\boldsymbol{\sigma}\mathbf{n}_\Gamma] = \operatorname{div}_\Gamma(\boldsymbol{\sigma}_\Gamma), \tag{7.61}$$

and an interface stress tensor $\boldsymbol{\sigma}_\Gamma$ such that $\boldsymbol{\sigma}_\Gamma = \boldsymbol{\sigma}_\Gamma\mathbf{P}$ holds. A variable surface tension coefficient corresponds to $\boldsymbol{\sigma}_\Gamma = \tau\mathbf{P}$, resulting in $\operatorname{div}_\Gamma(\boldsymbol{\sigma}_\Gamma) = -\tau\kappa\mathbf{n}_\Gamma + \nabla_\Gamma\tau$, cf. Remark 1.1.3. In the Boussinesq-Scriven model treated in Sect. 1.1.5 the interface stress tensor $\boldsymbol{\sigma}_\Gamma$ is as in (1.36). In the weak formulation, instead of the surface tension functional $f_\Gamma(\mathbf{v}) = -\tau\int_\Gamma \kappa\mathbf{n}\cdot\mathbf{v}\,ds$ we then have the generalization

$$f_\Gamma(\mathbf{v}) = \int_\Gamma \operatorname{div}_\Gamma(\boldsymbol{\sigma}_\Gamma)\cdot\mathbf{v}\,ds.$$

We can rewrite this using the partial integration identity (14.17), resulting in the general surface tension functional

$$f_\Gamma(\mathbf{v}) = -\int_\Gamma \operatorname{tr}(\boldsymbol{\sigma}_\Gamma\nabla_\Gamma\mathbf{v})\,ds = -\sum_{i=1}^{3}\int_\Gamma (e_i^T\boldsymbol{\sigma}_\Gamma)\nabla_\Gamma v_i\,ds, \tag{7.62}$$

with $\mathbf{v} = (v_1, v_2, v_3)^T$. We consider the case of a variable surface tension coefficient, i.e., $\boldsymbol{\sigma}_\Gamma = \tau\mathbf{P}$. For the discretization of the corresponding surface tension functional, as in the previous section we approximate Γ by Γ_h and $\mathbf{P}$ by $\tilde{\mathbf{P}}_h$. Thus we obtain the following generalization of (7.60):

$$\tilde{f}_{\Gamma_h}(\mathbf{v}_h) = -\int_{\Gamma_h} \tau \operatorname{tr}(\tilde{\mathbf{P}}_h \nabla_{\Gamma_h} \mathbf{v})\, ds = -\sum_{i=1}^{3} \int_{\Gamma_h} \tau \tilde{\mathbf{P}}_h e_i \cdot \nabla_{\Gamma_h} v_i \, ds. \tag{7.63}$$

Comparing (7.60) (constant τ) and (7.63) (variable τ) we observe that the only difference is that in case of a constant surface tension coefficient the term τ can be taken out of the integral. Below, in Sect. 7.7, we present an error analysis only for the case that τ is *constant*, in which the discretization (7.63) reduces to (7.60). If τ is a smooth function then an error analysis for the generalization (7.63) can be derived along the same lines as for the constant coefficient case presented in Sect. 7.7. Results of numerical experiments with this discrete variable surface tension functional are given in [184] and in Sect. 11.5.3.

7.7 Analysis of the Laplace-Beltrami discretization

In this section we derive a bound for

$$\sup_{\mathbf{v}_h \in \mathbf{V}_h} \frac{f_\Gamma(\mathbf{v}_h) - f_{\Gamma_h}(\mathbf{v}_h)}{\|\mathbf{v}_h\|_1}, \tag{7.64}$$

where f_{Γ_h} is the discretization of the surface tension force as in (7.54). We also derive a bound for this error measure with f_{Γ_h} replaced by $\tilde{f}_{\Gamma_h}$ as in (7.60). For $\mathbf{V}_h = V_h^3$ we take the finite element space of piecewise quadratics:

$$V_h = \{\, v \in C(\Omega) : v_{|T} \in \mathcal{P}_2 \quad \text{for all } \ T \in \mathcal{T}_h \,\}. \tag{7.65}$$

The choice of this finite element space is not essential in our analysis. The results also hold if for V_h we take another conforming piecewise polynomial finite element space.

7.7.1 Preliminaries

Properties of Γ and of Γ_h

We recall some notation and definitions from Sect. 7.3. The function d is the signed distance function

$$d : U \to \mathbb{R}, \qquad |d(x)| := \operatorname{dist}(x, \Gamma) \quad \text{for all } x \in U.$$

Thus Γ is the zero level set of d. We assume $d < 0$ on the interior of Γ (that is, in Ω_1) and $d > 0$ on the exterior. Note that $\mathbf{n}_\Gamma = \nabla d$ on Γ. We define $\mathbf{n}(x) := \nabla d(x)$ for all $x \in U$. Thus $\mathbf{n} = \mathbf{n}_\Gamma$ on Γ and $\|\mathbf{n}(x)\| = 1$ for all $x \in U$.

The Hessian of d is denoted by $\mathbf{H}$:

$$\mathbf{H}(x) = \nabla^2 d(x) \in \mathbb{R}^{3\times 3} \quad \text{for all } x \in U. \tag{7.66}$$

The eigenvalues of $\mathbf{H}(x)$ are denoted by $\kappa_1(x), \kappa_2(x)$ and 0. For $x \in \Gamma$ the eigenvalues $\kappa_i(x)$, $i = 1, 2$, are the *principal curvatures*, and $\kappa(x) = \kappa_1(x) + \kappa_2(x)$ is the *mean curvature*, cf. Chap. 14.

In the analysis we always assume that the following (technical) assumption is satisfied, namely that the neighborhood U of Γ is sufficiently small in the following sense. We assume that U is a strip of width $\delta > 0$ with

$$\delta^{-1} > \max_{i=1,2} \|\kappa_i(x)\|_{L^\infty(\Gamma)}. \tag{7.67}$$

We define the orthogonal projection:

$$\mathbf{P}(x) = \mathbf{I} - \mathbf{n}(x)\mathbf{n}(x)^T \quad \text{for } x \in U. \tag{7.68}$$

From $\nabla\big(\mathbf{n}(x)^T\mathbf{n}(x)\big) = 0$ it follows that $\big(\nabla\mathbf{n}(x)\big)\mathbf{n}(x) = \nabla^2 d(x)\mathbf{n}(x) = 0$ holds for all $x \in U$. Hence we obtain the following:

$$\mathbf{P}(x)\mathbf{H}(x) = \mathbf{H}(x)\mathbf{P}(x) = \mathbf{H}(x) \quad \text{for all } x \in U.$$

We introduce assumptions on the approximate interface Γ_h. We emphasize, that although we use the notation Γ_h, *this interface must not necessarily be constructed using the method explained in Sect.* 7.3. Our analysis below is presented in a more general setting. In Remark 7.7.3 we explain how the concrete interface construction that is discussed in Sect. 7.3 fits in this more general setting.

Let $\{\Gamma_h\}_{h>0}$ be a family of polygonal approximations of Γ. We assume that each Γ_h is contained in U and consists of a set $\mathcal{F}_h$ of *triangular faces*:

$$\Gamma_h = \bigcup_{F \in \mathcal{F}_h} F. \tag{7.69}$$

For $F_1, F_2 \in \mathcal{F}_h$ with $F_1 \neq F_2$ we assume that $F_1 \cap F_2$ is either empty or a common edge or a common vertex. The parameter h_Γ denotes the maximal diameter of the triangles in $\mathcal{F}_h$:

$$h_\Gamma = \max_{F \in \mathcal{F}_h} \operatorname{diam}(F).$$

By $\mathbf{n}_h(x)$ we denote the outward pointing unit normal on Γ_h. This normal is piecewise constant with possible discontinuities at the edges of the triangles in $\mathcal{F}_h$. We recall the discrete analogon of the orthogonal projection $\mathbf{P}$:

$$\mathbf{P}_h(x) := \mathbf{I} - \mathbf{n}_h(x)\mathbf{n}_h(x)^T \quad \text{for } x \in \Gamma_h,\ x \text{ not on an edge}.$$

The tangential derivative along Γ_h can be written as $\nabla_{\Gamma_h} g = \mathbf{P}_h \nabla g$.

Assumption 7.7.1 We need assumptions which guarantee that Γ_h is "sufficiently close" to Γ. Related to this we assume that $\Gamma_h \subset U$ and that the following holds:

$$|d(x)| \le c h_\Gamma^2 \quad \text{for all } x \in \Gamma_h, \tag{7.70a}$$

$$\operatorname{ess\,sup}_{x\in\Gamma_h} \|\mathbf{n}(x) - \mathbf{n}_h(x)\| \le \min\{c_0, c\, h_\Gamma\}, \quad \text{with } c_0 < \sqrt{2}. \tag{7.70b}$$

Here c, c_0 denote generic constants independent of h_Γ.

Remark 7.7.2 For a given $x \in \Gamma_h$, let θ be the angle between $\mathbf{n}(x)$ and $\mathbf{n}_h(x)$. Then $\cos\theta = \mathbf{n}(x)^T\mathbf{n}_h(x)$, $\sin\theta = \|\mathbf{P}\mathbf{n}_h(x)\|$ and $\|\mathbf{n}(x) - \mathbf{n}_h(x)\|^2 = 2(1-\cos\theta)$. Elementary manipulations show that the condition (7.70b) holds if and only if the following two conditions are satisfied:

$$\operatorname{ess\,inf}_{x\in\Gamma_h} \mathbf{n}(x)^T\mathbf{n}_h(x) \ge c > 0, \tag{7.71a}$$

$$\operatorname{ess\,sup}_{x\in\Gamma_h} \|\mathbf{P}(x)\mathbf{n}_h(x)\| \le c h_\Gamma. \tag{7.71b}$$

Remark 7.7.3 Related to these assumptions we note the following. In Theorem 7.3.1 it is shown that under certain (reasonable) assumptions the construction explained in Sect. 7.3 results in a family $\{\Gamma_h\}$ that satisfies the conditions (7.70).

Extensions

We introduce extensions that will be used in the analysis below. The techniques that we use are from the paper [83]. For proofs of certain results we will refer to that paper.

As in Sect. 7.3 we define a locally (in a neighborhood of Γ) orthogonal coordinate system by using the projection $\mathbf{p} : U \to \Gamma$:

$$\mathbf{p}(x) = x - d(x)\mathbf{n}(x) \quad \text{for all } x \in U.$$

We assume that the decomposition $x = \mathbf{p}(x) + d(x)\mathbf{n}(x)$ is unique for all $x \in U$. Note that

$$\mathbf{n}(x) = \mathbf{n}\big(\mathbf{p}(x)\big) \quad \text{for all } x \in U.$$

We use an extension operator defined as follows. For a (scalar) function v defined on Γ we define

$$v_\Gamma^e(x) := v\big(x - d(x)\mathbf{n}(x)\big) = v\big(\mathbf{p}(x)\big) \quad \text{for all } x \in U,$$

i.e., v is extended along normals on Γ. We will also need extensions of functions defined on Γ_h. This is done again by extending along normals $\mathbf{n}(x)$. For v defined on Γ_h we define, for $x \in \Gamma_h$,

$$v_{\Gamma_h}^e(x + \alpha\mathbf{n}(x)) := v(x) \quad \text{for all } \alpha \in \mathbb{R} \quad \text{with } x + \alpha\mathbf{n}(x) \in U. \tag{7.72}$$

The projection $\mathbf{p}$ and the extensions v_Γ^e, $v_{\Gamma_h}^e$ are illustrated in Fig. 7.8.
In the following two lemmas some properties of these extensions are given. Proofs are elementary and can be found in [83]. In the remainder we assume that Assumption 7.7.1 is satisfied.

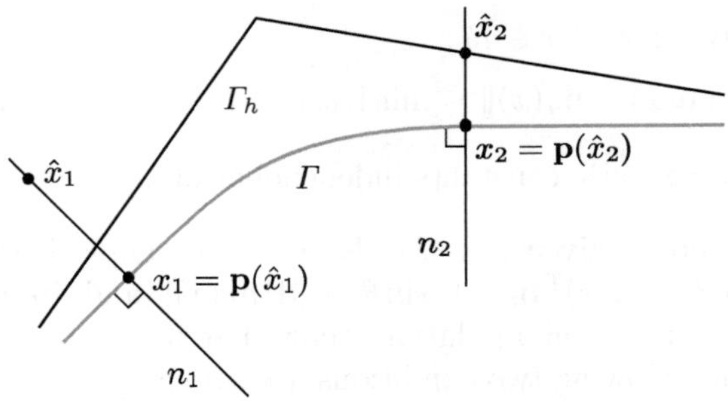

Fig. 7.8. Example of projection $\mathbf{p}$ and construction of extension operators. n_1 and n_2 are straight lines perpendicular to Γ. For v defined on Γ we have $v_\Gamma^e \equiv v(x_1)$ on n_1. For v_h defined on Γ_h we have $v_{\Gamma_h}^e \equiv v_h(\hat{x}_2)$ on n_2.

Lemma 7.7.4 *For v defined on Γ and sufficiently smooth the following holds:*

$$\nabla_{\Gamma_h} v_\Gamma^e(x) = \mathbf{P}_h(x)\big(\mathbf{I} - d(x)\mathbf{H}(x)\big)\mathbf{P}(x)\nabla_\Gamma v(\mathbf{p}(x)) \quad \text{a.e. on } \Gamma_h. \tag{7.73}$$

Proof. Given in Sect. 2.3 in [83]. □

In (7.73) (and also below) we have results "a.e. on Γ_h" because quantities (derivatives, $\mathbf{P}_h$, etc.) are not well-defined on the edges of the triangulation Γ_h.

Lemma 7.7.5 *For $x \in \Gamma_h$ (not on an edge) define*

$$\mu(x) = \big[\Pi_{i=1}^2 (1 - d(x)\kappa_i(x))\big]\mathbf{n}(x)^T \mathbf{n}_h(x), \tag{7.74}$$

$$\mathbf{A}(x) = \frac{1}{\mu(x)}\mathbf{P}(x)\big[\mathbf{I} - d(x)\mathbf{H}(x)\big]\mathbf{P}_h(x)\big[\mathbf{I} - d(x)\mathbf{H}(x)\big]\mathbf{P}(x). \tag{7.75}$$

Let $\mathbf{A}_{\Gamma_h}^e$ be the extension of $\mathbf{A}$ as in (7.72). The following identity holds for functions v and ψ that are defined on Γ_h and are sufficiently smooth:

$$\int_{\Gamma_h} \nabla_{\Gamma_h} v \cdot \nabla_{\Gamma_h} \psi \, ds = \int_\Gamma \mathbf{A}_{\Gamma_h}^e \nabla_\Gamma v_{\Gamma_h}^e \cdot \nabla_\Gamma \psi_{\Gamma_h}^e \, ds. \tag{7.76}$$

Proof. Given in Sect. 2.3 in [83]. □

Due to the assumptions in (7.71a) and (7.67) we have $\operatorname{ess\,inf}_{x\in\Gamma_h} \mu(x) > 0$ and thus $\mathbf{A}(x)$ is well defined and symmetric positive semi-definite.

A trace estimate

In the analysis of the discretization error in the next section we will need a bound for $\|\nabla_{\Gamma_h} v\|_{L^2(\Gamma_h)}$ in terms of $\|v\|_1$ for $v \in V_h$ (piecewise quadratics).

A possible approach is to apply an inverse inequality combined with a trace theorem, resulting in:

$$\|\nabla_{\Gamma_h} v\|_{L^2(\Gamma_h)} \leq c\, h_{\min}^{-1} \|v\|_{L^2(\Gamma_h)} \leq c\, h_{\min}^{-1} \|v\|_1 \quad \text{for all } v \in V_h. \tag{7.77}$$

This, however, turns out to be too crude. In order to be able to derive a better bound than the one in (7.77) we have to introduce some further assumptions which *relate the approximations* Γ_h of Γ *to the outer tetrahedral triangulation* $\mathcal{T}_h$ *that is used in the definition of the space* V_h, cf. (7.65).

Assumption 7.7.6 Let $\{\mathcal{T}_h\}$ be the family of tetrahedral triangulations that is used in the finite element space V_h. We assume that to each interface triangulation $\Gamma_h = \cup_{F \in \mathcal{F}_h} F$ there can be associated a set of tetrahedra $\mathcal{S}_h$ with the following properties:

$$\text{For each } F \in \mathcal{F}_h \text{ there is a corresponding } S_F \in \mathcal{S}_h \text{ with } F \subset S_F. \tag{7.78a}$$

$$\text{For } F_1, F_2 \in \mathcal{F}_h \text{ with } F_1 \neq F_2 \text{ we have } \mathrm{meas}_3(S_{F_1} \cap S_{F_2}) = 0. \tag{7.78b}$$

$$\text{The family } \{\mathcal{S}_h\}_{h>0} \text{ is shape-regular.} \tag{7.78c}$$

$$\exists\, c_0 > 0 : c_0 h \leq \mathrm{diam}(S_F) \leq ch \text{ for all } F \in \mathcal{F}_h, \text{ (quasi-uniformity).} \tag{7.78d}$$

$$\text{For each } S_F \in \mathcal{S}_h \text{ there is a tetrahedron } T \in \mathcal{T}_h \text{ such that } S_F \subset T. \tag{7.78e}$$

Note that the set of tetrahedra $\mathcal{S}_h$ has to be defined only close to the approximate interface Γ_h and that this set not necessarily forms a regular tetrahedral triangulation of Ω. Furthermore, it is *not* assumed that the family $\{\Gamma_h\}_{h>0}$ is shape-regular or quasi-uniform.

Remark 7.7.7 The construction in Sect. 7.3 is such that Assumption 7.7.6 is satisfied. We briefly explain this. Let $\mathcal{T}_h^\Gamma$ be the collection of tetrahedra that have a nonempty intersection with the zero level of the piecewise quadratic level set function and assume that $(\mathcal{T}_h^\Gamma)_{h>0}$ is quasi-uniform. Let $\mathcal{T}_{h'}^\Gamma$ be the triangulation obtained after one regular refinement of $\mathcal{T}_h^\Gamma$. Let Γ_h be as defined in (7.21). All $T \in \mathcal{T}_{h'}^\Gamma$ for which $\Gamma_T = T \cap \Gamma_h$ is a quadrilateral are further subdivided into two subtetrahedra such that for $T \cap \Gamma_h$ is always a triangle. The resulting triangulation is denoted by $\tilde{\mathcal{T}}_{h'}^\Gamma$. With this Γ_h and $\mathcal{S}_h = \tilde{\mathcal{T}}_{h'}^\Gamma$ the conditions formulated in Assumption 7.7.6 are satisfied.

In the following lemma we derive elementary properties of a standard affine mapping between a tetrahedron $S_F \in \mathcal{S}_h$ and the reference unit tetrahedron, which will be used in the proof of Theorem 7.7.9.

Lemma 7.7.8 *Assume that the family* $\{\Gamma_h\}_{h>0}$ *is such that Assumption 7.7.6 is satisfied. Take* $F \in \mathcal{F}_h$ *and the corresponding* $S_F \in \mathcal{S}_h$*. Let* $\hat{S}$ *be the reference unit tetrahedron and* $L(x) = \mathbf{J}x + \mathbf{b}$ *be an affine mapping such that* $L(\hat{S}) = S_F$*. Define* $\hat{F} := L^{-1}(F)$*. The following holds:*

$$\|\mathbf{J}\|^2 \frac{\mathrm{meas}_3(\hat{S})}{\mathrm{meas}_3(S_F)} \leq c\, h^{-1}, \tag{7.79}$$

$$\|\mathbf{J}^{-1}\|^2 \frac{\text{meas}_2(F)}{\text{meas}_2(\hat{F})} \leq c, \tag{7.80}$$

with constants c independent of F and h.

Proof. Let $\rho(S_F)$ be the diameter of the maximal ball contained in S_F and similarly for $\rho(\hat{S})$. From standard finite element theory we have

$$\|\mathbf{J}\| \leq \frac{\text{diam}(S_F)}{\rho(\hat{S})}, \quad \|\mathbf{J}^{-1}\| \leq \frac{\text{diam}(\hat{S})}{\rho(S_F)}.$$

Using (7.78c) and (7.78d) we then get

$$\|\mathbf{J}\|^2 \frac{\text{meas}_3(\hat{S})}{\text{meas}_3(S_F)} \leq c \frac{\text{diam}(S_F)^2}{\text{meas}_3(S_F)} \leq c\,\text{diam}(S_F)^{-1} \leq c\,h^{-1},$$

and thus the result in (7.79) holds.

The vertices of $\hat{F} = L^{-1}(F)$ are denoted by $\hat{V}_i$, $i = 1, 2, 3$. Let $\hat{V}_1\hat{V}_2$ be a longest edge of $\hat{F}$ and $\hat{M}$ the point on this edge such that $\hat{M}\hat{V}_3$ is perpendicular to $\hat{V}_1\hat{V}_2$. Define $V_i := L(\hat{V}_i)$, $i = 1, 2, 3$, and $M := L(\hat{M})$. Then V_i, $i = 1, 2, 3$, are the vertices of F and M lies on the edge V_1V_2. We then have

$$\begin{aligned}\text{meas}_2(\hat{F}) &= \frac{1}{2}\|\hat{V}_1 - \hat{V}_2\|\|\hat{V}_3 - \hat{M}\| = \frac{1}{2}\|\mathbf{J}^{-1}(V_1 - V_2)\|\|\mathbf{J}^{-1}(V_3 - M)\| \\ &\geq \frac{1}{2}\|\mathbf{J}\|^{-2}\|V_1 - V_2\|\|V_3 - M\| \geq c\,\frac{\rho(\hat{S})^2}{\text{diam}(S_F)^2}\,\text{meas}_2(F),\end{aligned}$$

with a constant $c > 0$. Thus we obtain

$$\|\mathbf{J}^{-1}\|^2 \frac{\text{meas}_2(F)}{\text{meas}_2(\hat{F})} \leq c\,\frac{\text{diam}(\hat{S})^2}{\rho(S_F)^2}\,\frac{\text{diam}(S_F)^2}{\rho(\hat{S})^2} \leq c,$$

which completes the proof. □

Theorem 7.7.9 *Assume that the family $\{\Gamma_h\}_{h>0}$ is such that Assumption 7.7.6 is satisfied. The following holds:*

$$\|\nabla_{\Gamma_h} v\|_{L^2(\Gamma_h)} \leq c\,h^{-\frac{1}{2}}\|v\|_1 \quad \textit{for all} \quad v \in V_h.$$

Proof. Note that

$$\|\nabla_{\Gamma_h} v\|^2_{L^2(\Gamma_h)} = \sum_{F \in \mathcal{F}_h} \|\nabla_F v\|^2_{L^2(F)}.$$

Take $F \in \mathcal{F}_h$ and let S_F be the associated tetrahedron as explained above. Let $\hat{S}$ be the reference unit tetrahedron and $L : \hat{S} \to S_T$ as in Lemma 7.7.8. Define $\hat{v} := v \circ L$. Using standard transformation rules and Lemma 7.7.8 we get

$$
\begin{aligned}
\|\nabla_F v\|_{L^2(F)}^2 &= \|\mathbf{P}_h \nabla v\|_{L^2(F)}^2 \le \|\nabla v\|_{L^2(F)}^2 = \sum_{|\alpha|=1} \|\partial^\alpha v\|_{L^2(F)}^2 \\
&\le c\,\|\mathbf{J}^{-1}\|^2 \sum_{|\alpha|=1} \|(\partial^\alpha \hat{v}) \circ L^{-1}\|_{L^2(F)}^2 \\
&\le c\,\|\mathbf{J}^{-1}\|^2 \frac{\mathrm{meas}_2(F)}{\mathrm{meas}_2(\hat{F})} \sum_{|\alpha|=1} \|\partial^\alpha \hat{v}\|_{L^2(\hat{F})}^2 \le c \sum_{|\alpha|=1} \|\partial^\alpha \hat{v}\|_{L^2(\hat{F})}^2 \\
&\le c \sum_{|\alpha|=1} \max_{x \in \hat{F}} \left|\partial^\alpha \hat{v}(x)\right|^2 \le c \sum_{|\alpha|=1} \max_{x \in \hat{S}} \left|\partial^\alpha \hat{v}(x)\right|^2,
\end{aligned}
$$

with a constant c independent of F. From (7.78e) it follows that $\hat{v}$ is a polynomial on $\hat{S}$ of maximal degree 2. On $\mathcal{P}_2^* := \{\, p \in \mathcal{P}_2 : p(0) = 0 \,\}$ we have, due to equivalence of norms:

$$
\sum_{|\alpha|=1} \max_{x \in \hat{S}} \left|\partial^\alpha \hat{v}(x)\right|^2 \le c \sum_{|\alpha|=1} \|\partial^\alpha \hat{v}\|_{L^2(\hat{S})}^2 \quad \text{for all } \hat{v} \in \mathcal{P}_2^*.
$$

Because, for $\hat{v} \in \mathcal{P}_2$ and $|\alpha| = 1$, $\partial^\alpha \hat{v}$ is independent of $\hat{v}(0)$, the same inequality holds for all $\hat{v} \in \mathcal{P}_2$. Thus we get

$$
\begin{aligned}
\|\nabla_F v\|_{L^2(F)}^2 &\le c \sum_{|\alpha|=1} \|\partial^\alpha \hat{v}\|_{L^2(\hat{S})}^2 \le c\,\|\mathbf{J}\|^2 \sum_{|\alpha|=1} \|(\partial^\alpha v) \circ L\|_{L^2(\hat{S})}^2 \\
&= c\,\|\mathbf{J}\|^2 \frac{\mathrm{meas}_3(\hat{S})}{\mathrm{meas}_3(S_F)} \sum_{|\alpha|=1} \|\partial^\alpha v\|_{L^2(S_F)}^2 \le c\,h^{-1} \|\nabla v\|_{L^2(S_F)}^2,
\end{aligned}
$$

with a constant c independent of F and h. Using (7.78b) we finally obtain

$$
\begin{aligned}
\|\nabla_{\Gamma_h} v\|_{L^2(\Gamma_h)}^2 &\le c\,h^{-1} \sum_{F \in \mathcal{F}_h} \|\nabla v\|_{L^2(S_F)}^2 \\
&\le c\,h^{-1} \int_\Omega (\nabla v)^2 \, dx \le c\,h^{-1} \|v\|_1^2,
\end{aligned}
$$

which proves the result. □

Remark 7.7.10 The analysis above also applies if instead of piecewise quadratics other piecewise polynomial finite element functions are used. Thus Theorem 7.7.9 also holds if for V_h we take another piecewise polynomial finite element space.

7.7.2 Error bounds for discrete surface tension functionals

In Sect. 7.6, for the surface tension functional

$$
f_\Gamma(\mathbf{v}_h) = -\tau \int_\Gamma \nabla_\Gamma \,\mathrm{id}_\Gamma \cdot \nabla_\Gamma \mathbf{v}_h \, ds = -\tau \sum_{i=1}^{3} \int_\Gamma \nabla_\Gamma (\mathrm{id}_\Gamma)_i \cdot \nabla_\Gamma (\mathbf{v}_h)_i \, ds
$$

we introduced the discretizations

$$f_{\Gamma_h}(\mathbf{v}_h) = -\tau \sum_{i=1}^{3} \int_{\Gamma_h} \nabla_{\Gamma_h}(\mathrm{id}_{\Gamma_h})_i \cdot \nabla_{\Gamma_h}(\mathbf{v}_h)_i \, ds, \tag{7.81}$$

$$\tilde{f}_{\Gamma_h}(\mathbf{v}_h) = -\tau \sum_{i=1}^{3} \int_{\Gamma_h} \tilde{\mathbf{P}}_h \nabla(\mathrm{id}_{\Gamma_h})_i \cdot \nabla_{\Gamma_h}(\mathbf{v}_h)_i \, ds. \tag{7.82}$$

In this section we derive error bounds for these discretizations. It suffices to consider only one term in this sum, say the i-th. We write id_Γ and v for the *scalar* functions $(\mathrm{id}_\Gamma)_i$ and $(\mathbf{v}_h)_i$, respectively, and id_{Γ_h} for $(\mathrm{id}_{\Gamma_h})_i$. With this notation we have

$$\nabla_\Gamma \mathrm{id}_\Gamma = \mathbf{P}\nabla \mathrm{id}_\Gamma = \mathbf{P}e_i \quad \text{on } \Gamma, \quad \nabla_{\Gamma_h} \mathrm{id}_{\Gamma_h} = \mathbf{P}_h \nabla \mathrm{id}_{\Gamma_h} = \mathbf{P}_h e_i \quad \text{on } \Gamma_h,$$
$$\tilde{\mathbf{P}}_h \nabla \mathrm{id}_{\Gamma_h} = \tilde{\mathbf{P}}_h e_i \quad \text{on } \Gamma_h,$$

where e_i denotes the i-th basis vector in $\mathbb{R}^3$. For the i-th term in these functionals we introduce the notation (ignoring the scaling with $-\tau$):

$$g(v) := \int_\Gamma \nabla_\Gamma \mathrm{id}_\Gamma \cdot \nabla_\Gamma v \, ds,$$
$$g_h(v) := \int_{\Gamma_h} \nabla_{\Gamma_h} \mathrm{id}_{\Gamma_h} \cdot \nabla_{\Gamma_h} v \, ds,$$
$$\tilde{g}_h(v) := \int_{\Gamma_h} \tilde{\mathbf{P}}_h \nabla \mathrm{id}_{\Gamma_h} \cdot \nabla_{\Gamma_h} v \, ds.$$

For the analysis it is convenient to introduce yet another functional:

$$\hat{g}_h(v) := \int_{\Gamma_h} \nabla_{\Gamma_h} \mathrm{id}_\Gamma^e \cdot \nabla_{\Gamma_h} v \, ds,$$

where id_Γ^e is the extension of id_Γ. Note that due to the occurrence of id_Γ the functional $\hat{g}_h(v)$ can not be used in practice. For the error $g(v) - g_h(v)$ we write $g(v) - g_h(v) = \big(g(v) - \hat{g}_h(v)\big) + \big(\hat{g}_h(v) - g_h(v)\big)$, derive bounds for $|g(v) - \hat{g}_h(v)|$ and $|\hat{g}_h(v) - g_h(v)|$ and then apply a triangle inequality. The same is done for the error $g(v) - \tilde{g}_h(v)$.

We start with the term $|g(v) - \hat{g}_h(v)|$. A bound for this is derived, based on the following splitting:

$$\begin{aligned}
&g(v) - \hat{g}_h(v) \\
&= \int_\Gamma \nabla_\Gamma \mathrm{id}_\Gamma \cdot \nabla_\Gamma v \, ds - \int_{\Gamma_h} \nabla_{\Gamma_h} \mathrm{id}_\Gamma^e \cdot \nabla_{\Gamma_h} v \, ds \\
&= \int_\Gamma \nabla_\Gamma \mathrm{id}_\Gamma \cdot \nabla_\Gamma v \, ds - \int_\Gamma \mathbf{A}_{\Gamma_h}^e \nabla_\Gamma \mathrm{id}_\Gamma \cdot \nabla_\Gamma v_{\Gamma_h}^e \, ds \qquad \text{(cf. (7.76))} \\
&= \int_\Gamma \nabla_\Gamma \mathrm{id}_\Gamma \cdot \nabla_\Gamma (v - v_{\Gamma_h}^e) \, ds + \int_\Gamma (\mathbf{I} - \mathbf{A}_{\Gamma_h}^e) \nabla_\Gamma \mathrm{id}_\Gamma \cdot \nabla_\Gamma v_{\Gamma_h}^e \, ds.
\end{aligned} \tag{7.83}$$

In the lemma below we give bounds for the two terms in (7.83).

Lemma 7.7.11 *Let Assumption 7.7.1 be satisfied. The following holds for all* $v \in V_h$*:*

$$\left| \int_\Gamma \nabla_\Gamma \operatorname{id}_\Gamma \cdot \nabla_\Gamma (v - v_{\Gamma_h}^e)\, ds \right| \leq c\, h_\Gamma \, \|v\|_{1,U}, \tag{7.84}$$

$$\left| \int_\Gamma (\mathbf{I} - \mathbf{A}_{\Gamma_h}^e) \nabla_\Gamma \operatorname{id}_\Gamma \cdot \nabla_\Gamma v_{\Gamma_h}^e \, ds \right| \leq c\, h_\Gamma^2 \, \|\nabla_{\Gamma_h} v\|_{L^2(\Gamma_h)}. \tag{7.85}$$

Proof. (7.84)–(7.85) are proved in Lemmas 4.1 and 4.3 in [129]. □

Using this we obtain a bound for the error $\|g - \hat{g}_h\|_{V_h'}$:

Theorem 7.7.12 *Let the Assumptions 7.7.1 and 7.7.6 be satisfied. The following holds:*

$$\sup_{v \in V_h} \frac{|g(v) - \hat{g}_h(v)|}{\|v\|_1} \leq c\, h_\Gamma. \tag{7.86}$$

Proof. The result in Lemma 7.7.11 implies

$$|g(v) - \hat{g}_h(v)| \leq c\, h_\Gamma \, \|v\|_{1,U} + c\, h_\Gamma^2 \, \|\nabla_{\Gamma_h} v\|_{L^2(\Gamma_h)} \quad \text{for all } v \in V_h.$$

From Theorem 7.7.9 we obtain $\|\nabla_{\Gamma_h} v\|_{L^2(\Gamma_h)} \leq c h_\Gamma^{-\frac{1}{2}} \|v\|_1$. Furthermore, $\|v\|_{1,U} \leq \|v\|_1$ holds. Thus the result in (7.86) holds. □

We now derive a bound for $|\hat{g}_h(v) - g_h(v)|$.

Lemma 7.7.13 *Let the Assumption 7.7.1 be satisfied. The following holds:*

$$|\hat{g}_h(v) - g_h(v)| \leq c h_\Gamma \|\nabla_{\Gamma_h} v\|_{L^2(\Gamma_h)} \quad \textit{for all} \;\; v \in V_h. \tag{7.87}$$

Proof. From Lemma 7.7.4 we get, for $x \in \Gamma_h$ (not on an edge),

$$\begin{aligned} \nabla_{\Gamma_h} \operatorname{id}_\Gamma^e(x) &= \mathbf{P}_h(x)\big(\mathbf{I} - d(x)\mathbf{H}(x)\big)\mathbf{P}(x) \nabla_\Gamma \operatorname{id}_\Gamma(\mathbf{p}(x)) \\ &= \mathbf{P}_h(x)\big(\mathbf{I} - d(x)\mathbf{H}(x)\big)\mathbf{P}(x) e_i. \end{aligned}$$

We also have $\nabla_{\Gamma_h} \operatorname{id}_{\Gamma_h} = \mathbf{P}_h \nabla \operatorname{id}_{\Gamma_h} = \mathbf{P}_h e_i$. Hence,

$$\begin{aligned} &\left| \int_{\Gamma_h} \big(\nabla_{\Gamma_h} \operatorname{id}_\Gamma^e - \nabla_{\Gamma_h} \operatorname{id}_{\Gamma_h}\big) \cdot \nabla_{\Gamma_h} v \, ds \right| \qquad (7.88) \\ &= \left| \int_{\Gamma_h} \big(\mathbf{P}_h(\mathbf{I} - d\mathbf{H})\mathbf{P} e_i - \mathbf{P}_h e_i\big) \cdot \nabla_{\Gamma_h} v \, ds \right| \\ &\leq c \operatorname{ess\,sup}_{x \in \Gamma_h} \|\mathbf{P}_h(x)\big(\mathbf{I} - d(x)\mathbf{H}(x)\big)\mathbf{P}(x) - \mathbf{P}_h(x)\| \, \|\nabla_{\Gamma_h} v\|_{L^2(\Gamma_h)} \\ &\leq c \operatorname{ess\,sup}_{x \in \Gamma_h} \big(\|\mathbf{P}_h(x)\big(\mathbf{I} - \mathbf{P}(x)\big)\| \qquad (7.89) \\ &\qquad + |d(x)| \, \|\mathbf{P}_h(x)\mathbf{H}(x)\mathbf{P}(x)\|\big) \, \|\nabla_{\Gamma_h} v\|_{L^2(\Gamma_h)}. \qquad (7.90) \end{aligned}$$

Note that $|d(x)| \le c\, h_\Gamma^2$ for $x \in \Gamma_h$, and

$$\operatorname{ess\,sup}_{x\in\Gamma_h} \|\mathbf{P}_h(x)\mathbf{H}(x)\mathbf{P}(x)\| \le \operatorname{ess\,sup}_{x\in\Gamma_h} \|\mathbf{H}(x)\| \le c.$$

For the term in (7.89) we have (dropping x in the notation):

$$\|\mathbf{P}_h(\mathbf{I}-\mathbf{P})\| = \|\mathbf{P}_h\mathbf{n}\mathbf{n}^T\| = \|\mathbf{P}_h\mathbf{n}\| = \|\mathbf{P}_h(\mathbf{n}-\mathbf{n}_h)\| \le \|\mathbf{n}-\mathbf{n}_h\| \le ch_\Gamma.$$

In the last inequality we used Assumption 7.7.1. Using these results in (7.89)-(7.90) and the definitions of $\hat{g}_h$, g_h, we get

$$|\hat{g}_h(v) - g_h(v)| \le ch_\Gamma \|\nabla_{\Gamma_h} v\|_{L^2(\Gamma_h)},$$

and thus the result is proved. □

This leads to a bound for the error $\|\hat{g}_h - g_h\|_{V_h'}$:

Theorem 7.7.14 *Let the Assumptions 7.7.1 and 7.7.6 be satisfied. The following holds:*

$$\sup_{v\in V_h} \frac{|\hat{g}_h(v) - g_h(v)|}{\|v\|_1} \le c\,\sqrt{h_\Gamma}. \tag{7.91}$$

Proof. The result follows from Lemma 7.7.13 and Theorem 7.7.9. □

As a direct consequence we obtain a discretization error bound for f_{Γ_h}:

Corollary 7.7.15 Let the Assumptions 7.7.1 and 7.7.6 be satisfied. For the surface tension force discretization f_{Γ_h} as defined in (7.81) the following holds:

$$\sup_{\mathbf{v}\in \mathbf{V}_h} \frac{|f_\Gamma(\mathbf{v}_h) - f_{\Gamma_h}(\mathbf{v}_h)|}{\|\mathbf{v}_h\|_1} \le \tau c\,\sqrt{h_\Gamma}.$$

Proof. It suffices to consider a bound for $\|g - g_h\|_{V_h'}$. From Theorem 7.7.12 and Theorem 7.7.14 it follows that

$$\|g-g_h\|_{V_h'} \le \|g-\hat{g}_h\|_{V_h'} + \|\hat{g}_h - g_h\|_{V_h'} \le ch_\Gamma + c\sqrt{h_\Gamma} \le c\sqrt{h_\Gamma},$$

which implies the error bound for f_{Γ_h}. □

An upper bound $\mathcal{O}(\sqrt{h_\Gamma})$ as in Corollary 7.7.15 for the error in the approximation of the localized force term may seem rather pessimistic, because Γ_h is an $\mathcal{O}(h_\Gamma^2)$ accurate approximation of Γ. Numerical experiments in Sect. 7.8 and results in [115], however, indicate that the bound is sharp.

Along the same lines as presented above for f_{Γ_h} we now derive an error bound for $\tilde{f}_{\Gamma_h}$. It suffices to consider $|g(v) - \tilde{g}_h(v)|$. We use the triangle inequality

$$|g(v) - \tilde{g}_h(v)| \le |g(v) - \hat{g}_h(v)| + |\hat{g}_h(v) - \tilde{g}_h(v)|.$$

The first term on the right-hand side is treated in Theorem 7.7.12. The next lemma gives a bound for the second term. In (7.92) we use the generalized normal $\tilde{\mathbf{n}}_h$ from (7.59).

Lemma 7.7.16 *Let Assumption 7.7.1 be satisfied. Furthermore, we assume that there exists* $p > 0$ *such that*

$$\|\mathbf{n}(x) - \tilde{\mathbf{n}}_h(x)\| \leq c\, h_\Gamma^p, \quad \text{for } x \in \Gamma_h. \tag{7.92}$$

Then the following holds:

$$|\hat{g}_h(v) - \tilde{g}_h(v)| \leq c h_\Gamma^{\min\{p,2\}} \|\nabla_{\Gamma_h} v\|_{L^2(\Gamma_h)} \quad \text{for all } \; v \in V_h.$$

Proof. We apply similar arguments as used in the proof of Lemma 7.7.13. We have

$$\nabla_{\Gamma_h} \operatorname{id}_\Gamma^e(x) = \mathbf{P}_h(x)\big(\mathbf{I} - d(x)\mathbf{H}(x)\big)\mathbf{P}(x) e_i$$

and $\tilde{\mathbf{P}}_h \nabla \operatorname{id}_{\Gamma_h} = \tilde{\mathbf{P}}_h e_i$ on Γ_h. Hence,

$$\left| \int_{\Gamma_h} (\nabla_{\Gamma_h} \operatorname{id}_\Gamma^e - \tilde{\mathbf{P}}_h \nabla \operatorname{id}_{\Gamma_h}) \cdot \nabla_{\Gamma_h} v \, ds \right| \tag{7.93}$$

$$= \left| \int_{\Gamma_h} (\mathbf{P}_h(\mathbf{I} - d\mathbf{H})\mathbf{P} e_i - \mathbf{P}_h \tilde{\mathbf{P}}_h e_i) \cdot \nabla_{\Gamma_h} v \, ds \right|$$

$$\leq c \operatorname{ess\,sup}_{x \in \Gamma_h} \|\mathbf{P}_h(x)\big(\mathbf{I} - d(x)\mathbf{H}(x)\big)\mathbf{P}(x) - \mathbf{P}_h(x)\tilde{\mathbf{P}}_h(x)\| \, \|\nabla_{\Gamma_h} v\|_{L^2(\Gamma_h)}$$

$$\leq c \operatorname{ess\,sup}_{x \in \Gamma_h} \big(\|\mathbf{P}_h(x)\big(\mathbf{P}(x) - \tilde{\mathbf{P}}_h(x)\big)\| \tag{7.94}$$

$$+ |d(x)| \, \|\mathbf{P}_h(x)\mathbf{H}(x)\mathbf{P}(x)\|\big) \, \|\nabla_{\Gamma_h} v\|_{L^2(\Gamma_h)}. \tag{7.95}$$

As in the proof of Lemma 7.7.13 we have $\operatorname{ess\,sup}_{x \in \Gamma_h} |d(x)| \|\mathbf{P}_h(x)\mathbf{H}(x)\mathbf{P}(x)\| \leq c h_\Gamma^2$. For the term in (7.94) we get (dropping x in the notation):

$$\|\mathbf{P}_h(\mathbf{P} - \tilde{\mathbf{P}}_h)\| \leq \|\mathbf{n}\mathbf{n}^T - \tilde{\mathbf{n}}_h \tilde{\mathbf{n}}_h^T\|$$

$$\leq \|(\mathbf{n} - \tilde{\mathbf{n}}_h)\mathbf{n}^T\| + \|\tilde{\mathbf{n}}_h(\mathbf{n} - \tilde{\mathbf{n}}_h)^T\| = 2\|\mathbf{n} - \tilde{\mathbf{n}}_h\| \leq c\, h_\Gamma^p.$$

Combination of these estimates proves the result. □

Remark 7.7.17 In Lemma 7.3.2 it is shown that for the approximate interface construction explained in Sect. 7.3 the assumption in (7.92) holds for $p \in (0, 2]$ if ϕ_h is an $\mathcal{O}(h_\Gamma^p)$ accurate (w.r.t $\|\cdot\|_{H^1_\infty}$) approximation of ϕ. For a piecewise quadratic level set approximation the optimal approximation quality is $\mathcal{O}(h_\Gamma^2)$, i.e., $p = 2$.

This leads to a bound for the error $\|\hat{g}_h - \tilde{g}_h\|_{V_h'}$:

Theorem 7.7.18 *Let the Assumptions 7.7.1, 7.7.6 and the one in* (7.92) *with* $p \geq 1\frac{1}{2}$ *be satisfied. The following holds:*

$$\sup_{v \in V_h} \frac{|\hat{g}_h(v) - \tilde{g}_h(v)|}{\|v\|_1} \leq c\, h_\Gamma. \tag{7.96}$$

Proof. The result follows from Lemma 7.7.16 and Theorem 7.7.9. □

As a direct consequence we obtain a discretization error bound for $\tilde{f}_{\Gamma_h}$:

Corollary 7.7.19 Let the assumptions as in Theorem 7.7.18 be satisfied. For the surface tension force discretization $\tilde{f}_{\Gamma_h}$ as defined in (7.82) the following holds:

$$\sup_{\mathbf{v}\in\mathbf{V}_h} \frac{|f_\Gamma(\mathbf{v}_h) - \tilde{f}_{\Gamma_h}(\mathbf{v}_h)|}{\|\mathbf{v}_h\|_1} \leq \tau c\, h_\Gamma.$$

Proof. It suffices to consider a bound for $\|g - \tilde{g}_h\|_{V_h'}$. From Theorem 7.7.12 and Theorem 7.7.18 it follows that

$$\|g - \tilde{g}_h\|_{V_h'} \leq \|g - \hat{g}_h\|_{V_h'} + \|\hat{g}_h - \tilde{g}_h\|_{V_h'} \leq c h_\Gamma,$$

which implies the error bound for $\tilde{f}_{\Gamma_h}$. □

This significant improvement ($\mathcal{O}(h_\Gamma)$ compared to the $\mathcal{O}(\sqrt{h_\Gamma})$ error bound for the functional f_{Γ_h}) is confirmed by numerical experiments in the next section.

7.8 Numerical experiments with the Laplace-Beltrami discretization

In this section we present results of a numerical experiment which indicates that the $\mathcal{O}(\sqrt{h})$ bound in Corollary 7.7.15 is sharp. Furthermore, for the improved approximation $\tilde{f}_{\Gamma_h}$ the $\mathcal{O}(h)$ bound will be confirmed numerically.

We consider the domain $\Omega := [-1,1]^3$ with $\Omega_1 := \{x \in \Omega : \|x\| < R\}$. In our experiments we take $R = \frac{1}{2}$.

For the discretization a uniform tetrahedral mesh $\mathcal{T}_0$ is used where the vertices form a $6 \times 6 \times 6$ lattice, hence $h_0 = \frac{1}{5}$. This coarse mesh $\mathcal{T}_0$ is locally refined in the vicinity of $\Gamma = \partial\Omega_1$. This repeated refinement process yields the gradually refined meshes $\mathcal{T}_1, \mathcal{T}_2, \ldots$ with *local* (i. e., close to the interface) mesh sizes $h_\Gamma = h_i = \frac{1}{5} \cdot 2^{-i}, i = 1, 2, \ldots$. Part of the tetrahedral triangulation $\mathcal{T}_4$ is shown in Fig. 7.9. The corresponding finite element spaces $\mathbf{V}_i := \mathbf{V}_{h_i} = (V_{h_i})^3$ consist of vector functions where each component is a continuous piecewise quadratic function on $\mathcal{T}_i$.

The interface $\Gamma = \partial\Omega_1$ is a sphere and thus the curvature $\kappa = \frac{2}{R}$ is constant. If we discretize the flow problem using $\mathbf{V}_i$ as discrete velocity space, we have to approximate the surface tension force

$$f_\Gamma(\mathbf{v}) = -\frac{2\tau}{R} \int_\Gamma \mathbf{n}_\Gamma \cdot \mathbf{v}\, ds = -\tau \int_\Gamma \nabla_\Gamma \operatorname{id}_\Gamma \cdot \nabla_\Gamma \mathbf{v}\, ds, \qquad \mathbf{v} \in \mathbf{V}_i. \quad (7.97)$$

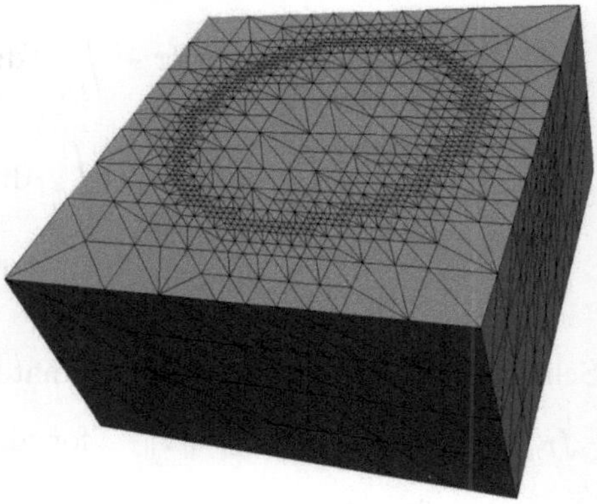

Fig. 7.9. Lower half part of the 4 times refined mesh $\mathcal{T}_4$.

To simplify notation, we take a fixed $i \geq 0$ and the corresponding local mesh size parameter is denoted by $h = h_i$. For the construction of an approximate interface Γ_h we use the approach described in Sect. 7.3, starting with ϕ equal to the exact signed distance function to Γ.

The discrete approximation of the surface tension force is

$$f_{\Gamma_h}(\mathbf{v}) = -\tau \int_{\Gamma_h} \nabla_{\Gamma_h} \mathrm{id}_{\Gamma_h} \cdot \nabla_{\Gamma_h} \mathbf{v} \, ds, \qquad \mathbf{v} \in \mathbf{V}_i.$$

We are interested in, cf. Corollary 7.7.15,

$$\|f_\Gamma - f_{\Gamma_h}\|_{\mathbf{V}_i'} := \sup_{\mathbf{v} \in \mathbf{V}_i} \frac{f_\Gamma(\mathbf{v}) - f_{\Gamma_h}(\mathbf{v})}{\|\mathbf{v}\|_1}. \tag{7.98}$$

The evaluation of $f_\Gamma(\mathbf{v})$, for $\mathbf{v} \in \mathbf{V}_i$, requires the computation of integrals on curved triangles or quadrilaterals $\Gamma \cap T$ where T is a tetrahedron from the triangulation $\mathcal{T}_i$. We are not able to compute these exactly. Therefore, we introduce an artificial force term which, in this model problem with a known constant curvature, is computable and sufficiently close to f_Γ.

Lemma 7.8.1 *For* $\mathbf{v} \in \mathbf{V} = H_0^1(\Omega)^3$ *define*

$$\hat{f}_{\Gamma_h}(\mathbf{v}) := -\frac{2\tau}{R} \int_{\Gamma_h} \mathbf{n}_h \cdot \mathbf{v} \, ds,$$

where $\mathbf{n}_h$ *is the piecewise constant outward unit normal on* Γ_h. *Then the following inequality holds:*

$$\|f_\Gamma - \hat{f}_{\Gamma_h}\|_{\mathbf{V}'} \leq ch. \tag{7.99}$$

Proof. Let $\Omega_{1,h} \subset \Omega$ be the domain enclosed by Γ_h, i.e., $\partial\Omega_{1,h} = \Gamma_h$. We define $D_h^+ := \Omega_1 \setminus \Omega_{1,h}$, $D_h^- := \Omega_{1,h} \setminus \Omega_1$ and $D_h := D_h^+ \cup D_h^-$. Due to Stokes' Theorem, for $\mathbf{v} \in \mathbf{V}$ we have

$$\begin{aligned}|f_\Gamma(\mathbf{v}) - \hat{f}_{\Gamma_h}(\mathbf{v})| &= \frac{2\tau}{R}\left|\int_{\Omega_1} \operatorname{div}\mathbf{v}\,dx - \int_{\Omega_{1,h}} \operatorname{div}\mathbf{v}\,dx\right| \\ &= \frac{2\tau}{R}\left|\int_{D_h^+} \operatorname{div}\mathbf{v}\,dx - \int_{D_h^-} \operatorname{div}\mathbf{v}\,dx\right| \\ &\le \frac{2\tau}{R}\int_{D_h} |\operatorname{div}\mathbf{v}|\,dx.\end{aligned}$$

Using the Cauchy-Schwarz inequality, we get the estimate

$$|f_\Gamma(\mathbf{v}) - \hat{f}_{\Gamma_h}(\mathbf{v})| \le c\sqrt{\mathrm{meas}_3(D_h)}\,\|\mathbf{v}\|_1 \quad \text{for all} \quad \mathbf{v} \in \mathbf{V},$$

which results in

$$\|f_\Gamma - \hat{f}_{\Gamma_h}\|_{\mathbf{V}'} \le c\sqrt{\mathrm{meas}_3(D_h)}. \tag{7.100}$$

Note that for the piecewise planar approximation Γ_h of the interface Γ we have $\mathrm{meas}_3(D_h) = \mathcal{O}(h^2)$ and thus (7.99) holds. □

From Lemma 7.8.1 we obtain $\|f_\Gamma - \hat{f}_{\Gamma_h}\|_{\mathbf{V}_j'} \le c\,h$ with a constant c independent of j. Thus we have

$$\|\hat{f}_{\Gamma_h} - f_{\Gamma_h}\|_{\mathbf{V}_i'} - ch \le \|f_\Gamma - f_{\Gamma_h}\|_{\mathbf{V}_i'} \le \|\hat{f}_{\Gamma_h} - f_{\Gamma_h}\|_{\mathbf{V}_i'} + ch. \tag{7.101}$$

The quantity $\|\hat{f}_{\Gamma_h} - f_{\Gamma_h}\|_{\mathbf{V}_i'}$ can be determined as follows. Since Γ_h is piecewise planar and $\mathbf{v} \in \mathbf{V}_i$ is a piecewise quadratic function, both $\hat{f}_{\Gamma_h}(\mathbf{v})$ and $f_{\Gamma_h}(\mathbf{v})$ can be computed exactly (up to machine accuracy) using suitable quadrature rules.

For the evaluation of the dual norm $\|\cdot\|_{\mathbf{V}_i'}$ we proceed as follows. Let $\{\boldsymbol{\xi}_j\}_{j=1,\dots,N}$ (with $N := \dim \mathbf{V}_i$) be the standard nodal basis in $\mathbf{V}_i$ and $J_{\mathbf{V}_i} : \mathbb{R}^N \to \mathbf{V}_i$ the isomorphism $J_{\mathbf{V}_i}\mathbf{x} = \sum_{k=1}^N x_k \boldsymbol{\xi}_k$. Let $\mathbf{M}_h$ be the mass matrix and $\mathbf{A}_h$ the discrete Laplacian:

$$\left.\begin{aligned}(\mathbf{M}_h)_{ij} &:= \int_\Omega \boldsymbol{\xi}_i \cdot \boldsymbol{\xi}_j\,dx, \\ (\mathbf{A}_h)_{ij} &:= \int_\Omega \nabla\boldsymbol{\xi}_i \cdot \nabla\boldsymbol{\xi}_j\,dx.\end{aligned}\right\} \quad 1 \le i,j \le N.$$

Define $\mathbf{C}_h = \mathbf{A}_h + \mathbf{M}_h$. Note that for $\mathbf{v} = J_{\mathbf{V}_i}\mathbf{x} \in \mathbf{V}_i$ we have $\|\mathbf{v}\|_1^2 = \langle \mathbf{C}_h\mathbf{x}, \mathbf{x}\rangle$. Take $e \in \mathbf{V}_i'$ and define $\mathbf{e} \in \mathbb{R}^N$ by $\mathbf{e}_j := e(\boldsymbol{\xi}_j)$, $j = 1, \dots, N$. Due to

$$\|e\|_{\mathbf{V}_i'} = \sup_{\mathbf{v}\in\mathbf{V}_i} \frac{|e(\mathbf{v})|}{\|\mathbf{v}\|_1} = \sup_{\mathbf{x}\in\mathbb{R}^N} \frac{|\sum_{j=1}^N x_j e(\boldsymbol{\xi}_j)|}{\sqrt{\langle \mathbf{C}_h\mathbf{x}, \mathbf{x}\rangle}}$$

we obtain

$$\|e\|_{\mathbf{V}_i'} = \sup_{\mathbf{x}\in\mathbb{R}^N} \frac{\langle \mathbf{x}, \mathbf{e}\rangle}{\sqrt{\langle \mathbf{C}_h\mathbf{x}, \mathbf{x}\rangle}} = \|\mathbf{C}_h^{-1/2}\mathbf{e}\| = \sqrt{\langle \mathbf{C}_h^{-1}\mathbf{e}, \mathbf{e}\rangle}. \tag{7.102}$$

Thus for the computation of $\|e\|_{\mathbf{V}_i'}$ we proceed in the following way:

1. Compute $\mathbf{e} = \left(e(\boldsymbol{\xi}_j)\right)_{j=1}^{N}$.
2. Solve the linear system $\mathbf{C}_h\,\mathbf{z} = \mathbf{e}$ up to machine accuracy.
3. Compute $\|e\|_{\mathbf{V}_i'} = \sqrt{\langle \mathbf{z},\, \mathbf{e}\rangle}$.

We applied this strategy to $e := \hat{f}_{\Gamma_h} - f_{\Gamma_h}$. The results are given in the second column in Table 7.8. The numerical order of convergence in the third column of this table clearly indicates an $\mathcal{O}(\sqrt{h})$ behavior. Due to (7.101) this implies the same $\mathcal{O}(\sqrt{h})$ convergence behavior for $\|f_\Gamma - f_{\Gamma_h}\|_{\mathbf{V}_i'}$. This indicates that the $\mathcal{O}(\sqrt{h})$ bound in Corollary 7.7.15 is sharp.

The same procedure can be applied with f_{Γ_h} replaced by the modified (improved) approximate surface tension force

$$\tilde{f}_{\Gamma_h}(\mathbf{v}) = -\tau \sum_{i=1}^{3} \int_{\Gamma_h} \tilde{\mathbf{P}}_h e_i \cdot \nabla_{\Gamma_h}(\mathbf{v})_i \, ds,$$

as defined in (7.82). This yields the results in the fourth column in Table 7.8. For this modification the numerical order of convergence is significantly better, namely at least first order in h. From (7.101) it follows that for $\|f_\Gamma - \tilde{f}_{\Gamma_h}\|_{\mathbf{V}_i'}$ we can expect $\mathcal{O}(h^p)$ with $p \geq 1$.

Summarizing, we conclude that the results of these numerical experiments confirm the theoretical $\mathcal{O}(\sqrt{h})$ error bound derived in the analysis in Sect. 7.7.2 and show that the modified approximation indeed leads to (much) better results.

Results of numerical experiments for a Stokes two-phase flow problem using both f_{Γ_h} and $\tilde{f}_{\Gamma_h}$ are presented in Sect. 7.10.3.

i	$\|\hat{f}_{\Gamma_h} - f_{\Gamma_h}\|_{\mathbf{V}_i'}$	order	$\|\hat{f}_{\Gamma_h} - \tilde{f}_{\Gamma_h}\|_{\mathbf{V}_i'}$	order
0	1.79 E-1	–	1.32 E-1	–
1	1.40 E-1	0.35	4.43 E-2	1.57
2	1.03 E-1	0.45	1.46 E-2	1.61
3	7.22 E-2	0.51	5.06 E-3	1.52
4	5.02 E-2	0.53	1.78 E-3	1.51

Table 7.8. Error norms and numerical order of convergence for different refinement levels.

7.9 XFEM discretization of the pressure

If surface tension forces are present the pressure is *discontinuous* across the interface Γ. We show that standard finite element spaces have poor approximation properties for such functions with a jump across Γ and introduce a so-called extended finite element space that is much better suited for discretization of the pressure variable. Most of the results presented in this section are from [128, 209].

In Sect. 7.9.1 we show, by means of a simple example, that if one uses standard finite element spaces for the discretization of a discontinuous function, then in general the approximation order (w.r.t. $\|\cdot\|_{L^2}$) is only $\mathcal{O}(\sqrt{h})$. In Sect. 7.9.2 we introduce *extended* finite element spaces, which are much better suited for the approximation of discontinuous functions. Some implementation issues related to XFEM are treated in Sect. 7.9.3. In Sect. 7.9.4 we present an analysis of the XFEM method. In Sect. 7.10 results of numerical experiments with this method are presented.

7.9.1 Approximation error for standard FE spaces

In this section we consider the approximation error

$$\inf_{q_h \in Q_h} \|q_h - p^*\|_{L^2}$$

for a few standard finite element spaces Q_h and explain why in general for a function p^* that is *discontinuous across* Γ_h one can expect no better bound for this approximation error than $c\sqrt{h}$. This serves as a motivation for an improved pressure finite element space as presented in Sect. 7.9.2. To explain the effect underlying the $\sqrt{h}$ behavior of the error bound we analyze a concrete two-dimensional example as illustrated in Fig. 7.10. We take $\Omega = (0,1)^2 \subset \mathbb{R}^2$ and define

$$\Omega_1 := \{\, x \in \Omega : x_1 \leq 1 - x_2 \,\}, \qquad \Omega_2 := \Omega \setminus \overline{\Omega}_1.$$

The interface Γ separating both subdomains from each other is given by

$$\Gamma = \{\, x \in \Omega : x_1 = 1 - x_2 \,\}.$$

A family of 2D triangulations $\{\mathcal{T}_h\}_{h>0}$ is constructed as follows. The starting triangulation $\mathcal{T}_0$ consists of two triangles, namely the ones with vertices $\{(0,0),(0,1),(1,1)\}$ and $\{(0,0),(1,0),(1,1)\}$. Then a global regular refinement strategy (connecting the midpoints of edges) is applied repeatedly. This results in a nested sequence of triangulations $\mathcal{T}_{h_k}$, $k = 1,2,\ldots$, with mesh size $h_k = 2^{-k}$. In Fig. 7.10 the triangulation $\mathcal{T}_{h_2}$ is shown. The set of triangles that contains the interface is given by (with $h := h_k$)

$$\mathcal{T}_h^\Gamma := \{\, T \in \mathcal{T}_h : \operatorname{meas}_1(T \cap \Gamma) > 0 \,\}.$$

In Fig. 7.10 the elements in $\mathcal{T}_{h_2}^\Gamma$ are colored gray.

For $h = h_k$ we consider the finite element spaces

$$\begin{aligned}
Q_h^0 &:= \{\, p : \Omega \to \mathbb{R} : p_{|T} \in \mathcal{P}_0 \quad \text{for all } T \in \mathcal{T}_h \,\} \quad \text{(piecewise constants)},\\
Q_h^{1,\mathrm{disc}} &:= \{\, p : \Omega \to \mathbb{R} : p_{|T} \in \mathcal{P}_1 \quad \text{for all } T \in \mathcal{T}_h \,\} \quad \text{(linear, discontinuous)},\\
Q_h^1 &:= \{\, p \in C(\Omega) : p_{|T} \in \mathcal{P}_1 \quad \text{for all } T \in \mathcal{T}_h \,\} \quad \text{(linear, continuous)}.
\end{aligned}$$

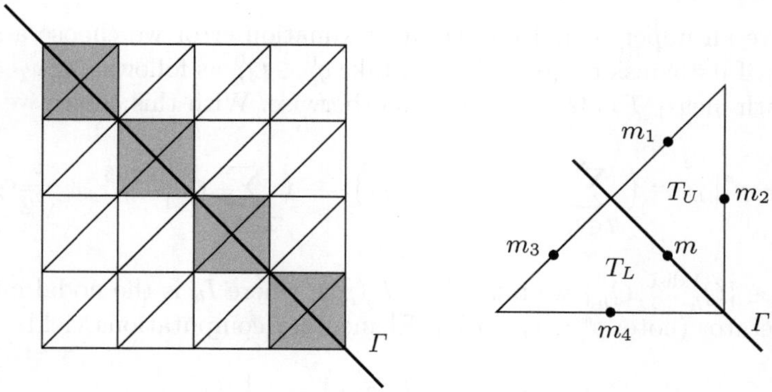

Fig. 7.10. Triangulation $\mathcal{T}_{h_2}$ and a triangle $T \in \mathcal{T}_{h_k}^{\Gamma}$.

Note that

$$Q_h^j \subset Q_h^{1,\mathrm{disc}} \quad \text{for } j = 0, 1. \tag{7.103}$$

We take p^* as follows: $p^*(x) = c_p > 0$ for all $x \in \Omega_1$, $p^*(x) = 0$ for all $x \in \Omega_2$. We study $\inf_{q_h \in Q_h} \|q_h - p^*\|_{L^2}$ for $Q_h \in \{Q_h^0, Q_h^{1,\mathrm{disc}}, Q_h^1\}$. For $Q_h = Q_h^{1,\mathrm{disc}}$ the identity

$$\inf_{q_h \in Q_h^{1,\mathrm{disc}}} \|q_h - p^*\|_{L^2}^2 = \sum_{T \in \mathcal{T}_h^{\Gamma}} \min_{q \in \mathcal{P}_1} \|q - p^*\|_{L^2(T)}^2$$

holds. Take $T \in \mathcal{T}_h^{\Gamma}$. Using a quadrature rule on triangles that is exact for all polynomials of degree two we get, cf. Fig. 7.10,

$$\begin{aligned}
\min_{q \in \mathcal{P}_1} \|q - p^*\|_{L^2(T)}^2 &= \min_{q \in \mathcal{P}_1} \Big(\int_{T_L} (q - c_p)^2 \, dx\, dy + \int_{T_U} q^2 \, dx\, dy \Big) \\
&= \frac{h^2}{12} \min_{q \in \mathcal{P}_1} \big((q(m_3) - c_p)^2 + (q(m_4) - c_p)^2 + (q(m) - c_p)^2 \\
&\qquad + q(m_1)^2 + q(m_2)^2 + q(m)^2 \big) \\
&\geq \frac{h^2}{12} \min_{q \in \mathcal{P}_1} \big((q(m) - c_p)^2 + q(m)^2 \big) = \frac{1}{24} c_p^2 h^2.
\end{aligned}$$

Thus we have

$$\inf_{q_h \in Q_h^{1,\mathrm{disc}}} \|q_h - p^*\|_{L^2} \geq \Big(\sum_{T \in \mathcal{T}_h^{\Gamma}} \frac{1}{24} c_p^2 h^2 \Big)^{\frac{1}{2}} = \big(\frac{2}{h} \frac{1}{24} c_p^2 h^2 \big)^{\frac{1}{2}} = \frac{1}{2\sqrt{3}} c_p \sqrt{h}.$$

Due to (7.103) this yields

$$\inf_{q_h \in Q_h} \|q_h - p^*\|_{L^2} \geq \frac{1}{2\sqrt{3}} c_p \sqrt{h} \quad \text{for } Q_h \in \{Q_h^0, Q_h^{1,\mathrm{disc}}, Q_h^1\}.$$

To derive an upper bound for the approximation error we choose a suitable $q_h \in Q_h$. First consider $Q_h = Q_h^0$ and take $q_h^0 \in Q_h^0$ as follows: $(q_h^0)_{|T} = c_p$ for all T with $\text{meas}_1(T \cap \Omega_1) > 0$, $q_h^0 = 0$ otherwise. With this choice we get

$$\|q_h^0 - p^*\|_{L^2} = \Big(\sum_{T \in \mathcal{T}_h^\Gamma} \|q_h^0 - p^*\|_{L^2(T)}^2 \Big)^{\frac{1}{2}} = \big(\sum_{T \in \mathcal{T}_h^\Gamma} c_p^2 \frac{1}{4} h^2 \big)^{\frac{1}{2}} = \frac{1}{\sqrt{2}} c_p \sqrt{h}.$$

For $Q_h \in \{Q_h^{1,\text{disc}}, Q_h^1\}$ we take $q_h^1 := I_h(p^*)$, where I_h is the nodal interpolation operator (note: $p^* = c_p$ on Γ). Elementary computations yield

$$\|q_h^1 - p^*\|_{L^2} = \big(\frac{1}{12} c_p^2 h\big)^{\frac{1}{2}} = \frac{1}{2\sqrt{3}} c_p \sqrt{h}.$$

Combination of these results yields

$$\frac{1}{2\sqrt{3}} c_p \sqrt{h} \leq \inf_{q_h \in Q_h} \|q_h - p^*\|_{L^2} \leq \frac{1}{\sqrt{2}} c_p \sqrt{h} \quad \text{for } Q_h \in \{Q_h^0, Q_h^{1,\text{disc}}, Q_h^1\}.$$

If instead of piecewise constants or piecewise linears we consider polynomials of higher degree, the approximation error still behaves like $\sqrt{h}$.

Similar examples, which have a $\sqrt{h}$ approximation error behavior, can be constructed using these finite element spaces on tetrahedral triangulations in 3D.

7.9.2 Extended finite element method (XFEM)

The analysis in the previous section, which is confirmed by numerical experiments in Sect. 7.10, leads to the conclusion that there is a need for an improved finite element space for the discretization of the pressure. In this section we introduce such a space which is based on an idea presented in [181, 30]. In these papers a so-called *extended finite element method* (XFEM) is introduced in the context of crack formations in structure mechanics which has good approximation properties for interface type of problems. A recent review on XFEM techniques is given in [113, 114]. XFEM belongs to the class of partition of unity methods (PUM) [18, 19].

Here we apply the XFEM method to two-phase flow problems by constructing a suitable extended pressure finite element space. In this section we explain the method. For $k \geq 1$ fixed we introduce the standard finite element space

$$Q_h = Q_h^k = \{\, q \in C(\Omega) : q|_T \in \mathcal{P}_k \quad \text{for all } T \in \mathcal{T}_h \,\}.$$

We explain the construction of the XFEM space for $k = 1$. This technique can easily be generalized to $k \geq 1$. Define the index set $\mathcal{J} = \{1, \dots, n\}$, where $n = \dim Q_h$ is the number of degrees of freedom. Let $\mathcal{B} := \{q_j\}_{j \in \mathcal{J}}$ be the nodal basis of Q_h, i. e. $q_j(x_i) = \delta_{i,j}$ for $i, j \in \mathcal{J}$ where $x_i \in \mathbb{R}^3$ denotes the vector of spatial coordinates of the i-th degree of freedom.

The idea of the XFEM method is to enrich the original finite element space Q_h by additional functions q_j^X for $j \in \mathcal{J}'$ where $\mathcal{J}' \subset \mathcal{J}$ is a given index set. An additional function q_j^X is constructed by multiplying the original nodal basis function q_j by a so called enrichment function Φ_j:

$$q_j^X(x) := q_j(x)\,\Phi_j(x). \tag{7.104}$$

This enrichment yields the extended finite element space

$$Q_h^X := Q_h \oplus \operatorname{span}\{\, q_j^X \,:\, j \in \mathcal{J}' \,\}.$$

This idea was introduced in [181] and further developed in [30] for different kinds of discontinuities (kinks, jumps), which may also intersect or branch. The choice of the enrichment function depends on the type of discontinuity. For representing jumps the Heaviside function is proposed to construct appropriate enrichment functions. Basis functions with kinks can be obtained by using the distance function as enrichment function [180].

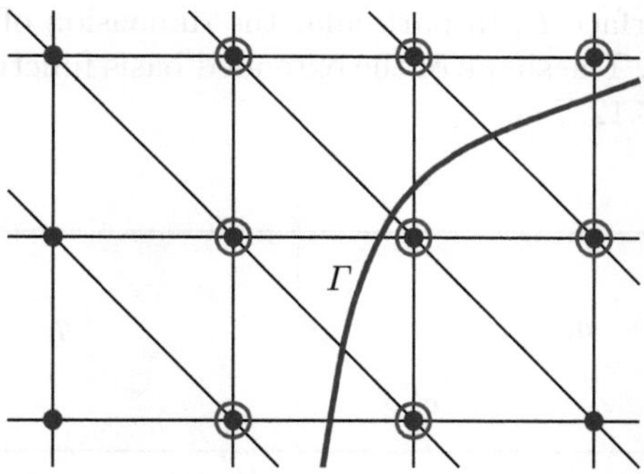

Fig. 7.11. Enrichment of P_1 finite elements in a 2D example. Dots represent degrees of freedom of original basis functions, circles indicate where additional functions are added in the vicinity of the interface Γ.

The index set of basis functions "close to the interface" is given by

$$\mathcal{J}_\Gamma := \{\, j \in \mathcal{J} \,:\, \operatorname{meas}_2(\Gamma \cap \operatorname{supp} q_j) > 0 \,\},$$

cf. Fig. 7.11 for a 2D example.

Let $\phi : \Omega \to \mathbb{R}$ be an indicator function such that ϕ is negative in Ω_1 and positive in Ω_2. For example the level set function could be used for ϕ. Let H be the Heaviside function and

$$H_\Gamma(x) := H(\phi(x)) = \begin{cases} 0 & x \in \Omega_1, \\ 1 & x \in \Omega_2. \end{cases}$$

Since we are interested in functions with a jump across the interface we define the enrichment function

$$\Phi_j^H(x) := H_\Gamma(x) - H_\Gamma(x_j), \quad j \in \mathcal{J}_\Gamma, \tag{7.105}$$

and a corresponding function

$$q_j^X := q_j \Phi_j^H, \quad j \in \mathcal{J}_\Gamma.$$

The second term in the definition of Φ_j^H is constant and may be omitted (as it doesn't introduce new functions in the function space), but ensures the nice property $q_j^X(x_i) = 0$ for all i, i. e., q_j^X vanishes in all degrees of freedom. As a consequence, we have $q_j^X \equiv 0$ in all T with $T \notin \mathcal{T}_h^\Gamma := \{T \in \mathcal{T}_h : \text{meas}_2(T \cap \Gamma) > 0\}$. In the following we will use the notation $q_j^\Gamma := q_j\, \Phi_j^H$ and the XFEM space is denoted by

$$Q_h^\Gamma := Q_h \oplus \text{span}\,\{\, q_j^\Gamma : j \in \mathcal{J}_\Gamma \,\}. \tag{7.106}$$

We emphasize that the extended finite element space Q_h^Γ depends on the location of the interface Γ. In particular the dimension of Q_h^Γ may change if the interface moves. The shape of the extended basis functions for the 1D case is sketched in Fig. 7.12.

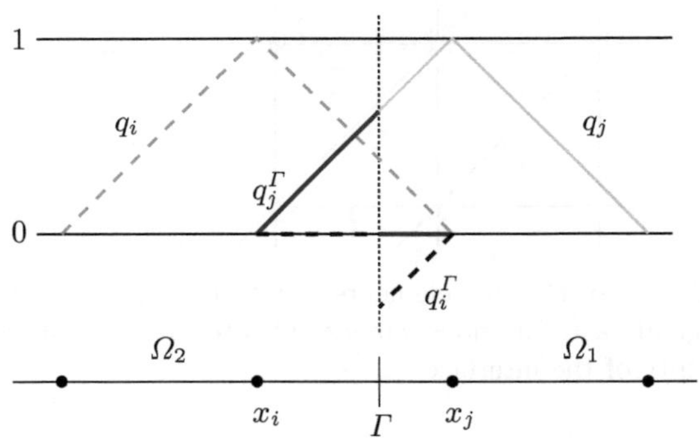

Fig. 7.12. Extended finite element basis functions q_i, q_i^Γ (dashed) and q_j, q_j^Γ (solid) for 1D case.

Remark 7.9.1 In [30] the XFEM is applied to problems from linear elasticity demonstrating the ability of the method to capture jumps and kinks. These discontinuities also branch or intersect in some of the examples, in this case more elaborate constructions of the enrichment functions are used.

In [66] the XFEM is also applied to a two-phase flow problem. In that paper discontinuous material properties ρ and μ, but *no surface tension forces* are taken into account. Thus there is no jump in pressure, but the velocity solution exhibits a kink (i.e., a discontinuity in the derivative) at the interface. For the pressure and the level set function standard finite element spaces are used. The velocity field is discretized with an extended finite element space enriched by $\mathbf{v}_j^X(x) = \mathbf{v}_j(x)\,|d(x)|$ to capture the kinks at the interface. The location of the interface is captured by a level set approach. The level set function is used as an approximate signed distance function.

A similar idea of space enrichment in the context of two-phase flow simulations is also suggested in [177].

The same finite element space Q_h^Γ is also used in the "unfitted finite element method" that is introduced in [24] for a class of elliptic interface problems.

7.9.3 Modifications and implementation issues

In this section we discuss a few practical issues related to the application of XFEM to non-stationary Navier-Stokes two-phase flow problems.

As Q_h^Γ depends on the location of the interface Γ, the space Q_h^Γ changes if the interface moves. Thus the discretization of $b(\cdot,\cdot)$ has to be updated each time when the level set function has changed. In a Navier-Stokes code solving non-stationary two-phase flow problems this is nothing special since the mass and stiffness matrices depend on discontinuous material properties like density and viscosity and thus have to be updated as well. What is special about the extended pressure finite element space is the fact that the dimension of Q_h^Γ may vary, i. e., some extended pressure unknowns may appear or disappear when the interface is moving. This has to be taken into account by a suitable interpolation procedure for the extended pressure unknowns.

Let Γ_h be a piecewise planar approximation of the interface Γ as described in Sect. 7.3. For practical reasons we do not consider Q_h^Γ but the space $Q_h^{\Gamma_h}$, which is the extended pressure finite element space described above but with Γ replaced by its approximation Γ_h. We discuss how in the discretization of a two-phase flow problem the construction of the discrete problem changes if instead of $(\mathbf{V}_h, Q_h)$ the pair $(\mathbf{V}_h, Q_h^{\Gamma_h})$ is used. For the velocity space $\mathbf{V}_h$ we use the standard space of piecewise quadratics. The use of another finite element space $Q_h^{\Gamma_h}$ (instead of standard piecewise linears) influences only the evaluation of $b(\cdot,\cdot)$.

For a basis function $\boldsymbol{\xi}_i \in \mathbf{V}_h$ and $j \in \mathcal{J}_\Gamma$ the evaluation of

$$b(\boldsymbol{\xi}_i, q_j^{\Gamma_h}) = -\sum_{T' \in \mathcal{T}_{h'}} \int_{T'} q_j^{\Gamma_h} \operatorname{div} \boldsymbol{\xi}_i \, dx$$

requires the computation of integrals with discontinuous integrands, as the extended pressure basis function $q_j^{\Gamma_h}$ has a jump across the interface. We sum over $T' \in \mathcal{T}_{h'}$ (and not $T \in \mathcal{T}_h$) because Γ_h is defined as in (7.21), i. e., Γ_h is piecewise planar corresponding to the refinement $\mathcal{T}_{h'}$ of $\mathcal{T}_h$. Let $T \in \mathcal{T}_h$

be a tetrahedron with $T \cap \operatorname{supp} q_j^{\Gamma_h} \neq \emptyset$ and $T' \in \mathcal{T}_{h'}$ with $T' \subset T$ a child tetrahedron created by regular refinement of T. Define

$$T_i' := T' \cap \Omega_{i,h}, \quad i = 1, 2.$$

Using the definition of $q_j^{\Gamma_h}$, cf. (7.104), (7.105), we get

$$\begin{aligned} \int_{T'} q_j^{\Gamma_h} \operatorname{div} \boldsymbol{\xi}_i \, dx &= \int_{T_2'} q_j \operatorname{div} \boldsymbol{\xi}_i \, dx - H_\Gamma(x_j) \int_{T'} q_j \operatorname{div} \boldsymbol{\xi}_i \, dx \\ &= \begin{cases} \int_{T_2'} q_j \operatorname{div} \boldsymbol{\xi}_i \, dx & \text{if } x_j \in \Omega_{1,h}, \\ -\int_{T_1'} q_j \operatorname{div} \boldsymbol{\xi}_i \, dx & \text{if } x_j \in \Omega_{2,h}. \end{cases} \end{aligned} \tag{7.107}$$

The integrands in the right-hand side of (7.107) are polynomial on the polyhedral subdomains T_1', T_2'. For the computation of the integral over T_i' we distinguish two cases. The face $T' \cap \Gamma_h$ is either a triangle or a quadrilateral. In the first case one of the sets T_1', T_2' is tetrahedral; without loss of generality let T_1' be tetrahedral. Then integration over T_2' can be computed by

$$\int_{T_2'} G(x) \, dx = \int_{T'} G(x) \, dx - \int_{T_1'} G(x) \, dx.$$

In the second case both T_1', T_2' are non-tetrahedral, but can each be subdivided into three sub-tetrahedra, cf. Fig. 7.13. In *all cases* the integration over T_i' can be reduced to *integration on tetrahedra*, for which standard quadrature rules can be applied.

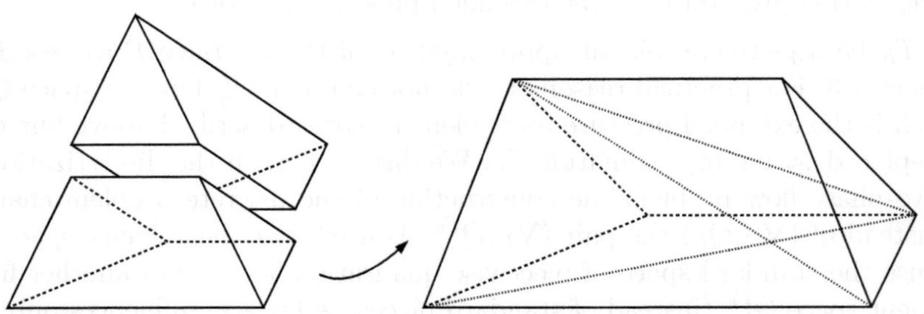

Fig. 7.13. Left: Parts of tetrahedron T' are non-tetrahedral, iff cutting face $T' \cap \Gamma_h$ is a quadrilateral. Right: Triangulation of the lower part into three tetrahedra.

Regarding stability, one has to treat carefully the situation where some extended basis functions q_j^Γ have a (very) "small" support. In such situations the resulting linear systems may become very ill-conditioned and the LBB-stability of the $(\mathbf{V}_h, Q_h^\Gamma)$ pair is questionable, cf. the numerical experiment in Sect. 7.10.3. One obvious possibility to deal with this instability problem is to skip the extended basis functions with relatively "small" contributions. What is meant by "small" will be specified now. In Sect. 7.9.4 we investigate which elements from the "added" space span $\{\, q_j^\Gamma : j \in \mathcal{J}_\Gamma \,\}$ can be deleted without loosing the optimal approximation quality of the extended finite element space. This leads to the following criterion, in which parameters $\tilde{c} > 0$ and $\alpha > 0$ are used. For $j \in \mathcal{J}_\Gamma$ we consider the following condition for the corresponding extended basis function q_j^Γ:

$$\|q_j^\Gamma\|_{l,T} \leq \tilde{c}\, h_T^\alpha \|q_j\|_{l,T} \qquad \text{for all } T \in \mathcal{T}_h^\Gamma. \tag{7.108}$$

Here $l \in \{0,1\}$ is the order of the Sobolev norm. We introduce the *reduced index set* $\tilde{\mathcal{J}}_\Gamma \subset \mathcal{J}_\Gamma$ by

$$\tilde{\mathcal{J}}_\Gamma := \{\, j \in \mathcal{J}_\Gamma : \text{(7.108) does } \textit{not} \text{ hold for } q_j^\Gamma \,\}$$

and the *reduced extended finite element space* $\tilde{Q}_h^\Gamma$

$$\tilde{Q}_h^\Gamma := Q_h \oplus \operatorname{span}\left\{\, q_j^\Gamma : j \in \tilde{\mathcal{J}}_\Gamma \,\right\}. \tag{7.109}$$

In other words, *only extended basis functions q_j^Γ are taken into account, for which* (7.108) *does not hold.* The criterion (7.108) quantifies what is meant by "small contributions". In this modified space $\tilde{Q}_h^\Gamma$ basis functions with very small supports are avoided and an approximation property of the following form can be shown to hold (Sect. 7.9.4):

$$\inf_{q \in \tilde{Q}_h^\Gamma} \|p - q\|_{l,\Omega_1 \cup \Omega_2} \leq c\big(h^{m-l} + h^{\alpha-l}\big)\|p\|_{m,\Omega_1 \cup \Omega_2}$$

for all $p \in H^m(\Omega_1 \cup \Omega_2)$ and integers l, m with $0 \leq l < m \leq 2$. Thus we maintain an optimal approximation error bound if in the criterion (7.108) we take $\alpha = m$. The choice of l and m depends on the norms in which the discretization error in the (pressure) variable p is measured. In our applications we use $l = 0$, $m = 2$, resulting in an optimal error bound $\mathcal{O}(h^2)$ for piecewise linear finite elements.

Numerical experiments, cf. Sect. 7.10.3, indicate that this reduction of the extended finite element has a significant influence on the LBB-stability of the $(\mathbf{V}_h, Q_h^\Gamma)$ finite element pair.

Remark 7.9.2 Because $\|q_j\|_{l,T} \sim c h_T^{\frac{3}{2}-l}$ for $l = 0, 1$, the condition (7.108) can be replaced by

$$\|q_j^\Gamma\|_{l,T} \leq \hat{c} h_T^{\alpha+\frac{3}{2}-l} \qquad \text{for all } T \in \mathcal{T}_h^\Gamma. \tag{7.110}$$

The constant $\hat{c}$ may differ from $\tilde{c}$ used in (7.108).

7.9.4 Analysis of XFEM

In this section we derive some properties of the XFEM method. We discuss the following topics: approximation quality, conditioning of a basis in the XFEM space and LBB-stability of the $(\mathbf{V}_h, Q_h^\Gamma)$ pair.

Approximation error bounds

For the approximation error bounds we consider the XFEM space Q_h^Γ with a given h-independent interface Γ. Note that in practice the space $Q_h^{\Gamma_h}$ is used; we do not consider this in the analysis, since it would lead to additional technical complications induced by the h-dependence of the interface.

For an integer $k \geq 0$ we define the space

$$H^k(\Omega_1 \cup \Omega_2) := \left\{ p \in L^2(\Omega) : p_{|\Omega_i} \in H^k(\Omega_i),\, i = 1, 2 \right\},$$

with the norm $\|p\|_{k,\Omega_1\cup\Omega_2}^2 := \|p\|_{k,\Omega_1}^2 + \|p\|_{k,\Omega_2}^2$. We need *restriction* operators $R_i : L^2(\Omega) \to L^2(\Omega)$, $i = 1, 2$

$$R_i v = \begin{cases} v_{|\Omega_i} & \text{on } \Omega_i \\ 0 & \text{on } \Omega \setminus \Omega_i \end{cases} \tag{7.111}$$

(in L^2 sense). The extended finite element space Q_h^Γ can also be characterized by the following property: $v \in Q_h^\Gamma$ if and only if there exist functions $v_1, v_2 \in Q_h$ such that $v|_{\Omega_i} = v_i|_{\Omega_i}$, $i = 1, 2$. In other words:

$$Q_h^{\Gamma_h} = R_1 Q_h \oplus R_2 Q_h. \tag{7.112}$$

We present an approximation error bound for the XFEM space:

Theorem 7.9.3 For integers l, m with $0 \leq l < m \leq 2$ the following holds:

$$\inf_{q \in Q_h^\Gamma} \|p - q\|_{l,\Omega_1\cup\Omega_2} \leq c\, h^{m-l} \|p\|_{m,\Omega_1\cup\Omega_2} \tag{7.113}$$

for all $p \in H^m(\Omega_1 \cup \Omega_2)$.

Proof. We use extension operators $\mathcal{E}_i^m : H^m(\Omega_i) \to H^m(\Omega)$, $i = 1, 2$, with $(\mathcal{E}_i^m w)_{|\Omega_i} = w$ and $\|\mathcal{E}_i^m w\|_m \leq c\|w\|_{m,\Omega_i}$, cf. [256]. For $m = 1, 2$, let $I_h^m : H^m(\Omega) \to Q_h$ be a (quasi-)interpolation operator such that $\|w - I_h^m w\|_l \leq c\, h^{m-l}\|w\|_m$ for all $w \in H^m(\Omega)$, $0 \leq l < m \leq 2$ (for example, nodal interpolation if $m = 2$). Let $m \in \{1, 2\}$ and $p \in H^m(\Omega_1 \cup \Omega_2)$ be given. Define $q^* \in Q_h^\Gamma$ by

$$q^* = R_1 I_h^m \mathcal{E}_1^m R_1 p + R_2 I_h^m \mathcal{E}_2^m R_2 p. \tag{7.114}$$

For this approximation we obtain

$$
\begin{aligned}
&\|p-q^*\|_{l,\Omega_1\cup\Omega_2}^2\\
&=\sum_{i=1}^{2}\|p-q^*\|_{l,\Omega_i}^2=\sum_{i=1}^{2}\|p-I_h^m\mathcal{E}_i^m R_i p\|_{l,\Omega_i}^2\\
&=\sum_{i=1}^{2}\|\mathcal{E}_i^m R_i p-I_h^m\mathcal{E}_i^m R_i p\|_{l,\Omega_i}^2\le\sum_{i=1}^{2}\|\mathcal{E}_i^m R_i p-I_h^m\mathcal{E}_i^m R_i p\|_l^2\\
&\le c\,h^{2(m-l)}\sum_{i=1}^{2}\|\mathcal{E}_i^m R_i p\|_m^2\le c\,h^{2(m-l)}\sum_{i=1}^{2}\|R_i p\|_{m,\Omega_i}^2\\
&=c\,h^{2(m-l)}\|p\|_{m,\Omega_1\cup\Omega_2}^2,
\end{aligned}
$$

which proves the result. □

Hence, the XFEM space has optimal approximation quality for piecewise smooth functions p, for example $\inf_{q_h\in Q_h^\Gamma}\|q_h-p\|_{L^2}\le ch^2$ if $p_{|\Omega_i}\in H^2(\Omega_i)$, $i=1,2$. Similar approximation results are given in [135]. In [209] also a result for the reduced XFEM space $\tilde{Q}_h^\Gamma$ is derived. In the analysis a global inverse inequality is used and therefore in the following theorem we use the assumption that the family of triangulations is quasi-uniform.

Theorem 7.9.4 Assume that the family $\{\mathcal{T}_h\}_{h>0}$ is quasi-uniform. For integers l, m with $0\le l<m\le 2$ the following holds, with $\tilde{Q}_h^\Gamma$ defined as in (7.109):

$$
\inf_{q\in\tilde{Q}_h^\Gamma}\|p-q\|_{l,\Omega_1\cup\Omega_2}\le c\left(h^{m-l}+h^{\alpha-l}\right)\|p\|_{m,\Omega_1\cup\Omega_2} \tag{7.115}
$$

for all $p\in H^m(\Omega_1\cup\Omega_2)$.

Proof. Theorem 4 in [209]. □

Properties of a basis in the XFEM space

In this section we derive properties of a basis in the space $Q_h^{\Gamma_h}$. Note that now we consider $Q_h^{\Gamma_h}$ (instead of Q_h^Γ). We first introduce some further notation. The restriction R_i, $i=1,2$, is as in (7.111), but with Ω_i replaced by $\Omega_{i,h}$ (recall: Γ_h defines the interface between $\Omega_{1,h}$ and $\Omega_{2,h}$). The nodal basis in Q_h is denoted by $\{q_j\}_{j\in\mathcal{J}}$, $\mathcal{J}=\{1,\ldots,n\}$. We introduce subsets of $\mathcal{J}$ for which the corresponding basis functions have a nonzero intersection with Γ_h:

$$
\begin{aligned}
\mathcal{J}_1^{\Gamma_h}&:=\{\,k\in\mathcal{J}:\,x_k\in\Omega_{2,h}\ \text{ and }\ \operatorname{supp}(q_k)\cap\Gamma_h\neq\emptyset\,\}\\
\mathcal{J}_2^{\Gamma_h}&:=\{\,k\in\mathcal{J}:\,x_k\in\Omega_{1,h}\ \text{ and }\ \operatorname{supp}(q_k)\cap\Gamma_h\neq\emptyset\,\}.
\end{aligned}
$$

Corresponding spaces are defined by

$$V_i^{\Gamma_h} := \operatorname{span}\left\{ R_i q_k : k \in \mathcal{J}_i^{\Gamma_h} \right\}, \quad i = 1, 2.$$

To avoid technical difficulties in the analysis, we make the (reasonable) assumption that $\operatorname{meas}_2(\Gamma_h \cap \partial T) = 0$ for all T, i.e., the interface Γ_h does not contain faces of the tetrahedra $T \in \mathcal{T}_h$. The extended finite element space Q_h^Γ can be represented as

$$Q_h^{\Gamma_h} = Q_h \oplus V_1^{\Gamma_h} \oplus V_2^{\Gamma_h}. \tag{7.116}$$

We will analyze the stability of the basis

$$\{q_k\}_{1 \le k \le n} \cup \{R_1 q_k\}_{k \in \mathcal{J}_1^{\Gamma_h}} \cup \{R_2 q_k\}_{k \in \mathcal{J}_2^{\Gamma_h}}. \tag{7.117}$$

We prove, cf. Theorem 7.9.7, that the *diagonally scaled mass matrix is uniformly* (w.r.t. h) *well-conditioned.* This holds independent of the size and the shape of the support of the basis functions $R_i q_k$. This immediately implies a similar result for the reduced XFEM space $\tilde{Q}_h^{\Gamma_h}$.

We first derive a strengthened Cauchy-Schwarz inequality between the spaces Q_h and $V_1^{\Gamma_h} \oplus V_2^{\Gamma_h}$. The collection of all vertices in the triangulation $\mathcal{T}_h$ is denoted by $\mathcal{V} := \{ x_k : k \in \mathcal{J} \}$. For each vertex $x \in \mathcal{V}$ let $\mathcal{T}(x)$ be the set of all tetrahedra in $\mathcal{T}_h$ that have x as a vertex. Define $\mathcal{T}_R = \{ T \in \mathcal{T}_h : T \cap \Gamma_h = \emptyset \}$. We introduce the assumption

$$\mathcal{T}(x) \cap \mathcal{T}_R \neq \emptyset \quad \text{for all } x \in \mathcal{V}. \tag{7.118}$$

For h sufficiently small this assumption is satisfied.

Lemma 7.9.5 *Assume that* (7.118) *holds. There exists a constant* $c_{CS} < 1$ *independent of* h *such that*

$$(v, w)_{L^2} \le c_{CS} \|v\|_{L^2} \|w\|_{L^2} \quad \textit{for all } v \in Q_h,\ w \in V_1^{\Gamma_h} \oplus V_2^{\Gamma_h}.$$

Proof. We use the notation $W = V_1^{\Gamma_h} \oplus V_2^{\Gamma_h}$. Let $P_W : L^2(\Omega) \to W$ be the L^2-orthogonal projection on W. Let $\mathcal{V}(T)$ denote the set of vertices of T. Transformation to a unit tetrahedron yields the norm equivalence

$$c_1 \|v\|_{L^2(T)}^2 \le |T| \sum_{x \in \mathcal{V}(T)} v(x)^2 \le c_2 \|v\|_{L^2(T)}^2 \tag{7.119}$$

for all $T \in \mathcal{T}_h$, $v \in Q_h$, with constants $c_1 > 0$ and c_2 independent of h. Due to (7.118) we have that for each $x \in \mathcal{V}(T)$ there exists a tetrahedron $\hat{T} \in \mathcal{T}(x) \cap \mathcal{T}_R$ with $x \in \mathcal{V}(\hat{T})$. Let $\mathcal{T}_R$ be as defined above and $\mathcal{T}_h^\Gamma := \mathcal{T}_h \setminus \mathcal{T}_R$ the set of all tetrahedra that have a nonzero intersection with Γ_h. We obtain for $v \in Q_h$ and $T \in \mathcal{T}_h^\Gamma$:

$$\begin{aligned} \|v\|_{L^2(T)}^2 &\le c\,|T| \sum_{x \in \mathcal{V}(T)} v(x)^2 \\ &\le c \sum_{x \in \mathcal{V}(T)} \sum_{\hat{T} \in \mathcal{T}(x) \cap \mathcal{T}_R} |\hat{T}| \sum_{y \in \mathcal{V}(\hat{T})} v(y)^2 \\ &\le c \sum_{x \in \mathcal{V}(T)} \|v\|_{L^2(\mathcal{T}(x) \cap \mathcal{T}_R)}^2. \end{aligned}$$

Hence,

$$\|v\|^2_{L^2(\mathcal{T}_h^\Gamma)} = \sum_{T\in\mathcal{T}_h^\Gamma} \|v\|^2_{L^2(T)} \le c\|v\|^2_{L^2(\mathcal{T}_R)}, \quad v \in Q_h,$$

holds with a constant c independent of h. This yields $\|v\|^2_{L^2} = \|v\|^2_{L^2(\mathcal{T}_h^\Gamma)} + \|v\|^2_{L^2(\mathcal{T}_R)} \le c\|v\|^2_{L^2(\mathcal{T}_R)}$ with c independent of h. Using this and $(P_W v)_{|\mathcal{T}_R} = 0$ we get, for $v \in Q_h$,

$$\|v - P_W v\|_{L^2} \ge \|v - P_W v\|_{L^2(\mathcal{T}_R)} = \|v\|_{L^2(\mathcal{T}_R)} \ge \hat{c}\|v\|_{L^2},$$

with a constant $\hat{c} > 0$ independent of h. Thus we get

$$\|P_W v\|^2_{L^2} = \|v\|^2_{L^2} - \|v - P_W v\|^2_{L^2} \le (1-\hat{c}^2)\|v\|^2_{L^2} =: c^2_{CS}\|v\|^2_{L^2}$$

for all $v \in Q_h$. Hence, for $v \in Q_h$, $w \in W$,

$$\begin{aligned}(v,w)_{L^2} &= (v, P_W w)_{L^2} = (P_W v, w)_{L^2} \le \|P_W v\|_{L^2}\|w\|_{L^2} \\ &\le c_{CS}\|v\|_{L^2}\|w\|_{L^2},\end{aligned}$$

which completes the proof. □

The spaces $V_1^{\Gamma_h}$ and $V_2^{\Gamma_h}$ are (due to disjoint supports of functions from these spaces) L^2-orthogonal. Thus we conclude that in the decomposition

$$Q_h^{\Gamma_h} = Q_h \oplus V_1^{\Gamma_h} \oplus V_2^{\Gamma_h}$$

we have a strengthened Cauchy-Schwarz inequality between Q_h and $V_1^{\Gamma_h} \oplus V_2^{\Gamma_h}$ and even orthogonality between $V_1^{\Gamma_h}$ and $V_2^{\Gamma_h}$.

For $v = w + w_1 + w_2 \in Q_h^{\Gamma_h}$, with $w \in Q_h$, $w_i \in V_i^{\Gamma_h}$, we have

$$\|v\|^2_{L^2} \le 2\big(\|w\|^2_{L^2} + \|w_1 + w_2\|^2_{L^2}\big) = 2\big(\|w\|^2_{L^2} + \|w_1\|^2_{L^2} + \|w_2\|^2_{L^2}\big)$$

and

$$\begin{aligned}\|v\|^2_{L^2} &= \|w\|^2_{L^2} + \|w_1 + w_2\|^2_{L^2} + 2(w, w_1 + w_2)_{L^2} \\ &\ge \|w\|^2_{L^2} + \|w_1 + w_2\|^2_{L^2} - 2c_{CS}\|w\|_{L^2}\|w_1 + w_2\|_{L^2} \\ &\ge (1 - c_{CS})\big(\|w\|^2_{L^2} + \|w_1\|^2_{L^2} + \|w_2\|^2_{L^2}\big).\end{aligned}$$

Hence we obtain

$$\begin{aligned}&(1 - c_{CS})\big(\|w\|^2_{L^2} + \|w_1\|^2_{L^2} + \|w_2\|^2_{L^2}\big) \\ &\le \|v\|^2_{L^2} \le 2\big(\|w\|^2_{L^2} + \|w_1\|^2_{L^2} + \|w_2\|^2_{L^2}\big).\end{aligned} \tag{7.120}$$

We now turn to the conditioning of the mass matrix. A function $v \in Q_h^{\Gamma_h}$ is represented in the basis $\{q_k\}_{1\le k\le n} \cup \{R_1 q_k\}_{k\in\mathcal{J}_1^{\Gamma_h}} \cup \{R_2 q_k\}_{k\in\mathcal{J}_2^{\Gamma_h}}$ as

$$v = \sum_{k=1}^{n} \alpha_k q_k + \sum_{i=1}^{2} \sum_{k \in \mathcal{J}_i^{\Gamma_h}} \beta_k^{(i)} R_i q_k =: w + w_1 + w_2, \tag{7.121}$$

where $w \in Q_h$, $w_i \in V_i^{\Gamma_h}$, $i = 1, 2$. It is well-known (cf. also (7.119)) that for $w = \sum_{k=1}^{n} \alpha_k q_k$ we have

$$c_1 \sum_{k=1}^{n} \alpha_k^2 \|q_k\|_{L^2}^2 \le \|w\|_{L^2}^2 \le c_2 \sum_{k=1}^{n} \alpha_k^2 \|q_k\|_{L^2}^2, \tag{7.122}$$

with constants $c_1 > 0$ and c_2 independent of h, i.e., the nodal basis $\{q_k\}_{1 \le k \le n}$ of Q_h is uniformly in h well-conditioned (w.r.t. $\|\cdot\|_{L^2}$). We prove a similar result for the basis $\{R_i q_k\}_{k \in \mathcal{J}_i^{\Gamma_h}}$ of $V_i^{\Gamma_h}$.

Lemma 7.9.6 *For $w_i = \sum_{k \in \mathcal{J}_i^{\Gamma_h}} \beta_k^{(i)} R_i q_k$, $i = 1, 2$, the following holds:*

$$\frac{\sqrt{2}-1}{2\sqrt{2}} \sum_{k \in \mathcal{J}_i^{\Gamma_h}} \big(\beta_k^{(i)}\big)^2 \|R_i q_k\|_{L^2}^2 \le \|w_i\|_{L^2}^2 \le 3 \sum_{k \in \mathcal{J}_i^{\Gamma_h}} \big(\beta_k^{(i)}\big)^2 \|R_i q_k\|_{L^2}^2. \tag{7.123}$$

Proof. It suffices to consider $i = 1$. We write $w_1 = \sum_{k \in \mathcal{J}_1^{\Gamma_h}} \beta_k R_1 q_k$. For each $T \in \mathcal{T}_h^{\Gamma} = \mathcal{T}_h \setminus \mathcal{T}_R$ there are at most 3 k-values in $\mathcal{J}_1^{\Gamma_h}$ with $(R_1 q_k)_{|T} \ne 0$ and thus

$$\begin{aligned} \|w_1\|_{L^2}^2 &= \sum_{T \in \mathcal{T}_h^{\Gamma}} \Big\| \sum_{k \in \mathcal{J}_1^{\Gamma_h}} \beta_k R_1 q_k \Big\|_{L^2(T)}^2 \le \sum_{T \in \mathcal{T}_h^{\Gamma}} \Big(\sum_{k \in \mathcal{J}_1^{\Gamma_h}} |\beta_k| \|R_1 q_k\|_{L^2(T)} \Big)^2 \\ &\le 3 \sum_{T \in \mathcal{T}_h^{\Gamma}} \sum_{k \in \mathcal{J}_1^{\Gamma_h}} |\beta_k|^2 \|R_1 q_k\|_{L^2(T)}^2 = 3 \sum_{k \in \mathcal{J}_1^{\Gamma_h}} |\beta_k|^2 \|R_1 q_k\|_{L^2}^2, \end{aligned}$$

which proves the upper bound in (7.123). A proof of the lower bound is given in Lemma 3 in [209]. □

Using the norm equivalences in (7.120), (7.122) and (7.123) we derive a spectral result for the mass matrix using standard arguments. Let $m = m_h := n + |\mathcal{J}_1^{\Gamma_h}| + |\mathcal{J}_2^{\Gamma_h}|$ be the dimension of $Q_h^{\Gamma_h}$ and $J : \mathbb{R}^m \to Q_h^{\Gamma_h}$ the isomorphism defined by (7.121):

$$J\mathbf{z} = J\big(\vec{\alpha}, \vec{\beta}^{(1)}, \vec{\beta}^{(2)}\big) = v.$$

The mass matrix $\mathbf{M} \in \mathbb{R}^{m \times m}$ is given by

$$\langle \mathbf{M}\mathbf{z}, \mathbf{z} \rangle = (J\mathbf{z}, J\mathbf{z})_{L^2} \quad \text{for all } \mathbf{z} \in \mathbb{R}^m.$$

Here $\langle \cdot, \cdot \rangle$ denotes the Euclidean scalar product.

Define $\mathrm{diag}(\mathbf{M}) =: \mathbf{D}_M$ with

$$\mathbf{D}_M = \begin{pmatrix} \mathbf{D} & & \emptyset \\ & \mathbf{D}_1 & \\ \emptyset & & \mathbf{D}_2 \end{pmatrix}, \quad \mathbf{D}_{k,k} = \|q_k\|_{L^2}^2, \quad 1 \le k \le n,$$

$$(\mathbf{D}_i)_{k,k} = \|R_i q_k\|_{L^2}^2, \quad k \in \mathcal{J}_i^{\Gamma_h}.$$

Theorem 7.9.7 *There are constants* $c_1 > 0$ *and* c_2 *independent of* h *such that*

$$c_1 \langle \mathbf{D}_M \mathbf{z}, \mathbf{z} \rangle \le \langle \mathbf{M}\mathbf{z}, \mathbf{z} \rangle \le c_2 \langle \mathbf{D}_M \mathbf{z}, \mathbf{z} \rangle \quad \textit{for all} \ \ \mathbf{z} \in \mathbb{R}^m.$$

Proof. From (7.120), (7.122) and (7.123) we get

$$\begin{aligned}
&\langle \mathbf{M}\mathbf{z}, \mathbf{z} \rangle = \|v\|_{L^2}^2 \le 2\big(\|w\|_{L^2}^2 + \|w_1\|_{L^2}^2 + \|w_2\|_{L^2}^2\big) \\
&\le 2\Big(c_2 \sum_{k=1}^n \alpha_k^2 \|q_k\|_{L^2}^2 + 3 \sum_{k \in \mathcal{J}_1^{\Gamma_h}} \big(\beta_k^{(1)}\big)^2 \|R_1 q_k\|_{L^2}^2 + 3 \sum_{k \in \mathcal{J}_2^{\Gamma_h}} \big(\beta_k^{(2)}\big)^2 \|R_2 q_k\|_{L^2}^2\Big) \\
&\le c\big(\langle \mathbf{D}\vec{\alpha}, \vec{\alpha} \rangle + \langle \mathbf{D}_1 \vec{\beta}^{(1)}, \vec{\beta}^{(1)} \rangle + \langle \mathbf{D}_2 \vec{\beta}^{(2)}, \vec{\beta}^{(2)} \rangle\big) = c\langle \mathbf{D}_M \mathbf{z}, \mathbf{z} \rangle,
\end{aligned}$$

with a constant c independent of h. Similarly, due to

$$\langle \mathbf{M}\mathbf{z}, \mathbf{z} \rangle = \|v\|_{L^2}^2 \ge (1 - c_{CS})\big(\|w\|_{L^2}^2 + \|w_1\|_{L^2}^2 + \|w_2\|_{L^2}^2\big),$$

and using the lower bounds in (7.122) and (7.123), we obtain $\langle \mathbf{M}\mathbf{z}, \mathbf{z} \rangle \ge c\langle \mathbf{D}_M \mathbf{z}, \mathbf{z} \rangle$ with a constant $c > 0$ independent of h. □

The result in this theorem proves that the matrix $\mathbf{D}_M^{-1}\mathbf{M}$ has a spectral condition number that is uniformly (w.r.t. h) bounded. Note that the constants in the spectral condition number bounds are also independent of the supports of the basis functions $R_i q_k$, $k \in \mathcal{J}_i^{\Gamma_h}$. In other words, a simple scaling is sufficient to control the stability (in L^2) of the basis functions with "very small" supports. Furthermore, we note that in the analysis we did *not* assume quasi-uniformity of the family of triangulations.

Corollary 7.9.8 Since the reduced extended finite element space $\tilde{Q}_h^{\Gamma_h}$ is spanned by a subset of the basis functions in (7.117), a similar L^2-stability result trivially holds for the basis in the space $\tilde{Q}_h^{\Gamma_h}$.

Remark 7.9.9 There are two canonical splittings of the XFEM space $Q_h^{\Gamma_h}$, namely the ones in (7.116) and in (7.112):

$$Q_h^{\Gamma_h} = Q_h \oplus V_1^{\Gamma_h} \oplus V_2^{\Gamma_h}, \quad Q_h^{\Gamma_h} = R_1 Q_h \oplus R_2 Q_h,$$

where R_i is the restriction operator as in (7.111), but now with respect to $\Omega_{i,h}$. In the analysis above we used the basis corresponding to the first splitting, cf. (7.117):

$$\{q_k\}_{1 \le k \le n} \cup \{R_1 q_k\}_{k \in \mathcal{J}_1^{\Gamma_h}} \cup \{R_2 q_k\}_{k \in \mathcal{J}_2^{\Gamma_h}}. \tag{7.124}$$

Let $\mathcal{J}_i$ be the index set of all k such that $\mathrm{supp}(q_k) \cap \Omega_{i,h} \neq \emptyset$ and define $V_i := \mathrm{span}\{\, q_k : k \in \mathcal{J}_i \,\}$. Note that $V_i \subset Q_h$ and $R_i Q_h = R_i V_i$ holds. The linear mapping $J : V_1 \times V_2 \to Q_h^{\Gamma_h}$

$$J(v_1, v_2) = R_1 v_1 + R_2 v_2 \tag{7.125}$$

is *bijective*. The second splitting induces the basis

$$\{R_1 q_k\}_{k \in \mathcal{J}_1} \cup \{R_2 q_k\}_{k \in \mathcal{J}_2}. \tag{7.126}$$

Related to the representations in these two bases we note the following. Take $v \in Q_h^{\Gamma_h}$ and let $v_i \in V_i$ be such that $v = R_1 v_1 + R_2 v_2$. For the representation in the basis (7.124) we have

$$v = \sum_{k=1}^{n} \alpha_k q_k + \sum_{i=1}^{2} \sum_{k \in \mathcal{J}_i^{\Gamma_h}} \beta_k^{(i)} R_i q_k,$$

with

$$\begin{aligned} \alpha_k &= v(x_k), \quad x_k \in \mathcal{V}, \quad k = 1, \ldots, n, \\ \beta_k^{(i)} &= v_i(x_k) - v(x_k), \quad k \in \mathcal{J}_i^{\Gamma_h}. \end{aligned}$$

For the representation in the basis (7.126) we have

$$v = \sum_{i=1}^{2} \sum_{k \in \mathcal{J}_i} \xi_k^{(i)} R_i q_k, \quad \text{with } \xi_k^{(i)} = v_i(x_k), \quad k \in \mathcal{J}_i.$$

LBB-stability

If the XFEM method is used for the discretization of the pressure variable in a two-phase flow problem, then the space $Q_h^{\Gamma_h}$ is combined with a finite element space for the velocity discretization. In this context the question of LBB-stability of the pair of spaces arises. As far as we know, this topic has not been investigated in the literature, yet. In our applications, for the velocity discretization we use the space $\mathbf{V}_h$ of piecewise quadratics. The Hood-Taylor pair $(\mathbf{V}_h, Q_h)$ is LBB-stable. If instead of Q_h we use the extended space $Q_h^{\Gamma_h}$ it is not known, whether the pair $(\mathbf{V}_h, Q_h^{\Gamma_h})$ is LBB-stable. Related to this we comment on results of a numerical experiment that are presented in Sect. 7.10.3. In this experiment it is observed that for the *reduced* XFEM space $\tilde{Q}_h^{\Gamma_h}$ the pair $(\mathbf{V}_h, \tilde{Q}_h^{\Gamma_h})$ has a (much) better LBB-stability property than the pair $(\mathbf{V}_h, Q_h^{\Gamma_h})$. Hence, at least in the model problem considered in Sect. 7.10.3, the concept of reduction of the original XFEM space appears to be important in view of LBB-stability. There is no theoretical analysis that explains this effect.

7.9.5 Numerical experiment with XFEM

In this experiment, for a given piecewise smooth function we compute best approximation errors for the spaces Q_h, Q_h^Γ and $\tilde{Q}_h^\Gamma$. The behavior of these approximation errors confirms the results of the theoretical analyses treated above.

We take $\Omega = (-1,1)^3$ and a planar interface $\Gamma = \{(x,y,z) \in \Omega : y+z = 0.05\}$ and $\Omega_1 = \{\, (x,y,z) \in \Omega : \; y+z < 0.05 \,\}$, $\Omega_2 = \Omega \setminus \Omega_1$. Let u be given by

$$u = \begin{cases} x^2+y^2+z^2 & \text{in } \Omega_1 \\ 3x^2+y^2+2z^2+2 & \text{in } \Omega_2. \end{cases}$$

We use a *uniform* triangulation of Ω with tetrahedra, resulting in a family $\{\mathcal{T}_{h_i}\}_{i \geq 0}$ with mesh size parameter $h = h_i = 2^{-i-1}$, $i = 0,1,2,\ldots$. The interface Γ and the triangulations are such that Γ is not aligned with the triangulation. Let Q_h be the space of continuous piecewise linear functions on $\mathcal{T}_h$ and Q_h^Γ, $\tilde{Q}_h^\Gamma$ the corresponding XFEM and reduced XFEM spaces, respectively. In the criterion (7.110) that is used in the construction of the space $\tilde{Q}_h^\Gamma$ the parameters l, α and $\hat{c}$ have to be chosen. We consider approximation errors in the L^2-norm and therefore we take $l = 0$ and $\alpha = 2$. We present results for different values of the cut-off parameter $\hat{c}$. Note that for $\hat{c} = 0$ we have $\tilde{Q}_h^\Gamma = Q_h^\Gamma$ (all discontinuous basis functions are kept) and for a sufficiently large $\hat{c}$ we have $\tilde{Q}_h^\Gamma = Q_h$ (all discontinuous basis functions are deleted). For $W_h \in \{Q_h, Q_h^\Gamma, \tilde{Q}_h^\Gamma\}$ we compute the best approximation of u in W_h, i.e. $u_h \in W_h$ such that

$$\|u - u_h\|_{L^2} = \inf_{w_h \in W_h} \|u - w_h\|_{L^2}.$$

Results for the approximation error $e_h := \|u - u_h\|_{L^2}$ are given in Table 7.9, Table 7.10. In the latter table we use the construction of the reduced space $\tilde{Q}_h^\Gamma$ based on the criterion (7.110) ($l = 0$, $\alpha = 2$) with different constants $\hat{c} = 10$, 1, 0.1. One-dimensional profiles of $u_h \in Q_h$ and $u_h \in Q_h^\Gamma$ are shown in Fig. 7.14.

# ref	$W_h = Q_h$	order	$W_h = Q_h^\Gamma$	order
0	1.60 E+0	-	1.44 E-1	-
1	1.20 E+0	0.41	3.71 E-2	1.96
2	8.88 E-1	0.43	9.37 E-3	1.99
3	6.27 E-1	0.50	2.35 E-3	1.99
4	4.52 E-1	0.47	5.89 E-4	2.00

Table 7.9. Approximation errors e_h for Q_h and Q_h^Γ.

The observed numerical order of convergence is consistent with the theoretically predicted improvement from $p = 0.5$ to $p = 2$. Furthermore, a good

ref#	$\hat{c} = 10$	order	$\hat{c} = 1$	order	$\hat{c} = 0.1$	order
0	1.60 E+0	-	1.60 E+0	-	1.77 E-1	-
1	1.20 E+0	0.41	2.69 E-1	2.57	4.01 E-2	2.14
2	8.88 E-1	0.43	4.72 E-2	2.51	9.37 E-3	2.10
3	1.37 E-2	6.01	8.98 E-3	2.39	2.35 E-3	1.99
4	2.60 E-3	2.40	5.89 E-4	3.93	5.89 E-4	2.00

Table 7.10. Approximation errors e_h for $\tilde{Q}_h^\Gamma$.

approximation quality appears to be not very sensitive with respect to the choice of the parameter $\hat{c}$.

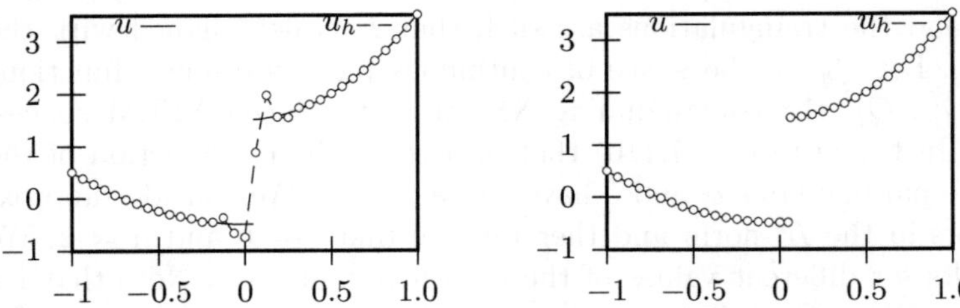

Fig. 7.14. 1D-profile of $u_h \in Q_h$ (left), $u_h \in Q_h^\Gamma$ (right) at $x = y = 0$, $h = 2^{-4}$.

The dimension of the space $\tilde{Q}_h^\Gamma$ depends on the value for $\hat{c}$. These dimensions corresponding to the spaces used in Tables 7.9 and 7.10 are given in Table 7.11.

# ref.	$\hat{c} = \infty$	$\hat{c} = 10$	$\hat{c} = 1$	$\hat{c} = 0.1$	$\hat{c} = 0$
0	125	125	125	186	205
1	729	729	872	954	1017
2	4913	4913	5730	6001	6001
3	35937	39008	39103	40161	40161
4	274625	290878	291005	291005	291005

Table 7.11. Dimension of the space $\tilde{Q}_h^\Gamma$.

Note that for not too small refinement levels i the dimension of the (modified) XFEM space is only slightly larger than that of the standard finite element space.

7.10 Numerical experiments for a Stokes problem

Due to the Laplace-Young law, typically the pressure has a jump across the interface, when surface tension forces are present ($\tau \neq 0$), cf. Remark 1.1.5 and Remark 7.10.1 below. In numerical simulations, this discontinuity and inadequate approximation of the localized surface force term often lead to strong unphysical oscillations of the velocity $\mathbf{u}_h$ at the interface, so called *spurious velocities* or *spurious currents*, cf. , e. g., [163, 111]. In this section we consider the relatively simple, but nevertheless interesting, test problem of a two-phase stationary Stokes problem (with $\mu_1 = \mu_2 = \mu$ in Ω) and investigate the discretization quality of the Laplace-Beltrami surface tension force approximation (Sect. 7.6) and of the extended finite element method for approximation of the pressure variable (Sect. 7.9.2). We will see that using the modified Laplace-Beltrami discretization $\tilde{f}_{\Gamma_h}$ and the XFEM space results in a significant reduction of the spurious velocities compared to the case where one uses f_{Γ_h} and the standard FEM space Q_h. We emphasize that these improved methods are *not* restricted to this simplified problem but apply to the general Navier-Stokes model as well.

7.10.1 A stationary Stokes test problem

For a given sufficiently smooth interface Γ, we introduce the following Stokes problem. For $\mathbf{V}_0 := H_0^1(\Omega)^3$, $Q := L_0^2(\Omega)$, find $(\mathbf{u}, p) \in \mathbf{V}_0 \times Q$ such that

$$\begin{aligned} a(\mathbf{u}, \mathbf{v}) + b(\mathbf{v}, p) &= (\rho \mathbf{g}, \mathbf{v})_{L^2} + f_\Gamma(\mathbf{v}) \qquad &\text{for all } \mathbf{v} \in \mathbf{V}_0, \\ b(\mathbf{u}, q) &= 0 &\text{for all } q \in Q, \end{aligned} \tag{7.127}$$

where

$$a(\mathbf{u}, \mathbf{v}) := \int_\Omega \mu \nabla \mathbf{u} \cdot \nabla \mathbf{v} \, d\mathbf{x}, \qquad b(\mathbf{v}, q) = -\int_\Omega q \operatorname{div} \mathbf{v} \, d\mathbf{x},$$

$$f_\Gamma(\mathbf{v}) := -\tau \int_\Gamma \kappa \mathbf{n} \cdot \mathbf{v} \, ds = -\tau \int_\Gamma \nabla_\Gamma \operatorname{id}_\Gamma \cdot \nabla_\Gamma \mathbf{v} \, ds,$$

with a viscosity $\mu > 0$ that is constant in Ω. Recall that $f_\Gamma \in \mathbf{V}_0'$. Well-posedness of this variational problem follows from the same arguments as used for the *one*-phase stationary Stokes problem in Sect. 2.2.2. Theorem 15.3.1 can be applied and yields well-posedness of the variational problem (7.127). The unique solution of this problem is denoted by $(\mathbf{u}^*, p^*) \in \mathbf{V}_0 \times Q$.

Remark 7.10.1 Assume that the domain Ω is convex. Then the problem (7.127) has a *smooth* velocity solution $\mathbf{u}^* \in \mathbf{V}_0 \cap H^2(\Omega)^3$ and a *piecewise smooth* pressure solution p with $p_{|\Omega_i} \in H^1(\Omega_i)$, $i = 1, 2$, which has a jump across Γ. These smoothness properties can be derived as follows. The curvature κ is assumed to be a smooth function (on Γ). Thus there exist

$\hat{p}_1 \in H^1(\Omega_1)$ such that $(\hat{p}_1)_{|\Gamma} = \kappa$ (in the sense of traces). Define $\hat{p} \in L^2(\Omega)$ by $\hat{p} = \hat{p}_1$ in Ω_1, $\hat{p} = 0$ on Ω_2. Note that for all $\mathbf{v} \in \mathbf{V}_0$,

$$\begin{aligned} f_\Gamma(\mathbf{v}) &= -\tau \int_\Gamma \kappa \mathbf{n}_\Gamma \cdot \mathbf{v}\, ds = -\tau \int_\Gamma \hat{p}_1 \mathbf{n}_\Gamma \cdot \mathbf{v}\, ds \\ &= -\tau \int_{\Omega_1} \hat{p}_1 \operatorname{div} \mathbf{v}\, d\mathbf{x} - \tau \int_{\Omega_1} \nabla \hat{p}_1 \cdot \mathbf{v}\, d\mathbf{x} \\ &= -\tau \int_\Omega \hat{p} \operatorname{div} \mathbf{v}\, d\mathbf{x} + \tau \int_\Omega \tilde{\mathbf{g}} \cdot \mathbf{v}\, d\mathbf{x}, \end{aligned}$$

with $\tilde{\mathbf{g}} \in L^2(\Omega)^3$ given by $\tilde{\mathbf{g}} = -\nabla \hat{p}_1$ in Ω_1, $\tilde{\mathbf{g}} = 0$ on Ω_2. Thus if $(\mathbf{u}^*, p^*)$ is the solution of (7.127) then $(\mathbf{u}^*, p^* - \tau \hat{p})$ satisfies the standard Stokes equations

$$\begin{aligned} a(\mathbf{u}^*, \mathbf{v}) + b(\mathbf{v}, p^* - \tau \hat{p}) &= (\rho \mathbf{g} + \tau \tilde{\mathbf{g}}, \mathbf{v})_{L^2} \quad \text{for all } \mathbf{v} \in \mathbf{V}_0, \\ b(\mathbf{u}^*, q) &= 0 \quad \text{for all } q \in Q. \end{aligned} \tag{7.128}$$

From regularity results on Stokes equations and the fact that Ω is convex we conclude that $\mathbf{u}^* \in H^2(\Omega)^3 \cap H_0^1(\Omega)^3$ and $p^* - \tau \hat{p} \in H^1(\Omega)$. Thus $[p^* - \tau \hat{p}]_\Gamma = 0$ (a.e. on Γ) holds, which implies

$$[p^*]_\Gamma = \tau [\hat{p}]_\Gamma = \tau \kappa,$$

i.e., p^* has a jump across Γ of the size $\tau\kappa$.

Example 7.10.2 (Static Droplet) A simple example that is used in the numerical experiments in Sect. 7.10.3 is the following. Let $\Omega := (-1,1)^3$ and Ω_1 a sphere with center at the origin and radius $r < 1$. We take $\mathbf{g} = 0$. In this case the curvature is constant, $\kappa = \frac{2}{r}$, and the solution of the Stokes problem (7.127) is given by $\mathbf{u}^* = 0$, $p^* = \tau \frac{2}{r} + c_0$ on Ω_1, $p^* = c_0$ on Ω_2 with a constant c_0 such that $\int_\Omega p^*\, dx = 0$.

Discretization error bounds

We assume that a piecewise planar surface Γ_h is known, which is close to the interface Γ in the sense of (7.70). The induced polyhedral approximations of the subdomains are $\Omega_{1,h} = \operatorname{int}(\Gamma_h)$ (region in the interior of Γ_h) and $\Omega_{2,h} = \Omega \setminus \overline{\Omega}_{1,h}$. Furthermore, we define the piecewise constant approximation of the density by $\rho_h = \rho_i$ on $\Omega_{i,h}$. We assume that for $\mathbf{v}_h \in \mathbf{V}_h$ the integrals in

$$(\rho_h \mathbf{g}, \mathbf{v}_h)_{L^2} = \rho_1 \int_{\Omega_{1,h}} \mathbf{g} \cdot \mathbf{v}_h\, dx + \rho_2 \int_{\Omega_{2,h}} \mathbf{g} \cdot \mathbf{v}_h\, dx$$

can be computed with high accuracy. This can be realized efficiently in our implementation because if one applies the standard finite element assembling strategy by using a loop over all tetrahedra $T \in \mathcal{T}_h$, then $T \cap \Omega_{i,h}$ is either empty or T or a relatively simple polygonal subdomain (due to the construction of Γ_h). For more details we refer to Sect. 7.9.3.

The discretization of (7.127) is as follows: determine $(\mathbf{u}_h, p_h) \in \mathbf{V}_h \times Q_h$ such that

$$\begin{aligned} a(\mathbf{u}_h, \mathbf{v}_h) + b(\mathbf{v}_h, p_h) &= (\rho_h \mathbf{g}, \mathbf{v}_h)_{L^2} + f_{\Gamma_h}(\mathbf{v}_h) \quad &&\text{for all } \mathbf{v}_h \in \mathbf{V}_h, \\ b(\mathbf{u}_h, q_h) &= 0 &&\text{for all } q_h \in Q_h. \end{aligned} \tag{7.129}$$

We do not restrict to a concrete pair of spaces $(\mathbf{V}_h, Q_h)$. For these spaces we (only) assume conformity $\mathbf{V}_h \subset \mathbf{V}_0$, $Q_h \subset Q$, and *LBB-stability* of the pair $(\mathbf{V}_h, Q_h)$. Approximations $f_{\Gamma_h}(\mathbf{v}_h)$ of $f_\Gamma(\mathbf{v}_h)$ are discussed in Sect. 7.6. Using standard finite element error analysis based on the Strang-lemma, cf. Sect. 15.4, we obtain the following discretization error bound.

Theorem 7.10.3 *Let* $(\mathbf{u}^*, p^*)$, $(\mathbf{u}_h, p_h)$ *be the solution of* (7.127) *and* (7.129), *respectively. Then the error bound*

$$\begin{aligned} \mu\|\mathbf{u}_h - \mathbf{u}^*\|_1 + \|p_h - p^*\|_{L^2} \le c\Big(\mu \inf_{\mathbf{v}_h \in \mathbf{V}_h} \|\mathbf{v}_h - \mathbf{u}^*\|_1 &+ \inf_{q_h \in Q_h} \|q_h - p^*\|_{L^2} \\ &+ \sup_{\mathbf{v}_h \in \mathbf{V}_h} \frac{|(\rho \mathbf{g}, \mathbf{v}_h)_{L^2} - (\rho_h \mathbf{g}, \mathbf{v}_h)_{L^2}|}{\|\mathbf{v}_h\|_1} \\ &+ \sup_{\mathbf{v}_h \in \mathbf{V}_h} \frac{|f_\Gamma(\mathbf{v}_h) - f_{\Gamma_h}(\mathbf{v}_h)|}{\|\mathbf{v}_h\|_1}\Big) \end{aligned} \tag{7.130}$$

holds with a constant c independent of h, μ and ρ.

Remark 7.10.4 Assume Ω to be convex. Then the problem (7.128) is H^2-regular and from a standard duality argument (and a scaling argument) it follows that

$$\|\mathbf{u}^* - \mathbf{u}_h\|_{L^2} \le ch\left(\|\mathbf{u}^* - \mathbf{u}_h\|_1 + \frac{1}{\mu}\|p^* - p_h\|_{L^2}\right)$$

holds, with a constant c independent of μ and h.

Corollary 7.10.5 Let $(\mathbf{u}^*, p^*)$, $(\mathbf{u}_h, p_h)$ be as in Theorem 7.10.3 and define

$$r_h := \sup_{\mathbf{v}_h \in \mathbf{V}_h} \frac{|(\rho \mathbf{g}, \mathbf{v}_h)_{L^2} - (\rho_h \mathbf{g}, \mathbf{v}_h)_{L^2}|}{\|\mathbf{v}_h\|_1} + \sup_{\mathbf{v}_h \in \mathbf{V}_h} \frac{|f_\Gamma(\mathbf{v}_h) - f_{\Gamma_h}(\mathbf{v}_h)|}{\|\mathbf{v}_h\|_1}.$$

The following holds:

$$\begin{aligned} \|\mathbf{u}_h - \mathbf{u}^*\|_1 &\le c\Big(\inf_{\mathbf{v}_h \in \mathbf{V}_h} \|\mathbf{v}_h - \mathbf{u}^*\|_1 + \frac{1}{\mu} \inf_{q_h \in Q_h} \|q_h - p^*\|_{L^2} + \frac{1}{\mu} r_h\Big), \\ \|\mathbf{u}_h - \mathbf{u}^*\|_{L^2} &\le ch\Big(\inf_{\mathbf{v}_h \in \mathbf{V}_h} \|\mathbf{v}_h - \mathbf{u}^*\|_1 + \frac{1}{\mu} \inf_{q_h \in Q_h} \|q_h - p^*\|_{L^2} + \frac{1}{\mu} r_h\Big), \\ \|p_h - p^*\|_{L^2} &\le c\Big(\mu \inf_{\mathbf{v}_h \in \mathbf{V}_h} \|\mathbf{v}_h - \mathbf{u}^*\|_1 + \inf_{q_h \in Q_h} \|q_h - p^*\|_{L^2} + r_h\Big), \end{aligned}$$

with constants c independent of h, μ and ρ. We observe that if $\mu \ll 1$ then in the velocity error we have an error amplification effect proportional to $\frac{1}{\mu}$. This effect does not occur in the discretization error for the pressure.

We comment on the terms occurring in the bound in (7.130). We start with the velocity approximation error term $\mu \inf_{\mathbf{v}_h \in \mathbf{V}_h} \|\mathbf{v}_h - \mathbf{u}^*\|_1$. Assume a situation in which the solution $\mathbf{u}^*$ of (7.127) is smooth: $\mathbf{u}^* \in H^2(\Omega)^3$. With standard finite element spaces $\mathbf{V}_h$ for the velocity (e.g., P_1 or P_2) we then obtain

$$\inf_{\mathbf{v}_h \in \mathbf{V}_h} \|\mathbf{v}_h - \mathbf{u}^*\|_1 \leq ch.$$

If $\mathbf{u}^* \in H^3(\Omega)^3$, then with quadratic finite elements the upper bound can be improved to ch^2.

Remark 7.10.6 Note that in Remark 7.10.1 the smoothness of $\mathbf{u}^*$ was shown under the assumption of equal viscosity in both phases, i. e., $\mu_1 = \mu_2$. If μ is discontinuous across Γ, then the normal derivative of $\mathbf{u}^*$ has a jump across Γ, which means that the velocity field $\mathbf{u}^*$ has a kink at Γ. If the grid is not aligned to the interface, then the approximation of such functions in standard finite element spaces $\mathbf{V}_h$ yields

$$\inf_{\mathbf{v}_h \in \mathbf{V}_h} \|\mathbf{v}_h - \mathbf{u}^*\|_1 \leq c\sqrt{h}.$$

In the case of large viscosity ratios $\max_{i=1,2} \mu_i / \min_{i=1,2} \mu_i$ (e. g., liquid-gas systems) the construction of specially adapted finite element spaces enabling first order convergence w.r.t. the H^1-norm is required, cf. [66, 166]. However, for liquid-liquid systems with small viscosity ratios the influence of this error source turns out to be rather small compared to the pressure approximation error (second term in (7.130)).

Related to the third term in (7.130) we note the following. Due to (7.70a) we get $|\operatorname{meas}_3(\Omega_i) - \operatorname{meas}_3(\Omega_{i,h})| \leq ch_\Gamma^2$, $i = 1, 2$, and using this we obtain

$$\begin{aligned} |(\rho \mathbf{g}, \mathbf{v}_h)_{L^2} - (\rho_h \mathbf{g}, \mathbf{v}_h)_{L^2}| &\leq \sum_{i=1}^{2} \rho_i \left| \int_{\Omega_i} \mathbf{g} \cdot \mathbf{v}_h \, dx - \int_{\Omega_{i,h}} \mathbf{g} \cdot \mathbf{v}_h \, dx \right| \\ &\leq c(\rho_1 + \rho_2)\, h_\Gamma \|\mathbf{v}_h\|_1, \end{aligned}$$

and thus an $\mathcal{O}(h_\Gamma)$ bound for the third term in (7.130).

The remaining two terms in (7.130) are less easy to handle. In Sect. 7.7 we treated the fourth term. It is shown that the approximation method based on the modified Laplace-Beltrami discretization $\tilde{f}_{\Gamma_h}$, cf. (7.60), results in a $\mathcal{O}(h_\Gamma)$ bound for this term whereas the Laplace-Beltrami approximation with f_{Γ_h}, cf. (7.54), only yields $\mathcal{O}(\sqrt{h_\Gamma})$.

The second term in (7.130) is treated in Sect. 7.9. It is shown that standard finite element spaces (e.g., P_0 or P_1) lead to a pressure discretization error $\inf_{q_h \in Q_h} \|q_h - p^*\|_{L^2} \sim \sqrt{h_\Gamma}$, and that for the extended finite element space (or its reduced variant) one has an L^2-error bound proportional to h_Γ^2.

Remark 7.10.7 Consider the problem as in Example 7.10.2. Then $\mathbf{u}^* = 0$, $\mathbf{g} = 0$ and the bound in (7.130) simplifies to

$$\mu \|\mathbf{u}_h\|_1 + \|p_h - p^*\|_{L^2} \le c\Big(\inf_{q_h \in Q_h} \|q_h - p^*\|_{L^2} + \sup_{\mathbf{v}_h \in \mathbf{V}_h} \frac{|f_\Gamma(\mathbf{v}_h) - f_{\Gamma_h}(\mathbf{v}_h)|}{\|\mathbf{v}_h\|_1} \Big). \tag{7.131}$$

In the following sections we consider the Galerkin discretization (7.129) of the Stokes problem with $\mathbf{g} = 0$ in the cube $\Omega = (-1,1)^3$. We assume constant viscosity $\mu = 1$. We will consider different interfaces Γ. The discrete problem is as follows: determine $\mathbf{v}_h \in \mathbf{V}_h$, $p_h \in Q_h$ such that

$$\begin{aligned} a(\mathbf{u}_h, \mathbf{v}_h) + b(\mathbf{v}_h, p_h) &= f_{\mathrm{SF},h}(\mathbf{v}_h) \quad \text{for all } \mathbf{v}_h \in \mathbf{V}_h, \\ b(\mathbf{u}_h, q_h) &= 0 \quad \text{for all } q_h \in Q_h, \end{aligned} \tag{7.132}$$

where $f_{\mathrm{SF},h} \in \mathbf{V}_h'$ are different approximations of f_Γ. We choose a uniform initial triangulation $\mathcal{T}_0$ of Ω where the vertices form a $5 \times 5 \times 5$ lattice and apply an adaptive refinement algorithm. Local refinement of the coarse mesh $\mathcal{T}_0$ in the vicinity of Γ yields the gradually refined meshes $\mathcal{T}_1, \mathcal{T}_2, \mathcal{T}_3, \mathcal{T}_4$ with local mesh sizes $h_\Gamma = h_i = 2^{-i-1}$, $i = 0, \ldots, 4$, at the interface. For the discretization of velocity we choose the standard finite element space of piecewise quadratics:

$$\mathbf{V}_h := \big\{ \mathbf{v} \in C(\Omega)^3 : \mathbf{v}|_T \in \mathcal{P}_2 \text{ for all } T \in \mathcal{T}_h, \;\; \mathbf{v}_{|\partial\Omega} = 0 \big\}.$$

We consider different choices for the pressure finite element space, namely piecewise constant or continuous piecewise linear elements, i. e., the spaces Q_h^0, Q_h^1 respectively, and the extended pressure space $Q_h^{\Gamma_h}$ introduced in Sect. 7.9.2. The discretization quality is quantified by computing norms of the errors

$$e_{\mathbf{u}} := \mathbf{u}^* - \mathbf{u}_h = -\mathbf{u}_h \quad \text{and} \quad e_p := p^* - p_h.$$

7.10.2 Test case A: Pressure jump at a planar interface

This simple test case is designed to examine interpolation errors of finite element spaces for the approximation of a discontinuous pressure variable. We consider two different interfaces Γ_1 and Γ_2, which are both planes. Γ_1 is defined by

$$\Gamma_1 = \{ x \in \Omega : x_3 = 0 \}.$$

In this case the two subdomains are given by $\Omega_1 := \{ x \in \Omega : x_3 < 0 \}$ and $\Omega_2 := \Omega \setminus \overline{\Omega}_1$, cf. Fig. 7.15. Interface Γ_2 is defined by

$$\Gamma_2 = \{ x \in \Omega : x_2 + x_3 = 1 \},$$

and the corresponding subdomains are $\Omega_1 := \{ x \in \Omega : x_2 + x_3 < 0 \}$ and $\Omega_2 := \Omega \setminus \overline{\Omega}_1$, cf. Fig. 7.17.

Since for these planar interfaces Γ_1, Γ_2 the curvature is zero we introduce an artificial surface force f_{ASF} given by

$$f_{\mathrm{ASF}}(\mathbf{v}) = -\sigma \int_{\Gamma} \mathbf{v} \cdot \mathbf{n} \, ds, \quad \mathbf{v} \in \mathbf{V},$$

with a constant $\sigma > 0$. Note that $f_{\mathrm{ASF}} \in \mathbf{V}'$. The unique solution of (7.127), with $f_\Gamma = f_{\mathrm{ASF}}$, is given by

$$\mathbf{u}^* = 0, \qquad p^* = \begin{cases} C + \sigma & \text{in } \Omega_1, \\ C & \text{in } \Omega_2. \end{cases}$$

Here C is a constant such that $\int_\Omega p^* \, dx = 0$. In the experiments below we use $\sigma = 1$. For both interfaces *the interface approximation* Γ_h *is exact*, i.e., $\Gamma_h = \Gamma$, allowing an *exact discretization* of the interfacial force, i.e., $f_{\mathrm{ASF},h} = f_{\mathrm{ASF}}$.

Due to $\mathbf{g} = 0$, $\mathbf{u}^* \in \mathbf{V}_h$ and the fact that $\|f_{\mathrm{ASF},h} - f_{\mathrm{ASF}}\|_{\mathbf{V}_h'} = 0$ the error bound (7.130) simplifies to

$$\mu \|e_{\mathbf{u}}\|_1 + \|e_p\|_{L^2} \leq c \inf_{q_h \in Q_h} \|p^* - q_h\|_{L^2}. \tag{7.133}$$

Thus the errors in velocity and pressure are solely controlled by the approximation quality of the finite element space Q_h.

The number of velocity and pressure unknowns for the grids $\mathcal{T}_0, \ldots, \mathcal{T}_4$ with different refinement levels are shown in Table 7.12. Note that $\dim Q_h^{\Gamma_h} > \dim Q_h^1$ due to the extended basis functions and that $\dim Q_h^0$ is even (much) larger.

interface	# ref.	$\dim \mathbf{V}_h$	$\dim Q_h^1$	$\dim Q_h^{\Gamma_h}$	$\dim Q_h^0$
$\Gamma = \Gamma_1$	0	1029	125	150	384
	1	6801	455	536	1984
	2	31197	1657	1946	8384
	3	131433	6235	7324	33984
	4	537717	24093	28318	136384
$\Gamma = \Gamma_2$	0	1029	125	190	384
	1	7749	543	768	2304
	2	42633	2313	3146	11556
	3	200469	9607	12808	52088
	4	871881	39229	51774	221796

Table 7.12. Dimensions of the finite element spaces for test case A.

We discuss the results obtained for the two cases $\Gamma = \Gamma_1$ and $\Gamma = \Gamma_2$.

Interface at $\Gamma = \Gamma_1$

For $\Gamma = \Gamma_1$, the interface Γ is *located at the element boundaries of tetrahedra* intersected by Γ, i.e., for each tetrahedron T intersecting Γ we have that $\Gamma \cap T$ is equal to a face of T.

In this special situation, the discontinuous pressure p^* can be represented exactly in the finite element space Q_h^0 of piecewise constants, thus the finite element solution $(\mathbf{u}_h, p_h) \in \mathbf{V}_h \times Q_h^0$ is equal to $(\mathbf{u}^*, p^*)$. This is confirmed by the numerical results: the exact solution $(\mathbf{u}^*, p^*)$ fulfills the discrete equations (up to rounding errors). The same holds for the extended finite element space $Q_h^{\Gamma_h}$.

For the space of continuous piecewise linear finite elements we have $p^* \notin Q_h^1$. The grid $\mathcal{T}_3$ and the corresponding pressure solution are shown in Figs. 7.15 and 7.16. The error norms for different grid refinement levels are shown in Table 7.13. The L^2-error of the pressure shows a decay of $\mathcal{O}(h^{1/2})$. This confirms the theoretical results for the approximation error $\min_{q \in Q_h^1} \|p^* - q_h\|_{L^2}$, cf. Sect. 7.9.1 and (7.133). The velocity error in the H^1-norm shows the same $\mathcal{O}(h^{1/2})$ behavior, whereas in the L^2-norm the error behaves like $\mathcal{O}(h^{3/2})$.

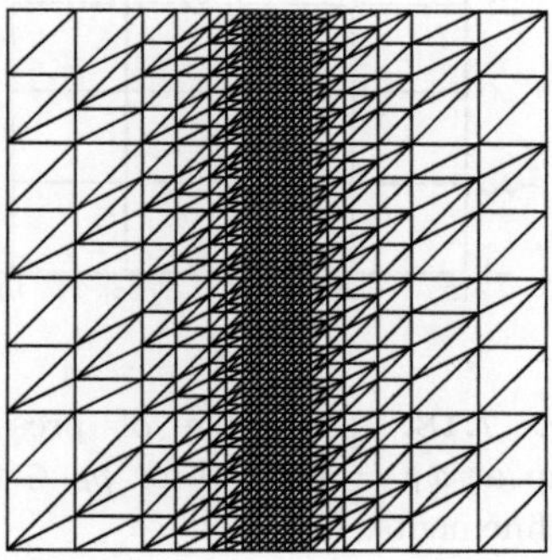

Fig. 7.15. Slice of grid at $x_1 = 0$ after 3 refinements for $\Gamma = \Gamma_1$.

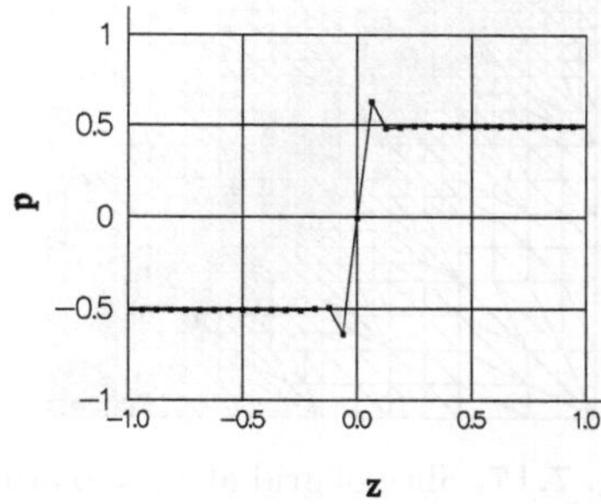

Fig. 7.16. 1D-profile of pressure jump at $x_1 = x_2 = 0$ for $p_h \in Q_h^1$. 3 refinements, $\Gamma = \Gamma_1$.

# ref.	$\|e_{\mathbf{u}}\|_{L^2}$	order	$\|e_{\mathbf{u}}\|_1$	order	$\|e_p\|_{L^2}$	order
0	4.26 E-2	–	4.26 E-1	–	5.32 E-1	–
1	1.85 E-2	1.2	3.41 E-1	0.32	3.78 E-1	0.49
2	7.09 E-3	1.38	2.55 E-1	0.42	2.68 E-1	0.5
3	2.60 E-3	1.45	1.85 E-1	0.46	1.90 E-1	0.5
4	9.37 E-4	1.47	1.33 E-1	0.48	1.34 E-1	0.5

Table 7.13. Errors for the $(\mathbf{V}_h, Q_h^1)$ finite element pair, $\Gamma = \Gamma_1$.

Interface at $\Gamma = \Gamma_2$

We now consider the case $\Gamma = \Gamma_2$. This problem corresponds to the 2D problem discussed in Sect. 7.9.1, cf. Fig. 7.10. Γ is chosen such that $\Gamma \cap F \neq F$ for all faces of the triangulations $\mathcal{T}_0, \mathcal{T}_1, \mathcal{T}_2, \mathcal{T}_3$. As a consequence, $p^* \notin Q_h^0$ and $p^* \notin Q_h^1$, but $p^* \in Q_h^{\Gamma_h}$. We checked that the finite element solution $(\mathbf{u}_h, p_h) \in \mathbf{V}_h \times Q_h^{\Gamma_h}$ is indeed equal to $(\mathbf{u}^*, p^*)$ (up to machine accuracy).

We first discuss results for P_1 finite elements. The grid $\mathcal{T}_3$, obtained after 3 times refinement, and the corresponding pressure solution for P_1 finite elements are shown in Figs. 7.17 and 7.18. The error norms for different grid refinement levels are shown in Table 7.14. The same convergence orders as for the case $\Gamma = \Gamma_1$ are obtained, cf. Table 7.13.

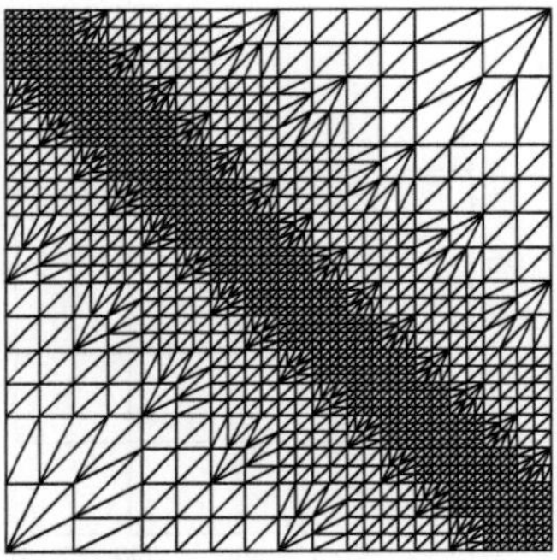

Fig. 7.17. Slice of grid at $x_1 = 0$ after 3 refinements for $\Gamma = \Gamma_2$.

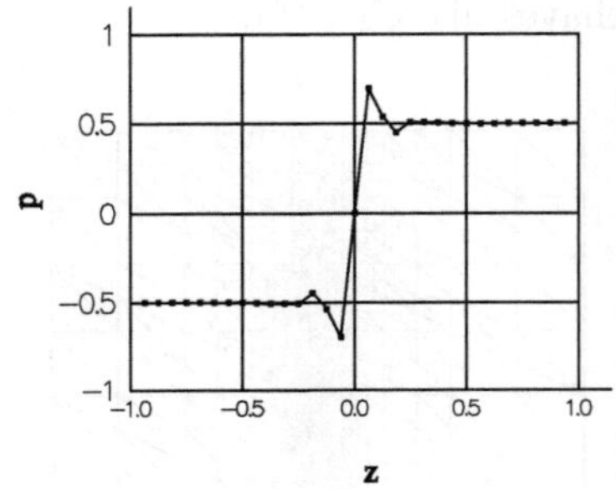

Fig. 7.18. 1D-profile of pressure jump at $x_1 = x_2 = 0$ for $p_h \in Q_h^1$. 3 refinements, $\Gamma = \Gamma_2$.

# ref.	$\lVert e_{\mathbf{u}} \rVert_{L^2}$	order	$\lVert e_{\mathbf{u}} \rVert_1$	order	$\lVert e_p \rVert_{L^2}$	order
0	2.53 E-2	–	2.56 E-1	–	5.44 E-1	–
1	1.24 E-2	1.02	2.25 E-1	0.18	3.99 E-1	0.45
2	5.03 E-3	1.31	1.75 E-1	0.36	2.88 E-1	0.47
3	1.89 E-3	1.41	1.29 E-1	0.44	2.06 E-1	0.48
4	6.88 E-4	1.46	9.35 E-2	0.47	1.46 E-1	0.49

Table 7.14. Errors for the $(\mathbf{V}_h, Q_h^1)$ finite element pair, $\Gamma = \Gamma_2$.

Results for the P_0 finite elements are shown in Table 7.15. Compared to P_1 finite elements, the errors are slightly larger but show similar convergence orders, i. e., $\mathcal{O}(h^{1/2})$ for the pressure L^2-error and velocity H^1-error, and $\mathcal{O}(h^{3/2})$ for the velocity L^2-error.

# ref.	$\|e_{\mathbf{u}}\|_{L^2}$	order	$\|e_{\mathbf{u}}\|_1$	order	$\|e_p\|_{L^2}$	order
0	3.98 E-2	–	3.49 E-1	–	7.30 E-1	–
1	1.64 E-2	1.28	2.75 E-1	0.35	4.89 E-1	0.58
2	6.14 E-3	1.41	2.04 E-1	0.43	3.35 E-1	0.54
3	2.22 E-3	1.47	1.48 E-1	0.46	2.34 E-1	0.52
4	7.92 E-4	1.49	1.06 E-1	0.48	1.65 E-1	0.51

Table 7.15. Errors for the $(\mathbf{V}_h, Q_h^0)$ finite element pair, $\Gamma = \Gamma_2$.

7.10.3 Test case B: Static droplet

In this test case (cf. Example 7.10.2) we consider a static droplet $\Omega_1 = \{ x \in \mathbb{R}^3 : \|x\| \leq r \}$ in the cube $\Omega = (-1,1)^3$ with $r = 2/3$. We assume that surface tension is present, i.e., $f_{\mathrm{SF}} = f_\Gamma$ with $\tau = 1$. This problem has the unique solution

$$\mathbf{u}^* = 0, \qquad p^* = \begin{cases} c_0 + \kappa & \text{in } \Omega_1, \\ c_0 & \text{in } \Omega_2. \end{cases}$$

Since $\kappa = 2/r$, the pressure jump is equal to $[p^*]_\Gamma = 3$. A 2D variant of this test case is presented in [111, 115, 226].

In this problem the errors in velocity and pressure are influenced by *two error sources*, namely the approximation error of the discontinuous pressure p^* in Q_h (as in test case A) and errors induced by the discretization of the surface force f_Γ, cf. (7.131).

The number of velocity and pressure unknowns for the grids $\mathcal{T}_0, \ldots, \mathcal{T}_4$ with different refinement levels are shown in Table 7.16. Note that $\dim Q_h^{\Gamma_h}$ is significantly larger than $\dim Q_h^1$, but that $\dim Q_h^{\Gamma_h} \ll \dim \mathbf{V}_h$.

#	test case B		
ref.	$\dim \mathbf{V}_h$	$\dim Q_h^1$	$\dim Q_h^{\Gamma_h}$
0	1029	125	176
1	5523	337	533
2	30297	1475	2295
3	139029	6127	9413
4	569787	24373	37355

Table 7.16. Dimensions of the finite element spaces for test case B.

We consider test case B for two different approximations of the surface tension functional f_Γ, namely the Laplace-Beltrami discretization f_{Γ_h} as in (7.54) and the modified Laplace-Beltrami discretization $\tilde{f}_{\Gamma_h}$ as in (7.60). For the pressure space we choose $Q_h = Q_h^1$ and $Q_h = Q_h^{\Gamma_h}$. We do not present results for the space Q_h^0 because these are similar to those for Q_h^1. Table 7.17 shows the decay of the pressure L^2-norm for the four different experiments. We observe poor $\mathcal{O}(h^{1/2})$ convergence in the cases where $p_h \in Q_h^1$ or when the surface tension force f_Γ is discretized by f_{Γ_h}. For the L^2 and H^1-norm of the velocity error the convergence orders are $\mathcal{O}(h^{3/2})$ and $\mathcal{O}(h^{1/2})$, respectively, which is similar to the results in test case A.

We emphasize that only for the *combination* of the extended pressure finite element space $Q_h^{\Gamma_h}$ with the improved approximation $\tilde{f}_{\Gamma_h}$ we achieve $\mathcal{O}(h^\alpha)$ convergence with $\alpha \geq 1$ for the pressure L^2-error. The velocity error in the H^1-norm shows a similar behavior (at least first order convergence), in the L^2-norm we even have second order convergence, cf. Table 7.18.

For the improved Laplace-Beltrami discretization $\tilde{f}_{\Gamma_h}$ the corresponding pressure solutions $p_h \in Q_h^1$ and $p_h \in Q_h^{\Gamma_h}$ are shown in Fig. 7.19. For the standard pressure space Q_h^1 we observe oscillations of the pressure at the interface inducing large spurious velocities in that region as shown in Fig. 7.20. For the XFEM pressure space $Q_h^{\Gamma_h}$ the pressure jump can be accurately resolved leading to a very large reduction of spurious velocities. In Fig. 7.21, on the left we illustrate these spurious velocities on the same scale as in Fig. 7.20 and on the right multiplied by a factor 20.

#	$\|e_p\|_{L^2}$ for $p_h \in Q_h^1$				$\|e_p\|_{L^2}$ for $p_h \in Q_h^{\Gamma_h}$			
ref.	f_{Γ_h}	order	$\tilde{f}_{\Gamma_h}$	order	f_{Γ_h}	order	$\tilde{f}_{\Gamma_h}$	order
0	1.60 E+0	–	1.60 E+0	–	3.12 E-1	–	1.64 E-1	–
1	1.07 E+0	0.57	1.07 E+0	0.57	1.00 E-1	1.64	4.97 E-2	1.73
2	8.23 E-1	0.38	8.23 E-1	0.38	6.24 E-2	0.68	1.66 E-2	1.58
3	5.80 E-1	0.51	5.80 E-1	0.51	4.28 E-2	0.54	7.16 E-3	1.22
4	4.13 E-1	0.49	4.13 E-1	0.49	2.95 E-2	0.54	2.83 E-3	1.34

Table 7.17. Pressure errors for the $(\mathbf{V}_h, Q_h^1)$ and $(\mathbf{V}_h, Q_h^\Gamma)$ finite element pair and different discretizations of f_Γ.

# ref.	$\|e_\mathbf{u}\|_{L^2}$	order	$\|e_\mathbf{u}\|_1$	order
0	7.16 E-3	–	1.10 E-1	–
1	1.57 E-3	2.19	4.26 E-2	1.37
2	3.25 E-4	2.28	1.70 E-2	1.33
3	8.57 E-5	1.92	7.43 E-3	1.19
4	1.75 E-5	2.29	2.40 E-3	1.63

Table 7.18. Errors and numerical order of convergence for the $(\mathbf{V}_h, Q_h^\Gamma)$ finite element pair and improved Laplace-Beltrami discretization $\tilde{f}_{\Gamma_h}$.

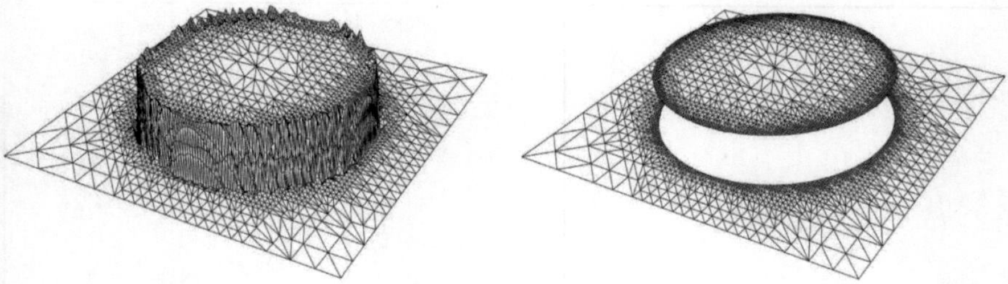

Fig. 7.19. Finite element pressure solution $p_h \in Q_h^1$ (left) and $p_h \in Q_h^{\Gamma_h}$ (right), visualized on slice at $x_3 = 0$.

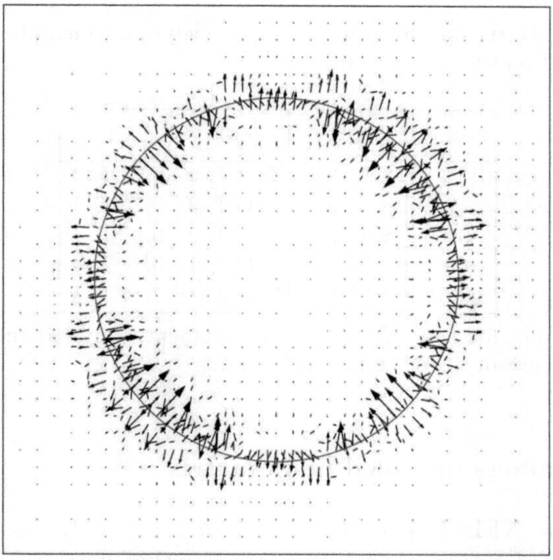

Fig. 7.20. Velocity $\mathbf{u}_h$ for the case $p_h \in Q_h^1$, visualized on slice at $x_3 = 0$.

μ-dependence of the errors

We repeated the computations of $(\mathbf{u}_h, p_h) \in \mathbf{V}_h \times Q_h^{\Gamma_h}$ for the improved Laplace-Beltrami discretization $\tilde{f}_{\Gamma_h}$ on the fixed grid $\mathcal{T}_3$ varying the viscosity μ. The errors are given in Table 7.19. We clearly observe that the velocity errors are proportional to μ^{-1} whereas the pressure error is independent of μ. This confirms the bound in (7.131).

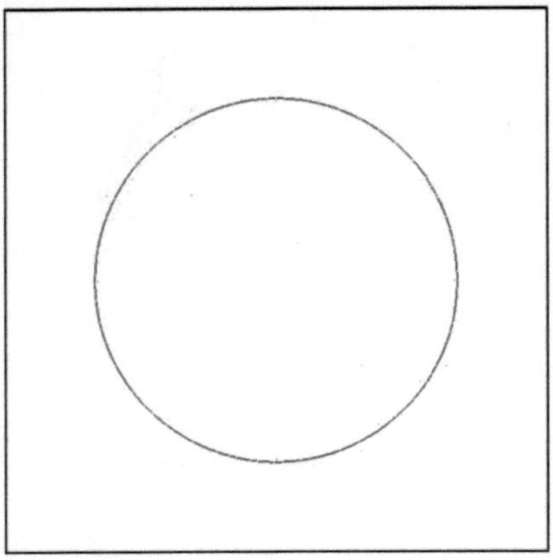

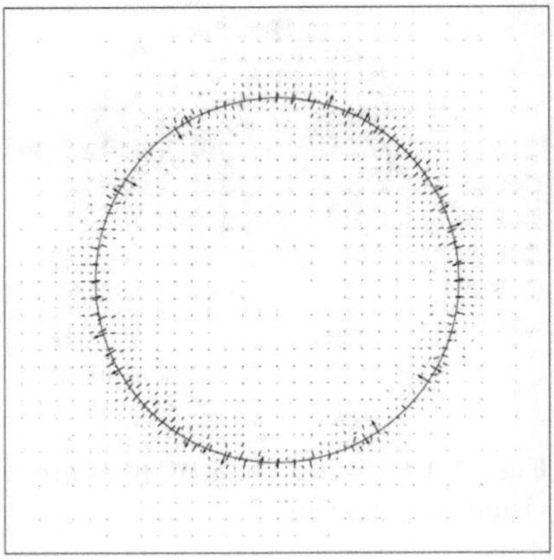

Fig. 7.21. Velocity $\mathbf{u}_h$ for the case $p_h \in Q_h^{\Gamma_h}$ (left) and magnified by a factor 20 (right), visualized on slice at $x_3 = 0$.

μ	$\|e_{\mathbf{u}}\|_{L^2}$	$\|e_{\mathbf{u}}\|_1$	$\|e_p\|_{L^2}$
10	8.62 E-6	7.51 E-4	8.71 E-3
1	8.57 E-5	7.43 E-3	7.16 E-3
0.1	8.58 E-4	7.44 E-2	6.87 E-3
0.01	8.57 E-3	7.44 E-1	6.88 E-3
0.001	8.57 E-2	7.43 E+0	7.16 E-3

Table 7.19. Errors for the $(\mathbf{V}_h, Q_h^\Gamma)$ finite element pair and improved Laplace-Beltrami discretization $\tilde{f}_{\Gamma_h}$ on $\mathcal{T}_3$ for different viscosities μ.

Condition numbers of scaled mass matrix

We consider the XFEM space $Q_h^{\Gamma_h}$, $h = h_i = 2^{-i-1}$, $i = 0, \ldots, 4$, used in the static droplet example from above. For this space we determined the mass matrix $\mathbf{M}_h$. With $\mathbf{D}_h := \mathrm{diag}(\mathbf{M}_h)$ we computed the spectral condition number of $\mathbf{D}_h^{-1}\mathbf{M}_h$, i.e., $\mathrm{cond}(\mathbf{D}_h^{-1}\mathbf{M}_h) = \lambda_{\max}(\mathbf{D}_h^{-1}\mathbf{M}_h)/\lambda_{\min}(\mathbf{D}_h^{-1}\mathbf{M}_h)$. For $h = h_i$, $i = 0, \ldots, 4$, the results are given in Table 7.20.

i	$\mathrm{cond}(\mathbf{D}_h^{-1}\mathbf{M}_h)$
0	16.16
1	11.24
2	12.08
3	12.93
4	12.98

Table 7.20. Spectral condition number of the scaled XFEM mass matrix.

These results clearly show the uniform boundedness of the spectral condition number of the scaled mass matrix, as proved in Theorem 7.9.7.

LBB-stability

In the theoretical analysis, in particular in Theorem 7.10.3, we assume that the pair of spaces that is used is LBB-stable. The standard P_2-P_1 Hood-Taylor pair $(\mathbf{V}_h, Q_h^1)$ is known to be LBB-stable. An obvious question is what happens with stability if for the pressure instead of Q_h^1 we take the (larger) space $\tilde{Q}_h^\Gamma$. We do not have a satisfactory theoretical analysis of this stability issue, yet. Here we present results of a numerical experiment for the static droplet example from above. We consider this problem with the discretization pair $(\mathbf{V}_h, \tilde{Q}_h^{\Gamma_h})$. The matrix representation of this discrete problem leads to a symmetric saddle point problem of the form

$$\mathbf{K}_h = \begin{pmatrix} \mathbf{A}_h & \mathbf{B}_h^T \\ \mathbf{B}_h & 0 \end{pmatrix}.$$

Recall that $h = h_i = 2^{-i-1}$, $i = 0, \ldots, 4$. The Schur complement matrix is given by $\mathbf{S}_h = \mathbf{B}_h \mathbf{A}_h^{-1} \mathbf{B}_h^T$. The LBB-constant for the $(\mathbf{V}_h, \tilde{Q}_h^{\Gamma_h})$ pair with $h = h_i$ is given by

$$C_{LBB}(i) = \inf_{p_h \in \tilde{Q}_h^{\Gamma_h,*}} \sup_{\mathbf{v} \in \mathbf{V}_h} \frac{(\operatorname{div} \mathbf{v}_h, p_h)_{L^2}}{\|\nabla \mathbf{v}_h\|_{L^2} \|p_h\|_{L^2}},$$

where $\tilde{Q}_h^{\Gamma_h,*}$ contains all functions from $\tilde{Q}_h^{\Gamma_h}$ that are L^2-orthogonal to the constant. Let $\mathbf{M}_h$ be the mass matrix in $\tilde{Q}_h^{\Gamma_h}$ and $m = m_i = \dim(\tilde{Q}_h^{\Gamma_h})$. Define $\mathbb{R}^{m,*} = \{\, \mathbf{y} \in \mathbb{R}^m : \langle \mathbf{y}, \mathbf{M}_h \mathbf{e} \rangle = 0 \,\}$, with $\mathbf{e} := (1, 1, \ldots, 1)^T$. The LBB constant can also be represented as follows, cf. (5.106),

$$C_{LBB}^2(i) = \inf_{\mathbf{y} \in \mathbb{R}^{m,*}} \frac{\langle \mathbf{S}_h \mathbf{y}, \mathbf{y} \rangle}{\langle \mathbf{M}_h \mathbf{y}, \mathbf{y} \rangle}, \tag{7.134}$$

and thus $C_{LBB}^2(i)$ is the smallest nonzero eigenvalue of $\mathbf{M}_h^{-1}\mathbf{S}_h$. Due to the fact that $\mathbf{M}_h$ is uniformly spectrally equivalent to its diagonal $\mathbf{D}_h$ we can instead consider the smallest nonzero eigenvalue of $\mathbf{D}_h^{-\frac{1}{2}} \mathbf{S}_h \mathbf{D}_h^{-\frac{1}{2}}$ which is denoted by $\lambda_{\min}^*(\mathbf{D}_h^{-1}\mathbf{S}_h)$. This eigenvalue can be approximated accurately using, for example, an inverse power iteration. In each iteration of this method the linear systems with matrix $\mathbf{D}_h^{-\frac{1}{2}} \mathbf{S}_h \mathbf{D}_h^{-\frac{1}{2}}$ can be solved using a CG method. We implemented this and computed (with sufficiently high accuracy) this smallest eigenvalue for several mesh sizes and for different values of the "cut-off" parameter $\hat{c}$ used in the definition of $\tilde{Q}_h^{\Gamma_h}$, cf. (7.110). The resulting values are presented in Table 7.21. Note that $\hat{c} = \infty$ corresponds to the space $Q_h = Q_h^1$. The rather irregular behavior in the columns in Table 7.21 may be caused by the fact that we compute the smallest nonzero eigenvalue of $\mathbf{D}_h^{-1}\mathbf{S}_h$ and not of $\mathbf{M}_h^{-1}\mathbf{S}_h$. We observe that for the full extended space $Q_h^{\Gamma_h}$, which coincides with $\tilde{Q}_h^{\Gamma_h}$ for $\hat{c} = 10^{-4}$, the LBB quantity $C_{LBB}^2(i)$ strongly deteriorates if i is increased. Hence, we conclude that with respect to LBB-stability it seems to be important (at least in this experiment) to use the *reduced* XFEM space $\tilde{Q}_h^{\Gamma_h}$ with a not too small parameter $\hat{c}$.

i	$\hat{c} = \infty$	$\hat{c} = 10$	$\hat{c} = 1$	$\hat{c} = 0.1$	$\hat{c} = 0.01$	$\hat{c} = 0.0001$
0	9.53 E-2	9.53 E-2	9.53 E-2	4.65 E-2	1.43 E-2	1.43 E-2
1	2.53 E-2	2.53 E-2	2.53 E-2	2.53 E-2	1.53 E-2	6.49 E-3
2	3.22 E-2	3.22 E-2	3.22 E-2	2.97 E-2	1.07 E-2	1.97 E-4
3	2.58 E-2	2.58 E-2	2.58 E-2	2.16 E-2	3.17 E-3	3.37 E-5
4	9.17 E-2	9.17 E-2	5.91 E-2	1.12 E-3	1.60 E-3	1.32 E-5

Table 7.21. Estimates of smallest nonzero eigenvalue of preconditioned Schur complement $\mathbf{D}_h^{-1}\mathbf{S}_h$.

7.11 Finite element discretization of two-phase flow problem

7.11.1 Spatial finite element discretization

In this section we combine the methods described in the previous sections to obtain a *semi*-discretization of a two-phase flow model. We recall the model given in (6.59): Find $\mathbf{u}(t) = \mathbf{u}(\cdot, t) \in \mathbf{V}_D$, $p(t) = p(\cdot, t) \in Q$, $\phi(t) = \phi(\cdot, t) \in W_{\mathbf{u},D}$ such that for almost all $t \in [0, T]$

$$\begin{aligned} m(\frac{\partial \mathbf{u}}{\partial t}, \mathbf{v}) &+ a(\mathbf{u}, \mathbf{v}) \\ + c(\mathbf{u}; \mathbf{u}, \mathbf{v}) + b(\mathbf{v}, p) &= (\rho \mathbf{g}, \mathbf{v})_{L^2} + f_\Gamma(\mathbf{v}) \quad \text{for all } \mathbf{v} \in \mathbf{V}_0, \\ b(\mathbf{u}, q) &= 0 \quad \text{for all } q \in Q, \end{aligned} \tag{7.135}$$

$$(\frac{\partial \phi}{\partial t}, v)_{L^2} + (\mathbf{u} \cdot \nabla \phi, v)_{L^2} = 0 \quad \text{for all } v \in L^2(\Omega),$$

together with initial conditions $\mathbf{u}(0) = \mathbf{u}_0$, $\phi(0) = \phi_0$ in Ω. The notation is as in Sect. 6.3:

$$\begin{aligned} \mathbf{V} &:= H^1(\Omega)^3, \\ \mathbf{V}_0 &:= \{ \mathbf{v} \in \mathbf{V} : \mathbf{v} = 0 \text{ on } \partial\Omega_D \}, \\ \mathbf{V}_D &:= \{ \mathbf{v} \in \mathbf{V} : \mathbf{v} = \mathbf{u}_D \text{ on } \partial\Omega_D \}, \\ Q &:= L_0^2(\Omega) = \left\{ q \in L^2(\Omega) : \int_\Omega q \, dx = 0 \right\}, \\ W_{\mathbf{u},D} &:= \{ w \in L^2(\Omega) : \mathbf{u} \cdot \nabla w \in L^2(\Omega), \ w_{|\partial\Omega_{in}} = \phi_D \}, \end{aligned}$$

and

$$m(\mathbf{u},\mathbf{v}) := \int_\Omega \rho\, \mathbf{u}\mathbf{v}\, dx,$$

$$a(\mathbf{u},\mathbf{v}) := \frac{1}{2}\int_\Omega \mu \operatorname{tr}\big(\mathbf{D}(\mathbf{u})\mathbf{D}(\mathbf{v})\big)\, dx,$$

$$b(\mathbf{v},q) := -\int_\Omega q \operatorname{div}\mathbf{v}\, dx,$$

$$c(\mathbf{u};\mathbf{v},\mathbf{w}) := \int_\Omega \rho(\mathbf{u}\cdot\nabla\mathbf{v})\mathbf{w}\, dx,$$

$$f_\Gamma(\mathbf{v}) := -\tau\int_\Gamma \kappa\mathbf{n}\cdot\mathbf{v}\, ds.$$

For the spatial discretization of this model we use the following methods:

- We construct *nested tetrahedral triangulations* $\{\mathcal{T}_h\}$ in the same way as for the one-phase flow problem, cf. Sect. 3.1.
- We apply streamline diffusion discretization of the level set equation, cf. Sect. 7.2.2, with piecewise *quadratic* finite elements. The space of piecewise quadratics is denoted by V_h.
- A polyhedral approximation Γ_h of Γ is constructed, as described in Sect. 7.3.
- For discretization of the velocity variable $\mathbf{u}$ we use the standard FE space of piecewise quadratics. The spaces are denoted by $\mathbf{V}_h$ ($\mathbf{v}_h = 0$ on $\partial\Omega_D$) and $\mathbf{V}_{D,h}$ ($\mathbf{v}_h$ =interpolation of $\mathbf{u}_D$ on $\partial\Omega_D$).
- Discretization of f_Γ by $\tilde{f}_{\Gamma_h}$ as explained in Sect. 7.6.
- For the discretization of the pressure variable p we use the extended finite element space $\tilde{Q}_h^{\Gamma_h}$, cf. Sect. 7.9.2. Default we use the variant in which new basis functions with "very small" support are deleted from the extended space (Sect. 7.9.3).

For the Galerkin discretization of the problem in (7.135) we proceed in the same way as for the one-phase Navier-Stokes equation. The semi-discretization reads as follows: Find $\mathbf{u}_h(t) \in \mathbf{V}_{D,h}$, $p_h(t) \in \tilde{Q}_h^{\Gamma_h}$ and $\phi_h(t) \in V_h(\phi_D)$ such that for $t \in [0,T]$:

$$\begin{aligned} m(\frac{\partial \mathbf{u}_h}{\partial t}(t),\mathbf{v}_h) + a(\mathbf{u}_h(t),\mathbf{v}_h) + c(\mathbf{u}_h(t);\mathbf{u}_h(t),\mathbf{v}_h) & \\ + b(\mathbf{v}_h,p_h(t)) = m(\mathbf{g},\mathbf{v}_h) + \tilde{f}_{\Gamma_h}(\mathbf{v}_h) \quad \forall\, \mathbf{v}_h \in \mathbf{V}_h, & \\ b(\mathbf{u}_h(t),q_h) = 0 \qquad \forall\, q_h \in Q_h^{\Gamma_h}, & \qquad (7.136) \\ \sum_{T\in\mathcal{T}_h} (\frac{\partial \phi_h}{\partial t}(t) + \mathbf{u}_h(t)\cdot\nabla\phi_h(t), v_h + \delta_T \mathbf{u}_h(t)\cdot\nabla v_h)_{L^2(T)} = 0 \;\; \forall\, v_h \in V_h. & \end{aligned}$$

Clearly, this is a method of lines approach. The finite element spaces $\mathbf{V}_h$ and V_h used for discretization of the velocity and of the level set function can be considered to be *independent* of t. The level set function trial space $V_h(\phi_D)$ depends on t if the inflow boundary data ϕ_D depend on t. If at a certain time $t = T_0 > 0$ the triangulation is adapted (local refinement and/or coarsening), the computed discrete solutions at $t = T_0$ are interpolated on the new triangulations. Then the next time interval $[T_0, T_1]$ can be treated by the method of lines with fixed discretization spaces $\mathbf{V}_h$ and V_h, that differ from those used on the previous interval $[0, T_0]$. If in the two-phase flow problem the interface is *stationary* then the pressure discretization space $\tilde{Q}_h^{\Gamma_h}$ can also be considered to be *independent* of t. In the more interesting case in which there is an evolving interface there is a strong *dependence* of $\tilde{Q}_h^{\Gamma_h}$ on t. In that case a method of lines discretization as in (7.136) induces difficulties regarding the time discretization and it is more natural to use a Rothe approach, cf. Remark 4.2.2. We come back to this issue in Sect. 8.1.2.

Let $\{\boldsymbol{\xi}_j\}_{1\leq j\leq N}$, $\{\psi_j\}_{1\leq j\leq K}$ and $\{\xi_j\}_{1\leq j\leq L}$ be (nodal) bases of $\mathbf{V}_h$, $\tilde{Q}_h^{\Gamma_h}$ and V_h, respectively. We emphasize again, that in case of an evolving interface we have $\tilde{Q}_h^{\Gamma_h} = \tilde{Q}_h^{\Gamma_h}(t)$ and thus in particular $K = K(t)$. The bases induce corresponding representations of the finite element functions in vector form. Functions $\mathbf{u}_h(t) \in \mathbf{V}_h$, $p_h(t) \in Q_h$ and $\phi_h(t) \in V_h$ can be represented as:

$$\mathbf{u}_h(t) = \sum_{j=1}^{N} u_j(t)\boldsymbol{\xi}_j, \quad \vec{\mathbf{u}}(t) := (u_1(t), \dots, u_N(t)),$$

$$p_h(t) = \sum_{j=1}^{K} p_j(t)\psi_j, \quad \vec{\mathbf{p}}(t) := (p_1(t), \dots, p_K(t)),$$

$$\phi_h(t) = \sum_{j=1}^{L} \phi_j(t)\xi_j + b_h(t), \quad \vec{\boldsymbol{\phi}}(t) := (\phi_1(t), \dots, \phi_L(t)),$$

with $b_h(t) \in V_h(\phi_D)$ such that $b_h(t)(x) = \phi_D(x,t)$ for all $x \in \mathcal{V}(\partial\Omega_{in})$ and $b_h(t)(x) = 0$ for all other vertices x, cf. (7.15). For $\phi_h \in V_h(\phi_D)$ and $\mathbf{u}_h \in \mathbf{V}_h$ (or $\mathbf{V}_{D,h}$) we introduce the following (mass and stiffness) matrices:

$$\mathbf{M}(\phi_h) \in \mathbb{R}^{N\times N}, \quad \mathbf{M}(\phi_h)_{ij} = \int_\Omega \rho(\phi_h)\,\boldsymbol{\xi}_i \cdot \boldsymbol{\xi}_j \, dx,$$
$$\mathbf{A}(\phi_h) \in \mathbb{R}^{N\times N}, \quad \mathbf{A}(\phi_h)_{ij} = \frac{1}{2}\int_\Omega \mu(\phi_h)\,\mathrm{tr}\left(\mathbf{D}(\boldsymbol{\xi}_i)\mathbf{D}(\boldsymbol{\xi}_j)\right) dx,$$
$$\mathbf{B}(\phi_h) \in \mathbb{R}^{K\times N}, \quad \mathbf{B}(\phi_h)_{ij} = -\int_\Omega \psi_i \,\mathrm{div}\, \boldsymbol{\xi}_j \, dx,$$
$$\mathbf{N}(\phi_h, \mathbf{u}_h) \in \mathbb{R}^{N\times N}, \quad \mathbf{N}(\phi_h, \mathbf{u}_h)_{ij} = \int_\Omega \rho(\phi_h)\,(\mathbf{u}_h \cdot \nabla \boldsymbol{\xi}_j) \cdot \boldsymbol{\xi}_i \, dx,$$
$$\mathbf{E}(\mathbf{u}_h) \in \mathbb{R}^{L\times L}, \quad \mathbf{E}(\mathbf{u}_h)_{ij} = \sum_{T\in\mathcal{T}_h}\int_T \xi_j(\xi_i + \delta_T \mathbf{u}_h \cdot \nabla \xi_i)\, dx,$$
$$\mathbf{H}(\mathbf{u}_h) \in \mathbb{R}^{L\times L}, \quad \mathbf{H}(\mathbf{u}_h)_{ij} = \sum_{T\in\mathcal{T}_h}\int_T (\mathbf{u}_h \cdot \nabla \xi_j)(\xi_i + \delta_T \mathbf{u}_h \cdot \nabla \xi_i)\, dx.$$

We also need the following vectors:

$$\vec{\mathbf{g}}(\phi_h) \in \mathbb{R}^N, \quad \vec{\mathbf{g}}(\phi_h)_i = \int_\Omega \rho(\phi_h)\,\mathbf{g}\cdot\boldsymbol{\xi}_i \, dx,$$
$$\vec{\mathbf{f}}_{\Gamma_h}(\phi_h) \in \mathbb{R}^N, \quad \vec{\mathbf{f}}_{\Gamma_h}(\phi_h)_i = \tilde{f}_{\Gamma_h}(\boldsymbol{\xi}_i),$$
$$\mathbf{b}(\mathbf{u}_h) \in \mathbb{R}^L, \quad \mathbf{b}(\mathbf{u}_h)_i = \sum_{T\in\mathcal{T}_h}\int_T \left(\frac{\partial b_h}{\partial t} + \mathbf{u}_h \cdot \nabla b_h\right)(\xi_i + \delta_T \mathbf{u}_h \cdot \nabla \xi_i)\, dx.$$

Below we write $\mathbf{M}(\vec{\boldsymbol{\phi}}(t)) := \mathbf{M}(\phi_h)$, and similarly for other matrices and vectors. Using these notations we obtain the following equivalent formulation of the coupled system of ordinary differential equations (7.136), where for simplicity we assumed $\mathbf{u}_D = 0$: Find $\vec{\mathbf{u}}(t) \in \mathbb{R}^N$, $\vec{\mathbf{p}}(t) \in \mathbb{R}^K$ and $\vec{\boldsymbol{\phi}}(t) \in \mathbb{R}^L$ such that for all $t \in [0, T]$

$$\mathbf{M}(\vec{\boldsymbol{\phi}}(t))\frac{d\vec{\mathbf{u}}}{dt}(t) + \mathbf{A}(\vec{\boldsymbol{\phi}}(t))\vec{\mathbf{u}}(t) + \mathbf{N}(\vec{\boldsymbol{\phi}}(t), \vec{\mathbf{u}}(t))\vec{\mathbf{u}}(t) + \mathbf{B}(\vec{\boldsymbol{\phi}}(t))^T\vec{\mathbf{p}}(t) = \vec{\mathbf{g}}(\vec{\boldsymbol{\phi}}(t)) + \vec{\mathbf{f}}_{\Gamma_h}(\vec{\boldsymbol{\phi}}(t)), \tag{7.137a}$$
$$\mathbf{B}(\vec{\boldsymbol{\phi}}(t))\vec{\mathbf{u}}(t) = 0, \tag{7.137b}$$
$$\mathbf{E}(\vec{\mathbf{u}}(t))\frac{d\vec{\boldsymbol{\phi}}}{dt}(t) + \mathbf{H}(\vec{\mathbf{u}}(t))\phi(t) = -\mathbf{b}(\vec{\mathbf{u}}(t)). \tag{7.137c}$$

In addition we have initial conditions for $\vec{\mathbf{u}}$ and $\vec{\boldsymbol{\phi}}$.

7.11.2 Numerical experiment with a two-phase flow problem

In Sect. 1.3.1 we presented simulation results of a rising butanol droplet in water, which is a system with a rather small surface tension coefficient

$\tau = 1.63 \cdot 10^{-3}\, N/m$. In this section we consider a similar rising droplet example, but now for a toluene-water system, where the surface tension coefficient is about 20 times larger. Hence, compared to the butanol-water system the numerical simulation of the fluid dynamics in the toluene-water system is (much) more challenging for the applied numerical methods. Below we compare the numerical results obtained by applying the reduced XFEM pressure space $\tilde{Q}_h^{\Gamma_h}$ and the standard FEM pressure space Q_h^1 of piecewise linears.

We use the standard two-phase model described in (7.135). Consider a single toluene droplet with an initial spherical shape with radius $r = 10^{-3}\, m$ inside a rectangular tank $\Omega = [0, 12 \cdot 10^{-3}] \times [0, 30 \cdot 10^{-3}] \times [0, 12 \cdot 10^{-3}]\, m^3$ filled with water, cf. Fig. 1.8. The material properties of this two-phase system are given in Table 7.22. Note that the properties of water slightly differ from those in Table 1.1 which is due to the fact that in the real experiment the water was saturated with toluene at an equilibrium state to avoid any mass transfer between the droplet and the ambient phase. Gravitation acts in negative x_2-direction, i. e., $\mathbf{g} = (0, -9.81, 0)\, m/s^2$. Initially at rest ($\mathbf{u}_0 = 0\, m/s$) the bubble starts to rise in x_2-direction due to buoyancy effects.

quantity (unit)	toluene	water
ρ (kg/m^3)	867.5	998.8
μ $(kg/m\,s)$	$5.96 \cdot 10^{-4}$	$1.029 \cdot 10^{-3}$
τ (N/m)	$34.31 \cdot 10^{-3}$	

Table 7.22. Material properties of the system toluene/water.

For the initial triangulation $\mathcal{T}_0$ the domain Ω is subdivided into $4 \times 10 \times 4$ sub-cubes each consisting of 6 tetrahedra. Then the grid is refined four times in the vicinity of the interface Γ. As time evolves the grid is adapted to the moving interface. The velocity space $\mathbf{V}_h$ consists of piecewise quadratics and the pressure is either discretized using the reduced XFEM space $\tilde{Q}_h^{\Gamma_h}$ with $\tilde{c} = 1$ or the standard finite element space Q_h^1 consisting of piecewise linears. The surface tension force term is discretized using the modified Laplace-Beltrami discretization $\tilde{f}_{\Gamma_h}$ as in (7.60). The level set function is discretized by piecewise quadratics and streamline-diffusion stabilization. A re-initialization $\mathrm{ReInit}(\phi_h)$ is performed as defined in (7.44) with $c = 10$. Mass conservation is forced in each time step as described in Sect. 7.4.2. For time discretization the decoupled implicit Euler scheme is applied with $\Delta t = 5 \cdot 10^{-4}$, cf. (8.23).

Figure 7.22 shows the initial shape of the droplet and the droplet shapes after 10 time steps for the cases $Q_h = \tilde{Q}_h^{\Gamma_h}$ and $Q_h = Q_h^1$, respectively. While the interface is smooth using the extended pressure finite element space, it shows many "spikes" in the case of the standard pressure space. These spikes are of course non-physical and only caused by numerical oscillations at the interface,

Fig. 7.22. Initial droplet shape (left) and after 10 time steps for the XFEM case (middle) and the standard FEM case (right).

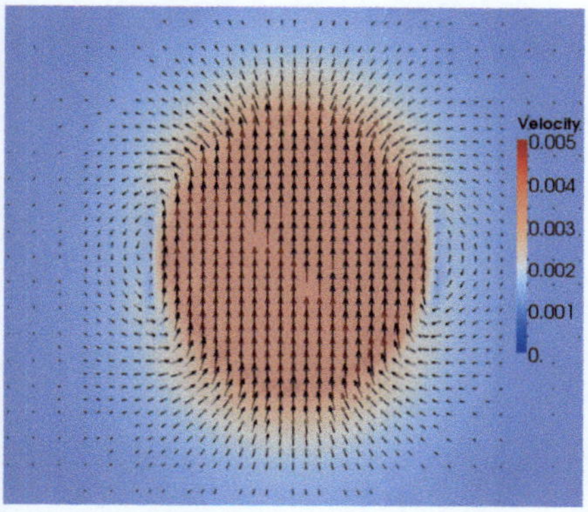

Fig. 7.23. Velocity field at interface for the XFEM case.

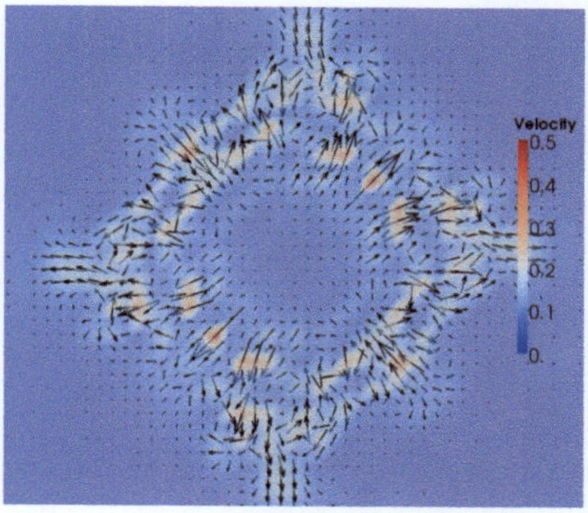

Fig. 7.24. Velocity field at interface for the standard FEM case.

so-called *spurious velocities*, which are shown in Fig. 7.24. The velocity field for the XFEM case $Q_h = \tilde{Q}_h^{\Gamma_h}$ is smooth showing the characteristic vortices, cf. Fig. 7.23. Note that the scaling of the color coding in both figures is very different, with a maximum velocity of $5 \cdot 10^{-3}\, m$ for the extended pressure space compared to $5 \cdot 10^{-1}\, m$ for the standard pressure space. These results clearly show, that for this realistic two-phase flow example the standard pressure space Q_h^1 is not suitable, whereas the (reduced) extended pressure space $\tilde{Q}_h^{\Gamma_h}$ yields satisfactory results.

8

Time integration

For the two-phase flow problem, the time discretization is based on a generalization of the θ-scheme given in Sect. 4.2 for the one-phase flow Navier-Stokes equations. This generalized method is not found in the literature and therefore we describe its derivation in detail. The need for a generalization has two reasons. Firstly, opposite to the one-phase flow problem the mass matrix $\mathbf{M}$ is no longer constant but may vary in time. Secondly, if in the discretization the XFEM space is used then the matrix $\mathbf{B}$ is in general also time dependent (due to the dynamics of the interface). In Sect. 8.1 below we present the derivation of a generalized θ-schema. In Sect. 8.2 we give a variant of an implicit Euler method in which there is in each time step a decoupling of the the unknowns $(\vec{\mathbf{u}}, \vec{\mathbf{p}})$ and $\vec{\phi}$.

8.1 A generalized θ-scheme

In this section we derive a generalized θ-scheme for the two-phase flow discrete problem in (7.137). We distinguish two cases: In the first case we allow $\mathbf{M}$ to be time dependent but assume that $\mathbf{B}$ does *not* depend on t. In the second case we allow both $\mathbf{M}$ and $\mathbf{B}$ to be time dependent. The resulting schemes turn out to be very similar.

8.1.1 Case I: B independent of time

We first consider the Navier-Stokes part in (7.137) and assume that the matrix $\mathbf{B}$ *does not depend on t*. This holds if the pressure discretization space does not depend on t, e.g. if we use Q_h instead of $\tilde{Q}_h^\Gamma$ or if the interface is stationary. For the derivation of a generalized θ-scheme we use the same approach as in Sect. 4.2. We introduce the notation

$$\mathbf{G}(\vec{\mathbf{u}}, \vec{\phi}, \vec{\mathbf{g}}, \vec{\mathbf{f}}_{\Gamma_h}) = \vec{\mathbf{g}}(\vec{\phi}(t)) + \vec{\mathbf{f}}_{\Gamma_h}(\vec{\phi}(t)) - \mathbf{A}(\vec{\phi}(t))\vec{\mathbf{u}}(t) - \mathbf{N}(\vec{\phi}(t), \vec{\mathbf{u}}(t))\vec{\mathbf{u}}(t).$$

Then the Navier-Stokes equations can be written as

$$\begin{aligned}\mathbf{M}(\vec{\phi}(t))\frac{d\vec{\mathbf{u}}}{dt}(t) + \mathbf{B}^T\vec{\mathbf{p}}(t) &= \mathbf{G}(\vec{\mathbf{u}},\vec{\phi},\vec{\mathbf{g}},\vec{\mathbf{f}}_{\Gamma_h}) \\ \mathbf{B}\vec{\mathbf{u}}(t) &= 0,\end{aligned} \tag{8.1}$$

or, equivalently,

$$\begin{aligned}\frac{d\vec{\mathbf{u}}}{dt}(t) + \mathbf{M}(\vec{\phi}(t))^{-1}\mathbf{B}^T\vec{\mathbf{p}}(t) &= \mathbf{M}(\vec{\phi}(t))^{-1}\mathbf{G}(\vec{\mathbf{u}},\vec{\phi},\vec{\mathbf{g}},\vec{\mathbf{f}}_{\Gamma_h}) \\ \mathbf{B}\vec{\mathbf{u}}(t) &= 0.\end{aligned} \tag{8.2}$$

To obtain a system of ODEs we use the same approach as in Sect. 4.2 and eliminate the algebraic equation $\mathbf{B}\vec{\mathbf{u}}(t) = 0$ and the (Lagrange multiplier) $\vec{\mathbf{p}}(t)$. In the notation we suppress the dependence on $\vec{\phi}, \vec{\mathbf{g}}$ and $\vec{\mathbf{f}}_{\Gamma_h}$ and write $\mathbf{M}(t) = \mathbf{M}(\vec{\phi}(t))$, $\mathbf{G}(\vec{\mathbf{u}},t) = \mathbf{G}(\vec{\mathbf{u}},\vec{\phi},\vec{\mathbf{g}},\vec{\mathbf{f}}_{\Gamma_h})$. From $\mathbf{B}\frac{d\vec{\mathbf{u}}}{dt}(t) = 0$ (here we used that $\mathbf{B}$ does not depend on t) and substitution of $\frac{d\vec{\mathbf{u}}}{dt}(t)$ from the first equation we obtain

$$\mathbf{S}(t)\vec{\mathbf{p}}(t) = \mathbf{B}\mathbf{M}(t)^{-1}\mathbf{G}(\vec{\mathbf{u}},t), \quad \mathbf{S}(t) := \mathbf{B}\mathbf{M}(t)^{-1}\mathbf{B}^T. \tag{8.3}$$

The matrix $\mathbf{B}^T$ has full rank and thus $\mathbf{S}(t)$ is invertible (on the subspace of FE pressure functions with $(p_h,1)_{L^2} = 0$). Using (8.3) we can eliminate $\vec{\mathbf{p}}(t)$ from the first equation in (8.2) resulting in

$$\begin{aligned}\frac{d\vec{\mathbf{u}}}{dt}(t) &= \big[\mathbf{I} - \mathbf{M}(t)^{-1}\mathbf{B}^T\mathbf{S}(t)^{-1}\mathbf{B}\big]\mathbf{M}(t)^{-1}\mathbf{G}(\vec{\mathbf{u}},t) \\ &=: \mathbf{P}(t)\mathbf{M}(t)^{-1}\mathbf{G}(\vec{\mathbf{u}},t).\end{aligned} \tag{8.4}$$

The projection $\mathbf{P}(t) = \mathbf{I} - \mathbf{M}(t)^{-1}\mathbf{B}^T\mathbf{S}(t)^{-1}\mathbf{B}$ satisfies $\mathbf{B}\mathbf{P}(t) = 0$. Hence, if $\mathbf{B}\vec{\mathbf{u}}(0) = 0$ then the solution $\vec{\mathbf{u}}(t)$ of the ordinary differential equation (8.4) remains in the subspace $\mathrm{Ker}(\mathbf{B})$. To this system of ODEs the θ-scheme is applied, resulting in the discretization

$$\begin{aligned}\frac{\vec{\mathbf{u}}^{n+1} - \vec{\mathbf{u}}^n}{\Delta t} &= \theta\mathbf{P}(t_{n+1})\mathbf{M}(t_{n+1})^{-1}\mathbf{G}(\vec{\mathbf{u}}^{n+1},t_{n+1}) \\ &\quad + (1-\theta)\mathbf{P}(t_n)\mathbf{M}(t_n)^{-1}\mathbf{G}(\vec{\mathbf{u}}^n,t_n).\end{aligned} \tag{8.5}$$

We assume that for each n this system has a unique solution $\vec{\mathbf{u}}^{n+1}$ (which is the case for Δt sufficiently small). If $\mathbf{B}\vec{\mathbf{u}}^0 = 0$ then $\mathbf{B}\vec{\mathbf{u}}^n = 0$ for all $n \geq 1$. For the implementation of this method it is convenient to eliminate the projection $\mathbf{P}$ by introducing a suitable Lagrange multiplier. Define $\vec{\mathbf{p}}^k := \mathbf{S}(t_k)^{-1}\mathbf{B}\mathbf{M}(t_k)^{-1}\mathbf{G}(\vec{\mathbf{u}}^k,t_k)$. Then (8.5) takes the form

$$\begin{aligned}\frac{\vec{\mathbf{u}}^{n+1} - \vec{\mathbf{u}}^n}{\Delta t} &= \theta\mathbf{M}(t_{n+1})^{-1}\big(\mathbf{G}(\vec{\mathbf{u}}^{n+1},t_{n+1}) - \mathbf{B}^T\vec{\mathbf{p}}^{n+1}\big) \\ &\quad + (1-\theta)\mathbf{M}(t_n)^{-1}\big(\mathbf{G}(\vec{\mathbf{u}}^n,t_n) - \mathbf{B}^T\vec{\mathbf{p}}^n\big).\end{aligned} \tag{8.6}$$

Assume that $\vec{\mathbf{u}}^0$ is such that $\mathbf{B}\vec{\mathbf{u}}^0 = 0$. The sequence $(\vec{\mathbf{u}}^n)_{n\geq 0}$ defined by the θ-scheme (8.5) satisfies (8.6) and also $\mathbf{B}\vec{\mathbf{u}}^n = 0$ for all n. We use $\vec{\mathbf{p}}^k$ as a Lagrange multiplier to enforce $\mathbf{B}\vec{\mathbf{u}}^k = 0$ as follows. Given $\vec{\mathbf{u}}^0$ with $\mathbf{B}\vec{\mathbf{u}}^0 = 0$ let $\vec{\mathbf{p}}^0$ be given by

$$\vec{\mathbf{p}}^0 = \mathbf{S}(t_0)^{-1}\mathbf{B}\mathbf{M}(t_0)^{-1}\mathbf{G}(\vec{\mathbf{u}}^0, t_0), \tag{8.7}$$

and for $n \geq 0$ let $\vec{\mathbf{u}}^{n+1}$, $\vec{\mathbf{p}}^{n+1}$ be such that

$$\begin{aligned} \frac{\vec{\mathbf{u}}^{n+1} - \vec{\mathbf{u}}^n}{\Delta t} &= \theta \mathbf{M}(t_{n+1})^{-1}\big(\mathbf{G}(\vec{\mathbf{u}}^{n+1}, t_{n+1}) - \mathbf{B}^T\vec{\mathbf{p}}^{n+1}\big) \\ &\quad + (1-\theta)\mathbf{M}(t_n)^{-1}\big(\mathbf{G}(\vec{\mathbf{u}}^n, t_n) - \mathbf{B}^T\vec{\mathbf{p}}^n\big) \\ \mathbf{B}\vec{\mathbf{u}}^{n+1} &= 0 \end{aligned} \tag{8.8}$$

holds. Note that in this saddle point type system the projection $\mathbf{P}$ is *not* used. Due to (8.6) this system has a solution. If we assume that for each n the saddle point problem (8.8) has a unique solution (which is true for Δt sufficiently small) then this yields the solution of the θ-scheme in (8.5).

For $\theta = 0$ the method in (8.8) corresponds to the explicit Euler method applied to (8.4), which is not very useful due to its poor stability properties. We consider $\theta \neq 0$. In this case, in (8.8) inverses of both $\mathbf{M}(t_{n+1})$ and $\mathbf{M}(t_n)$ occur. The latter can be avoided by introducing an additional variable, leading to a more convenient (but equivalent) formulation of (8.8). This is done as follows. Define

$$\vec{\mathbf{z}}^k := \mathbf{M}(t_k)^{-1}\big(\mathbf{G}(\vec{\mathbf{u}}^k, t_k) - \mathbf{B}^T\vec{\mathbf{p}}^k\big), \quad k \geq 0,$$

i.e.,

$$\begin{aligned} \mathbf{M}(t_0)\vec{\mathbf{z}}^0 &= \mathbf{G}(\vec{\mathbf{u}}^0, t_0) - \mathbf{B}^T\vec{\mathbf{p}}^0 \\ \theta\vec{\mathbf{z}}^{k+1} &= \frac{\vec{\mathbf{u}}^{k+1} - \vec{\mathbf{u}}^k}{\Delta t} - (1-\theta)\vec{\mathbf{z}}^k, \quad k \geq 0. \end{aligned}$$

Using this, (8.8) can be reformulated as

$$\begin{aligned} \mathbf{M}(t_{n+1})\frac{\vec{\mathbf{u}}^{n+1} - \vec{\mathbf{u}}^n}{\Delta t} + \theta\mathbf{B}^T\vec{\mathbf{p}}^{n+1} &= \theta\mathbf{G}(\vec{\mathbf{u}}^{n+1}, t_{n+1}) + (1-\theta)\mathbf{M}(t_{n+1})\vec{\mathbf{z}}^n \\ \mathbf{B}\vec{\mathbf{u}}^{n+1} &= 0 \\ \theta\vec{\mathbf{z}}^{n+1} &= \frac{\vec{\mathbf{u}}^{n+1} - \vec{\mathbf{u}}^n}{\Delta t} - (1-\theta)\vec{\mathbf{z}}^n, \end{aligned}$$

for $n \geq 0$ and a starting value $\vec{\mathbf{z}}^0$ as defined above. Besides the fact that in this reformulated version the matrix $\mathbf{M}(t_n)$ is avoided it also has further advantages in view of implementation. The operator $\mathbf{G}(\vec{\mathbf{u}}^n, t_n)$ in (8.8) contains matrices (e.g. $\mathbf{A}$) and vectors (e.g. $\vec{\mathbf{f}}_{\Gamma_h}$) that have to be computed or stored from the previous time step. This is not needed in the reformulated version.

Application of the θ-scheme to the level set equation (7.137c) results in the discretization given in (7.17).

Combining these results and inserting the notation for $\mathbf{G}$ we obtain, for $\theta \neq 0$, the following coupled nonlinear system for $(\vec{\mathbf{u}}^n, \vec{\mathbf{p}}^n, \vec{\phi}^n) \to (\vec{\mathbf{u}}^{n+1}, \vec{\mathbf{p}}^{n+1}, \vec{\phi}^{n+1})$:

Given $\vec{\mathbf{u}}^0$, $\vec{\phi}^0$, determine $\vec{\mathbf{p}}^0$, $\vec{\mathbf{z}}^0$ and $\vec{\mathbf{w}}^0$ as follows:

$$\begin{aligned}
\mathbf{G}(\vec{\mathbf{u}}^0, \vec{\phi}^0, \vec{\mathbf{g}}, \vec{\mathbf{f}}_{\Gamma_h}) &= \vec{\mathbf{g}}(\vec{\phi}^0) + \vec{\mathbf{f}}_{\Gamma_h}(\vec{\phi}^0) - \mathbf{A}(\vec{\phi}^0)\vec{\mathbf{u}}^0 - \mathbf{N}(\vec{\phi}^0, \vec{\mathbf{u}}^0)\vec{\mathbf{u}}^0 \\
\mathbf{B}\mathbf{M}(\vec{\phi}^0)^{-1}\mathbf{B}^T\vec{\mathbf{p}}^0 &= \mathbf{B}\mathbf{M}(\vec{\phi}^0)^{-1}\mathbf{G}(\vec{\mathbf{u}}^0, \vec{\phi}^0, \vec{\mathbf{g}}, \vec{\mathbf{f}}_{\Gamma_h}) \\
\mathbf{M}(\vec{\phi}^0)\vec{\mathbf{z}}^0 &= \mathbf{G}(\vec{\mathbf{u}}^0, \vec{\phi}^0, \vec{\mathbf{g}}, \vec{\mathbf{f}}_{\Gamma_h}) - \mathbf{B}^T\vec{\mathbf{p}}^0 \\
\mathbf{E}(\vec{\mathbf{u}}^0)\vec{\mathbf{w}}^0 &= -\mathbf{H}(\vec{\mathbf{u}}^0)\vec{\phi}^0 - \mathbf{b}(\vec{\mathbf{u}}^0).
\end{aligned} \tag{8.9}$$

For $n \geq 0$:

$$\begin{aligned}
&\mathbf{M}(\vec{\phi}^{n+1})\frac{\vec{\mathbf{u}}^{n+1}}{\Delta t} + \theta\big[\mathbf{A}(\vec{\phi}^{n+1}) + \mathbf{N}(\vec{\phi}^{n+1}, \vec{\mathbf{u}}^{n+1})\big]\vec{\mathbf{u}}^{n+1} + \theta\mathbf{B}^T\vec{\mathbf{p}}^{n+1} \\
&\quad = \mathbf{M}(\vec{\phi}^{n+1})\frac{\vec{\mathbf{u}}^n}{\Delta t} + \theta[\vec{\mathbf{g}}(\vec{\phi}^{n+1}) + \vec{\mathbf{f}}_{\Gamma_h}(\vec{\phi}^{n+1})] + (1-\theta)\mathbf{M}(\vec{\phi}^{n+1})\vec{\mathbf{z}}^n, \\
&\mathbf{B}\vec{\mathbf{u}}^{n+1} = 0, \\
&\mathbf{E}(\vec{\mathbf{u}}^{n+1})\frac{\vec{\phi}^{n+1}}{\Delta t} + \theta\big(\mathbf{H}(\vec{\mathbf{u}}^{n+1})\vec{\phi}^{n+1} + \mathbf{b}(\vec{\mathbf{u}}^{n+1})\big) \\
&\quad = \mathbf{E}(\vec{\mathbf{u}}^{n+1})\frac{\vec{\phi}^n}{\Delta t} + (1-\theta)\mathbf{E}(\vec{\mathbf{u}}^{n+1})\vec{\mathbf{w}}^n, \\
&\theta\vec{\mathbf{z}}^{n+1} = \frac{\vec{\mathbf{u}}^{n+1} - \vec{\mathbf{u}}^n}{\Delta t} - (1-\theta)\vec{\mathbf{z}}^n, \\
&\theta\vec{\mathbf{w}}^{n+1} = \frac{\vec{\phi}^{n+1} - \vec{\phi}^n}{\Delta t} - (1-\theta)\vec{\mathbf{w}}^n.
\end{aligned} \tag{8.10}$$

Remark 8.1.1 The derivation above shows that the scheme in (8.9)-(8.10) is a reformulation of the θ-scheme applied to the system of ODEs in (8.4). The latter is A-stable and first order accurate for $\theta \in (0, 1]$ and second order accurate for $\theta = \frac{1}{2}$. This θ-scheme (8.9)-(8.10) can be used to derive a fractional-step θ-scheme, similar to the method for the one-phase Navier-Stokes problem in Sect. 4.3, cf. Remark 4.3.1.

Special case: implicit Euler method

For the case $\theta = 1$ the scheme takes a much simpler form. In particular the sequences for $\vec{\mathbf{z}}^n$ and $\vec{\mathbf{w}}^n$ are avoided and one does not need a starting value for the pressure $\vec{\mathbf{p}}^0$, cf. (8.9). The resulting method is as follows:

Given $\vec{\mathbf{u}}^0$, $\vec{\phi}^{\,0}$, determine for $n \geq 0$:

$$\begin{aligned}
\mathbf{M}(\vec{\phi}^{\,n+1})\frac{\vec{\mathbf{u}}^{n+1}}{\Delta t} + \big[\mathbf{A}(\vec{\phi}^{\,n+1}) + \mathbf{N}(\vec{\phi}^{\,n+1}\,\vec{\mathbf{u}}^{n+1})\big]\vec{\mathbf{u}}^{n+1} + \mathbf{B}^T\vec{\mathbf{p}}^{n+1}& \\
= \mathbf{M}(\vec{\phi}^{\,n+1})\frac{\vec{\mathbf{u}}^{n}}{\Delta t} + \vec{\mathbf{g}}(\vec{\phi}^{\,n+1}) + \vec{\mathbf{f}}_{\Gamma_h}(\vec{\phi}^{\,n+1})&, \\
\mathbf{B}\vec{\mathbf{u}}^{n+1} = 0&, \\
\mathbf{E}(\vec{\mathbf{u}}^{n+1})\frac{\vec{\phi}^{\,n+1}}{\Delta t} + \mathbf{H}(\vec{\mathbf{u}}^{n+1})\vec{\phi}^{\,n+1} + \mathbf{b}(\vec{\mathbf{u}}^{n+1}) = \mathbf{E}(\vec{\mathbf{u}}^{n+1})\frac{\vec{\phi}^{\,n}}{\Delta t}&.
\end{aligned} \tag{8.11}$$

A variant of the θ-scheme

The θ-scheme presented above has many variants. We consider one particular variant that is of interest for the following two reasons. Firstly, in this method we will not need the sequence $\vec{\mathbf{z}}^n$ and secondly, this method can be generalized to the case with a time dependent $\mathbf{B}$, cf. Sect. 8.1.2. Instead of the θ-scheme in (8.5) we now use the variant:

$$\frac{\vec{\mathbf{u}}^{n+1} - \vec{\mathbf{u}}^n}{\Delta t} = \mathbf{P}(t_{n+\frac{1}{2}})\mathbf{M}(t_{n+\frac{1}{2}})^{-1}\big(\theta\mathbf{G}(\vec{\mathbf{u}}^{n+1}, t_{n+1}) + (1-\theta)\mathbf{G}(\vec{\mathbf{u}}^n, t_n)\big). \tag{8.12}$$

This method has the same consistency order and stability properties as the one in (8.5). We assume that for each n this system has a unique solution $\vec{\mathbf{u}}^{n+1}$ (which is the case for Δt sufficiently small). If $\mathbf{B}\vec{\mathbf{u}}^0 = 0$ then $\mathbf{B}\vec{\mathbf{u}}^n = 0$ for all $n \geq 1$. Again we introduce a suitable Lagrange multiplier, which is slightly different from the one used above:

$$\vec{\mathbf{p}}^{k+1} := \mathbf{S}(t_{k+\frac{1}{2}})^{-1}\mathbf{B}\mathbf{M}(t_{k+\frac{1}{2}})^{-1}\big(\theta\mathbf{G}(\vec{\mathbf{u}}^{k+1}, t_{k+1}) + (1-\theta)\mathbf{G}(\vec{\mathbf{u}}^k, t_k)\big),\ k \geq 0.$$

If $\mathbf{M}$ does *not* depend on t, then this multiplier is the same as the one used in the θ-scheme for a one-phase Navier-Stokes equation, as described in Sect. 4.2. Using this, (8.12) can be reformulated in the following equivalent form. Given $\vec{\mathbf{u}}^0$ with $\mathbf{B}\vec{\mathbf{u}}^0 = 0$, for $n \geq 0$ let $\vec{\mathbf{u}}^{n+1}$, $\vec{\mathbf{p}}^{n+1}$ be such that

$$\begin{aligned}
\mathbf{M}(t_{n+\frac{1}{2}})\frac{\vec{\mathbf{u}}^{n+1} - \vec{\mathbf{u}}^n}{\Delta t} &= \theta\mathbf{G}(\vec{\mathbf{u}}^{n+1}, t_{n+1}) \\
&\quad + (1-\theta)\mathbf{G}(\vec{\mathbf{u}}^n, t_n) - \mathbf{B}^T\vec{\mathbf{p}}^{n+1} \\
\mathbf{B}\vec{\mathbf{u}}^{n+1} &= 0.
\end{aligned} \tag{8.13}$$

Note that this is slightly different from the method in (8.8). In particular the pressure is needed only at the new time level $n+1$. Furthermore, only *one* mass matrix $\mathbf{M}(t_{n+\frac{1}{2}})$ occurs. Due to this we can avoid the variable $\vec{\mathbf{z}}^k$. In principle, the same idea can be applied to avoid the help sequence $\vec{\mathbf{w}}^k$ in the discretization of the level set equation. This is left to the reader. In practice it may be more convenient to use

$$\hat{\mathbf{M}}(t_{n+\frac{1}{2}}) := \frac{1}{2}\big(\mathbf{M}(t_{n+1}) + \mathbf{M}(t_n)\big)$$

instead of $\mathbf{M}(t_{n+\frac{1}{2}})$. The derivation above then still applies, provided in $\mathbf{P}(t_{n+\frac{1}{2}})$ the matrix $\mathbf{M}(t_{n+\frac{1}{2}})$ is replaced by $\hat{\mathbf{M}}(t_{n+\frac{1}{2}})$. Thus, instead of (8.9)-(8.10) we obtain, for $\theta \neq 0$, the following method:

Given $\vec{\phi}^{\,0}$, determine $\vec{\mathbf{w}}^0$ as follows:

$$\mathbf{E}(\vec{\mathbf{u}}^0)\vec{\mathbf{w}}^0 = -\mathbf{H}(\vec{\mathbf{u}}^0)\vec{\phi}^{\,0} - \mathbf{b}(\vec{\mathbf{u}}^0). \tag{8.14}$$

Given $\vec{\mathbf{u}}^0$, $\vec{\mathbf{w}}^0$, for $n \geq 0$:

$$\begin{aligned}
\hat{\mathbf{M}}(\vec{\phi}^{\,n+\frac{1}{2}})\frac{\vec{\mathbf{u}}^{n+1}}{\Delta t} &+ \theta\big[\mathbf{A}(\vec{\phi}^{\,n+1}) + \mathbf{N}(\vec{\phi}^{\,n+1}, \vec{\mathbf{u}}^{n+1})\big]\vec{\mathbf{u}}^{n+1} + \mathbf{B}^T\vec{\mathbf{p}}^{n+1} \\
&= \hat{\mathbf{M}}(\vec{\phi}^{\,n+\frac{1}{2}})\frac{\vec{\mathbf{u}}^{n}}{\Delta t} - (1-\theta)\big[\mathbf{A}(\vec{\phi}^{\,n}) + \mathbf{N}(\vec{\phi}^{\,n}, \vec{\mathbf{u}}^{n})\big]\vec{\mathbf{u}}^{n} \\
&\quad +\theta[\vec{\mathbf{g}}(\vec{\phi}^{\,n+1}) + \vec{\mathbf{f}}_{\Gamma_h}(\vec{\phi}^{\,n+1})] + (1-\theta)[\vec{\mathbf{g}}(\vec{\phi}^{\,n}) + \vec{\mathbf{f}}_{\Gamma_h}(\vec{\phi}^{\,n})], \\
\mathbf{B}\vec{\mathbf{u}}^{n+1} &= 0, \\
\mathbf{E}(\vec{\mathbf{u}}^{n+1})\frac{\vec{\phi}^{\,n+1}}{\Delta t} &+ \theta\big(\mathbf{H}(\vec{\mathbf{u}}^{n+1})\vec{\phi}^{\,n+1} + \mathbf{b}(\vec{\mathbf{u}}^{n+1})\big) \\
&= \mathbf{E}(\vec{\mathbf{u}}^{n+1})\frac{\vec{\phi}^{\,n}}{\Delta t} + (1-\theta)\mathbf{E}(\vec{\mathbf{u}}^{n+1})\vec{\mathbf{w}}^n, \\
\theta\vec{\mathbf{w}}^{n+1} &= \frac{\vec{\phi}^{\,n+1} - \vec{\phi}^{\,n}}{\Delta t} - (1-\theta)\vec{\mathbf{w}}^n.
\end{aligned} \tag{8.15}$$

We use the notation $\hat{\mathbf{M}}(\vec{\phi}^{\,n+\frac{1}{2}}) := \frac{1}{2}\big(\mathbf{M}(\vec{\phi}^{\,n+1}) + \mathbf{M}(\vec{\phi}^{\,n})\big)$. Compared to the scheme (8.9)-(8.10), in (8.14)-(8.15) we do not use the variable $\vec{\mathbf{z}}^k$ and we do *not* have to determine $\vec{\mathbf{p}}^0$: the scheme is of the form $(\vec{\mathbf{u}}^n, \vec{\phi}^{\,n}, \vec{\mathbf{w}}^n) \to (\vec{\mathbf{u}}^{n+1}, \vec{\mathbf{p}}^{n+1}, \vec{\phi}^{\,n+1}, \vec{\mathbf{w}}^{n+1})$. For $\theta = 1$ this method is very similar to the implicit Euler scheme in (8.11). The only difference lies in the use of $\hat{\mathbf{M}}(\vec{\phi}^{\,n+\frac{1}{2}})$ instead of $\mathbf{M}(\vec{\phi}^{\,n+1})$.

8.1.2 Case II: B may depend on time

In this section we allow both $\mathbf{M}$ and $\mathbf{B}$ to be time dependent. In our application this time dependence occurs if we use XFEM finite element spaces for pressure discretization. Then the function $t \to \mathbf{B}(t)$ is in general *not* smooth. It can even happen that for certain t_0 the dimension of $\mathbf{B}(t_0)$ is different from the dimension of $\mathbf{B}(t_0 + \epsilon)$ with $\epsilon > 0$, arbitrarily small. Due to this, it is not clear how to derive a space-time discrete problem, based on the method of lines. As already indicated in Remark 4.2.2, a Rothe approach may be more appropriate in this situation. In this section, for the general case in which both $\mathbf{M}$ and $\mathbf{B}$ may be time dependent, we derive a fully discrete problem based on the Rothe technique. For the special case that $\mathbf{B}$ does not depend on t the resulting method will turn out to be the same as the one in (8.14)-(8.15).

In the Rothe method, starting from the variational problem in (7.135) we first apply a time discretization followed by a space discretization. It suffices to consider the Navier-Stokes part in (7.135). For simplicity we only consider the case with homogeneous Dirichlet boundary conditions for velocity, i.e., $\mathbf{V}_D = \mathbf{V}_0$. Recall that $\mathbf{V}_{\text{div}} := \{\, \mathbf{v} \in \mathbf{V}_0 : \operatorname{div} \mathbf{v} = 0 \,\}$. We introduce the notation

$$\gamma(\mathbf{u}, \mathbf{v}, t) = \gamma(\mathbf{u}, \mathbf{v}, \phi, \mathbf{g}, f_\Gamma) := (\rho \mathbf{g}, \mathbf{v})_{L^2} + f_\Gamma(\mathbf{v}) - a(\mathbf{u}, \mathbf{v}) - c(\mathbf{u}; \mathbf{u}, \mathbf{v}),$$

which is the continuous analogon of the vector $\mathbf{G}(\vec{\mathbf{u}}, \vec{\boldsymbol{\phi}}, \vec{\mathbf{g}}, \vec{\mathbf{f}}_{\Gamma_h})$ used above. The projected version of the Navier-Stokes equations in (7.135) is as follows, cf. Sect. 2.2.3: find $\mathbf{u}(t) = \mathbf{u}(\cdot, t) \in \mathbf{V}_{\text{div}}$ such that

$$m(\frac{\partial \mathbf{u}}{\partial t}, \mathbf{v}) = \gamma(\mathbf{u}, \mathbf{v}, t) \quad \text{for all } \; \mathbf{v} \in \mathbf{V}_{\text{div}}. \tag{8.16}$$

For the time discretization of this problem we consider the following variant of the θ-scheme: Given $\mathbf{u}^0 = \mathbf{u}_0 \in \mathbf{V}_{\text{div}}$, for $n \geq 0$, $\mathbf{u}^{n+1} \in \mathbf{V}_{\text{div}}$ is determined by

$$\begin{aligned} &\int_\Omega \hat{\rho}(t_{n+\frac{1}{2}}) \frac{\mathbf{u}^{n+1} - \mathbf{u}^n}{\Delta t} \mathbf{v} \, dx \\ &= \theta \gamma(\mathbf{u}^{n+1}, \mathbf{v}, t_{n+1}) + (1 - \theta) \gamma(\mathbf{u}^n, \mathbf{v}, t_n) \quad \text{for all } \; \mathbf{v} \in \mathbf{V}_{\text{div}}, \end{aligned} \tag{8.17}$$

with

$$\hat{\rho}(t_{n+\frac{1}{2}}) := \frac{1}{2} \big[\rho\big(\phi(t_n)\big) + \rho\big(\phi(t_{n+1})\big) \big].$$

Instead of $\hat{\rho}(t_{n+\frac{1}{2}})$ we could also use $\rho\big(\phi(t_{n+\frac{1}{2}})\big)$. We introduce a pressure variable $p \in L_0^2(\Omega)$, i.e. instead of (8.17) we consider the following problem: determine $\mathbf{u}^{n+1} \in \mathbf{V}_0$, $p \in L_0^2(\Omega)$ such that

$$\begin{aligned} &\int_\Omega \hat{\rho}(t_{n+\frac{1}{2}}) \frac{\mathbf{u}^{n+1} - \mathbf{u}^n}{\Delta t} \mathbf{v} \, dx + b(\mathbf{v}, p) \\ &= \theta \gamma(\mathbf{u}^{n+1}, \mathbf{v}, t_{n+1}) + (1 - \theta) \gamma(\mathbf{u}^n, \mathbf{v}, t_n) \quad \text{for all } \; \mathbf{v} \in \mathbf{V}_0, \\ &b(\mathbf{u}^{n+1}, q) = 0 \quad \text{for all } \; q \in L_0^2(\Omega). \end{aligned} \tag{8.18}$$

This is a nonlinear Oseen type of problem, which, for Δt sufficiently small, has a unique solution $(\mathbf{u}^{n+1}, p)$. The velocity solution $\mathbf{u}^{n+1}$ solves (8.17), too. The method in (8.18) defines the time-discretization of the Navier-Stokes part in (7.135). To obtain a fully discrete problem the saddle point problem can be discretized by a Galerkin method in space. For this we use, for example, the pair $(\mathbf{V}_h, \tilde{Q}_h^\Gamma)$, i.e., piecewise quadratics for velocity and the (reduced) XFEM space for pressure. For the latter we use the space at the time $t = t_{n+1}$, denoted by $\tilde{Q}_h^\Gamma = \tilde{Q}_h^\Gamma(t_{n+1})$. Given an approximation $\mathbf{u}_h^n$ (not necessarily in $\mathbf{V}_h$) of $\mathbf{u}^n$, the time step $t_n \to t_{n+1}$ in (8.18) is then discretized in space as follows: determine $\mathbf{u}_h^{n+1} \in \mathbf{V}_h$, $p_h \in \tilde{Q}_h^\Gamma(t_{n+1})$ such that

$$\begin{aligned}
&\int_\Omega \hat{\rho}(t_{n+\frac{1}{2}})\frac{\mathbf{u}_h^{n+1} - \mathbf{u}_h^n}{\Delta t}\mathbf{v}_h \, dx + b(\mathbf{v}_h, p_h) \\
&= \theta\gamma(\mathbf{u}_h^{n+1}, \mathbf{v}_h, t_{n+1}) + (1-\theta)\gamma(\mathbf{u}_h^n, \mathbf{v}, t_n) \quad \text{for all } \mathbf{v}_h \in \mathbf{V}_h, \\
&b(\mathbf{u}_h^{n+1}, q_h) = 0 \quad \text{for all } q_h \in \tilde{Q}_h^\Gamma(t_{n+1}).
\end{aligned} \tag{8.19}$$

Again, in practice we use $\Gamma_h(t)$ instead of $\Gamma(t)$. Note that in the construction of $\tilde{Q}_h^{\Gamma_h}(t_{n+1})$ we need $\Gamma_h(t_{n+1})$ and thus this space is defined *implicitly* only. In the numerical realization of this fully discrete problem we need an iterative decoupling strategy between the level set equation (evolution of $\Gamma(t)$) and the Navier-Stokes problem (8.19).
If we represent this fully discrete problem using the standard bases in $\mathbf{V}_h$, $\tilde{Q}_h^{\Gamma_h}(t_{n+1})$ we obtain the following method:

Given $\vec{\phi}^{\,0}$, determine $\vec{\mathbf{w}}^0$ as follows:

$$\mathbf{E}(\vec{\mathbf{u}}^0)\vec{\mathbf{w}}^0 = -\mathbf{H}(\vec{\mathbf{u}}^0)\vec{\phi}^{\,0} - \mathbf{b}(\vec{\mathbf{u}}^0). \tag{8.20}$$

Given $\vec{\mathbf{u}}^0$, $\vec{\mathbf{w}}^0$, for $n \geq 0$:

$$\begin{aligned}
&\hat{\mathbf{M}}(\vec{\phi}^{\,n+\frac{1}{2}})\frac{\vec{\mathbf{u}}^{n+1}}{\Delta t} + \theta\big[\mathbf{A}(\vec{\phi}^{\,n+1}) + \mathbf{N}(\vec{\phi}^{\,n+1}, \vec{\mathbf{u}}^{n+1})\big]\vec{\mathbf{u}}^{n+1} + \mathbf{B}(\vec{\phi}^{\,n+1})^T\vec{\mathbf{p}}^{n+1} \\
&\quad = \hat{\mathbf{M}}(\vec{\phi}^{\,n+\frac{1}{2}})\frac{\vec{\mathbf{u}}^n}{\Delta t} - (1-\theta)\big[\mathbf{A}(\vec{\phi}^{\,n}) + \mathbf{N}(\vec{\phi}^{\,n}, \vec{\mathbf{u}}^n)\big]\vec{\mathbf{u}}^n \\
&\quad\quad + \theta[\vec{\mathbf{g}}(\vec{\phi}^{\,n+1}) + \vec{\mathbf{f}}_{\Gamma_h}(\vec{\phi}^{\,n+1})] + (1-\theta)[\vec{\mathbf{g}}(\vec{\phi}^{\,n}) + \vec{\mathbf{f}}_{\Gamma_h}(\vec{\phi}^{\,n})], \\
&\mathbf{B}(\vec{\phi}^{\,n+1})\vec{\mathbf{u}}^{n+1} = 0, \\
&\mathbf{E}(\vec{\mathbf{u}}^{n+1})\frac{\vec{\phi}^{\,n+1}}{\Delta t} + \theta\big(\mathbf{H}(\vec{\mathbf{u}}^{n+1})\vec{\phi}^{\,n+1} + \mathbf{b}(\vec{\mathbf{u}}^{n+1})\big) \\
&\quad = \mathbf{E}(\vec{\mathbf{u}}^{n+1})\frac{\vec{\phi}^{\,n}}{\Delta t} + (1-\theta)\mathbf{E}(\vec{\mathbf{u}}^{n+1})\vec{\mathbf{w}}^n, \\
&\theta\vec{\mathbf{w}}^{n+1} = \frac{\vec{\phi}^{\,n+1} - \vec{\phi}^{\,n}}{\Delta t} - (1-\theta)\vec{\mathbf{w}}^n.
\end{aligned} \tag{8.21}$$

Note that this is a generalization of the method in (8.14)-(8.15): the only difference is that we use $\mathbf{B}(\vec{\phi}^{\,n+1})$ instead of $\mathbf{B}$. In the derivation above, there is some freedom concerning the choice of the pressure discretization space. The pressure p in (8.18) is an approximation of $\theta p(t_{n+1}) + (1-\theta)p(t_n)$, cf. Sect. 4.2. Therefore, in the discretization, instead of replacing p by $p_h \in \tilde{Q}_h^\Gamma(t_{n+1})$, as is done in (8.19), it may be better to replace p by $\theta p_h^{n+1} + (1-\theta)p_h^n$, with $p_h^n \in \tilde{Q}_h^\Gamma(t_n)$ the pressure discretization from the previous time step and $p_h^{n+1} \in \tilde{Q}_h^\Gamma(t_{n+1})$ the unknown Lagrange multiplier. In the formulation (8.21), instead of $\mathbf{B}(\vec{\phi}^{\,n+1})^T\vec{\mathbf{p}}^{n+1}$ one then obtains

$$\theta\mathbf{B}(\vec{\phi}^{\,n+1})^T\vec{\mathbf{p}}^{n+1} + (1-\theta)\mathbf{B}(\vec{\phi}^{\,n})^T\vec{\mathbf{p}}^{n}.$$

In this method one needs a starting value for the pressure $\vec{\mathbf{p}}^0$, which can be determined as in (8.9). For $\theta = \frac{1}{2}$ another option is to use, instead of the two spaces $\tilde{Q}_h^\Gamma(t_n)$ and $\tilde{Q}_h^\Gamma(t_{n+1})$, the space $\tilde{Q}_h^\Gamma(t_{n+\frac{1}{2}})$, which is the XFEM space corresponding to the zero level of $\phi(t_n) + \phi(t_{n+1})$. This variant leads to a method as (8.21) in which the matrices $\mathbf{B}(\vec{\phi}^{\,n+1})$ are replaced by $\mathbf{B}(\vec{\phi}^{\,n+\frac{1}{2}})$, with $\vec{\phi}^{\,n+\frac{1}{2}} := \frac{1}{2}(\vec{\phi}^{\,n} + \vec{\phi}^{\,n+1})$.

The θ-schemes treated above can be used to derive a fractional-step θ-scheme, similar to the method for the one-phase Navier-Stokes problem in Sect. 4.3, cf. Remark 4.3.1.

Concerning (error) analyses of the time discretization methods treated in this section we note the following. For the (less interesting) case of a stationary interface the method of lines approach can be applied and the resulting method (8.9)-(8.10) is a reformulation of the θ-scheme applied to a system of ODEs. Therefore one may expect this method to have accuracy and stability properties as discussed in Sect. 4.1 for the θ-scheme. For problems with an evolving interface we derived, based on a Rothe approach, the method (8.20)-(8.21). Its (accuracy and stability) properties concerning time discretization are not clear. In Sect. 8.3 we present results of a numerical experiment with this scheme. To our knowledge a systematic analysis of time discretization methods for two-phase incompressible flow problems has not been performed, yet.

For the case of an evolving interface it is natural to use space-time finite elements for the discretization of the two-phase Navier-Stokes equations. In such a setting a space-time XFEM space for the discretization of the pressure can be used which avoids the problems caused the fact that the XFEM pressure space $\tilde{Q}_h^\Gamma(t)$ used above is time dependent. We explain this space-time XFEM space in Sect. 11.5.2. The application of this space-time approach to the two-phase flow problem is not treated here.

8.2 An implicit Euler method with decoupling

We present a variant of an implicit Euler method which is particularly attractive due to its simplicity. We present this method for the general case, i.e. both $\mathbf{M}$ and $\mathbf{B}$ may be time dependent. We consider a semi-implicit variant of the method in (8.17), in the sense that the $\mathbf{u}$ part is treated implicitly but other time dependent parts are treated explicitly. More precisely, we propose the following variant: Given $\mathbf{u}^0 = \mathbf{u}_0 \in \mathbf{V}_{\mathrm{div}}$, for $n \geq 0$, $\mathbf{u}^{n+1} \in \mathbf{V}_{\mathrm{div}}$ is determined by

$$\begin{aligned} &\int_\Omega \rho(t_n)\frac{\mathbf{u}^{n+1}-\mathbf{u}^n}{\Delta t}\mathbf{v}\,dx \\ &= \theta\gamma(\mathbf{u}^{n+1},\mathbf{v},t_n) + (1-\theta)\gamma(\mathbf{u}^n,\mathbf{v},t_n) \quad \text{for all } \mathbf{v} \in \mathbf{V}_{\mathrm{div}}. \end{aligned} \tag{8.22}$$

As above we can introduce a pressure variable $p \in L_0^2(\Omega)$, resulting in the Oseen problem: determine $\mathbf{u}^{n+1} \in \mathbf{V}_0$, $p \in L_0^2(\Omega)$ such that

$$\begin{aligned} &\int_\Omega \rho(t_n)\frac{\mathbf{u}^{n+1}-\mathbf{u}^n}{\Delta t}\mathbf{v}\,dx + b(\mathbf{v},p) \\ &\quad = \theta\gamma(\mathbf{u}^{n+1},\mathbf{v},t_n) + (1-\theta)\gamma(\mathbf{u}^n,\mathbf{v},t_n) \quad \text{for all } \mathbf{v} \in \mathbf{V}_0, \\ &b(\mathbf{u}^{n+1},q) = 0 \quad \text{for all } q \in L_0^2(\Omega). \end{aligned}$$

We discretize this saddle point problem using a Galerkin approach with spaces $\mathbf{V}_h$, $\tilde{Q}_h^\Gamma(t_n)$ (and *not* $\tilde{Q}_h^\Gamma(t_{n+1})$). The same semi-implicit treatment can be applied to the level set equation and thus we get the following time integration scheme for the coupled problem:

Given $\vec{\mathbf{u}}^0$, $\vec{\boldsymbol{\phi}}^0$, determine for $n \geq 0$:

$$\begin{aligned} \mathbf{M}(\vec{\boldsymbol{\phi}}^n)\frac{\vec{\mathbf{u}}^{n+1}}{\Delta t} &+ \left[\mathbf{A}(\vec{\boldsymbol{\phi}}^n) + \mathbf{N}(\vec{\boldsymbol{\phi}}^n, \vec{\mathbf{u}}^{n+1})\right]\vec{\mathbf{u}}^{n+1} + \mathbf{B}(\vec{\boldsymbol{\phi}}^n)^T\vec{\mathbf{p}}^{n+1} \\ &= \mathbf{M}(\vec{\boldsymbol{\phi}}^n)\frac{\vec{\mathbf{u}}^n}{\Delta t} + \vec{\mathbf{g}}(\vec{\boldsymbol{\phi}}^n) + \vec{\mathbf{f}}_{\Gamma_h}(\vec{\boldsymbol{\phi}}^n), \\ \mathbf{B}(\vec{\boldsymbol{\phi}}^n)\vec{\mathbf{u}}^{n+1} &= 0, \\ \mathbf{E}(\vec{\mathbf{u}}^n)\frac{\vec{\boldsymbol{\phi}}^{n+1}}{\Delta t} &+ \mathbf{H}(\vec{\mathbf{u}}^n)\vec{\boldsymbol{\phi}}^{n+1} + \mathbf{b}(\vec{\mathbf{u}}^n) = \mathbf{E}(\vec{\mathbf{u}}^n)\frac{\vec{\boldsymbol{\phi}}^n}{\Delta t}. \end{aligned} \tag{8.23}$$

This scheme is similar to the one in (8.11), but now the mass matrices $\mathbf{M}$ and $\mathbf{E}$ are treated explicitly (i.e. evaluated at t_n instead of t_{n+1}). In the Navier-Stokes equations the level set function is evaluated at t_n, and in the level set equation the velocity is evaluated at t_n. Due to this, per time step there is a *decoupling* between the Navier-Stokes and level set equation. Of course this strong gain in simplicity is accompanied by a loss of accuracy.

8.3 Numerical experiments

In this section we present results of a numerical experiment with different time integration schemes for a two-phase *Stokes* problem. As a test case we consider a rising droplet $\Omega_1(t)$ inside a cuboid domain $\Omega = (0,1)^2 \times (0,2)$.

For $t = 0$ the droplet is spherical with center point $x_c = (0.5, 0.5, 0.5)$ and radius 0.25, i. e., $\Omega_1(0) = \{ x \in \Omega : \|x - x_c\| < 0.25 \}$. The initial conditions are given by $\mathbf{u}(0) = 0$ and $\phi(0)$ equal to the signed distance function for $\Omega_1(0)$. Homogeneous Dirichlet boundary conditions $\mathbf{u} = 0$ are prescribed on the whole boundary $\partial\Omega$. The material parameters are chosen as follows: densities $\rho_1 = 1$, $\rho_2 = 10$, dynamic viscosities $\mu_1 = 2$, $\mu_2 = 1$ and surface tension coefficient $\tau = 0.1$. The gravitational force $\mathbf{g} = (0, 0, -10)$ acts in negative x_3-direction, hence the droplet is rising in positive x_3-direction due to buoyancy effects. Figure 8.1 shows the position of the rising droplet for $t = 0, 0.5, 1$, respectively.

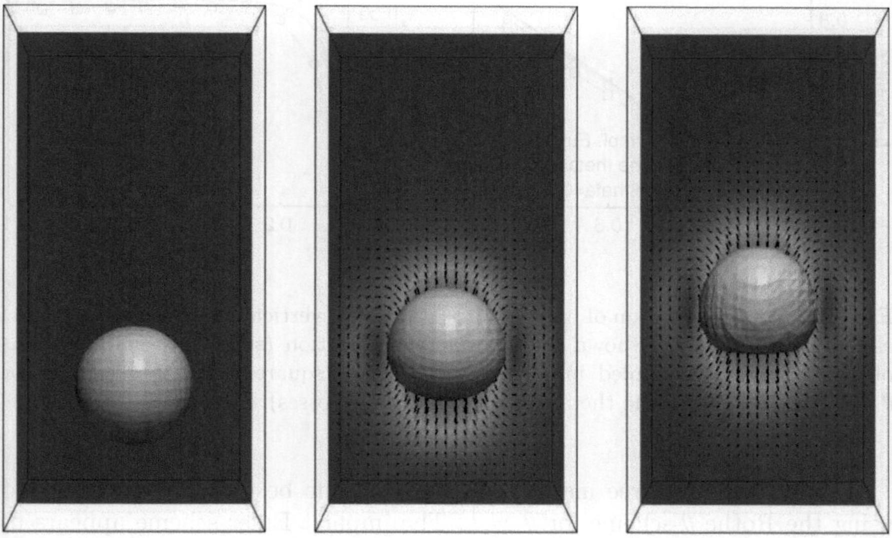

Fig. 8.1. Rising droplet for $t = 0$, 0.5, 1 (from left to right). Shown is the is the interface $\Gamma(t)$ and the velocity $\mathbf{u}$ on a cut plane $x_2 = 0.5$.

We used a time-*in*dependent triangulation (no adaptivity) obtained by a uniform subdivision of Ω in $12 \times 12 \times 24$ subcubes, each of which is subdivided into 6 tetrahedra. For spatial discretization we use piecewise quadratics for velocity and a reduced XFEM space for pressure, i.e. the pair $\mathbf{V}_h \times \tilde{Q}_h^\Gamma$. Note that the XFEM space $\tilde{Q}_h^\Gamma$ depends on time since the interface evolves in time.

We consider three time integration schemes, namely the decoupled implicit Euler scheme in (8.23) and the Rothe θ-scheme (8.20)–(8.21) for $\theta = 1$ and $\theta = \frac{1}{2}$, respectively. The different schemes are applied for time step sizes $\Delta t = \frac{1}{n_t}$ with $n_t = 10, 20, 40, 80, 160$. The mass conservation strategy described in Sect. 7.4.2 is applied in each time step, but no reparametrization of the level set function is performed, cf. Sect. 7.4.1. The computed approximations $\vec{\mathbf{u}}^{n_t}$ of $\mathbf{u}(1)$ are compared to a reference solution $\vec{\mathbf{u}}^{\mathrm{ref}}$ which is computed by the scheme (8.20)–(8.21) with $\theta = \frac{1}{2}$ and time step size $\Delta t = 10^{-3}$. The barycenter and the rise velocity of the droplet $\Omega_1(t)$ are defined by

$$\bar{x}(t) = |\Omega_1(t)|^{-1} \int_{\Omega_1(t)} x \, dx, \quad \bar{\mathbf{u}}(t) = |\Omega_1(t)|^{-1} \int_{\Omega_1(t)} \mathbf{u} \, dx.$$

The third (=vertical) component of these quantities is plotted in Fig. 8.2 as a function of time t. The results of the three time discretization schemes with $\Delta t = 0.1$ are shown, as well as the results for the reference solution.

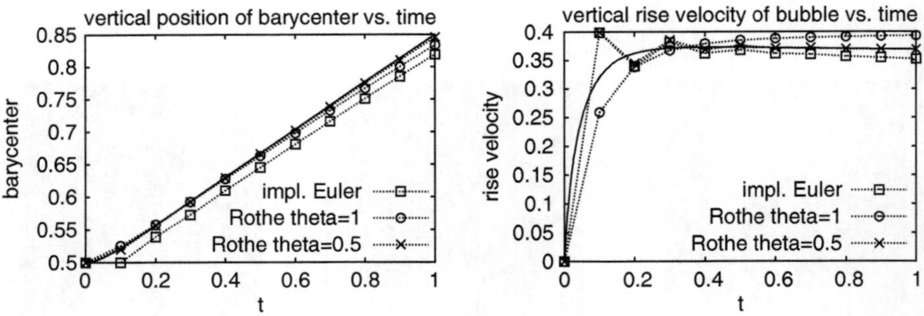

Fig. 8.2. Vertical position of barycenter $\bar{x}_3$ (left) and vertical rise velocity $\bar{\mathbf{u}}_3$ (right) as a function of time. Shown are the reference solution (solid line) and solutions obtained by the decoupled implicit Euler scheme (squares), Rothe θ-scheme for $\theta = 1$ (circles) and Rothe theta-scheme for $\theta = \frac{1}{2}$ (crosses) for $n_t = 10$ time steps.

Comparing the three methods we see that the best results are obtained using the Rothe θ-scheme for $\theta = \frac{1}{2}$. The implicit Euler scheme appears to have a time lag for the position of the barycenter. Further experiments show that this time lag is of the order of magnitude of Δt. This time lag effect is probably due to the (too) strong decoupling that is used in this scheme. Comparing the results for the rise velocity of the droplet, the Euler scheme leads to an under-estimation whereas the Rothe θ-scheme for $\theta = 1$ over-estimates this value. Again the Rothe θ-scheme for $\theta = \frac{1}{2}$ shows a very good match with the reference solution.
Tables 8.1–8.3 show the convergence rates for the the three time discretization schemes.

The results show first order convergence, both in the L^2- and H^1-norm, for the decoupled implicit Euler scheme and the Rothe θ-scheme with $\theta = 1$. For the Rothe θ-scheme with $\theta = \frac{1}{2}$ the observed order fluctuates between first and second order. The reason for this behavior is not understood, yet, but is probably related to the pressure XFEM space. When using the standard piecewise linear finite element space for the pressure, second order convergence is observed.

n_t	$\|\vec{\mathbf{u}}^{\text{ref}} - \vec{\mathbf{u}}^{n_t}\|_{L^2}$	order	$\|\vec{\mathbf{u}}^{\text{ref}} - \vec{\mathbf{u}}^{n_t}\|_1$	order
10	1.57 E-2	—	2.48 E-1	—
20	7.65 E-3	1.04	1.27 E-1	0.96
40	3.79 E-3	1.01	6.67 E-2	0.93
80	1.88 E-3	1.01	3.23 E-2	1.05
160	9.34 E-4	1.01	1.66 E-2	0.96

Table 8.1. Convergence behavior of the decoupled implicit Euler scheme (8.23) w.r.t. time step size.

n_t	$\|\vec{\mathbf{u}}^{\text{ref}} - \vec{\mathbf{u}}^{n_t}\|_{L^2}$	order	$\|\vec{\mathbf{u}}^{\text{ref}} - \vec{\mathbf{u}}^{n_t}\|_1$	order
10	1.33 E-2	—	1.77 E-1	—
20	6.90 E-3	0.95	9.59 E-2	0.88
40	3.53 E-3	0.96	5.02 E-2	0.93
80	1.81 E-3	0.97	2.61 E-2	0.94
160	9.19 E-4	0.98	1.34 E-2	0.97

Table 8.2. Convergence behavior of the Rothe θ-scheme (8.20)–(8.21) for $\theta = 1$ w.r.t. time step size.

n_t	$\|\vec{\mathbf{u}}^{\text{ref}} - \vec{\mathbf{u}}^{n_t}\|_{L^2}$	order	$\|\vec{\mathbf{u}}^{\text{ref}} - \vec{\mathbf{u}}^{n_t}\|_1$	order
10	2.24 E-3	—	7.57 E-2	—
20	6.53 E-4	1.78	3.81 E-2	0.99
40	2.63 E-4	1.31	2.05 E-2	0.89
80	9.21 E-5	1.52	8.00 E-3	1.36
160	2.42 E-5	1.93	2.41 E-3	1.73

Table 8.3. Convergence behavior of the Rothe θ-scheme (8.20)–(8.21) for $\theta = \frac{1}{2}$ w.r.t. time step size.

9

Iterative solvers

In this chapter we address the issue of iterative solvers for the coupled nonlinear system of equations that arises in each time step of the implicit time integration methods treated in Chap. 8.

9.1 Decoupling and linearization

In a two-phase flow problem, besides the nonlinear system for the unknowns $\vec{\mathbf{u}}, \vec{\mathbf{p}}$, we also have, in each time step, the nonlinear coupling between the flow variable $\vec{\mathbf{u}}$ and the level set function $\vec{\phi}$. One usually applies an iterative strategy in which per iteration the unknown $\vec{\mathbf{u}}$ is decoupled from $\vec{\phi}$. To explain this in more detail we consider a time step in the θ-scheme (8.10). Other time integration methods can be treated in a very similar manner. The nonlinear system for $\vec{\mathbf{u}}^{n+1}, \vec{\mathbf{p}}^{n+1}, \vec{\phi}^{\,n+1}$ in (8.10) is of the form

$$\begin{cases} [\frac{1}{\Delta t}\mathbf{M} + \theta\mathbf{A}](\vec{\phi}^{\,n+1})\vec{\mathbf{u}}^{n+1} + \theta\mathbf{N}(\vec{\phi}^{\,n+1}, \vec{\mathbf{u}}^{n+1})\vec{\mathbf{u}}^{n+1} + \theta\mathbf{B}^T\vec{\mathbf{p}}^{n+1} \\ = \theta[\vec{\mathbf{g}} + \vec{\mathbf{f}}_{\Gamma_h}](\vec{\phi}^{\,n+1}) + \mathbf{M}(\vec{\phi}^{\,n+1})(\frac{\vec{\mathbf{u}}^n}{\Delta t} + (1-\theta)\vec{\mathbf{z}}^n) \\ \mathbf{B}\vec{\mathbf{u}}^{n+1} = 0 \\ [\frac{1}{\Delta t}\mathbf{E} + \theta\mathbf{H}](\vec{\mathbf{u}}^{n+1})\vec{\phi}^{\,n+1} = \mathbf{E}(\vec{\mathbf{u}}^{n+1})(\frac{\vec{\phi}^{\,n}}{\Delta t} + (1-\theta)\vec{\mathbf{w}}^n) - \theta\mathbf{b}(\vec{\mathbf{u}}^{n+1}). \end{cases}$$

If one uses the XFEM approach for pressure discretization then *the matrix* $\mathbf{B}$ *depends on the position of the interface and has to be replaced by* $\mathbf{B}(\vec{\phi}^{\,n+1})$, cf. (8.20)-(8.21). The nonlinear coupling between $\vec{\mathbf{u}}^{n+1}$ and $\vec{\phi}^{\,n+1}$ can be treated by several decoupling strategies. We treat a few methods. For this we first introduce some notation that simplifies the presentation. At $t = t_n$ the values of $\vec{\mathbf{u}}^n$, $\vec{\mathbf{w}}^n$, $\vec{\mathbf{z}}^n$ and $\vec{\phi}^{\,n}$ are known. The unknowns $\vec{\mathbf{u}}^{n+1}$, $\vec{\mathbf{p}}^{n+1}$ and $\vec{\phi}^{\,n+1}$ are denoted by $\mathbf{x}$, $\mathbf{y}$ and ϕ, respectively. We also define

$$\mathbf{f}_1 := \frac{\vec{\mathbf{u}}^n}{\Delta t} + (1-\theta)\vec{\mathbf{z}}^n, \quad \mathbf{f}_2 := \frac{\vec{\phi}^{\,n}}{\Delta t} + (1-\theta)\vec{\mathbf{w}}^n.$$

Using this notation the nonlinear system can be written as follows:

$$
\begin{aligned}
[\tfrac{1}{\Delta t}\mathbf{M} + \theta\mathbf{A}](\boldsymbol{\phi})\mathbf{x} + \theta\mathbf{N}(\boldsymbol{\phi},\mathbf{x})\mathbf{x} + \theta\mathbf{B}^T\mathbf{y} &= \theta[\vec{\mathbf{g}} + \vec{\mathbf{f}}_{\Gamma_h}](\boldsymbol{\phi}) + \mathbf{M}(\boldsymbol{\phi})\mathbf{f}_1 \\
\mathbf{B}\mathbf{x} &= 0 \\
[\tfrac{1}{\Delta t}\mathbf{E} + \theta\mathbf{H}](\mathbf{x})\boldsymbol{\phi} &= \mathbf{E}(\mathbf{x})\mathbf{f}_2 - \theta\mathbf{b}(\mathbf{x}).
\end{aligned}
\tag{9.1}
$$

The $(\mathbf{x},\mathbf{y})$ unknowns can be decoupled from the level set unknown $\boldsymbol{\phi}$ by a simple block Gauss-Seidel approach in which the level set equation is solved for the level set unknowns and the Navier-Stokes equations for the velocity and pressure unknowns:

Decoupling by a block Gauss-Seidel iteration

This method is as follows: Initialize $\mathbf{x}$ and $\boldsymbol{\phi}$ with the values from the previous time step, i.e., $\mathbf{x}^0 = \vec{\mathbf{u}}^n$, $\boldsymbol{\phi}^0 = \vec{\boldsymbol{\phi}}^n$. Iterate for $k = 0, 1, \ldots$

- Compute the level set vector $\boldsymbol{\phi}^{k+1}$ from the *linear* system

$$
\big[\tfrac{1}{\Delta t}\mathbf{E} + \theta\mathbf{H}\big](\mathbf{x}^k)\boldsymbol{\phi}^{k+1} = \mathbf{E}(\mathbf{x}^k)\mathbf{f}_2 - \theta\mathbf{b}(\mathbf{x}^k). \tag{9.2}
$$

- Solve the following equations for $(\mathbf{x}^{k+1}, \mathbf{y}^{k+1})$:

$$
\begin{aligned}
&\big[\tfrac{1}{\Delta t}\mathbf{M} + \theta\mathbf{A}\big](\boldsymbol{\phi}^{k+1})\mathbf{x}^{k+1} + \theta\mathbf{N}(\boldsymbol{\phi}^{k+1},\mathbf{x}^{k+1})\mathbf{x}^{k+1} + \theta\mathbf{B}^T\mathbf{y}^{k+1} \\
&= \theta\big[\vec{\mathbf{g}} + \vec{\mathbf{f}}_{\Gamma_h}\big](\boldsymbol{\phi}^{k+1}) + \mathbf{M}(\boldsymbol{\phi}^{k+1})\mathbf{f}_1 \\
&\mathbf{B}\mathbf{x}^{k+1} = 0.
\end{aligned}
\tag{9.3}
$$

The nonlinear system (9.3) is very similar to the discrete Navier-Stokes equations of a one-phase flow problem. Note, however, that the matrices $\mathbf{M}$, $\mathbf{A}$, $\mathbf{N}$ in (9.3) depend on the viscosity and density, which in case of a two-phase problem are *not* constant in the whole domain. Furthermore, the matrix $\mathbf{B}$ is different from the one-phase flow case if for the two-phase flow problem we use the extended finite element space for pressure discretization. This nonlinear system has, after rescaling of the first equation by a factor $\frac{1}{\theta}$, the form

$$
\begin{aligned}
(\frac{1}{\theta\Delta t}\mathbf{M} + \mathbf{A})\mathbf{x} + \mathbf{N}(\mathbf{x})\mathbf{x} + \mathbf{B}^T\mathbf{y} &= \mathbf{b} \\
\mathbf{B}\mathbf{x} &= \mathbf{c},
\end{aligned}
\tag{9.4}
$$

and can be solved using the defect correction method explained in Sect. 5.1.

In the level set method one applies re-initialization after each or a few time step(s). In our approach we also apply a volume correction technique, cf. Sect. 7.4.2. Such corrections of the level set function may lead to inconsistencies. To explain this more precisely, we introduce some notation related to the re-initialization and volume correction methods. Given a (piecewise quadratic) finite element function ϕ_h, we assume that we have a re-initialization operator $\phi_h \to \mathrm{ReInit}(\phi_h)$, cf. Remark 7.4.4. Let $\mathbf{R}$ be the vector representation of this operator, i.e. if $\boldsymbol{\phi}$ is the vector representation of ϕ_h, then $\mathbf{R}(\boldsymbol{\phi})$ is the vector representation of $\mathrm{ReInit}(\phi_h)$. Similarly, the vector representation of the mass correction method in Sect. 7.4.2 is denoted by $\mathbf{V}(\boldsymbol{\phi})$. The block Gauss-Seidel method, based on the subproblems (9.2)-(9.3), is represented in a compact form as follows: Take $\mathbf{x}^0 = \vec{\mathbf{u}}^n$, $\boldsymbol{\phi}^0 = \vec{\boldsymbol{\phi}}^n$; iterate for $k = 0, 1, \ldots$

$$\text{Determine } \boldsymbol{\phi}^{k+1} \text{ such that } \quad \mathbf{F}_1(\boldsymbol{\phi}^{k+1}, \mathbf{x}^k) = 0. \tag{9.5a}$$

$$\text{Determine } \mathbf{x}^{k+1} \text{ such that } \quad \mathbf{F}_2(\mathbf{x}^{k+1}, \boldsymbol{\phi}^{k+1}) = 0. \tag{9.5b}$$

In (9.5b) the (pressure) variable $\mathbf{y}^{k+1}$ from (9.3) is part of the operator $\mathbf{F}_2$, since it plays no role in the coupling between the level set and Navier-Stokes equations. Based on a stopping criterion this iteration is terminated and as approximations for velocity and level set vectors at $t = t_{n+1}$ one takes: $(\vec{\mathbf{u}}^{n+1}, \vec{\boldsymbol{\phi}}^{n+1}) = (\mathbf{x}^{k+1}, \boldsymbol{\phi}^{k+1})$. If *after* completion of this time step one applies a re-initialization $\mathbf{R}(\vec{\boldsymbol{\phi}}^{n+1})$ and/or volume correction $\mathbf{V}(\vec{\boldsymbol{\phi}}^{n+1})$ this results in a modification $\vec{\boldsymbol{\psi}}^{n+1} \neq \vec{\boldsymbol{\phi}}^{n+1}$ and the pair $(\vec{\mathbf{u}}^{n+1}, \vec{\boldsymbol{\psi}}^{n+1})$ will in general *not* satisfy the discrete Navier-Stokes equations in (9.5b). This *inconsistency* can be eliminated by solving (9.5b) once more with $\boldsymbol{\phi}^{k+1}$ replaced by $\vec{\boldsymbol{\psi}}^{n+1}$, resulting in a new velocity vector $\vec{\mathbf{u}}^{n+1}$. This, however, requires an additional solve of the problem (9.5b) (which is expensive) and the corrected pair $(\vec{\mathbf{u}}^{n+1}, \vec{\boldsymbol{\psi}}^{n+1})$ may now have a large residual $\mathbf{F}_1(\vec{\boldsymbol{\psi}}^{n+1}, \vec{\mathbf{u}}^{n+1})$ in the level set equation (9.5a). Since the equation in (9.5a) describes the discrete evolution of the level set function and the re-initialization and volume corrections result in modifications of the level set function, a natural alternative is to incorporate the correction operators into the block Gauss-Seidel iteration. We describe one particular strategy that is used in our simulations. After solving the discrete level set equation in (9.5a) we first apply a re-initialization step $\mathbf{R}(\boldsymbol{\phi}^{k+1})$, then a volume correction $\mathbf{V}\big(\mathbf{R}(\boldsymbol{\phi}^{k+1})\big)$ and then use this level set vector (instead of $\boldsymbol{\phi}^{k+1}$) in (9.5b). Hence, the block Gauss-Seidel iteration now takes the following form:

$$\text{Determine } \boldsymbol{\phi}^{k+1} \text{ such that } \quad \mathbf{F}_1(\boldsymbol{\phi}^{k+1}, \mathbf{x}^k) = 0, \tag{9.6a}$$

$$\text{Determine } \mathbf{x}^{k+1} \text{ such that } \quad \mathbf{F}_2\big(\mathbf{x}^{k+1}, \mathbf{V}(\mathbf{R}(\boldsymbol{\phi}^{k+1}))\big) = 0. \tag{9.6b}$$

This method is consistent in the sense that the pair $\big(\mathbf{x}^{k+1}, \mathbf{V}(\mathbf{R}(\boldsymbol{\phi}^{k+1}))\big)$ satisfies the discrete Navier-Stokes equations for all k. Furthermore, error control is relatively easy since it can be based on the size of the residuals of the equations (9.6). For $k+1$ sufficiently large, such that a stopping criterion is satisfied, we take as approximations for velocity and the level set function at $t = t_{n+1}$: $(\vec{\mathbf{u}}^{n+1}, \vec{\boldsymbol{\phi}}^{n+1}) = \big(\mathbf{x}^{k+1}, \mathbf{V}(\mathbf{R}(\boldsymbol{\phi}^{k+1}))\big)$. The method in (9.6) is feasible in practice, since the volume correction operator $\mathbf{V}$ is inexpensive (compared to other operations in the evaluation of $\mathbf{F}_2$) and for the re-initialization $\mathbf{R}$ one "often" has $\mathbf{R}(\boldsymbol{\phi}^{k+1}) = \boldsymbol{\phi}^{k+1}$, as the Fast Marching Method is applied only if the size of the gradient of the level set function represented by ϕ^{k+1} is too small or too large, cf. Remark 7.4.4. The iterations in (9.6a) and (9.6b) define solution operators:

$$\phi^{k+1} =: \mathbf{S}_1(\mathbf{x}^k), \quad \mathbf{x}^{k+1} =: \mathbf{S}_2(\phi^{k+1}), \tag{9.7}$$

and the method (9.6) can be rewritten as

$$\mathbf{x}^{k+1} = \mathbf{S}_2(\mathbf{S}_1(\mathbf{x}^k)), \quad \mathbf{x}^0 = \vec{\mathbf{u}}^n. \tag{9.8}$$

Let $\mathbf{x}^*$ be a fixed point, i.e. $\mathbf{x}^* = \mathbf{S}_2(\mathbf{S}_1(\mathbf{x}^*))$ and define $\phi^* := \mathbf{S}_1(\mathbf{x}^*)$. Due to the perturbations introduced by the re-initialization $\mathbf{R}$ and volume correction $\mathbf{V}$, the pair $(\mathbf{x}^*, \phi^*)$ in general does not solve the coupled system (9.1). However, $(\mathbf{x}^*, \phi^*)$ is a solution of the level set equation in (9.1) and the Navier-Stokes equations in (9.1) are solved by the pair $(\mathbf{x}, \phi) = \big(\mathbf{x}^*, \mathbf{V}(\mathbf{R}(\phi^*))\big)$. Due to the fact that the operators $\mathbf{F}_1$ and $\mathbf{F}_2$ are highly nonlinear, the fixed point method (9.8) can have a (very) slow convergence and acceleration is desirable. Many convergence acceleration techniques are known, cf. [86]. Below we discuss two strategies. The first one, treated in detail below, uses a special structure in our coupled problem. It is based on a linearization of the surface tension force term. This results in a modified nonlinear operator $\tilde{\mathbf{F}}_2$ instead of $\mathbf{F}_2$ in (9.6b), with corresponding solution operator $\tilde{\mathbf{S}}_2$. Instead of the fixed point method (9.8) we get a modified iteration

$$\mathbf{x}^{k+1} = \tilde{\mathbf{S}}_2(\mathbf{S}_1(\mathbf{x}^k)), \quad \mathbf{x}^0 = \vec{\mathbf{u}}^n, \tag{9.9}$$

which, due to the choice of $\tilde{\mathbf{F}}_2$ has better contraction properties. The modification is such that (9.9) has the same fixed point as (9.8).

The second convergence acceleration strategy is an application of the general Broyden acceleration method for nonlinear problems.

Convergence acceleration by linearization of the surface tension force

There is a strong coupling between the Navier-Stokes and level set equations through the surface tension force term $\vec{\mathbf{f}}_{\Gamma_h}(\phi)$ in (9.1). A modification of the block Gauss-Seidel method (9.6), which has better convergence properties, can be obtained by a suitable linearization of this term. We explain the basic idea behind this approach by restricting to a strongly simplified problem, in which, however, the (coupling through the) surface tension force still has essentially the same form as for the general class of two-phase flow problems considered above. We take a non-stationary *Stokes* flow with *constant viscosity and density* in the whole domain: $\mu_1 = \mu_2 = 1$, $\rho_1 = \rho_2 = 1$. For the surface tension force discretization we use, instead of $\tilde{f}_{\Gamma_h}$, the simpler functional in (7.54):

$$f_{\Gamma_h}(\mathbf{v}_h) := -\tau \int_{\Gamma_h} \nabla_{\Gamma_h} \mathrm{id}_{\Gamma_h} \cdot \nabla_{\Gamma_h} \mathbf{v}_h \, ds, \qquad \mathbf{v}_h \in \mathbf{V}_h.$$

Discretization in time is done by the implicit Euler method. Since in this simplified case the mass and stiffness matrices do not depend on $\boldsymbol{\phi}$ and the convection term $\mathbf{N}(\boldsymbol{\phi}, \mathbf{x})$ vanishes, the Navier-Stokes part of the coupled problem (9.1) simplifies to

$$\begin{aligned} \tilde{\mathbf{A}}\mathbf{x} + \mathbf{B}^T\mathbf{y} &= \vec{\mathbf{f}}_{\Gamma_h}(\boldsymbol{\phi}) + \mathbf{b}, \\ \mathbf{B}\mathbf{x} &= 0, \\ \text{with } \mathbf{b} = \vec{\mathbf{g}} + \mathbf{M}\mathbf{f}_1, &\quad \tilde{\mathbf{A}} = \frac{1}{\Delta t}\mathbf{M} + \mathbf{A}. \end{aligned} \tag{9.10}$$

Note that $\tilde{\mathbf{A}}$ is symmetric positive definite. In (9.10) the coupling with the level set function is only through the surface tension force term $\vec{\mathbf{f}}_{\Gamma_h}(\boldsymbol{\phi})$. The solution pair $(\mathbf{x}, \boldsymbol{\phi})$ yields the velocity and level set functions at time $t = t_{n+1}$: $\mathbf{x} = \vec{\mathbf{u}}^{n+1}$, $\boldsymbol{\phi} = \vec{\boldsymbol{\phi}}^{n+1}$. The i-th component of the surface tension force vector is given by

$$\vec{\mathbf{f}}_{\Gamma_h}(\boldsymbol{\phi})_i = -\tau \int_{\Gamma_h(t_{n+1})} \nabla_{\Gamma_h(t_{n+1})} \mathrm{id}_{\Gamma_h(t_{n+1})} \cdot \nabla_{\Gamma_h(t_{n+1})} \boldsymbol{\xi}_i \, ds, \tag{9.11}$$

with $\boldsymbol{\xi}_i$ the i-th nodal basis function in $\mathbf{V}_h$ and $\Gamma_h(t_{n+1})$ the interface approximation corresponding to $\boldsymbol{\phi} = \vec{\boldsymbol{\phi}}^{n+1}$. The latter is obtained through an implicit Euler time discretization of the level set equation, which is the third equation in (9.1) with $\theta = 1$. We assume that there is no re-initialization or volume correction. Denoting by $P_h : \mathbb{R}^m \to \mathbf{V}_h$ the finite element isomorphism $P_h\mathbf{v} = \sum_{i=1}^m v_i \boldsymbol{\xi}_i$, we have

$$\Gamma_h(t_{n+1}) \approx \{\, x + \Delta t (P_h \vec{\mathbf{u}}^{n+1})(x) : x \in \Gamma_h(t_n) \,\} =: \Gamma_h(\vec{\mathbf{u}}^{n+1}) = \Gamma_h(\mathbf{x}).$$

Note that the notation $\Gamma_h(\mathbf{x})$ differs from $\Gamma(t_{n+1})$. The former denotes the approximation of $\Gamma(t_{n+1})$ that is obtained by transporting the known interface approximation $\Gamma(t_n)$ over a time Δt in the direction of the velocity field $\mathbf{x}$. Using the result in (9.11) and the approximation $\Gamma_h(t_{n+1}) \approx \Gamma_h(\mathbf{x})$ we can eliminate $\boldsymbol{\phi}$ from (9.10) and instead consider the following very similar problem

$$\begin{aligned} \tilde{\mathbf{A}}\mathbf{x} + \mathbf{B}^T\mathbf{y} &= \mathbf{f}(\mathbf{x}) + \mathbf{b}, \\ \mathbf{B}\mathbf{x} &= 0, \end{aligned} \tag{9.12}$$

with

$$\mathbf{f}(\mathbf{x})_i := -\tau \int_{\Gamma_h(\mathbf{x})} \nabla_{\Gamma_h(\mathbf{x})} \mathrm{id}_{\Gamma_h(\mathbf{x})} \cdot \nabla_{\Gamma_h(\mathbf{x})} \boldsymbol{\xi}_i \, ds, \quad 1 \le i \le m. \tag{9.13}$$

The problem (9.12) is nonlinear due to the nonlinearity of $\mathbf{f}$. An obvious iterative method for solving this problem is given by: $\mathbf{x}^0 := \vec{\mathbf{u}}^n$, for $k \geq 0$ solve

$$\begin{aligned} \tilde{\mathbf{A}}\mathbf{x}^{k+1} + \mathbf{B}^T\mathbf{y}^{k+1} &= \mathbf{f}(\mathbf{x}^k) + \mathbf{b}, \\ \mathbf{B}\mathbf{x}^{k+1} &= 0. \end{aligned} \tag{9.14}$$

A better method is obtained if we replace $\mathrm{id}_{\Gamma_h(\mathbf{x})}$ in (9.13) by a better approximation than $\mathrm{id}_{\Gamma_h(\mathbf{x}^k)}$, namely $\mathrm{id}_{\Gamma_h(\mathbf{x}^{k+1})}$. For a description of this modified method we introduce some further notation. For $\mathbf{v} \in \mathbb{R}^m$ we define the velocity field

$$\mathrm{id}_{\Gamma_h(\mathbf{v})} : x \to x + \Delta t(P_h\mathbf{v})(x), \quad x \in \Omega.$$

The restriction of this velocity field to $\Gamma_h(t_n)$ results in a shifted interface

$$\Gamma_h(\mathbf{v}) := \left\{ \mathrm{id}_{\Gamma_h(\mathbf{v})}(x) : x \in \Gamma_h(t_n) \right\},$$

which is consistent with the notation $\Gamma_h(\mathbf{x})$ introduced above. For $\mathbf{v}$, $\mathbf{w} \in \mathbb{R}^m$ we have

$$\mathrm{id}_{\Gamma_h(\mathbf{v})} = \mathrm{id}_{\Gamma_h(\mathbf{w})} + \Delta t P_h(\mathbf{v} - \mathbf{w}),$$

i.e., $\mathbf{v} \to \mathrm{id}_{\Gamma_h(\mathbf{v})}$ is an *affine* mapping. We introduce the generalization of (9.13) given by

$$\mathbf{F}(\mathbf{v}, \mathbf{w})_i := -\tau \int_{\Gamma_h(\mathbf{v})} \nabla_{\Gamma_h(\mathbf{v})} \mathrm{id}_{\Gamma_h(\mathbf{w})} \cdot \nabla_{\Gamma_h(\mathbf{v})} \boldsymbol{\xi}_i \, ds, \quad 1 \leq i \leq m,$$

which is well-defined since $\mathrm{id}_{\Gamma_h(\mathbf{v})}$, $\mathbf{v} \in \mathbb{R}^m$, is a velocity field on Ω. The affine mapping $\mathbf{w} \to \mathbf{F}(\mathbf{v}, \mathbf{w})$ satisfies

$$\begin{aligned} &\mathbf{F}(\mathbf{v}, \mathbf{w}_1) = \mathbf{F}(\mathbf{v}, \mathbf{w}_2) - \Delta t \, \mathbf{L}_\mathbf{v}(\mathbf{w}_1 - \mathbf{w}_2), \\ &(\mathbf{L}_\mathbf{v})_{ij} := \tau \int_{\Gamma_h(\mathbf{v})} \nabla_{\Gamma_h(\mathbf{v})} \boldsymbol{\xi}_j \cdot \nabla_{\Gamma_h(\mathbf{v})} \boldsymbol{\xi}_i \, ds, \quad 1 \leq i, j \leq m. \end{aligned} \tag{9.15}$$

The matrix $\mathbf{L}_\mathbf{v}$ is symmetric positive semi-definite. As mentioned above, instead of $\mathrm{id}_{\Gamma_h(\mathbf{x}^k)}$ we use $\mathrm{id}_{\Gamma_h(\mathbf{x}^{k+1})}$ in the surface tension force term. This results in the iterative method

$$\begin{aligned} \tilde{\mathbf{A}}\mathbf{x}^{k+1} + \mathbf{B}^T\mathbf{y}^{k+1} &= \mathbf{F}(\mathbf{x}^k, \mathbf{x}^{k+1}) + \mathbf{b}, \\ \mathbf{B}\mathbf{x}^{k+1} &= 0. \end{aligned}$$

Using (9.15) we reformulate this method and thus obtain the following alternative iterative method for solving (9.12): take $\mathbf{x}^0 := \vec{\mathbf{u}}^n$, for $k \geq 0$ solve

$$\begin{aligned} (\tilde{\mathbf{A}} + \Delta t \mathbf{L}_{\mathbf{x}^k})\mathbf{x}^{k+1} + \mathbf{B}^T\mathbf{y}^{k+1} &= \mathbf{f}(\mathbf{x}^k) + \Delta t \mathbf{L}_{\mathbf{x}^k}\mathbf{x}^k + \mathbf{b}, \\ \mathbf{B}\mathbf{x}^{k+1} &= 0. \end{aligned} \tag{9.16}$$

In the following theorem we give convergence results for the two methods in (9.14) and (9.16). We use the Euclidean norm $\| \cdot \|$ on $\mathbb{R}^m$.

Theorem 9.1.1 *Let* $\mathbf{x}$ *be solution of* (9.12). *Assume that on the neighborhood* $B = \{\hat{\mathbf{x}} \in \mathbb{R}^m : \|\hat{\mathbf{x}} - \mathbf{x}\| \leq r\}$, *with a given* $r > 0$, *we have*

$$\begin{aligned} \|\mathbf{F}(\mathbf{x}_1, \mathbf{z}) - \mathbf{F}(\mathbf{x}_2, \mathbf{z})\| &\leq L_1 \|\mathbf{x}_1 - \mathbf{x}_2\| \quad \textit{for all } \mathbf{x}_1, \mathbf{x}_2, \mathbf{z} \in B, \\ \|\mathbf{L}_{\hat{\mathbf{x}}}\| &\leq L_2 \quad \textit{for all } \hat{\mathbf{x}} \in B. \end{aligned}$$

Define $\mathbf{e}^i = \mathbf{x} - \mathbf{x}^i$, $i = 0, 1, 2, \ldots$. *If* $\mathbf{x}^k \in B$ *then the following holds:*

$$\|\mathbf{e}^{k+1}\| \leq (L_1 + \Delta t L_2) \|\tilde{\mathbf{A}}^{-1}\| \|\mathbf{e}^k\| \quad \textit{for the method in } (9.14), \tag{9.17}$$

$$\|\mathbf{e}^{k+1}\| \leq L_1 \|(\tilde{\mathbf{A}} + \Delta t \mathbf{L}_{\mathbf{x}^k})^{-1}\| \|\mathbf{e}^k\| \quad \textit{for the method in } (9.16). \tag{9.18}$$

Proof. First we consider a saddle point problem of the form

$$\begin{aligned} \mathbf{C}\mathbf{e} + \mathbf{B}^T \mathbf{y} &= \mathbf{d} \\ \mathbf{B}\mathbf{e} &= 0, \end{aligned}$$

with a symmetric positive definite matrix $\mathbf{C}$. A straightforward computation yields

$$\mathbf{e} = \mathbf{C}^{-\frac{1}{2}} (\mathbf{I} - \hat{\mathbf{B}}^T (\hat{\mathbf{B}} \hat{\mathbf{B}}^T)^{-1} \hat{\mathbf{B}}) \mathbf{C}^{-\frac{1}{2}} \mathbf{d}, \quad \hat{\mathbf{B}} := \mathbf{B}\mathbf{C}^{-\frac{1}{2}}.$$

Hence, using $\|\mathbf{I} - \hat{\mathbf{B}}^T (\hat{\mathbf{B}} \hat{\mathbf{B}}^T)^{-1} \hat{\mathbf{B}}\| = 1$, we get

$$\|\mathbf{e}\| \leq \|\mathbf{C}^{-1}\| \|\mathbf{d}\|.$$

We use the identity

$$\begin{aligned} \mathbf{f}(\mathbf{x}) - \mathbf{f}(\mathbf{x}^k) &= \mathbf{F}(\mathbf{x}, \mathbf{x}) - \mathbf{F}(\mathbf{x}^k, \mathbf{x}) + \mathbf{F}(\mathbf{x}^k, \mathbf{x}) - \mathbf{F}(\mathbf{x}^k, \mathbf{x}^k) \\ &= \mathbf{F}(\mathbf{x}, \mathbf{x}) - \mathbf{F}(\mathbf{x}^k, \mathbf{x}) - \Delta t \mathbf{L}_{\mathbf{x}^k} (\mathbf{x} - \mathbf{x}^k). \end{aligned} \tag{9.19}$$

For the method in (9.14) we obtain

$$\begin{aligned} \tilde{\mathbf{A}} \mathbf{e}^{k+1} + \mathbf{B}^T (\mathbf{y} - \mathbf{y}^{k+1}) &= \mathbf{f}(\mathbf{x}) - \mathbf{f}(\mathbf{x}^k), \\ \mathbf{B}\mathbf{e}^{k+1} &= 0. \end{aligned}$$

Thus $\|\mathbf{e}^{k+1}\| \leq \|\tilde{\mathbf{A}}^{-1}\| \|\mathbf{f}(\mathbf{x}) - \mathbf{f}(\mathbf{x}^k)\|$ holds. Using (9.19) we get

$$\|\mathbf{f}(\mathbf{x}) - \mathbf{f}(\mathbf{x}^k)\| \leq (L_1 + \Delta t L_2) \|\mathbf{x} - \mathbf{x}^k\|$$

and thus the result (9.17) holds. For the method (9.16) we obtain

$$\begin{aligned} (\tilde{\mathbf{A}} + \Delta t \mathbf{L}_{\mathbf{x}^k}) \mathbf{e}^{k+1} + \mathbf{B}^T (\mathbf{y} - \mathbf{y}^{k+1}) &= \mathbf{f}(\mathbf{x}) - \mathbf{f}(\mathbf{x}^k) + \Delta t \mathbf{L}_{\mathbf{x}^k} (\mathbf{x} - \mathbf{x}^k), \\ \mathbf{B}\mathbf{e}^{k+1} &= 0. \end{aligned}$$

From (9.19) it follows that

$$\mathbf{f}(\mathbf{x}) - \mathbf{f}(\mathbf{x}^k) + \Delta t \mathbf{L}_{\mathbf{x}^k} (\mathbf{x} - \mathbf{x}^k) = \mathbf{F}(\mathbf{x}, \mathbf{x}) - \mathbf{F}(\mathbf{x}^k, \mathbf{x}).$$

Thus we get

$$\begin{aligned}\|\mathbf{e}^{k+1}\| &\leq \|(\tilde{\mathbf{A}} + \Delta t\mathbf{L}_{\mathbf{x}^k})^{-1}\|\|\mathbf{f}(\mathbf{x}) - \mathbf{f}(\mathbf{x}^k) + \Delta t\mathbf{L}_{\mathbf{x}^k}(\mathbf{x} - \mathbf{x}^k)\| \\ &= \|(\tilde{\mathbf{A}} + \Delta t\mathbf{L}_{\mathbf{x}^k})^{-1}\|\|\mathbf{F}(\mathbf{x},\mathbf{x}) - \mathbf{F}(\mathbf{x}^k,\mathbf{x})\| \\ &\leq L_1\|(\tilde{\mathbf{A}} + \Delta t\mathbf{L}_{\mathbf{x}^k})^{-1}\|\|\mathbf{x} - \mathbf{x}^k\|.\end{aligned}$$

Hence, the result (9.17) holds. □

From the result in this theorem we see that the linearization of the surface tension force applied in the method (9.16) has two positive effects. Firstly, the matrix $\tilde{\mathbf{A}}$ is further "stabilized" by adding the symmetric positive semi-definite matrix $\Delta t\,\mathbf{L}_{\mathbf{x}^k}$. Secondly, instead of the sum of the Lipschitz constants $L_1 + \Delta t L_2$, only the constant L_1 occurs in the error reduction bound. From (9.15) it can be seen that the matrix $\mathbf{L}_{\mathbf{x}^k}$ corresponds to a discrete diffusion type operator (which is ill conditioned). The error term $\mathbf{L}_{\mathbf{x}^k}(\mathbf{x}-\mathbf{x}^k)$ is treated *explicitly* ($\mathbf{x}$ replaced by $\mathbf{x}^k$) in the method (9.14), whereas it is treated *implicitly* ($\mathbf{x}$ replaced by $\mathbf{x}^{k+1}$) in the method (9.16). Furthermore, from

$$\|(\tilde{\mathbf{A}} + \Delta t\mathbf{L}_{\mathbf{x}^k})^{-1}\| \leq \|\tilde{\mathbf{A}}^{-1}\| = \|(\frac{1}{\Delta t}\mathbf{M} + \mathbf{A})^{-1}\| \leq \Delta t\|\mathbf{M}^{-1}\|$$

it follows that for Δt sufficiently small both methods converge.

Remark 9.1.2 The approach used in the iterative method (9.16) is very similar to the semi-implicit time discretization of the curvature used in [23]. In that paper a space-time finite element discretization of the Navier-Stokes equations with a free capillary surface $\Gamma(t)$ is analyzed. An accurate and stable method is obtained by a Laplace-Beltrami variational approximation of the curvature, in which the tangential derivatives are evaluated on the surface parametrization X^{n+1} at the new time level $t = t_{n+1}$, whereas the domain of integration is $\Gamma_h(t_n)$, cf. Remark 3 in [23]. A similar semi-implicit method is often used in the discretization of mean curvature flows, cf. [78].

We now address how the surface tension linearization approach, explained and analyzed for the simplified Stokes problem (9.12), can be applied to the general Navier-Stokes case. For discretization of the surface tension force we use the improved functional in (7.60):

$$\tilde{f}_{\Gamma_h}(\mathbf{v}_h) = -\tau \int_{\Gamma_h} \tilde{\mathbf{P}}_h \nabla\,\mathrm{id}_{\Gamma_h} \cdot \nabla_{\Gamma_h} \mathbf{v}_h \, ds.$$

Let $\boldsymbol{\phi}$ be a given level set vector and $\Gamma_h(\boldsymbol{\phi})$ the corresponding interface approximation. For linearization of the surface tension, instead of the matrix $\mathbf{L}_{\hat{\mathbf{x}}}$ in (9.15) we now use

$$\mathbf{L}(\boldsymbol{\phi})_{ij} = \tau \int_{\Gamma_h(\boldsymbol{\phi})} \tilde{\mathbf{P}}_h \nabla \boldsymbol{\xi}_j \cdot \nabla_{\Gamma_h(\boldsymbol{\phi})} \boldsymbol{\xi}_i \, ds.$$

Note that due to $\tilde{\mathbf{P}}_h$ the matrix $\mathbf{L}(\phi)$ is in general non-symmetric. Instead of (9.3) we use the following iteration, which is a generalization of (9.16). Solve for $(\mathbf{x}^{k+1}, \mathbf{y}^{k+1})$:

$$\begin{aligned}
&\left[\tfrac{1}{\Delta t}\mathbf{M} + \theta\mathbf{A} + \theta\Delta t\mathbf{L}\right](\phi^{k+1})\mathbf{x}^{k+1} + \theta\mathbf{N}(\phi^{k+1}, \mathbf{x}^{k+1})\mathbf{x}^{k+1} + \theta\mathbf{B}^T\mathbf{y}^{k+1} \\
&= \theta\left[\vec{\mathbf{g}} + \vec{\mathbf{f}}_{\Gamma_h}\right](\phi^{k+1}) + \theta\Delta t\mathbf{L}(\phi^{k+1})\mathbf{x}^k + \mathbf{M}(\phi^{k+1})\mathbf{f}_1 \qquad (9.20)\\
&\mathbf{B}\mathbf{x}^{k+1} = 0.
\end{aligned}$$

The coupling of this with the level set iteration in (9.2) and with re-initialization and volume correction results in a method of the form (cf. (9.6)):

$$\begin{aligned}
&\text{Determine } \phi^{k+1} \text{ such that } \quad \mathbf{F}_1(\phi^{k+1}, \mathbf{x}^k) = 0. \qquad (9.21a)\\
&\text{Determine } \mathbf{x}^{k+1} \text{ such that } \quad \tilde{\mathbf{F}}_2\big(\mathbf{x}^{k+1}, \mathbf{V}(\mathbf{R}(\phi^{k+1}))\big) = 0. \qquad (9.21b)
\end{aligned}$$

Here $\tilde{\mathbf{F}}_2\big(\mathbf{x}^{k+1}, \mathbf{V}(\mathbf{R}(\phi^{k+1}))\big) = 0$ is a compact representation of the iteration in (9.20), with ϕ^{k+1} replaced by $\mathbf{V}(\mathbf{R}(\phi^{k+1}))$. The iterations in (9.21) define solution operators:

$$\phi^{k+1} =: \mathbf{S}_1(\mathbf{x}^k), \quad \mathbf{x}^{k+1} =: \tilde{\mathbf{S}}_2(\phi^{k+1}),$$

and the method (9.21) can be represented as

$$\mathbf{x}^{k+1} = \tilde{\mathbf{S}}_2(\mathbf{S}_1(\mathbf{x}^k)). \qquad (9.22)$$

For a fixed point of this iteration, say $\mathbf{x}^*$, we have $\mathbf{x}^{k+1} = \mathbf{x}^k = \mathbf{x}^*$ and the terms $\mathbf{L}(\phi^{k+1})\mathbf{x}^{k+1}$ and $\mathbf{L}(\phi^{k+1})\mathbf{x}^k$ in (9.20) cancel. Hence, at the fixed point $\tilde{\mathbf{F}}_2 = \mathbf{F}_2$ holds and $\mathbf{x}^*$ is a fixed point of (9.8), too.

Remark 9.1.3 The method (9.6) and its improved version (9.21) are decoupling iterations for the scheme in (8.10). It is straightforward to derive similar methods for other schemes treated in Chap. 8.

Convergence acceleration by a Broyden method

We apply the Broyden method, which is very easy to implement and turns out to result in a significant improvement of efficiency. We explain this method for the fixed point method in (9.22). The same approach can be applied to (9.6) or other fixed point iterations.
A fixed point $\mathbf{x}^*$, i.e. $\mathbf{x}^* = \tilde{\mathbf{S}}_2(\mathbf{S}_1(\mathbf{x}^*))$ is a zero of $\mathbf{S} := \mathbf{I} - \tilde{\mathbf{S}}_2 \circ \mathbf{S}_1$:

$$\mathbf{S}(\mathbf{x}^*) = 0. \qquad (9.23)$$

Note that an evaluation of $\mathbf{S}(\mathbf{x})$ requires (only) one evaluation of $\tilde{\mathbf{S}}_2 \circ \mathbf{S}_1$, i.e., one iteration of the method in (9.21). The method (9.22) is the fixed point iteration for solving $(\mathbf{I} - \mathbf{S})(\mathbf{x}^*) = \mathbf{x}^*$. One may expect faster convergence if a Newton type of method is applied for solving the problem (9.23). Jacobians of $\mathbf{S}$, however, are not available and therefore we consider a Broyden method, which requires only evaluations of $\mathbf{S}(\mathbf{x})$. We briefly explain the so-called contravariant version of this method. For an extensive treatment and convergence analysis we refer to [86]. Let $\mathbf{x}^0$ be given and $\mathbf{J}_0$ a given approximation of the Jacobian $\frac{d\mathbf{S}}{d\mathbf{x}}(\mathbf{x}^0)$ such that the system $\mathbf{J}_0\mathbf{y} = \mathbf{b}$ can be solved with low costs. In applications one often takes $\mathbf{J}_0 = \mathbf{I}$. The Broyden method for computing an approximate solution of (9.23) is defined as follows:

$$\mathbf{x}^{k+1} = \mathbf{x}^k - \mathbf{J}_k^{-1}\mathbf{S}(\mathbf{x}^k), \quad k = 0, 1, 2, \dots,$$
$$\mathbf{J}_k^{-1} := \mathbf{J}_{k-1}^{-1}\Big(\mathbf{I} - \frac{\mathbf{S}(\mathbf{x}^k)\Delta\mathbf{S}_k^T}{\|\Delta\mathbf{S}_k\|^2}\Big), \quad k = 1, 2, , \dots,$$

with $\|\cdot\|$ the Euclidean norm and $\Delta\mathbf{S}_k := \mathbf{S}(\mathbf{x}^k) - \mathbf{S}(\mathbf{x}^{k-1})$. We assume $\Delta\mathbf{S}_k \neq 0$. The approximate Jacobian $\mathbf{J}_k$ is such that, for $k \geq 1$,

$$\mathbf{J}_k^{-1}\mathbf{z} = \mathbf{J}_{k-1}^{-1}\mathbf{z} \quad \text{for all } \mathbf{z} \text{ with } \mathbf{z}^T\Delta\mathbf{S}_k = 0,$$
$$\mathbf{J}_k^{-1}\Delta\mathbf{S}_k = \mathbf{x}^k - \mathbf{x}^{k-1}.$$

The latter is the *secant property* $\mathbf{S}(\mathbf{x}_k) - \mathbf{S}(\mathbf{x}_{k-1}) = \mathbf{J}_k(\mathbf{x}^k - \mathbf{x}^{k-1})$ of the approximate Jacobian $\mathbf{J}_k$. For the implementation of this method we introduce the help sequence

$$\mathbf{v}^{k,k} = \mathbf{S}(\mathbf{x}^k), \quad \mathbf{v}^{k,\ell-1} = \Big(\mathbf{I} - \frac{\mathbf{S}(\mathbf{x}^\ell)\Delta\mathbf{S}_\ell^T}{\|\Delta\mathbf{S}_\ell\|^2}\Big)\mathbf{v}^{k,\ell}, \quad \ell = k, k-1, \dots, 1. \quad (9.24)$$

The update $\mathbf{J}_k^{-1}\mathbf{S}(\mathbf{x}^k)$ needed in the Broyden method satisfies

$$\mathbf{J}_k^{-1}\mathbf{S}(\mathbf{x}^k) = \mathbf{J}_k^{-1}\mathbf{v}^{k,k} = \mathbf{J}_{k-1}^{-1}\mathbf{v}^{k,k-1} = \ldots = \mathbf{J}_0^{-1}\mathbf{v}^{k,0}.$$

Hence, for computing this update one has to determine $\mathbf{S}(\mathbf{x}^k)$ and recursively the vectors $\mathbf{v}^{k,k-1}, \dots, \mathbf{v}^{k,0}$. The latter can be realized with low costs if the vectors $\mathbf{S}(\mathbf{x}^\ell)$, $\Delta\mathbf{S}_\ell$, $1 \leq \ell \leq k$, are stored, cf. (9.24). Finally a system $\mathbf{J}_0\mathbf{y} = \mathbf{v}^{k,0}$ has to be solved. In [86] it is shown that, under certain assumptions, the rate of convergence of this method is superlinear. In our applications we observed that (with $\mathbf{J}_0 = \mathbf{I}$) this method leads to a significant convergence acceleration of the fixed point iteration (9.22), with low additional arithmetic costs.

9.1.1 Numerical experiment

In this section we consider the *Stokes* model of a curvature-driven flow and use it as a starting point to compare the different convergence acceleration schemes described in Sect. 9.1. A similar curvature-driven flow problem is presented in [22].

Consider an initially ellipsoidal droplet $\Omega_1 \subset \Omega = (-1, 1)^3$ with center point $(0, 0, 0)$ and diameter 1.2 in x_1-direction and diameter 0.8 in x_2- and

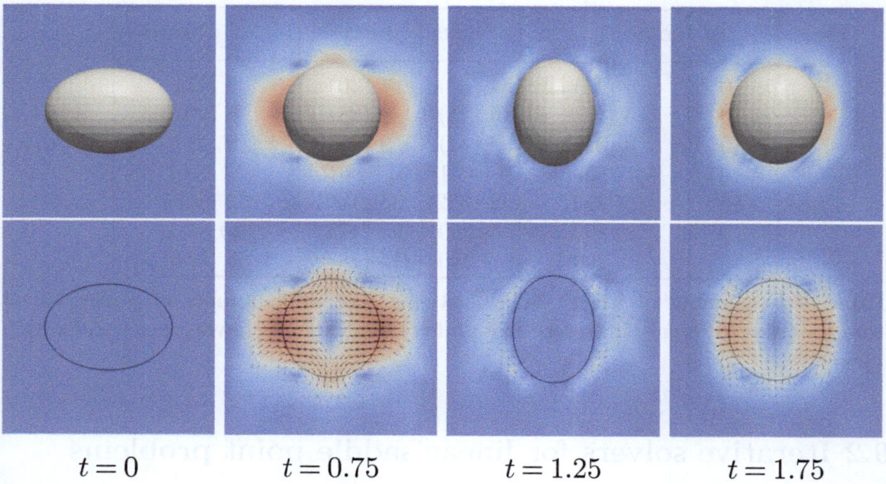

Fig. 9.1. Curvature-driven flow: droplet shape and velocity field for different times.

x_3-direction, respectively. The droplet is initially at rest ($\mathbf{u}_0 = 0$). The material properties are chosen as $\mu_1 = 0.1$, $\mu_2 = 0.01$, $\rho_1 = 10$, $\rho_2 = 1$. We assume surface tension with $\tau = 1$ and no gravity, i. e., $\mathbf{g} = 0$. Due to its ellipsoidal shape the droplet has a non-constant curvature and the droplet starts to deform due to surface tension. The tips with relatively large curvature are moving towards the center whereas the regions with small curvature are moving outwards, leading to an oscillation of the droplet shape, cf. Fig. 9.1. This oscillation has a certain damping, leading to a spherical shape for $t \to \infty$.

Taking the discrete solution for $t = 0.75$ as initial condition for $\mathbf{x}$ and ϕ, i. e., $(\mathbf{x}^0, \phi^0) = (\vec{\mathbf{u}}, \vec{\phi})|_{t=0.75}$, one time step of the (coupled) implicit Euler scheme is performed. The coupled system of equations is solved using the block Gauss-Seidel iteration (9.5), reducing the residual below a tolerance of 10^{-8}. The number of fixed point iterations and the total number of Stokes solver iterations are reported in Table 9.1 for different time step sizes Δt and for the case with or without convergence acceleration by the linearization of the surface tension force (LinST) and/or the Broyden method. For small time steps $\Delta t < 10^{-2}$ all methods perform equally well. This is due to the good starting values provided by the discrete solution for $t = 0.75$ which are very close to the solution $(\mathbf{x}^*, \phi^*) = (\vec{\mathbf{u}}, \vec{\phi})|_{t=0.75+\Delta t}$ of the coupled system. For $t \geq 10^{-2}$ the method without acceleration fails to converge within 250 fixed point iterations which was the maximum iteration number chosen in this experiment. The accelerated methods increase the robustness of the solver. In particular, for large time steps the linearization of the surface tension force turns out to be essential.

Δt	no acceleration	Broyden	LinST	LinST+Broyden
10^{-3}	4 (18)	4 (20)	4 (20)	4 (17)
$2 \cdot 10^{-3}$	5 (23)	5 (23)	5 (28)	5 (29)
$5 \cdot 10^{-3}$	9 (48)	8 (42)	9 (45)	8 (41)
10^{-2}	> 250	14 (79)	17 (70)	11 (56)
$2 \cdot 10^{-2}$	> 250	92 (352)	35 (140)	62 (150)
$5 \cdot 10^{-2}$	> 250	> 250	95 (471)	120 (481)
10^{-1}	> 250	> 250	214 (1578)	66 (656)

Table 9.1. Number of fixed point iterations (and total number of Stokes solver iterations) for different time step sizes and combination of convergence acceleration schemes.

9.2 Iterative solvers for linear saddle point problems

The decoupling strategies result in saddle point problems of the form

$$\mathbf{K}\begin{pmatrix}\mathbf{x}\\ \mathbf{y}\end{pmatrix} := \begin{pmatrix}\tilde{\mathbf{A}} & \mathbf{B}^T\\ \mathbf{B} & 0\end{pmatrix}\begin{pmatrix}\mathbf{x}\\ \mathbf{y}\end{pmatrix} = \begin{pmatrix}\mathbf{f}_1\\ \mathbf{f}_2\end{pmatrix}, \qquad \tilde{\mathbf{A}} := \mathbf{A} + \mathbf{N}(\mathbf{x}^{\mathrm{old}}) + \beta\mathbf{M}, \tag{9.25}$$

that have a very similar structure as the ones in Sect. 5.3. We can apply iterative solvers as explained in Chap. 5. If $\mathbf{N} = 0$ then this saddle point problem is *symmetric* and methods like preconditioned MINRES or inexact Uzawa, cf. Sect. 5.2, are applicable. For the general discrete Oseen problem some iterative methods (e.g., preconditioned GCR) are treated in Sect. 5.3. A key issue in all these iterative solvers is an appropriate preconditioning of $\tilde{\mathbf{A}}$ and of the Schur complement $\mathbf{S} = \mathbf{B}\tilde{\mathbf{A}}^{-1}\mathbf{B}^T$. The matrix $\tilde{\mathbf{A}}$ can be preconditioned by a multigrid solver or by a preconditioned Krylov subspace method. The problem of how to precondition the Schur complement can be very difficult, in particular if in the two-phase flow problem we have (very) large jumps in the viscosity and/or density values across the interface. Below we discuss a few Schur complement preconditioning methods.

9.2.1 Schur complement preconditioners

In Sect. 5.4.3 we treated Schur complement preconditioners for *one*-phase flow problems. We considered three problem classes: stationary Stokes-, generalized Stokes and Oseen problems. For the first two classes "optimal" (i.e. robust w.r.t. variation in the mesh size and in the time step) Schur complement preconditioners are known, namely the mass matrix preconditioner in Theorem 5.4.27 and the Cahouet-Chabard preconditioner in (5.108), respectively. Fairly complete theoretical analyses are available which show the optimality of these preconditioners. These analyses are essentially applications of the abstract results in Sect. 15.5. For the case of the Oseen problem an "optimal" (i.e. robust with respect to variation in the mesh size, the time step and in the convection/diffusion ratio) is not known, yet. We briefly addressed two possible preconditioners: if the Oseen problem is diffusion-dominated it makes sense to use the same Cahouet-Chabard preconditioner as for the generalized Stokes case, otherwise the so-called $BFBt$ preconditioner might be a good option.

In this section we treat the issue of Schur complement preconditioning for the case of a *two*-phase flow problem. If the jumps in viscosity and density are small, as is the case in many liquid-liquid systems, then the linear systems that arise in the discrete linearized (Navier-)Stokes equations are similar to those that occur in a one-phase flow problem. Hence, in that case the preconditioners discussed above for the one-phase flow problem can be expected to perform well. If on the other hand the jumps in density and/or viscosity are large (e.g. in liquid-gas systems) then one needs suitably modified preconditioners. In that case the problem of finding "optimal" Schur complement preconditioners is largely unsolved. This problem is hard, since one then wants to have a feasible preconditioner such that the preconditioned Schur complement is well-conditioned independent of many parameters, namely the mesh size, the time step, the convection/diffusion ratio, the density ratio and the viscosity ratio. Below we present some first results on Schur complement preconditioning for two-phase flow problems. As for the one-phase flow problem in Sect. 5.4.3 we consider three classes of problems with increasing complexity: stationary Stokes- ((9.25) with $\mathbf{N} = 0$ and $\beta = 0$), generalized Stokes ((9.25) with $\mathbf{N} = 0$) and Oseen problems (general case (9.25)).

Stationary Stokes problem

For the derivation of an appropriate Schur complete preconditioner for the discrete stationary Stokes problem we first derive a preconditioner for the continuous Schur complement operator, cf. also Sect. 15.5.3. We consider the following model problem of a stationary Stokes problem with a discontinuous piecewise constant viscosity coefficient $\mu = \mu_i$ in Ω_i, $i = 1, 2$. We consider the three-dimensional case, i.e. $\Omega \subset \mathbb{R}^3$. Without loss of generality we can assume that Ω_1 is the subdomain with the larger viscosity coefficient and by suitable rescaling we can assume $\mu_1 = 1$, i.e we take

$$\mu = \begin{cases} 1 & \text{in } \Omega_1 \\ \mu_2 & \text{in } \Omega_2, \quad \mu_2 \in (0, 1]. \end{cases}$$

The model Stokes problem that we consider reads: determine $\mathbf{u}$ and p such that

$$\begin{aligned} -\operatorname{div}(\mu(x)\nabla \mathbf{u}) + \nabla p &= \mathbf{g} \quad \text{in } \Omega_i, \\ \operatorname{div} \mathbf{u} &= 0 \quad \text{in } \Omega_i, \\ [\mathbf{u}] = 0, \quad [(-p\mathbf{I} + \mu \nabla \mathbf{u})\mathbf{n}] &= \tilde{\mathbf{g}} \quad \text{on } \Gamma = \partial\Omega_1 \cap \partial\Omega_2, \\ \mathbf{u} &= 0 \quad \text{on } \partial\Omega. \end{aligned}$$

The function $\tilde{\mathbf{g}}$ is assumed to be given and is used to model a localized surface tension force. For the variational formulation of this problem it turns out to be convenient to use the space

$$M := \left\{ p \in L^2(\Omega) : \int_\Omega \mu^{-1} p(x)\, dx = 0 \right\} \tag{9.26}$$

instead of the standard pressure space $L_0^2(\Omega)$. Note that $p \in M$ iff $p + c \in L_0^2(\Omega)$ with a constant $c = c(p)$ given by $c = \frac{\mu_2 - 1}{|\Omega|} \int_{\Omega_2} p(x)\, dx$. For the velocity we use $\mathbf{V} = H_0^1(\Omega)^3$ and on $\mathbf{V}$ we introduce the bilinear form

$$a_\mu(\mathbf{u}, \mathbf{v}) = (\mu \nabla \mathbf{u}, \nabla \mathbf{v})_{L^2} = \sum_{i=1}^{3} \int_\Omega \mu \nabla u_i \cdot \nabla v_i\, dx.$$

The variational problem is as follows: given $\mathbf{f} \in \mathbf{V}'$ find $(\mathbf{u}, p) \in \mathbf{V} \times M$ such that

$$\begin{aligned} a_\mu(\mathbf{u}, \mathbf{v}) + b(\mathbf{v}, p) &= \mathbf{f}(\mathbf{v}) \quad \text{for } \mathbf{v} \in \mathbf{V}, \\ b(\mathbf{u}, q) &= 0 \qquad \text{for } q \in M. \end{aligned} \tag{9.27}$$

The functional $\mathbf{f}(\mathbf{v}) := (\mathbf{g}, \mathbf{v})_{L^2} + \int_\Gamma \tilde{\mathbf{g}} \cdot \mathbf{v}\, ds$ takes both the (exterior) force $\mathbf{g}$ and the localized force $\tilde{\mathbf{g}}$ into account. The bilinear form $a_\mu(\cdot, \cdot)$ defines a scalar product on $\mathbf{V}$. We use the norm induced by this scalar product:

$$\|\mathbf{u}\|_V := a_\mu(\mathbf{u}, \mathbf{u})^{\frac{1}{2}} \qquad \text{for } \mathbf{u} \in \mathbf{V}. \tag{9.28}$$

On $L^2(\Omega)$ (and thus also on M), besides the L^2 scalar product we will also use a weighted L^2 scalar product:

$$(p, q)_M := \int_\Omega \mu^{-1} p\, q\, dx = (\mu^{-1} p, q)_{L^2} \qquad \text{for } p, q \in M, \tag{9.29}$$

and $\|p\|_M := (p, p)_M^{\frac{1}{2}}$. The norms $\|\cdot\|_V$ and $\|\cdot\|_M$, which both depend on μ, are such that continuity and an inf-sup property can be shown to hold, with constants *independent* of the value of the viscosity coefficient μ:

Theorem 9.2.1 *The following holds:*

$$|b(\mathbf{u}, p)| \leq \sqrt{3} \|\mathbf{u}\|_V \|p\|_M \quad \textit{for all } \mathbf{u} \in \mathbf{V},\ p \in M.$$

There exists a constant $\gamma_b > 0$ *independent of* μ *such that*

$$\sup_{\mathbf{u} \in \mathbf{V}} \frac{b(\mathbf{u}, p)}{\|\mathbf{u}\|_V} \geq \gamma_b \|p\|_M \quad \textit{for all } p \in M.$$

Proof. A proof of these results is given in [193]. □

Using standard arguments, cf. Sect. 15.5, one obtains the following robust (i.e. uniformly w.r.t. variation in μ) preconditioner of the Schur complement. We identify $L^2(\Omega)$ with its dual, i.e. in particular for $g \in M'$ we have $g(p) = (g, p)_{L^2}$ for all $p \in M$. The Schur complement mapping $S : M \to M$ corresponding to the problem (9.27) and its induced norm on M are given by

$$\|p\|_S = (Sp, p)_{L^2}^{\frac{1}{2}} = \sup_{\mathbf{v} \in \mathbf{V}} \frac{b(\mathbf{v}, p)}{\|\mathbf{v}\|_V}.$$

Corollary 9.2.2 Define $I_\mu : L^2(\Omega) \to L^2(\Omega)$ by $(I_\mu p, q)_{L^2} = (p, q)_M$. Then the following holds:

$$\gamma_b^2 (I_\mu p, p)_{L^2} \le (Sp, p)_{L^2} \le 3(I_\mu p, p)_{L^2} \quad \text{for all } p \in M. \tag{9.30}$$

Proof. Follows with the same arguments as in Remark 15.5.1. □

Hence, the (simple) operator I_μ, which is the *scaled identity* given by $I_\mu p = \mu^{-1} p$, can be used as a preconditioner for the Schur complement S, and on M the spectral condition number of $I_\mu^{-1} S$ is bounded by $3\gamma_b^{-2}$, which is independent of μ. In this sense I_μ is a *robust preconditioner* of S.

In [193] it is shown, that an analogon holds for the *discrete* case. We briefly explain this discrete Schur complement preconditioner, more details can be found in [193]. Assume that we use finite element subspaces $\mathbf{V}_h \subset \mathbf{V}$, $M_h \subset M$. We consider the Galerkin discretization of the Stokes problem (9.27): find $(\mathbf{u}_h, p_h) \in \mathbf{V}_h \times M_h$ such that

$$\begin{aligned} a_\mu(\mathbf{u}_h, \mathbf{v}_h) + b(\mathbf{v}_h, p_h) &= \mathbf{f}(\mathbf{v}_h) \quad \text{for } \mathbf{v}_h \in \mathbf{V}_h, \\ b(\mathbf{u}_h, q_h) &= 0 \qquad\quad \text{for } q_h \in M_h. \end{aligned} \tag{9.31}$$

The matrix representation of this discrete problem has a form as in (9.25), with $\mathbf{N} = 0$ and $\beta = 0$. The Schur complement matrix is given by $\mathbf{S} = \mathbf{B}\mathbf{A}^{-1}\mathbf{B}^T$. In practice the finite element space M_h is constructed by taking a finite element space, denoted by M_h^+ (e.g. $M_h^+ = \mathbb{X}_h^{k-1}$ or the P_1-XFEM space), and then adding the orthogonality condition:

$$M_h = \{\, p_h \in M_h^+ : (p_h, 1)_M = 0 \,\}.$$

Let $(\psi_i)_{1 \le i \le K}$ be the standard (nodal) basis in M_h^+ and $\mathbf{M}_\mu$ the mass matrix corresponding to $(\cdot,\cdot)_M$:

$$(\mathbf{M}_\mu)_{ij} = (\psi_i, \psi_j)_M = \int_\Omega \mu^{-1} \psi_i \psi_j \, dx, \quad 1 \le i, j \le K. \tag{9.32}$$

Hence, compared to the standard mass matrix in the pressure finite element space, there is a *scaling* with μ^{-1}. The vector representation of the subspace $M_h \subset M_h^+$ of functions $p_h \in M_h^+$ that satisfy the orthogonality condition

$(p_h, 1)_M = 0$ is given by $\left(\mathbf{M}_\mu(1,\ldots,1)\right)^\perp$. On that space the Schur complement matrix $\mathbf{S}$ is nonsingular. In [193] it is proved that if the pair $(\mathbf{V}_h, M_h)$ is LBB stable then the following holds:

$$\hat{\beta}^2\langle \mathbf{M}_\mu \mathbf{y}, \mathbf{y}\rangle \leq \langle \mathbf{S}\mathbf{y}, \mathbf{y}\rangle \leq 3\langle \mathbf{M}_\mu \mathbf{y}, \mathbf{y}\rangle \quad \text{for all } \mathbf{y} \in \left(\mathbf{M}_\mu(1,\ldots,1)\right)^\perp, \tag{9.33}$$

which is a discrete analogon of the result in (9.30). The constant $\hat{\beta} > 0$ is the LBB constant for the pair $(\mathbf{V}_h, M_h)$. In the analysis (cf. Theorem 6 in [193]), apart from the LBB-stability assumption we need a further technical assumption, which is satisfied for standard finite element spaces, and we need that the triangulations used in the finite element spaces are *fitted* to the interface. For our applications the latter assumption is *not* realistic. It turns out, however, that for our applications the scaled mass matrix $\mathbf{M}_\mu$ yields a good preconditioner for the Schur complement (in case of a Stokes problem) even if this assumption is not satisfied.

Remark 9.2.3 In [192] it is shown that similar results hold if instead of M one uses the standard pressure space $L_0^2(\Omega)$ with norm $\|\cdot\|_M$.

Generalized Stokes problem

We now treat the issue of Schur complement preconditioning of a problem as in in (9.25), with $\mathbf{N} = 0$. For the case of "small" jumps in viscosity and density it makes sense to use the same preconditioners as in the *one*-phase flow problem, thus in particular for the generalized Stokes problem this is the Cahouet-Chabard preconditioner in (5.108). We introduce a generalization of that preconditioner for the case of a generalized Stokes problem with discontinuous viscosity and density coefficients. This preconditioner has a certain robustness property with respect to the size of the jumps in the viscosity and density coefficients. The results that we present are from [193, 189]. We apply the general abstract analysis presented in Sect. 15.5. As for the stationary Stokes problem treated above we first derive the preconditioner for the *continuous* Schur complement operator and then briefly address a discrete analogon. We consider the following simplified generalized Stokes interface problem: Find $\mathbf{u}$ and p such that

$$-\operatorname{div}(\mu(x)\mathbf{D}(\mathbf{u})) + \xi\rho(x)\mathbf{u} + \nabla p = \mathbf{g} \quad \text{in } \Omega_i, \tag{9.34a}$$

$$\operatorname{div}\mathbf{u} = 0 \quad \text{in } \Omega_i, \qquad i = 1, 2, \tag{9.34b}$$

$$[\mathbf{u}] = 0, \ [\boldsymbol{\sigma}\mathbf{n}] = \tilde{\mathbf{g}} \quad \text{on } \Gamma = \partial\Omega_1 \cap \partial\Omega_2, \tag{9.34c}$$

$$\mathbf{u} = 0 \quad \text{on } \partial\Omega. \tag{9.34d}$$

We use standard notations: $\boldsymbol{\sigma} = -p\mathbf{I} + \mu\mathbf{D}(\mathbf{u})$, $\mathbf{D}(\mathbf{u}) = \nabla\mathbf{u} + \nabla\mathbf{u}^T$.

The function $\tilde{\mathbf{g}}$ is given and models surface tension effects. We assume piecewise constant viscosity and density. Suitable scaling can be used to ensure that viscosity and density are equal to one in Ω_1. Hence, we assume

$$\mu = \begin{cases} 1 & \text{in } \Omega_1, \\ \mu_2 > 0 & \text{in } \Omega_2, \end{cases} \qquad \rho = \begin{cases} 1 & \text{in } \Omega_1, \\ \rho_2 > 0 & \text{in } \Omega_2. \end{cases} \tag{9.35}$$

The weak formulation leads to a saddle point problem of generalized Stokes type: determine $\mathbf{u} \in \mathbf{V} = H_0^1(\Omega)^3$, $p \in L_0^2(\Omega)$ such that

$$\begin{aligned} a_\mu(\mathbf{u}, \mathbf{v}) + \xi c_\rho(\mathbf{u}, \mathbf{v}) + b(\mathbf{v}, p) &= \mathbf{f}(\mathbf{v}) \quad \text{for all } \mathbf{v} \in \mathbf{V}, \\ b(\mathbf{u}, q) &= 0 \quad \text{for all } q \in L_0^2(\Omega), \end{aligned} \tag{9.36}$$

with

$$a_\mu(\mathbf{u}, \mathbf{v}) := \frac{1}{2} \int_\Omega \mu \operatorname{tr}\big(\mathbf{D}(\mathbf{u})\mathbf{D}(\mathbf{v})\big)\, dx\ , \quad c_\rho(\mathbf{u}, \mathbf{v}) := (\rho \mathbf{u}, \mathbf{v})_{L^2},$$

$$b(\mathbf{v}, p) := -(p, \operatorname{div} \mathbf{v})_{L^2}, \quad \mathbf{f}(\mathbf{v}) := (\mathbf{g}, \mathbf{v})_{L^2} + \int_\Gamma \tilde{\mathbf{g}} \cdot \mathbf{v}\, ds.$$

We introduce the following Hilbert spaces:

$$\begin{aligned} H_1 &= \{\ \mathbf{v} \in \mathbf{V}, \text{ with } \|\mathbf{v}\|_{H_1}^2 := \frac{1}{2} \int_\Omega \mu \operatorname{tr}\big((\mathbf{D}(\mathbf{v})^2\big)\, dx\ \}, \\ H_2 &= \{\ \mathbf{v} \in L^2(\Omega)^3, \text{ with } \|\mathbf{v}\|_{H_2} := \|\rho^{\frac{1}{2}} \mathbf{v}\|_{L^2}\ \}. \end{aligned}$$

Due to Korn's inequality $\|\cdot\|_{H_1}$ defines a norm on $\mathbf{V}$. Related to this norm we need the following *uniform* (w.r.t. μ) equivalence result. Assume that one of the following conditions is satisfied:

$$\operatorname{meas}(\partial\Omega_i \cap \partial\Omega) > 0 \quad \text{for } \ i = 1, 2, \tag{9.37a}$$

$$\operatorname{meas}(\partial\Omega_2 \cap \partial\Omega) > 0 \quad \text{and } \ \mu_2 \geq C > 0. \tag{9.37b}$$

Then there exists a constant $\tilde{c} > 0$ *independent* of μ (but depending on C from (9.37b)) such that

$$\tilde{c} \|\mu^{\frac{1}{2}} \nabla \mathbf{v}\|_{L^2} \leq \|\mathbf{v}\|_{H_1} \leq \|\mu^{\frac{1}{2}} \nabla \mathbf{v}\|_{L^2} \quad \text{for all } \ \mathbf{v} \in H_1. \tag{9.38}$$

This result follows from Lemma 5 in [193] (there, instead of condition (9.37b) the condition "$\operatorname{meas}(\partial\Omega_1 \cap \partial\Omega) > 0$ and $\mu_2 \leq C$" is used; the arguments in the proof, however, still apply if the condition (9.37b) is used). Note that in view of our applications (stationary droplet problem) the condition (9.37b) is reasonable.

We use the same pressure space M as in the stationary Stokes problem treated above:

$$M = \big\{ p \in L^2(\Omega) : (p, 1)_M = 0, \text{ with } \ \|p\|_M^2 := (\mu^{-1} p, p)_{L^2} \big\}.$$

The scalar product corresponding to $\|\cdot\|_M$ is denoted by $(\cdot, \cdot)_M$. These bilinear forms and spaces satisfy the Assumptions (15.27a)-(15.27d) that are the basis of the general analysis presented in Sect. 15.5.

Lemma 9.2.4 *Assume that* (9.37a) *or* (9.37b) *is satisfied. The following holds with constants* $\Gamma_b, \gamma_b > 0$ *independent of* μ *and* ρ:

$$\|\mathbf{u}\|_{H_1}^2 = a_\mu(\mathbf{u},\mathbf{u}), \quad a_\mu(\mathbf{u},\mathbf{v}) \le \|\mathbf{u}\|_{H_1}\|\mathbf{v}\|_{H_1}, \quad \mathbf{u},\mathbf{v} \in H_1, \tag{9.39a}$$

$$\|\mathbf{u}\|_{H_2}^2 = c_\rho(\mathbf{u},\mathbf{u}), \quad c_\rho(\mathbf{u},\mathbf{v}) \le \|\mathbf{u}\|_{H_2}\|\mathbf{v}\|_{H_2}, \quad \mathbf{u},\mathbf{v} \in H_2, \tag{9.39b}$$

$$|b(\mathbf{u},p)| \le \Gamma_b \|\mathbf{u}\|_{H_1}\|p\|_M, \quad \mathbf{u} \in H_1,\ p \in M, \tag{9.39c}$$

$$\gamma_b \|p\|_M \le \sup_{\mathbf{u}\in H_1} \frac{b(\mathbf{u},p)}{\|\mathbf{u}\|_{H_1}}, \quad p \in M. \tag{9.39d}$$

Proof. The results in (9.39a)-(9.39b) are direct consequences of $\|\mathbf{u}\|_{H_1}^2 = a_\mu(\mathbf{u},\mathbf{u})$ and $\|\mathbf{u}\|_{H_2}^2 = c_\rho(\mathbf{u},\mathbf{u})$. The results in (9.39c)-(9.39d) follow from (9.38) and Theorem 9.2.1. □

We apply the general abstract results derived in Sect. 15.5 with the spaces H_1, H_2, M defined above, with $\hat{a}(\cdot,\cdot) = a_\mu(\cdot,\cdot)$, $\hat{c}(\cdot,\cdot) = c_\rho(\cdot,\cdot)$, $\hat{b}(\cdot,\cdot) = b(\cdot,\cdot)$ and with $\tau = \xi$. The application is along the same lines as treated for the generalized Stokes problem (without jumps in viscosity and density) in Sect. 15.5.3.

We identify $L^2(\Omega)$ with its dual and then have

$$\begin{aligned} H_1' &= \{\ \mathbf{f} \in \mathbf{V}',\ \text{with } \|\mathbf{f}\|_{H_1'} = \sup_{\mathbf{v}\in\mathbf{V}} \frac{\mathbf{f}(\mathbf{v})}{\|\mathbf{v}\|_{H_1}}\ \},\\ H_2' &= \{\ \mathbf{v} \in L^2(\Omega)^3, \text{with } \|\mathbf{v}\|_{H_2'} = \|\rho^{-\frac12}\mathbf{v}\|_{L^2}\ \},\\ M' &= \{\ p \in M,\ \text{with } \|p\|_{M'} = \|\mu^{\frac12} p\|_{L^2}\ \}. \end{aligned}$$

The norm $\|\cdot\|_{H_2}$ is equivalent to the standard L^2-norm. Hence, the space $W = \{p \in M : Bp = \nabla p \in H_2'\}$ is, apart from the different orthogonalization condition $(p,1)_M$ that is used in M, the same as the one for the generalized Stokes problem in Sect. 15.5.3:

$$W = H^1(\Omega) \cap M, \quad \text{with norm } \|p\|_W = \|\rho^{-\frac12}\nabla p\|_{L^2}. \tag{9.40}$$

The Schur complement operator $S_{\mu,\rho} : M \to M$ corresponding to the generalized Stokes interface problem (9.36) is characterized by

$$(S_{\mu,\rho}p,p)_{L^2}^{\frac12} = \sup_{\mathbf{v}\in\mathbf{V}} \frac{(p, \operatorname{div}\mathbf{v})_{L^2}}{\big(a_\mu(\mathbf{v},\mathbf{v}) + \xi c_\rho(\mathbf{v},\mathbf{v})\big)^{\frac12}}, \quad p \in M. \tag{9.41}$$

We take the preconditioner given in Theorem 15.5.10:

$$(\tilde{S}_{\mu,\rho}p,p)_{L^2}^{\frac12} = \|p\|_{M+\xi^{-1}W} = \inf_{q\in W}\big(\|p-q\|_M + \xi^{-1}\|\rho^{-\frac12}\nabla q\|_{L^2}^2\big)^{\frac12}. \tag{9.42}$$

This preconditioner can be characterized using a Neumann solution operator by applying a similar approach as in Theorem 15.5.14. We can apply the general analysis of Sect. 15.5 to derive a uniform spectral bound $S_{\mu,\rho} \le c\tilde{S}_{\mu,\rho}$. This is summarized in the following theorem.

Theorem 9.2.5 *Assume that one of the conditions* (9.37a) *or* (9.37b) *is satisfied. Denote by* $-\Delta_\rho^{-1}: M \to H^1(\Omega) \cap M$ *the solution operator of the following Neumann interface problem: Given* $f \in M$*, find* $p = -\Delta_\rho^{-1} f \in H^1(\Omega) \cap M$ *such that*

$$(\rho^{-1}\nabla p, \nabla q)_{L^2} = (f, q)_{L^2}, \quad \textit{for all} \;\; q \in H^1(\Omega) \cap M.$$

The scaled identity $I_\mu : L^2(\Omega) \to L^2(\Omega)$ *is defined by* $(I_\mu p, q)_{L^2} = (p, q)_M$ *for all* $p, q \in L^2(\Omega)$*. Then*

$$\tilde{S}_{\mu,\rho}^{-1} = I_\mu^{-1} - \xi \Delta_\rho^{-1} \tag{9.43}$$

holds, and for all $p \in M$

$$(S_{\mu,\rho} p, p)_{L^2} \le c(\tilde{S}_{\mu,\rho} p, p)_{L^2}$$

holds, with a constant c *independent of* ξ*,* μ *and* ρ*.*

Proof. The proof is based on applying Theorem 15.5.6. A complete analysis is given in [189]. □

The preconditioner $\tilde{S}_{\mu,\rho}$ in (9.43) is a natural generalization of the Cahouet-Chabard preconditioner derived in Sect. 15.5.3. In the case of jumps in viscosity and/or density we now have a scaling (with μ^{-1}) in the identity operator I_μ and a scaling (with ρ^{-1}) in the Neumann operator Δ_ρ.

In Theorem 9.2.5 we have a spectral inequality $S_{\mu,\rho} \le c\tilde{S}_{\mu,\rho}$ that is uniform with respect to both the parameter ξ ($\sim \frac{1}{\Delta t}$) and the jumps in the coefficients μ, ρ *without using any regularity assumptions.* To derive a spectral inequality $\tilde{S}_{\mu,\rho} \le cS_{\mu,\rho}$ we need (at least in our analysis) regularity results for a stationary Stokes interface problem of the form

$$-\operatorname{div}\left(\mu(x)\mathbf{D}(\mathbf{u})\right) + \nabla p = \mathbf{g} \quad \text{in } \Omega_i, \tag{9.44a}$$

$$\operatorname{div}\mathbf{u} = 0 \quad \text{in } \Omega_i, \quad i = 1, 2, \tag{9.44b}$$

$$[\mathbf{u}] = 0, \; [\boldsymbol{\sigma}\mathbf{n}] = \tilde{\mathbf{g}} \quad \text{on } \Gamma, \tag{9.44c}$$

$$\mathbf{u} = 0 \quad \text{on } \Omega. \tag{9.44d}$$

Similarly to the Stokes case in Sect. 15.5.3, verifying Assumption 15.5.7 is based on regularity properties of this problem. This important issue is largely unsolved. The following result is found in the literature (see [223]): If the interface $\Gamma = \partial\Omega_1 \cap \partial\Omega_2$ is sufficiently smooth and has no common points with $\partial\Omega$ and $\mathbf{g} \in L^2(\Omega)^3$ then a solution $(\mathbf{u}, p)$ of (9.44a)–(9.44d) belongs to $H^2(\Omega_i)^3 \times H^1(\Omega_i)$, $i = 1, 2$. However, in these results and in other analyses known in the literature the dependence of constants in the a-priori estimates on μ is not known. Due to this we are not able to prove a result $\tilde{S}_{\mu,\rho} \le cS_{\mu,\rho}$ that is uniform both with respect to ξ and the jumps in μ, ρ. Below we present a result in which the spectral inequality is *uniform with respect to* ξ *only.*

Theorem 9.2.6 *Assume that one of the conditions* (9.37a) *or* (9.37b) *is satisfied and that the domain* $\Omega \subset \mathbb{R}^3$ *is such that the Stokes problem* (15.53) *is* H^2*-regular. Let* $\tilde{S}_{\mu,\rho}$ *be the preconditioner from* (9.43). *There exists a constant* c *independent of* ξ *such that for all* $p \in M$

$$(\tilde{S}_{\mu,\rho}p, p)_{L^2} \leq c(S_{\mu,\rho}p, p)_{L^2}$$

holds.

Proof. We refer to [189]. □

We briefly address a *discrete analogon* of the preconditioner $\tilde{S}_{\mu,\rho}$. Assume that we use finite element subspaces $\mathbf{V}_h \subset \mathbf{V}$, $M_h \subset M$. We consider the Galerkin discretization of the Stokes problem (9.36): find $(\mathbf{u}_h, p_h) \in \mathbf{V}_h \times M_h$ such that

$$\begin{aligned} a_\mu(\mathbf{u}_h, \mathbf{v}_h) + \xi c_\rho(\mathbf{u}_h, \mathbf{v}_h) + b(\mathbf{v}_h, p_h) &= \mathbf{f}(\mathbf{v}_h) \quad \text{for } \mathbf{v}_h \in \mathbf{V}_h, \\ b(\mathbf{u}_h, q_h) &= 0 \qquad\quad \text{for } q_h \in M_h. \end{aligned} \tag{9.45}$$

The matrix representation of this discrete problem has a form as in (9.25), with $\tilde{\mathbf{A}} = \mathbf{A} + \xi\mathbf{M}$. The Schur complement matrix is given by $\mathbf{S} = \mathbf{B}\tilde{\mathbf{A}}^{-1}\mathbf{B}^T$. Let $\mathbf{M}_\mu$ be the scaled mass matrix in the discrete pressure space as in (9.32), i.e.

$$(\mathbf{M}_\mu)_{ij} = (\psi_i, \psi_j)_M = \int_\Omega \mu^{-1}\psi_i\psi_j \, dx, \quad 1 \leq i, j \leq K.$$

We now distinguish two cases: $M_h \subset H^1(\Omega)$ (e.g. $M_h = \mathbb{X}_h^1$) and $M_h \not\subset H^1(\Omega)$ (e.g. XFEM pressure space). In the former case we can discretize the Neumann problem used in Theorem 9.2.5 in the space M_h, resulting in a discrete approximation of Δ_ρ^{-1}. Let $-\Delta_{\rho,h}^{-1} : M_h \to M_h$, $-\Delta_{\rho,h}^{-1} f_h = p_h$ be the solution operator of the following discrete Neumann problem: Find $p_h \in M_h$ such that

$$(\rho^{-1}\nabla p_h, \nabla q_h)_{L^2} = (f_h, q_h)_{L^2} \quad \text{for all } q_h \in M_h.$$

Let $\mathbf{A}_\rho^{-1}$ be the matrix representation of this solution operator. Then the discrete analogon of $\tilde{S}_{\mu,\rho}^{-1}$ in (9.43) is given by

$$\mathbf{Q}_S^{-1} = \tilde{\mathbf{S}}_{\mu,\rho}^{-1} := \mathbf{M}_\mu^{-1} + \xi\mathbf{A}_\rho^{-1}. \tag{9.46}$$

In [189] this preconditioner is used in numerical experiments with a Hood-Taylor P_2-P_1 finite element pair and turns out to have very good robustness properties w.r.t. variation in h, ξ, and the jumps in μ, ρ.

We now address the case $M_h \not\subset H^1(\Omega)$. Due to this nonconformity, a Galerkin discretization of the Neumann problem in the space M_h is not possible. Instead one can use the form of the Schur complement preconditioner as in (15.50). The discrete analogon is given by

$$\mathbf{Q}_S^{-1} = \tilde{\mathbf{S}}_{\mu,\rho}^{-1} := \mathbf{M}_\mu^{-1} + \xi(\mathbf{B}\mathbf{M}^{-1}\mathbf{B}^T)^{-1}. \tag{9.47}$$

Analogous to $\mathbf{A}_\rho$, for the operator $\mathbf{B}\mathbf{M}^{-1}\mathbf{B}^T$ there is a weighting with the density, namely through the weighted velocity mass matrix $\mathbf{M}$, cf. the definition of $\mathbf{M}$ in Sect. 7.11.1.

Remark 9.2.7 In [189] it is shown that similar results hold if instead of M one uses the standard pressure space $L_0^2(\Omega)$ with norm $\|\cdot\|_M$.

Oseen problem

We briefly address Schur complement preconditioning for the general Oseen problem (9.25). If the Oseen problem is diffusion dominated, the modified Cahouet-Chabard preconditioner treated above, cf. (9.46) or (9.47), is an obvious possibility. For stronger convection the following alternative may be better. In Sect. 5.4.3 we explained the $BFBt$ Schur complement preconditioner for a one-phase Oseen problem. Its definition is completely algebraic (i.e., based on the given mass and stiffness matrices) and can immediately be extended to the case of a two-phase Oseen problem. We recall the definition of this preconditioner. Let $\mathbf{M}^*$ be the velocity mass matrix (*without* a weighting with ρ) and $\mathbf{M}_1 := \operatorname{diag}(\mathbf{M}^*)$. The $BFBt$ preconditioner, cf. (5.109), is given by

$$\mathbf{Q}_S^{-1} = (\mathbf{B}\mathbf{M}_1^{-1}\mathbf{B}^T)^{-1}\,\mathbf{B}\,\mathbf{M}_1^{-1}\,\tilde{\mathbf{A}}\,\mathbf{M}_1^{-1}\,\mathbf{B}^T\,(\mathbf{B}\mathbf{M}_1^{-1}\mathbf{B}^T)^{-1}. \tag{9.48}$$

It turns out, that in the case of discretization with the XFEM method the matrix $\mathbf{B}\mathbf{M}_1^{-1}\mathbf{B}^T$ is in general extremely ill-conditioned due to large differences in scaling of the basis functions. Hence, systems with this matrix are very hard to solve. It is much better to use this preconditioner in a different form, in which this poor scaling is avoided. For this we introduce $\tilde{\mathbf{B}} = \mathbf{M}_{p,1}^{-\frac{1}{2}}\mathbf{B}\mathbf{M}_1^{-\frac{1}{2}}$, with $\mathbf{M}_p$ the mass matrix in the pressure space (without scaling) and $\mathbf{M}_{p,1} := \operatorname{diag}(\mathbf{M}_p)$. Thus $\tilde{\mathbf{B}}$ is a rescaled version of the discrete divergence matrix $\mathbf{B}$. The $BFBt$ Schur complement preconditioner $\mathbf{Q}_S^{-1}$ can also be represented as

$$\mathbf{Q}_S^{-1} = \mathbf{M}_{p,1}^{-\frac{1}{2}}(\tilde{\mathbf{B}}\tilde{\mathbf{B}}^T)^{-1}\,\tilde{\mathbf{B}}\,\mathbf{M}_1^{-\frac{1}{2}}\,\tilde{\mathbf{A}}\,\mathbf{M}_1^{-\frac{1}{2}}\,\tilde{\mathbf{B}}^T\,(\tilde{\mathbf{B}}\,\tilde{\mathbf{B}}^T)^{-1}\,\mathbf{M}_{p,1}^{-\frac{1}{2}}. \tag{9.49}$$

In this representation we have to solve linear systems with the matrix $\tilde{\mathbf{B}}\,\tilde{\mathbf{B}}^T$ instead of with $\mathbf{B}\mathbf{M}_1^{-1}\mathbf{B}^T$. Due to better scaling, systems with $\tilde{\mathbf{B}}\,\tilde{\mathbf{B}}^T$ are much easier to solve than those with $\mathbf{B}\mathbf{M}_1^{-1}\mathbf{B}^T$.

9.3 Numerical experiments

We reconsider the static droplet test case described in Sect. 7.10.3, where a static droplet $\Omega_1 = \{\, x \in \mathbb{R}^3 : \|x\| \le r \,\}$ is located inside the cube $\Omega = (-1,1)^3$ with $r = 2/3$. Assuming $\mathbf{g} = 0$ and $\tau = 1$, i.e., no gravitational force but only surface tension is acting, the solution of the stationary two-phase Stokes problem is analytically known: the velocity vanishes in the whole domain, i. e., $\mathbf{u}^* = 0$, and the pressure is piecewise constant in the two phases with a jump $[p]_\Gamma = \tau\kappa = 3$ across the interface Γ, cf. Fig. 7.19.

For spatial discretization a coarse grid $\mathcal{T}_0$ consisting of $4 \times 4 \times 4$ subcubes each consisting of 6 tetrahedra is created. This coarse grid is successively refined in the vicinity of the interface yielding a sequence of tetrahedral grids $\mathcal{T}_i$, $i = 0, \ldots, 3$ with grid sizes $h_i = 2^{-i-1}$, $i = 0, \ldots, 3$. See Fig. 7.9 for an illustration of a very similar adaptive mesh. We use piecewise quadratic finite elements for the velocity space $\mathbf{V}_h$ and either piecewise linears ($Q_h = Q_h^1$) or the reduced XFEM discretization ($Q_h = \tilde{Q}_h^{\Gamma_h}$, cf. Sect. 7.9) for the pressure space Q_h.

The discretization leads to an algebraic saddle point system of the form

$$\mathbf{K}\begin{pmatrix}\mathbf{x}\\ \mathbf{y}\end{pmatrix} := \begin{pmatrix}\tilde{\mathbf{A}} & \mathbf{B}^T\\ \mathbf{B} & 0\end{pmatrix}\begin{pmatrix}\mathbf{x}\\ \mathbf{y}\end{pmatrix} = \begin{pmatrix}\mathbf{f}_1\\ \mathbf{f}_2\end{pmatrix}, \qquad \tilde{\mathbf{A}} := \mathbf{A} + \xi\mathbf{M}. \tag{9.50}$$

In the following sections we will analyze the iterative solution of this system for the stationary Stokes case ($\xi = 0$) and the generalized Stokes case ($\xi > 0$), cf. Sects. 9.3.1 and 9.3.2, respectively. The inexact Uzawa method is used in both cases for the iterative solution of (9.50) applying appropriate preconditioners $\mathbf{Q}_A$ and $\mathbf{Q}_S$. The starting vector is chosen as $(\vec{\mathbf{u}}_0, \vec{p}_0) = (0, 0)$. The iteration is stopped if a reduction of the Euclidean norm of the starting residual by a factor of 10^6 is achieved, i. e., $\|\mathbf{r}^k\| \leq 10^{-6}\|\mathbf{r}^0\|$. To measure the arithmetic costs which are dominated by the application of the preconditioners, the number of evaluations of $\mathbf{Q}_A^{-1}$ are counted. Note that the number of $\mathbf{Q}_S^{-1}$ and $\mathbf{Q}_A^{-1}$ evaluations are almost the same, cf. Remark 5.2.11, so we will not report the numbers of $\mathbf{Q}_S^{-1}$ evaluations in the following.

9.3.1 Stationary Stokes case

We first consider the stationary Stokes case where $\tilde{\mathbf{A}} = \mathbf{A}$ in (9.50). The preconditioners are chosen as follows. The Schur complement is preconditioned by $\mathbf{Q}_S^{-1} = \mathbf{M}_\mu^{-1}$. For $\mathbf{Q}_A$ we compared a Krylov-based approach to a multigrid approach, i. e., for $\mathbf{Q}_A^{-1}$ we use an SSOR-preconditioned CG method with zero initial guess reducing the Euclidean norm of the residual by a factor of 100 or, alternatively, apply 5 multigrid V-cycle iterations to $\tilde{\mathbf{A}}$. Computations are performed for different refinement levels and for fixed $\mu_2 = 1$, varying the viscosity μ_1 of the droplet. The number of $\mathbf{Q}_A^{-1}$ evaluations are reported in Table 9.2 for the standard FEM space $Q_h = Q_h^1$ and in Tables 9.3 and 9.4 for the reduced XFEM space $Q_h = \tilde{Q}_h^{\Gamma_h}$.

We first discuss the results for the standard pressure space $Q_h = Q_h^1$ given in Table 9.2. For the Krylov-based preconditioner $\mathbf{Q}_A$ the results are robust w. r. t. variation in the grid size h. Furthermore, robustness w. r. t. the viscosity ratio μ_1/μ_2 is observed for moderate ratios up to $1 : 100$. There is a slight increase of the number of $\mathbf{Q}_A^{-1}$ evaluations for the “extreme” case

$Q_h = Q_h^1$	Krylov-based $\mathbf{Q}_A^{-1}$			multigrid-based $\mathbf{Q}_A^{-1}$		
	$h=1/4$	$h=1/8$	$h=1/16$	$h=1/4$	$h=1/8$	$h=1/16$
$\mu_1/\mu_2 = 1$	36	34	36	42	38	36
$\mu_1/\mu_2 = 10^{-1}$	40	41	46	44	51	46
$\mu_1/\mu_2 = 10^{-2}$	40	40	47	77	79	97
$\mu_1/\mu_2 = 10^{-3}$	68	64	69	294	413	480

Table 9.2. Stationary Stokes case, standard FEM pressure space: number of $\mathbf{Q}_A^{-1}$ evaluations for different grid sizes h and viscosity ratios μ_1/μ_2, using a multigrid or Krylov-based preconditioner $\mathbf{Q}_A^{-1}$.

$Q_h = \tilde{Q}_h^{\Gamma_h}$,	Krylov-based $\mathbf{Q}_A^{-1}$			multigrid-based $\mathbf{Q}_A^{-1}$		
$\tilde{c} = 1$	$h=1/4$	$h=1/8$	$h=1/16$	$h=1/4$	$h=1/8$	$h=1/16$
$\mu_1/\mu_2 = 1$	36	34	36	42	39	37
$\mu_1/\mu_2 = 10^{-1}$	40	35	48	44	46	55
$\mu_1/\mu_2 = 10^{-2}$	40	48	71	77	81	139
$\mu_1/\mu_2 = 10^{-3}$	68	68	184	294	459	860

Table 9.3. Stationary Stokes case, XFEM pressure space, $\tilde{c} = 1$: number of $\mathbf{Q}_A^{-1}$ evaluations for different grid sizes h and viscosity ratios μ_1/μ_2, using a multigrid or Krylov-based preconditioner $\mathbf{Q}_A^{-1}$.

$Q_h = \tilde{Q}_h^{\Gamma_h}$,	Krylov-based $\mathbf{Q}_A^{-1}$			multigrid-based $\mathbf{Q}_A^{-1}$		
$\tilde{c} = 0.1$	$h=1/4$	$h=1/8$	$h=1/16$	$h=1/4$	$h=1/8$	$h=1/16$
$\mu_1/\mu_2 = 1$	38	34	37	43	42	43
$\mu_1/\mu_2 = 10^{-1}$	41	46	64	45	56	79
$\mu_1/\mu_2 = 10^{-2}$	51	80	131	100	144	197
$\mu_1/\mu_2 = 10^{-3}$	119	206	266	351	511	1338

Table 9.4. Stationary Stokes case, XFEM pressure space, $\tilde{c} = 0.1$: number of $\mathbf{Q}_A^{-1}$ evaluations for different grid sizes h and viscosity ratios μ_1/μ_2, using a multigrid or Krylov-based preconditioner $\mathbf{Q}_A^{-1}$.

$\mu_1/\mu_2 = 10^{-3}$, but compared to the case $\mu_1/\mu_2 = 10^{-2}$ the numbers are less than doubled.

Analyzing the results for the multigrid-based preconditioner $\mathbf{Q}_A$ we again observe a moderate robustness w. r. t. variation in h, but a strong dependence on μ_1/μ_2, especially for the "extreme" case $\mu_1/\mu_2 = 10^{-3}$, which will be discussed below.

We now turn to the XFEM case, i. e., $Q_h = \tilde{Q}_h^{\Gamma}$. The results for the reduced XFEM space with $\tilde{c} = 1$, cf. (7.108), are given in Table 9.3. For $h = 1/4, 1/8$ the results are comparable to the standard FEM case, but for $h = 1/16$ the effect of the smaller LBB constant becomes visible in terms of larger iteration numbers. For smaller $\tilde{c}$ the LBB constant deteriorates even more, which is confirmed by numerical experiments using a reduced XFEM space with $\tilde{c} = 0.1$, cf. Table 9.4. For this case there is a strong dependence of the number of $\mathbf{Q}_A^{-1}$ iterations on the viscosity ratio μ_1/μ_2.

We draw the following conclusions:

- The results show that for large viscosity ratios the *standard* multigrid method does not yield a satisfactory preconditioner $\mathbf{Q}_A$. This can be traced back to the fact that for discrete diffusion problems the multigrid method is known to be robust w. r. t. variation in h, but not w. r. t. the size of the jumps in the diffusion coefficient. The deterioration in the rate of convergence for problems with large jumps is known from numerical experiments in the literature but also reflected in theoretical analyses. For example, in [258] it is shown that using a multigrid V-cycle as preconditioner $\mathbf{Q}_A$ for the diffusion matrix $\mathbf{A}$ the condition number of the preconditioned system can be bounded by

$$\operatorname{cond}_2(\mathbf{Q}_A^{-1}\mathbf{A}) \leq C \min\left\{h^{-1}, \frac{\max \mu_i}{\min \mu_i}\right\}$$

  with a constant C independent of h, μ. This bound predicts a worse preconditioning quality for an increasing viscosity ratio, in particular for small h. A significant improvement can be obtained if modified multigrid solvers are used, for example, an *algebraic multigrid* method or a multigrid method with so-called *matrix dependent* prolongation and restriction operators.
- From the reasonably constant $\mathbf{Q}_A^{-1}$-evaluation counts in Table 9.2, in case of a Krylov-based $\mathbf{Q}_A^{-1}$, we conclude that $\mathbf{Q}_S^{-1} = \mathbf{M}_\mu^{-1}$ is a robust Schur complement preconditioner if in the discretization we use the standard FEM pressure space Q_h.
- Consider the discretization with the XFEM pressure space $\tilde{Q}_h^{\Gamma_h}$. The results in Table 9.3 and Table 9.4, for the case with a Krylov-based $\mathbf{Q}_A^{-1}$, indicate that the Schur complement preconditioner $\mathbf{M}_\mu^{-1}$ is still fairly robust, *provided* one does not take the value of the cut-off parameter $\tilde{c}$ too small. A theoretical analysis that explains the dependence of the quality of the Schur complement preconditioner on $\tilde{c}$ is not known.

Hence, even for this relatively simple stationary Stokes problem with a discontinuous viscosity coefficient, discretized with XFEM, the issue of robust Schur complement preconditioning needs further analysis.

9.3.2 Generalized Stokes case

In this section we treat the generalized Stokes case with $\tilde{\mathbf{A}} = \mathbf{A} + \xi\mathbf{M}$, $\xi > 0$, in (9.50), stemming from some time discretization where $\xi \sim (\Delta t)^{-1}$. The preconditioners are chosen as follows. The Schur complement is preconditioned by $\mathbf{Q}_S$ given in (9.47), but replacing the velocity mass matrix $\mathbf{M}$ in the Schur complement by its diagonal $\operatorname{diag}(\mathbf{M})$. For $\mathbf{Q}_A^{-1}$ we use an SSOR-preconditioned CG method with zero initial guess reducing the Euclidean norm of the residual by a factor of 100. Computations are performed for different refinement levels, different values of ξ and for fixed material properties $\rho_2 = \mu_2 = 1$ of the ambient phase, varying the viscosity μ_1 and density ρ_1 of the droplet. The parameter ranges considered in the numerical experiments are chosen as follows:

spatial discretization:	$h = 2^{-i}$	for $i = 2, 3, 4$,
temporal discretization:	$\xi = (\Delta t)^{-1} = 10^j$	for $j = 0, 1, 2, 3$,
droplet density:	$\rho_1 = 10^{-k}\rho_2$	for $k = 0, 1, 2, 3$,
droplet viscosity:	$\mu_1 = 10^{-l}\mu_2$	for $l = 0, 1, 2, 3$.

We first consider the standard FEM case, i. e., $Q_h = Q_h^1$. For fixed discretization parameters $h = 1/8$, $\xi = 100$ and varying material parameters $\mu_1 = 10^{-i}\mu_2$, $i = 0, 1, 2, 3$ and $\rho_1 = 10^{-j}\rho_2$, $j = 0, 1, 2, 3$, the number of $\mathbf{Q}_A^{-1}$ evaluations are reported in Table 9.5. The obtained results indicate robustness w. r. t. viscosity and density ratios in the whole range considered in the experiments. Now we fix the droplet material properties $\mu_1 = 10^{-2}\mu_2$ and $\rho_1 = 10^{-3}\rho_2$, varying the discretization parameters $h = 1/4, 1/8, 1/16$ and $\xi = 10^j$, $j = 0, 1, 2, 3$. The corresponding numbers of $\mathbf{Q}_A^{-1}$ evaluations can be found in the left part of Table 9.7. We observe only a mild dependence of the iteration numbers on the values of h and ξ, where the iteration numbers increase for smaller grid sizes h and larger time step sizes Δt. Varying all four parameters, the maximal iteration number observed was 72 for the case $h = 1/16$, $\xi = 1$, $\rho_1 = 10^{-1}\rho_2$, $\mu_1 = 10^{-3}\mu_2$. Overall, for the considered parameter range of h, ξ, ρ_1, μ_1 the iterative method turns out to be quite robust.

$h = 1/8$, $\xi = 10^2$	$\rho_1/\rho_2 = 1$	$\rho_1/\rho_2 = 10^{-1}$	$\rho_1/\rho_2 = 10^{-2}$	$\rho_1/\rho_2 = 10^{-3}$
$\mu_1/\mu_2 = 1$	19	19	19	19
$\mu_1/\mu_2 = 10^{-1}$	19	20	23	23
$\mu_1/\mu_2 = 10^{-2}$	29	25	26	26
$\mu_1/\mu_2 = 10^{-3}$	39	43	42	39

Table 9.5. Generalized Stokes case, standard FEM pressure space: number of $\mathbf{Q}_A^{-1}$ evaluations for different density ratios ρ_1/ρ_2 and viscosity ratios μ_1/μ_2.

The experiments were repeated for the XFEM case, i. e., taking the reduced XFEM space $Q_h = \tilde{Q}_h^\Gamma$ with $\tilde{c} = 1$ as pressure space. Table 9.6 shows the number of $\mathbf{Q}_A^{-1}$ iterations for fixed discretization parameters $h = 1/8$, $\xi = 100$, varying the material parameters ρ_1, μ_1 of the droplet. The results are comparable to those obtained for the standard FEM pressure space, cf. Table 9.5. Fixing $\rho_1 = 10^{-3}\rho_2$, $\mu_1 = 10^{-2}\mu_2$ and varying the discretization parameters h, ξ, the obtained iteration numbers are reported in the right part of Table 9.7. For the grid sizes $h = 1/4, 1/8$ the results are comparable to the standard FEM case, cf. left part of Table 9.7. As in the stationary

Stokes case, for the finest mesh with grid size $h = 1/16$ we see larger iteration numbers which can again be explained by the deteriorating LBB constant of the $\mathbf{V}_h \times \tilde{Q}_h^{\Gamma}$ finite element pair. Varying all four parameters, the maximal iteration number observed was 163 for the "extreme" case $h = 1/16$, $\xi = 1$, $\rho_1 = 10^{-2}\rho_2$, $\mu_1 = 10^{-3}\mu_2$. The result gets worse when we take a smaller $\tilde{c}$ in the cut-off criterion (7.108).

$h = 1/8$, $\xi = 10^2$	$\rho_1/\rho_2 = 1$	$\rho_1/\rho_2 = 10^{-1}$	$\rho_1/\rho_2 = 10^{-2}$	$\rho_1/\rho_2 = 10^{-3}$
$\mu_1/\mu_2 = 1$	19	22	19	19
$\mu_1/\mu_2 = 10^{-1}$	24	22	26	26
$\mu_1/\mu_2 = 10^{-2}$	29	30	32	32
$\mu_1/\mu_2 = 10^{-3}$	41	42	41	48

Table 9.6. Generalized Stokes case, XFEM pressure space, $\tilde{c} = 1$: number of $\mathbf{Q}_A^{-1}$ evaluations for different density ratios ρ_1/ρ_2 and viscosity ratios μ_1/μ_2.

$\mu_1/\mu_2 = 10^{-2}$,	$Q_h = Q_h^1$			$Q_h = \tilde{Q}_h^{\Gamma_h}$, $\tilde{c} = 1$		
$\rho_1/\rho_2 = 10^{-3}$	$h = 1/4$	$h = 1/8$	$h = 1/16$	$h = 1/4$	$h = 1/8$	$h = 1/16$
$\xi = 1$	34	39	41	34	48	66
$\xi = 10^1$	28	34	44	28	39	64
$\xi = 10^2$	27	26	33	27	32	55
$\xi = 10^3$	27	33	29	27	35	37

Table 9.7. Generalized Stokes case: number of $\mathbf{Q}_A^{-1}$ evaluations for different grid sizes h and values of ξ, applying a standard FEM or reduced XFEM discretization for the pressure.

We repeated the experiments for the case that the material properties of the ambient phase are chosen as $\mu_2 = 10^{-3}$, $\rho_2 = 10^3$, which are roughly the material properties of water, allowing for a more realistic scenario than $\mu_2 = \rho_2 = 1$. It turns out that in this case we have a very robust behavior. This is due to the fact that $\tilde{\mathbf{A}} = \mathbf{A} + \xi\mathbf{M}$ is strongly dominated by the μ-independent term $\xi\mathbf{M}$ because of the large ratio ρ/μ of density to dynamic viscosity. We illustrate the robustness by considering an important concrete example. For fixed material properties $\mu_1 = 10^{-2}\mu_2$, $\rho_1 = 10^{-3}\rho_2$ (i. e., an air-like fluid) the discretization parameters h, ξ are varied. This is in an interesting case from the practical point of view, since the considered two-phase system is close to that of an air bubble in water, where the jumps of density and dynamic viscosity are rather large. The results are reported in Table 9.8 for the standard FEM case ($Q_h = Q_h^1$) and the XFEM case ($Q_h = \tilde{Q}_h^{\Gamma_h}$, $\tilde{c} = 1$). For both cases the preconditioner shows robustness w. r. t. the discretization parameters h, ξ. Comparing the FEM and XFEM case one observes an increase of the iteration numbers for small h. As before, this effect becomes stronger the smaller the cut-off parameter $\tilde{c}$ is chosen.

"water-air"	$Q_h = Q_h^1$			$Q_h = \tilde{Q}_h^{\Gamma_h}$, $\tilde{c} = 1$		
	$h = 1/4$	$h = 1/8$	$h = 1/16$	$h = 1/4$	$h = 1/8$	$h = 1/16$
$\xi = 1$	28	31	30	28	35	48
$\xi = 10^1$	23	27	26	23	30	42
$\xi = 10^2$	23	27	26	23	30	42
$\xi = 10^3$	23	27	26	23	30	42

Table 9.8. Generalized Stokes case, "water-air" system: number of $\mathbf{Q}_A^{-1}$ evaluations for different grid sizes h and values of ξ, applying a standard FEM or reduced XFEM discretization for the pressure.

Part III

Mass transport

10

Mathematical model

10.1 Introduction

We consider the model for transport of a dissolved species as given in (1.24). In (1.24) the unknown quantity (concentration) is denoted by $c = c(x,t)$, the velocity field by $\mathbf{u}$ and the diffusion coefficient by D. In this and the next chapter we use a different notation for these quantities: the unknown function (concentration) is denoted by $u(x,t)$ (instead of c), the velocity field by $\mathbf{w}$ (instead of $\mathbf{u}$) and the diffusion coefficient by α. In this notation the mass transport equation, in strong formulation, is as follows:

$$\frac{\partial u}{\partial t} + \mathbf{w} \cdot \nabla u - \operatorname{div}(\alpha \nabla u) = f \quad \text{in } \Omega_i(t),\ i = 1,2,\ t \in [0,T], \tag{10.1a}$$

$$[\alpha \nabla u \cdot \mathbf{n}]_\Gamma = 0, \tag{10.1b}$$

$$[\beta u]_\Gamma = 0, \tag{10.1c}$$

$$u(\cdot,0) = u_0 \quad \text{in } \Omega_i(t),\ i = 1,2, \tag{10.1d}$$

$$u(\cdot,t) = 0 \quad \text{on } \partial\Omega,\ t \in [0,T]. \tag{10.1e}$$

We assume that $\Omega_1 = \Omega_1(t)$, $t \in [0,T]$, is *given*, with $\partial\Omega_1$ sufficiently smooth and $\partial\Omega_1 \cap \partial\Omega = \emptyset$, i.e. one phase is completely surrounded by the other one. A typical example is a droplet surrounded by another fluid. $\mathbf{n}$ denotes the unit normal at Γ pointing from Ω_1 into Ω_2. In (10.1a) we have standard parabolic convection-diffusion equations, which are coupled by the interface conditions in (10.1b) and (10.1c). The diffusion coefficient $\alpha = \alpha(x,t)$ is assumed to be piecewise constant:

$$\alpha = \alpha_i > 0 \quad \text{in } \Omega_i(t).$$

In general we have $\alpha_1 \neq \alpha_2$. The interface condition in (10.1b) results from the conservation of mass principle. The condition in (10.1c) is the so-called *Henry condition*. In this condition the coefficient $\beta = \beta(x,t)$ is strictly positive and piecewise constant:

$$\beta = \beta_i > 0 \quad \text{in } \Omega_i(t).$$

In general we have $\beta_1 \neq \beta_2$, in which case the solution u is *discontinuous across the interface.* We assume that for the function u_0 in the initial condition (10.1d) the conditions in (10.1b), (10.1c) are satisfied. For simplicity we only consider homogeneous Dirichlet boundary conditions in (10.1e).

As noted above, the interface $\Gamma = \Gamma(t)$ is assumed to be given and to be sufficiently smooth. Also $\mathbf{w} = \mathbf{w}(x,t)$ is assumed to be a given sufficiently smooth velocity field. Clearly, in the setting of a two-phase flow problem the interface and the velocity field result from the Navier-Stokes equations which model the fluid dynamics. If in the two-phase flow system there is no significant influence of the concentration u on fluid dynamics, it is reasonable to assume that in the transport problem (10.1) the interface Γ and velocity field $\mathbf{w}$ are given quantities. In certain other cases, for example if there is a strong dependence of the surface tension coefficient τ on the concentration u, this assumption can be unrealistic.

In this chapter we present suitable weak formulations of the mass transport model (10.1). These formulations are used in the derivation of Galerkin finite element discretizations in Chap. 11. Although the mass transport model (10.1) consists of (relatively simple) convection-diffusion equations in the subdomains, its numerical treatment requires special finite element techniques, since the diffusion coefficient is discontinuous across the interface (which is not aligned with the triangulation) and the solution u has to satisfy a jump condition across the interface.

In this chapter we distinguish the following two cases:

- Firstly, in Sect. 10.2 we treat the special case in which both the interface and the velocity field are assumed to be *stationary*, i.e., independent of t. A model example is a droplet at a stationary position is a stationary flow field. For this case a weak formulation easily follows from results known in the literature.
- The second, more general, and from a practical point point of view more interesting, case of a *non-stationary* interface and a time-dependent velocity field is treated in Sect. 10.3. This general case requires a more elaborate analysis.

The distinction of these two cases is also useful for the analysis of the finite element methods treated in Chap. 11.

Remark 10.1.1 The discontinuity of u across the interface can be avoided by introducing transformed quantities $\tilde{u} := \beta u$, $\tilde{\alpha} := \alpha/\beta$, $\tilde{\mathbf{w}} := \mathbf{w}/\beta$. Then (10.1a)-(10.1c) can be reformulated as

$$\beta^{-1}\frac{\partial \tilde{u}}{\partial t} + \tilde{\mathbf{w}} \cdot \nabla \tilde{u} - \operatorname{div}(\tilde{\alpha}\nabla \tilde{u}) = f \quad \text{in } \Omega_i,\ i = 1,2,\ t \in [0,T], \tag{10.2a}$$

$$[\tilde{\alpha}\nabla \tilde{u} \cdot \mathbf{n}]_\Gamma = 0, \tag{10.2b}$$

$$[\tilde{u}]_\Gamma = 0. \tag{10.2c}$$

In this formulation we have continuity of $\tilde{u}$ across Γ but, compared to (10.1a), a discontinuous subdomain dependent scaling factor β^{-1} in front of the time derivative, which causes difficulties.

We will consider the model in the formulation (10.1a)-(10.1e), which compared to (10.2) is closer to physics.

10.2 Weak formulation: stationary interface

In this section, based on results known in the literature on parabolic equations, we derive a weak formulation for the transport problem in (10.1). For this known theory to be applicable we have to assume that the interface and the velocity field are stationary, i.e., Γ and $\mathbf{w}$ do not depend on t, cf. also Remark 10.2.5. Due to the fact that the underlying problem is a two-phase flow with two incompressible immiscible phases it is reasonable to make the following assumptions about the velocity field $\mathbf{w} = \mathbf{w}(x)$:

$$\operatorname{div} \mathbf{w} = 0 \quad \text{in } \Omega_i,\ i = 1, 2, \quad \mathbf{w} \cdot \mathbf{n} = 0 \quad \text{on } \Gamma, \quad \|\mathbf{w}\|_{L^\infty(\Omega)} \leq c < \infty. \tag{10.3}$$

In the remainder of this section we assume that (10.3) holds.

For a weak formulation we introduce suitable Hilbert spaces. First we define the space of functions for which all weak first derivatives exist on both Ω_1 and Ω_2 and which in addition are zero (in trace sense) on $\partial\Omega$. In the literature this space is usually denoted by $H_0^1(\Omega_1 \cup \Omega_2)$:

$$H_0^1(\Omega_1 \cup \Omega_2) := \{\, v \in L^2(\Omega) : v_{|\Omega_i} \in H^1(\Omega_i),\ i = 1, 2,\ v_{|\partial\Omega} = 0 \,\}.$$

For $v \in H_0^1(\Omega_1 \cup \Omega_2)$ we write $v_i := v_{|\Omega_i}$, $i = 1, 2$. Furthermore

$$H := L^2(\Omega), \quad V := \{\, v \in H_0^1(\Omega_1 \cup \Omega_2) : [\beta v]_\Gamma = 0 \,\}. \tag{10.4}$$

Note:

$$v \in V \ \Leftrightarrow \beta v \in H_0^1(\Omega). \tag{10.5}$$

On H we use the scalar product

$$(u, v)_0 := (\beta u, v)_{L^2} = \int_\Omega \beta u v \, dx,$$

which clearly is equivalent to the standard scalar product on $L^2(\Omega)$. The corresponding norm is denoted by $\|\cdot\|_0$. For $u, v \in H^1(\Omega_i)$ we define $(u, v)_{1,\Omega_i} := \beta_i \int_{\Omega_i} \nabla u_i \cdot \nabla v_i \, dx$ and furthermore

$$(u, v)_{1,\Omega_1 \cup \Omega_2} := (u, v)_{1,\Omega_1} + (u, v)_{1,\Omega_2}, \quad u, v \in V.$$

The corresponding norm is denoted by $|\cdot|_{1,\Omega_1 \cup \Omega_2}$. This norm is equivalent to

$$\left(\|\cdot\|_0^2 + |\cdot|_{1,\Omega_1 \cup \Omega_2}^2\right)^{\frac{1}{2}} =: \|\cdot\|_{1,\Omega_1 \cup \Omega_2}.$$

We emphasize that the norms $\|\cdot\|_0$ and $\|\cdot\|_{1,\Omega_1\cup\Omega_2}$ depend on β. The space $(V,(\cdot,\cdot)_{1,\Omega_1\cup\Omega_2})$ is a Hilbert space. We obtain a Gelfand triple $V \hookrightarrow H \equiv H' \hookrightarrow V'$, with dense and continuous embeddings $\hookrightarrow$. In the following the same spaces $L^2(0,T;V)$, $C([0,T];H)$ as in Sect. 2.2.3 are used, cf. Sect. 2.2.1 for a definition.

The bilinear form

$$a(u,v) := (\alpha u, v)_{1,\Omega_1\cup\Omega_2} + (\mathbf{w}\cdot\nabla u, v)_0, \quad u,v \in V, \tag{10.6}$$

is continuous on V and using (10.3) we get, for $u \in V$,

$$\begin{aligned}
&(\mathbf{w}\cdot\nabla u, u)_0 \\
&= \sum_{i=1,2} \beta_i \int_{\Omega_i} \mathbf{w}\cdot\nabla u_i\, u_i\, dx \\
&= \int_\Gamma \mathbf{w}\cdot\mathbf{n}[\beta u^2]_\Gamma\, ds - \sum_{i=1,2}\beta_i \int_{\Omega_i} \operatorname{div}\mathbf{w}\, u_i^2\, dx - (\mathbf{w}\cdot\nabla u, u)_0 \\
&= -(\mathbf{w}\cdot\nabla u, u)_0.
\end{aligned} \tag{10.7}$$

Hence, $(\mathbf{w}\cdot\nabla u, u)_0 = 0$ holds. This yields ellipticity of $a(\cdot,\cdot)$:

$$a(u,u) \geq \min_{i=1,2} \alpha_i\, |u|^2_{1,\Omega_1\cup\Omega_2} \quad \text{for all } u \in V. \tag{10.8}$$

Now consider the following weak formulation of (10.1a)-(10.1e). Given $f \in V'$, $u_0 \in H$, determine $u \in W^1(0,T;V) := \{\, v \in L^2(0,T;V) : v' \in L^2(0,T;V')\,\}$ such that

$$u(0) = u_0, \quad \frac{d}{dt}(u(t),v)_0 + a(u,v) = f(v) \quad \text{for all } v \in V. \tag{10.9}$$

Here $\frac{d}{dt}(u(t),v)_0 = u'(v)$ corresponds to a weak derivative $u' \in L^2(0,T;V')$ as explained in Lemma 2.2.6. Hence, due to (2.28) $u \in C([0,T];H)$ holds and thus the initial condition $u = u_0$ is well-defined. From Theorem 2.2.7 it follows that the weak formulation (10.9) *has a unique solution.*

Remark 10.2.1 This existence and uniqueness result still holds (cf. [238, 106] and Remark 2.2.8) if instead of ellipticity of the bilinear form $a(\cdot,\cdot)$, cf. (10.8), one has the weaker property

$$a(u,u) \geq c_0|u|^2_{1,\Omega_1\cup\Omega_2} - c_1\|u\|_0^2 \quad \text{for all } u \in V,$$

with constants $c_0 > 0$ and c_1 independent of u. Using $|(\mathbf{w}\cdot\nabla u, u)_0| \leq c|u|_{1,\Omega_1\cup\Omega_2}\|u\|_0$ it easily follows that this property holds *without* using the first two assumptions in (10.3). We introduce these assumptions because they simplify the presentation of the analysis for the continuous problem and we need them in our analysis of the Nitsche-XFEM method in Sect. 11.2.

The weak derivative $u' \in L^2(0,T;V')$ in (10.9), which satisfies $u'(v) = \frac{d}{dt}(u(t),v)_0$ for $v \in V$, can be replaced by a more regular one if we assume some regularity of the data f and u_0. Related to this regularity issue we first consider the *stationary* problem: for $f \in H$,

$$\text{find } u \in V \text{ such that } a(u,v) = (f,v)_0 \quad \text{for all } v \in V. \tag{10.10}$$

We assume that the unique solution u of this problem satisfies $u_i \in H^2(\Omega_i)$, $i = 1,2$ and

$$\|u\|_{2,\Omega_1\cup\Omega_2} := \left(\|u\|^2_{1,\Omega_1\cup\Omega_2} + |u|^2_{2,\Omega_1\cup\Omega_2}\right)^{\frac{1}{2}} \leq c\,\|f\|_0 \tag{10.11}$$

holds, with a constant c independent of f. For the stationary problem it is no restriction to assume $\beta_1 = \beta_2$, since the general case can be reduced to that by a transformation as in (10.2). For the symmetric case $\mathbf{w} = 0$, this regularity result is given in [65]. For the general case such regularity results are derived in Chap. 3 of [162] (cf. also [161]). Using this regularity assumption it follows from Theorem II.3.2 in [236] that the following holds:

Theorem 10.2.2 *Assume that* (10.11) *is satisfied. Take*

$$f \in H, \quad u_0 \in V_{\text{reg}} := \left\{ v \in V : v_i \in H^2(\Omega_i), \quad i = 1,2 \right\}. \tag{10.12}$$

Then the unique solution $u \in W^1(0,T;V)$ *of* (10.9) *satisfies* $u \in C([0,T];V_{\text{reg}})$ *and its weak derivative* $u' =: \frac{du}{dt}$ *has the regularity property*

$$\frac{du}{dt} \in L^2(0,T;V) \cap C([0,T];H). \tag{10.13}$$

Hence u *satisfies, for almost all* $t \in (0,T)$*:*

$$(\frac{du}{dt}, v)_0 + a(u,v) = (f,v)_0 \quad \textit{for all } v \in V. \tag{10.14}$$

We now show that the variational problem (10.14) is indeed a correct weak formulation of the problem (10.1a)-(10.1e).

Lemma 10.2.3 *Take* $f \in H$, $u_0 \in V_{\text{reg}}$. *Assume that* (10.1a)-(10.1e) *has a solution* $u(x,t)$ *which is sufficiently smooth such that for* $u : t \to u(\cdot,t)$ *we have* $u \in C([0,T];V_{\text{reg}})$ *and* $\frac{du}{dt} \in L^2(0,T;H)$. *This* u *solves the variational problem* (10.14).
Conversely, if $u \in C([0,T];V_{\text{reg}})$ *with* $u(0) = u_0$ *solves the variational problem* (10.14) *then* u *satisfies* (10.1a) *in a weak* $L^2(\Omega_i)$ *sense and* (10.1b), (10.1c), (10.1e) *in trace sense.*

Proof. Take $u \in C([0,T];V_{\text{reg}})$ with $\frac{du}{dt} \in L^2(0,T;H)$, and $v \in V$. Using $[\beta v]_\Gamma = 0$ and the notation $\{w\}_\Gamma := \frac{1}{2}\left((w_1)_{|\Gamma} + (w_2)_{|\Gamma}\right)$ for the average of a function $w \in H^1(\Omega_1 \cup \Omega_2)$ we get

$$[\alpha\nabla u\cdot\mathbf{n}\beta v]_\Gamma = [\alpha\nabla u\cdot\mathbf{n}]_\Gamma\,\{\beta v\}_\Gamma + \{\alpha\nabla u\cdot\mathbf{n}\}_\Gamma\,[\beta v]_\Gamma = [\alpha\nabla u\cdot\mathbf{n}]_\Gamma\,\{\beta v\}_\Gamma.$$

Using this we obtain

$$\begin{aligned}(\frac{du}{dt},v)_0 + a(u,v) &= (\frac{du}{dt},v)_0 + (\mathbf{w}\cdot\nabla u, v)_0\\ &\quad - \sum_{i=1,2}\int_{\Omega_i}\operatorname{div}(\alpha_i\nabla u_i)\beta_i v_i\,dx + \int_\Gamma [\alpha\nabla u\cdot\mathbf{n}\beta v]_\Gamma\,ds\\ &= \sum_{i=1,2}\int_{\Omega_i}(\frac{du_i}{dt} + \mathbf{w}\cdot\nabla u_i - \operatorname{div}(\alpha_i\nabla u_i))\beta_i v_i\,dx\\ &\quad + \int_\Gamma [\alpha\nabla u\cdot\mathbf{n}]_\Gamma\,\{\beta v\}_\Gamma\,ds.\end{aligned} \tag{10.15}$$

If u satisfies (10.1a), (10.1b) we thus obtain

$$(\frac{du}{dt},v)_0 + a(u,v) = (f,v)_0 \quad \text{for all } \; v\in V,$$

i.e., (10.14) holds. Conversely, if $u \in C([0,T];V_{\text{reg}})$ with $u(0) = u_0$ solves the variational problem (10.14) we obtain for $u_i(t,x) := u(t)(x)_{|\Omega_i}$

$$\begin{aligned}&\sum_{i=1,2}\int_{\Omega_i}(\frac{\partial u_i}{\partial t} + \mathbf{w}\cdot\nabla u_i - \operatorname{div}(\alpha_i\nabla u_i) - f)\beta_i v_i\,dx\\ &\quad + \int_\Gamma [\alpha\nabla u\cdot\mathbf{n}]_\Gamma\,\{\beta v\}_\Gamma\,ds = 0\end{aligned}$$

for all $v \in V$. This implies that $\frac{\partial u_i}{\partial t} + \mathbf{w}\cdot\nabla u_i - \operatorname{div}(\alpha_i\nabla u_i) = f$ in $L^2(\Omega_i)$ sense and $[\alpha\nabla u\cdot\mathbf{n}]_\Gamma = 0$ in trace sense. The properties in (10.1c) and (10.1e) hold due to $u \in V$. □

For the result in (10.15) it is essential that we multiply the equation (10.1a) by βv and not by v. This explains why in the scalar products $(\cdot,\cdot)_0$ and $(\cdot,\cdot)_{1,\Omega_1\cup\Omega_2}$ we use the weighting with the (piecewise constant) function β.

Remark 10.2.4 Using (10.5) the weighting with β in the scalar products in (10.14) can be eliminated, resulting in the following equivalent variational equation:

$$(\frac{du}{dt},v)_{L^2} + (\alpha\nabla u,\nabla v)_{L^2} + (\mathbf{w}\cdot\nabla u, v)_{L^2} = (f,v)_{L^2} \quad \forall\, v\in H_0^1(\Omega). \tag{10.16}$$

In the finite element discretization in Chap. 11 we prefer the formulation in (10.14), since it uses the same space V both as solution space and as test space. In (10.16) we have V and $H_0^1(\Omega)$ as solution and test space, respectively.

Remark 10.2.5 With additional technical manipulations, the analysis presented in this section can be generalized such that it also covers the case of a time dependent (sufficiently smooth) velocity field $\mathbf{w} = \mathbf{w}(x,t)$. Then the bilinear form $a(\cdot,\cdot)$ in (10.6) depends on t, i.e., we have $a(t;u,v)$. This bilinear form is continuous and elliptic uniformly in $t \in [0,T]$ and an analysis as in e.g. [256], Sect. 26, can be applied. This analysis does *not* apply to the case of a non-stationary interface.

10.3 Weak formulation: non-stationary interface

In this section we derive a weak formulation of the transport problem (10.1) for the general case that $\Gamma = \Gamma(t)$ is time-dependent and the velocity field $\mathbf{w} = \mathbf{w}(x,t)$ may depend on t. We assume that $\Gamma(t)$ is sufficiently smooth for all $t \in [0,T]$. In the analysis presented in the previous section it is essential for the formulation of the parabolic mass transport problem in the space $L^2(0,T;V)$ that the space V *does not depend on* t. If, however, the interface is non-stationary, then $H_0^1(\Omega_1 \cup \Omega_2)$, and thus also V, is time dependent. Due to this, the analysis of the previous section is not applicable in case $\Gamma = \Gamma(t)$. Instead, a *space-time* variational formulation should be used. Although the transport problem (10.1) is relatively simple, since it consists of two coupled parabolic problems, we are not aware of any literature in which a rigorous analysis of an appropriate weak formulation of this problem for the case of a non-stationary interface is given. Below we present such an analysis.

For the velocity field we assume that for almost all $t \in [0,T]$:

$$\mathbf{w}(\cdot,t) \in H^1(\Omega)^3, \quad \|\mathbf{w}(\cdot,t)\|_{L^\infty(\Omega)} \le c < \infty, \quad \operatorname{div}\mathbf{w} = 0 \ \text{ in } \Omega.$$

Furthermore, we assume that the interface $\Gamma(t)$ is *transported by the velocity field* $\mathbf{w}$.

We consider the 3D case, i.e. $\Omega \subset \mathbb{R}^3$. The analysis applies, with only minor modifications to the general case $\Omega \subset \mathbb{R}^d$. For simplicity we assume that $\Omega_1(t)$ is connected and completely contained in Ω. i.e., $\partial\Omega_1(t) \cap \partial\Omega = \emptyset$, $\Gamma(t) = \partial\Omega_1(t)$. In the subsections below we first introduce suitable space-time Sobolev spaces and derive some properties of these spaces (Sect. 10.3.1), then we introduce a space-time weak formulation (Sect. 10.3.2) and finally prove well-posedness of this weak formulation (Sect. 10.3.3).

10.3.1 Preliminaries

The spaces and techniques treated in this section can be found in papers on parabolic problems in so-called noncylindrical domains, which means that the spatial domain in which the problem is formulated depends on t. The first extensive treatment of this topic is given in [167]. Below we use some results from this paper.

The remainder of this section is somewhat technical. For the readers convenience we outline the main results. We first introduce spaces on the space-time domain $Q_T := \Omega \times (0,T)$. The spaces V_β and W_β introduced in (10.18), (10.22) are generalizations of the spaces $L^2(0,T;V)$ and $W^1(0,T;V)$ used for the stationary interface case in the previous section. These spaces V_β, W_β are used in the space-time weak formulation presented in Sect. 10.3.2. Dense subspaces of piecewise smooth functions $\mathcal{V}_\beta \subset V_\beta$, cf. (10.20), and $\mathcal{W}_\beta \subset W_\beta$, cf. (10.21), are introduced. An important difference between these two subspaces is that functions from $\mathcal{V}_\beta$ are zero on the whole boundary of Q_T, whereas functions from $\mathcal{W}_\beta$ are in general zero only on $\partial\Omega \times [0,T]$. An alternative definition of the space W_β, which is useful for the analysis of well-posedness in Sect. 10.3.3, is derived in Proposition 10.3.2. For initial conditions to be well-defined we can use the embedding property $W_\beta \hookrightarrow C([0,T];L^2(\Omega))$ proved in Lemma 10.3.4. In the analysis of well-posedness we need two partial integration identities that hold in the space W_β, namely the ones derived in Corollary 10.3.5 and Lemma 10.3.6.

The (open) space-time cylinder is denoted by $Q_T := \Omega \times (0,T) \subset \mathbb{R}^4$, and the space-time interface is given by $\Gamma_* := \{\, (x,t) \in Q_T : x \in \Gamma(t),\ t \in (0,T) \,\}$. The space-time cylinder is split in subdomains

$$Q_i := \{\, (x,t) \in Q_T : x \in \Omega_i(t),\ t \in (0,T) \,\}, \quad i = 1,2.$$

For $v \in L^2(Q_T)$ we define $v_i := v_{|Q_i}$. We introduce a scalar product in which, on Q_i, we take first derivatives with respect to all *spatial* variables, but *no derivative* with respect to t:

$$\|u\|^2_{H^{1,0}(Q_i)} := \sum_{j=1}^{3} \Big\| \frac{\partial u}{\partial x_j} \Big\|^2_{L^2(Q_i)} + \|u\|^2_{L^2(Q_i)}, \quad u \in C^1(\overline{Q}_i). \tag{10.17}$$

The induced Hilbert space is given by

$$H^{1,0}(Q_i) := \overline{C^1(\overline{Q}_i)}^{\|\cdot\|_{H^{1,0}(Q_i)}} = \Big\{\, u \in L^2(Q_i) : \frac{\partial u}{\partial x_j} \in L^2(Q_i),\ j = 1,2,3 \,\Big\}.$$

From [167] it follows that under mild assumptions on Γ_* there exist bounded linear trace operators

$$\gamma_i : H^{1,0}(Q_i) \to L^2(\Gamma_*), \quad i = 1,2,$$
$$\gamma_\Omega : H^{1,0}(Q_2) \to L^2(\partial\Omega \times (0,T)).$$

In the remainder we assume that such bounded linear trace operators exist. Then for $u \in L^2(Q_T)$ with $u_i \in H^{1,0}(Q_i)$, $i = 1,2$, the operators $u_{|\partial\Omega} = \gamma_\Omega u$ and $[u] = [u]_{\Gamma_*} = \gamma_1 u - \gamma_2 u$ are well-defined. The first space that plays an important role in the space-time weak formulation of the mass transport equation is the following analogon of the space V in (10.4):

$$V_\beta := \{\, u \in L^2(Q_T) : u_i \in H^{1,0}(Q_i),\ i = 1,2,\ u_{|\partial\Omega} = 0,\ [\beta u]_{\Gamma_*} = 0 \,\}. \tag{10.18}$$

This space is equipped with the norm

$$\|u\|_V^2 := \sum_{i=1}^{2} \|u_i\|_{H^{1,0}(Q_i)}^2, \quad u \in V_\beta.$$

This space can also be characterized as follows. Let

$$H_0^{1,0}(Q_T) := \overline{C_0^1(Q_T)}^{\|\cdot\|_{H^{1,0}(Q_T)}},$$

with $\|\cdot\|_{H^{1,0}(Q_T)}$ as in (10.17) but with Q_i replaced by Q_T, be the space-time analogon of $H_0^1(\Omega)$ (i.e., no derivatives w.r.t. t). Then

$$v \in V_\beta \;\Leftrightarrow\; \beta v \in H_0^{1,0}(Q_T) \tag{10.19}$$

holds. Using this we obtain as alternative characterization of V_β:

$$V_\beta = \overline{\mathcal{V}_\beta}^{\|\cdot\|_V}, \quad \text{with } \mathcal{V}_\beta := \left\{\beta^{-1}\phi : \phi \in C_0^1(Q_T)\right\}. \tag{10.20}$$

We introduce a subspace of V_β of functions for which a suitable weak time derivative is well-defined. This space is an analogon for the noncylindrical case of the space $W^1(0,T;V)$ introduced in Sect. 2.2.3. We need the dual space of $H_0^{1,0}(Q_T)$, denoted by

$$H^{-1,0}(Q_T) := H_0^{1,0}(Q_T)'.$$

For $u \in V_\beta$ the distributional time derivative $\frac{\partial u}{\partial t}$ is the linear functional given by

$$\frac{\partial u}{\partial t}(\phi) := -\int_{Q_T} u \frac{\partial \phi}{\partial t}\, dx\, dt, \quad \phi \in C_0^1(Q_T).$$

Define $\mathcal{W}_\beta \subset V_\beta$ by

$$\mathcal{W}_\beta := \left\{\beta^{-1}\psi : \psi \in C^1(\overline{Q}_T),\ \psi_{|\partial\Omega} = 0\right\}. \tag{10.21}$$

Note that opposite to $\mathcal{V}_\beta$ in (10.20), a function from $\mathcal{W}_\beta$ is not necessarily equal to zero on all of ∂Q_T.

Lemma 10.3.1 *For $\psi \in \mathcal{W}_\beta$ we have $\frac{\partial \psi}{\partial t} \in H^{-1,0}(Q_T)$.*

Proof. Take $\psi \in \mathcal{W}_\beta$, $\phi \in C_0^1(Q_T)$. Then $\psi_i \in C^1(\overline{Q}_i)$, $i = 1, 2$, and

$$\begin{aligned}\frac{\partial \psi}{\partial t}(\phi) &= -\int_{Q_T} \psi \frac{\partial \phi}{\partial t}\, dx\, dt \\ &= \sum_{i=1}^{2} \left(-\int_{\Gamma_*} \gamma_i \psi_i\, \phi n_4\, ds\, dt + \int_{Q_i} \frac{\partial \psi_i}{\partial t} \phi\, dx\, dt \right),\end{aligned}$$

where $n_4 = \hat{\mathbf{n}}_4$ is the fourth component (i. e., the temporal direction) of the unit normal $\hat{\mathbf{n}}$ at Γ_*. Using the bounds for the trace operators γ_i and Cauchy-Schwarz inequalities we obtain

$$|\frac{\partial \psi}{\partial t}(\phi)| \leq c\|\phi\|_{H^{1,0}(Q_T)}.$$

Using a density argument it follows that $\frac{\partial \psi}{\partial t} \in H_0^{1,0}(Q_T)' = H^{-1,0}(Q_T)$ holds. □

We introduce the closure of $\mathcal{W}_\beta$ in V_β w.r.t. the topology induced by $\|\cdot\|_V^2 + \|\frac{\partial}{\partial t} \cdot \|_{H^{-1,0}(Q_T)}^2$:

$$W_\beta := \overline{\mathcal{W}_\beta}^{\|\cdot\|_W}, \quad \text{with } \|v\|_W^2 := \|v\|_V^2 + \|\frac{\partial v}{\partial t}\|_{H^{-1,0}(Q_T)}^2. \tag{10.22}$$

The space W_β is contained in $\tilde{W}_\beta := \{ v \in V_\beta : \frac{\partial v}{\partial t} \in H^{-1,0}(Q_T) \}$. We claim that $W_\beta = \tilde{W}_\beta$ holds. This claim is formulated in the following proposition, for which we only give a sketch of a proof, based on [7]. In the remainder we use this proposition only in Lemma 10.3.10 and Theorem 10.3.11.

Proposition 10.3.2 *Assume that for $i = 1, 2$, there are bounded C^1 bijections $\Phi_i : \Omega_i(0) \times (0,T) \to Q_i$ such that $\Omega_i(t) = \{ \Phi_i(\tilde{x}, t) : \tilde{x} \in \Omega_i(0) \}$, $0 < t < T$. For W_β as in (10.22) the following holds:*

$$W_\beta = \left\{ v \in V_\beta : \frac{\partial v}{\partial t} \in H^{-1,0}(Q_T) \right\}. \tag{10.23}$$

Proof. First we consider the case of a stationary interface, i.e., $\Gamma(t)$ does not depend on t and thus $\beta(x,t) = \beta(x)$ (the subdomains Q_i are cylindrical). We write $\mathcal{H} = H_0^1(\Omega)$. The spaces $L^2(0,T;\mathcal{H})$ and $L^2(0,T;\mathcal{H}') = L^2(0,T;\mathcal{H})'$ can be identified with $H_0^{1,0}(Q_T)$ and $H^{-1,0}(Q_T)$, respectively. Define

$$X := \left\{ u \in L^2(0,T;\mathcal{H}) : \frac{d(\beta^{-1}u)}{dt} \in L^2(0,T;\mathcal{H}') \right\}, \tag{10.24}$$

with $\frac{dw}{dt} = w'$ the weak time derivative as in Sect. 2.2.3. On X we use the norm

$$\|u\|_X^2 = \|u\|_{L^2(0,T;\mathcal{H})}^2 + \|\frac{d(\beta^{-1}u)}{dt}\|_{L^2(0,T;\mathcal{H}')}^2.$$

The space on the right-hand side in (10.23) is denoted by $\tilde{W}_\beta$. The equivalence $v \in \tilde{W}_\beta \Leftrightarrow \beta v \in X$ holds and $\|\cdot\|_W$ and $\|\beta \cdot\|_X$ are equivalent norms on $\tilde{W}_\beta$. The scaling with β^{-1} in (10.24) is not essential for properties of the space X, since it can be incorporated as a weighting factor in the $L^2(\Omega)$ scalar product. From the literature, e.g. [256], it follows that $C^\infty(0,T;\mathcal{H})$ is dense in X. Using the density of $C_0^\infty(\Omega)$ in $\mathcal{H}$ it follows that $\mathcal{W} := \{ \psi \in C^1(Q_T) : \psi_{|\partial\Omega} = 0 \}$ is dense in X. Take $v \in \tilde{W}_\beta$. Then $\beta v \in X$ and there is a sequence (ψ_m) in

$\mathcal{W}$ with $\lim_{m\to\infty} \|\psi_m - \beta v\|_X = 0$. This implies $\lim_{m\to\infty} \|\beta^{-1}\psi_m - v\|_W = 0$ for the sequence $(\beta^{-1}\psi_m)$ from $\mathcal{W}_\beta$. Thus $\mathcal{W}_\beta$ is dense in $\tilde{W}_\beta$, i.e. $W_\beta = \overline{\mathcal{W}_\beta}^{\|\cdot\|_W} = \tilde{W}_\beta$ holds, which proves the claim for the case of a stationary interface.

We now treat the general case. Define the cylindrical subdomain $\tilde{Q}_i = \Omega_i(0) \times (0,T)$, $i = 1,2$. A function $v \in V_\beta$ can be represented in transformed variables $(\tilde{x}, t)$ by $\tilde{v}(\tilde{x},t) := v(\Phi_i(\tilde{x},t),t)$, $(\tilde{x},t) \in \tilde{Q}_i$. Due to the smoothness assumption on the bijection Φ_i, the Sobolev norm of the transformed function $\|\tilde{v}\|_{\tilde{V}}^2 := \sum_{i=1}^2 \|\tilde{v}_i\|_{H^{1,0}(\tilde{Q}_i)}$ is equivalent to $\|v\|_V^2$, cf. [8] Sect. 3.34. Furthermore, $\|\frac{\partial \tilde{v}}{\partial t}\|_{H^{-1,0}(\tilde{Q}_T)}$ is equivalent to $\|\frac{\partial v}{\partial t}\|_{H^{-1,0}(Q_T)}$. Using these norm equivalences and the density result for the special case of a stationary interface (cylindrical subdomains) the density result for the general case can be proved. □

In the variational formulation of the transport problem presented in Sect. 10.3.2 below the spaces V_β and W_β play an important role. For the analysis in Sect. 10.3.3 we need some further properties, which are derived in the remainder of this section.

As mentioned above, we assume that the interface $\Gamma(t)$ is transported by the velocity field $\mathbf{w}(x,t)$. From this it follows that

$$\hat{\mathbf{n}} \cdot \begin{pmatrix} \mathbf{w} \\ 1 \end{pmatrix} = 0 \tag{10.25}$$

holds, with $\hat{\mathbf{n}} \in \mathbb{R}^4$ the unit normal at Γ_*, pointing outward from Q_1. For this normal the identity

$$\hat{\mathbf{n}} = \nu \begin{pmatrix} \mathbf{n}_\Gamma \\ -\mathbf{w} \cdot \mathbf{n}_\Gamma \end{pmatrix}, \quad \nu := \frac{1}{\sqrt{1 + (\mathbf{w} \cdot \mathbf{n}_\Gamma)^2}} \tag{10.26}$$

holds, where $\mathbf{n}_\Gamma \in \mathbb{R}^3$ is the unit normal at Γ.

Lemma 10.3.3 *For all $\psi \in \mathcal{W}_\beta$ the identity*

$$\begin{aligned} &\frac{\partial \psi}{\partial t}(\beta\psi) - \int_{Q_T} \psi \mathbf{w} \cdot \nabla(\beta\psi)\, dx\, dt \\ &= \frac{1}{2}\|\beta\psi(\cdot,T)\|_{L^2(\Omega)}^2 - \frac{1}{2}\|\beta\psi(\cdot,0)\|_{L^2(\Omega)}^2 \end{aligned} \tag{10.27}$$

holds.

Proof. Take $\psi \in \mathcal{W}_\beta$, $\phi \in \mathcal{V}_\beta$. Then $\beta\phi \in C_0^1(Q_T)$ holds. From partial integration and using (10.25) we obtain, with the notation $\hat{\mathbf{n}} =: \begin{pmatrix} \hat{\mathbf{n}}_x \\ n_4 \end{pmatrix}$,

$$\begin{aligned}\frac{\partial\psi}{\partial t}(\beta\phi) - \int_{Q_T} \psi\mathbf{w}\cdot\nabla(\beta\phi)\,dx\,dt &= -\int_{Q_T} \psi\frac{\partial(\beta\phi)}{\partial t} + \psi\mathbf{w}\cdot\nabla(\beta\phi)\,dx\,dt \\ &= \sum_{i=1}^{2}\int_{Q_i}(\frac{\partial\psi_i}{\partial t} + \mathbf{w}\cdot\nabla\psi_i)\beta\phi\,dx\,dt - \int_{\Gamma_*}[\psi]\beta\phi(n_4+\hat{\mathbf{n}}_x\cdot\mathbf{w})\,ds\,dt \quad (10.28) \\ &= \sum_{i=1}^{2}\int_{Q_i}(\frac{\partial\psi_i}{\partial t} + \mathbf{w}\cdot\nabla\psi_i)\beta\phi\,dx\,dt.\end{aligned}$$

From a continuity argument it follows that

$$\frac{\partial\psi}{\partial t}(\beta v) - \int_{Q_T} \psi\mathbf{w}\cdot\nabla(\beta v)\,dx\,dt = \sum_{i=1}^{2}\int_{Q_i}(\frac{\partial\psi_i}{\partial t} + \mathbf{w}\cdot\nabla\psi_i)\beta v\,dx\,dt \quad (10.29)$$

holds for all $v \in V_\beta$. In particular it holds for $v = \psi \in \mathcal{W}_\beta \subset V_\beta$. Taking $v = \psi$ in (10.29) and applying partial integration again we get

$$\begin{aligned}&\sum_{i=1}^{2}\int_{Q_i}(\frac{\partial\psi_i}{\partial t} + \mathbf{w}\cdot\nabla\psi_i)\beta\psi\,dx\,dt \\ &= \int_\Omega \beta\psi(\cdot,T)^2\,dx - \int_\Omega \beta\psi(\cdot,0)^2\,dx - \sum_{i=1}^{2}\int_{Q_i}(\frac{\partial\psi_i}{\partial t} + \mathbf{w}\cdot\nabla\psi_i)\beta\psi\,dx\,dt,\end{aligned}$$

since the boundary integral $\int_{\Gamma_*}$ vanishes, cf. (10.28). Using this in (10.29), with $v = \psi$, the result is proved. □

In the next lemma we derive a generalization of the embedding property $W^1(0,T;V) \hookrightarrow C([0,T];H)$ given in (2.28).

Lemma 10.3.4 *There is a continuous embedding $W_\beta \hookrightarrow C([0,T];L^2(\Omega))$.*

Proof. It suffices to prove

$$\sup_{t\in[0,T]} \|\psi(\cdot,t)\|_{L^2(\Omega)} \leq c\|\psi\|_W \quad \text{for all} \quad \psi \in \mathcal{W}_\beta.$$

Take $\psi \in \mathcal{W}_\beta$. From Lemma 10.3.3 we have

$$\begin{aligned}&\frac{\partial\psi}{\partial t}(\beta\psi) - \int_{Q_T} \psi\mathbf{w}\cdot\nabla(\beta\psi)\,dx\,dt \\ &= \frac{1}{2}\|\beta\psi(\cdot,T)\|^2_{L^2(\Omega)} - \frac{1}{2}\|\beta\psi(\cdot,0)\|^2_{L^2(\Omega)}.\end{aligned} \quad (10.30)$$

We take $t_0 \in [\frac{1}{2}T, T]$. Let $\mathcal{W}_\beta(Q_{t_0})$ be as in (10.21), but with T replaced by t_0. Note that $\psi \in \mathcal{W}_\beta = \mathcal{W}_\beta(Q_T)$ implies $\psi \in \mathcal{W}_\beta(Q_{t_0})$. The identity (10.30) holds with Q_T replaced by Q_{t_0}. For the time derivative we then have $\frac{\partial\psi}{\partial t} \in H^{-1,0}(Q_{t_0})$ and

$$\|\frac{\partial \psi}{\partial t}\|_{H^{-1,0}(Q_{t_0})} \leq \|\frac{\partial \psi}{\partial t}\|_{H^{-1,0}(Q_T)} \tag{10.31}$$

holds. Let $\theta = \theta(t)$ be a smooth function with $\theta(t) \in [0,1]$ for all t, $\theta(0) = 0$, $\theta(t) = 1$ for all $t \in [\frac{1}{2}T, T]$. The result (10.30) holds with Q_T replaced by Q_{t_0} and, since $\theta\psi \in \mathcal{W}_\beta$, with ψ replaced by $\theta\psi$. Using this and $\theta(t_0) = 1$, $\theta(0) = 0$ we get, with $c_0 := 2\max\{\beta_1^{-2}, \beta_2^{-2}\}$,

$$\begin{aligned}\|\psi(\cdot,t_0)\|_{L^2(\Omega)}^2 &\leq c_0\frac{1}{2}\|\beta\theta(t_0)\psi(\cdot,t_0)\|_{L^2(\Omega)}^2 - c_0\frac{1}{2}\|\beta\theta(0)\psi(\cdot,0)\|_{L^2(\Omega)}^2 \\ &= c_0\frac{\partial(\theta\psi)}{\partial t}(\beta\theta\psi) - c_0\int_{Q_{t_0}} \theta^2\psi\mathbf{w}\cdot\nabla(\beta\psi)\,dx\,dt.\end{aligned}$$

For the second term on the right-hand side we have

$$|\int_{Q_{t_0}} \theta^2\psi\mathbf{w}\cdot\nabla(\beta\psi)\,dx\,dt| \leq c\|\psi\|_{L^2(Q_{t_0})}\|\nabla(\beta\psi)\|_{L^2(Q_{t_0})} \leq c\|\psi\|_V^2.$$

For the first term we get

$$\begin{aligned}|\frac{\partial(\theta\psi)}{\partial t}(\beta\theta\psi)| &\leq \|\frac{\partial(\theta\psi)}{\partial t}\|_{H^{-1,0}(Q_{t_0})}\|\theta\beta\psi\|_{H^{1,0}(Q_{t_0})} \\ &\leq c\|\frac{\partial(\theta\psi)}{\partial t}\|_{H^{-1,0}(Q_{t_0})}\|\psi\|_V.\end{aligned} \tag{10.32}$$

We consider the term $\|\frac{\partial(\theta\psi)}{\partial t}\|_{H^{-1,0}(Q_{t_0})}$ in (10.32). For $\phi \in C_0^1(Q_{t_0})$ we have

$$\frac{\partial(\theta\psi)}{\partial t}(\phi) = -\int_{Q_{t_0}} \theta\psi\frac{\partial\phi}{\partial t}\,dx\,dt = -\int_{Q_{t_0}} \psi\frac{\partial(\theta\phi)}{\partial t}\,dx\,dt + \int_{Q_{t_0}} \theta'\psi\phi\,dx\,dt.$$

Hence, using $\theta\phi \in C_0^1(Q_{t_0})$ we obtain

$$\begin{aligned}|\frac{\partial(\theta\psi)}{\partial t}(\phi)| &\leq \|\frac{\partial\psi}{\partial t}\|_{H^{-1,0}(Q_{t_0})}\|\theta\phi\|_{H^{1,0}(Q_{t_0})} + c\|\psi\|_V\|\phi\|_{H^{1,0}(Q_{t_0})} \\ &\leq c(\|\frac{\partial\psi}{\partial t}\|_{H^{-1,0}(Q_{t_0})} + \|\psi\|_V)\|\phi\|_{H^{1,0}(Q_{t_0})}.\end{aligned}$$

Using (10.31) this yields

$$\|\frac{\partial(\theta\psi)}{\partial t}\|_{H^{-1,0}(Q_{t_0})} \leq c(\|\frac{\partial\psi}{\partial t}\|_{H^{-1,0}(Q_T)} + \|\psi\|_V).$$

Using this in (10.32) we obtain

$$|\frac{\partial(\theta\psi)}{\partial t}(\beta\theta\psi)| \leq c(\|\frac{\partial\psi}{\partial t}\|_{H^{-1,0}(Q_T)}^2 + \|\psi\|_V^2),$$

and combing these results yields

$$\sup_{t\in[\frac{1}{2}T,T]} \|\psi(\cdot,t)\|_{L^2(\Omega)} \leq c\|\psi\|_W.$$

The same bound can be derived for $t_0 \in [0, \frac{1}{2}T]$ by replacing t_0 by $T - t_0$ and applying the same arguments. □

Corollary 10.3.5 The identity (10.27) holds for all $\psi \in W_\beta$.

Proof. This follows from the density of $\mathcal{W}_\beta$ in W_β with respect to $\|\cdot\|_W$, continuity of the bilinear forms on the left-hand side in (10.27) w.r.t. $\|\cdot\|_W$ and the inequality $\|\beta\psi(\cdot,t)\|_{L^2(\Omega)} \leq c\|\psi\|_W$ for all $\psi \in W_\beta$. □

Lemma 10.3.6 *For all $u, v \in W_\beta$ the following holds:*

$$\frac{\partial u}{\partial t}(\beta v) + \frac{\partial v}{\partial t}(\beta u) = \int_\Omega (\beta uv)_{|t=T}\, dx - \int_\Omega (\beta uv)_{|t=0}\, dx - \int_{\Gamma_*} [\beta uv] n_4 \, ds\, dt,$$

with $n_4 = \hat{\mathbf{n}}_4$ the fourth component of the unit normal at Γ_*.

Proof. Take $\psi \in \mathcal{W}_\beta$. Define

$$l_\psi(v) := \int_{Q_T} \frac{\partial(\beta\psi)}{\partial t} v \, dx\, dt - \int_{\Gamma_*} [\beta\psi v] n_4 \, ds, \quad v \in V_\beta.$$

From the definition of the distributional derivative and a density argument it follows that $\frac{\partial \psi}{\partial t}(\beta v) = l_\psi(v)$ for all $v \in V_\beta$. For $\phi \in \mathcal{W}_\beta$ we have

$$l_\psi(\phi) = -l_\phi(\psi) + \int_\Omega (\beta\psi\phi)_{|t=T}\, dx - \int_\Omega (\beta\psi\phi)_{|t=0}\, dx - \int_{\Gamma_*} [\beta\psi\phi] n_4 \, ds.$$

This yields

$$\frac{\partial \psi}{\partial t}(\beta\phi) + \frac{\partial \phi}{\partial t}(\beta\psi) = \int_\Omega (\beta\psi\phi)_{|t=T}\, dx - \int_\Omega (\beta\psi\phi)_{|t=0}\, dx - \int_{\Gamma_*} [\beta\psi\phi] n_4 \, ds$$

for all $\psi, \phi \in \mathcal{W}_\beta$. By a density argument this even holds for $\psi, \phi \in W_\beta$. □

10.3.2 Space-time weak formulation

In this section we introduce a weak formulation of the mass transport problem (10.1). We restrict to the case with an initial condition $u_0 = 0$. The case $u_0 \neq 0$ can be treated by a shift argument.

For $u \in W_\beta$, due to the result in Lemma 10.3.4 the function $u(\cdot, 0) \in L^2(\Omega)$ is well-defined. We introduce the subspace of W_β of functions with initial data equal to zero:

$$W_{\beta,0} := \{ u \in W_\beta : u(\cdot, 0) = 0 \quad \text{in } \Omega \}.$$

The space $(W_{\beta,0}, \|\cdot\|_W)$ is a Hilbert space. We introduce the following space-time weak formulation of the mass transport equation:

Determine $u \in W_{\beta,0}$ such that

$$\frac{\partial u}{\partial t}(v) - \int_{Q_T} u\mathbf{w} \cdot \nabla v \, dx\, dt + \sum_{i=1}^{2} \int_{Q_i} \alpha_i \nabla u_i \cdot \nabla v \, dx\, dt = \int_{Q_T} f v \, dx\, dt \quad (10.33)$$

for all $v \in H_0^{1,0}(Q_T)$.

Remark 10.3.7 The formulation in (10.33) generalizes the one for the stationary interface case given in (10.16). Due to the property (10.19) the test space $H_0^{1,0}(Q_T)$ can be replaced by V_β. For the trial and test space we then have the nice embedding relation $W_{\beta,0} \subset V_\beta$. Using the test space V_β the variational equation (10.33) takes the form

$$\frac{\partial u}{\partial t}(\beta v) - \int_{Q_T} u\mathbf{w} \cdot \nabla(\beta v)\, dx\, dt + \sum_{i=1}^{2} \int_{Q_i} \alpha_i \beta_i \nabla u_i \cdot \nabla v_i \, dx\, dt = \int_{Q_T} \beta f v \, dx\, dt$$

for all $v \in V_\beta$. This generalizes the problem in (10.14).

We show that this problem is consistent with the strong formulation in (10.1):

Lemma 10.3.8 *Assume that the weak formulation* (10.33) *has a solution* $u \in W_{\beta,0}$ *that is sufficiently smooth, namely* $u \in H^1(Q_i)$, $\frac{\partial u}{\partial x_j} \in H^{1,0}(Q_i)$, *for* $j = 1,2,3$ *and* $i = 1,2$. *Then* u *satisfies* (10.1a)–(10.1e) (*in* L^2*-sense*), *with* $u_0 = 0$.

Proof. Due to $u \in W_{\beta,0}$ the properties (10.1c)–(10.1e), with $u_0 = 0$, hold for u. In (10.33) we take $v = \phi \in C_0^1(Q_T)$. Using (10.19), partial integration on the subdomains Q_i and the property (10.25) we get

$$\frac{\partial u}{\partial t}(\phi) - \int_{Q_T} u\mathbf{w} \cdot \nabla \phi \, dx\, dt = \sum_{i=1}^{2} \int_{Q_i} (\frac{\partial u_i}{\partial t} + \mathbf{w} \cdot \nabla u_i) \phi \, dx\, dt,$$

and

$$\sum_{i=1}^{2} \int_{Q_i} \alpha_i \nabla u_i \cdot \nabla \phi \, dx\, dt = -\sum_{i=2}^{2} \int_{Q_i} \operatorname{div}(\alpha_i \nabla u_i) \phi \, dx\, dt + \int_{\Gamma_*} \nu [\alpha \nabla u \cdot \mathbf{n}]_\Gamma \phi \, ds,$$

with $\mathbf{n} \in \mathbb{R}^3$ the unit normal at $\Gamma(t)$ and ν a suitable scaling parameter. Hence, we obtain

$$\sum_{i=1}^{2} \int_{Q_i} \left(\frac{\partial u_i}{\partial t} + \mathbf{w} \cdot \nabla u_i - \operatorname{div}(\alpha_i \nabla u_i) - f \right) \phi \, dx\, dt + \int_{\Gamma_*} \nu [\alpha \nabla u \cdot \mathbf{n}]_\Gamma \phi \, ds\, dt = 0$$

for all $\phi \in C_0^1(Q_T)$. Thus (10.1a) and (10.1b) are satisfied. □

10.3.3 Well-posedness of the space-time weak formulation

In this section we prove that the space-time variational problem (10.33) has a unique solution. We assume that the right-hand side $f \in L^2(Q_T)$. For the analysis we apply Theorem 15.1.1. As Hilbert spaces we use $H_1 = W_{\beta,0}$, $H_2 = H_0^{1,0}(Q_T)$. For notational convenience we write $H = H_0^{1,0}(Q_T)$, and thus $\|\cdot\|_{H_0^{1,0}(Q_T)} = \|\cdot\|_{H^{1,0}(Q_T)} = \|\cdot\|_H$. For the bilinear form on the left-hand side in (10.33) we use the notation $a : W_{\beta,0} \times H \to \mathbb{R}$, and this bilinear is split into two parts:

$$a(u,v) := \frac{\partial u}{\partial t}(v) - \int_{Q_T} u\mathbf{w}\cdot\nabla v\,dx\,dt + \sum_{i=1}^{2}\int_{Q_i}\alpha_i\nabla u_i\cdot\nabla v\,dx\,dt$$
$$=: a_1(u,v) + a_2(u,v),$$

with the elliptic part

$$a_2(u,v) = \sum_{i=1}^{2}\int_{Q_i}\alpha_i\nabla u_i\cdot\nabla v\,dx\,dt.$$

Introduce $f(v) := \int_{Q_T} fv\,dx\,dt$. Using this notation the weak formulation takes the form:

$$\text{determine } u\in W_{\beta,0} \text{ such that } \; a(u,v) = f(v) \quad \text{for all} \;\; v\in H. \qquad (10.34)$$

The bilinear form $a(\cdot,\cdot)$ is *continuous* on $W_{\beta,0}\times H$. The problem (10.34) is well-posed iff the following two conditions are satisfied, cf. Theorem 15.1.1:

$$\exists\,\varepsilon>0: \quad \inf_{u\in W_{\beta,0}}\sup_{v\in H}\frac{a(u,v)}{\|u\|_W\|v\|_H} \geq \varepsilon \qquad (\text{"inf-sup condition"}), \qquad (10.35a)$$

$$[\,a(u,v)=0 \;\text{ for all } u\in W_{\beta,0}] \;\text{ implies } \; v=0. \qquad (10.35b)$$

Lemma 10.3.9 *The inf-sup condition* (10.35a) *is fulfilled.*

Proof. Take $u\in W_{\beta,0}$. Note that $W_{\beta,0}\subset W_\beta\subset V_\beta$ and thus $\beta u\in H$, cf. (10.19). Define $g_u := \frac{\partial u}{\partial t}(\cdot)\in H'$. Recall that $\|u\|_W^2 = \|u\|_V^2 + \|g_u\|_{H'}^2$. For $v,w\in H$ we have $a_2(v,w) = \int_{Q_T}\alpha\nabla v\cdot\nabla w\,dx\,dt$, hence using a Friedrichs inequality we conclude that t he bilinear form $a_2(\cdot,\cdot)$ is continuous and elliptic on $H\times H$:

$$\exists\,\delta>0: \quad \delta\,\|v\|_H^2 \leq a_2(v,v) \quad \text{for all} \;\; v\in H.$$

Let $z\in H$ be such that

$$a_2(z,v) = g_u(v) \quad \text{for all} \;\; v\in H.$$

Using

$$\|g_u\|_{H'} = \sup_{v\in H}\frac{a_2(z,v)}{\|v\|_H} \leq c\|z\|_H$$

it follows that there is a constant $\xi>0$ independent of u such that

$$\xi\|g_u\|_{H'}^2 \leq \delta\|z\|_H^2 \leq g_u(z) \leq \|g_u\|_{H'}\|z\|_H. \qquad (10.36)$$

Take $v := z + \mu\beta u \in H$, with a fixed, sufficiently large $\mu > 0$. Using (10.36) we obtain

$$\|v\|_H \leq \|z\|_H + \mu\|\beta u\|_H \leq c_\mu(\|g_u\|_{H'} + \|u\|_V) \leq \frac{1}{2}c_\mu\|u\|_W. \tag{10.37}$$

Substitution of this $v \in H$ in the bilinear form yields

$$\begin{aligned}
a(u,v) &= \frac{\partial u}{\partial t}(z+\mu\beta u) - \int_{Q_T} u\mathbf{w}\cdot\nabla(z+\mu\beta u)\;dxdt + a_2(u, z+\mu\beta u) \\
&= g_u(z) + \mu\left[\frac{\partial u}{\partial t}(\beta u) - \int_{Q_T} u\mathbf{w}\cdot\nabla(\beta u)\,dx\,dt\right] \\
&\quad - \int_{Q_T} u\mathbf{w}\cdot\nabla z\,dx\,dt + a_2(u,z) + \mu a_2(u,\beta u) \\
&\geq g_u(z) - \int_{Q_T} u\mathbf{w}\cdot\nabla z\;dxdt + a_2(u,z) + \mu a_2(u,\beta u),
\end{aligned}$$

where in the last inequality we used Corollary 10.3.5 and $u(x,0) = 0$, since $u \in W_{\beta,0}$. Now note that

$$\Big|\int_{Q_T} u\mathbf{w}\cdot\nabla z\,dx\,dt + a_2(u,z)\Big| \leq \tilde{c}\|u\|_V\|z\|_H$$

holds with $\tilde{c}$ independent of u. For the term $a_2(u,\beta u)$ we have

$$\begin{aligned}
a_2(u,\beta u) &\geq \min\{\beta_1^{-1},\beta_2^{-1}\}\;a_2(\beta u,\beta u) \geq \min\{\beta_1^{-1},\beta_2^{-1}\}\;\delta\|\beta u\|_H^2 \\
&\geq \delta\min\{\beta_1^{-1},\beta_2^{-1}\}\;\min\{\beta_1^2,\beta_2^2\}\;\|u\|_V^2 =: \hat{c}\|u\|_V^2.
\end{aligned} \tag{10.38}$$

Hence, for μ sufficiently large, independent of u, we get

$$\begin{aligned}
a(u,v) &\geq g_u(z) - (\tilde{c}\delta^{-\frac{1}{2}}\|u\|_V)(\delta^{\frac{1}{2}}\|z\|_H) + \mu\hat{c}\|u\|_V^2 \\
&\geq g_u(z) - \frac{1}{2}\delta\|z\|_H^2 + \big(\mu\hat{c} - \frac{1}{2}\tilde{c}^2\delta^{-1}\big)\|u\|_V^2 \\
&\geq \frac{1}{2}g_u(z) + \big(\mu\hat{c} - \frac{1}{2}\tilde{c}^2\delta^{-1}\big)\|u\|_V^2 \\
&\geq \frac{1}{2}\xi(\|g_u\|_{H'}^2 + \|u\|_V^2) = \frac{1}{2}\xi\|u\|_W^2.
\end{aligned}$$

Combining this with (10.37) we obtain

$$a(u,v) \geq \varepsilon\;\|u\|_W\|v\|_H,$$

with $\varepsilon > 0$ independent of u, which proves the inf-sup property. □

We now consider the second condition (10.35b). In the proof of the next lemma we use Proposition 10.3.2.

Lemma 10.3.10 *Condition* (10.35b) *is fulfilled.*

Proof. Let $v \in H$ be such that $a(u,v) = 0$ for all $u \in W_{\beta,0}$. Define $\hat{v} := \frac{1}{\beta} v \in V_\beta$. Introduce, for $u, v \in H^{1,0}(Q_1 \cup Q_2)$,

$$d(u,v) := \sum_{i=1}^{2} \int_{Q_i} -u_i \mathbf{w} \cdot \nabla v_i + \alpha_i \nabla u_i \cdot \nabla v_i \, dx\, dt,$$

hence, $\frac{\partial u}{\partial t}(\beta \hat{v}) + d(u, \beta \hat{v}) = 0$ for all $u \in W_{\beta,0}$. For arbitrary $u \in \mathcal{V}_\beta \subset W_{\beta,0}$ we have

$$\frac{\partial u}{\partial t}(\beta \hat{v}) = \int_{Q_T} \frac{\partial (\beta u)}{\partial t} \hat{v} \, dx\, dt - \int_{\Gamma_*} [\beta u \hat{v}] n_4 \, ds = -\frac{\partial \hat{v}}{\partial t}(\beta u) - \int_{\Gamma_*} [\beta u \hat{v}] n_4 \, ds,$$

and thus

$$\begin{aligned} \frac{\partial \hat{v}}{\partial t}(\beta u) &= -\frac{\partial u}{\partial t}(\beta \hat{v}) - \int_{\Gamma_*} [\beta u \hat{v}] n_4 \, ds = d(u, \beta \hat{v}) - \int_{\Gamma_*} [\beta u \hat{v}] n_4 \, ds \\ &= d(\beta u, \hat{v}) - \int_{\Gamma_*} [\beta u \hat{v}] n_4 \, ds. \end{aligned}$$

The linear functional $w \to d(w, \hat{v}) - \int_{\Gamma_*} [w \hat{v}] n_4 \, ds$ is bounded on H. Thus it follows that $\frac{\partial \hat{v}}{\partial t}(\cdot) \in H'$, i.e., $\hat{v} \in W_\beta$ and, by a density argument,

$$\frac{\partial \hat{v}}{\partial t}(\beta u) - d(u, \beta \hat{v}) + \int_{\Gamma_*} [\beta u \hat{v}] n_4 \, ds = 0 \quad \text{for all} \quad u \in V_\beta. \tag{10.39}$$

Using $-d(u, \beta \hat{v}) = \frac{\partial u}{\partial t}(\beta \hat{v})$ for all $u \in W_{\beta,0}$ and Lemma 10.3.6 we get

$$0 = \frac{\partial \hat{v}}{\partial t}(\beta u) + \frac{\partial u}{\partial t}(\beta \hat{v}) + \int_{\Gamma_*} [\beta u \hat{v}] n_4 \, ds = \int_\Omega (\beta u \hat{v})_{t=T} \, dx \quad \text{for all} \quad u \in W_{\beta,0}.$$

This implies $\hat{v}(\cdot, T) = 0$. In equation (10.39) we take $u = \hat{v} \in W_\beta$, apply partial integration and use Corollary 10.3.5, resulting in

$$\begin{aligned} 0 &= \frac{\partial \hat{v}}{\partial t}(\beta \hat{v}) - d(\hat{v}, \beta \hat{v}) + \int_{\Gamma_*} [\beta \hat{v}^2] n_4 \, ds \\ &= \frac{\partial \hat{v}}{\partial t}(\beta \hat{v}) - \int_{Q_T} \beta \hat{v} \mathbf{w} \cdot \nabla \hat{v} \, dx\, dt + \int_{\Gamma_*} [\beta \hat{v}^2] n_4 \, ds - a_2(\hat{v}, \beta \hat{v}) \\ &= -\frac{1}{2} \|\beta \hat{v}(\cdot, 0)\|^2_{L^2(\Omega)} - a_2(\hat{v}, \beta \hat{v}). \end{aligned}$$

Using ellipticity of $a_2(\cdot, \beta \cdot)$, cf. (10.38), this implies $\hat{v} = 0$ and thus $v = 0$. □.

Theorem 10.3.11 *For each $f \in L^2(Q_T)$ the space-time variational problem* (10.33) *has a unique solution $u \in W_{\beta,0}$ and $\|u\|_W \le c \|f\|_{L^2(Q_T)}$ holds with a constant c independent of f.*

Proof. This follows from Theorem 15.1.1 and the Lemmas 10.3.9 and 10.3.10. □

11

Finite element discretization

In this chapter we discuss a finite element discretization method for the mass transport problem in (10.1). Compared to the usual convection-diffusion problems there are two issues that make this problem more complicated. Firstly, one has to deal with the Henry interface condition in (10.1c) and secondly, due to this condition the solution u is *discontinuous* across the interface. In our applications, due to the level set technique for capturing the interface, the triangulation is *not* fitted to Γ.

As in Chap. 10 we distinguish two cases. In the Sects. 11.1–11.4 we restrict ourselves to the case of a *stationary* interface and velocity field, as in Sect. 10.2. In Sect. 11.5 we address the finite element discretization for the general case of the mass transport problem with a *non*-stationary interface and velocity field.

We restrict ourselves to the *diffusion dominated* case. If the mass transport problem is *convection dominated* one needs additional stabilization of the finite element methods treated in this chapter. One possibility would be to use the streamline-diffusion stabilization approach as explained in Sect. 7.2.2. However, we do not treat such stabilized versions.

11.1 Nitsche-XFEM method

In this section we derive a special finite element method that is very suitable for the discretization of the mass transport problem (10.1). For this derivation and the error analysis (in the Sects. 11.2 and 11.3) we restrict ourselves to the case in which both the interface Γ and the velocity filed $\mathbf{w}$ are assumed to be independent of t. Generalization of this method to the case of a non-stationary Γ and $\mathbf{w}$ is discussed in Sect. 11.5.

For a conforming finite element discretization of the variational problem (10.14) one needs finite element functions ϕ that satisfy the interface condition $[\beta\phi] = 0$, which is very inconvenient, in particular if the interface Γ crosses the elements. An alternative approach is to use a technique, due to Nitsche [186], in which the XFEM space is used and the interface condition $[\beta u] = 0$ is enforced, in a weak sense, by modifying the bilinear form. An overview of applications of Nitsche's method to interface problems is given in [137].

We present the Nitsche-XFEM method for the 3D case, along the same lines as in [135, 136]. Let $\{\mathcal{T}_h\}_{h>0}$ be a regular family of tetrahedral triangulations of Ω. A triangulation $\mathcal{T}_h$ consists of tetrahedra T, with $h_T := \mathrm{diam}(T)$ and $h := \max\{\, h_T : T \in \mathcal{T}_h \,\}$. For any tetrahedron $T \in \mathcal{T}_h$ let $T_i := T \cap \Omega_i$ be the part of T in Ω_i. For any T with $T \cap \Gamma \neq \emptyset$ we define $\Gamma_T := T \cap \Gamma$. We introduce the finite element space

$$Q_h^\Gamma := \left\{\, v \in H_0^1(\Omega_1 \cup \Omega_2) : \, v_{|T_i} \ \text{ is linear for all } \ T \in \mathcal{T}_h, \ i = 1,2 \,\right\}. \quad (11.1)$$

In the variational formulation of the continuous problem (10.14) the space $V := \left\{\, v \in H_0^1(\Omega_1 \cup \Omega_2) : \, [\beta v]_\Gamma = 0 \,\right\}$ is used. Note that $Q_h^\Gamma \subset H_0^1(\Omega_1 \cup \Omega_2)$, but $Q_h^\Gamma \not\subset V$, since the Henry interface condition $[\beta v_h] = 0$ does not necessarily hold for $v_h \in Q_h^\Gamma$.

Remark 11.1.1 The space Q_h^Γ defined above is almost the same as the extended finite element space (XFEM) treated in Sect. 7.9.2, cf. (7.106) and (7.116). The only difference is that in (11.1) we restrict to finite element functions with homogeneous Dirichlet boundary conditions on $\partial\Omega$, whereas in Sect. 7.9.2 there are no essential boundary conditions in the XFEM space. For implementation issues we refer to Sects. 7.9.2 and 7.9.3. In practice we use this space with a (piecewise planar) approximation Γ_h of Γ. It follows from Theorem 7.9.3 that this finite element space has optimal approximation properties for functions $u \in H^m(\Omega_1 \cup \Omega_2)$, $m = 1, 2$.

Define

$$(\kappa_i)_{|T} = \frac{|T_i|}{|T|}, \qquad T \in \mathcal{T}_h, \ i = 1, 2,$$

hence, $\kappa_1 + \kappa_2 = 1$. For v sufficiently smooth such that $(v_i)_{|\Gamma}$, $i = 1, 2$, are well-defined, we define the weighted average

$$\{v\} := \kappa_1 (v_1)_{|\Gamma} + \kappa_2 (v_2)_{|\Gamma}. \quad (11.2)$$

For the average and jump operators the following identity holds for all f, g such that these operators are well-defined:

$$[fg] = \{f\}[g] + [f]\{g\} - (\kappa_1 - \kappa_2)[f][g]. \quad (11.3)$$

Let $(f, g)_\Gamma := \int_\Gamma fg \, ds$ be the $L^2(\Gamma)$ scalar product. We introduce the bilinear form

$$\begin{aligned} a_h(u,v) := {} & (\alpha u, v)_{1,\Omega_1 \cup \Omega_2} + (\mathbf{w} \cdot \nabla u, v)_0 - ([\beta u], \{\alpha \nabla v \cdot \mathbf{n}\})_\Gamma \\ & - (\{\alpha \nabla u \cdot \mathbf{n}\}, [\beta v])_\Gamma + \lambda h^{-1}([\beta u], [\beta v])_\Gamma, \end{aligned} \quad (11.4)$$

with $\lambda > 0$ a parameter. This bilinear form is well-defined on the space Q_h^Γ but also on

$$W_{\text{reg}} := \{ v \in H_0^1(\Omega_1 \cup \Omega_2) : v_i \in H^2(\Omega_i),\ i = 1,2 \}.$$

The space W_{reg} is larger than the space V_{reg} in (10.12). The interface condition $[\beta v] = 0$ is fulfilled for all $v \in V_{\text{reg}}$ but *not necessarily* for $v \in W_{\text{reg}}$. Using this bilinear form we define a method of lines discretization of (10.14).

Let $\hat{u}_0 \in Q_h^\Gamma$ be an approximation of u_0. For $t \in [0,T]$ let $u_h(t) \in Q_h^\Gamma$ be such that $u_h(0) = \hat{u}_0$ and

$$\left(\frac{du_h}{dt}, v_h\right)_0 + a_h(u_h, v_h) = (f, v_h)_0 \quad \text{for all} \quad v_h \in Q_h^\Gamma. \tag{11.5}$$

In practice one uses Γ_h instead of Γ. Using the (nodal) basis $(q_j)_{j\in\mathcal{J}} \cup (q_j^\Gamma)_{j\in\mathcal{J}_\Gamma}$ in Q_h^Γ, cf. (7.106), the function $u_h = u_h(t)$ can be represented as

$$u_h(t) = \sum_{j\in\mathcal{J}} u_j(t) q_j + \sum_{j\in\mathcal{J}_\Gamma} u_j^\Gamma(t) q_j^\Gamma$$

and the problem (11.5) can be written as a system of coupled ordinary differential equations for the time dependent coefficients $u_j(t)$, $j \in \mathcal{J}$, and $u_j^\Gamma(t)$, $j \in \mathcal{J}_\Gamma$. Note that since we assumed Γ to be stationary, $m := |\mathcal{J}| + |\mathcal{J}_\Gamma|$ is independent of t. In general this does not hold for a non-stationary Γ. For the matrix-vector representation of the system of ordinary differential equations we introduce some notation. Without loss of generality we assume that $\mathcal{J}_\Gamma = \{1, \ldots, |\mathcal{J}_\Gamma|\}$. Denoting the extended basis functions q_j^Γ, $j \in \mathcal{J}_\Gamma$ by $q_{|\mathcal{J}|+j} := q_j^\Gamma$ and the corresponding coefficients by $u_{|\mathcal{J}|+j} := u_j^\Gamma$, the mass and stiffness matrix are given by

$$\mathbf{M}_{ij} = (q_j, q_i)_0, \quad \mathbf{A}_{ij} = a_h(q_j, q_i), \quad 1 \le i, j \le m.$$

Furthermore, $\vec{\mathbf{b}}_i = (f, q_i)_0$, $1 \le i \le m$, and $\vec{\mathbf{u}}_0$ denotes the vector representation of the initial condition $\hat{u}_0$. For simplicity we assume f, and hence also $\vec{\mathbf{b}}$, to be independent of t. Then the system of ordinary differential equations takes the form

$$\begin{aligned} \mathbf{M}\frac{d\vec{\mathbf{u}}}{dt}(t) + \mathbf{A}\vec{\mathbf{u}}(t) &= \vec{\mathbf{b}}, \\ \vec{\mathbf{u}}(0) &= \vec{\mathbf{u}}_0. \end{aligned} \tag{11.6}$$

As an example, for the time discretization we consider the θ-scheme:

$$\begin{aligned} \vec{\mathbf{u}}^0 &= \vec{\mathbf{u}}_0, \\ \frac{1}{\Delta t}\mathbf{M}(\vec{\mathbf{u}}^{n+1} - \vec{\mathbf{u}}^n) + \theta\mathbf{A}\vec{\mathbf{u}}^{n+1} + (1-\theta)\mathbf{A}\vec{\mathbf{u}}^n &= \vec{\mathbf{b}}, \quad n = 0, 1, \ldots. \end{aligned} \tag{11.7}$$

Below we analyze the errors in these space and time discretization methods. In Sect. 11.2 we analyze the spatial discretization error in the Nitsche method (11.5). In Sect. 11.3 we analyze the error after space *and* time discretization. In Sect. 11.4 we present results of a few numerical experiments with the Nitsche-XFEM method.

Remark 11.1.2 We discuss an alternative equivalent representation of the Nitsche-XFEM discretization in (11.5) which is better suitable for a generalization of the Nitsche-XFEM method to a space-time discretization, as discussed in Sect. 11.5.2. This formulation uses the subspaces $V_i \subset Q_h$, $i = 1,2$, and the isomorphism between $V_1 \times V_2$ and Q_h^Γ (or $Q_h^{\Gamma_h}$) as explained in Remark 7.9.9. We introduce some notation. In the bilinear form $a_h(\cdot,\cdot)$ in (11.4) the first two terms correspond to the bilinear form $a(\cdot,\cdot)$ of the continuous problem, cf. (10.6), and the three interface terms come from the Nitsche method. The bilinear form $a(\cdot,\cdot)$ is split according to the subdomains:

$$a_i(u,v) := \int_{\Omega_i} \beta_i \alpha_i \nabla u \cdot \nabla v \, dx + \int_{\Omega_i} \beta_i \mathbf{w} \cdot \nabla u v \, dx, \quad u,v \in H^1(\Omega_i), \quad i = 1,2.$$

Note that β_i, α_i are constant on Ω_i. For a pair of functions (w_1,w_2) the jump and average are given by

$$[(w_1,w_2)] := (w_1 - w_2)_{|\Gamma}, \qquad \{(w_1,w_2)\} := (\kappa_1 w_1 + \kappa_2 w_2)_{|\Gamma}.$$

We define the Nitsche interface functional

$$\begin{aligned} N_\Gamma\big((u_1,u_2),(v_1,v_2)\big) := &- \int_\Gamma [(\beta_1 u_1, \beta_2 u_2)]\{\alpha_1 \nabla v_1 \cdot \mathbf{n}, \alpha_2 \nabla v_2 \cdot \mathbf{n}\} \, ds \\ &- \int_\Gamma \{\alpha_1 \nabla u_1 \cdot \mathbf{n}, \alpha_2 \nabla u_2 \cdot \mathbf{n}\}[(\beta_1 v_1, \beta_2 v_2)] \, ds \\ &+ \lambda h^{-1} \int_\Gamma [(\beta_1 u_1, \beta_2 u_2)][(\beta_1 v_1, \beta_2 v_2)], \quad u_i, v_i \in V_i. \end{aligned}$$

The Nitsche-XFEM method can be reformulated in the space $V_1 \times V_2$ as follows. Take $(\hat{u}_1, \hat{u}_2) \in V_1 \times V_2$ such that $R_1\hat{u}_1 + R_2\hat{u}_1$ is an approximation of the initial condition u_0. Determine $(u_1(t), u_2(t)) \in V_1 \times V_2$ such that $(u_1(0), u_2(0)) = (\hat{u}_1, \hat{u}_2)$ and for $t \in [0,T]$:

$$\begin{aligned} &\sum_{i=1}^{2} \big[(\beta_i \frac{du_i}{dt}, v_i)_{L^2(\Omega_i)} + a_i(u_i,v_i)\big] + N_\Gamma\big((u_1,u_2),(v_1,v_2)\big) \\ &= \sum_{i=1}^{2} (\beta_i f, v_i)_{L^2(\Omega_i)} \quad \text{for all} \quad (v_1,v_2) \in V_1 \times V_2. \end{aligned} \tag{11.8}$$

In this formulation instead of the XFEM space Q_h^Γ we use the subspaces $V_i \subset Q_h$ of the standard space Q_h of linear finite elements. In (11.8) we do not use integrals over Ω but only over Ω_i, $i = 1,2$. For the implementation of this formulation in the space $V_1 \times V_2$ it is natural to use the basis as in (7.126).

11.2 Analysis of the Nitsche-XFEM method

In this section we present an error analysis of the method of lines discretization given in (11.5), which is based on the analyses presented in [135, 136]. We assume that the velocity field $\mathbf{w}$ is such that the conditions in (10.3) are satisfied. Related to the regular family of triangulations $\{\mathcal{T}_h\}$ we formulate the same assumptions as in [135, 136]:

Assumption 11.2.1 Consider a $T \in \mathcal{T}_h$ with $T \cap \Gamma \neq \emptyset$. We assume that the intersection of the interface Γ with ∂T is either a connected curve or a face of T. For the first case, take a plane through three of the points of the intersection of Γ and the edges of T and let $\Gamma_{T,h}$ be the intersection of T and this plane. We assume that, in suitable local coordinates (ξ, η, θ), $\Gamma_T = \Gamma \cap T$ can be parametrized by a function on $\Gamma_{T,h}$:

$$\Gamma_T = \{\, (\xi, \eta, \theta) : (\xi, \eta, 0) \in \Gamma_{T,h}, \; \theta = g(\xi, \eta) \,\}.$$

The assumptions formulated in 11.2.1 are satisfied on sufficiently fine meshes. We start the analysis with a consistency result:

Lemma 11.2.2 *Let $u = u(t) \in V_{\mathrm{reg}}$ be the solution defined in Theorem 10.2.2. Then $u(t)$ satisfies*

$$(\frac{du}{dt}, v_h)_0 + a_h(u, v_h) = (f, v_h)_0 \quad \textit{for all} \;\; v_h \in Q_h^\Gamma, \;\; t \in [0, T]. \qquad (11.9)$$

Proof. From Lemma 10.2.3 we have that $u = u(t)$ satisfies $[\alpha \nabla u \cdot \mathbf{n}] = 0$, $[\beta u] = 0$. Using this and (11.3) we obtain, for $v_h \in Q_h^\Gamma$:

$$\begin{aligned}
&- \sum_{i=1,2} \int_{\Omega_i} \operatorname{div}(\alpha_i \nabla u_i) \beta v_h \, dx + (\mathbf{w} \cdot \nabla u, v_h)_0 \\
&= - \int_\Gamma [\alpha \nabla u \cdot \mathbf{n} \beta v_h] \, ds + (\alpha u, v_h)_{1, \Omega_1 \cup \Omega_2} + (\mathbf{w} \cdot \nabla u, v_h)_0 \\
&= -(\{\alpha \nabla u \cdot \mathbf{n}\}, [\beta v_h])_\Gamma + (\alpha u, v_h)_{1, \Omega_1 \cup \Omega_2} + (\mathbf{w} \cdot \nabla u, v_h)_0 = a_h(u, v_h).
\end{aligned}$$

Furthermore, u solves (10.1a) (in the sense as in Lemma 10.2.3). Multiplication of (10.1a) by βv_h and integration over Ω results in

$$\begin{aligned}
(f, v_h)_0 &= (\frac{du}{dt}, v_h)_0 + (\mathbf{w} \cdot \nabla u, v_h)_0 - \sum_{i=1,2} \int_{\Omega_i} \operatorname{div}(\alpha_i \nabla u_i) \beta v_h \, dx \\
&= (\frac{du}{dt}, v_h)_0 + a_h(u, v_h),
\end{aligned}$$

and thus the consistency result holds. □

For the error analysis we introduce suitable norms, as in [135]. Let $\mathcal{T}_h^\Gamma$ denote the set of all tetrahedra that are intersected by Γ. We define

$$\|v\|_{1/2,h,\Gamma}^2 := \sum_{T\in\mathcal{T}_h^\Gamma} h_T^{-1}\|v\|_{L^2(\Gamma_T)}^2, \tag{11.10}$$

$$\|v\|_{-1/2,h,\Gamma}^2 := \sum_{T\in\mathcal{T}_h^\Gamma} h_T\|v\|_{L^2(\Gamma_T)}^2, \tag{11.11}$$

$$|\!|\!|v|\!|\!|^2 := |v|_{1,\Omega_1\cup\Omega_2}^2 + \|\{\nabla v\cdot\mathbf{n}\}\|_{-1/2,h,\Gamma}^2 + \|[\beta v]\|_{1/2,h,\Gamma}^2. \tag{11.12}$$

Different from [135] we have a scaling with β in the terms $|v|_{1,\Omega_1\cup\Omega_2}$ and $\|[\beta v]\|_{1/2,h,\Gamma}$. The bilinear form $a_h(\cdot,\cdot)$ has the following continuity and ellipticity properties with respect to the norm $|\!|\!|\cdot|\!|\!|$.

Lemma 11.2.3 *There exist constants c_1, $c_2 > 0$ such that for λ sufficiently large (independent of h) the following holds:*

$$|a_h(u,v)| \le c_1|\!|\!|u|\!|\!|\,|\!|\!|v|\!|\!| \quad \text{for all} \quad u,v\in Q_h^\Gamma + W_{\text{reg}}, \tag{11.13}$$

$$a_h(v_h,v_h) \ge c_2|\!|\!|v_h|\!|\!|^2 \quad \text{for all} \quad v_h\in Q_h^\Gamma. \tag{11.14}$$

Proof. First note that $|(f,g)_\Gamma| \le \|f\|_{1/2,h,\Gamma}\|g\|_{-1/2,h,\Gamma}$ holds. Take $u,v\in Q_h^\Gamma + W_{\text{reg}}$. Using the Cauchy-Schwarz inequality and the definitions of the norms we obtain

$$\begin{aligned}
&|a_h(u,v)| \\
&\le c\,|u|_{1,\Omega_1\cup\Omega_2}|v|_{1,\Omega_1\cup\Omega_2} + c\,|u|_{1,\Omega_1\cup\Omega_2}\|v\|_0 \\
&\quad + \|[\beta u]\|_{1/2,h,\Gamma}\|\{\alpha\nabla v\cdot\mathbf{n}\}\|_{-1/2,h,\Gamma} + \|\{\alpha\nabla u\cdot\mathbf{n}\}\|_{-1/2,h,\Gamma}\|[\beta v]\|_{1/2,h,\Gamma} \\
&\quad + \lambda\|[\beta u]\|_{1/2,h,\Gamma}\|[\beta v]\|_{1/2,h,\Gamma} \le c\,|\!|\!|u|\!|\!|\,|\!|\!|v|\!|\!|,
\end{aligned}$$

which proves the continuity. Using the assumptions (10.3) we obtain for $v_h\in Q_h^\Gamma$, cf. (10.7), $(\mathbf{w}\cdot\nabla v_h, v_h)_0 = 0$. Hence, using $h_T\le h$ we get, with $c_0 = \min_{i=1,2}\alpha_i$,

$$\begin{aligned}
&a_h(v_h,v_h) \\
&\ge |\alpha^{\frac12}v_h|_{1,\Omega_1\cup\Omega_2}^2 - 2|(\{\alpha\nabla v_h\cdot\mathbf{n}\},[\beta v_h])_\Gamma| + \lambda\|[\beta v_h]\|_{1/2,h,\Gamma}^2 \\
&\ge c_0|v_h|_{1,\Omega_1\cup\Omega_2}^2 - 2\|\{\alpha\nabla v_h\cdot\mathbf{n}\}\|_{-1/2,h,\Gamma}\|[\beta v_h]\|_{1/2,h,\Gamma} + \lambda\|[\beta v_h]\|_{1/2,h,\Gamma}^2 \\
&\ge c_0|v_h|_{1,\Omega_1\cup\Omega_2}^2 - \delta\|\{\alpha\nabla v_h\cdot\mathbf{n}\}\|_{-1/2,h,\Gamma}^2 + (\lambda-\delta^{-1})\|[\beta v_h]\|_{1/2,h,\Gamma}^2
\end{aligned} \tag{11.15}$$

for arbitrary $\delta > 0$. The term $\|\{\alpha\nabla v_h\cdot\mathbf{n}\}\|_{-1/2,h,\Gamma}$ can be estimated as follows. For a given tetrahedron T and $v_h\in Q_h^\Gamma$, $\|\nabla(v_h)_i(x)\|$ (where $\|\cdot\|$ is the Euclidean norm on $\mathbb{R}^3$) is constant for $x\in T_i$, say equal to $\hat{c}_i$. Hence,

$$\begin{aligned}
h_T\kappa_i^2\|\alpha_i\nabla(v_h)_i\cdot\mathbf{n}\|_{L^2(\Gamma_T)}^2 &\leq h_T\kappa_i^2|\Gamma_T|\alpha_i^2\hat{c}_i^2 \leq ch_T\kappa_i^2\frac{|\Gamma_T|}{|T_i|}\|\beta_i^{\frac{1}{2}}\nabla(v_h)_i\|_{L^2(T_i)}^2 \\
&= ch_T\frac{|T_i||\Gamma_T|}{|T|^2}\|\beta_i^{\frac{1}{2}}\nabla(v_h)_i\|_{L^2(T_i)}^2 \\
&\leq c\|\beta_i^{\frac{1}{2}}\nabla(v_h)_i\|_{L^2(T_i)}^2,
\end{aligned}$$

and thus

$$\begin{aligned}
\|\{\alpha\nabla v_h\cdot\mathbf{n}\}\|_{-1/2,h,\Gamma}^2 &\leq 2\sum_{i=1}^2\kappa_i^2\|\alpha_i\nabla(v_h)_i\cdot\mathbf{n}\|_{-1/2,h,\Gamma}^2 \\
&= 2\sum_{i=1}^2\sum_{T\in\mathcal{T}_h^\Gamma}h_T\kappa_i^2\|\alpha_i\nabla(v_h)_i\cdot\mathbf{n}\|_{L^2(\Gamma_T)}^2 \\
&\leq c\sum_{i=1}^2\sum_{T\in\mathcal{T}_h^\Gamma}\|\beta_i^{\frac{1}{2}}\nabla(v_h)_i\|_{L^2(T_i)}^2 \\
&\leq c\sum_{i=1}^2\|\beta_i^{\frac{1}{2}}\nabla(v_h)_i\|_{L^2(\Omega_i)}^2 = c\,|v_h|_{1,\Omega_1\cup\Omega_2}^2.
\end{aligned}$$

With this constant c and for $\lambda \geq \frac{c_0}{1+2c}+\frac{1+2c}{c_0}$, taking $\delta=\frac{c_0}{1+2c}$ in (11.15) we get

$$\begin{aligned}
&a_h(v_h,v_h) \\
&\geq c_0|v_h|_{1,\Omega_1\cup\Omega_2}^2-\frac{c_0}{1+2c}\|\{\alpha\nabla v_h\cdot\mathbf{n}\}\|_{-1/2,h,\Gamma}^2+(\lambda-\frac{1+2c}{c_0})\|[\beta v_h]\|_{1/2,h,\Gamma}^2 \\
&\geq c_0|v_h|_{1,\Omega_1\cup\Omega_2}^2-\frac{2c_0}{1+2c}\|\{\alpha\nabla v_h\cdot\mathbf{n}\}\|_{-1/2,h,\Gamma}^2+\frac{c_0}{1+2c}\|\{\alpha\nabla v_h\cdot\mathbf{n}\}\|_{-1/2,h,\Gamma}^2 \\
&+\frac{c_0}{1+2c}\|[\beta v_h]\|_{1/2,h,\Gamma}^2 \\
&\geq\frac{c_0}{1+2c}|v_h|_{1,\Omega_1\cup\Omega_2}^2+\frac{c_0}{1+2c}\|\{\alpha\nabla v_h\cdot\mathbf{n}\}\|_{-1/2,h,\Gamma}^2+\frac{c_0}{1+2c}\|[\beta v_h]\|_{1/2,h,\Gamma}^2 \\
&=\frac{c_0}{1+2c}|\!|\!|v_h|\!|\!|^2,
\end{aligned}$$

and thus the ellipticity result in (11.14). □

Below, in Theorem 11.2.4, we derive an approximation error bound for the XFEM space Q_h^Γ with respect to the norm $|\!|\!|\cdot|\!|\!|$. The analysis is essentially the same as in [135, 136]. In Sect. 7.9.4, Theorem 7.9.3, we derived such a bound in the (more standard) norms $\|\cdot\|_{L^2}$ and $\|\cdot\|_1$. As in (7.114) we can define an interpolation operator $I_h^*: W_{\mathrm{reg}}\to Q_h^\Gamma$ of the form:

$$I_h^*v = R_1I_h^2\mathcal{E}_1^2R_1v+R_2I_h^2\mathcal{E}_2^2R_2v,$$

with R_i restriction operators, $R_iv:=v_{|\Omega_i}$, $\mathcal{E}_i^2: H^2(\Omega_i)\to H^2(\Omega)$ bounded extension operators (that preserve homogeneous Dirichlet boundary conditions on $\partial\Omega$) and I_h^2 the standard nodal interpolation operator on $H^2(\Omega)\cap H_0^1(\Omega)$. For these operators the following error bounds hold:

$$\begin{aligned} &\|\mathcal{E}_i^2 v - I_h^2\mathcal{E}_i^2 v\|_{L^2(T)} \leq c h_T^m \|\mathcal{E}_i^2 v\|_{H^m(T)}, \quad m=1,2, \quad v\in W_{\rm reg}, \\ &\|\mathcal{E}_i^2 v - I_h^2\mathcal{E}_i^2 v\|_{H^1(T)} \leq c h_T \|\mathcal{E}_i^2 v\|_{H^2(T)}, \quad v\in W_{\rm reg}, \end{aligned} \tag{11.16}$$

with a constant c independent of v and of $T\in\mathcal{T}_h$. Furthermore, we will use in the analysis below the following trace inequality that is known in the literature:

$$\|w\|_{L^2(\partial T)}^2 \leq c\big(h_T^{-1}\|w\|_{L^2(T)}^2 + h_T\|w\|_{H^1(T)}^2\big), \quad \text{for all } \ w\in H^1(T), \tag{11.17}$$

with a constant c independent of w and of $T\in\mathcal{T}_h$. This result can be obtained by using an affine transformation of T to the unit tetrahedron $\hat{T}$ and applying $\|w\|_{L^2(\partial\hat{T})}^2 \leq c\|w\|_{L^2(\hat{T})}\|w\|_{H^1(\hat{T})}$ for all $w\in H^1(\hat{T})$, cf. e.g. Theorem 1.6.6 in [53].

Theorem 11.2.4 *Let $I_h^*: W_{\rm reg}\to Q_h^\Gamma$ be the interpolation operator defined above. There exists a constant c such that*

$$\|v - I_h^* v\| \leq c\, h \|v\|_{2,\Omega_1\cup\Omega_2} \quad \textit{for all} \ \ v\in W_{\rm reg} \tag{11.18}$$

holds.

Proof. Take $T\in\mathcal{T}_h^\Gamma$, i.e., $\Gamma_T := T\cap\Gamma \neq \emptyset$. Due to Assumption 11.2.1 the intersection of Γ_T with ∂T is either a connected curve or a face. In the latter case it is trivial, that the estimate (11.17) also holds with ∂T replaced by Γ_T. We now show that also in the former case, the result (11.17) with ∂T replaced by Γ_T holds. We use the local coordinates $z := (\xi,\eta,\theta)$ as defined in Assumption 11.2.1. Define the level set function $\phi(\xi,\eta,\theta) = \theta - g(\xi,\eta)$. From Assumption 11.2.1 it follows that Γ_T is the zero level of ϕ. Take a fixed $i\in\{1,2\}$ and $T_i := \Omega_i\cap T$. Let $\mathbf{n}(z) = (n_1(z), n_2(z), n_3(z))^T$ be the unit outward pointing normal on ∂T_i. Note that $\Gamma_T\subset\partial T_i$. For $w\in H^1(T)$ the following holds:

$$\begin{aligned} 2\int_{T_i} w\frac{\partial w}{\partial\theta}\,dz &= \int_{T_i} {\rm div}_z \begin{pmatrix} 0\\ 0\\ w^2\end{pmatrix} dz = \int_{\partial T_i} \mathbf{n}\cdot \begin{pmatrix} 0\\ 0\\ w^2\end{pmatrix} ds = \int_{\partial T_i} n_3(s) w^2(s)\,ds \\ &= \int_{\Gamma_T} n_3(s)w^2(s)\,ds + \int_{\partial T_i\setminus\Gamma_T} n_3(s)w^2(s)\,ds. \end{aligned} \tag{11.19}$$

For $z\in\Gamma_T$ we have $\mathbf{n}(z) = \pm\frac{\nabla_z\phi(z)}{\|\nabla_z\phi(z)\|}$ and from $\nabla_z\phi(z) = \begin{pmatrix} -\nabla_{(\xi,\eta)}g(\xi,\eta)\\ 1\end{pmatrix}$ it follows that

$$n_3(z) = \big(\|\nabla_{(\xi,\eta)}g(\xi,\eta)\|^2 + 1\big)^{-\frac{1}{2}}, \quad z\in\Gamma_T.$$

From smoothness properties of Γ it follows that $1 \leq n_3(z)^{-1} \leq c$ with c independent of T. Using this and the identity in (11.19) we get, for arbitrary $w \in H^1(T)$:

$$\begin{aligned}\int_{\Gamma_T} w^2\,ds &\leq c\int_{\Gamma_T} n_3\,w^2\,ds \leq c\|w\|_{L^2(T_i)}\|w\|_{H^1(T_i)} + c\int_{\partial T_i\setminus\Gamma_T} w^2\,ds\\ &\leq c\|w\|_{L^2(T)}\|w\|_{H^1(T)} + c\int_{\partial T} w^2\,ds\\ &\leq ch_T^{-1}\|w\|^2_{L^2(T)} + h_T\|w\|^2_{H^1(T)},\end{aligned}$$

where in the last inequality we used (11.17). Thus we have shown that

$$\|w\|^2_{L^2(\Gamma_T)} \leq c\big(h_T^{-1}\|w\|^2_{L^2(T)} + h_T\|w\|^2_{H^1(T)}\big) \quad \text{for all} \quad w \in H^1(T) \tag{11.20}$$

holds, with c independent of w, T *and* of Γ_T.

We take $v \in W_{\mathrm{reg}}$ and write $e_h := v - I_h^* v$. We have

$$|||e_h|||^2 = |e_h|^2_{1,\Omega_1\cup\Omega_2} + \|\{\nabla e_h\cdot\mathbf{n}\}\|^2_{-1/2,h,\Gamma} + \|[\beta e_h]\|^2_{1/2,h,\Gamma}. \tag{11.21}$$

For the first term it easily follows, as in the proof of Theorem 7.9.3, that

$$|e_h|^2_{1,\Omega_1\cup\Omega_2} \leq ch^2\|v\|^2_{2,\Omega_1\cup\Omega_2}$$

holds. We now consider the third term on the right-hand side in (11.21):

$$\|[\beta e_h]\|^2_{1/2,h,\Gamma} = \sum_{T\in\mathcal{T}_h^\Gamma} h_T^{-1}\|[\beta e_h]\|^2_{L^2(\Gamma_T)}.$$

For $T \in \mathcal{T}_h^\Gamma$ we have (with $v_i = R_i v = v_{|\Omega_i}$)

$$\begin{aligned}h_T^{-1}\|[\beta e_h]\|^2_{L^2(\Gamma_T)} &\leq c\sum_{i=1}^2 h_T^{-1}\|R_i e_h\|^2_{L^2(\Gamma_T)} = c\sum_{i=1}^2 h_T^{-1}\|v_i - R_i I_h^* v\|^2_{L^2(\Gamma_T)}\\ &= c\sum_{i=1}^2 h_T^{-1}\|\mathcal{E}_i^2 v_i - I_h^2\mathcal{E}_i^2 v_i\|^2_{L^2(\Gamma_T)}\\ &\leq c\sum_{i=1}^2\Big(h_T^{-2}\|\mathcal{E}_i^2 v_i - I_h^2\mathcal{E}_i^2 v_i\|^2_{L^2(T)} + \|\mathcal{E}_i^2 v_i - I_h^2\mathcal{E}_i^2 v_i\|^2_{H^1(T)}\Big),\end{aligned}$$

where in the last inequality we used (11.20). Using the local error bounds for the standard nodal interpolation operator I_h^2 given in (11.16) and summing over $T \in \mathcal{T}_h^\Gamma$ we obtain

$$\begin{aligned}\|[\beta e_h]\|^2_{1/2,h,\Gamma} &\leq c\sum_{T\in\mathcal{T}_h^\Gamma}\sum_{i=1}^2 h_T^2\|\mathcal{E}_i^2 v_i\|^2_{H^2(T)}\\ &\leq ch^2\sum_{i=1}^2\|\mathcal{E}_i^2 v_i\|_{H^2(\Omega)} \leq ch^2\|v\|^2_{2,\Omega_1\cup\Omega_2}.\end{aligned}$$

We finally apply similar arguments to derive a bound for the second term on the right-hand side in (11.21):

$$\|\{\nabla e_h \cdot \mathbf{n}\}\|^2_{-1/2,h,\Gamma} = \sum_{T \in \mathcal{T}_h^\Gamma} h_T \|\{\nabla e_h \cdot \mathbf{n}\}\|^2_{L^2(\Gamma_T)}.$$

For $T \in \mathcal{T}_h^\Gamma$ we have, with $w_i := \nabla(\mathcal{E}_i^2 v_i - I_h^2 \mathcal{E}_i^2 v_i) \cdot \mathbf{n}$:

$$\begin{aligned}
h_T \|\{\nabla e_h \cdot \mathbf{n}\}\|^2_{L^2(\Gamma_T)} &\leq \sum_{i=1}^{2} h_T \|\nabla(v_i - R_i I_h^2 \mathcal{E}_i^2 v_i) \cdot \mathbf{n}\|^2_{L^2(\Gamma_T)} \\
&= \sum_{i=1}^{2} h_T \|w_i\|^2_{L^2(\Gamma_T)} \leq c \sum_{i=1}^{2} \left(\|w_i\|^2_{L^2(T)} + h_T^2 \|w_i\|^2_{H^1(T)} \right) \\
&\leq c \sum_{i=1}^{2} \left(\|\mathcal{E}_i^2 v_i - I_h^2 \mathcal{E}_i^2 v_i\|^2_{H^1(T)} + h_T^2 \|\mathcal{E}_i^2 v_i\|^2_{H^2(T)} \right) \\
&\leq c \sum_{i=1}^{2} h_T^2 \|\mathcal{E}_i^2 v_i\|^2_{H^2(T)}.
\end{aligned}$$

Summing over $T \in \mathcal{T}_h^\Gamma$ we obtain

$$\|\{\nabla e_h \cdot \mathbf{n}\}\|^2_{-1/2,h,\Gamma} \leq c \sum_{T \in \mathcal{T}_h^\Gamma} \sum_{i=1}^{2} h_T^2 \|\mathcal{E}_i^2 v_i\|^2_{H^2(T)} \leq c h^2 \|v\|^2_{2,\Omega_1 \cup \Omega_2},$$

which completes the proof. □

In the error analysis we use the elliptic projection $R_h : W_{\mathrm{reg}} + Q_h^\Gamma \to Q_h^\Gamma$, defined by

$$a_h(R_h v, w_h) = a_h(v, w_h) \quad \text{for all} \quad w_h \in Q_h^\Gamma.$$

In the following two lemmas we derive error bounds for this projection.

Lemma 11.2.5 *The following holds:*

$$|\!|\!| R_h v - v |\!|\!| \leq c\, h \|v\|_{2,\Omega_1 \cup \Omega_2} \quad \textit{for all} \quad v \in W_{\mathrm{reg}}.$$

Proof. For $v \in W_{\mathrm{reg}}$ define $\chi_h := R_h v - I_h^* v \in Q_h^\Gamma$. Using Lemma 11.2.3 and Theorem 11.2.4 we get, with $c_2 > 0$:

$$\begin{aligned}
c_2 |\!|\!| \chi_h |\!|\!|^2 &\leq a_h(\chi_h, \chi_h) = a_h(R_h v - I_h^* v, \chi_h) \\
&= a_h(v - I_h^* v, \chi_h) \leq c_1 |\!|\!| v - I_h^* v |\!|\!| \, |\!|\!| \chi_h |\!|\!| \leq c h \|v\|_{2,\Omega_1 \cup \Omega_2} |\!|\!| \chi_h |\!|\!|.
\end{aligned}$$

Hence, $|\!|\!| \chi_h |\!|\!| \leq c\, h \|v\|_{2,\Omega_1 \cup \Omega_2}$ holds and thus

$$|\!|\!| R_h v - v |\!|\!| \leq |\!|\!| \chi_h |\!|\!| + |\!|\!| v - I_h^* v |\!|\!| \leq c\, h \|v\|_{2,\Omega_1 \cup \Omega_2}$$

holds. □

In the next lemma we derive an L^2-norm error bound based on a standard duality argument. For this we need an H^2-regularity property for the stationary problem.

Lemma 11.2.6 *Assume that the H^2-regularity property* (10.11) *is valid. The following holds:*

$$\|R_h v - v\|_0 \leq c\,h^2 \|v\|_{2,\Omega_1\cup\Omega_2} \quad \textit{for all} \;\; v \in W_{\mathrm{reg}}.$$

Proof. For $v \in W_{\mathrm{reg}}$ define $e_h := R_h v - v \in Q_h^\Gamma + W_{\mathrm{reg}}$. Introduce the bilinear form

$$\tilde{a}(u,v) = (\alpha u, v)_{1,\Omega_1\cup\Omega_2} - (\mathbf{w}\cdot\nabla u, v)_0, \quad u,v \in H_0^1(\Omega_1\cup\Omega_2).$$

Using $\mathbf{w}\cdot\mathbf{n} = 0$ on Γ and $\operatorname{div}\mathbf{w} = 0$ we obtain by partial integration on the subdomains Ω_i, $-(\mathbf{w}\cdot\nabla u, v)_0 = (\mathbf{w}\cdot\nabla v, u)_0$ and thus $\tilde{a}(u,v) = a(v,u)$ for $u,v \in H_0^1(\Omega_1\cup\Omega_2)$. Let $\tilde{u} \in V = \{\, v \in H_0^1(\Omega_1\cup\Omega_2) : [\beta v]_\Gamma = 0 \,\}$ be the unique solution of

$$\tilde{a}(\tilde{u}, v) = (e_h, v)_0 \quad \text{for all} \;\; v \in V.$$

This dual problem has the same regularity properties as the one in (10.10), i.e., $\tilde{u} \in H^2(\Omega_1\cup\Omega_2)$ and

$$\|\tilde{u}\|_{2,\Omega_1\cup\Omega_2} \leq c\|e_h\|_0,$$

with a constant c independent of e_h. Using this regularity property, combined with $[\beta\tilde{u}] = 0$ (since $\tilde{u} \in V$) it follows that $\tilde{u}$ solves the following problem:

$$-\operatorname{div}(\alpha\nabla\tilde{u}) - \mathbf{w}\cdot\nabla\tilde{u} = e_h \quad \text{in} \;\; \Omega_i, \; i = 1,2, \;\; (\text{in } L^2 \text{ sense}), \tag{11.22a}$$

$$[\alpha\nabla\tilde{u}\cdot\mathbf{n}]_\Gamma = 0, \tag{11.22b}$$

$$[\beta\tilde{u}]_\Gamma = 0. \tag{11.22c}$$

Multiplication of (11.22a) with βe_h, integration over Ω and applying partial integration on the subdomains Ω_i we obtain, using (11.22b),(11.22c):

$$\begin{aligned}
(e_h, e_h)_0 &= (\alpha\tilde{u}, e_h)_{1,\Omega_1\cup\Omega_2} - (\mathbf{w}\cdot\nabla\tilde{u}, e_h)_0 - \int_\Gamma [\alpha\nabla\tilde{u}\cdot\mathbf{n}\beta e_h]\,ds \\
&= (\alpha e_h, \tilde{u})_{1,\Omega_1\cup\Omega_2} + (\mathbf{w}\cdot\nabla e_h, \tilde{u})_0 - ([\beta e_h], \{\alpha\nabla\tilde{u}\cdot\mathbf{n}\})_\Gamma \\
&\quad - (\{\alpha\nabla e_h\cdot\mathbf{n}\}, [\beta\tilde{u}])_\Gamma + \lambda h^{-1}([\beta e_h], [\beta\tilde{u}])_\Gamma \\
&= a_h(e_h, \tilde{u}).
\end{aligned}$$

Using this in combination with Lemma 11.2.5 and Theorem 11.2.4 we get

$$\begin{aligned}
(e_h, e_h)_0 &= a_h(e_h, \tilde{u}) = a_h(e_h, \tilde{u} - I_h^*\tilde{u}) \leq c_1 |\!|\!| e_h |\!|\!| \, |\!|\!| \tilde{u} - I_h^*\tilde{u} |\!|\!| \\
&\leq c\,h^2 \|v\|_{2,\Omega_1\cup\Omega_2} \|\tilde{u}\|_{2,\Omega_1\cup\Omega_2} \leq c\,h^2 \|v\|_{2,\Omega_1\cup\Omega_2} \|e_h\|_0,
\end{aligned}$$

which completes the proof. □

We now derive an error bound for the semi-discretization by the Nitsche-XFEM method in (11.5). We require that the solution $u = u(t) \in V_{\mathrm{reg}}$ as defined in Theorem 10.2.2 has sufficient regularity, in particular $\frac{du}{dt} \in L^1(0,T;W_{\mathrm{reg}})$. The analysis uses standard arguments as in, for example, [238] or the proof of Theorem 3.2.14.

Theorem 11.2.7 *Assume that the regularity property* (10.11) *is valid. Let* $u = u(t) \in V_{\mathrm{reg}}$ *be the solution as in Theorem 10.2.2 and* $u_h = u_h(t) \in Q_h^\Gamma$ *the solution of* (11.5) *with* $u_h(0) = \hat{u}_0$. *The following holds for* $t \in [0,T]$*:*

$$\|u_h(t) - u(t)\|_0 \leq \|\hat{u}_0 - R_h u_0\|_0 + c\,h^2\{\|u_0\|_{2,\Omega_1\cup\Omega_2} + \int_0^t \left\|\frac{du}{dt}\right\|_{2,\Omega_1\cup\Omega_2} d\tau\}.$$

Proof. Introduce the splitting $u_h(t) - u(t) = \theta(t) + \rho(t)$, with $\theta := u_h - R_h u$, $\rho := R_h u - u$. From Lemma 11.2.6 we have

$$\begin{aligned}\|\rho(t)\|_0 &\leq ch^2\|u(t)\|_{2,\Omega_1\cup\Omega_2} \\ &\leq c\,h^2\big(\|u_0\|_{2,\Omega_1\cup\Omega_2} + \int_0^t \left\|\frac{du}{dt}\right\|_{2,\Omega_1\cup\Omega_2} d\tau\big).\end{aligned} \tag{11.23}$$

For $\theta = \theta(t) \in Q_h^\Gamma$ we have, using Lemma 11.2.2:

$$\begin{aligned}\|\theta\|_0 \frac{d}{dt}\|\theta\|_0 &= \frac{1}{2}\frac{d}{dt}\|\theta\|_0^2 = \big(\frac{d\theta}{dt},\theta\big)_0 \leq \big(\frac{d\theta}{dt},\theta\big)_0 + a_h(\theta,\theta) \\ &= \big(\frac{du_h}{dt},\theta\big)_0 + a_h(u_h,\theta) - \big(\frac{dR_h u}{dt},\theta\big)_0 - a_h(R_h u,\theta) \\ &= (f,\theta)_0 - a_h(u,\theta) - \big(\frac{dR_h u}{dt},\theta\big)_0 \\ &= \big(\frac{du}{dt},\theta\big)_0 - \big(\frac{dR_h u}{dt},\theta\big)_0 = \big(w - R_h w,\theta\big)_0,\end{aligned}$$

with $w = \frac{du}{dt}$. We assumed sufficient regularity, in particular $w \in W_{\mathrm{reg}}$. Using Lemma 11.2.6 we get

$$\big(w - R_h w,\theta\big)_0 \leq c\,h^2 \left\|\frac{du}{dt}\right\|_{2,\Omega_1\cup\Omega_2} \|\theta\|_0.$$

Thus we have

$$\frac{d}{dt}\|\theta\|_0 \leq c\,h^2 \left\|\frac{du}{dt}\right\|_{2,\Omega_1\cup\Omega_2}.$$

Integration over $[0,t]$ and using $\|\theta(0)\|_0 = \|\hat{u}_0 - R_h u_0\|_0$ proves the desired result. □

Related to the error term $\|\hat{u}_0 - R_h u_0\|_0$ in Theorem 11.2.7 we note the following. If we assume $u_0 \in W_{\mathrm{reg}}$ and for the approximation of this initial condition we take $\hat{u}_0 := I_h^* u_0 \in Q_h^\Gamma$, then we get

$$\begin{aligned} \|\hat{u}_0 - R_h u_0\|_0 &= \|I_h^* u_0 - R_h u_0\|_0 \\ &\leq \|I_h^* u_0 - u_0\|_0 + \|R_h u_0 - u_0\|_0 \leq ch^2 \|u_0\|_{2,\Omega_1 \cup \Omega_2}. \end{aligned} \tag{11.24}$$

Hence, we conclude that for the semi-discretization of the mass transport problem with the Nitsche-XFEM method we obtain an *optimal* L^2-error bound, namely of the form ch^2.

11.3 Time discretization

The semi-discretization (11.5), resulting from the Nitsche-XFEM method, can be combined with standard time discretization methods. For example, the θ-scheme takes the following form. For $n = 0, 1, \ldots, N-1$, with $N\Delta t = T$, set $u_h^0 := \hat{u}_0 \in Q_h^\Gamma$, and determine $u_h^{n+1} \in Q_h^\Gamma$ such that for all $v_h \in Q_h^\Gamma$

$$\left(\frac{u_h^{n+1} - u_h^n}{\Delta t}, v_h \right)_0 + \theta a_h(u_h^{n+1}, v_h) + (1-\theta) a_h(u_h^n, v_h) = (f, v_h)_0 \tag{11.25}$$

holds. For simplicity we assume that f does not depend on t. The matrix representation of this problem is given in (11.7). In practice very often either $\theta = 0.5$ (Crank-Nicolson) or $\theta = 1$ (implicit Euler) is used. The error analysis of this *full (i.e. space and time) discretization* method can be performed using standard arguments, as in [238]. For completeness we derive an error bound for the implicit Euler method. Again we require that the solution $u = u(t) \in V_{\mathrm{reg}}$ as defined in Theorem 10.2.2 has sufficient regularity, in particular $\frac{du}{dt} \in L^1(0,T; W_{\mathrm{reg}})$ and $\frac{d^2u}{dt^2} \in L^1(0,T; L^2(\Omega))$. Furthermore, we again assume that the stationary problem (10.10) has the regularity property (10.11).

Theorem 11.3.1 *Let $u = u(t) \in V_{\mathrm{reg}}$ be the solution defined in Theorem 10.2.2 and $u_h^n \in Q_h^\Gamma$, $n = 0, 1, \ldots, N$ the solution of the θ-scheme (11.25) for $\theta = 1$. The following holds:*

$$\begin{aligned} &\|u_h^n - u(t_n)\|_0 \leq \|\hat{u}_0 - R_h u_0\|_0 \\ &+ ch^2 \Big\{ \|u_0\|_{2,\Omega_1\cup\Omega_2} + \int_0^{t_n} \left\| \frac{du}{dt} \right\|_{2,\Omega_1\cup\Omega_2} d\tau \Big\} + \Delta t \int_0^{t_n} \left\| \frac{d^2u}{dt^2} \right\|_0 d\tau. \end{aligned}$$

Proof. We use the splitting $u_h^n - u(t_n) = \big(u_h^n - R_h u(t_n)\big) + \big(R_h u(t_n) - u(t_n)\big) =: \theta^n + \rho^n$. For $\|\rho^n\|_0 = \|\rho(t_n)\|_0$ we have a bound as in (11.23). For the backward difference quotient we introduce the notation $\bar{\partial} w^n := (w^n - w^{n-1})/\Delta t$. Using the definition of u_h^n in (11.25), the definition of the semi-discretization in (11.5) and the consistency result in Lemma 11.2.2 we obtain

$$\begin{aligned}(\bar\partial\theta^n, v_h)_0 + a_h(\theta^n, v_h) &= \frac{1}{\Delta t}(u_h^n - u_h^{n-1}, v_h)_0 + a_h(u_h^n, v_h) \\ &\quad - (\bar\partial R_h u(t_n), v_h)_0 - a_h(R_h u(t_n), v_h) \\ &= (f, v_h)_0 - a_h(u(t_n), v_h) - (\bar\partial R_h u(t_n), v_h)_0 \\ &= \Big(\frac{du(t_n)}{dt}, v_h\Big)_0 - (R_h\bar\partial u(t_n), v_h)_0 =: (\omega^n, v_h)_0,\end{aligned}$$

with

$$\omega^n = \frac{du(t_n)}{dt} - R_h\bar\partial u(t_n) = \big[(I - R_h)\bar\partial u(t_n)\big] - \left[\bar\partial u(t_n) - \frac{du(t_n)}{dt}\right] =: \omega_1^n - \omega_2^n.$$

Taking $v_h = \theta^n \in Q_h^\Gamma$ and using $a_h(\theta^n, \theta^n) \geq 0$ we get

$$\|\theta^n\|_0^2 - (\theta^{n-1}, \theta^n) \leq \Delta t \|\omega^n\|_0 \|\theta^n\|_0.$$

Hence,

$$\|\theta^n\|_0 \leq \|\theta^{n-1}\|_0 + \Delta t\|\omega^n\|_0,$$

and

$$\begin{aligned}\|\theta^n\|_0 &\leq \|\theta^0\|_0 + \Delta t \sum_{j=1}^n \|\omega^j\|_0 \\ &\leq \|\hat u_0 - R_h u_0\|_0 + \Delta t \sum_{j=1}^n \|\omega_1^j\|_0 + \Delta t \sum_{j=1}^n \|\omega_2^j\|_0.\end{aligned} \tag{11.26}$$

For $\|\omega_1^j\|_0$ we obtain with Lemma 11.2.6

$$\begin{aligned}\|\omega_1^j\|_0 &= \left\| \frac{1}{\Delta t}(I - R_h) \int_{t_{j-1}}^{t_j} \frac{du}{dt}\, d\tau \right\|_0 \leq \frac{1}{\Delta t} \int_{t_{j-1}}^{t_j} \left\| (I - R_h)\frac{du}{dt} \right\|_0 d\tau \\ &\leq c\frac{h^2}{\Delta t} \int_{t_{j-1}}^{t_j} \left\| \frac{du}{dt} \right\|_{2,\Omega_1\cup\Omega_2} d\tau,\end{aligned}$$

and thus

$$\Delta t \sum_{j=1}^n \|\omega_1^j\|_0 \leq ch^2 \int_0^{t_n} \left\| \frac{du}{dt} \right\|_{2,\Omega_1\cup\Omega_2} d\tau. \tag{11.27}$$

For ω_2^j we have

$$\Delta t\, \omega_2^j = u(t_j) - u(t_{j-1}) - \Delta t \frac{du(t_j)}{dt} = -\int_{t_{j-1}}^{t_j} (\tau - t_{j-1}) \frac{\partial^2 u(\tau)}{\partial t^2}\, d\tau,$$

and thus

$$\Delta t \sum_{j=1}^{n} \|\omega_2^j\|_0 \leq \sum_{j=1}^{n} \int_{t_{j-1}}^{t_j} (\tau - t_{j-1}) \left\| \frac{\partial^2 u}{\partial t^2} \right\|_0 d\tau \leq \Delta t \int_0^{t_n} \left\| \frac{\partial^2 u}{\partial t^2} \right\|_0 d\tau. \quad (11.28)$$

Using the results from (11.27), (11.28) in (11.26) in combination with the bound for $\|\rho^n\|_0$ from (11.23) we obtain the result. □

If in the initial condition we take $\hat{u}_0 = I_h^* u_0$ we can use the error bound in (11.24) and thus for the full Nitsche-XFEM-Euler discretization we obtain an optimal error bound of the form $c(h^2 + \Delta t)$.

Remark 11.3.2 We comment on the iterative solution of the linear system that arises in each time step of the fully discrete problem (11.7). This system is of the form

$$\hat{\mathbf{A}}\mathbf{x} = \mathbf{b}, \quad \hat{\mathbf{A}} := \frac{1}{\Delta t}\mathbf{M} + \theta \mathbf{A}. \quad (11.29)$$

In Sect. 7.9.4 properties of the XFEM mass matrix **M** are derived, in particular the result in Theorem 7.9.7 implies that the diagonally scaled mass matrix is well-conditioned. Hence, for "small" Δt ($\Delta t < h^2$) a Krylov subspace method with a simple (e.g. Gauss-Seidel or ILU) preconditioner can be expected to be an efficient iterative solver for the system (11.29). For "large" time steps Δt one needs a solver that can deal efficiently with the poorly conditioned matrix **A**. This topic has not been studied in the literature, yet. For problems in which the mass transport problem is diffusion dominated, an efficient preconditioner of **A** may be developed based on the following observation. Since diffusion is assumed to be dominant, instead of **A** we can consider its symmetric part $\frac{1}{2}(\mathbf{A} + \mathbf{A}^T)$. From the results in Lemma 11.2.3 it follows that it suffices to develop a preconditioner that is spectrally equivalent to the symmetric positive definite matrix that represents the (energy) scalar product given in (11.12).

11.4 Numerical experiments

In Subsection 11.4.1 we present two three-dimensional test problems of the form (10.1a)-(10.1e) on $\Omega = (0,1)^3$ with different interfaces Γ. In the first problem we take a simple planar interface. Thus we avoid errors due to numerical interface approximation. In the second experiment we consider a cylindrical interface, which has to be approximated on the unfitted tetrahedral meshes that we use. In both cases, the exact solution u is smooth, known and satisfies the interface conditions (10.1b)-(10.1c). The velocity field $\mathbf{w}$ does not depend on t and satisfies the assumptions (10.3).

11.4.1 Test problems

Case 1: Planar interface

The domain $\Omega = (0,1)^3$ is subdivided into the two subdomains $\Omega_1 := \{ (x,y,z) \in \Omega : z < 0.34113 \}$ and $\Omega_2 := \Omega \setminus \overline{\Omega}_1$, which are separated by the interface $\Gamma := \{ (x,y,z) \in \Omega : z = 0.34113 \}$. The position of the interface is chosen to avoid grid matching.

We choose the coefficients $(\alpha_1, \alpha_2) = (1,2)$, $(\beta_1, \beta_2) = (2,1)$ and a stationary velocity $\mathbf{w} := (y(1-z), x, 0)^T$. Note that $\operatorname{div} \mathbf{w} = 0$ in Ω and $\mathbf{w} \cdot \mathbf{n} = 0$ on Γ. The right-hand side f is taken such that the exact solution is given by

$$u(x,y,z,t) := \begin{cases} \mathrm{e}^{-t} \cos(\pi x) \cos(2\pi y) a z (z+b) & \text{for } x \text{ in } \Omega_1, \\ \mathrm{e}^{-t} \cos(\pi x) \cos(2\pi y) z(z-1) & \text{for } x \text{ in } \Omega_2, \end{cases} \tag{11.30}$$

where the parameters a and b are determined from the interface conditions (10.1b)-(10.1c). We take homogeneous Dirichlet boundary conditions on the boundary parts $z = 0$ and $z = 1$ and homogeneous Neumann boundary conditions on the remaining part of the boundary.

Case 2: Cylindrical interface

In this experiment, the domain $\Omega = (-\frac{1}{2}, \frac{1}{2})^3$ is subdivided into

$$\Omega_1 = \{ (x,y,z) \in \Omega : x^2 + y^2 < R^2 \}, \quad \Omega_2 = \Omega \setminus \Omega_1,$$

with $R = 0.1$. We take a constant velocity field $\mathbf{w} = (0,0,1)^T$ and coefficients $(\alpha_1, \alpha_2) = (1,5)$, $(\beta_1, \beta_2) = (2,1)$. The velocity field satisfies $\operatorname{div} \mathbf{w} = 0$ in Ω and $\mathbf{w} \cdot \mathbf{n} = 0$ on Γ. The exact solution is given by

$$u(x,y,z,t) := \begin{cases} e^{-t}\big(\alpha_2(x^2+y^2-R^2) + \beta_2\big) & \text{in } \Omega_1, \\ e^{-t}\big(\alpha_1(x^2+y^2-R^2) + \beta_1\big) & \text{in } \Omega_2. \end{cases} \tag{11.31}$$

We take homogeneous Neumann conditions on the boundary parts $z = 0$ and $z = 1$. On the remaining part of the boundary the values of u are used as inhomogeneous Dirichlet conditions.

11.4.2 Numerical results

We start with a brief discussion of some implementation issues related to the Nitsche-XFEM discretization as in (11.5) (semi-discretization), (11.7) (full discretization). For the finite element spaces and the bilinear form $a_h(\cdot,\cdot)$ we need an accurate approximation of the interface Γ. For the first test problem this is easy. For the second one we use the approach discussed in Sect. 7.3, which we briefly recall. Let d be the signed distance function to Γ, for which it is easy to give an explicit formula using cylindrical coordinates. A *piecewise planar* approximation Γ_h of Γ is determined by computing an approximation of the zero level of d as follows. Corresponding to the given triangulation $\mathcal{T}_h$ we introduce one further regular refinement, denoted by $\mathcal{T}_h'$. Let $I(d)$ be the continuous piecewise *linear* function on $\mathcal{T}_h'$ which interpolates d at the vertices of each tetrahedron in $\mathcal{T}_h'$. Then Γ_h is defined as

$$\Gamma_h := \{ x \in \Omega : I(d)(x) = 0 \}. \tag{11.32}$$

Thus we obtain polygonal subdomains $\Omega_{1,h}$ and $\Omega_{2,h}$ as approximations for Ω_1 and Ω_2. For a tetrahedron $T \in \mathcal{T}_h$ with $T \cap \Gamma_h \neq \emptyset$, let $T' \in \mathcal{T}_h'$ be one of the 8 regular children of T. We introduce the notations $T_i' := T' \cap \Omega_{i,h}$, $i = 1, 2$, and $\Gamma_{h,T'} = T' \cap \Gamma_h$, the restrictions of the subdomains and interface to T'. Then T_i' is either a tetrahedron or a prism, which can be subdivided into 3 subtetrahedra, cf. Fig 7.13. The planar segment $\Gamma_{h,T'}$ is either a triangle or a quadrilateral, which can be subdivided into 2 triangles. Thus integrals over T_i' or $\Gamma_{h,T'}$ of smooth functions can be easily determined with high accuracy using standard (Gauss) quadrature.

In the implementation of (11.7) the basis q_j, $j \in \mathcal{J}$, $q_j^{\Gamma_h}$, $j \in \mathcal{J}_\Gamma$ of $Q_h^{\Gamma_h}$, as explained in Sect. 7.9.2 is used. Due to the *discontinuity* of the basis functions $q_i^{\Gamma_h}$ across Γ_h, one has to be careful in computing quantities like $(q_i^{\Gamma_h}, \psi_h)_0$ and $(q_i^{\Gamma_h}, \psi_h)_{1,\Omega_1 \cup \Omega_2}$, with $\psi_h \in \{q_j, q_j^{\Gamma_h}\}$ using quadrature. The calculation of these integrals is done elementwise by summing the contributions from all tetrahedra in $\mathcal{T}_h$. Let x_i, x_j be two vertices of a tetrahedron T that intersects the approximate interface, i.e., $T \cap \Gamma_h \neq \emptyset$; assume that $x_i \in \Omega_{1,h}$. A local integral on T is assembled over all 8 children T' of T as follows. Using $supp(q_i^{\Gamma_h}) \cap T' = T_2'$ we obtain

$$\int_{T'} \beta q_i^{\Gamma_h} \psi_h dx = \int_{T_2'} \beta_2 \, q_i^{\Gamma_h} \, \psi_h dx = \int_{T_2'} \beta_2 \, q_i \, \psi_h dx. \tag{11.33}$$

Since β_2 is constant on T_2' and q_i and ψ_h are linear functions on T_2' the latter integral is easy to determine. The other volume integrals in $a_h(q_i^\Gamma, \psi_h)$ can be computed in a similar way. In our test problems a Gauss quadrature rule of degree three on a tetrahedron is sufficient to compute the volume integrals in the bilinear form $a_h(\cdot,\cdot)$, with Γ replaced by Γ_h, exactly. The scalar product $(f, \psi_h)_0$ in the right-hand side of (11.5) is approximated with high accuracy (using the same assembling process as described above) by a Gauss quadrature rule of degree five on tetrahedra. The interfacial integrals over Γ_h (instead of Γ) in $a_h(q_i^{\Gamma_h}, \psi_h)$ are approximated by summing the local integrals on each planar segment $\Gamma_{h,T'}$ of Γ_h using a Gauss quadrature rule of degree five on a triangle. Below we present results for the two test problems.

Case 1: Planar interface

For the spatial discretization, we first create a uniform grid with mesh size $h = h_i = 2^{-i-1}$, with $i = 2, 3, 4$. Starting from this uniform grid the elements near the interface are refined two times further, i. e., the local mesh size close to the interface is $h_{\Gamma,i} = \frac{1}{4} h_i$. For the case $i = 4$ this results in a problem with 1 293 754 tetrahedra and 226 087 unknowns. The approximation of the initial value $\hat{u}_0$ is chosen as $I_h^*(u(\cdot, 0))$, with I_h^* the interpolation operator as in Theorem 11.2.4. For the parameter λ in the bilinear form $a_h(\cdot,\cdot)$ we take the value $\lambda = 100$. This choice is based on numerical experiments. It turns out that the error behavior is not very sensitive with respect to the choice of the parameter value. The results are essentially the same for all $10^1 \leq \lambda \leq 10^3$.

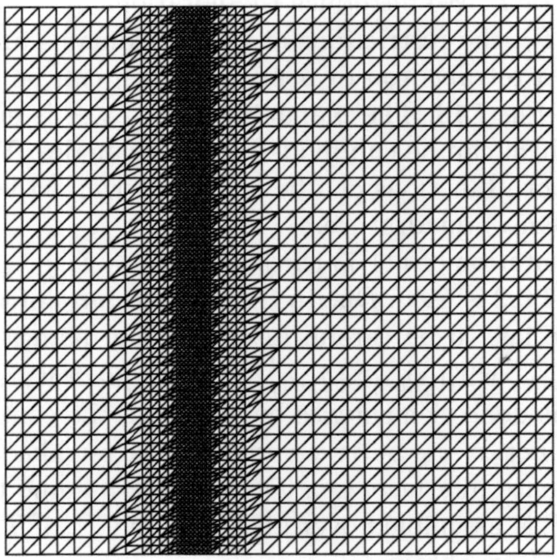

Fig. 11.1. Test 1: A slice of the tetrahedral mesh at $x = 0.25$, for the case $i = 3$.

The semi-discretization $u_h(t)$ is not known. We computed an accurate approximation of $u_h(t)$ using the implicit Euler time-stepping scheme with a time step size Δt which is sufficiently small (in our experiments: $\Delta t = 10^{-4}$) such that the error due to the time discretization is negligible compared to the space discretization error. The resulting reference solution is denoted by $u_h^*(t)$. In Table 11.1, the errors $\|u_h^*(T) - u(T)\|_{L^2}$ at $T = 0.15$ are displayed. These results are consistent with the theoretical bound $\mathcal{O}(h^2)$ given in Theorem 11.2.7.

i	$\|u_h^*(T) - u(T)\|_{L^2}$	order
2	7.39 E-3	-
3	2.02 E-3	1.87
4	5.23 E-4	1.95

Table 11.1. Case 1: Spatial discretization error at $T = 0.15$.

The exact solution satisfies $[\beta u]_\Gamma = 0$. In the Nitsche discretization this interface condition is satisfied only approximately. For a *stationary* elliptic problem it is shown in [135] that for the discretization u_h the error in this interface condition is bounded by $\|[\beta u_h]\|_{L^2(\Gamma)} \leq c h^{\frac{3}{2}} \|u\|_{2,\Omega_1 \cup \Omega_2}$. For the parabolic case a theoretical bound for this error quantity is not known. We computed the errors $\|[\beta u_h^*]\|_{L^2(\Gamma)}$ for our problem; the results are given in Table 11.2.

As expected, the interface condition (10.1c) is satisfied only approximately. The error $\|[\beta u_h^*]\|_{L^2(\Gamma)}$ seems to behave like $\mathcal{O}(h)$.

i	$\|[\beta u_h^*(T)]\|_{L^2(\Gamma)}$	order
2	1.57 E-4	-
3	7.98 E-5	0.97
4	3.90 E-5	1.03

Table 11.2. Case 1: Error in the Henry interface condition at $T = 0.15$.

The numerical solution for $i = 3$ at $T = 0.15$ on the cross section $x = 0.25$ is shown in Fig. 11.2.

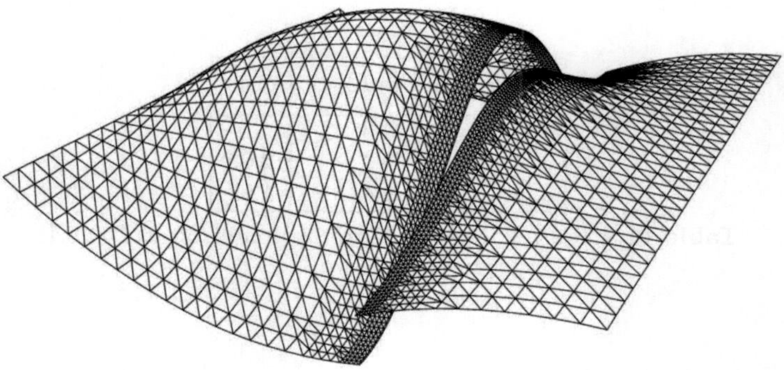

Fig. 11.2. Case 1: Numerical solution at $T = 0.15$ on the cross section $x = 0.25$.

Now we study the time discretization error bound for the implicit Euler method in Theorem 11.3.1. We use the fixed mesh size $h = h_3 = \frac{1}{16}$ as described above and compute a reference solution with $\Delta t = 10^{-4}$ in the time interval $[0, T]$, $T = 0.2$, which is denoted by $u_h^*(t)$. The Euler discretization, i.e. (11.7) (or, equivalently, (11.25)) with $\theta = 1$ and time step $\Delta t = \frac{T}{n}$ results in approximations $u_h^n(T)$ of $u_h^*(T)$. For the cases $n = 5, 10, 20$ the temporal errors in the L^2-norm, i.e. $\|u_h^n(0.2) - u_h^*(0.2)\|_{L^2}$, are given in Table 11.3. We observe the expected first order of convergence in Δt.

Case 2: Cylindrical interface

For the spatial discretization we proceed as in case 1. A difference is that we now have an approximation Γ_h of Γ, cf. the explanation given above. We use uniform grids with mesh size $h = h_i = \frac{1}{5}2^{-i}$, $i = 1, 2, 3$, and refine the elements

n	$\|u_h^n - u_h^*(0.2)\|_{L^2}$	order
5	1.25 E-5	-
10	6.09 E-6	1.04
20	3.01 E-6	1.02

Table 11.3. Case 1: Time discretization error at $T = 0.2$.

near the interface (only) one time further, due to memory limitation. For the case $i = 3$, the grid already contains 1 043 040 tetrahedra which leads to a problem with 166 059 unknowns. The implicit Euler method with $\Delta t = 10^{-4}$ is used to compute the reference solution $u_h^*(t)$ for $t \in [0, 0.1]$. The error $\|u_h^*(T) - u(T)\|_{L^2}$ at $T = 0.1$ is given in Table 11.4. The error reduction is in good agreement with the $\mathcal{O}(h^2)$ error bound derived in Theorem 11.2.7. The $L^2(\Gamma_h)$-norm of the jump $[\beta u_h^n]$ at the approximated interface Γ_h again appears to have a numerical convergence order 1, cf. Table 11.5.

i	$\|u_h^*(T) - u(T)\|_{L^2}$	order
1	3.27 E-3	-
2	8.07 E-4	2.02
3	2.06 E-4	1.97

Table 11.4. Case 2: Spatial discretization error at $T = 0.1$.

i	$\|[\beta u_h^n]\|_{L^2(\Gamma_h)}$	order
1	1.03 E-3	-
2	5.01 E-4	1.04
3	2.20 E-4	1.19

Table 11.5. Case 2: Error in the Henry interface condition at $T = 0.1$.

Figure 11.3 shows the numerical solution on the cross section $z = 0.5$ at $T = 0.1$.

11.5 Discretization in case of a non-stationary interface

In this section we treat the discretization of the mass transport problem with a *non*-stationary interface. A first possibility, that is explained in Sect. 11.5.1, is to apply a Rothe approach to the strong formulation in (10.1) and combine this with a Nitsche-XFEM space discretization. This works well if one uses the implicit Euler time discretization. It is, however, not clear how in this

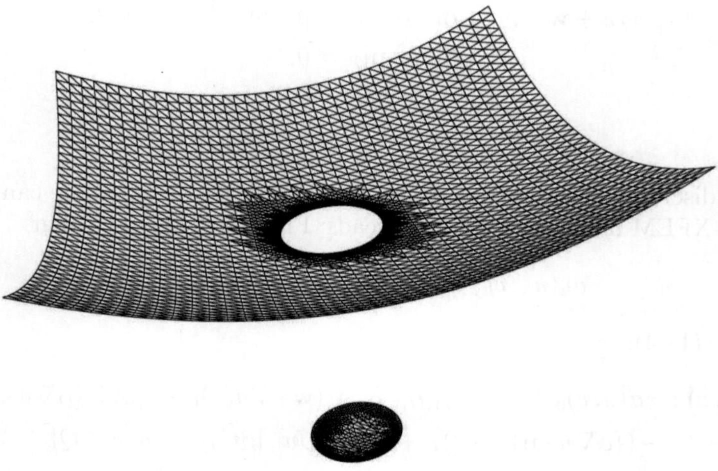

Fig. 11.3. Test 2: Numerical solution at $T = 0.1$ on the cross section $z = 0$.

Rothe approach one can obtain a (Crank-Nicolson type) higher order time discretization. It is more natural to base the discretization on the space-time weak formulation derived in Sect. 10.3. This leads to space-time finite element methods that are treated in Sect. 11.5.2.

11.5.1 Rothe's method combined with Nitsche-XFEM

Assume that $\Gamma = \Gamma(t)$ is known for $t \in [0, T]$ and an initialization $u^0(x) := u_0(x)$, $x \in \Omega$, is given. An implicit Euler time discretization applied to the transport problem (10.1) results in the following sequence of stationary problems: For $n \geq 0$, determine $u^{n+1} = u^{n+1}(x)$, $x \in \Omega_1(t_{n+1}) \cup \Omega_2(t_{n+1})$ such that, for $i = 1, 2$,

$$\frac{u^{n+1}}{\Delta t} + \mathbf{w}^{n+1} \cdot \nabla u^{n+1} - \operatorname{div}(\alpha \nabla u^{n+1}) = f + \frac{u^n}{\Delta t} \quad \text{in } \Omega_i(t_{n+1}), \tag{11.34a}$$

$$[\alpha \nabla u^{n+1} \cdot \mathbf{n}]_{\Gamma(t_{n+1})} = 0, \tag{11.34b}$$

$$[\beta u^{n+1}]_{\Gamma(t_{n+1})} = 0, \tag{11.34c}$$

$$u^{n+1} = 0 \quad \text{on } \partial\Omega. \tag{11.34d}$$

We used the notation $\mathbf{w}^{n+1} := \mathbf{w}(\cdot, t_{n+1})$. For a fixed n we write $\Omega_i(t_{n+1}) =: \Omega_i$, $\Gamma(t_{n+1}) =: \Gamma$, $u^{n+1} =: u$, $\mathbf{w}^{n+1} =: \mathbf{w}$ and $\sigma := \frac{1}{\Delta t}$. Thus in each time step we have a *stationary* problem of the form

$$\begin{aligned}\sigma u + \mathbf{w} \cdot \nabla u - \operatorname{div}(\alpha \nabla u) &= g \quad \text{in } \Omega_i, \ i = 1, 2, \\ [\alpha \nabla u \cdot \mathbf{n}]_\Gamma &= 0, \\ [\beta u]_\Gamma &= 0, \\ u &= 0 \quad \text{on } \partial \Omega.\end{aligned}$$

For the discretization of the weak formulation of this problem one can use the Nitsche-XFEM method, which now reads: Find $u_h \in Q_h^\Gamma$ such that

$$a_h(u_h, v_h) = (g, v_h)_0 \quad \text{for all } v_h \in Q_h^\Gamma, \tag{11.35}$$

with, cf. (11.4):

$$\begin{aligned} a_h(u, v) := \sigma(u, v)_0 + (\alpha u, v)_{1,\Omega_1 \cup \Omega_2} + (\mathbf{w} \cdot \nabla u, v)_0 - ([\beta u], \{\alpha \nabla v \cdot \mathbf{n}\})_\Gamma \\ - (\{\alpha \nabla u \cdot \mathbf{n}\}, [\beta v])_\Gamma + \lambda h^{-1}([\beta u], [\beta v])_\Gamma, \quad u, v \in Q_h^\Gamma. \end{aligned}$$

An analysis of the spatial discretization in (11.35) is given in [184]. There, in the problem related norm $\|\!|\cdot|\!\|$, cf. (11.12), an error bound of the form

$$\|\!| u - u_h |\!\| \le ch \|u\|_{2,\Omega_1 \cup \Omega_2}$$

is proved. The constant c in this error bound depends on $\sigma = \frac{1}{\Delta t}$. An error analysis of the full, i.e. space and time, discretization is not known. In numerical experiments on test problems this method has discretization errors in the L^2-norm that are of order h^2 w.r.t. space discretization, cf. the results in the following remark.

Remark 11.5.1 We present results of a numerical experiment with the Rothe-Nitsche-XFEM method explained above. We consider the *non-stationary* transport problem (10.1) in the unit cube $\Omega = (0,1)^3$ and with $\Omega_1(0)$ a sphere of radius $R = 0.2$ centered at the barycenter of Ω. This sphere is moving with a constant velocity $\mathbf{w} = (0,1,0)^T$, i.e. $\Omega_1(t) = \Omega_1(0) + t\mathbf{w}$. Let $d(x,t)$ be the distance from the point $x \in \Omega$ to the center of $\Omega_1(t)$. We take the piecewise quadratic solution

$$u(x,t) := \begin{cases} \alpha_2\big(d(x,t)^2 - R^2\big) + 0.1\beta_2 & \text{in } \Omega_1, \\ \alpha_1\big(d(x,t)^2 - R^2\big) + 0.1\beta_1 & \text{in } \Omega_2, \end{cases} \tag{11.36}$$

with coefficients $(\alpha_1, \alpha_2) := (1, 5)$, $(\beta_1, \beta_2) := (2, 1)$. The values of u on $\partial\Omega$ are used as inhomogeneous Dirichlet conditions. For a discussion of implementation aspects, in particular of the piecewise planar approximation $\Gamma_h(t)$ of $\Gamma(t)$ we refer to Sect. 11.4.2. We discretize the problem first in time using the implicit Euler method with the time step size $\Delta t = 10^{-4}$. The resulting convection-diffusion-reaction problem of the form (11.34) is discretized with the Nitsche method (11.35). We first create a uniform grid with mesh size $h = h_i = 2^{-i-1}$, $i = 2, 3, 4$, then locally refine the elements near the interface two times further. After 1000 time steps, we obtain the approximation u_h^{1000} of

$u(0.1)$. The errors $\|u_h^{1000} - u(0.1)\|_{L^2}$ for $i = 2, 3, 4$ are displayed in Table 11.6. The results show second order convergence. Note that $\|u_h^{1000} - u(0.1)\|_{L^2}$ contains both time and spatial discretization errors, but the time step is sufficiently small such that the spatial discretization error is dominant.

i	$\|u_h^{1000} - u(0.1)\|_{L^2}$	order
2	7.90 E-3	-
3	2.00 E-3	1.98
4	5.00 E-4	2.00

Table 11.6. Discretization error at $T = 0.1$.

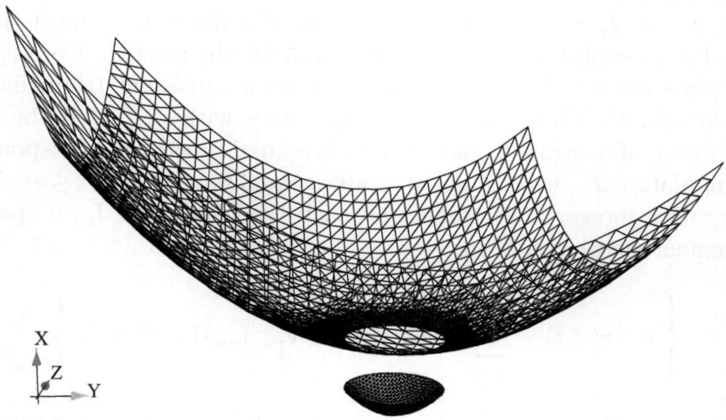

Fig. 11.4. Numerical solution for $h = 1/16$ at $T = 0.1$ on the cross section $x = 0.5$.

Since the interface evolves the XFEM space $Q_h^{\Gamma_h}(t)$ is in general time-dependent. Hence a reference solution is not available and we cannot compute the order of convergence for the time discretization.

It is not clear how the Euler time discretization (11.34) can be improved based on a Crank-Nicolson time discretization, because in that case both $\mathrm{div}(\alpha\nabla u^{n+1})$ and $\mathrm{div}(\alpha\nabla u^n)$ occur in the differential equation. The function u^{n+1} is discontinuous across $\Gamma(t_{n+1})$, whereas u^n is discontinuous across $\Gamma(t_n)$. It is not clear how to treat this discrepancy. This difficulty does not occur if instead of the Rothe approach a space-time discretization method is applied. This topic is discussed in the next section.

11.5.2 Nitsche-XFEM space-time discretization

We first introduce the basic idea of a space-time finite element method for a parabolic problem, cf. for example [153, 105, 238]. We then combine this approach with the Nitsche technique for handling the Henry interface condition in case of a non-stationary interface.

We consider the model parabolic problem

$$\begin{aligned} \frac{\partial u}{\partial t} - \Delta u &= f \quad \text{in } \Omega,\, t \in [0,T], \\ u(\cdot,0) &= u_0 \quad \text{in } \Omega, \\ u(\cdot,t) &= 0 \quad \text{on } \partial\Omega. \end{aligned} \tag{11.37}$$

For simplicity we assume f to be independent of t. We use a partitioning of the time domain $0 = t_0 < t_1 < \ldots < t_N = T$, with a fixed time step size $\Delta t = T/N$, i.e. $t_j = j\Delta t$. This assumption of a fixed time step is made to simplify the presentation, but is *not* essential for the method. Corresponding to each time interval $I_n := (t_{n-1}, t_n)$ we have a consistent triangulation $\mathcal{T}_n$ of the domain Ω. This triangulation may vary with n. Let V_n be a finite element space of continuous piecewise polynomial functions corresponding to the triangulation $\mathcal{T}_n$, with boundary values equal to zero. For $1 \leq n \leq N$ and a nonnegative integer k we define, on each time slab $\Omega \times I_n$, a space-time finite element space as follows:

$$V_{kn} := \left\{ v : v(x,t) = \sum_{j=0}^{k} t^j \phi_j(x),\ \phi_j \in V_n,\ (x,t) \in \Omega \times I_n \right\}, \tag{11.38}$$

for $1 \leq n \leq N$. The corresponding space-time discretization of (11.37) reads: Determine u_h such that for all $n = 1, 2, \ldots, N$, $(u_h)_{|\Omega \times I_n} \in V_{kn}$ and

$$\begin{aligned} &\int_{t_{n-1}}^{t_n} \Big(\frac{\partial u_h}{\partial t}, v_h\Big)_{L^2} + (\nabla u_h, \nabla v_h)_{L^2}\, dt + ([u_h]^{n-1}, v_h^{n-1,+})_{L^2} \\ &= \int_{t_{n-1}}^{t_n} (f, v_h)_{L^2}\, dt \quad \text{for all } v_h \in V_{kn}, \end{aligned} \tag{11.39}$$

where $(\cdot,\cdot)_{L^2} = (\cdot,\cdot)_{L^2(\Omega)}$,

$$[w_h]^n = w_h^{n,+} - w_h^{n,-}, \quad w_h^{n,+(-)} = \lim_{s \to 0+(-)} w_h(\cdot, t_n + s),$$

and $u_h^{0,-} \in Q_0$ an approximation of the initial data u_0. For an analysis of this discretization method we refer to the literature, e.g. [238].

We consider two important special cases, namely $k = 0$, $k = 1$. If $k = 0$ then $v_h \in V_{kn}$ does not depend on t. Define $u_h^n(x) := u_h(x,t)$, $t \in I_n$. The method (11.39) for determining $u_h^n \in V_n$ reduces to the implicit Euler scheme:

$$\frac{1}{\Delta t}(u_h^n - u_h^{n-1}, v_h)_{L^2} + (\nabla u_h^n, \nabla v_h)_{L^2} = (f, v_h)_{L^2} \quad \text{for all } v_h \in V_n.$$

We now consider $k = 1$. Then on $\Omega \times I_n$ the function u_h^n can be represented as $u_h^n(x,t) = \hat{u}_h^n(x) + \frac{1}{\Delta t}(t - t_{n-1})\tilde{u}_h^n(x)$, with $\hat{u}_h^n$, $\tilde{u}_h^n \in V_n$. These unknown functions are determined by the coupled system

$$(\hat{u}_h^n + \tilde{u}_h^n, v_h)_{L^2} + \Delta t\big(\nabla \hat{u}_h^n + \frac{1}{2}\nabla \tilde{u}_h^n, \nabla v_h\big)_{L^2} = (u_h^{n-1,-}, v_h)_{L^2} + \Delta t(f, v_h)_{L^2},$$
$$\frac{1}{2}(\tilde{u}_h^n, v_h)_{L^2} + \Delta t\big(\frac{1}{2}\nabla \hat{u}_h^n + \frac{1}{3}\nabla \tilde{u}_h^n, \nabla v_h\big)_{L^2} = \frac{1}{2}\Delta t(f, v_h)_{L^2},$$

for all $v_h \in V_n$, cf. [238].

Remark 11.5.2 Let $\hat{\mathbf{u}}^n$ and $\tilde{\mathbf{u}}^n$ be the vector representations of $\hat{u}_h^n$ and $\tilde{u}_h^n$, respectively, and $\mathbf{u}^n = \hat{\mathbf{u}}^n + \tilde{\mathbf{u}}^n$ the vector representation of $u_h^n(\cdot, t_n)$. The mass and stiffness matrices in V_n are denoted by $\mathbf{M}$ and $\mathbf{A}$. Then the scheme for $k = 1$ can be rewritten in matrix-vector form as follows:

$$\begin{aligned}\mathbf{M}(\hat{\mathbf{u}}^n + \tilde{\mathbf{u}}^n) + \Delta t\mathbf{A}(\hat{\mathbf{u}}^n + \frac{1}{2}\tilde{\mathbf{u}}^n) &= \mathbf{M}\mathbf{u}^{n-1} + \mathbf{b},\\ \frac{1}{2}\mathbf{M}\tilde{\mathbf{u}}^n + \Delta t\mathbf{A}(\frac{1}{2}\hat{\mathbf{u}}^n + \frac{1}{3}\tilde{\mathbf{u}}^n) &= \frac{1}{2}\mathbf{b}.\end{aligned} \tag{11.40}$$

For a *fixed* triangulation, $\mathcal{T}_n = \mathcal{T}$ for all n, this is a time discretization scheme for the semi-discrete problem (in vector notation)

$$\mathbf{M}\frac{d\mathbf{u}}{dt} + \mathbf{A}\mathbf{u} = \mathbf{b}, \quad \text{i.e.} \quad \frac{d\mathbf{u}}{dt} + \mathbf{M}^{-1}\mathbf{A}\mathbf{u} = \mathbf{M}^{-1}\mathbf{b},$$

and $\mathbf{u}^n$ is an approximation of $\mathbf{u}(t_n)$. The spectrum of $\mathbf{M}^{-1}\mathbf{A}$ is contained in $(0, \infty)$. Using a transformation to the eigenvector basis of $\mathbf{M}^{-1}\mathbf{A}$, for the analysis of accuracy and stability of the scheme (11.40) we consider the simple scalar model problem $\frac{du}{dt} - \lambda u = 0$, with an arbitrary $\lambda \in (-\infty, 0)$. The solution is given by $u(t) = u(0)e^{\lambda t}$. This model test problem is the same as the one considered in Sect. 4.1, cf. (4.10). The scheme (11.40) applied to this test problem is given by

$$\begin{aligned}&\hat{u}^n + \tilde{u}^n - \Delta t\lambda\big(\hat{u}^n + \frac{1}{2}\tilde{u}^n\big) = u^{n-1}\\ &\frac{1}{2}\tilde{u}^n - \Delta t\lambda\big(\frac{1}{2}\hat{u}^n + \frac{1}{3}\tilde{u}^n\big) = 0, \quad u^n = \hat{u}^n + \tilde{u}^n.\end{aligned}$$

This can be rewritten as

$$u^n = g_{ST}(\Delta t\lambda)u^{n-1}, \quad \text{with} \quad g_{ST}(z) := \frac{1 + \frac{1}{3}z}{1 - \frac{2}{3}z + \frac{1}{6}z^2}.$$

The stability function $g_{ST}(z)$ (cf. Sect. 4.1) is a so-called Padé approximation of e^z. From Taylor expansion it follows that $e^z = g_{ST}(z) + \mathcal{O}(z^4)$, $(z \to 0)$, hence the method has consistency order 3. The method has the optimal smoothing property $\lim_{z\to-\infty} g_{ST}(z) = 0$ and the stability property

$$|g_{ST}(z)| \leq 1 \quad \text{for all} \quad z \in (-\infty, 0].$$

This is an optimal stability property if one considers only real eigenvalues $\lambda \in (-\infty, 0]$ in the test problem. For general *complex* $z \in \mathbb{C}$, with $\mathrm{Re}(z) \leq 0$, the inequality $|g(z)| \leq 1$ does not necessarily hold, i.e., the method is not A-stable.

In the finite element method introduced above both the trial and test functions are *continuous* in space and *discontinuous* in time. This discontinuity is at the interface between time slabs. Alternatively one can use trial functions that are *continuous* in space *and* time combined with test functions that are *continuous* in space and *discontinuous* in time. Another variant has been developed in which trial and test functions are *discontinuous* in space *and* time. We do not treat such alternative methods here. We only consider the continuous in space and discontinuous in time variant explained above and show how an XFEM space-time version can be derived. An XFEM space-time technique for a class of spatially *one*-dimensional hyperbolic problems is presented in [67, 68]. We combine the space-time method described above with the Nitsche approach for handling the Henry interface condition, resulting in a rather general finite element discretization technique applicable to the space-time weak formulation of the mass transport problem given in Sect. 10.3.2.

We use notations as in Sect. 10.3 and introduce the XFEM method along the same lines as in Sect. 7.9.2 but now with the space-time subdomains $Q_i \subset \mathbb{R}^4$ instead of the spatial subdomains $\Omega_i \subset \mathbb{R}^3$. We take a fixed time slab $\Omega \times I_n$. Let V_n be the (simplicial) finite element space introduced above and $\{q_j\}_{j\in\mathcal{J}}$ the nodal basis in this space. For $k \geq 1$, let $\xi_0, \ldots, \xi_k$ be the Lagrange basis in $\mathcal{P}_k([t_{n-1}, t_n])$, corresponding to equidistant node points, which include the interval end points t_{n-1}, t_n. For $k = 0$ we choose the basis function $\xi_0 := 1$ of $\mathcal{P}_0([t_{n-1}, t_n])$. In practice one typically uses small values of k ($k \leq 2$), in which case the Lagrange basis is well-conditioned. To each q_j, $j \in \mathcal{J}$, there correspond $k+1$ space-time functions

$$q_{j,\ell}(x,t) := q_j(x)\xi_\ell(t), \quad \ell = 0, \ldots, k, \quad (x,t) \in \Omega \times I_n,$$

and $q_{j,\ell} = 0$ otherwise. These form a basis of V_{kn}:

$$V_{kn} = \mathrm{span}\{\, q_{j,\ell} : j \in \mathcal{J},\ 0 \leq \ell \leq k \,\}.$$

For example, for $k = 0$ we have

$$q_{j,0}(x,t) = q_j(x), \quad (x,t) \in \Omega \times I_n,$$

and for $k = 1$:

$$q_{j,0}(x,t) = \frac{1}{\Delta t}(t_n - t)q_j(x), \quad q_{j,1}(x,t) = \frac{1}{\Delta t}(t - t_{n-1})q_j(x), \quad (x,t) \in \Omega \times I_n.$$

Using $\overline{\text{supp}(\xi_\ell)} = [t_{n-1}, t_n]$, we obtain $\overline{\text{supp}(q_{j,\ell})} = \overline{\text{supp}(q_j)} \times [t_{n-1}, t_n]$, *in*dependent of ℓ. The index set of basis functions "close to the interface" is given by

$$\mathcal{J}_{\Gamma_*} := \{\, j \in \mathcal{J} : \text{meas}_3\left(\Gamma_* \cap \text{supp}(q_{j,0})\right) > 0 \,\},$$

cf. Fig. 11.5 for a 1D example. Let H_{Γ_*} be the Heaviside function correspond-

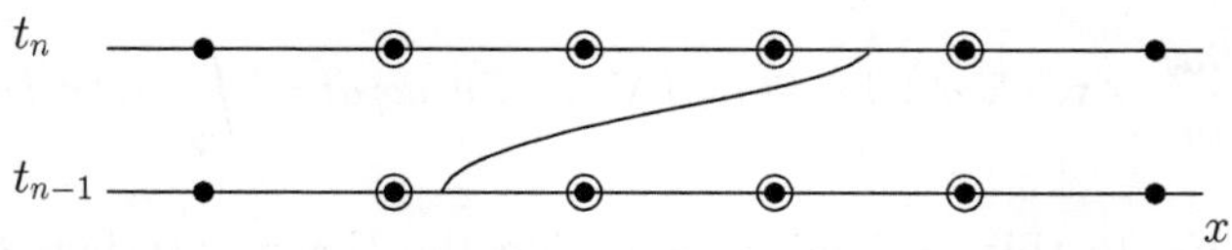

Fig. 11.5. Enrichment index set for 1D example. Dots represent degrees of freedom of original basis functions, circles indicate where additional functions are added in the vicinity of the interface Γ_*.

ing to Q_2:

$$H_{\Gamma_*}(x,t) := \begin{cases} 0 & \text{if } \ (x,t) \in Q_1, \\ 1 & \text{if } \ (x,t) \in Q_2. \end{cases}$$

For each space-time node index (j, ℓ), $j \in \mathcal{J}_{\Gamma_*}$, $0 \leq \ell \leq k$, a corresponding enrichment function is given by

$$\Phi^H_{j,\ell}(x,t) := \begin{cases} H_{\Gamma_*}(x,t) & \text{if } \ k = 0 \\ H_{\Gamma_*}(x,t) - H_{\Gamma_*}(x_j, \hat{t}_\ell) & \text{if } \ k > 0, \end{cases} \tag{11.41}$$

with $\hat{t}_\ell \in I_n$ the Lagrange node for which $\xi_\ell(\hat{t}_\ell) = 1$ holds. The additional basis functions are defined as follows:

$$q^{\Gamma_*}_{j,\ell} := q_{j,\ell}\Phi^H_{j,\ell}, \quad j \in \mathcal{J}_{\Gamma_*}, \ 0 \leq \ell \leq k.$$

The term $H_{\Gamma_*}(x_j, \hat{t}_\ell)$ in the definition of Φ^H_j is constant and may be omitted (as it doesn't introduce new functions in the function space), but ensures that $q^{\Gamma_*}_{j,\ell}(x_j, \hat{t}_\ell) = 0$ holds in all space-time grid points $(x_j, \hat{t}_\ell)$ if $k > 0$. The space-time XFEM space on the time slab $\Omega \times I_n$ is given by

$$V^{\Gamma_*}_{kn} := V_{kn} \oplus \text{span}\left\{ q^{\Gamma_*}_{j,\ell} : j \in \mathcal{J}_{\Gamma_*}, \ 0 \leq \ell \leq k \right\}. \tag{11.42}$$

Note that for $v \in \oplus^N_{n=1} V^{\Gamma_*}_{kn}$ we have $v_i := v_{|Q_i} \in H^{1,0}(Q_i)$ for $i = 1, 2$, and $v_{|\partial\Omega} = 0$. We use $\oplus^N_{n=1} V^{\Gamma_*}_{kn}$ for a space-time discretization of the variational formulation (10.33). Similar to (11.39) the jumps between time slabs $\Omega \times I_{n-1}$

and $\Omega \times I_n$ are controlled by a term of the form $([u_h]^{n-1}, \cdot)_{L^2}$ and as in Sect. 11.1 the Henry interface condition $[\beta u]_{\Gamma_*} = 0$ is satisfied in a weak sense by adding suitable interface jump terms to the bilinear form. We use the weak formulation of the continuous problem as explained in Remark 10.3.7 and, for the derivation of the method, assume that the solution u has the smoothness property $u_i \in H^1(Q_i)$, $i = 1, 2$, which implies that the weak derivative $\frac{\partial u_i}{\partial t}$ is well-defined in the usual sense on Q_i. Using partial integration, cf. (10.29), it follows that the weak solution $u \in W_{\beta,0} \subset V_\beta$ satisfies

$$\sum_{i=1}^{2} \int_{Q_i} (\frac{\partial u_i}{\partial t} + \mathbf{w} \cdot \nabla u_i)\beta_i v_i + \alpha_i \beta_i \nabla u_i \cdot \nabla v_i \, dx \, dt = \int_{Q_T} \beta f v \, dx \, dt \quad (11.43)$$

for all $v \in V_\beta$. In this problem formulation the Henry interface condition is incorporated as an essential condition in the space V_β. As in Sect. 11.1, in the discrete problem we “eliminate” this condition from the trial and test space and instead add interface terms to the bilinear form. We introduce the notation $Q_i^n := Q_i \cap (\Omega \times I_n) = \{ (x,t) \in Q_i : t \in I_n \}$, $i = 1, 2$, $Q_T^n := \Omega \times I_n$ and $\Gamma_*^n := \{ (x,t) \in \Gamma_* : t \in I_n \}$. For a test function $v \in V_{kn}^{\Gamma_*}$ (extended by zero outside of $\Omega \times I_n$) the integrals $\int_{Q_i}$ and $\int_{Q_T}$ in (11.43) can be replaced by integrals $\int_{Q_i^n}$ and $\int_{Q_T^n}$, respectively. Taking such a test function and applying partial integration results in

$$\begin{aligned}\sum_{i=1}^{2} \int_{Q_i^n} \alpha_i \beta_i \nabla u_i \cdot \nabla v_i \, dx \, dt = &- \sum_{i=1}^{2} \int_{Q_i^n} \operatorname{div}(\alpha_i \nabla u_i) \beta_i v_i \, dx \, dt \\ &+ \int_{\Gamma_*^n} \nu [\alpha \nabla u \cdot \mathbf{n}_\Gamma \beta v] \, ds,\end{aligned}$$

with $\mathbf{n}_\Gamma \in \mathbb{R}^3$ the unit normal at Γ and $\nu = (1 + \mathbf{w} \cdot \mathbf{n}_\Gamma)^{-\frac{1}{2}}$. We introduce an averaging operator across Γ_* given by $\{v\} := \frac{1}{2}(v_1)_{|\Gamma_*} + \frac{1}{2}(v_2)_{|\Gamma_*}$.

Remark 11.5.3 Note that in the averaging we do not use a weighting. The weighting used in the average introduced in (11.2) is essential for the analysis presented in Sect. 11.2. In numerical experiments, however, we observe that in general the results do not deteriorate significantly if instead of the weighted average the standard unweighted one is used. For the introduction of the space-time Nitsche method we restrict, for simplicity, to an unweighted average.

For u with $[\alpha \nabla u \cdot \mathbf{n}_\Gamma] = 0$ we have $[\alpha \nabla u \cdot \mathbf{n}_\Gamma \beta v] = \{\alpha \nabla u \cdot \mathbf{n}_\Gamma\}[\beta v]$. Thus in view of consistency the interface term $-\int_{\Gamma_*^n} \nu \{\alpha \nabla u \cdot \mathbf{n}_\Gamma\}[\beta v] \, ds$ is added to the space-time bilinear form. For symmetry reasons (not essential) we also add $-\int_{\Gamma_*^n} \nu [\beta u]\{\alpha \nabla v \cdot \mathbf{n}_\Gamma\} \, ds$, and for stability we add $\lambda h^{-1} \int_{\Gamma_*^n} [\beta u][\beta v] \, ds$, with a parameter $\lambda > 0$. In order to present the discrete problem in a compact form we introduce some further notation:

$$a_i^n(u,v) := \int_{Q_i^n} (\frac{\partial u_i}{\partial t} + \mathbf{w} \cdot \nabla u_i)\beta_i v_i + \alpha_i \beta_i \nabla u_i \cdot \nabla v_i \, dx \, dt, \quad i = 1, 2,$$

$$N_{\Gamma_*}^n(u,v) := -\int_{\Gamma_*^n} \nu \{\alpha \nabla u \cdot \mathbf{n}_\Gamma\}[\beta v] \, ds - \int_{\Gamma_*^n} \nu [\beta u]\{\alpha \nabla v \cdot \mathbf{n}_\Gamma\} \, ds$$
$$+ \lambda h^{-1} \int_{\Gamma_*^n} [\beta u][\beta v] \, ds.$$

The resulting *space-time Nitsche-XFEM* discretization reads as follows:

Determine u_h such that for all $n = 1, \ldots, N$, $(u_h)_{|\Omega \times I_n} \in V_{kn}^{\Gamma_*}$ and

$$\sum_{i=1}^{2} a_i^n(u_h, v_h) + ([u_h]^{n-1}, \beta v_h^{n-1,+})_{L^2} + N_{\Gamma_*}^n(u_h, v_h) = \int_{t_{n-1}}^{t_n} (f, \beta v_h)_{L^2} \, dt \tag{11.44}$$

for all $v_h \in V_{kn}^{\Gamma_*}$.

Remark 11.5.4 Note that in this discretization *the same* space $V_{kn}^{\Gamma_*}$ is used for the trial and test functions. This is similar to the space-time discretization of the heat equation in (11.39), where one space V_{kn} is used both as trial and test space on the time slab $\Omega \times I_n$. As discussed above, for the heat equation there are alternative possibilities for the finite element spaces. One popular choice is to use trial functions that are *globally continuous in time.* An advantage of this choice is that the jump term $[u_h]^{n-1}$ vanishes, due to continuity of u_h. A well-posed discretization of the heat equation is obtained if for $k \geq 1$ we take trial functions $u_h \in C(\Omega \times [0,T])$ with $(u_h)_{|\Omega \times I_n} \in V_{kn}$ and test functions $v_h \in V_{(k-1)n}$. Note that in this variant the trial functions are polynomials of degree k w.r.t. time whereas the test functions are polynomials of degree $k-1$ w.r.t. time. One easily verifies that the number of degrees of freedom for such a trial function u_h (on the time slab $\Omega \times I_n$) equals the dimension of the test space $V_{(k-1)n}$.

The reason for using a "discontinuous in time" trial and test space in the XFEM method (11.44) is that it is not obvious how a well-posed variant with a continuous in time trial space can be defined. We indicate which difficulty arises. Consider $k = 1$ and a situation as in Fig. 11.5. The XFEM extension yields 8 additional degrees of freedom in the trial function u_h on the time slab $\Omega \times I_n$, corresponding to the basis functions $q_{j,\ell}^{\Gamma_*}$, $j \in \mathcal{J}_{\Gamma_*}$, $\ell = 0, 1$. These additional XFEM basis function are by construction *continuous* on Q_1 and on Q_2. Therefore a continuity condition for u_h at $t = t_{n-1}$ is automatically satisfied by these additional basis functions. The XFEM test space, with $k = 0$, i.e. $V_{0,n}^{\Gamma_*}$, contains (only) 4 additional XFEM basis functions. Therefore there is a mismatch between the degrees of freedom in the trial space and the dimension of the test space.

The term $([u_h]^{n-1}, \beta v_h^{n-1,+})_{L^2}$ controls the discontinuity at $t = t_{n-1}$ and thus one should take $\beta = \beta(\cdot, t_{n-1})$. The initial condition u_0 is approximated by

$u_h^{0,-} \in Q_h^\Gamma$ (XFEM space corresponding to $\Gamma(0)$). An important difference between this variational problem and the one in Sect. 11.1 is that for a nonstationary interface the XFEM space Q_h^Γ used in Sect. 11.1 varies with t, due to $\Gamma = \Gamma(t)$, whereas the space $V_{kn}^{\Gamma_*}$ used on the time slab $\Omega \times I_n$ in (11.44) does not depend on $t \in I_n$. The formulation in (11.44) yields a general framework for the development of concrete feasible methods. One issue is that in the specification of the space $V_{kn}^{\Gamma_*}$ one has to choose a finite element space V_n (for discretization of the spatial variable) and a value for k. Another important numerical aspect, hidden in the general formulation in (11.44), is that one needs space-time quadrature for computing mass and stiffness matrices. For this quadrature one needs an approximation of Γ_*^n, i.e. of $\Gamma(t)$ for $t \in I_n$. The accuracy of the quadrature depends on the quality of this approximation.

From the general formulation in (11.44) we derive a concrete feasible method. For the approximation of $\Gamma(t)$, $t \in I_n$, we take $\Gamma(t) \approx \Gamma(t_n) =: \Gamma_n$, and thus Γ_*^n is approximated by $\Gamma_n \times I_n$. This results in a cylindrical approximation $\Omega_i(t_n) \times I_n$ of the space-time subdomain Q_i^n. In addition we choose $k = 0$, i.e. the functions u_h, v_h do not depend on $t \in I_n$. Define $u_h^n(x) := u_h(x,t)$, $t \in I_n$. On the time slab $\Omega \times I_n$ the space-time XFEM space $V_{0n}^{\Gamma_*}$ is replaced by the time-independent XFEM space $Q_h^{\Gamma_n}$. The approximate interface Γ_n is stationary in I_n and thus we take $\nu = 1$. Furthermore, $\mathbf{w}(x,t)$ and $f(x,t)$ are replaced by $\mathbf{w}(x,t_n) =: \mathbf{w}^n(x)$ and $f(x,t_n) =: f^n(x)$, respectively. Then the scheme (11.44) reads: Determine $u_h^n \in Q_h^{\Gamma_n}$ such that

$$\begin{aligned}
&\frac{1}{\Delta t}(u_h^n - u_h^{n-1}, \beta v_h)_{L^2(\Omega)} + \sum_{i=1}^{2} \int_{\Omega_i(t_n)} \beta_i \mathbf{w}^n \cdot \nabla u_h^n v_h + \alpha_i \beta_i \nabla u_h^n \cdot \nabla v_h \, dx \\
&- \big(\{\alpha \nabla u_h^n \cdot \mathbf{n}_{\Gamma_n}\}, [\beta v_h]\big)_{L^2(\Gamma_n)} - \big([\beta u_h^n], \{\alpha \nabla v_h \cdot \mathbf{n}_{\Gamma_n}\}\big)_{L^2(\Gamma_n)} \qquad (11.45) \\
&+ \lambda h^{-1}\big([\beta u_h^n], [\beta v_h]\big)_{L^2(\Gamma_n)} = (f^n, \beta v_h)_{L^2(\Omega)} \quad \text{for all } v_h \in Q_h^{\Gamma_n}.
\end{aligned}$$

This scheme is (almost) the same as the one resulting from the Rothe method treated in Sect. 11.5.1, cf. (11.35). Since in our applications $\Gamma_n = \Gamma(t_n)$ is usually not known, we replace it by a (piecewise planar) approximation. The implementation of this method requires similar (quadrature) routines as the ones used in the Nitsche-XFEM method, cf. Sect. 11.4.2. Related to this quadrature we note the following. In the numerical evaluation of the term $(u_h^n - u_h^{n-1}, \beta v_h)_{L^2(\Omega)} = (\beta(\cdot,t_{n-1})u_h^n - \beta(\cdot,t_{n-1})u_h^{n-1}, v_h)_{L^2(\Omega)}$ discontinuities both across Γ_{n-1} (of $\beta(\cdot,t_{n-1})$ and u_h^{n-1}) and across Γ_n (of u_h^n and v_h) have to be taken into account. In the quadrature for the term $(\beta(\cdot,t_{n-1})u_h^{n-1}, v_h)_{L^2(\Omega)}$ it may be reasonable to neglect the discontinuity across Γ_{n-1}, since, due to the Henry condition, the jump $[\beta(\cdot,t_{n-1})u_h^{n-1}]_{\Gamma_{n-1}}$ is relatively small.

Along the same lines a more accurate, and from an implementational point of view (even) more challenging, method can be derived if we keep the interface approximation $\Gamma(t) \approx \Gamma_n$, $t \in I_n$, but instead of $k = 0$ use $k = 1$, resulting in a space-time XFEM finite element space $V_{1n}^{\Gamma_n}$.

The discretization accuracy is also improved if a better approximation of Γ_*^n is used. Essentially this boils down to "interpolating" between the interfaces $\Gamma(t_{n-1})$ and $\Gamma(t_n)$ (or their approximations). For the spatially 1D case this is simple, because the interface is a point. In two- or three dimensions it is not obvious how to obtain an accurate interpolation in an efficient way. A relatively simple (but not very accurate) strategy is to use the piecewise constant (in time) approximation $\Gamma(t_{n-1})$ for $t \in [t_{n-1}, t_{n-1} + \frac{1}{2}\Delta t]$, $\Gamma(t_n)$ for $t \in (t_{n-1} + \frac{1}{2}\Delta t, t_n]$.

A systematic investigation of these or other variants of the space-time Nitsche-XFEM discretization is not available in the literature, yet.

11.5.3 Numerical experiment: mass transport coupled with fluid dynamics

The three dimensional numerical simulation of two-phase fluid-dynamics coupled with mass transport is a very challenging task. Only in the recent literature one can find a few publications in which results of such simulations are presented, e.g. [44, 43, 197, 249, 200].

Combining the numerical methods for discretization of the mass transport problem with the numerical methods for the simulation of the fluid dynamics treated in Part II, one has all ingredients for the numerical simulation of a two-phase flow problem in which fluid-dynamics is coupled with mass transport. In [184] these methods are combined. In this section we present a few results from [184] to illustrate the effects that mass transport and Marangoni effects can have on the fluid dynamics.

For the fluid-dynamics we use the model (1.19)–(1.21), with a variable surface tension coefficient τ:

$$\begin{cases} \rho_i(\dfrac{\partial \mathbf{u}}{\partial t} + (\mathbf{u}\cdot\nabla)\mathbf{u}) = \operatorname{div}\boldsymbol{\sigma}_i + \rho_i\mathbf{g} \\ \operatorname{div}\mathbf{u} = 0 \end{cases} \quad \text{in } \Omega_i, \quad i = 1, 2,$$

$$[\boldsymbol{\sigma}\mathbf{n}_\Gamma] = -\tau\kappa\mathbf{n}_\Gamma + \nabla_\Gamma\tau, \ [\mathbf{u}] = 0 \quad \text{on } \Gamma,$$
$$V_\Gamma = \mathbf{u}\cdot\mathbf{n}_\Gamma \quad \text{on } \Gamma.$$

This is coupled with the level set equation

$$\frac{\partial\phi}{\partial t} + \mathbf{u}\cdot\nabla\phi = 0 \quad \text{in } \Omega$$

for capturing the interface and with the model (1.24) for mass transport:

$$\frac{\partial c}{\partial t} + \mathbf{u}\cdot\nabla c = \operatorname{div}(\alpha_i\nabla c) \quad \text{in } \Omega_i, \quad i = 1, 2,$$
$$[\alpha\nabla c\cdot\mathbf{n}]_\Gamma = 0 \quad \text{on } \Gamma,$$
$$[\beta c]_\Gamma = 0 \quad \text{on } \Gamma.$$

We use homogeneous Dirichlet and Neumann boundary conditions on $\partial\Omega$ for $\mathbf{u}$ and c, respectively. Suitable initial conditions for $\mathbf{u}$, c and ϕ are given below. For the numerical simulation of the fluid dynamics model we use the methods discussed in Part II. We summarize a few key components. For discretization of velocity and pressure we use the P_2-XFEM pair. The level set equation is discretized using quadratic finite elements and streamline diffusion stabilization. For the approximation of the interface we use the method explained in Sect. 7.3. The variable surface tension force is discretized using the discrete Laplace-Beltrami functional $\tilde{f}_{\Gamma_h}$, as explained in Sect. 7.6.1. For time discretization the simple (but less accurate) implicit Euler method given in Sect. 8.2 is applied. The mass transport equation is spatially discretized using the Nitsche-XFEM method explained in this chapter. We use the Rothe approach presented in Sect. 11.5.1.

In the experiment described below we consider a dependence of the surface tension coefficient on the concentration c, i.e. $\tau = \tau(c)$. Due to this, a concentration gradient along the interface leads to a variable surface tension, which induces so-called *Marangoni convection*. Hence, there is a strong coupling between the fluid-dynamics and mass transport models. In the numerical simulation we used, in each time step, a simple fixed point iterative strategy. For given c the fluid dynamics part is solved and then the computed velocity field $\mathbf{u}$ is used in the mass transport problem, resulting in an update of the concentration c. This is repeated until the update is smaller than a given tolerance.

We performed simulations with the *n-butanol - water - succinic acid* system, in which an n-butanol droplet is rising in water due to gravity. The computational domain is $\Omega = [0, 0.02] \times [0, 0.04] \times [0, 0.02]\,\mathrm{m}^3$. The dispersed phase (droplet) and continuous phase are contained in the subdomains Ω_1 and Ω_2, respectively. At $t = 0$, the droplet is at rest and has a spherical shape, with a diameter of $2 \cdot 10^{-3}\,\mathrm{m}$ and centered at $(0.01, 0.01, 0.01)\,\mathrm{m}$. This defines the initial conditions for $\mathbf{u}$ and ϕ.

The size and the initial position of the droplet are chosen such that its dynamics can be considered to be independent of wall effects. The right-hand side function $\mathbf{g}$ is taken as $\mathbf{g} = (0, -g, 0)$, with $g = 9.81\,\mathrm{m/s}^2$. The viscosity μ and density ρ for each phase at 20°C are given in Table 11.7.

butanol-water		
	$\mu\,[Pa \cdot s]$	$\rho\,[kg/m^3]$
Ω_1	3.28 E-3	845
Ω_2	1.39 E-3	987

Table 11.7. Material properties.

The concentrations in Ω_1 and Ω_2 are denoted by c_1 and c_2, respectively. At $t = 0$, the water phase contains a solute of succinic acid with initial concentration $c_2(0) > 0$ while the dispersed phase is clean. The initial condition for the concentration is

$$c(x, 0) = \begin{cases} 0 & \text{in } \Omega_1, \\ c_2(0) & \text{in } \Omega_2. \end{cases}$$

In general, this initial condition doesn't satisfy the Henry condition.

The Henry interface condition for the system at 20°C corresponds to the weighting with $\beta = (1, 1.2143)$. The diffusion coefficients for this system are given by $\alpha_r = (2.29 \cdot 10^{-10}, 5.83 \cdot 10^{-10})\,\mathrm{m}^2/\mathrm{s}$, which are extremely small values. For typical flow fields the mass transport equation is strongly convection dominated. On the meshes we use, the cell Peclet number is of the order of magnitude 10^6. For such a case the Nitsche-XFEM space discretization explained in Sects. 11.1, 11.5.1 is *not stable.* A suitable stabilization for the Nitsche-XFEM method is not available, yet. Moreover, the very thin concentration boundary layers require a very high resolution close to the interface, which leads to high computational costs. As a first step towards an efficient numerical simulation of this very demanding two-phase flow problem, we consider a *simplified case* in which the instability and high boundary layer resolution is avoided by taking artificial *much larger diffusion coefficients*, namely $\alpha_F = 10^5 \alpha_r$.

In [178], based on the experimental data for different systems, the surface tension coefficient τ for the system that we consider is modeled by

$$\tau = C_0 + C_1 x_C + C_2 x_C^2 \quad [10^{-3}\,\mathrm{N/m}], \tag{11.46}$$

with $C_0 := 1.625$, $C_1 = -28.08$, $C_2 = 222.8$, and $x_C = c_2|_\Gamma$ is the restriction to the interface of the concentration of the solute in the continuous phase. Due to the limited solubility of succinic acid in water, which is 5.8% at 20°C, the model (11.46) for the surface tension coefficient τ is valid only for x_C in the interval $[0, 0.058]$, cf. Fig. 11.6. In simulations with a constant τ, we take $\tau = C_0 = 1.625 \cdot 10^{-3}\,\mathrm{N/m}$.

A multilevel tetrahedral triangulation is used for the spatial discretization. For the initial triangulation, we partition the domain into $10 \times 20 \times 10$ cubes and then each of them is subdivided into six tetrahedra. The grid is then refined three times further near the interface, which results in a smallest mesh size of $2.5 \cdot 10^{-4}$m. When the interface moves, the grid is adaptively refined and coarsened. In Fig. 11.7 the grid is illustrated on the cross section $x = 0.01$ at $t = 0$ and $t = 0.4\,\mathrm{s}$.

In the simulations, we consider different initial concentrations, namely $c_2(0) = 1\%, 2.5\%$ and 5%. We illustrate the following effects: occurrence of concentration boundary layers and a significant influence of a *variable* surface tension coefficient on both the rise velocity and the shape of the droplet.

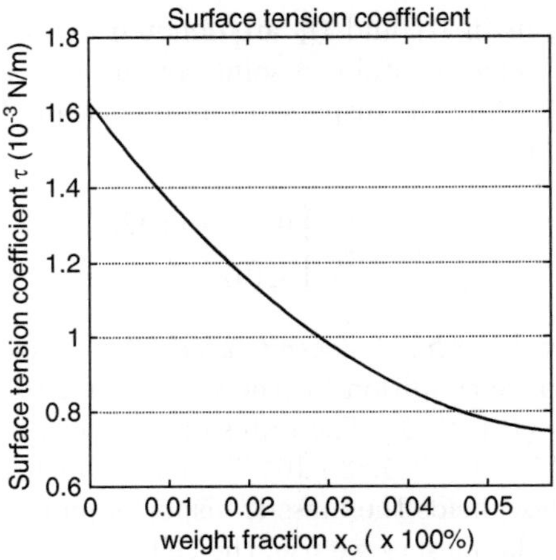

Fig. 11.6. Surface tension coefficient τ vs. x_C; model (11.46).

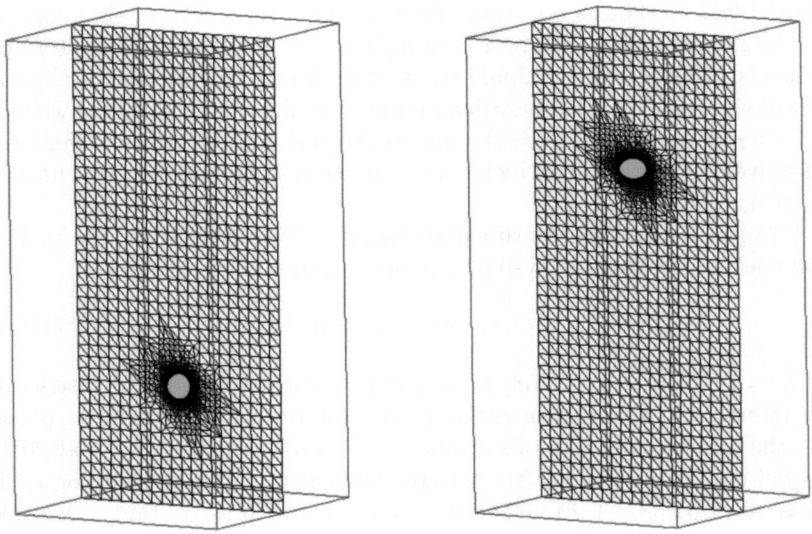

Fig. 11.7. Part of the grid on the cross section $x = 0.01$ at $t = 0$ (left) and $t = 0.4$ (right).

Concentration boundary layers

We take $c_2(0) = 1\%$ and a variable τ as in (11.46). The concentrations inside and outside the droplet at different times are displayed on the cross section $x = 0.01$ in Fig. 11.8.

Due to $\beta_1 < \beta_2$, the Henry condition requires $c_1(x,t) > c_2(x,t)$ for $x \in \Gamma$ and $t > 0$. For $t = 0$ we have $0 = c_1(x,0) < c_2(x,0)$ for $x \in \Gamma$, and thus the initial condition does *not* satisfy the Henry condition. This inconsistency in the initial condition causes a very small relaxation time in which there is a

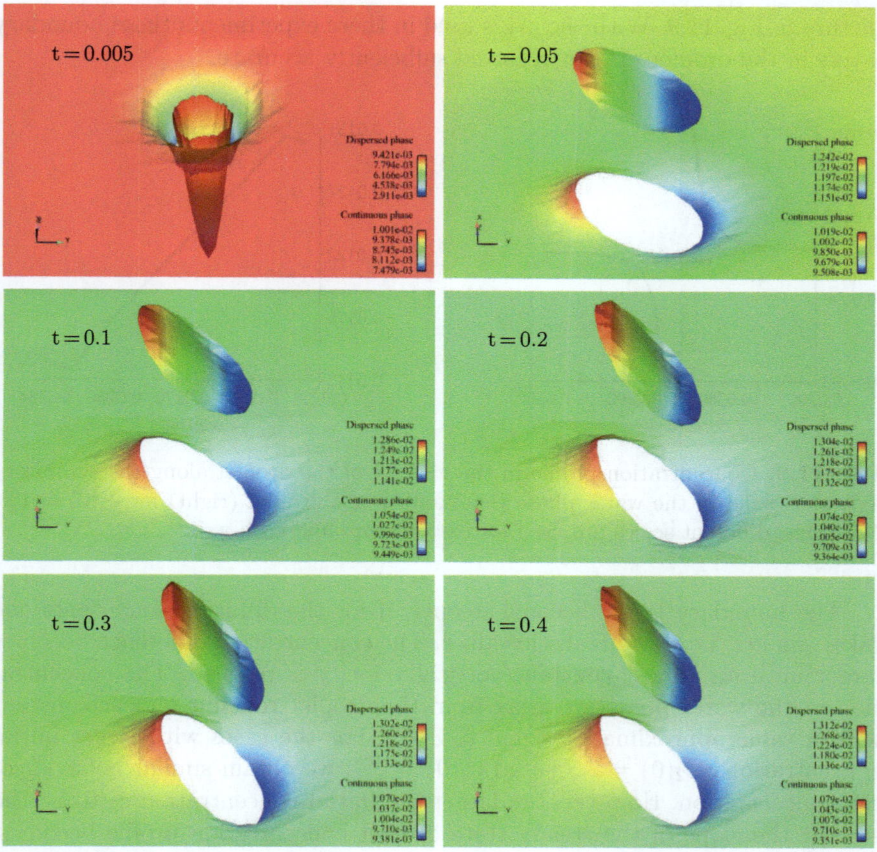

Fig. 11.8. Concentration distribution in the droplet region on the cross section $x = 0.01$ for $c_2(0) = 1\%$.

rapid change in the solution, cf. the result at $t = 0.005$ in Fig. 11.8. At the initial stage, the velocity in the droplet is very low and the mass transport is mainly due to diffusion. At the interface the concentration satisfies the Henry condition instantaneously, but in the middle of the droplet, the concentration is still very low due to the small diffusivity. After a short time ($t = 0.05\,\mathrm{s}$), as the droplet is accelerated, the role of the convection becomes larger. Then the concentration profile in the droplet is almost linear, cf. Fig. 11.9. The maximal and minimal values are attained at the lower and upper part of the droplet, respectively. At the interface boundary layers appear. In the left picture in Fig. 11.9 one observes the concentration boundary layers in the water phase, which can also be seen (less clearly) in Fig. 11.8. Due to the interface condition $[\alpha \nabla c \cdot \mathbf{n}]_\Gamma = 0$ the normal derivative at Γ must have the same sign in the water phase and in the droplet. This explains the nonmonotonic concentration profile in the droplet close to the interface, as can be observed in the right picture in Fig. 11.9. With the grids used in these experiments these boundary layers in the droplet are not resolved sufficiently accurate.

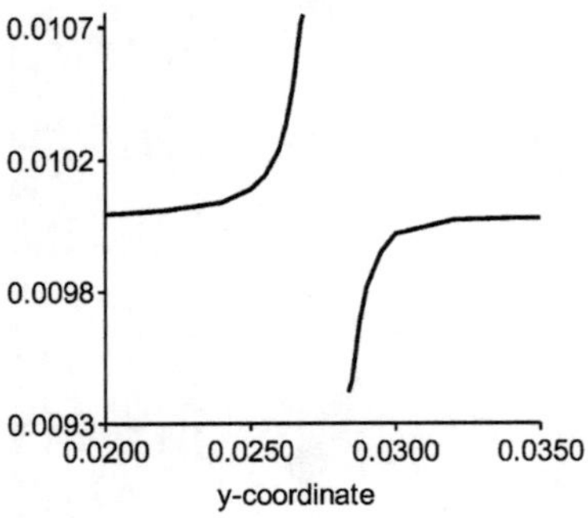

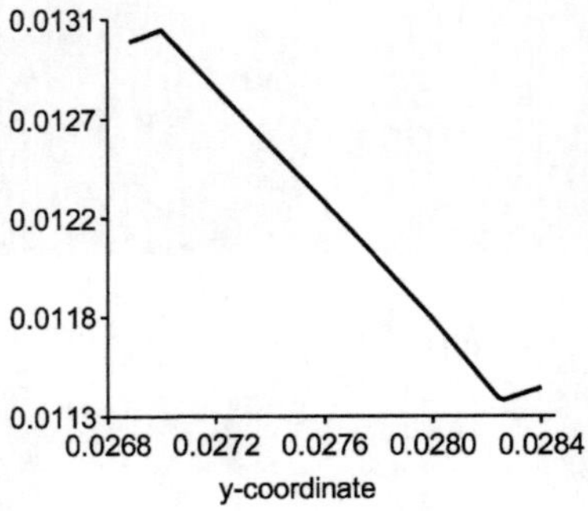

Fig. 11.9. Concentration profile in the vicinity of the droplet along the symmetry axis at $t = 0.4$ in the water phase (left) and in the droplet (right) for $c_2(0) = 1\%$. Note that different scales are used for each picture and each axis.

The boundary layers become steeper, if for the diffusion coefficients one takes smaller, i.e. more realistic values. The concentration difference becomes larger until an almost constant vertical velocity is reached. The concentration in the water phase far away from the droplet remains almost constant with a value approximately equal to $c_2(0)$. For the cases with larger initial concentrations, $c_2(0) = 2.5\%$ and $c_2(0) = 5\%$, we obtain similar behavior of the concentration. However, the larger the initial concentration $c_2(0)$ is, the steeper the slope of the steady state droplet concentration profile becomes. Furthermore, the concentration profiles are very similar if instead of the variable surface tension coefficient we use the constant value $\tau = C_0$.

Influence of variable surface tension coefficient on rise velocity

Although the variable surface tension coefficient appears to have little effect on the concentration profile in the droplet it does *have a significant effect on the dynamics of the droplet.* In Fig. 11.10, the vertical velocity of the droplet with respect to different initial concentrations is plotted over time. Note that for constant τ, the dynamics of the droplet are independent of the mass transport. After the same initial phase, the droplet rise velocity with variable τ is lower than the one with constant τ due to Marangoni effects. Note that the variable surface tension $c \to \tau(c)$ is monotonically decreasing for $c \in [0, 5.8\%]$. At the lower part of the interface, the higher concentration results in a lower value of the surface tension coefficient. Marangoni convection occurs, which retards the motion of the droplet. The larger the initial concentration $c_2(0)$ is, the stronger the Marangoni convection becomes. As a consequence, a lower value of the terminal droplet velocity is obtained. This effect is also observed in other literature, e.g. [200, 250].

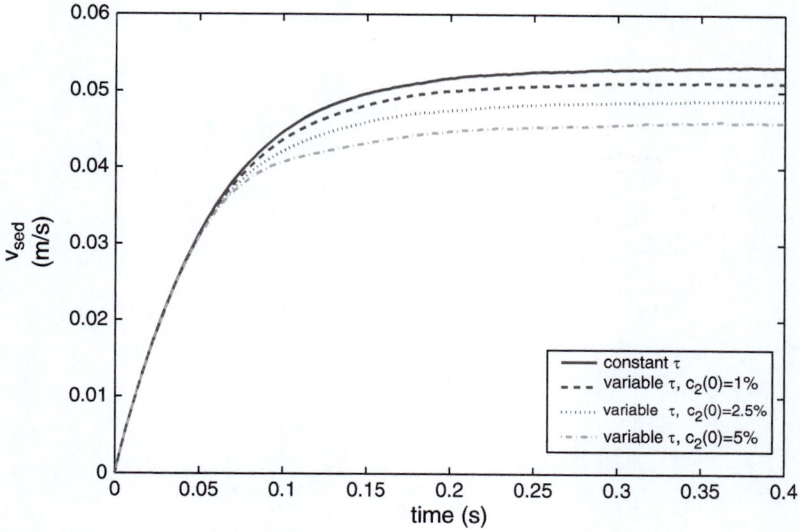

Fig. 11.10. Rise velocity of the droplet.

Influence of variable surface tension coefficient on shape of the droplet

In case of a *variable* surface tension coefficient the droplet becomes flatter for higher initial concentrations, which induces a lower rise velocity. In Fig. 11.11 we show, for a constant and for a variable surface tension coefficient, the shape of the droplet and vector fields of the relative velocity $\mathbf{v}_{rel} := \mathbf{v} - \mathbf{v}_{sed}$ with respect to the barycenter of the droplet (i. e., velocity in a Lagrangian reference system attached to the droplet) at $t = 0.4$ s for the case $c_2(0) = 5\%$. Inspection of the results shows that, for $c_2(0) = 5\%$, the vertical diameter changes from 1.57 mm (constant surface tension) to 1.30 mm (variable surface tension).

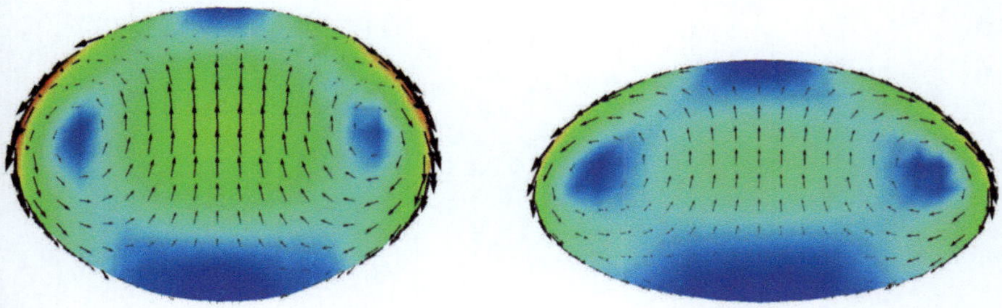

Fig. 11.11. Velocity field in Lagrangian reference system at steady state: constant τ (left) and variable τ (right), $c_2(0) = 5\%$.

Part IV

Surfactant transport

12

Mathematical model

We recall the model for transport of surfactants, derived in Sect. 1.1.4. In Sect. 1.1.4 the concentration of the surfactant is denoted by $S(x,t)$, the velocity field by $\mathbf{u}$ and the diffusion coefficient by D_Γ. For simplicity we assume D_Γ to be constant. By a rescaling we can take $D_\Gamma = 1$. In this and the next chapter we use a different notation for the unknown concentration and for the velocity field, namely $u = u(x,t)$ and $\mathbf{w} = \mathbf{w}(x,t)$, respectively. With this notation the convection-diffusion equation for the unknown concentration takes the following form, cf. (1.25):

$$\dot{u} + u \,\mathrm{div}_\Gamma \mathbf{w} = \mathrm{div}_\Gamma \nabla_\Gamma u, \tag{12.1}$$

where $\dot{u}$ denotes the material derivative of $u(x,t)$. Using the definitions of the material derivative and of the Laplace-Beltrami operator this equation can be written as

$$\frac{\partial u}{\partial t} + \mathbf{w} \cdot \nabla u + u \,\mathrm{div}_\Gamma \mathbf{w} - \Delta_\Gamma u = 0. \tag{12.2}$$

Remark 12.0.5 In (12.2) we do not model ad- or desorption effects. These would lead to a source term f of the form $f = f(u_\Omega)$, where u_Ω denotes the concentration of u in Ω_i ($i = 1$ or $i = 2$) evaluated at the interface Γ. Suitable models for f are hard to derive. In the remainder we restrict ourselves to (12.2), i.e., we do not treat models that include ad- or desorption effects.

In this chapter we consider weak formulations of the convection-diffusion problem (12.2). These weak formulations form the basis for the finite element methods treated in Chap. 13. We distinguish two cases, namely surfactant transport on a *stationary* or on a *non-stationary* interface.

12.1 Surfactant transport on a stationary interface

In this section we assume that the interface Γ is stationary, sufficiently smooth, bounded and $\partial\Gamma = \emptyset$, i.e. Γ does not have a boundary. Before we turn to the weak formulation of the convection-diffusion problem (12.2) we first consider the so-called *Laplace-Beltrami equation*, which is an important model problem that is often used in the literature. This equation models a pure diffusion process on a given sufficiently smooth surface (in our case, interface) Γ. It reads as follows: For a given f determine u such that

$$-\Delta_\Gamma u = f \quad \text{on } \Gamma.$$

Since $-\int_\Gamma \Delta_\Gamma u \, ds = \int_\Gamma \nabla_\Gamma u \cdot \nabla_\Gamma 1 \, ds = 0$ we introduce the assumption $\int_\Gamma f \, ds = 0$. A well-posed weak formulation is as follows: For given $f \in L_0^2(\Gamma) := \{ v \in L^2(\Gamma) : \int_\Gamma v \, ds = 0 \}$, determine $u \in H_*^1(\Gamma)$, with $H_*^1(\Gamma) := \{ v \in H^1(\Gamma) : \int_\Gamma v \, ds = 0 \}$, such that

$$\int_\Gamma \nabla_\Gamma u \cdot \nabla_\Gamma v \, ds = \int_\Gamma f v \, ds \qquad \text{for all } v \in H_*^1(\Gamma). \tag{12.3}$$

The bilinear form $(u,v) \to \int_\Gamma \nabla_\Gamma u \nabla_\Gamma v \, ds$ is continuous and elliptic on $H_*^1(\Gamma)$ and thus, using the Lax-Milgram lemma, it follows that this weak formulation has a unique solution u. Furthermore, it can be shown that this solution has the regularity property $u \in H^2(\Gamma)$ and $\|u\|_{H^2(\Gamma)} \le c \|f\|_{L^2(\Gamma)}$, with a constant c independent of f, cf. [92].

We now consider a weak formulation of the convection-diffusion problem in (12.2). We use the results presented in the abstract Hilbert space setting in Sect. 2.2.3, in particular Theorem 2.2.7. We apply the abstract results with the Hilbert spaces $V = H_*^1(\Gamma)$, $H = L_0^2(\Gamma)$. In case of a stationary interface we have $\mathbf{w} \cdot \mathbf{n} = 0$ and thus $\mathbf{w} \cdot \nabla u = \mathbf{w} \cdot \nabla_\Gamma u$. Using this we can rewrite (12.2) as

$$\frac{\partial u}{\partial t} + \mathrm{div}_\Gamma(\mathbf{w}u) - \Delta_\Gamma u = 0. \tag{12.4}$$

We introduce the bilinear form

$$\hat{a}(u,v) = \int_\Gamma \nabla_\Gamma u \cdot \nabla_\Gamma v + \mathrm{div}_\Gamma(\mathbf{w}u)\, v \, ds, \quad u, v \in H_*^1(\Gamma).$$

A sufficiently smooth solution u of (12.4), if it exists, satisfies $\int_\Gamma \frac{\partial u}{\partial t} v \, ds + \hat{a}(u,v) = 0$ for all $v \in H_*^1(\Gamma)$. We introduce the following weak formulation of (12.4), cf. (2.29) in Sect. 2.2.3:

> Find $u \in W^1(0,T; H_*^1(\Gamma))$ such that
>
> $$\begin{aligned} &\frac{d}{dt}(u(t),v)_{L^2(\Gamma)} + \hat{a}(u(t),v) = 0 \quad \text{for all } v \in H_*^1(\Gamma),\ t \in (0,T), \\ &u(0) = u_0. \end{aligned} \tag{12.5}$$

The space $W^1(0,T;H^1_*(\Gamma))$ is as defined in Sect. 2.2.3, i.e. $W^1(0,T;H^1_*(\Gamma)) = \{\, v \in L^2(0,T;H^1_*(\Gamma)) : v' \in L^2(0,T;H^1_*(\Gamma)') \text{ exists} \,\}$, where v' denotes the weak time derivative of v. This space has the following continuous embedding property, cf. (2.28),

$$W^1(0,T;H^1_*(\Gamma)) \hookrightarrow C([0,T];L^2_0(\Gamma)). \tag{12.6}$$

Based on Theorem 2.2.7 and Remark 2.2.7 we derive the following well-posedness result:

Theorem 12.1.1 *Assume* $\mathbf{w} \in H^1(\Gamma)^3$ *and* $\|\operatorname{div}\mathbf{w}\|_{L^\infty(\Gamma)} \leq c$. *For each* $u_0 \in L^2_0(\Gamma)$ *there exists a unique solution* u *of* (12.5) *and the linear mapping* $u_0 \to u$ *is continuous from* $L^2_0(\Gamma)$ *into* $W^1(0,T;H^1_*(\Gamma))$.

Proof. We apply Theorem 2.2.7 with the ellipticity condition replaced by the Garding inequality, cf. Remark 2.2.8. Using the smoothness assumption on the velocity field $\mathbf{w}$ it follows that $\hat{a}(\cdot,\cdot)$ is continuous on $H^1_*(\Gamma) \times H^1_*(\Gamma)$. Using the Green formula (14.15) and $\mathbf{w}\cdot\mathbf{n} = 0$ we get

$$\int_\Gamma \operatorname{div}_\Gamma(\mathbf{w}u)\, u\, ds = -\int_\Gamma u\mathbf{w}\cdot\nabla_\Gamma u\, ds = -\int_\Gamma \operatorname{div}_\Gamma(\mathbf{w}u)\, u\, ds + \int_\Gamma \operatorname{div}_\Gamma \mathbf{w}\, u^2\, ds,$$

and thus $\int_\Gamma \operatorname{div}_\Gamma(\mathbf{w}u)\, u\, ds = \frac{1}{2}\int_\Gamma \operatorname{div}_\Gamma \mathbf{w}\, u^2\, ds$. Using this we obtain with suitable constants $\gamma_V > 0$ and γ_H:

$$\begin{aligned} \hat{a}(u,u) &= \|\nabla_\Gamma u\|^2_{L^2(\Gamma)} + \frac{1}{2}\int_\Gamma \operatorname{div}_\Gamma \mathbf{w}\, u^2\, ds \\ &\geq \gamma_V \|u\|^2_{H^1_*(\Gamma)} - \gamma_H \|u\|^2_{L^2(\Gamma)} \quad \text{for all } \; u \in H^1_*(\Gamma). \end{aligned}$$

Hence $\hat{a}(\cdot,\cdot)$ satisfies the Garding condition. □

In Chap. 13 we treat finite element discretizations of the variational problems in (12.3) and (12.5).

12.2 Surfactant transport on a non-stationary interface

We now consider the case in which the interface may vary in time. We assume that for all $t \in [0,T]$ the interface $\Gamma(t)$ is *sufficiently smooth.* Precise sufficient smoothness conditions on $\Gamma(t)$ are formulated in Sect. 2 in [94]. Below we consider two weak formulations of the convection-diffusion equation (12.2). The first one, which is introduced in [94], is a variational problem *in space only,* in which the test space depends on time. The second one is a *space-time* variational problem. We introduce both, because as we will see in Chap. 13, these two formulations induce different finite element discretization approaches with their own merits.

The first weak formulation is taken from [94]. The smoothness assumptions on $\Gamma(t)$ are such that the space-time interface

$$\Gamma_* := \cup_{t\in[0,T]}\Gamma(t) \times \{t\}$$

is a three-dimensional hypersurface in $\mathbb{R}^4$. On this hypersurface one can define the corresponding Sobolev space of all functions for which all weak first derivatives exist. As usual, this space is denoted by $H^1(\Gamma_*)$. On $H^1(\Gamma_*)$ the norm equivalence

$$\|u\|^2_{H^1(\Gamma_*)} \sim \|u\|^2_{L^2(\Gamma_*)} + \|\nabla_\Gamma u\|^2_{L^2(\Gamma_*)} + \|\dot{u}\|^2_{L^2(\Gamma_*)}, \quad u \in H^1(\Gamma_*),$$

holds, with $\dot{u} = \frac{\partial u}{\partial t} + \mathbf{w}\cdot\nabla u$ the material derivative. In the analysis below we need a smoothness property of the velocity field $\mathbf{w}$. In the remainder of this section we assume

$$\mathbf{w} \in H^1(\Gamma_*)^3, \quad \|\mathrm{div}_\Gamma \mathbf{w}\|_{L^\infty(\Gamma_*)} < \infty.$$

We introduce the following weak formulation of (12.2):

Find $u \in H^1(\Gamma_*)$ such that for almost all $t \in (0,T)$:

$$\begin{aligned} &\int_{\Gamma(t)} \dot{u}v + uv\,\mathrm{div}_\Gamma\mathbf{w} + \nabla_\Gamma u\cdot\nabla_\Gamma v\,ds = 0 \quad \forall\; v(\cdot,t) \in H^1\big(\Gamma(t)\big), \\ &u(\cdot,0) = u_0. \end{aligned} \tag{12.7}$$

Clearly, a strong solution of (12.2) is a solution of (12.7). The following result is proved in [94].

Theorem 12.2.1 *Assume $u_0 \in H^1(\Gamma(0))$. Then there exists a unique solution of the variational problem* (12.7).

Proof. For a proof we refer to [94]. □

In the variational formulation (12.7) we use different trial and test spaces, namely $H^1(\Gamma_*)$ and $H^1\big(\Gamma(t)\big)$, respectively. Note, however, that for any $v \in H^1(\Gamma_*)$ we have $v(\cdot,t) \in H^1\big(\Gamma(t)\big)$ for almost all $t \in [0,T]$. This is due to the fact that from $\int_0^T\int_{\Gamma(t)} \nabla_\Gamma v\cdot\nabla_\Gamma v + v^2\,ds\,dt < \infty$ it follows (Fubini's theorem) that $\int_{\Gamma(t)} \nabla_\Gamma v\cdot\nabla_\Gamma v + v^2\,ds < \infty$ for almost all $t \in [0,T]$.

For a space-time variational formulation we introduce a test space consisting of functions in $L^2(\Gamma_*)$ for which the weak *spatial* first derivatives exist, i.e.

$$H^{1,0}(\Gamma_*) := \left\{ v \in L^2(\Gamma_*) : \|\nabla_\Gamma v\|_{L^2(\Gamma_*)} < \infty \right\},$$

which is a Hilbert space w.r.t. the norm

$$\|v\|^2_{H^{1,0}(\Gamma_*)} = \|v\|^2_{L^2(\Gamma_*)} + \|\nabla_\Gamma v\|^2_{L^2(\Gamma_*)}.$$

The weak formulation is as follows:

Find $u \in H^1(\Gamma_*)$ such that:

$$\int_0^T \int_{\Gamma(t)} \dot{u}v + uv\,\mathrm{div}_\Gamma \mathbf{w} + \nabla_\Gamma u \cdot \nabla_\Gamma v \, ds\, dt = 0 \quad \forall\, v \in H^{1,0}(\Gamma_*), \qquad u(\cdot,0) = u_0. \tag{12.8}$$

This formulation is a well-posed problem:

Theorem 12.2.2 *Assume $u_0 \in H^1(\Gamma(0))$. Then there exists a unique solution of the variational problem* (12.8).

Proof. On $H^1(\Gamma_*) \times H^{1,0}(\Gamma_*)$ we introduce the bilinear form

$$a(u,v) := \int_0^T \int_{\Gamma(t)} \dot{u}v + uv\,\mathrm{div}_\Gamma \mathbf{w} + \nabla_\Gamma u \cdot \nabla_\Gamma v \, ds\, dt,$$

which is continuous. Let $u \in H^1(\Gamma_*)$ be the solution of (12.7). Integration of the identity in (12.7) over $t \in [0,T]$ results in

$$a(u,v) = 0 \quad \text{for all} \quad v \in C^\infty(\overline{G}_T).$$

Since $C^\infty(\overline{G}_T)$ is dense in $H^{1,0}(\Gamma_*)$ is follows that u is a solution of (12.8). We now prove uniqueness, using a Gronwall argument. A corollary of the Gronwall lemma is as follows (cf. lemma 29.3 in [256]): if $f \in C([0,T])$, $f(t) \geq 0$ for all $t \in [0,T]$, and there exists a constant C such that

$$f(\tau) \leq C \int_0^\tau f(t)\, dt \quad \text{for all} \quad \tau \in [0,T],$$

this implies that $f(t) = 0$ for all $t \in [0,T]$. Let $w \in H^1(\Gamma_*)$ be a solution of (12.8) with $w(\cdot,0) = u_0 = 0$. Define $f(t) := \int_{\Gamma(t)} w(s,t)^2\, ds$. A variant of the embedding property (12.6) yields that f is continuous on $[0,T]$. Clearly $f(0) = 0$, $f \geq 0$ on $[0,T]$. Take a $\tau \in (0,T]$. By the differentiation rule (14.21b) we obtain

$$\begin{aligned} f(\tau) = f(\tau) - f(0) &= \int_0^\tau \frac{d}{dt} \int_{\Gamma(t)} w^2 \, ds\, dt \\ &= \int_0^\tau \int_{\Gamma(t)} 2\dot{w}w + w^2 \mathrm{div}_\Gamma \mathbf{w}\, ds\, dt. \end{aligned}$$

We use (12.8) with a test function v given by $v(\cdot,t) = w(\cdot,t)$ for $t \leq \tau$, $v(\cdot,t) = 0$ otherwise. Then we obtain

$$\begin{aligned} \int_0^\tau \int_{\Gamma(t)} \dot{w}w \, ds\, dt &= -\int_0^\tau \int_{\Gamma(t)} w^2 \mathrm{div}_\Gamma \mathbf{w}\, ds\, dt - \int_0^\tau \int_{\Gamma(t)} \nabla_\Gamma w \cdot \nabla_\Gamma w \, ds\, dt \\ &\leq -\int_0^\tau \int_{\Gamma(t)} w^2 \mathrm{div}_\Gamma \mathbf{w}\, ds\, dt. \end{aligned}$$

Hence, we get

$$f(\tau) \leq \|\mathrm{div}_\Gamma \mathbf{w}\|_{L^\infty(\Gamma_*)} \int_0^\tau f(t)\, dt.$$

Application of the Gronwall corollary yields $f(t) = \int_{\Gamma(t)} w(x,t)^2\, ds = 0$ for all $t \in [0,T]$. Hence $w = 0$ a.e. on Γ_*, which implies uniqueness. □

We compare the weak formulations in (12.7) and (12.8). In both formulations we use the same trial space $H^1(\Gamma_*)$. In (12.7) we have, for each $t \in (0,T)$, a variational formulation and a corresponding test space on the interface $\Gamma(t)$. In (12.8) the variational formulation and the corresponding test space are on the space-time domain Γ_*. This difference in the variational problems leads to (very) different finite element discretization methods, treated in Chap. 13.

13

Finite element methods for surfactant transport equations

In this chapter we treat different finite element approaches for interfacial transport problems as in (12.2). We first present a short overview of important classes of methods and then, in the following sections, treat some of these methods in more detail.

Recall that for the (numerical) treatment of the interface in Sect. 6.2 we distinguished between *Lagrangian interface tracking* and *Eulerian interface capturing* approaches. In the former the interface is explicitly represented (or approximated using an interface triangulation) and then tracked along characteristics. In the latter approach, instead of the interface, one tracks some phase indicator function. In numerics this leads to, for example, the VOF and level set techniques. These different (numerical) interface representation methods induce different finite element approaches for solving partial differential equations on the interface. In the literature, for discretization of partial differential equations on a surface (or interface) one can find finite element methods based on Lagrangian interface tracking and methods based on Eulerian interface capturing. The Lagrangian methods use finite element spaces on regular (triangular) triangulations Γ_h that approximate the interface Γ. These methods were first developed and analyzed for the case of a *stationary* interface and only recently Lagrangian finite element methods for the case of a *non-stationary* interface have been investigated. In Sect. 13.1 we discuss these methods in more detail. Eulerian interface capturing methods use an indicator function (e.g., level set function) to represent the interface and for discretization of a partial differential equation on the interface one uses finite element spaces corresponding to the (tetrahedral) triangulation on which this indicator function is discretized. Note that in general this triangulation is *in*dependent of the location of the interface. As for the Lagrangian case, these Eulerian methods were first developed for a stationary interface and afterwards extended to problems with a non-stationary interface. Finite element methods based on Eulerian interface capturing are treated in Sect. 13.2.

13.1 Finite element methods based on Lagrangian interface tracking

In this section we first discuss methods for the case of a *stationary* interface Γ and then treat methods for *non-stationary* Γ.

Stationary interface

The paper [92] contains the first analysis of a finite element method for discretizing an elliptic equation (Laplace-Beltrami equation) on a stationary surface. We outline the main ideas of this method and its analysis. To simplify the presentation, we only consider the 3D case. Let Γ be a sufficiently smooth two-dimensional surface without boundary, embedded in $\mathbb{R}^3$. We consider the Laplace-Beltrami equation (12.3) on Γ. The surface Γ is approximated by a *regular* family $\{\Gamma_h\}$ of triangulations. Each triangulation Γ_h is consistent (no hanging nodes) and it is assumed that all vertices in the triangulation lie on Γ. The space of scalar functions that are continuous on Γ_h and linear on each triangle in the triangulation Γ_h is denoted by V_h. The discretization of the Laplace-Beltrami equation is as follows:

determine $u_h \in V_h \cap H^1_*(\Gamma_h)$ such that

$$\int_{\Gamma_h} \nabla_{\Gamma_h} u_h \cdot \nabla_{\Gamma_h} v_h \, ds = \int_{\Gamma_h} f_h v_h \, ds \qquad \text{for all } v_h \in V_h \cap H^1_*(\Gamma_h). \quad (13.1)$$

Here $H^1_*(\Gamma_h) = \left\{ v \in H^1(\Gamma_h) : \int_{\Gamma_h} v \, ds = 0 \right\}$, ∇_{Γ_h} is the tangential derivative corresponding to Γ_h and f_h is a suitable extension of f, as explained below. For the definition of this extension and the discretization error analysis we need a suitable local coordinate system, which is the same as the one used in the Sects. 7.3.1, 7.7.1. On a neighborhood U of Γ we introduce the signed distance function $d : U \to \mathbb{R}$, $|d(x)| := \text{dist}(x, \Gamma)$ for all $x \in U$. Thus Γ is the zero level set of d. We assume $d < 0$ on the interior of Γ and $d > 0$ on the exterior. Note that $\mathbf{n}_\Gamma = \nabla d$ on Γ. We define $\mathbf{n}(x) := \nabla d(x)$ for all $x \in U$. Thus $\mathbf{n} = \mathbf{n}_\Gamma$ on Γ and $\|\mathbf{n}(x)\| = 1$ for all $x \in U$. We assume $\Gamma_h \subset U$ for all $\Gamma_h \in \{\Gamma_h\}$. We introduce a locally orthogonal coordinate system by using the projection $\mathbf{p} : U \to \Gamma$:

$$\mathbf{p}(x) = x - d(x)\mathbf{n}(x) \quad \text{for all } \; x \in U.$$

We assume that the decomposition $x = \mathbf{p}(x) + d(x)\mathbf{n}(x)$ is unique for all $x \in U$. Note that

$$\mathbf{n}(x) = \mathbf{n}(\mathbf{p}(x)) \quad \text{for all } \; x \in U.$$

Using these ingredients one can define "suitable" extensions. For $f \in L^2(\Gamma)$ we define its extension f^e as follows:

$$f^e(x) := f(\mathbf{p}(x)) \quad \text{for} \;\; x \in U.$$

For an illustration of the projection $\mathbf{p}$ and the construction of the extension f^e we refer to Fig. 7.8.

Analogous to the right-hand side f in (12.3) its lifted version $f_h \in L^2(\Gamma_h)$ in (13.1) should have zero mean value. Therefore we define:

$$f_h := f^e - |\Gamma_h|^{-1} \int_{\Gamma_h} f^e \, ds \tag{13.2}$$

with $|\Gamma_h| := \int_{\Gamma_h} 1 \, ds$. Using the Lax-Milgram lemma it follows that the discrete problem (13.1) with f_h as defined in (13.2) *has a unique solution* u_h. For the discretization error analysis one has to compare the discrete solution u_h, which is defined on Γ_h, with the solution u of the Laplace-Beltrami equation, which is defined on Γ. This can be done by introducing a lift U_h of u_h:

$$U_h(\mathbf{p}(x)) := u_h(x), \quad x \in \Gamma_h. \tag{13.3}$$

The following result is from [92].

Theorem 13.1.1 *Let u be the solution of the Laplace-Beltrami equation* (12.3) *and u_h its discrete approximation as in* (13.1), *with f_h as in* (13.2). *Let U_h be the lift of u_h given in* (13.3). *Then*

$$\|u - U_h\|_{L^2(\Gamma)} + h\|u - U_h\|_{H^1(\Gamma)} \leq c\, h^2$$

holds, with a constant c independent of h.

Proof. A proof is given in Lemma 6, Lemma 7 in [92]. □

Given the interface triangulation Γ_h the discretization (13.1) is easy to implement. The signed distance function d plays an important role in the error analysis but is of minor importance for the implementation of the method. In (13.1) it is used only in constructing the extension f_h of f and thus it can be avoided if a sufficiently accurate approximation f_h of f can be constructed using another technique.

The method described above has recently been extended from *linear* to *higher order* finite elements in [82]. To obtain higher-order convergence it is generally necessary to approximate Γ to higher order in addition to employing higher-order finite element spaces. In the implementation of the higher-order finite element method explicit knowledge of the signed distance function is essential, cf. [82].

An *adaptive* finite element discretization method for the Laplace-Beltrami equation, based on linear finite elements and suitable a-posteriori error estimators is treated in [83].

In [95] the linear finite element method for the elliptic Laplace-Beltrami equation on a stationary interface as described above is extended to a *parabolic* problem, on a *stationary* interface Γ, of the form

$$\int_\Gamma \frac{\partial u}{\partial t} v + \nabla_\Gamma u \cdot \nabla_\Gamma v \, ds = 0 \quad \text{for all} \quad v \in H^1(\Gamma), \quad t \in (0, T],$$

with initial condition $u(\cdot, 0) = u_0$ on Γ. This problem can be discretized using a standard method of lines technique as follows. Let V_h be the space of continuous linear finite elements on the triangulation Γ_h, as used in the discretization of the Laplace-Beltrami equation described above. Let $\{\psi_i\}_{1 \leq i \leq N}$ be the standard nodal basis in this space. For approximating the solution $u = u(x,t)$, $(x,t) \in \Gamma \times [0,T]$, of the parabolic problem we use the ansatz $u_h(x,t) = \sum_{j=1}^N u_j(t)\, \psi_j(x)$, with $(x,t) \in \Gamma_h \times [0,T]$. As test functions we use $v = \psi_i$, $i = 1, \ldots, N$. If we write $\vec{\mathbf{u}} = \vec{\mathbf{u}}(t) = (u_1, \ldots, u_N)^T$ we thus obtain the semi-discrete problem

$$\mathbf{M}\frac{d\vec{\mathbf{u}}}{dt}(t) + \mathbf{S}\vec{\mathbf{u}}(t) = 0, \quad t \in (0, T], \quad \vec{\mathbf{u}}(0) = \vec{\mathbf{u}}_0, \tag{13.4}$$

with $\vec{\mathbf{u}}_0$ the vector representation of an approximation $u_{0,h} \in V_h$ of the initial data u_0. The mass and stiffness matrices are given by

$$\mathbf{M}_{ij} = \int_{\Gamma_h} \psi_i \psi_j \, ds, \quad \mathbf{S}_{ij} = \int_{\Gamma_h} \nabla_{\Gamma_h} \psi_i \cdot \nabla_{\Gamma_h} \psi_j \, ds, \quad 1 \leq i, j \leq N.$$

The matrices $\mathbf{M}$ and $\mathbf{S}$ are symmetric positive definite and symmetric positive semidefinite, respectively. Optimal error bounds for the semi-discrete problem (13.4) are proved in [95]. A full space and time discrete problem is obtained by combining the semi-discretization with standard time discretization methods. Extensions of this method to other types of parabolic problems and to surfaces with boundary ($\partial\Gamma \neq \emptyset$) are presented in [95].

Non-stationary interface

In [94] the Lagrangian finite element technique explained above is generalized to problems with a *non*-stationary (or "evolving") interface. A similar technique was first introduced for the special case of the mean curvature flow problem in [93]. We explain the main ideas from [94].

The interface (or surface) $\Gamma(t)$, with $\partial\Gamma(t) = \emptyset$, is assumed to be smoothly evolving, with normal velocity $\mathbf{w} \cdot \mathbf{n}_\Gamma$. Starting point is the weak formulation of the transport problem in (12.7). Let $\{\Gamma_h\}$ be a regular family of consistent triangulations of $\Gamma(0)$. Define $\Gamma_h(0) := \Gamma_h$. The vertices of $\Gamma_h(0)$ are denoted by $X_j(0)$, $j = 1, \ldots, N$. We assume that $\Gamma_h(0)$ interpolates $\Gamma(0)$, i.e., all vertices $X_j(0)$ lie on $\Gamma(0)$. A crucial point in this finite element method is that it is purely Lagrangian, in the sense that *the vertices are transported with the flow field* $\mathbf{w}$, i.e., $\dot{X}_j(t) = \mathbf{w}(X_j(t), t)$ for all j and $t \in [0, T]$. This grid movement induces corresponding interpolating triangulations $\Gamma_h(t)$ of $\Gamma(t)$. Note that depending on the velocity field $\mathbf{w}$ and the deformation of $\Gamma(t)$, for $t > 0$ the triangulations $\Gamma_h(t)$ may become (strongly) distorted. A possible remedy then is to retriangulate the interface $\Gamma(t_0)$ for a suitable t_0. This requires interpolation operations between $\Gamma_h(t_0)$ and this retriangulation. Below we assume that T is sufficiently small such that for $t \in [0,T]$ the triangulations $\Gamma_h(t)$ remain sufficiently shape regular.

Corresponding to each vertex $X_j(t)$ of $\Gamma_h(t)$ there is a nodal basis function $\psi_j(\cdot,t) : \Gamma_h(t) \to \mathbb{R}$ such that $\psi_j(\cdot,t)$ is continuous on $\Gamma_h(t)$, linear on each triangle in $\Gamma_h(t)$ and $\psi_j(X_i(t),t)$ equals one if $i=j$ and zero otherwise. By construction these nodal basis functions have the property

$$\dot{\psi}_j = 0 \quad \text{on} \quad \Gamma_h(t), \quad j = 1,\ldots,N. \tag{13.5}$$

The space of continuous piecewise linears is given by

$$V_h(t) := \operatorname{span}\{\, \psi_j(\cdot,t) \,:\, 1 \le j \le N \,\}.$$

Let $u_{0,h} \in V_h(0)$ be an approximation of the initial data u_0. The following is a discretization of the problem (12.7):

Find $u_h(\cdot,t) = \sum_{j=1}^N u_j(t)\psi_j(\cdot,t) \in V_h(t)$ such that for all $t \in (0,T)$:

$$\begin{aligned} \int_{\Gamma_h(t)} \dot{u}_h v_h + u_h v_h \operatorname{div}_{\Gamma_h} \mathbf{w} + \nabla_{\Gamma_h} u_h \cdot \nabla_{\Gamma_h} v_h \, ds = 0 \;\; \forall\, v_h \in V_h(t), \\ u_h(\cdot,0) = u_{0,h}. \end{aligned} \tag{13.6}$$

In [94] an error analysis of this semi-discretization is presented. We outline a main result. For $u_h(\cdot,t) \in V_h(t)$, which is defined on $\Gamma_h(t)$, let $U_h(\cdot,t)$ be its lift to $\Gamma(t)$, as in (13.3),

$$U_h(\mathbf{p}(x),t) := u_h(x,t), \quad x \in \Gamma_h(t). \tag{13.7}$$

Theorem 13.1.2 *Let u be the solution of* (12.7)*, which is assumed to be sufficiently smooth, and u_h the discrete solution from* (13.6) *with $u_{0,h}$ the nodal interpolation of u_0. Let U_h be the lift of u_h as in* (13.7)*. Then*

$$\sup_{t\in[0,T]} \|u(\cdot,t) - U_h(\cdot,t)\|^2_{L^2(\Gamma(t))} + \int_0^T \|\nabla_\Gamma\big(u(\cdot,t) - U_h(\cdot,t)\big)\|^2_{L^2(\Gamma(t))}\, dt \le ch^2$$

holds.

Proof. A proof is given in Theorem 6.2 in [94]. □

The error bound is optimal with respect to the norm $\left(\int_0^T \|\nabla_\Gamma \cdot\|^2_{L^2(\Gamma(t))}\, dt\right)^{\frac{1}{2}}$ but suboptimal with respect to the norm $\sup_{t\in[0,T]} \|u(\cdot,t) - U_h(\cdot,t)\|_{L^2(\Gamma(t))}$. Results of numerical experiments in [94] indicate an (optimal) error behavior $\sup_{t\in[0,T]} \|u(\cdot,t) - U_h(\cdot,t)\|_{L^2(\Gamma(t))} \sim ch^2$. This optimal error bound is proved in the paper [98].

We now show that using the property (13.5) the discretization of the time variable can be realized by a simple method. In (13.6) it suffices to restrict the trial space to the set of basis functions $v_h = \psi_i$, $i = 1, \ldots, N$. For these basis functions we have $\dot{\psi}_i = 0$ and thus, using the Reynolds formula (14.21b) we get

$$\int_{\Gamma_h(t)} \dot{u}_h \psi_i + u_h \psi_i \operatorname{div}_{\Gamma_h} \mathbf{w} \, ds = \int_{\Gamma_h(t)} \dot{(u_h \psi_i)} + u_h \psi_i \operatorname{div}_{\Gamma_h} \mathbf{w} \, ds$$
$$= \frac{d}{dt} \int_{\Gamma_h(t)} u_h \psi_i \, ds.$$

Hence, instead of (13.6) we can determine $u_h(\cdot, t) = \sum_{j=1}^{N} u_j(t) \psi_j(\cdot, t) \in V_h(t)$ such that $u_h(\cdot, 0) = u_{0,h}$ and

$$\frac{d}{dt} \int_{\Gamma_h(t)} u_h \psi_i \, ds + \int_{\Gamma_h(t)} \nabla_{\Gamma_h} u_h \cdot \nabla_{\Gamma_h} \psi_i \, ds = 0, \quad \text{for } i = 1, \ldots, N.$$

Define $\vec{\mathbf{u}} = \vec{\mathbf{u}}(t) = (u_1, \ldots, u_N)^T$ and the time-dependent mass and stiffness matrices $\mathbf{M}(t)$, $\mathbf{S}(t)$ by

$$\mathbf{M}(t)_{ij} = \int_{\Gamma_h(t)} \psi_j(s,t) \psi_i(s,t) \, ds, \; \mathbf{S}(t)_{ij} = \int_{\Gamma_h(t)} \nabla_{\Gamma_h} \psi_j(s,t) \cdot \nabla_{\Gamma_h} \psi_i(s,t) \, ds.$$

For the unknown coefficient vector $\vec{\mathbf{u}}(t)$ we then obtain the system of ordinary differential equations

$$\frac{d}{dt}\big(\mathbf{M}(t)\vec{\mathbf{u}}(t)\big) + \mathbf{S}(t)\vec{\mathbf{u}}(t) = 0, \quad t \in [0, T].$$

For all $t \in [0, T]$ the matrices $\mathbf{M}(t)$ and $\mathbf{S}(t)$ are symmetric positive definite and symmetric positive semidefinite, respectively. A time discretization is straightforward. The implicit Euler method, for example, leads to

$$\mathbf{M}(t_{n+1})\vec{\mathbf{u}}^{n+1} + \Delta t \mathbf{S}(t_{n+1})\vec{\mathbf{u}}^{n+1} = \mathbf{M}(t_n)\vec{\mathbf{u}}^n, \quad n = 0, 1, \ldots,$$

with $\vec{\mathbf{u}}^0$ the vector representation of $u_{h,0}$. This method has a very simple structure. The main difficulty is hidden in the Lagrangian movement of the grid points $X_j(t)$ of $\Gamma_h(t)$. As noted above, retriangulation procedures and corresponding interpolation operators may be required if the triangulation $\Gamma_h(t)$ becomes too distorted. Clearly, this method is based on an interface tracking approach and thus it encounters severe difficulties in case of topological singularities (e.g., droplet collision).

13.2 Finite element methods based on Eulerian interface capturing

A general framework for the numerical treatment of partial differential equations on implicit surfaces based on *finite difference* methods was proposed in [36]. In that paper finite difference approximations on rectangular grids *in*dependent of an implicit static surface are considered. In [259] these Eulerian finite difference techniques are extended to problems with moving interfaces. We do not treat these finite difference methods, but restrict ourselves to Eulerian finite element discretizations.

13.2.1 An extension-based Eulerian finite element method

In [79] an Eulerian finite element method for discretizing the Laplace-Beltrami equation on a *stationary* surface Γ is presented. We outline the main idea of this method. Let the two-dimensional surface Γ be contained in a three-dimensional domain Ω and $\phi : \Omega \to \mathbb{R}$ be a level set function whose zero level set is Γ. Assume that $\nabla\phi \neq 0$ on Ω. Define the tangential derivative ∇_ϕ as follows:

$$\mathbf{n} := \frac{\nabla\phi}{\|\nabla\phi\|}, \quad \mathbf{P} := \mathbf{I} - \mathbf{n}\mathbf{n}^T, \quad \nabla_\phi u := \mathbf{P}\nabla u.$$

Note that on Γ this tangential derivative equals ∇_Γ. Below, to avoid technical difficulties due to the fact that $\Delta_\Gamma c = 0$ for a function c that is constant on Γ, we consider the variant of the Laplace-Beltrami equation in which a zero order term is added, i.e.: determine $u \in H^1(\Gamma)$ such that

$$\int_\Gamma \nabla_\Gamma u \cdot \nabla_\Gamma v + uv \, ds = \int_\Gamma fv \, ds \quad \text{for all} \quad v \in H^1(\Gamma). \tag{13.8}$$

Based on the level set function there is a natural extension of this Laplace-Beltrami equation to the domain Ω. Instead of this variational problem on Γ one considers the variational problem to find $u \in H^1_\phi(\Omega)$ such that

$$\int_\Omega (\nabla_\phi u \cdot \nabla_\phi v + uv) \|\nabla\phi\| \, dx = \int_\Omega f^e v \|\nabla\phi\| \, dx \quad \text{for all} \quad v \in H^1_\phi(\Omega), \tag{13.9}$$

with f^e an extension of the data f and a suitable Sobolev space $H^1_\phi(\Omega)$, cf. below.

Remark 13.2.1 The derivation of the extended equation (13.9) from (13.8) is essentially an application of the following *co-area formula*. Assume that the level set function $\phi : \overline{\Omega} \to \mathbb{R}$ is Lipschitz continuous. Define $\phi_{\min} := \min\{\, \phi(x) : x \in \overline{\Omega} \,\}$, $\phi_{\max} := \max\{\, \phi(x) : x \in \overline{\Omega} \,\}$ and let the level sets in Ω, i.e. $\Gamma_r := \{\, x \in \Omega : \phi(x) = r \,\}$ with $r \in [\phi_{\min}, \phi_{\max}]$, be two-dimensional hypersurfaces in $\mathbb{R}^3$. Then the co-area formula

$$\int_\Omega g \,\|\nabla\phi\| \, dx = \int_{-\infty}^{\infty} \int_{\phi^{-1}(r)} g \, ds dr = \int_{\phi_{\min}}^{\phi_{\max}} \int_{\Gamma_r} g \, ds dr$$

holds for functions g that are integrable on Ω. Using a suitable extension of f the problem (13.8) is not only considered on $\Gamma = \Gamma_0$ but on all Γ_r, $r \in [\phi_{\min}, \phi_{\max}]$. Integration over the level set values r then yields

$$\int_{\phi_{\min}}^{\phi_{\max}} \int_{\Gamma_r} \nabla_\phi u \cdot \nabla_\phi v + uv \, ds dr = \int_{\phi_{\min}}^{\phi_{\max}} \int_{\Gamma_r} f^e v \, ds dr,$$

for suitable test functions v (e.g. $v \in C^1(\overline{\Omega})$). Application of the co-area formula results in the extended problem (13.9).

Note that due to the tangential derivative the problem (13.9) is degenerated in the sense that there is *no diffusion in the normal direction.* This is also reflected in the strong formulation of the equation corresponding to (13.9), given by

$$-\operatorname{div}\left(\|\nabla\phi\|\mathbf{P}\nabla u\right) + \|\nabla\phi\| u = \|\nabla\phi\| f^e \quad \text{in } \Omega,$$

and which has, due to $\mathbf{Pn} = 0$, a singular diffusion tensor $\|\nabla\phi\|\mathbf{P}$. Therefore, instead of the standard Sobolev space $H^1(\Omega)$ one has to use other function spaces, which are ϕ-dependent, e.g. $H^1_\phi(\Omega) := \left\{ v \in L^2(\Omega) : \nabla_\phi u \in L^2(\Omega)^3 \right\}$. If the level sets of ϕ intersect the boundary $\partial\Omega$ one has to formulate appropriate artificial boundary conditions for u to make the problem well-posed. Natural boundary conditions (leading to well-posedness) are automatically satisfied if one uses a domain Ω of the form

$$\Omega = \left\{ x \in \mathbb{R}^3 : c_0 < \phi(x) < c_1 \right\}, \quad \text{with } c_0 < 0 < c_1. \tag{13.10}$$

Appropriate function spaces, well-posedness and regularity properties are studied in [79]. In [79, 58] it is proved that the restriction $u_{|\Gamma}$ of the solution u of (13.9) solves the Laplace-Beltrami problem (13.8) on Γ.

For the finite element discretization one starts with a triangulation of Ω, denoted by $\mathcal{T}_h$. A banded computational domain D_h is defined as follows. Let $I(\phi)$ be the piecewise linear nodal interpolation of ϕ on the triangulation $\mathcal{T}_h$. Define

$$D_h = \{ x \in \Omega : |I(\phi)(x)| < c_0 h \},$$

with a sufficiently large constant c_0 such that $T \subset D_h$ for all tetrahedra $T \in \mathcal{T}_h$ that have a nonzero intersection with Γ. A local triangulation is defined by

$$\mathcal{T}_h^\Gamma := \{ T \in \mathcal{T}_h : T \cap D_h \neq \emptyset \}.$$

Let V_h^Γ be the space of continuous piecewise linear finite elements on the local triangulation $\mathcal{T}_h^\Gamma$. The discrete problem is formulated on the h-banded computational domain D_h:

determine $u_h \in V_h^\Gamma$ such that for all $v_h \in V_h^\Gamma$

$$\int_{D_h} \left(\nabla_{\phi_h} u_h \cdot \nabla_{\phi_h} v_h + u_h v_h\right) \|\nabla I(\phi)\| \, dx = \int_{D_h} f^e v_h \|\nabla I(\phi)\| \, dx \tag{13.11}$$

holds.

Here $\nabla_{\phi_h} u_h := \mathbf{P}_h \nabla u_h$, with $\mathbf{P}_h = \mathbf{I} - \mathbf{n}_h \mathbf{n}_h^T$, $\mathbf{n}_h = \nabla I(\phi)/\|\nabla I(\phi)\|$. Note that in this discretization one needs a suitable extension f^e of the data f. An optimal discretization error bound is given in the following theorem.

Theorem 13.2.2 *Assume that the right-hand side f and the solution u of* (13.9) *are sufficiently smooth. Let u_h be the solution of* (13.11)*. Then*

$$\|u - u_h\|_{H^1(\Gamma)} \leq ch$$

holds.

Proof. A proof is given in Theorem 4.2 in [79]. There the smoothness assumptions for f and u are specified. □

Discretization error bounds in the $\|\cdot\|_{L^2(\Gamma)}$ norm are not known. Numerical experiments in [79] suggest a close to second order convergence in this norm.

Related to the implementation of the discrete problem (13.11) we note the following. The boundary of the computational domain D_h is piecewise planar (triangles or quadrilaterals), similar to the approximate interface Γ_h treated in Sect. 7.3. This boundary cuts tetrahedra $T \in \mathcal{T}_h^\Gamma$ and thus for computing the entries of the stiffness matrix special quadrature techniques are needed, which are very similar to those used in the implementation of the XFEM method as discussed in Sect. 7.9.3, cf. Fig. 7.13.

The Eulerian finite element method (13.11) for the elliptic Laplace-Beltrami equation on a stationary surface Γ has a canonical analogon (based on the method of lines) for *parabolic* problems on a stationary surface. This method is presented in [96]. In that paper for the computational domain, instead of the h-narrow band D_h defined above, the whole domain Ω is used. For this Eulerian finite element method for parabolic problems, no theoretical bounds for the (semi-)discretization error are known.

A further generalization of this method to transport problems on *non-stationary* interfaces (evolving surfaces) is given in [97]. We outline the main idea. Consider the surfactant transport problem (12.2), or its weak formulation in (12.7) and assume that the smooth interface $\Gamma(t)$ is evolving with a smooth velocity $\mathbf{w}$. A first step is a suitable extension of this transport problem to a fixed domain $\Omega \subset \mathbb{R}^3$ which contains $\Gamma(t)$, $0 \leq t \leq T$. Assuming $\mathbf{w} \cdot \mathbf{n}_{\partial\Omega} = 0$, this leads to a variational problem:

$$\frac{d}{dt}\int_\Omega uv\|\nabla\phi\|\,dx + \int_\Omega \nabla_\phi u \cdot \nabla_\phi v\|\nabla\phi\|\,dx = \int_\Omega u\dot{v}\|\nabla\phi\|\,dx \qquad (13.12)$$

for all v from a suitable set of test functions. This parabolic problem in Ω can be shown to be a consistent extension of the surfactant transport problem (12.2) on Γ. For an Eulerian discretization we introduce a time-*in*dependent triangulation $\mathcal{T}_h$ of Ω. Let V_h be the corresponding finite element space of continuous piecewise linears. For $v_h \in V_h$ we have $\dot{v}_h = \mathbf{w} \cdot \nabla v_h$. We obtain the following semi-discretization of (13.12):

determine $u_h(\cdot,t) \in V_h$ such that

$$\begin{aligned} &\frac{d}{dt}\int_\Omega u_h v_h \|\nabla\phi\|\,dx + \int_\Omega \nabla_\phi u_h \cdot \nabla_\phi v_h \|\nabla\phi\|\,dx \\ &= \int_\Omega u_h \mathbf{w}\cdot\nabla v_h \|\nabla\phi\|\,dx, \quad \text{for all } v_h \in V_h,\ t\in[0,T]. \end{aligned} \tag{13.13}$$

Of course, this has to be supplemented by a suitable initial condition for $u_h(\cdot,0)$. A discretization error analysis of this method is not known.

For the matrix-vector representation of this semi-discrete problem we introduce the standard nodal basis in V_h, denoted by ψ_i, $i = 1,\ldots,N$. The semi-discrete solution is written as $u_h(\cdot,t) = \sum_{j=1}^N u_j(t)\psi_j$. Define $\vec{\mathbf{u}} = \vec{\mathbf{u}}(t) = (u_1,\ldots,u_N)^T$ and the time-dependent matrices

$$\begin{aligned} \mathbf{M}(t)_{ij} &= \int_\Omega \psi_j(x)\psi_i(x)\|\nabla\phi(x,t)\|\,dx, \\ \mathbf{S}(t)_{ij} &= \int_\Omega \nabla_\phi\psi_j(x)\cdot\nabla_\phi\psi_i(x)\|\nabla\phi(x,t)\|\,dx, \\ \mathbf{C}(t)_{ij} &= \int_\Omega \psi_j(x)\mathbf{w}(x,t)\cdot\nabla\psi_i(x)\|\nabla\phi(x,t)\|\,dx. \end{aligned}$$

For the unknown coefficient vector $\vec{\mathbf{u}}(t)$ we then obtain the system of ordinary differential equations

$$\frac{d}{dt}\big(\mathbf{M}(t)\vec{\mathbf{u}}(t)\big) + \mathbf{S}(t)\vec{\mathbf{u}}(t) = \mathbf{C}(t)\vec{\mathbf{u}}(t), \quad t\in[0,T]. \tag{13.14}$$

Note that in this spatial semi-discretization we have unknowns $u_j(t)$ corresponding to *all* grid points in Ω and not (as in (13.11)) in a small h-banded domain around Γ. The Eulerian nature of this method is reflected in the fact that the domain of integration Ω and the basis functions ψ_i, ψ_j used in the definition of the matrices $\mathbf{M}(t)$, $\mathbf{S}(t)$ and $\mathbf{C}(t)$ do *not* depend on t. Application of a time discretization to (13.14) is straightforward. After an implicit time-discretization one obtains, in each time step, a linear system. In the development of an efficient iterative solver for this linear system one has to take into account that the extended problem (13.12) has a strongly anisotropic diffusion.

The Eulerian finite element methods treated above, cf. (13.11) and (13.13), are based on an *extension* of the partial differential equation given on the interface Γ to a larger three-dimensional domain Ω that contains Γ. For the discretization one uses finite element spaces on a given time independent triangulation $\mathcal{T}_h$ of Ω (or a local triangulation $\mathcal{T}_h^\Gamma$). In the subsections below we treat *an alternative Eulerian approach in which the extension procedure is avoided.* The main idea is to use the finite element spaces corresponding to $\mathcal{T}_h$

and restrict these to the interface Γ (or its approximation Γ_h). For the case of a stationary interface the method is explained in detail in Sect. 13.2.2. This approach turns out to be extremely easy to implement if finite element methods for two-phase flow problems as treated in Chap. 7 are already available, cf. Remark 13.2.6. Results of numerical experiments and a discretization error analysis of this method are given in the Sects. 13.2.3 and 13.2.4, respectively. The generalization of this Eulerian approach to the case of a *non*-stationary interface is treated in Sect. 13.2.5 and results of numerical experiments with this method are presented in Sect. 13.2.6.

13.2.2 Eulerian surface finite element method for a stationary interface

In this section we present an Eulerian finite element discretization of the Laplace-Beltrami problem (12.3) on a *stationary* interface Γ, with $\partial\Gamma = \emptyset$. This method is introduced in [195]. The method, which we explain using the Laplace-Beltrami problem, has a straightforward extension to the surfactant transport equation, cf. Remark 13.2.8 below.

Let $\{\mathcal{T}_h\}_{h>0}$ be a regular family of tetrahedral triangulations of a *fixed* domain $\Omega \subset \mathbb{R}^3$ that contains Γ. Take $\mathcal{T}_h \in \{\mathcal{T}_h\}_{h>0}$. We need an approximation Γ_h of Γ and assume that this approximate interface (or surface) has the following properties. We assume that Γ_h is a $C^{0,1}$ surface without boundary and that Γ_h can be partitioned in planar segments, triangles or quadrilaterals, consistent with the outer triangulation $\mathcal{T}_h$. This can be formally defined as follows. For any tetrahedron $S_F \in \mathcal{T}_h$ such that $\mathrm{meas}_2(S_F \cap \Gamma_h) > 0$ define $F = S_F \cap \Gamma_h$. We assume that each F is *planar*, i.e., either a triangle or a quadrilateral. Thus, Γ_h can be decomposed as

$$\Gamma_h = \bigcup_{F \in \mathcal{F}_h} F, \qquad (13.15)$$

where $\mathcal{F}_h$ is the set of all triangles or quadrilaterals F such that $F = S_F \cap \Gamma_h$ for some tetrahedron $S_F \in \mathcal{T}_h$. Note that if F coincides with a face of an element in $\mathcal{T}_h$ then the corresponding S_F is not unique. In this case, we choose one arbitrary but fixed tetrahedron S_F, which has F as a face.

Remark 13.2.3 The construction of Γ_h as described in Sect. 7.3 satisfies the assumptions made above, cf. Fig. 7.5.

For discretization of the problem (12.3) *we use a finite element space induced by the continuous linear finite elements on* $\mathcal{T}_h$. This is done as follows. We define a subdomain that contains Γ_h:

$$\omega_h := \bigcup_{F \in \mathcal{F}_h} S_F, \qquad (13.16)$$

and introduce the finite element space

$$V_h := \{ v_h \in C(\omega_h) : v_{h|S_F} \in \mathcal{P}_1 \ \text{ for all } \ F \in \mathcal{F}_h \}. \tag{13.17}$$

This space induces the following space on Γ_h:

$$V_h^{\Gamma_h} := \{ \psi_h \in H^1(\Gamma_h) : \exists \ v_h \in V_h : \psi_h = v_h|_{\Gamma_h} \}. \tag{13.18}$$

The spaces V_h and $V_h^{\Gamma_h}$ are called *outer* and *surface* finite element space, respectively. The surface space is used for a Galerkin discretization of (12.3):

determine $u_h \in V_h^{\Gamma_h}$ with $\int_{\Gamma_h} u_h \, ds = 0$ such that

$$\int_{\Gamma_h} \nabla_{\Gamma_h} u_h \cdot \nabla_{\Gamma_h} \psi_h \, ds = \int_{\Gamma_h} f_h \psi_h \, ds \qquad \text{for all } \ \psi_h \in V_h^{\Gamma_h}, \tag{13.19}$$

with f_h a suitable extension of f such that $\int_{\Gamma_h} f_h \, ds = 0$, cf. (13.2). Due to the Lax-Milgram lemma this problem has a unique solution u_h. In the discretization method (13.19) we restricted to *linear* finite elements. This restriction is not essential; the method has a straightforward extension to higher order finite elements.

In Sect. 13.2.3 we present results of numerical experiments that indicate that the discretization method has optimal convergence rates. In Sect. 13.2.4 we present a discretization error analysis of this method showing that under reasonable assumptions we indeed have optimal error bounds.

In the remarks below we give some comments related to this approach.

Remark 13.2.4 (Shape irregularity) The family $\{\mathcal{T}_h\}_{h>0}$ of tetrahedral triangulations of Ω is *shape-regular* but the family $\{\mathcal{F}_h\}_{h>0}$ of interface triangulations is in general *not shape-regular*. In our numerical experiments, cf. Sect. 13.2.3, $\mathcal{F}_h$ contains a significant number of strongly deteriorated triangles that have very small angles. Moreover, neighboring triangles can have very different areas, cf. Fig. 13.4 for an illustration. As we will prove in Sect. 13.2.4, optimal discretization bounds hold if $\{\mathcal{T}_h\}_{h>0}$ is shape-regular; for $\{\mathcal{F}_h\}_{h>0}$ shape-regularity is *not* required.

Remark 13.2.5 (Related finite element space) Each quadrilateral in $\mathcal{F}_h$ can be subdivided into two triangles. Let $\tilde{\mathcal{F}}_h$ be the induced set consisting of *only* triangles and such that $\cup_{F \in \tilde{\mathcal{F}}_h} F = \Gamma_h$. Define

$$W_h^{\Gamma_h} := \left\{ \psi_h \in C(\Gamma_h) : \psi_h|_F \in \mathcal{P}_1 \quad \text{for all } \ F \in \tilde{\mathcal{F}}_h \right\}. \tag{13.20}$$

The space $W_h^{\Gamma_h}$ is the space of continuous functions that are piecewise linear on the triangles of Γ_h. This space, on a *shape-regular* interface triangulation, is used in the Lagrangian finite element method discussed in Sect. 13.1. Clearly $V_h^{\Gamma_h} \subset W_h^{\Gamma_h}$ holds. In general, however, $V_h^{\Gamma_h} \neq W_h^{\Gamma_h}$ holds. This follows, for example, from the fact that corresponding to a quadrilateral segment F one

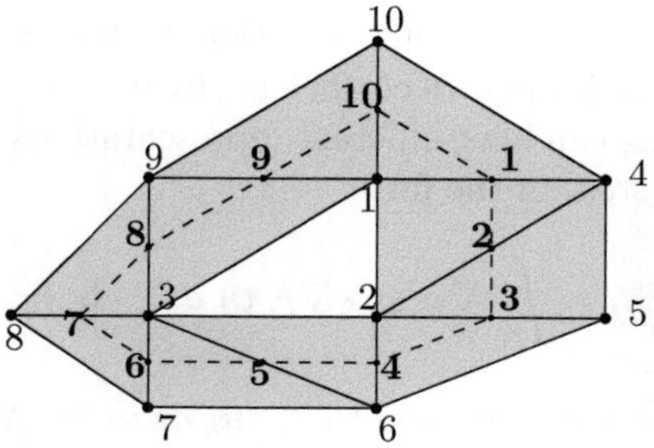

Fig. 13.1. Example with $V_h^{\Gamma_h} \neq W_h^{\Gamma_h}$.

has 3 degrees of freedom (representation of a linear function) in $V_h^{\Gamma_h}$ and 4 degrees of freedom (linear functions on the two triangles in which F is subdivided) in $W_h^{\Gamma_h}$. A 2D example with $V_h^{\Gamma_h} \neq W_h^{\Gamma_h}$ is given in Fig. 13.1.

In this example ω_h consists of 10 triangles (shaded). The nodal basis functions corresponding to these triangles are denoted by $\{\psi_i\}_{1\leq i\leq 10}$. The line segments of the interface Γ_h (represented by dashed lines) intersect midpoints of edges of the triangles. The space $W_h^{\Gamma_h}$ consists of piecewise linears on Γ_h and is spanned by the 1D nodal basis functions at the intersection points labeled by boldface $\mathbf{1},\ldots,\mathbf{10}$. Clearly $\dim(W_h^{\Gamma_h}) = 10$. In this example we have $\dim(V_h^{\Gamma_h}) = 9$. For the piecewise linear function $v_h = \sum_{i=1}^{10} \alpha_i\psi_i$ with $\alpha_i = -1$ for $i = 1, 2, 3$ and $\alpha_i = 1$ for $i = 4, \ldots, 10$ we have $(v_h)_{|\Gamma_h} = 0$.

Remark 13.2.6 (Implementation) Let $(v_i)_{1\leq i\leq m}$ be the collection of all vertices of all tetrahedra in ω_h and ψ_i the nodal linear finite element basis function in the outer finite element space V_h corresponding to v_i. Then $V_h^{\Gamma_h}$ is spanned by the functions $\psi_i|_{\Gamma_h}$, $1 \leq i \leq m$. These functions are *not* necessarily linear independent, cf. Remark 13.2.5. In the implementation of this method one has to compute integrals of the form

$$\int_F \nabla_{\Gamma_h}\psi_j \cdot \nabla_{\Gamma_h}\psi_i \, ds, \quad \int_F f_h\psi_i \, ds \quad \text{for} \quad F \in \mathcal{F}_h. \tag{13.21}$$

The domain F is either a triangle or a quadrilateral. The first integral can be computed exactly. For the second one standard quadrature rules can be applied. The algorithmic components for computing these quantities are *almost exactly the same* as the ones used in the computation of the discrete surface tension functional $\tilde{f}_{\Gamma_h}$ (or f_{Γ_h}) treated in Sect. 7.6. In other words, once the implementation of this surface tension force discretization is available, the quantities in (13.21) can be determined with only very little additional implementation effort.

Remark 13.2.7 (Linear system) Let ψ_i, $i = 1, \ldots, m$, be the nodal basis functions of V_h as in Remark 13.2.6. In general, the set $\{\, \psi_i|_{\Gamma_h} : 1 \leq i \leq m \,\}$ is not a basis, but only a generating system of the finite element space $V_h^{\Gamma_h}$. This follows already from the example $v_h = \sum_{i=1}^{10} \alpha_i\psi_i$ at the end of Remark 13.2.5.

Let $u_h = \sum_{i=1}^{m} u_i \psi_{i|\Gamma_h}$ be the unique solution of the discrete problem (13.19) and denote the vector of unknown coefficients by $\vec{\mathbf{u}} = (u_1, \ldots, u_m)^T$. In general the unique solution u_h can have different representations $\vec{\mathbf{u}}$. The discretization results in a linear system of the form

$$\mathbf{S}\vec{\mathbf{u}} = \vec{\mathbf{b}}, \quad \mathbf{S}_{ij} = \int_{\Gamma_h} \nabla_{\Gamma_h}\psi_j \cdot \nabla_{\Gamma_h}\psi_i \, ds, \quad \vec{\mathbf{b}}_i = \int_{\Gamma_h} f_h \psi_i \, ds,$$

with the additional constraint (due to $\int_{\Gamma_h} u_h \, ds = 0$) $\langle \mathbf{M}\vec{\mathbf{u}}, \mathbf{e}\rangle = 0$, where $\mathbf{M}$ the mass matrix and $\mathbf{e} = (1, 1, \ldots, 1)^T$. The stiffness matrix $\mathbf{S}$ is symmetric positive semidefinite. Every solution $\mathbf{u}$ of this linear system yields the desired unique discrete solution u_h. Due to the fact that a nonstandard generating system is used the question of conditioning of the stiffness and mass matrix arises. This topic is studied in [194]. We consider an introductory example, which illustrates some interesting properties and then briefly address some general conditioning results. We consider a strongly simplified situation of a one-dimensional interface $\Gamma = [0, 1]$ embedded in $\mathbb{R}^2$. The outer finite element space is based on a uniform triangulation $\mathcal{T}_h$ as illustrated in Fig. 13.2. The number of vertices is denoted by m ($m = 11$ in Fig. 13.2) and $h := \frac{2}{m-3}$ is a measure for the mesh size of the triangulation. The interface $\Gamma = [0, 1]$ is located in the middle between the upper and lower line of the outer triangulation.

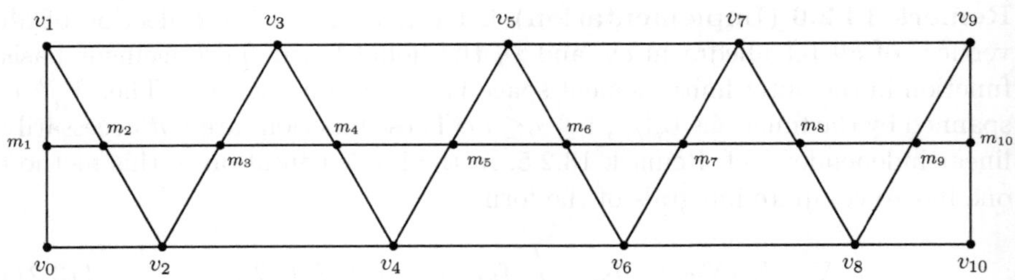

Fig. 13.2. Example with a uniform triangulation.

The nodal basis function corresponding to v_i is denoted by ψ_i, $i = 0, 1, \ldots, m-1$. We represent $u_h \in V_h^\Gamma$ as $u_h = \sum_{i=0}^{m-1} u_i \psi_{i|\Gamma}$. The vector representation is given by $\vec{\mathbf{u}} = (u_0, u_1, \ldots, u_{m-1})^T \in \mathbb{R}^m$. The mass and stiffness matrix are defined by

$$\langle \mathbf{M}\vec{\mathbf{u}}, \vec{\mathbf{u}}\rangle = \int_0^1 u_h(x)^2 \, dx, \quad \langle \mathbf{S}\vec{\mathbf{u}}, \vec{\mathbf{u}}\rangle = \int_0^1 u_h'(x)^2 \, dx.$$

Let $m_i = \frac{v_{i-1}+v_i}{2} \in \Gamma$, $i = 1, \ldots, m-2$ denote the mid-vertices of the edges of the outer triangulation that are crossing Γ. Now note that

$$\int_0^1 u_h(x)^2\,dx = \sum_{i=1}^{m-2}\int_{m_i}^{m_{i+1}} u_h(x)^2\,dx$$
$$\sim h\sum_{i=1}^{m-2}\left(u_h(m_i)^2+u_h(m_{i+1})^2\right)\sim h\sum_{i=1}^{m-1}u_h(m_i)^2$$
$$=\frac{h}{4}\sum_{i=1}^{m-1}\left(u_h(v_{i-1})+u_h(v_i)\right)^2=\frac{h}{4}\sum_{i=1}^{m-1}\left(u_{i-1}+u_i\right)^2=\frac{h}{4}\langle \mathbf{L}\vec{\mathbf{u}},\mathbf{L}\vec{\mathbf{u}}\rangle,$$

with

$$\mathbf{L}=\begin{pmatrix} 1 & 1 & & & \\ & 1 & 1 & \emptyset & \\ & \emptyset & \ddots & \ddots & \\ & & & 1 & 1 \end{pmatrix}\in\mathbb{R}^{(m-1)\times m}.$$

Thus the diagonally scaled mass matrix is spectrally equivalent to $\mathbf{L}^T\mathbf{L}$. The matrix $\mathbf{L}^T\mathbf{L}$ has one zero eigenvalue $\lambda_1=0$, with corresponding eigenvector $(1,-1,1,-1,\ldots)^T$. The smallest nonzero eigenvalue is $\lambda_2\sim h^2$, and thus for the effective condition number we obtain $\frac{\lambda_{\max}}{\lambda_2}\sim h^{-2}$.

For the stiffness matrix we obtain the following:

$$\int_0^1 u_h'(x)^2\,dx=\sum_{i=1}^{m-2}\int_{m_i}^{m_{i+1}}u_h'(x)^2\,dx\sim h\sum_{i=1}^{m-2}\left(\frac{u_h(m_{i+1})-u_h(m_i)}{m_{i+1}-m_i}\right)^2$$
$$\sim\frac{1}{h}\sum_{i=1}^{m-2}\left(\left(u_h(v_i)+u_h(v_{i+1})\right)-\left(u_h(v_{i-1})+u_h(v_i)\right)\right)^2$$
$$=\frac{1}{h}\sum_{i=1}^{m-2}\left(u_{i+1}-u_{i-1}\right)^2=\frac{1}{h}\langle\hat{\mathbf{L}}\vec{\mathbf{u}},\hat{\mathbf{L}}\vec{\mathbf{u}}\rangle,$$

with

$$\hat{\mathbf{L}}=\begin{pmatrix} -1 & 0 & 1 & & & \emptyset \\ & -1 & 0 & 1 & & \\ & & \ddots & \ddots & \ddots & \\ \emptyset & & & -1 & 0 & 1 \end{pmatrix}\in\mathbb{R}^{(m-2)\times m}.$$

Thus the diagonally scaled stiffness matrix is spectrally equivalent to $\hat{\mathbf{L}}^T\hat{\mathbf{L}}$. The matrix $\hat{\mathbf{L}}^T\hat{\mathbf{L}}$ has two zero eigenvalues $\lambda_1=\lambda_2=0$, with corresponding eigenvectors $(1,-1,1,-1,\ldots)^T$, $(1,1,\ldots,1)^T$. The smallest nonzero eigenvalue is $\lambda_3\sim h^2$, and thus for the effective condition number we obtain $\frac{\lambda_{\max}}{\lambda_3}\sim h^{-2}$. Summarizing, in this simple example, both the mass and stiffness matrix are singular and for both matrices the effective condition number behaves like h^{-2}. The zero eigenvalue of $\mathbf{M}$ and one of the zero eigenvalues of $\mathbf{S}$ are caused by the fact that we use a generating system of m functions $\psi_{i|\Gamma}$, $0\le i\le m-1$, to span the trace space V_h^Γ of dimension $m-1$. The second zero eigenvalue of $\mathbf{S}$ is caused by the singularity of the bilinear form: $\int_0^1 u_h'(x)^2\,ds = 0$ if $u_h(x) = \text{constant}$.

We now consider a case as illustrated in Fig. 13.3 in which there is one vertex v_k ($k = 4$ in Fig. 13.3) for which $\operatorname{dist}(v_k, \Gamma) = \epsilon = \delta\frac{h}{2}$, with $\delta \in (0,1]$ and k such that $\frac{m}{k}$ is a fixed number if the mesh is refined.

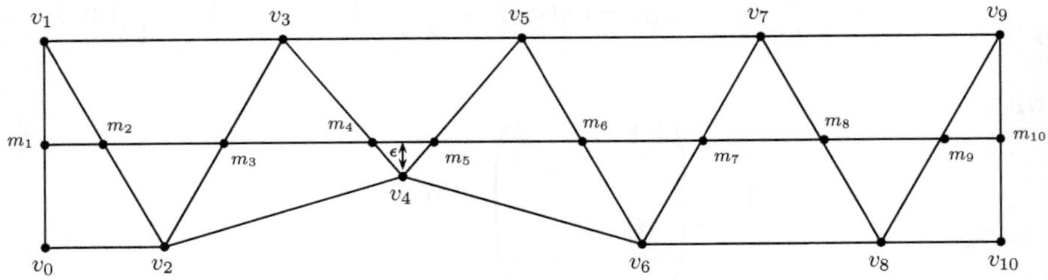

Fig. 13.3. Degenerated case with a small distance.

For this case we obtain:

$$\begin{aligned}\int_0^1 u_h(x)^2\,dx &\sim h\sum_{i=1}^{m-1} u_h(m_i)^2 \sim h\sum_{i=1}^{k-1}\big(u_{i-1}+u_i\big)^2 \\ &\quad + h(\delta u_{k-1}+u_k)^2 + h(u_k+\delta u_{k+1})^2 + h\sum_{i=k+2}^{m-1}\big(u_{i-1}+u_i\big)^2 \\ &\sim h\langle \tilde{\mathbf{L}}\vec{\mathbf{u}}, \tilde{\mathbf{L}}\vec{\mathbf{u}}\rangle,\end{aligned}$$

with

$$\tilde{\mathbf{L}} = \begin{pmatrix} 1 & 1 & & & & & \\ & \ddots & \ddots & & \emptyset & & \\ & & 1 & 1 & & & \\ & & & \delta & 1 & & \\ & & & & 1 & \delta & \\ & & \emptyset & & & 1 & 1 \\ & & & & & & \ddots \;\ddots \end{pmatrix} \in \mathbb{R}^{(m-1)\times m}.$$

Thus the diagonally scaled mass matrix is spectrally equivalent to $\tilde{\mathbf{L}}^T\tilde{\mathbf{L}}$. A straightforward (MATLAB) calculation yields that this matrix has one zero eigenvalue $\lambda_1 = 0$ and for $\delta \leq \sqrt{h}$ the first nonzero eigenvalue is of size $\lambda_2 \sim h\delta^2$. Hence for the effective condition number of the scaled mass matrix we obtain $\frac{\lambda_{\max}}{\lambda_2} \sim h^{-1}\delta^{-2}$ and thus a strong growth in the effective condition number for $\delta \downarrow 0$. This example indicates that the dependence of the effective spectral condition number on the distances of the vertices of the outer triangulation to Γ is a delicate issue.

We now briefly address results in a more general setting. In [194] one can find results of numerical experiments in two- and three-dimensional examples which indicate that in the 3D case *both for the diagonally scaled mass and stiffness matrix* (effective) spectral condition numbers behave like $O(h^{-2})$ and in the 2D case the behavior of these condition numbers is $O(h^{-3})$ and $O(h^{-2})$, respectively. Some of these experimental results for the 3D case are given in Sect. 13.2.3. Here h denotes the mesh size of the *outer* triangulation, which (in the analysis) is assumed to be quasi-uniform in a small neighborhood of the interface. In [194] an analysis for the *two*-dimensional case is given which proves these conditioning properties (up to an additional logarithmic term $|\ln h|$) under certain assumptions on the distribution of the vertices near the interface.

Remark 13.2.8 For the case of a stationary interface (surface) the finite element method explained above has an obvious generalization to parabolic problems on the interface. We describe the application to the surfactant transport equation. A further (less obvious) generalization of this Eulerian surface finite element method to partial differential equations on a *non*-stationary interface is treated in Sect. 13.2.5 below.

Recall the weak formulation of the surfactant transport problem in (12.5): Find $u \in W^1(0,T; H^1_*(\Gamma))$ such that $u(0) = u_0$ and, for all $t \in (0,T)$,

$$\frac{d}{dt}(u(t),v)_{L^2(\Gamma)} + (\nabla_\Gamma u, \nabla_\Gamma v)_{L^2(\Gamma)} + (\mathrm{div}_\Gamma(\mathbf{w}u), v)_{L^2(\Gamma)} = 0 \quad \forall\, v \in H^1_*(\Gamma).$$

Let $V_h^{\Gamma_h}$ be the surface finite element space from above and $u_{0,h} \in V_h^{\Gamma_h} \cap H^1_*(\Gamma_h)$ an approximation of the function u_0 in the initial condition. The semi-discretization is as follows: For $t \in [0,T]$ determine $u_h(t) = u_h(\cdot,t) \in V_h^{\Gamma_h}$ with $u_h(0) = u_{h,0}$ and

$$\Big(\frac{du_h}{dt}, v_h\Big)_{L^2(\Gamma_h)} + (\nabla_{\Gamma_h} u_h, \nabla_{\Gamma_h} v_h)_{L^2(\Gamma_h)} + (\mathrm{div}_{\Gamma_h}(\mathbf{w}u_h), v_h)_{L^2(\Gamma_h)} = 0$$

for all $v_h \in V_h^{\Gamma_h}$. This ordinary differential equation for $u_h(t)$ can be combined with standard time discretization schemes.

13.2.3 Numerical experiments

In this section we present results of a few numerical experiments. As a test problem we consider the Laplace-Beltrami equation

$$-\Delta_\Gamma u = f \quad \text{on } \Gamma,$$

with $\Gamma = \{x \in \mathbb{R}^3 \mid \|x\|_2 = 1\}$ and $\Omega = (-2,2)^3$. This example is taken from [79]. The source term f is taken such that the solution is given by

$$u(x) = \frac{12}{\|x\|^3}\left(3x_1^2 x_2 - x_2^3\right), \quad x = (x_1, x_2, x_3) \in \Omega.$$

Using the representation of u in spherical coordinates one can verify that u is an eigenfunction of $-\Delta_\Gamma$:

$$u(r,\varphi,\theta) = 12\sin(3\varphi)\sin^3\theta, \qquad -\Delta_\Gamma u = 12u =: f(r,\varphi,\theta). \tag{13.22}$$

The right-hand side f satisfies the compatibility condition $\int_\Gamma f\,ds = 0$, likewise does u. Note that u and f are constant along normals at Γ.

A family $\{\mathcal{T}_l\}_{l\geq 0}$ of tetrahedral triangulations of Ω is constructed as follows. We triangulate Ω by starting with a uniform subdivision into 48 tetrahedra with mesh size $h_0 = 2$. Then we apply an adaptive red-green refinement algorithm, cf. Sect. 3.1, in which in each refinement step the tetrahedra that contain Γ are refined such that on level $l = 1, 2, \ldots$, we have

$$h_T \leq 2^{-l} h_0 \quad \text{for all } T \in \mathcal{T}_l \quad \text{with} \quad T \cap \Gamma \neq \emptyset.$$

The family $\{\mathcal{T}_l\}_{l\geq 0}$ is consistent and shape-regular. The interface Γ is the zero level of $\phi(x) := \|x\|^2 - 1$. Let I be the standard nodal interpolation operator on $\mathcal{T}_l$. The discrete interface is given by $\Gamma_{h_l} := \{\, x \in \Omega : I(\phi)(x) = 0 \,\}$. Let $\{\psi_i\}_{1\leq i\leq m}$ be the nodal basis functions corresponding to the vertices of the tetrahedra in ω_h, cf. (13.16). The restrictions of these functions to Γ_h span the finite element space V_h^Γ. The entries $\int_{\Gamma_h} \nabla_{\Gamma_h}\psi_i \cdot \nabla_{\Gamma_h}\psi_j\,ds$ of the stiffness matrix are computed within machine accuracy. For the right-hand side of the Galerkin discretization (13.19) we need an extension f_h of f. In order to be consistent with the theoretical analysis in the next section we take the constant extension of f along the normals at Γ, i.e., we take $f_h(r,\varphi,\theta) = f(1,\varphi,\theta) + c_h$, with $f(r,\varphi,\theta)$ as in (13.22) and c_h such that $\int_{\Gamma_h} f_h\,ds = 0$. For the computation of the integrals $\int_F f_h\psi_h\,ds$ with $F \in \mathcal{F}_h$ we use a quadrature-rule that is exact up to order five. The computed solution u_h is normalized such that $\int_{\Gamma_h} u_h\,ds = 0$.

The discretization errors in the $L^2(\Gamma_h)$ norm are given in Table 13.1. The extension u^e of u is given by $u^e(r,\varphi,\theta) := u(1,\varphi,\theta)$, cf. (13.22).

level l	$\|u^e - u_h\|_{L^2(\Gamma_h)}$	factor
1	4.42 E-1	–
2	1.15 E-1	3.85
3	2.30 E-2	3.88
4	7.30 E-3	4.06
5	1.87 E-3	3.91
6	4.63 E-4	4.03
7	1.16 E-4	4.00

Table 13.1. Discretization errors and error reduction.

These results clearly indicate an h^2 error behavior, which will be confirmed by the theoretical analysis in Sect. 13.2.4. To illustrate the fact that in this approach the triangulation of the approximate manifold Γ_h is strongly shape-irregular we show a part of this triangulation in Fig. 13.4. The discrete solution is visualized in Fig. 13.5.

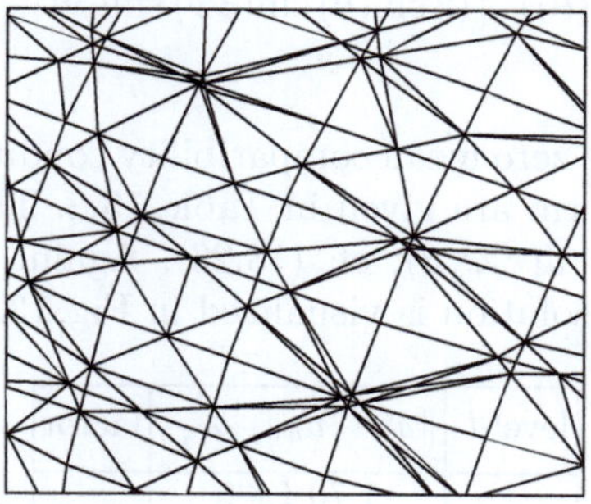

Fig. 13.4. Details of the induced triangulation of Γ_h.

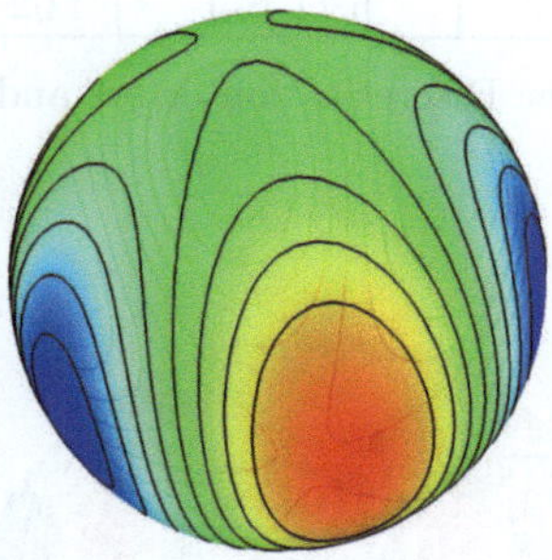

Fig. 13.5. Level lines of the discrete solution u_h.

To demonstrate the flexibility of the method with respect to the shape of Γ we repeat the previous experiment but now with a torus instead of the unit sphere. We take $\Gamma = \left\{ x \in \Omega : r^2 = x_3^2 + (\sqrt{x_1^2 + x_2^2} - R)^2 \right\}$, with $R = 1$, $r = 0.6$, and $\Omega = (-2, 2)^3$. In the coordinate system (ρ, φ, θ), with

$$x = R \begin{pmatrix} \cos\varphi \\ \sin\varphi \\ 0 \end{pmatrix} + \rho \begin{pmatrix} \cos\varphi\cos\theta \\ \sin\varphi\cos\theta \\ \sin\theta \end{pmatrix},$$

the ρ-direction is normal to Γ, $\frac{\partial x}{\partial \rho} \perp \Gamma$ for $x \in \Gamma$. Thus, the following solution u and corresponding right-hand side f are constant in the normal direction:

$$\begin{aligned} u(x) &= \sin(3\varphi)\cos(3\theta+\varphi), \\ f(x) &= r^{-2}(9\sin(3\varphi)\cos(3\theta+\varphi)) \\ &\quad - (R + r\cos(\theta)^{-2}(-10\sin(3\varphi)\cos(3\theta+\varphi) - 6\cos(3\varphi)\sin(3\theta+\varphi)) \\ &\quad - (r(R+r\cos(\theta))^{-1}(3\sin(\theta)\sin(3\varphi)\sin(3\theta+\varphi)). \end{aligned} \tag{13.23}$$

Both u and f satisfy the zero mean compatibility condition. The discretization errors in the $L^2(\Gamma_h)$-norm are given in Table 13.2. The extension u^e of u is given by $u^e(\rho,\varphi,\theta) := u(r,\varphi,\theta)$, cf. (13.23). Again, we observe a h^2 error behavior. The discrete solution is visualized in Fig. 13.6.

level l	$\|u^e - u_h\|_{L^2(\Gamma_h)}$	factor
1	1.70 E+0	–
2	5.30 E-1	3.21
3	1.40 E-1	3.77
4	3.63 E-2	3.86
5	9.32 E-3	3.90
6	2.30 E-3	4.05
7	5.71 E-4	4.02

Table 13.2. Torus: Discretization errors and error reduction.

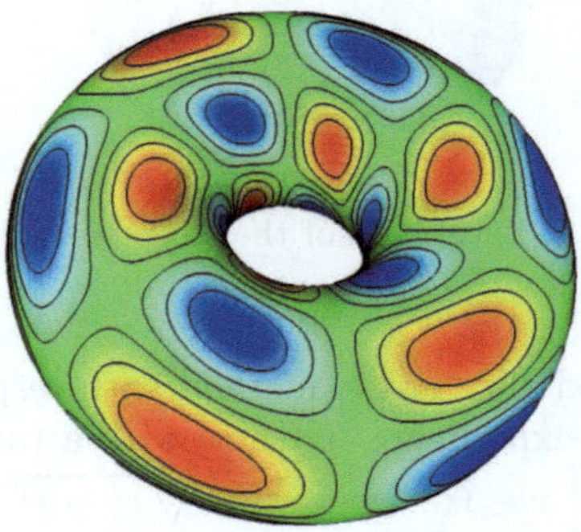

Fig. 13.6. Torus: Level lines of the discrete solution u_h.

Numerical results on conditioning

For the example with the unit sphere we computed the spectrum of the *scaled* mass and stiffness matrices. The mass matrix $\mathbf{M}$ and stiffness matrix $\mathbf{S}$ have entries

$$\mathbf{M}_{ij} = \int_{\Gamma_h} \psi_i \psi_j \, ds, \quad \mathbf{S}_{ij} = \int_{\Gamma_h} \nabla_{\Gamma_h} \psi_i \cdot \nabla_{\Gamma_h} \psi_j \, ds, \quad 1 \leq i, j \leq m.$$

Define $\mathbf{D}_M := \mathrm{diag}(\mathbf{M})$, $\mathbf{D}_S := \mathrm{diag}(\mathbf{S})$ and the scaled matrices

$$\tilde{\mathbf{M}} := \mathbf{D}_M^{-\frac{1}{2}} \mathbf{M} \mathbf{D}_M^{-\frac{1}{2}}, \; \tilde{\mathbf{S}} := \mathbf{D}_S^{-\frac{1}{2}} \mathbf{S} \mathbf{D}_S^{-\frac{1}{2}}.$$

For different refinement levels we computed the largest and smallest eigenvalues of $\tilde{\mathbf{M}}$ and $\tilde{\mathbf{S}}$. We use an ordering $\lambda_1 \leq \lambda_2 \leq \ldots \leq \lambda_m$. The results are given in Table 13.3 and Table 13.4.

level l	m	factor	λ_1	λ_2	λ_m	λ_m/λ_2	factor
1	112	-	3.8 E-17	2.61 E-2	2.86	109	-
2	472	4.2	4.0 E-17	5.80 E-3	2.83	488	4.5
3	1922	4.1	0	1.20 E-3	2.83	2358	4.8
4	7646	4.0	0	2.90 E-4	2.83	9759	4.1

Table 13.3. Eigenvalues of scaled mass matrix $\tilde{\mathbf{M}}$.

level l	m	factor	λ_1	λ_2	λ_3	λ_m	λ_m/λ_3	factor
1	112	-	0	0	5.50 E-2	2.17	39.5	-
2	472	4.2	0	0	1.30 E-2	2.26	174	4.4
3	1922	4.1	0	0	2.80 E-3	2.47	882	5.0
4	7646	4.0	0	0	6.90 E-4	2.61	3783	4.3

Table 13.4. Eigenvalues of scaled stiffness matrix $\tilde{\mathbf{S}}$.

These results show that for the scaled mass matrix there is one eigenvalue very close to or equal to zero and for the effective condition number we have $\frac{\lambda_m}{\lambda_2} \sim m \sim h_l^{-2}$. For the scaled stiffness matrix we observe that there are two eigenvalues close to or equal to zero and an effective condition number $\frac{\lambda_m}{\lambda_3} \sim m \sim h_l^{-2}$. In Fig. 13.7 for both matrices the eigenvalues λ_k are shown for $k \geq k_{\min}$, with $k_{\min} = 2$ for the scaled mass matrix and $k_{\min} = 3$ for the scaled stiffness matrix. These results indicate a relation of the form

$$\log(\lambda_k) \approx \log(\lambda_m) + \log\left(\frac{k}{m}\right), \quad k = k_{\min}, \ldots, m,$$

and $\log(\lambda_m) = \mathcal{O}(1)$. Hence, in particular

$$\lambda_{k_{\min}} \approx c \, \frac{k_{\min}}{m}, \quad \lambda_m = \mathcal{O}(1),$$

which is consistent with the results in the Tables 13.3 and 13.4.

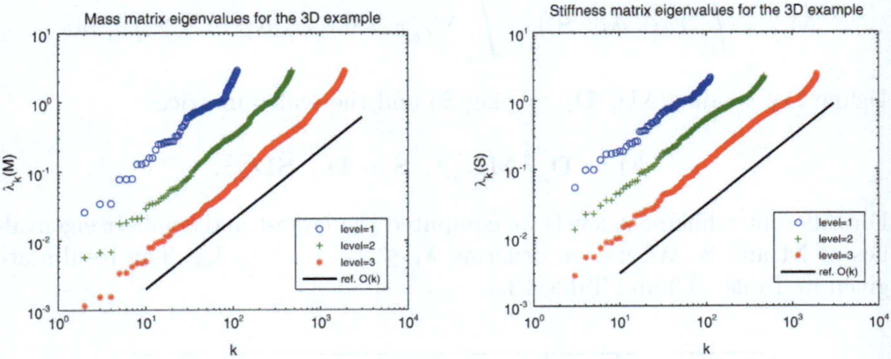

Fig. 13.7. Eigenvalue distributions for scaled mass matrix $\tilde{\mathbf{M}}$ (left) and for scaled stiffness matrix $\tilde{\mathbf{S}}$ (right) for the 3D example.

13.2.4 Discretization error analysis

In this section for the discrete problem (13.19) we derive discretization error bounds, both in the H^1- and the L^2-norm on Γ_h. These results are from [195]. We first collect some preliminaries, then derive approximation error bounds and finally present discretization error bounds.

Preliminaries

We will need a Poincaré type inequality that is given in the following lemma.

Lemma 13.2.9 *Consider a bounded domain $\Omega \subset \mathbb{R}^n$ and a subdomain $S \subset \Omega$. Assume that Ω is such that the following Poincaré inequality is valid:*

$$\|f\|_{L^2(\Omega)} \leq C_\Omega \|\nabla f\|_{L^2(\Omega)} \quad \text{for all } f \in H^1(\Omega) \text{ with } \int_\Omega f\,dx = 0. \tag{13.24}$$

Then for any $f \in H^1(\Omega)$ the following estimate holds:

$$\|f\|^2_{L^2(\Omega)} \leq \frac{|\Omega|}{|S|}\left(2\|f\|^2_{L^2(S)} + 3C^2_\Omega\|\nabla f\|^2_{L^2(\Omega)}\right). \tag{13.25}$$

Proof. We introduce the projectors $\Pi_k : H^1(\Omega) \to \mathbb{R}$, $k = 1, 2$,

$$\Pi_1 f := |\Omega|^{-1}\int_\Omega f\,dx, \qquad \Pi_2 f := |S|^{-1}\int_S f\,dx.$$

Since $\|(I - \Pi_1)f\|^2_{L^2(\Omega)} = \|f\|^2_{L^2(\Omega)} - |\Omega||\Pi_1 f|^2$, the Poincaré inequality (13.24) can be rewritten in the equivalent form

$$\|f\|^2_{L^2(\Omega)} \leq |\Omega||\Pi_1 f|^2 + C^2_\Omega\|\nabla f\|^2_{L^2(\Omega)} \quad \text{for all } f \in H^1(\Omega). \tag{13.26}$$

For any $f \in H^1(\Omega)$, with $\Pi_1 f = 0$, using the Cauchy-Schwarz and Poincaré inequalities we get

$$\begin{aligned} |\Pi_2 f| &= |S|^{-1} \left| \int_S f \, dx \right| \leq |S|^{-\frac{1}{2}} \|f\|_{L^2(S)} \\ &\leq |S|^{-\frac{1}{2}} \|f\|_{L^2(\Omega)} \leq C_\Omega |S|^{-\frac{1}{2}} \|\nabla f\|_{L^2(\Omega)}. \end{aligned} \tag{13.27}$$

Define $M := C_\Omega |S|^{-\frac{1}{2}}$. Note that for $f \in H^1(\Omega)$ we have $\Pi_1(I - \Pi_1)f = 0$ and thus from (13.27) we obtain

$$|(\Pi_2 - \Pi_1)f| = |\Pi_2(I - \Pi_1)f| \leq M \|\nabla(I - \Pi_1)f\|_{L^2(\Omega)} = M \|\nabla f\|_{L^2(\Omega)}.$$

Hence, for any $f \in H^1(\Omega)$ we have

$$\begin{aligned} |\Pi_1 f|^2 &\leq 2|\Pi_2 f|^2 + 2|(\Pi_2 - \Pi_1)f|^2 \\ &\leq 2|\Pi_2 f|^2 + 2M^2 \|\nabla f\|^2_{L^2(\Omega)} \\ &\leq 2|S|^{-1} \|f\|^2_{L^2(S)} + 2M^2 \|\nabla f\|^2_{L^2(\Omega)}. \end{aligned} \tag{13.28}$$

Estimates (13.26) and (13.28) imply:

$$\begin{aligned} \|f\|^2_{L^2(\Omega)} &\leq \max\{|\Omega|, C_\Omega^2 M^{-2}\} \left(|\Pi_1 f|^2 + M^2 \|\nabla f\|^2_{L^2(\Omega)} \right) \\ &= |\Omega| \left(|\Pi_1 f|^2 + M^2 \|\nabla f\|^2_{L^2(\Omega)} \right) \\ &\leq |\Omega| \left(2|S|^{-1} \|f\|^2_{L^2(S)} + 3M^2 \|\nabla f\|^2_{L^2(\Omega)} \right) \\ &= |\Omega||S|^{-1} \left(2\|f\|^2_{L^2(S)} + 3C_\Omega^2 \|\nabla f\|^2_{L^2(\Omega)} \right), \end{aligned}$$

which proves the inequality in (13.25). □

Remark 13.2.10 In the analysis below we shall apply Lemma 13.2.9 for the case of a *convex* domain Ω. For convex domains the following upper bound is well-known [201] for the Poincaré constant:

$$C_\Omega \leq \frac{\mathrm{diam}(\Omega)}{\pi}. \tag{13.29}$$

We define a neighborhood of Γ:

$$U = \left\{ x \in \mathbb{R}^3 : \mathrm{dist}(x, \Gamma) < c \right\},$$

with c sufficiently small and assume that $\Gamma_h \subset U$. Let $d : U \to \mathbb{R}$ be the signed distance function. Its Hessian is denoted by $\mathbf{H}(x) = D^2 d(x)$, and has eigenvalues denoted by $\kappa_1(x), \kappa_2(x)$, and 0. On U we use normals, projections and extensions that are the same as the ones used previously, for example in the Sects. 7.7 and 13.1:

$$\begin{aligned}
\mathbf{n}(x) &= \nabla d(x),\\
\mathbf{P}(x) &= \mathbf{I} - \mathbf{n}(x)\mathbf{n}(x)^T,\\
\mathbf{p}(x) &= x - d(x)\mathbf{n}(x),\\
v^e(x) &= v(\mathbf{p}(x)) \quad (\text{for } v \text{ defined on } \Gamma).
\end{aligned}$$

We assume that the decomposition $x = \mathbf{p}(x) + d(x)\mathbf{n}(x)$ is unique for all $x \in U$. Note that $\mathbf{n}(x) = \mathbf{n}(\mathbf{p}(x))$ for all $x \in U$. We define a discrete analogon of the orthogonal projection $\mathbf{P}$:

$$\mathbf{P}_h(x) := \mathbf{I} - \mathbf{n}_h(x)\mathbf{n}_h(x)^T \quad \text{for } \; x \in \Gamma_h, \; x \text{ not on an edge}.$$

Here $\mathbf{n}_h(x)$ denotes the (outward pointing) normal at $x \in \Gamma_h$. The tangential derivative along Γ_h can be written as $\nabla_{\Gamma_h} g(x) = \mathbf{P}_h(x)\nabla g(x)$ for $x \in \Gamma_h$ (not on an edge).

In the analysis we use techniques from [83, 92], for example, the formula

$$\nabla v^e(x) = (\mathbf{I} - d(x)\mathbf{H}(x))\nabla_\Gamma v(\mathbf{p}(x)) \quad \text{a.e. on } \; U \tag{13.30}$$

(cf. Sect. 2.3 in [83]), which implies

$$\nabla_{\Gamma_h} v^e(x) = \mathbf{P}_h(x)\big(\mathbf{I} - d(x)\mathbf{H}(x)\big)\nabla_\Gamma v(\mathbf{p}(x)) \quad \text{a.e. on } \; \Gamma_h. \tag{13.31}$$

Furthermore, for v sufficiently smooth and $|\nu| = 2$, the inequality

$$|D^\nu v^e(x)| \leq c\Big(\sum_{|\nu|=2} |\mathrm{D}_\Gamma^\nu v(\mathbf{p}(x))| + \|\nabla_\Gamma v(\mathbf{p}(x))\|\Big) \quad \text{a.e. on } \; U \tag{13.32}$$

holds, cf. Lemma 3 in [92]. We define an h-neighborhood of Γ:

$$U_h = \left\{ x \in \mathbb{R}^3 : \operatorname{dist}(x, \Gamma) < c_1 h \right\},$$

and assume that h is sufficiently small, such that $\omega_h \subset U_h \subset U$ and

$$5c_1 h < \left(\max_{i=1,2} \|\kappa_i\|_{L^\infty(\Gamma)}\right)^{-1}. \tag{13.33}$$

From (2.5) in [83] we have the following formula for the principal curvatures κ_i:

$$\kappa_i(x) = \frac{\kappa_i(\mathbf{p}(x))}{1 + d(x)\kappa_i(\mathbf{p}(x))} \quad \text{for } x \in U. \tag{13.34}$$

Hence, from (13.33) and (13.34) it follows that

$$\|d\|_{L^\infty(U_h)} \max_{i=1,2} \|\kappa_i\|_{L^\infty(U_h)} \leq \frac{1}{4} \tag{13.35}$$

holds. In the remainder we assume that

$$\operatorname{ess\,sup}_{x\in\Gamma_h} |d(x)| \leq c_0 h^2, \tag{13.36}$$

$$\operatorname{ess\,sup}_{x\in\Gamma_h} \|\mathbf{n}(x) - \mathbf{n}_h(x)\| \leq \tilde{c}_0 h \tag{13.37}$$

holds. These are reasonable assumptions, cf. Theorem 7.3.1.

Lemma 13.2.11 *There are constants $c_1 > 0$ and c_2 independent of h such that for all $u \in H^2(\Gamma)$ the following inequalities hold:*

$$c_1 \|u^e\|_{L^2(U_h)} \leq \sqrt{h}\|u\|_{L^2(\Gamma)} \leq c_2 \|u^e\|_{L^2(U_h)}, \tag{13.38a}$$

$$c_1 \|\nabla u^e\|_{L^2(U_h)} \leq \sqrt{h}\|\nabla_\Gamma u\|_{L^2(\Gamma)} \leq c_2 \|\nabla u^e\|_{L^2(U_h)}, \tag{13.38b}$$

$$\|D^\nu u^e\|_{L^2(U_h)} \leq c_2 \sqrt{h}\|u\|_{H^2(\Gamma)}, \quad |\nu| = 2. \tag{13.38c}$$

Proof. Define

$$\mu(x) := \big(1 - d(x)\kappa_1(x)\big)\big(1 - d(x)\kappa_2(x)\big), \quad x \in U_h.$$

From (2.20), (2.23) in [83] we have

$$\mu(x) dx = dr\, ds(\mathbf{p}(x)), \quad x \in U,$$

where dx is the volume measure in U_h, ds the surface measure on Γ, and r the local coordinate at $x \in \Gamma$ in the direction $\mathbf{n}(\mathbf{p}(x)) = \mathbf{n}(x)$. Using (13.35) we get

$$\frac{9}{16} \leq \mu(x) \leq \frac{25}{16} \quad \text{for all } x \in U_h. \tag{13.39}$$

Using the local coordinate representation $x = (\mathbf{p}(x), r)$, for $x \in U$, we have

$$\begin{aligned}\int_{U_h} u^e(x)^2 \mu(x)\, dx &= \int_{-c_1 h}^{c_1 h} \int_\Gamma \left[u^e(\mathbf{p}(x), r)\right]^2\, ds(\mathbf{p}(x)) dr \\ &= \int_{-c_1 h}^{c_1 h} \int_\Gamma \left[u(\mathbf{p}(x), 0)\right]^2\, ds(\mathbf{p}(x))\, dr = 2c_1 h \|u\|^2_{L^2(\Gamma)}.\end{aligned}$$

Combining this with (13.39) yields the result in (13.38a).

From (13.30) we have that $u^e \in H^1(U_h)$. Note that

$$\int_{U_h} \left[\nabla u^e(x)\right]^2 \mu(x)\, dx = \int_{-c_1 h}^{c_1 h} \int_\Gamma \left[(\mathbf{I} - d(x)\mathbf{H}(x))\nabla_\Gamma u(\mathbf{p}(x), 0)\right]^2 ds(\mathbf{p}(x))\, dr.$$

Using this in combination with $\|d(x)\mathbf{H}(x)\| \leq \frac{1}{4}$ for all $x \in U_h$ (cf. (13.35)) and the bounds in (13.39) we obtain the result in (13.38b). Finally, using similar arguments and the bound in (13.32) one can derive the bound in (13.38c). □

Approximation error bounds

Let $I_h : C(\overline{\omega_h}) \to V_h$ be the nodal interpolation operator. We use the approximation property of the linear finite element space V_h: For $v \in H^2(\omega_h)$

$$\|v - I_h v\|_{H^k(\omega_h)} \leq C\, h^{2-k} \|v\|_{H^2(\omega_h)}, \quad k = 0, 1. \tag{13.40}$$

A consequence of this approximation result is given in the following lemma.

Lemma 13.2.12 *For $u \in H^2(\Gamma)$ and $k = 0,1$ we have*

$$\|u^e - I_h u^e\|_{H^k(\omega_h)} \le C\, h^{\frac{5}{2}-k}\|u\|_{H^2(\Gamma)}. \tag{13.41}$$

Proof. From (13.40) and (13.38b) we obtain

$$\begin{aligned}\|u^e - I_h u^e\|_{H^k(\omega_h)} &\le C\, h^{2-k}\|u^e\|_{H^2(\omega_h)} \le C\, h^{2-k}\|u^e\|_{H^2(U_h)}\\ &\le C\, h^{\frac{5}{2}-k}\|u\|_{H^2(\Gamma)},\end{aligned}$$

which proves the result. □

The following two lemmas play a crucial role in the analysis. In both lemmas we use a "pull back" strategy based on Lemma 13.2.9. For this we introduce a special local coordinate system as follows. For a subdomain $\omega \subset \mathbb{R}^3$ let $\rho(\omega)$ be the diameter of the largest ball that is contained in ω. Take an arbitrary planar segment F of Γ_h, i.e., $F \in \mathcal{F}_h$. Let $S_F \in \mathcal{T}_h$ be the tetrahedron such that $\Gamma_h \cap S_F = F$. There exists a planar extension F^e of F such that $F^e \subset U$, F^e is convex, $\mathbf{p}(S_F) \subset \mathbf{p}(F^e)$, and

$$\mathrm{diam}(F^e) \simeq \rho(F^e) \simeq h. \tag{13.42}$$

The existence of such a planar extension is discussed in Remark 8 in [195]. This extension F^e is used to define a coordinate system in the neighborhood $N_F := \{\, x \in U : \mathbf{p}(x) \in \mathbf{p}(F^e)\,\}$. Note that $S_F \subset N_F$. Every $x \in N_F$ has a unique decomposition of the form

$$x = s + \tilde{d}(x)\mathbf{n}(x), \quad \text{with } \; s \in F^e, \quad \tilde{d}(x) := \pm\|s - x\|. \tag{13.43}$$

The sign of $\tilde{d}(x)$ is determined by taking into account on which side of the plane F^e the point x lies. Note that $\tilde{d}$ is a signed distance, along the normal $\mathbf{n}(x)$, to the planar segment F^e. The representation in this coordinate system is denoted by Φ, i.e., $\Phi(x) = (s(x), \tilde{d}(x))$. This coordinate system is illustrated, for the 2D case, in Fig. 13.8.

For $x \in F^e$ we thus have $\Phi(x) = (s(x), 0)$. Due to the shape-regularity of $\mathcal{T}_h$ there exists, in the Φ-coordinate system, a cylinder B_F that has the following properties:

$$B_F = F_b^e \times [d_0, d_1] \subset S_F, \quad F_b^e \subset F^e, \quad |F_b^e| \simeq h^2, \quad d_1 - d_0 \simeq h. \tag{13.44}$$

This coordinate system and the cylinder $B_F \subset S_F$ are used in the analysis below.

Lemma 13.2.13 *Let v_h be a linear function on N_F and $u \in H^2(\Gamma)$. There exists a constant c independent of v_h, u, and F such that the following inequality holds:*

$$\|\nabla_{\Gamma_h}(u^e - v_h)\|_{L^2(F^e)} \le c h^{-\frac{1}{2}}\|\nabla(u^e - v_h)\|_{L^2(S_F)} + h\|u\|_{H^2(\mathbf{p}(F^e))}. \tag{13.45}$$

Here ∇_{Γ_h} denotes the projection of the gradient on F^e.

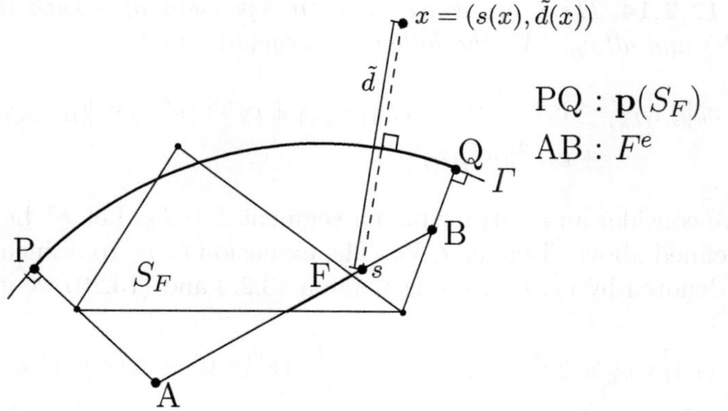

Fig. 13.8. 2D illustration of coordinate system.

Proof. Using Lemma 13.2.9, (13.29), and (13.32) we obtain

$$\begin{aligned}\|\nabla_{\Gamma_h}(u^e - v_h)\|^2_{L^2(F^e)} &\le c\|\nabla_{\Gamma_h}(u^e - v_h)\|^2_{L^2(F^e_b)} + ch^2\|\nabla^2_{\Gamma_h} u^e\|^2_{L^2(F^e)} \\ &\le c\|\nabla_{\Gamma_h}(u^e - v_h)\|^2_{L^2(F^e_b)} + ch^2\|u\|^2_{H^2(\mathbf{p}(F^e))}. \quad (13.46)\end{aligned}$$

We consider the first term in (13.46). We write $\nabla v_h =: c_F$ and use the notation $x = (s(x), \tilde{d}(x)) =: (s, y)$ in the Φ-coordinate system. From (13.30) we have

$$\nabla_\Gamma u(\mathbf{p}(x)) = \nabla u^e(s, y) + d(x)\mathbf{H}(x)\nabla_\Gamma u(\mathbf{p}(x)).$$

Using this and (13.31) we obtain

$$\begin{aligned}&\|\nabla_{\Gamma_h}(u^e - v_h)\|^2_{L^2(F^e_b)} \\ &= \|\nabla_{\Gamma_h} u^e - \mathbf{P}_h c_F\|^2_{L^2(F^e_b)} \\ &\le 2\|\mathbf{P}_h(\nabla_\Gamma u) \circ \mathbf{p} - \mathbf{P}_h c_F\|^2_{L^2(F^e_b)} + 2\|\mathbf{P}_h d\mathbf{H}(\nabla_\Gamma u) \circ \mathbf{p}\|^2_{L^2(F^e_b)} \\ &\le c\|(\nabla_\Gamma u) \circ \mathbf{p} - c_F\|^2_{L^2(F^e_b)} + ch^2\|u\|^2_{H^1(\mathbf{p}(S_F))} \\ &= c\int_{F^e_b} \|\nabla_\Gamma u(\mathbf{p}(s,0)) - c_F\|^2\, ds + ch^2\|u\|^2_{H^1(\mathbf{p}(S_F))} \\ &\le ch^{-1}\int_{d_0}^{d_1}\int_{F^e_b} \|\nabla_\Gamma u(\mathbf{p}(s,0)) - c_F\|^2\, ds\, dy + ch^2\|u\|^2_{H^1(\mathbf{p}(S_F))} \\ &\le ch^{-1}\int_{d_0}^{d_1}\int_{F^e_b} \|\nabla u^e(\mathbf{p}(s,y)) - c_F\|^2\, ds\, dy + ch^2\|u\|^2_{H^1(\mathbf{p}(S_F))} \\ &\le ch^{-1}\|\nabla(u^e - v_h)\|^2_{L^2(B_F)} + ch^2\|u\|^2_{H^1(\mathbf{p}(S_F))} \\ &\le ch^{-1}\|\nabla(u^e - v_h)\|^2_{L^2(S_F)} + ch^2\|u\|^2_{H^1(\mathbf{p}(S_F))}.\end{aligned}$$

Combining this with the result in (13.46) completes the proof. □

Lemma 13.2.14 *There are constants c_i independent of h such that for all $u \in H^2(\Gamma)$ and all $v_h \in V_h$ the following inequality holds:*

$$\begin{aligned} \|u^e - v_h\|_{L^2(\Gamma_h)} \leq\, & c_1 h^{-\frac{1}{2}} \|u^e - v_h\|_{L^2(\omega_h)} + c_2 h^{\frac{1}{2}} \|u^e - v_h\|_{H^1(\omega_h)} \\ & + c_3 h^2 \|u\|_{H^2(\Gamma)}. \end{aligned} \tag{13.47}$$

Proof. We consider an arbitrary planar segment $F \in \Gamma_h$. Let F^e be its extension as defined above. Take $v_h \in V_h$. The extension of v_h to a linear function on F^e is denoted by v_h, too. Using Lemma 13.2.9 and (13.29) we get:

$$\begin{aligned} \|u^e - v_h\|^2_{L^2(F)} \leq \|u^e - v_h\|^2_{L^2(F^e)} &= \int_{F^e} (u^e(s,0) - v_h(s,0))^2 \, ds \\ &\leq c \int_{F^e_b} (u^e(s,0) - v_h(s,0))^2 \, ds \\ &+ ch^2 \int_{F^e} \|\nabla_{\Gamma_h} (u^e(s,0) - v_h(s,0))\|^2 \, ds. \end{aligned} \tag{13.48}$$

We consider the first term on the right-hand side of (13.48). For a linear function g and $0 \leq \delta_0 < \delta_1$ we have $g(\delta_i)^2 \leq \frac{6}{\delta_1 - \delta_0} \int_{\delta_0}^{\delta_1} g(t)^2 \, dt$ for $i = 0, 1$, and $g(0) = g(\delta_0)\frac{\delta_1}{\delta_1 - \delta_0} - g(\delta_1)\frac{\delta_0}{\delta_1 - \delta_0}$. Hence, $|g(0)| \leq \frac{2\delta_1}{\delta_1 - \delta_0} \max_{i=0,1} |g(\delta_i)|$ and thus

$$g(0)^2 \leq 24 \left(\frac{\delta_1}{\delta_1 - \delta_0} \right)^2 \frac{1}{\delta_1 - \delta_0} \int_{\delta_0}^{\delta_1} g(t)^2 \, dt \tag{13.49}$$

holds. Without loss of generality we can assume that d_0, d_1 from (13.44) satisfy $0 \leq d_0 < d_1$. Furthermore, we have $\frac{d_i}{d_1 - d_0} \leq c$ for $i = 1, 2$, with c independent of h. Using this and (13.49) applied to the linear function $y \to c + v_h(s,y)$ we obtain

$$\begin{aligned} & \int_{F^e_b} (u^e(s,0) - v_h(s,0))^2 \, ds \leq ch^{-1} \int_{F^e_b} \int_{d_0}^{d_1} (u^e(s,0) - v_h(s,y))^2 \, dy \, ds \\ &= ch^{-1} \int_{F^e_b} \int_{d_0}^{d_1} (u^e(s,y) - v_h(s,y))^2 \, dy \, ds = ch^{-1} \|u^e - v_h\|^2_{L^2(B_F)} \\ &\leq ch^{-1} \|u^e - v_h\|^2_{L^2(S_F)}. \end{aligned}$$

For the second term on the right-hand side of (13.48) we apply Lemma 13.2.13, and thus we get

$$\|u^e - v_h\|^2_{L^2(F)} \leq ch^{-1} \|u^e - v_h\|^2_{L^2(S_F)} + ch \|\nabla(u^e - v_h)\|^2_{L^2(S_F)} + ch^4 \|u\|^2_{H^2(\mathbf{p}(F^e))}.$$

Summation over all $F \in \mathcal{F}_h$ gives (13.47). □

Lemma 13.2.15 *There are constants c_1, c_2 independent of h such that for all $u \in H^2(\Gamma)$ and all $v_h \in V_h$ the following inequality holds:*

$$\|u^e - v_h\|_{H^1(\Gamma_h)} \le c_1 h^{-\frac{1}{2}} \|u^e - v_h\|_{H^1(\omega_h)} + c_2 h \|u\|_{H^2(\Gamma)}. \tag{13.50}$$

Proof. Take $u \in H^2(\Gamma)$ and $v_h \in V_h$. By definition of the H^1-norm on Γ_h we get

$$\|u^e - v_h\|^2_{H^1(\Gamma_h)} = \|u^e - v_h\|^2_{L^2(\Gamma_h)} + \|\nabla_{\Gamma_h}(u^e - v_h)\|^2_{L^2(\Gamma_h)}.$$

For the first term on the right-hand side we can apply Lemma 13.2.14, and use

$$h^{-\frac{1}{2}} \|u^e - v_h\|_{L^2(\omega_h)} + c_2 h^{\frac{1}{2}} \|u^e - v_h\|_{H^1(\omega_h)} \le c h^{-\frac{1}{2}} \|u^e - v_h\|_{H^1(\omega_h)}.$$

We now consider the second term

$$\|\nabla_{\Gamma_h}(u^e - v_h)\|^2_{L^2(\Gamma_h)} = \sum_{F \in \mathcal{F}_h} \|\nabla_{\Gamma_h}(u^e - v_h)\|^2_{L^2(F)}.$$

Take a $F \in \mathcal{F}_h$ and extend v_h linearly outside F. This extension is denoted by v_h, too. Using Lemma 13.2.13 we get

$$\begin{aligned}\|\nabla_{\Gamma_h}(u^e - v_h)\|^2_{L^2(F)} &\le \|\nabla_{\Gamma_h}(u^e - v_h)\|^2_{L^2(F^e)} \\ &\le c h^{-1} \|\nabla(u^e - v_h)\|^2_{L^2(S_F)} + h^2 \|u\|^2_{H^2(\mathbf{p}(F^e))}.\end{aligned}$$

Summation over $F \in \mathcal{F}_h$ yields

$$\|\nabla_{\Gamma_h}(u^e - v_h)\|^2_{L^2(\Gamma_h)} \le c h^{-1} \|u^e - v_h\|^2_{H^1(\omega_h)} + c h^2 \|u\|^2_{H^2(\Gamma)},$$

and thus the proof is completed. □

As a direct consequence of the previous two lemmas we obtain the following main theorem.

Theorem 13.2.16 *For each $u \in H^2(\Gamma)$ the following hold:*

$$\inf_{v_h \in V_h^\Gamma} \|u^e - v_h\|_{L^2(\Gamma_h)} \le \|u^e - (I_h u^e)_{|\Gamma_h}\|_{L^2(\Gamma_h)} \le C h^2 \|u\|_{H^2(\Gamma)}, \tag{13.51}$$

$$\inf_{v_h \in V_h^\Gamma} \|u^e - v_h\|_{H^1(\Gamma_h)} \le \|u^e - (I_h u^e)_{|\Gamma_h}\|_{H^1(\Gamma_h)} \le C h \|u\|_{H^2(\Gamma)}, \tag{13.52}$$

with a constant C independent of u and h.

Proof. Combine the results in the Lemmas 13.2.14 and 13.2.15 with the result in Lemma 13.2.12. □

Finite element discretization error bounds

Using the approximation error bounds derived in Theorem 13.2.16 fairly standard arguments lead to optimal discretization error bounds. Below we present two main results (from [195]).

For $x \in \Gamma_h$ define

$$\mu_h(x) = (1 - d(x)\kappa_1(x))(1 - d(x)\kappa_2(x))\mathbf{n}(x)^T\mathbf{n}_h(x).$$

The integral transformation formula

$$\mu_h(x)ds_h(x) = ds(\mathbf{p}(x)), \quad x \in \Gamma_h \tag{13.53}$$

holds, where $ds_h(x)$ and $ds(\mathbf{p}(x))$ are the surface measures on Γ_h and Γ, respectively, cf. (2.20) in [83]. From

$$\|\mathbf{n}(x) - \mathbf{n}_h(x)\|^2 = 2\big(1 - \mathbf{n}(x)^T\mathbf{n}_h(x)\big),$$

the assumption in (13.37) and $|d(x)| \leq ch^2$, $|\kappa_i(x)| \leq c$ we obtain

$$\text{ess sup}_{x\in\Gamma_h} |1 - \mu_h(x)| \leq ch^2, \tag{13.54}$$

with a constant c independent of h.

Theorem 13.2.17 *Let $u \in H^2(\Gamma)$ be the solution of (12.3), $u_h \in V_h^\Gamma$ the solution of (13.19), with $f_h = f^e - c_f$, where c_f is such that $\int_{\Gamma_h} f_h \, ds = 0$. The following discretization error bound holds:*

$$\|\nabla_{\Gamma_h}(u^e - u_h)\|_{L^2(\Gamma_h)} \leq c\, h\, \|f\|_{L^2(\Gamma)}, \tag{13.55}$$

with a constant c independent of f and h.

Proof. We only present the main idea of the proof given in [195]. It uses the relation

$$\int_\Gamma \nabla_\Gamma u \cdot \nabla_\Gamma v \, ds = \int_{\Gamma_h} \mathbf{A}_h \nabla_{\Gamma_h} u^e \cdot \nabla_{\Gamma_h} v^e \, ds_h \quad \text{for all} \;\; v \in H^1(\Gamma),$$

with a matrix $\mathbf{A}_h$ that is close to the identity, in the sense that

$$\|\mathbf{P}_h(\mathbf{I} - \mathbf{A}_h)\| \leq ch^2 \tag{13.56}$$

can be shown to hold. Define

$$c_f := |\Gamma_h|^{-1} \int_{\Gamma_h} f^e \, ds_h, \quad \delta_f := (1 - \mu_h)f^e - c_f.$$

The following perturbed Galerkin orthogonality property holds: for arbitrary $\psi_h \in V_h^\Gamma$ we have

$$\begin{aligned}\int_{\Gamma_h} \nabla_{\Gamma_h}(u^e - u_h)\cdot\nabla_{\Gamma_h}\psi_h\,ds_h \\ = \int_{\Gamma_h} (\mathbf{I}-\mathbf{A}_h)\nabla_{\Gamma_h}u^e\cdot\nabla_{\Gamma_h}\psi_h\,ds_h - \int_{\Gamma_h}\delta_f\psi_h\,ds_h \\ = \int_{\Gamma_h}\mathbf{P}_h(\mathbf{I}-\mathbf{A}_h)\nabla_{\Gamma_h}u^e\cdot\nabla_{\Gamma_h}\psi_h\,ds_h - \int_{\Gamma_h}\delta_f\psi_h\,ds_h.\end{aligned} \tag{13.57}$$

Using

$$\|\delta_f\|_{L^2(\Gamma_h)} \le ch^2\|f\|_{L^2(\Gamma)}$$

and (13.56) the terms on the right-hand side can be shown to be "small". The proof uses a standard error analysis, based on the approximation error bounds derived in Theorem 13.2.16 and in which instead of the usual Galerkin orthogonality property of the finite element solution the relation (13.57) is used. □

Remark 13.2.18 We indicate how the error bound (13.55) in $H^1(\Gamma_h)$ yields a similar bound in $H^1(\Gamma)$. For this we need the extension of functions defined on Γ_h along the normals $\mathbf{n}$ on Γ: For $v \in C(\Gamma_h)$ we define, for $x \in \Gamma_h$,

$$v^{e,h}(x+\alpha\mathbf{n}(x)) := v(x) \quad \text{for all } \alpha\in\mathbb{R}, \text{ with } x+\alpha\mathbf{n}(x)\in U. \tag{13.58}$$

The following holds (cf. [83], Lemma 3.3 in [129]):

$$\|\nabla_\Gamma v^{e,h}\|_{L^2(\Gamma)} \le c\|\nabla_{\Gamma_h} v\|_{L^2(\Gamma_h)} \quad \text{for all } v\in H^1(\Gamma_h)\cap C(\Gamma_h).$$

Using this for the error $v = u^e - u_h$ and noting that $(u^e)^{e,h} = u$ on Γ the bound (13.55) yields

$$\|\nabla_\Gamma(u - u_h^{e,h})\|_{L^2(\Gamma)} \le ch\|f\|_{L^2(\Gamma)},$$

i.e., an optimal error bound in $H^1(\Gamma)$.

We now present an L^2-norm discretization error bound.

Theorem 13.2.19 *Let u and u_h be as in Theorem 13.2.17. The error bound*

$$\|u^e - u_h\|_{L^2(\Gamma_h)} \le c\,h^2\,\|f\|_{L^2(\Gamma)} \tag{13.59}$$

holds, with a constant c independent of f and h.

Proof. We only sketch the main points. A detailed proof is given in [195]. A perturbed duality argument is used. Denote $e_h := (u^e - u_h)|_{\Gamma_h}$ and let e_h^l be the lift of e_h on Γ, as in (13.58), and $c_e := |\Gamma|^{-1}\int_\Gamma e_h^l\,ds$. Consider the problem: Find $w\in H^1(\Gamma)$, with $\int_\Gamma w\,ds = 0$, such that

$$\int_\Gamma \nabla_\Gamma w\cdot\nabla_\Gamma v\,ds = \int_\Gamma (e_h^l - c_e)v\,ds \qquad \text{for all } v\in H^1(\Gamma). \tag{13.60}$$

The solution w satisfies $w \in H^2(\Gamma)$ and $\|w\|_{H^2(\Gamma)} \leq c\|e_h^l\|_{L^2(\Gamma)/\mathbb{R}}$, with $\|e_h^l\|_{L^2(\Gamma)/\mathbb{R}} := \|e_h^l - c_e\|_{L^2(\Gamma)}$. Using (13.60), the H^2-regularity bound, the approximation properties in Theorem 13.2.16, the discretization error bound in Theorem 13.2.17 and perturbation arguments, we obtain

$$\|e_h^l\|_{L^2(\Gamma)/\mathbb{R}} \leq ch^2\|f\|_{L^2(\Gamma)}.$$

A further perturbation argument yields

$$\|e_h\|_{L^2(\Gamma_h)} \leq c\|e_h^l\|_{L^2(\Gamma)} \leq c\big(\|e_h^l\|_{L^2(\Gamma)/\mathbb{R}} + |c_e|\big) \leq ch^2\|f\|_{L^2(\Gamma)},$$

and thus the result holds. □

13.2.5 Eulerian space-time surface finite element method for a non-stationary interface

In this section we explain how the surface finite element approach introduced in Sect. 13.2.2, cf. (13.19), can be generalized to problems with a *non-stationary* interface. This generalization is based on a natural idea, similar to the one used in Sect. 11.5.2 for the derivation of the space-time XFEM method, namely to use a space-time variational approach and suitable space-time finite element spaces. We explain the method for the surfactant transport equation, but it can also be applied (with obvious modifications) to other elliptic or parabolic equations on the interface that are formulated in a space-time variational form.

Let $Q_T = \Omega \times (0,T) \subset \mathbb{R}^4$ be the space-time cylinder. The space-time interface $\Gamma_* = \cup_{t\in(0,T)}\Gamma(t) \times \{t\} \subset Q_T$ is assumed to be a three-dimensional hypersurface. The interface $\Gamma(t)$ is transported by the (smooth) velocity field $\mathbf{w}(x,t)$. We recall the weak space-time variational formulation of the surfactant transport equation (12.8): Find $u \in H^1(\Gamma_*)$ with $u(\cdot,0) = u_0$ such that

$$(\dot{u} + u\,\mathrm{div}_\Gamma \mathbf{w}, v)_{L^2(\Gamma_*)} + (\nabla_\Gamma u, \nabla_\Gamma v)_{L^2(\Gamma_*)} = 0 \quad \forall\, v \in H^{1,0}(\Gamma_*). \qquad (13.61)$$

For the discretization of this problem we use the same trace technique as in Sect. 13.2.2: First we introduce standard space-time finite element spaces on the cylinder Q_T (outer domain) and then we use the restriction of these spaces to Γ_* for a Galerkin discretization of (13.61). The basic idea of the space-time finite element method is explained in Sect. 11.5.2; we use notation as in that section. As noted in Sect. 11.5.2 there are different variants of space-time methods, e.g., one can use trial functions that are continuous w.r.t. time or trial functions that may be discontinuous in time. In Sect. 11.5.2 we used the latter variant, cf. Remark 11.5.4. Here we choose the former variant since it leads to a somewhat simplified presentation due to the fact that the jump terms between time slabs vanish. This choice, however, is not essential for the method presented below.

We use a partitioning of the time interval $0 = t_0 < \ldots < t_N = T$. Corresponding to each time interval $I_n = [t_{n-1}, t_n)$ we assume a consistent triangulation $\mathcal{T}_n$ of the spatial domain Ω. To simplify the presentation we assume uniform time steps $\Delta t = T/N$, and a triangulation independent of n, denoted by $\mathcal{T}_h$. These assumptions are not essential. The time slabs are denoted by $S_n = \Omega \times I_n$, $1 \leq n \leq N$. Let V_h be a given finite element space of continuous piecewise polynomial (e.g., piecewise linear) functions corresponding to the triangulation $\mathcal{T}_h$. We introduce the space of piecewise space-time polynomials

$$V_k^{\mathrm{ST}} := \left\{ v(x,t) = \sum_{j=0}^{k} t^j \phi_j(x) : \phi_j \in V_h \right\}, \quad k \geq 0. \tag{13.62}$$

The test functions are allowed to be discontinuous between time slabs. Hence we introduce

$$W_k := \left\{ v : Q_T \to \mathbb{R} : v_{|S_n} \in V_k^{\mathrm{ST}} \quad \text{for } 1 \leq n \leq N \right\}, \quad k \geq 0.$$

Note that $v \in W_k$ implies $v_{|\Gamma_*} \in H^{1,0}(\Gamma_*)$. The corresponding *trace* space is denoted by $W_k^{\Gamma_*} := \left\{ v_{|\Gamma_*} : v \in W_k \right\}$. Using this test space there is no global coupling in time direction and thus a time marching procedure can be applied. This discrete time marching is obtained by choosing test functions that are equal to zero except on one time slab. To make this more precise we need some further notation. Let Γ_*^n be the part of the space-time interface Γ_* that is contained in the time slab S_n, i.e. $\Gamma_*^n = \cup_{t \in I_n} \Gamma(t) \times \{t\}$. The test space $W_k^{\Gamma_*}$ is a direct sum of spaces corresponding to the time intervals I_n, $1 \leq n \leq N$:

$$W_k^{\Gamma_*} = \bigoplus_{n=1}^{N} V_k^{\Gamma_*^n}, \quad \text{with } V_k^{\Gamma_*^n} := \left\{ v_{|\Gamma_*^n} : v \in V_k^{\mathrm{ST}} \right\}.$$

We now turn to the trial functions. As mentioned above, we consider trial functions that are continuous with respect to t. By this we mean continuity on the space-time interface Γ_*. To enforce this continuity, in our method we need a well-defined local extension of functions from the finite element trace space, which we now introduce. Let $V_h^{\Gamma(t_n)}$ be the finite element trace space corresponding to $\Gamma(t_n)$, i.e. $v_h \in V_h^{\Gamma(t_n)}$ iff there exists $w_h \in V_h$ such that $v_h = (w_h)_{|\Gamma(t_n)}$. Clearly, this w_h is *not* unique. For a given n, $1 \leq n \leq N$, let $\omega_n \subset \Omega$ be the union of all tetrahedra which have a nonzero intersection with $\Gamma(t)$ for some $t \in I_n$,

$$\omega_n = \cup \left\{ T \in \mathcal{T}_h : \mathrm{meas}_2(T \cap \Gamma(t)) > 0 \quad \text{for a } t \in I_n \right\}, \tag{13.63}$$

and $V_h(\omega_n) := \left\{ v_{|\omega_n} : v \in V_h \right\}$ the space of outer finite element functions restricted to the subdomain ω_n. We *assume* a well-defined extension operator

$$E_n : V_h^{\Gamma(t_{n-1})} \to V_h(\omega_n). \tag{13.64}$$

A natural choice for this operator (which we use in our experiments in Sect. 13.2.6) is as follows. Let $\omega_n^* \subset \omega_n$ be the subdomain consisting of all tetrahedra which have a nonzero intersection with $\Gamma(t_{n-1})$, i.e.,

$$\omega_n^* = \cup\{\, T \in \mathcal{T}_h : \operatorname{meas}_2(T \cap \Gamma(t_{n-1})) > 0 \,\}. \tag{13.65}$$

In general, i.e. except for special situations with exceptional geometric constellations, for each $v_h \in V^{\Gamma(t_{n-1})}$ there is a *unique* $w_h \in V_h(\omega_n^*)$ such that $v_h = (w_h)_{|\Gamma(t_{n-1})}$. The extension $E_n v_h$ is defined by extending this unique w_h by zero values at all vertices not in ω_h^*.

Taking test functions $v_h \in V_{k-1}^{\Gamma_*^n}$ we obtain the following time marching procedure, with $u_h^0 \in V_h^{\Gamma(0)}$ an approximation of the initial data u_0. For $n = 1, \ldots, N$: Determine $u_h = u_h^n \in V_k^{\Gamma_*^n}$ such that

$$E_n u_h(\cdot, t_{n-1}) = E_n u_h^{n-1}(\cdot, t_{n-1}) \quad \text{on } \omega_n, \tag{13.66}$$

and

$$(\dot{u}_h + u_h \operatorname{div}_\Gamma \mathbf{w}, v_h)_{L^2(\Gamma_*^n)} + (\nabla_\Gamma u_h, \nabla_\Gamma v_h)_{L^2(\Gamma_*^n)} = 0 \tag{13.67}$$

for all $v_h \in V_{k-1}^{\Gamma_*^n}$. The role of the extension operator E_n in (13.66) will become clear from the analysis below.

A systematic theoretical error analysis of this discretization method is not known, yet. Some analysis and results of numerical experiments with this method can be found in [123], cf. also Sect. 13.2.6 below. Note that for both the trial and test functions the trace space $V_m^{\Gamma_*^n}$, of space-time polynomials $v \in V_m^{\mathrm{ST}}$ restricted to Γ_*^n, is used. The degree m with respect to time is taken one larger for the trial function space than for the test function space, since the trial function must satisfy the condition (13.66).

For the implementation and analysis of the discretization (13.66)-(13.67) it is convenient to reformulate the problem in a slightly different form, based on rewriting the first term in (13.67) as follows:

$$\begin{aligned}
&(\dot{u}_h + u_h \operatorname{div}_\Gamma \mathbf{w}, v_h)_{L^2(\Gamma_*^n)} \\
&= \int_{t_{n-1}}^{t_n} \int_{\Gamma(t)} \dot{u}_h v_h + u_h v_h \operatorname{div}_\Gamma \mathbf{w} \, ds \, dt \\
&= \int_{t_{n-1}}^{t_n} \int_{\Gamma(t)} (\dot{\overline{u_h v_h}}) + u_h v_h \operatorname{div}_\Gamma \mathbf{w} - u_h \dot{v}_h \, ds \, dt \\
&= \int_{t_{n-1}}^{t_n} \frac{d}{dt} \int_{\Gamma(t)} u_h v_h \, ds \, dt - (u_h, \dot{v}_h)_{L^2(\Gamma_*^n)} \\
&= (u_h, v_h)_{L^2(\Gamma(t_n))} - (u_h, v_h)_{L^2(\Gamma(t_{n-1}))} - (u_h, \dot{v}_h)_{L^2(\Gamma_*^n)}.
\end{aligned}$$

This leads to the following *space-time Galerkin discretization* of the surfactant transport problem (13.61).

For $n = 1, \ldots, N$, determine $u_h = u_h^n \in V_k^{\Gamma_*^n}$ such that

$$E_n u_h(\cdot, t_{n-1}) = E_n u_h^{n-1}(\cdot, t_{n-1}) \quad \text{on } \omega_n, \tag{13.68}$$

and

$$\begin{aligned}(u_h, v_h)_{L^2(\Gamma(t_n))} &- (u_h, \dot{v}_h)_{L^2(\Gamma_*^n)} + (\nabla_\Gamma u_h, \nabla_\Gamma v_h)_{L^2(\Gamma_*^n)} \\ &= (u_h^{n-1}, v_h)_{L^2(\Gamma(t_{n-1}))} \quad \text{for all } v_h \in V_{k-1}^{\Gamma_*^n}.\end{aligned} \tag{13.69}$$

This formulation has two advantages compared to the one in (13.67). Firstly, in (13.67) the time derivative $\frac{\partial}{\partial t}$ is applied to a trial function u_h, whereas in (13.69) it is applied to a test function v_h. The trial functions u_h and test functions v_h are of degree k and $k-1$, respectively, with respect to time. Due to this the quadrature w.r.t. time can be simpler for the term $(u_h, \dot{v}_h)_{L^2(\Gamma_*^n)}$ than for the term $(\dot{u}_h, v_h)_{L^2(\Gamma_*^n)}$, cf. Remark 13.2.21 below. A second (minor) advantage is that in (13.69) the space-time term $(u_h \operatorname{div}_\Gamma \mathbf{w}, v_h)_{L^2(\Gamma_*^n)}$ does not occur.

A further reformulation is obtained by "inserting" the continuity condition (13.68) into a suitable representation of the trial function u_h. For $k \geq 1$, let $\ell_0, \ldots, \ell_k \in \mathcal{P}_k$ be the Lagrange basis on $[t_{n-1}, t_n]$, corresponding to equidistant nodes. The function ℓ_0 corresponds to the node t_{n-1}, hence $\ell_0(t_{n-1}) = 1$, $\ell_0(t_n) = 0$, $\ell_j(t_{n-1}) = 0$ for all $1 \leq j \leq k$. We also introduce $\ell(t) := \frac{t - t_{n-1}}{\Delta t}$. The space V_k as in (13.62) can be represented as $V_k = \left\{ \sum_{j=0}^k \ell_j \phi_j : \phi_j \in V_h \right\}$. Using this it follows that

$$V_k = \{ \ell_0 \phi_0 + \ell \psi : \phi_0 \in V_h, \ \psi \in V_{k-1} \}, \quad k \geq 1.$$

Hence, a trial function $u_h \in V_k^{\Gamma_*^n}$ can be represented as the restriction to Γ_*^n of a function

$$u_h(x,t) = \ell_0(t)\phi_0(x) + \ell(t)\psi(x,t), \quad \psi \in V_{k-1}. \tag{13.70}$$

Since $\Gamma_*^n \subset I_n \times \omega_n$, it follows that $(u_h)_{|\Gamma_*^n}$ is known if both $\psi_{|\Gamma_*^n}$ and $(\phi_0)_{|\omega_n}$ are known. Inserting $t = t_{n-1}$ in (13.70) we see that $\phi_0(x)$, $x \in \omega_n$, is determined by the continuity condition (13.68). This identification of $(\phi_0)_{|\omega_n}$ by means of the continuity condition relies on the fact that in (13.68) we prescribe values on ω_n (not only on $\Gamma(t_{n-1})$). This explains why we introduced the extension operator E_n.

For a *given* function $\ell_0\phi_0$ on Γ_*^n, the function $\psi \in V_{k-1}^{\Gamma_*^n}$ can be determined using the equation (13.69): Let $\psi \in V_{k-1}^{\Gamma_*^n}$ be such that

$$\begin{aligned}&(\psi, v_h)_{L^2(\Gamma(t_n))} - (\ell\psi, \dot{v}_h)_{L^2(\Gamma_*^n)} + (\ell\nabla_\Gamma\psi, \nabla_\Gamma v_h)_{L^2(\Gamma_*^n)} \\ &= (u_h^{n-1}, v_h)_{L^2(\Gamma(t_{n-1}))} + (\ell_0\phi_0, \dot{v}_h)_{L^2(\Gamma_*^n)} - (\ell_0\nabla_\Gamma\phi_0, \nabla_\Gamma v_h)_{L^2(\Gamma_*^n)}\end{aligned} \tag{13.71}$$

for all $v_h \in V_{k-1}^{\Gamma_*^n}$. Note that in this variational problem the *trial and test space are the same.* The bilinear form used in this variational formulation is denoted by

$$a(w,w) = (w,v)_{L^2(\Gamma(t_n))} - (\ell w, \dot{v})_{L^2(\Gamma_*^n)} + (\ell \nabla_\Gamma w, \nabla_\Gamma v)_{L^2(\Gamma_*^n)}, \qquad (13.72)$$

and can be shown to be elliptic on $V_{k-1}^{\Gamma_*^n}$, provided a mild condition on Δt is satisfied:

Lemma 13.2.20 *Assume that* $\Delta t \| \operatorname{div}_\Gamma \mathbf{w} \|_{L^\infty(\Gamma_*)} \le c_0$. *Then*

$$a(v,v) \ge \frac{1-c_0}{2\Delta t} \|v\|^2_{L^2(\Gamma_*^n)} \quad \textit{for all} \;\; v \in V_{k-1}^{\Gamma_*^n}$$

holds.

Proof. Note that $(\dot{\ell v^2}) = \frac{1}{\Delta t} v^2 + 2\ell v \dot{v}$ holds. Using this we get

$$\begin{aligned}
&(\ell v, \dot{v})_{L^2(\Gamma_*^n)} \\
&= \int_{t_{n-1}}^{t_n} \int_{\Gamma(t)} \ell v \dot{v} \, ds\, dt = \frac{1}{2} \int_{t_{n-1}}^{t_n} \int_{\Gamma(t)} (\dot{\ell v^2})\, ds\, dt - \frac{1}{2\Delta t} \|v\|^2_{L^2(\Gamma_*^n)} \\
&= \frac{1}{2} \int_{t_{n-1}}^{t_n} \int_{\Gamma(t)} (\dot{\ell v^2}) + \ell v^2 \operatorname{div}_\Gamma \mathbf{w} \, ds\, dt \\
&\quad - \frac{1}{2} (\ell v \operatorname{div}_\Gamma \mathbf{w}, v)_{L^2(\Gamma_*^n)} - \frac{1}{2\Delta t} \|v\|^2_{L^2(\Gamma_*^n)} \\
&= \frac{1}{2} \int_{t_{n-1}}^{t_n} \frac{d}{dt} \int_{\Gamma(t)} \ell v^2 \, ds\, dt - \frac{1}{2} (\ell v \operatorname{div}_\Gamma \mathbf{w}, v)_{L^2(\Gamma_*^n)} - \frac{1}{2\Delta t} \|v\|^2_{L^2(\Gamma_*^n)} \\
&\le \frac{1}{2} \|v\|^2_{L^2(\Gamma(t_n))} - \frac{1}{2\Delta t} \|v\|^2_{L^2(\Gamma_*^n)} \big(1 - \Delta t \| \operatorname{div}_\Gamma \mathbf{w} \|_{L^\infty(\Gamma_*)}\big).
\end{aligned}$$

Hence,

$$\begin{aligned}
a(v,v) &= (v,v)_{L^2(\Gamma(t_n))} - (\ell v, \dot{v})_{L^2(\Gamma_*^n)} + (\ell \nabla_\Gamma v, \nabla_\Gamma v)_{L^2(\Gamma_*^n)} \\
&\ge \|v\|^2_{L^2(\Gamma(t_n))} - \frac{1}{2} \|v\|^2_{L^2(\Gamma(t_n))} + \frac{1-c_0}{2\Delta t} \|v\|^2_{L^2(\Gamma_*^n)} \\
&\ge \frac{1-c_0}{2\Delta t} \|v\|^2_{L^2(\Gamma_*^n)},
\end{aligned}$$

and thus the result holds. □

We assume that Δt is sufficiently small such that for c_0 as in Lemma 13.2.20 we have $c_0 < 1$. The function ϕ_0 is determined by the continuity condition (13.68). Due to ellipticity of $a(\cdot,\cdot)$, a unique ψ is determined by (13.71) and thus $(u_h)_{|\Gamma_*^n}$ is known. Summarizing, we obtain the following *well-defined* space-time discretization method for the surfactant transport equation:

Take $u_h^0 \in V_h^{\Gamma(0)}$. For $1 \leq n \leq N$, determine $\psi \in V_{k-1}^{\Gamma_*^n}$ such that

$$\begin{aligned} a(\psi, v_h) = (u_h^{n-1}, v_h)_{L^2(\Gamma(t_{n-1}))} + (\ell_0 E_n u_h^{n-1}, \dot{v}_h)_{L^2(\Gamma_*^n)} \\ - (\ell_0 \nabla_\Gamma E_n u_h^{n-1}, \nabla_\Gamma v_h)_{L^2(\Gamma_*^n)} \quad \text{for all } \; v_h \in V_{k-1}^{\Gamma_*^n}. \end{aligned} \tag{13.73}$$

The discrete approximation at $t = t_n$ is given by $u_h^n := \psi(\cdot, t_n)_{|\Gamma(t_n)}$.

This method is a natural extension to problems with a non-stationary interface of the Eulerian interface method treated in the Sects. 13.2.2-13.2.4. A discretization error analysis of this generalization is not known, yet.

For the implementation of a space-time method as in (13.73) an important issue is the use of suitable quadrature routines, which we briefly address. A detailed study is given in [123]. Integrals over the space-time interface part Γ_*^n occur for which numerical integration is required. Clearly, if the approximation order of the space V_k, cf. (13.62), is increased more accurate quadrature is required. Here we restrict to the simplest case $k = 1$, for which the trapezoidal quadrature rule for approximation of the time integral seems to be sufficiently accurate. We apply the trapezoidal rule to the time integrals in $(\cdot, \cdot)_{L^2(\Gamma_*^n)}$ that occur in (13.73), i.e.:

$$\begin{aligned} &- (\ell \psi, \dot{v}_h)_{L^2(\Gamma_*^n)} + (\ell \nabla_\Gamma \psi, \nabla_\Gamma v_h)_{L^2(\Gamma_*^n)} \\ &\approx -\frac{\Delta t}{2}(\psi, \dot{v}_h)_{L^2(\Gamma(t_n))} + \frac{\Delta t}{2}(\nabla_\Gamma \psi, \nabla_\Gamma v_h)_{L^2(\Gamma(t_n))}, \end{aligned}$$

and

$$\begin{aligned} &(\ell_0 E_n u_h^{n-1}, \dot{v}_h)_{L^2(\Gamma_*^n)} - (\ell_0 \nabla_\Gamma E_n u_h^{n-1}, \nabla_\Gamma v_h)_{L^2(\Gamma_*^n)} \\ &\approx \frac{\Delta t}{2}(u_h^{n-1}, \dot{v}_h)_{L^2(\Gamma(t_{n-1}))} - \frac{\Delta t}{2}(\nabla_\Gamma u_h^{n-1}, \nabla_\Gamma v_h)_{L^2(\Gamma(t_{n-1}))}. \end{aligned}$$

Note that in the quadrature approximation the extension operator E_n is not needed. For $k = 1$ the test functions v_h are constant with respect to t and thus $\dot{v} = \mathbf{w} \cdot \nabla v_h$. This results in the following Crank-Nicolson type discrete problem: Determine $u_h^n \in V_h^{\Gamma(t_n)}$ such that

$$\begin{aligned} &(u_h^n, v_h)_{L^2(\Gamma(t_n))} - \frac{\Delta t}{2}(u_h^n, \mathbf{w} \cdot \nabla v_h)_{L^2(\Gamma(t_n))} + \frac{\Delta t}{2}(\nabla_\Gamma u_h^n, \nabla_\Gamma v_h)_{L^2(\Gamma(t_n))} \\ &= (u_h^{n-1}, v_h)_{L^2(\Gamma(t_{n-1}))} + \frac{\Delta t}{2}(u_h^{n-1}, \mathbf{w} \cdot \nabla v_h)_{L^2(\Gamma(t_{n-1}))} \\ &- \frac{\Delta t}{2}(\nabla_\Gamma u_h^{n-1}, \nabla_\Gamma v_h)_{L^2(\Gamma(t_{n-1}))} \quad \text{for all } \; v_h \in V_h^{\Gamma(t_n)}. \end{aligned} \tag{13.74}$$

In practice, the discrete problem (13.74) is used with the exact interface $\Gamma(t_j)$ replaced by a (piecewise planar) approximation $\Gamma_h(t_j)$. Well-posedness of the variational problem (13.74) is studied in [123].

Remark 13.2.21 If instead of (13.69) one uses the equivalent formulation in (13.67), one has to apply quadrature to $(\dot{u}_h, v_h)_{L^2(\Gamma_*^n)}$. Consider the time derivative of the unknown part $\ell\psi$ of u_h, cf. (13.70), i.e. $\frac{\partial}{\partial t}(\ell\psi) = \frac{1}{\Delta t}\psi$ (since, for $k = 1$, ψ is constant in t). Using the trapezoidal rule as approximation of the time integral in $(\psi, v_h)_{L^2(\Gamma_*^n)}$ results in terms

$$\int_{\Gamma(t_{n-1})} \psi v_h \, ds, \quad \int_{\Gamma(t_n)} \psi v_h \, ds.$$

Hence, one has to evaluate $\psi \in V_0^{\Gamma_*^n}$ both on the "old" interface $\Gamma(t_{n-1})$ and on the "new" interface $\Gamma(t_n)$. This is avoided if the formulation (13.69) is used, cf. (13.74).

Remark 13.2.22 A less nice property of the space-time method in (13.73) is that it depends on the extension operator E_n. The following variant does not need such an operator. We restrict to the case $k = 1$. The method introduced above is based on the representation $u_h = \ell_0\phi_0 + \ell\psi$, with $\phi_0, \psi \in V_h(\omega_n)$. Hence $t \to u_h(x,t)$ is linear for all $x \in \omega_n$. The function ϕ_0 is determined by the continuity condition (13.68), and the function ψ by the variational problem (13.73). We introduce a variant in which trial functions u_h are used such that $t \to u_h(x,t)$ is linear only for $x \in \omega_n^*$, with $\omega_n^* \subset \omega_n$ as in (13.65). The nodal basis functions, corresponding to nodes x_j in ω_n, are denoted by ξ_j, i.e., $V_h(\omega_n) = \text{span}\{\, \xi_j : x_j \in \omega_n \,\}$. This induces a splitting

$$V_h(\omega_n) = \text{span}\{\, \xi_j : x_j \in \omega_n^* \,\} \oplus \text{span}\{\, \xi_j : x_j \in \omega_n \setminus \omega_n^* \,\} =: V_h^* \oplus V_h^r.$$

The corresponding splitting of $\psi \in V_h(\omega_n)$ is denoted by $\psi = \psi^* + \psi^r$, with $\psi^* \in V_h^*$, $\psi^r \in V_h^r$. We consider trial functions of the form

$$u_h = \ell_0\phi_0 + \ell\psi^* + \psi^r, \quad \phi_0 \in V_h^*, \, \psi = \psi^* + \psi^r \in V_h(\omega_n).$$

The function ϕ_0 can be determined from the data $u_h^{n-1}(x)$, $x \in \Gamma(t_{n-1})$, *without* using an extension operator. The function ψ is determined by a variational problem similar to the one in (13.73).

13.2.6 Numerical experiments

We present results of some numerical experiments with the method treated in the previous section. The outer domain, given by $\Omega = (-2,2)^3$, contains a spherical non-stationary interface

$$\Gamma(t) = \partial B_1(0) + t\mathbf{w}, \quad \mathbf{w} = (0,0,10)^T, \quad t \in [0,T], \; T = 0.05,$$

with $\partial B_1(0)$ the sphere with radius 1 centered at the origin. On this interface we consider the surfactant transport equation

$$\dot{u} + u \operatorname{div}_\Gamma \mathbf{w} - \Delta_\Gamma u = 0 \quad \text{on } \Gamma(t),$$

$$u(x,0) = -\frac{13}{8}\sqrt{35/\pi}\frac{\|x\|^2}{12+\|x\|^2}(3x_1^2x_2 - x_2^3) \quad \text{on } \Gamma(0).$$

The signed distance function to $\Gamma(t)$ is denoted by $\phi(x,t)$. Hence, $\Gamma(t)$ is the zero level of $\phi(\cdot,t)$. A family of uniform tetrahedral triangulations $\mathcal{T}_{h_l}$ of Ω is used, with mesh size parameter $h_l = h_0 2^{-(l+1)}$, $h_0 := 4$, $2 \leq l \leq 5$. On a triangulation $\mathcal{T}_h$ the exact interface $\Gamma(t)$ is approximated by a piecewise planar approximation $\Gamma_h(t)$ as explained in Sect. 7.3, i.e. $\Gamma_h(t)$ is obtained as the zero level of the piecewise linear interpolation of the level set function $\phi(x,t)$ on the refined grid $\mathcal{T}_{\frac{1}{2}h}$. For the space-time finite element discretization of the surfactant equation on $\Gamma_h(t)$ we consider the lowest order case, i.e. $k=1$, and use the trapezoidal rule for approximating the time integrals. This results in the discrete problem (13.74), with $\Gamma(t_j)$ replaced by $\Gamma_h(t_j)$.

First we consider, for a fixed mesh size $h = h_l$, the rate of convergence with respect to $\Delta t = T/N$. It appears that for $\Delta t \downarrow 0$, i.e. $N \to \infty$, the discrete approximation $u_h^N(\cdot,T)$ converges towards a limit solution. A reference solution that approximates this limit solution is computed with $N = 2^{10}$ and denoted by $u_h^{\mathrm{ref}}(\cdot,T)$. Note that this reference solution depends on $h = h_l$. For $l = 2,3,4$, we computed the discrete solution $u_{h_l}^N(\cdot,T)$ for different time steps $\Delta t = T/N$, $N < 1024$, and determined the corresponding error measure

$$e_l^{\Delta t} := \|u_{h_l}^N(\cdot,T) - u_{h_l}^{\mathrm{ref}}(\cdot,T)\|_{L^2(\Gamma_{h_l}(T))}.$$

The results are shown in Fig. 13.9.

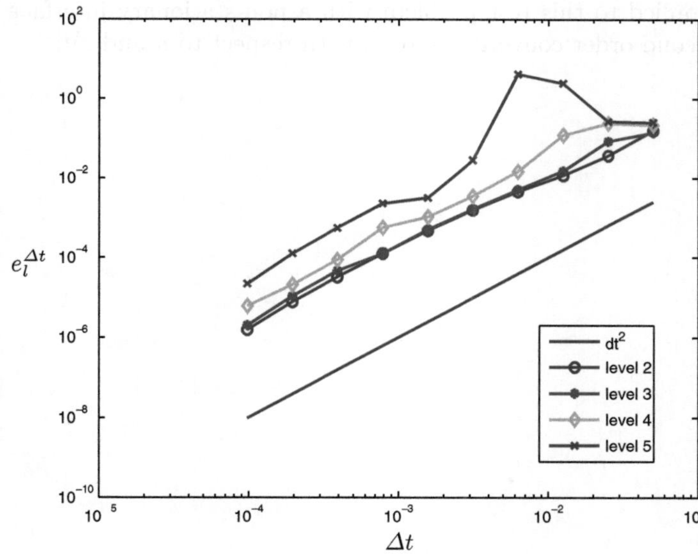

Fig. 13.9. Error $e_l^{\Delta t}$ as a function of Δt, for $l = 2,3,4$.

The results show a $\mathcal{O}(\Delta t^2)$ convergence behavior. To give an indication of the space-time error behavior we computed a reference solution on level $l = 5$ and with $N = 1024$, denoted by $u^{\text{ref}}(\cdot, T)$. For different Δt and $l = 2, 3, 4$, the discrete solution $u_{h_l}^N(\cdot, T)$ is determined and compared to this reference solution. This comparison is done by lifting the discrete solution $u_{h_l}^N(\cdot, T)$ from $\Gamma_{h_l}(T)$ to $\Gamma_{h_5}(T)$, resulting in $\hat{u}_{h_l}^N(\cdot, T)$ and computing the error measure

$$\hat{e}_l^{\Delta t} := \|\hat{u}_{h_l}^N(\cdot, T) - u^{\text{ref}}(\cdot, T)\|_{L^2(\Gamma_{h_5}(T))}.$$

This lift can be determined in a natural way, since $u_{h_l}^N(\cdot, T)$ is the restriction of a finite element function, say w_{h_l}, from the outer finite element space V_{h_l}. For the lift we take the restriction of this w_{h_l} to $\Gamma_{h_5}(T)$. Results are given in Table 13.5. From the results it turns out that we should take $\Delta t \ll h$ to balance the spatial and time discretization errors.

l	h_l	Δt	$\hat{e}_l^{\Delta t}$
2	2.50 E-1	1.25 E-2	1.80 E-1
3	1.25 E-1	6.25 E-3	4.42 E-2
4	6.25 E-2	3.13 E-3	1.01 E-2

Table 13.5. Discretization error $\hat{e}_l^{\Delta t}$.

These results indicate that for the space-time Crank-Nicolson type of method applied to this test problem with a non-stationary interface we can achieve second order convergence both with respect to h and Δt.

Part V

Appendix

14

Appendix A: Results from differential geometry

14.1 Results for a stationary surface

We assume $\Gamma \subset \mathbb{R}^d$ to be an oriented C^2-hypersurface, i.e., for $x^* \in \Gamma$ there exists an open set $U_{x^*} \subset \mathbb{R}^d$ with $x^* \in U_{x^*}$ and a scalar function $\psi \in C^2(U_{x^*})$ such that

$$U_{x^*} \cap \Gamma = \{ x \in U_{x^*} : \psi(x) = 0 \}, \text{ and } \nabla\psi(x) \neq 0 \text{ for all } x \in U_{x^*} \cap \Gamma. \tag{14.1}$$

Moreover, for such a hypersurface Γ there exists an open neighborhood $U \supset \Gamma$ and a vector function $\mathbf{n} \in C^1(U)^d$ such that $\|\mathbf{n}(x)\|_2 = 1$ and for all $x \in \Gamma$ the vector $\mathbf{n}(x)$ is orthogonal to the tangent plane at x (i.e., the plane orthogonal to $\nabla\psi(x)$).

For $f \in C^1(U)$ we define the *tangential derivative*

$$\nabla_\Gamma f(x) := \big(\mathbf{I} - \mathbf{n}(x)\mathbf{n}(x)^T\big)\nabla f(x) =: \mathbf{P}(x)\nabla f(x), \quad x \in \Gamma,$$

with the orthogonal projection $\mathbf{P}(x) = \mathbf{I} - \mathbf{n}(x)\mathbf{n}(x)^T$. The values of $\nabla_\Gamma f(x)$ depend only on values $f(x)$, $x \in \Gamma$. For $\mathbf{f} \in C^1(U)^d$ the *tangential divergence* is defined as follows. We use the notation $\nabla_i = \frac{\partial}{\partial x_i}$.

$$\begin{aligned} \mathrm{div}_\Gamma\, \mathbf{f} := \nabla_\Gamma^T \mathbf{f} &= \sum_{i=1}^d (\mathbf{P}\nabla)_i f_i = \sum_{i=1}^d \nabla_i f_i - \sum_{i=1}^d n_i \mathbf{n}^T(\nabla f_i) \\ &= \mathrm{div}\, \mathbf{f} - \mathbf{n}^T \nabla\mathbf{f}\, \mathbf{n}. \end{aligned}$$

Recall that $\nabla\mathbf{f} := \big(\nabla f_1 \ldots \nabla f_d\big)$. Hence,

$$\mathrm{div}_\Gamma\, \mathbf{f}(x) = \mathrm{div}\, \mathbf{f}(x) - \mathbf{n}(x)^T \nabla\mathbf{f}(x)\, \mathbf{n}(x), \quad x \in \Gamma. \tag{14.2}$$

It can be shown that the value of $\operatorname{div}_\Gamma \mathbf{f}(x),\ x \in \Gamma$, depends only on values $\mathbf{f}(x),\ x \in \Gamma$. Another useful representation is

$$\operatorname{div}_\Gamma \mathbf{f}(x) = \operatorname{tr}\big(\nabla_\Gamma \mathbf{f}(x)\big).$$

Using these definitions we derive some elementary and useful results. For a detailed treatment we refer to the literature, e.g. [78, 15]. From $\mathbf{n}(x)^T\mathbf{n}(x) = 1$ we obtain $\nabla\big(\mathbf{n}(x)^T\mathbf{n}(x)\big) = 0$, which implies

$$\big(\nabla \mathbf{n}(x)\big)\mathbf{n}(x) = 0, \quad x \in U. \tag{14.3}$$

Using this and $\mathbf{f} = \mathbf{n}$ in (14.2) yields

$$\operatorname{div}_\Gamma \mathbf{n}(x) = \operatorname{div} \mathbf{n}(x), \quad x \in \Gamma. \tag{14.4}$$

Let $\phi \in C^1(U)$. Using (14.2), (14.3), (14.4) we get

$$\operatorname{div}_\Gamma \big(\phi(x)\mathbf{n}(x)\big) = \phi(x) \operatorname{div}_\Gamma \mathbf{n}(x), \quad x \in \Gamma. \tag{14.5}$$

The matrix $\mathbf{H}(x) := \mathbf{P}(x)\nabla\mathbf{n}(x) = \nabla_\Gamma \mathbf{n}(x) \in \mathbb{R}^{d\times d}$, $x \in \Gamma$, is symmetric, cf. Remark 14.1.1, and its entries depend only on values of $\mathbf{n}(x),\ x \in \Gamma$. From (14.3) it follows that $\mathbf{H}$ has an eigenvalue 0. The other real eigenvalues $\kappa_1(x), \ldots, \kappa_{d-1}(x)$ are called *principal curvatures* at $x \in \Gamma$ and

$$\kappa(x) := \sum_{i=1}^{d-1} \kappa_i(x) = \operatorname{tr}\mathbf{H}(x), \quad x \in \Gamma, \tag{14.6}$$

is called the *mean curvature* at x (in the literature also another definition $\kappa(x) = \frac{1}{d-1}\sum_{i=1}^{d-1}\kappa_i(x)$ is used).

Remark 14.1.1 In the special case that ψ in (14.1) is a signed distance function $d : U \to \mathbb{R}$, one can choose $\mathbf{n} = \nabla d$. Then (14.3) implies that the Hessian $\tilde{\mathbf{H}} := \nabla^2 d$, which is symmetric, satisfies $\tilde{\mathbf{H}}\mathbf{n}(x) = 0$. Hence $\mathbf{n}(x)^T\tilde{\mathbf{H}} = 0$ and thus $\mathbf{H}(x) = \mathbf{P}(x)\nabla\mathbf{n}(x) = \mathbf{P}(x)\tilde{\mathbf{H}}(x) = \tilde{\mathbf{H}}(x)$ holds, i.e., $\mathbf{H}$ is symmetric and the $\kappa_i(x)$ are the eigenvalues of the Hessian $\nabla^2 d(x)$, $x \in \Gamma$.

From $\operatorname{tr}\mathbf{H}(x) = \operatorname{div}\mathbf{n}(x)$ and (14.4) it follows that

$$\boxed{\kappa(x) = \operatorname{div}\mathbf{n}(x) = \operatorname{div}_\Gamma \mathbf{n}(x), \quad x \in \Gamma,} \tag{14.7}$$

holds. For $\operatorname{id}_\Gamma(x) := x = (x_1, \ldots, x_d)^T$ we have $\nabla(\operatorname{id}_\Gamma)_j = e_j$ (j-th basis vector in $\mathbb{R}^d$) and thus, using (14.5) we get for $1 \leq j \leq d$,

$$\operatorname{div}_\Gamma \nabla_\Gamma (\operatorname{id}_\Gamma)_j = \operatorname{div}_\Gamma(\mathbf{P}e_j) = -\operatorname{div}_\Gamma(\mathbf{n}n_j) = -n_j \operatorname{div}_\Gamma \mathbf{n} = -\kappa\, n_j. \tag{14.8}$$

With $\Delta_\Gamma := \operatorname{div}_\Gamma \nabla_\Gamma$ and $\Delta_\Gamma \operatorname{id}_\Gamma = \big(\Delta_\Gamma(\operatorname{id}_\Gamma)_1, \ldots, \Delta_\Gamma(\operatorname{id}_\Gamma)_d\big)^T$ we thus obtain the following *Laplace-Beltrami characterization of the mean curvature*:

$$-\Delta_\Gamma \operatorname{id}_\Gamma(x) = \kappa(x)\mathbf{n}(x), \quad x \in \Gamma. \tag{14.9}$$

From (14.8) we also obtain

$$\operatorname{div}_\Gamma \mathbf{P} = -\begin{pmatrix} \operatorname{div}_\Gamma(n_1\mathbf{n}) \\ \operatorname{div}_\Gamma(n_2\mathbf{n}) \\ \operatorname{div}_\Gamma(n_3\mathbf{n}) \end{pmatrix} = -\kappa\mathbf{n}. \tag{14.10}$$

We now in addition assume that Γ is the boundary of an open bounded set $\Omega_1 \subset \mathbb{R}^d$, hence Γ has no boundary. The following partial integration rule is known in the literature, e.g. [120]. For $g \in C^1(U)$ and $1 \le i \le d$ we have

$$\int_\Gamma (\nabla_\Gamma)_i g \, ds = \int_\Gamma g \, \kappa \, n_i \, ds. \tag{14.11}$$

Note that no minus sign occurs. Applying this with g replaced by $f(\nabla_\Gamma)_i g$ results in

$$\int_\Gamma (\nabla_\Gamma)_i f \, (\nabla_\Gamma)_i g + f(\nabla_\Gamma)_i^2 g \, ds = \int_\Gamma f \, (\nabla_\Gamma)_i g \, \kappa \, n_i \, ds.$$

Summing over $i = 1, \dots, d$ and with $\sum_{i=1}^d (\nabla_\Gamma)_i g n_i = \mathbf{n}^T \nabla_\Gamma g = \mathbf{n}^T \mathbf{P} \nabla g = 0$ we obtain the Green's formula

$$\int_\Gamma \nabla_\Gamma f \cdot \nabla_\Gamma g \, ds = -\int_\Gamma f \, \Delta_\Gamma g \, ds. \tag{14.12}$$

For vector functions $\mathbf{f}, \mathbf{g}$ (sufficiently smooth) this yields

$$\int_\Gamma \nabla_\Gamma \mathbf{f} \cdot \nabla_\Gamma \mathbf{g} \, ds = -\int_\Gamma \mathbf{f} \cdot \Delta_\Gamma \mathbf{g} \, ds, \tag{14.13}$$

with $\nabla_\Gamma \mathbf{f} \cdot \nabla_\Gamma \mathbf{g} := \sum_{i=1}^d \nabla_\Gamma f_i \cdot \nabla_\Gamma g_i$, $\Delta_\Gamma \mathbf{g} := \left(\Delta_\Gamma g_1, \dots, \Delta_\Gamma g_d\right)^T$. Using these results we obtain the following integral identities involving the curvature.

Lemma 14.1.2 *Assume that Γ is sufficiently smooth. For $\mathbf{f} \in H^1(U)^d$ the following holds:*

$$\int_\Gamma \kappa \mathbf{n} \cdot \mathbf{f} \, ds = \int_\Gamma \nabla_\Gamma \mathbf{f} \cdot \nabla_\Gamma \operatorname{id}_\Gamma \, ds \tag{14.14a}$$

$$= \int_\Gamma \operatorname{tr}\left(\mathbf{P}\nabla \mathbf{f}\right) ds. \tag{14.14b}$$

Proof. We take $\mathbf{f} \in C^1(U)^d$. Using $\mathbf{g} = \mathrm{id}_\Gamma$ in (14.13) and the result in (14.9) we obtain the result in (14.14a). Let e_i be the i-th basis vector in $\mathbb{R}^d$. From $\nabla_\Gamma(\mathrm{id}_\Gamma)_i = \mathbf{P}\nabla(\mathrm{id}_\Gamma)_i = \mathbf{P}e_i$ and $\mathbf{P}^2 = \mathbf{P} = \mathbf{P}^T$ it follows that

$$\begin{aligned}\nabla_\Gamma \mathbf{f} \cdot \nabla_\Gamma \,\mathrm{id}_\Gamma &= \sum_{i=1}^d \nabla_\Gamma f_i \cdot \nabla_\Gamma(\mathrm{id}_\Gamma)_i = \sum_{i=1}^d (\mathbf{P}\nabla f_i)\cdot(\mathbf{P}e_i)\\ &= \sum_{i=1}^d (\mathbf{P}\nabla f_i)\cdot e_i = \mathrm{tr}(\mathbf{P}\nabla\mathbf{f}),\end{aligned}$$

and thus the result in (14.14b) holds. From a density argument it follows that these results also hold for $\mathbf{f} \in H^1(U)^d$. □

From (14.11) with g replaced by $f_i g$, $i = 1, \ldots, d$, and summation of i we obtain

$$\int_\Gamma g\,\mathrm{div}_\Gamma \mathbf{f} + \mathbf{f}\cdot\nabla_\Gamma g \, ds = \int_\Gamma \kappa\,\mathbf{f}\cdot\mathbf{n}\,g\,ds, \tag{14.15}$$

and thus, due to $\mathbf{Pn} = 0$,

$$\int_\Gamma g\,\mathrm{div}_\Gamma(\mathbf{Pf})\,ds = -\int_\Gamma \mathbf{f}\cdot\nabla_\Gamma g\,ds, \quad \int_\Gamma \mathrm{div}_\Gamma(\mathbf{Pf})\,ds = 0. \tag{14.16}$$

From the first result in (14.16) it follows that for $\mathbf{G} \in \mathbb{R}^{3\times 3}$

$$\int_\Gamma \mathrm{div}_\Gamma(\mathbf{GP})\cdot\mathbf{g}\,ds = -\int_\Gamma \mathrm{tr}(\mathbf{G}\nabla_\Gamma\mathbf{g})\,ds \tag{14.17}$$

holds. We finally discuss a generalization of the partial integration rule in (14.11) for a smooth hypersurface *with* boundary. Let $\gamma \subset \Gamma$ be a hypersurface with boundary $\partial\gamma$. The unit outer normal to $\partial\gamma$ that is tangential to Γ is denoted by $\boldsymbol{\mu} = (\mu_1, \ldots, \mu_d)$. The following holds:

$$\int_\gamma (\nabla_\Gamma)_i g\,ds = \int_\gamma g\,\kappa\,n_i\,ds + \int_{\partial\gamma} g\,\mu_i\,d\tilde{s}, \quad 1 \le i \le d,$$

with $d\tilde{s}$ the surface measure on $\partial\gamma$. As immediate corollaries of this we obtain

$$\int_\gamma \nabla_\Gamma g\,ds = \int_\gamma g\,\kappa\,\mathbf{n}\,ds + \int_{\partial\gamma} g\,\boldsymbol{\mu}\,d\tilde{s}, \tag{14.18}$$

and, for all (smooth) tangential vector functions $\mathbf{g}$, i.e., with $\mathbf{g}(x)\cdot\mathbf{n}(x) = 0$ for all $x \in \gamma$,

$$\int_\gamma \mathrm{div}_\Gamma\,\mathbf{g}\,ds = \int_{\partial\gamma} \mathbf{g}\cdot\boldsymbol{\mu}\,d\tilde{s}.$$

The latter implies

$$\int_\gamma \mathrm{div}_\Gamma(\mathbf{GP})\,ds = \int_{\partial\gamma} \mathbf{G}\boldsymbol{\mu}\,d\tilde{s}, \tag{14.19}$$

for $\mathbf{G} \in \mathbb{R}^{3\times 3}$.

14.2 Results for an evolving surface

Below we derive some useful identities for *moving* hypersurfaces $\Gamma(t)$. For each $t \in [0,T]$ we assume that $\Gamma(t)$ has the properties formulated above (compact, oriented C^2-hypersurface). We assume that the evolution $t \to \Gamma(t)$ is sufficiently smooth, cf. [78] for more details. Let $\mathbf{u}(x,t)$ be a C^1 vector field with which $\Gamma(t)$ is transported, i.e. $\frac{d}{dt}\Gamma(t) = \mathbf{u}(\Gamma(t),t)$ holds for $t \in [0,T]$ and $\Gamma(0)$ is given. Let $V(x,t)$ be the *normal velocity* of $\Gamma(t)$ at $x \in \Gamma(t)$: $V = \mathbf{u}\cdot\mathbf{n}$.

Remark 14.2.1 If $\Gamma(t)$ is characterized as the zero level of the signed distance function $d(\cdot,t) : U \to \mathbb{R}$, with $d < 0$ in $U \cap \Omega_1$ then for the normal velocity we have $V(x,t) = -\frac{\partial d}{\partial t}(x,t)$ for $x \in \Gamma(t)$.

We recall the fundamental Reynolds' transport theorem:

$$\begin{aligned}\frac{d}{dt}\int_{\Omega_1(t)} f(x,t)\,dx &= \int_{\Omega_1(t)} \dot{f} + f\operatorname{div}\mathbf{u}\,dx \\ &= \int_{\Omega_1(t)} \frac{\partial f}{\partial t} + \mathbf{u}\cdot\nabla f + f\operatorname{div}\mathbf{u}\,dx.\end{aligned} \tag{14.20}$$

Lemma 14.2.2 *Assume that $g(x,t)$ is continuously differentiable w.r.t. both x and t in a neighborhood of $\cup_{0<t<T}\Gamma(t)$. Assume $\Gamma(t) = \partial\Omega_1(t)$ with $\Omega_1(t) \subset \mathbb{R}^d$ open and bounded. The following holds, where we use the material derivative $\dot{g} = \frac{\partial g}{\partial t} + \mathbf{u}\cdot\nabla g$:*

$$\frac{d}{dt}\int_{\Omega_1(t)} g\,dx = \int_{\Omega_1(t)} \frac{\partial g}{\partial t}\,dx + \int_{\Gamma(t)} gV\,ds, \tag{14.21a}$$

$$\frac{d}{dt}\int_{\Gamma(t)} g\,ds = \int_{\Gamma(t)} \dot{g} + g\operatorname{div}_\Gamma\mathbf{u}\,ds \tag{14.21b}$$

$$= \int_{\Gamma(t)} \frac{\partial g}{\partial t} + V\big(\mathbf{n}\cdot\nabla g + \kappa\, g\big)\,ds. \tag{14.21c}$$

$$= \int_{\Gamma(t)} \frac{\partial g}{\partial t} + V\mathbf{n}\cdot\nabla g + g\operatorname{div}_\Gamma(V\mathbf{n})\,ds. \tag{14.21d}$$

Proof. The result in (14.21a) follows from (14.20) and

$$\int_{\Omega_1(t)} \mathbf{u}\cdot\nabla g + g\operatorname{div}\mathbf{u}\,dx = \int_{\Omega_1(t)} \operatorname{div}(g\mathbf{u})\,ds = \int_{\Gamma(t)} g\mathbf{u}\cdot\mathbf{n}\,ds.$$

Note that (14.21b) is an analogon for $\Gamma(t)$ of the first result in (14.20). For a proof of (14.21b) we refer to the literature, e.g. [78, 46]. Using (14.15) we get

$$\begin{aligned}\int_{\Gamma(t)} \dot{g} + g \operatorname{div}_\Gamma \mathbf{u}\, ds &= \int_{\Gamma(t)} \frac{\partial g}{\partial t} + \mathbf{u}\cdot\nabla g - \mathbf{u}\cdot\nabla_\Gamma g + V\kappa\, g\, ds \\ &= \int_{\Gamma(t)} \frac{\partial g}{\partial t} + \mathbf{u}\cdot(\mathbf{I}-\mathbf{P})\nabla g + V\kappa\, g\, ds \\ &= \int_{\Gamma(t)} \frac{\partial g}{\partial t} + V\mathbf{n}\cdot\nabla g + V\kappa\, g\, ds,\end{aligned}$$

which proves the result (14.21c). From

$$\operatorname{div}_\Gamma(V\mathbf{n}) = V \operatorname{div}_\Gamma \mathbf{n} = V\kappa$$

it follows that (14.21c) equals (14.21d). □

Let $\mathbf{w} := V\mathbf{n}$ be the normal velocity field, which completely determines the evolution of $\Gamma(t)$. Introduce the material derivative corresponding to this flow field $\mathbf{w}$:

$$\ddot{g} := \frac{\partial g}{\partial t} + \mathbf{w}\cdot\nabla g.$$

Then (14.21d) can be written as

$$\frac{d}{dt}\int_{\Gamma(t)} g\, ds = \int_{\Gamma(t)} \ddot{g} + g \operatorname{div}_\Gamma \mathbf{w}\, ds,$$

which is the analogon for $\Gamma(t)$ of the Reynolds theorem in (14.20) but now corresponding to the normal velocity field $\mathbf{w}$ instead of $\mathbf{u}$.

Remark 14.2.3 We note that for the result in (14.21b) to hold, it is not necessary to assume that $\Gamma(t)$ is the boundary of a domain $\Omega_1(t)$. In [46] it is shown that the result also holds, for $t \in (t_0 - \delta, t_0 + \delta)$ with $\delta > 0$ sufficiently small, if $\Gamma(t_0)$ is a compact C^2-hypersurface in Ω (and thus $\partial\Gamma(t_0)$ may be nonempty) and the velocity field $\mathbf{u}$ is C^1 on $\Omega \times [t_0 - \delta, t_0 + \delta]$.

15

Appendix B: Variational formulations in Hilbert spaces

In this appendix we collect some results on the well-posedness of variational problems in Hilbert spaces. These results are known in the literature. For most of the proofs we refer to the literature.

15.1 Variational problems and Galerkin discretization

We start with a remark on notation: In this appendix, for elements from a Hilbert space we use boldface notation (e.g., $\mathbf{u}$), elements from the dual space (i.e., bounded linear functionals) are denoted by f, g, etc., and for linear operators between spaces we use capitals (e.g., L).

Let H_1 and H_2 be Hilbert spaces. A bilinear form $k : H_1 \times H_2 \to \mathbb{R}$ is *continuous* if there is a constant M such that for all $\mathbf{x} \in H_1$, $\mathbf{y} \in H_2$:

$$|k(\mathbf{x},\mathbf{y})| \leq M\|\mathbf{x}\|_{H_1}\|\mathbf{y}\|_{H_2}. \tag{15.1}$$

For a continuous bilinear form $k : H_1 \times H_2 \to \mathbb{R}$ we define its norm by $\|k\| = \sup\{\,|k(\mathbf{x},\mathbf{y})| : \|\mathbf{x}\|_{H_1} = 1,\ \|\mathbf{y}\|_{H_2} = 1\,\}$. A fundamental result is given in the following theorem:

Theorem 15.1.1 *Let H_1, H_2 be Hilbert spaces and $k: H_1 \times H_2 \to \mathbb{R}$ be a continuous bilinear form. For $f \in H_2'$ consider the variational problem:*

$$\text{find} \quad \mathbf{u} \in H_1 \quad \text{such that} \quad k(\mathbf{u}, \mathbf{v}) = f(\mathbf{v}) \quad \text{for all} \quad \mathbf{v} \in H_2. \tag{15.2}$$

The following two statements are equivalent*:*

1. *For arbitrary $f \in H_2'$ the problem* (15.2) *has a unique solution $\mathbf{u} \in H_1$ and $\|\mathbf{u}\|_{H_1} \leq c\|f\|_{H_2'}$ holds with a constant c independent of f.*
2. *The conditions* (15.3) *and* (15.4) *hold:*

$$\exists\, \varepsilon > 0 : \quad \sup_{\mathbf{v} \in H_2} \frac{k(\mathbf{u}, \mathbf{v})}{\|\mathbf{v}\|_{H_2}} \geq \varepsilon \|\mathbf{u}\|_{H_1} \quad \text{for all} \quad \mathbf{u} \in H_1, \tag{15.3}$$

$$\forall\, \mathbf{v} \in H_2,\ \mathbf{v} \neq 0,\ \exists\ \mathbf{u} \in H_1 : \quad k(\mathbf{u}, \mathbf{v}) \neq 0. \tag{15.4}$$

Moreover, for the constants c and ε one can take $c = \frac{1}{\varepsilon}$.

A proof of this result can be found in e.g. [106].

Remark 15.1.2 The condition (15.4) can also be formulated as follows:

$$\left[\mathbf{v} \in H_2 \ \text{ such that } \ k(\mathbf{u}, \mathbf{v}) = 0 \ \text{ for all } \ \mathbf{u} \in H_1\right] \ \Rightarrow \ \mathbf{v} = 0.$$

The condition (15.3) is equivalent to

$$\exists\, \varepsilon > 0 : \quad \inf_{\mathbf{u} \in H_1 \setminus \{0\}} \sup_{\mathbf{v} \in H_2} \frac{k(\mathbf{u}, \mathbf{v})}{\|\mathbf{u}\|_{H_1} \|\mathbf{v}\|_{H_2}} \geq \varepsilon, \tag{15.5}$$

and is often called the *inf-sup condition.* In the *finite* dimensional case with $\dim(H_1) = \dim(H_2) < \infty$ this condition implies the result in (15.4) and thus is *necessary and sufficient* for existence and uniqueness.

The Galerkin discretization of the problem (15.2) is based on the following simple idea. We assume *finite dimensional subspaces* $H_{1,h} \subset H_1$, $H_{2,h} \subset H_2$ (note: in concrete cases the index h will correspond to some mesh size parameter) and consider the finite dimensional variational problem

$$\text{find } \mathbf{u}_h \in H_{1,h} \text{ such that } k(\mathbf{u}_h, \mathbf{v}_h) = f(\mathbf{v}_h) \text{ for all } \mathbf{v}_h \in H_{2,h}. \tag{15.6}$$

This problem is called a *Galerkin discretization* of (15.2) (in $H_{1,h} \times H_{2,h}$). We now discuss the well-posedness of this Galerkin-discretization. First note that the continuity of $k : H_{1,h} \times H_{2,h} \to \mathbb{R}$ follows from (15.1). From Theorem 15.1.1 it follows that we need the conditions (15.3) and (15.4) with H_i replaced by $H_{i,h}$, $i = 1, 2$. However, because $H_{i,h}$ is finite dimensional we only need (15.3) since this implies (15.4). Thus we formulate the following (discrete) inf-sup condition in the space $H_{1,h} \times H_{2,h}$:

$$\exists\, \varepsilon_h > 0 : \quad \sup_{\mathbf{v}_h \in H_{2,h}} \frac{k(\mathbf{u}_h, \mathbf{v}_h)}{\|\mathbf{v}_h\|_{H_2}} \geq \varepsilon_h \|\mathbf{u}_h\|_{H_1} \quad \text{for all} \quad \mathbf{u}_h \in H_{1,h}. \tag{15.7}$$

We prove a fundamental result in which the *discretization* error $\|\mathbf{u}-\mathbf{u}_h\|_{H_1}$ is bounded by an *approximation* error $\inf_{\mathbf{v}_h\in H_{1,h}} \|\mathbf{u}-\mathbf{v}_h\|_{H_1}$. In the literature this result is often called "Céa's lemma".

Theorem 15.1.3 (Céa's lemma.) *Let H_1, H_2 be Hilbert spaces and $k : H_1 \times H_2 \to \mathbb{R}$ be a bilinear form. Assume that* (15.1), (15.3), (15.4), (15.7) *hold. Then the variational problem* (15.2) *and its Galerkin discretization* (15.6) *have unique solutions* $\mathbf{u}$ *and* $\mathbf{u}_h$*, respectively. Furthermore, the inequality*

$$\|\mathbf{u}-\mathbf{u}_h\|_{H_1} \le \Big(1+\frac{M}{\varepsilon_h}\Big) \inf_{\mathbf{v}_h\in H_{1,h}} \|\mathbf{u}-\mathbf{v}_h\|_{H_1} \tag{15.8}$$

holds.

Proof. The existence and uniqueness of $\mathbf{u}$ and $\mathbf{u}_h$ follow from Theorem 15.1.1 and the fact that in the finite dimensional case (15.3) implies (15.4). From (15.2) and (15.6) it follows that

$$k(\mathbf{u}-\mathbf{u}_h, \mathbf{v}_h) = 0 \quad \text{for all} \quad \mathbf{v}_h \in H_{2,h}. \tag{15.9}$$

For arbitrary $\mathbf{v}_h \in H_{1,h}$ we have, due to (15.7), (15.9), (15.1):

$$\begin{aligned}\|\mathbf{v}_h-\mathbf{u}_h\|_{H_1} &\le \frac{1}{\varepsilon_h} \sup_{\mathbf{w}_h\in H_{2,h}} \frac{k(\mathbf{v}_h-\mathbf{u}_h,\mathbf{w}_h)}{\|\mathbf{w}_h\|_{H_2}}\\ &= \frac{1}{\varepsilon_h} \sup_{\mathbf{w}_h\in H_{2,h}} \frac{k(\mathbf{v}_h-\mathbf{u},\mathbf{w}_h)}{\|\mathbf{w}_h\|_{H_2}} \le \frac{M}{\varepsilon_h}\|\mathbf{v}_h-\mathbf{u}\|_{H_1}.\end{aligned}$$

From this and the triangle inequality

$$\|\mathbf{u}-\mathbf{u}_h\|_{H_1} \le \|\mathbf{u}-\mathbf{v}_h\|_{H_1} + \|\mathbf{v}_h-\mathbf{u}_h\|_{H_1} \quad \text{for all} \quad \mathbf{v}_h \in H_{1,h}$$

the result follows. □

15.2 Application to elliptic problems

In this section we apply the results from Sect. 15.1 in the special case $H_1 = H_2 =: H$ and with a bilinear form $k : H \times H \to \mathbb{R}$ that is assumed to be *H-elliptic*, i.e., there exists a constant $\gamma > 0$ such

$$k(\mathbf{u},\mathbf{u}) \ge \gamma\|\mathbf{u}\|_H^2 \quad \text{for all} \quad \mathbf{u} \in H.$$

From this property it follows that the conditions (15.3), (15.4) and (15.7) are satisfied with $\varepsilon = \varepsilon_h = \gamma$. Thus, as an immediate consequence of Theorem 15.1.1 we obtain the following famous result, often called "Lax-Milgram lemma".

Theorem 15.2.1 (Lax-Milgram lemma) *Let H be a Hilbert space and $k : H \times H \to \mathbb{R}$ a continuous H-elliptic bilinear form with ellipticity constant γ. Then for every $f \in H'$ there exists a unique $\mathbf{u} \in H$ such that*

$$k(\mathbf{u}, \mathbf{v}) = f(\mathbf{v}) \quad \text{for all} \quad \mathbf{v} \in H. \tag{15.10}$$

Furthermore, the inequality $\|\mathbf{u}\|_H \leq \frac{1}{\gamma}\|f\|_{H'}$ holds.

If the bilinear form is in addition *symmetric*, i.e., $k(\mathbf{u}, \mathbf{v}) = k(\mathbf{v}, \mathbf{u})$ for all $\mathbf{u}, \mathbf{v} \in H$, then there is a natural correspondence between the variational problem (15.10) and a minimization problem:

Theorem 15.2.2 *Let H be a Hilbert space and $k : H \times H \to \mathbb{R}$ a continuous H-elliptic* symmetric *bilinear form. For $f \in H'$ let $\mathbf{u} \in H$ be the unique solution of the variational problem* (15.10). *Then $\mathbf{u}$ is the unique minimizer of the functional*

$$J(\mathbf{v}) := \frac{1}{2}k(\mathbf{v}, \mathbf{v}) - f(\mathbf{v}). \tag{15.11}$$

Proof. From the Lax-Milgram lemma it follows that the variational problem (15.10) has a unique solution $\mathbf{u} \in H$. For arbitrary $\mathbf{z} \in H$, $\mathbf{z} \neq 0$, we have, with ellipticity constant $\gamma > 0$:

$$\begin{aligned} J(\mathbf{u}+\mathbf{z}) &= \frac{1}{2}k(\mathbf{u}+\mathbf{z}, \mathbf{u}+\mathbf{z}) - f(\mathbf{u}+\mathbf{z}) \\ &= \frac{1}{2}k(\mathbf{u}, \mathbf{u}) - f(\mathbf{u}) + k(\mathbf{u}, \mathbf{z}) - f(\mathbf{z}) + \frac{1}{2}k(\mathbf{z}, \mathbf{z}) \\ &= J(\mathbf{u}) + \frac{1}{2}k(\mathbf{z}, \mathbf{z}) \geq J(\mathbf{u}) + \frac{1}{2}\gamma\|\mathbf{z}\|_H^2 > J(\mathbf{u}). \end{aligned}$$

This proves the desired result. □

In the elliptic case one can improve the discretization error bound in Céa's lemma. First we give a result in which the term $1 + \frac{M}{\varepsilon_h} = 1 + \frac{M}{\gamma}$ is replaced by $\frac{M}{\gamma}$.

Theorem 15.2.3 *Consider the problem* (15.10) *and its Galerkin discretization in the subspace $H_h \subset H$. Assume that the conditions as in Theorem 15.2.1 are satisfied. Then the variational problem* (15.10) *and its Galerkin discretization have unique solutions $\mathbf{u}$ and $\mathbf{u}_h$, respectively. Furthermore, the inequality*

$$\|\mathbf{u} - \mathbf{u}_h\|_H \leq \frac{M}{\gamma} \inf_{\mathbf{v}_h \in H_h} \|\mathbf{u} - \mathbf{v}_h\|_H \tag{15.12}$$

holds.

Proof. Theorem 15.2.1 can be applied both to (15.10) and its Galerkin discretization. Thus we conclude that unique solutions $\mathbf{u}$ of (15.10) and $\mathbf{u}_h$ of the Galerkin discretization exist. Using $k(\mathbf{u}-\mathbf{u}_h,\mathbf{v}_h)=0$ for all $\mathbf{v}_h \in H_h$ and the ellipticity and continuity properties, we get for arbitrary $\mathbf{v}_h \in H_h$:

$$\begin{aligned}\|\mathbf{u}-\mathbf{u}_h\|_H^2 &\leq \frac{1}{\gamma}k(\mathbf{u}-\mathbf{u}_h,\mathbf{u}-\mathbf{u}_h) = \frac{1}{\gamma}k(\mathbf{u}-\mathbf{u}_h,\mathbf{u}-\mathbf{v}_h)\\ &\leq \frac{M}{\gamma}\|\mathbf{u}-\mathbf{u}_h\|_H\|\mathbf{u}-\mathbf{v}_h\|_H.\end{aligned}$$

Hence the inequality in (15.12) holds. □

An improvement of the bound in (15.12) can be obtained if $k(\cdot,\cdot)$ is symmetric:

Theorem 15.2.4 *Assume that the conditions as in Theorem 15.2.3 are satisfied. If in addition the bilinear form $k(\cdot,\cdot)$ is* symmetric, *the inequality*

$$\|\mathbf{u}-\mathbf{u}_h\|_H \leq \sqrt{\frac{M}{\gamma}}\inf_{\mathbf{v}_h\in H_h}\|\mathbf{u}-\mathbf{v}_h\|_H \tag{15.13}$$

holds.

Proof. Introduce the norm $|\!|\!|v|\!|\!| := k(\mathbf{v},\mathbf{v})^{\frac{1}{2}}$ on H. Note that

$$\sqrt{\gamma}\|\mathbf{v}\|_H \leq |\!|\!|v|\!|\!| \leq \sqrt{M}\|\mathbf{v}\|_H \quad \text{for all} \quad \mathbf{v}\in H.$$

The space $(H, |\!|\!|\cdot|\!|\!|)$ is a Hilbert space and due to $|\!|\!|\mathbf{v}|\!|\!|^2 = k(\mathbf{v},\mathbf{v})$, $k(\mathbf{u},\mathbf{v}) \leq |\!|\!|\mathbf{u}|\!|\!|\,|\!|\!|\mathbf{v}|\!|\!|$ the bilinear form has ellipticity constant and continuity constant w.r.t. the norm $|\!|\!|\cdot|\!|\!|$ both equal to 1. Application of Theorem 15.2.3 in the space $(H, |\!|\!|\cdot|\!|\!|)$ yields

$$|\!|\!|\mathbf{u}-\mathbf{u}_h|\!|\!| \leq \inf_{\mathbf{v}_h\in H_h}|\!|\!|\mathbf{u}-\mathbf{v}_h|\!|\!|,$$

and thus we obtain

$$\begin{aligned}\|\mathbf{u}-\mathbf{u}_h\|_H &\leq \frac{1}{\sqrt{\gamma}}|\!|\!|\mathbf{u}-\mathbf{u}_h|\!|\!| \leq \frac{1}{\sqrt{\gamma}}\inf_{\mathbf{v}_h\in H_h}|\!|\!|\mathbf{u}-\mathbf{v}_h|\!|\!|\\ &\leq \sqrt{\frac{M}{\gamma}}\inf_{\mathbf{v}_h\in H_h}\|\mathbf{u}-\mathbf{v}_h\|_H,\end{aligned}$$

which completes the proof. □

15.3 Application to saddle point problems

We introduce an abstract saddle point problem. Let V and M be Hilbert spaces and

$$\hat{a}: V\times V\to\mathbb{R}, \quad \hat{b}: V\times M\to\mathbb{R},$$

be continuous bilinear forms. For $f_1 \in V'$, $f_2 \in M'$ we define the following variational problem: find $(\boldsymbol{\phi}, \boldsymbol{\lambda}) \in V \times M$ such that

$$\hat{a}(\boldsymbol{\phi}, \boldsymbol{\psi}) + \hat{b}(\boldsymbol{\psi}, \boldsymbol{\lambda}) = f_1(\boldsymbol{\psi}) \quad \text{for all} \quad \boldsymbol{\psi} \in V \tag{15.14a}$$

$$\hat{b}(\boldsymbol{\phi}, \boldsymbol{\mu}) = f_2(\boldsymbol{\mu}) \quad \text{for all} \quad \boldsymbol{\mu} \in M. \tag{15.14b}$$

This variational problem can be put in the general framework of Sect. 15.1 as follows. Define $H := V \times M$ and

$$\begin{aligned} &k : H \times H \to \mathbb{R}, \qquad k(\mathbf{u}, \mathbf{v}) := \hat{a}(\boldsymbol{\phi}, \boldsymbol{\psi}) + \hat{b}(\boldsymbol{\phi}, \boldsymbol{\mu}) + \hat{b}(\boldsymbol{\psi}, \boldsymbol{\lambda}), \\ &\text{with} \quad \mathbf{u} := (\boldsymbol{\phi}, \boldsymbol{\lambda}), \ \mathbf{v} := (\boldsymbol{\psi}, \boldsymbol{\mu}). \end{aligned} \tag{15.15}$$

On H we use the product norm $\|\mathbf{u}\|_H^2 = \|\boldsymbol{\phi}\|_V^2 + \|\boldsymbol{\lambda}\|_M^2$, for $\mathbf{u} = (\boldsymbol{\phi}, \boldsymbol{\lambda}) \in H$. If we define $f \in H' = V' \times M'$ by $f(\boldsymbol{\phi}, \boldsymbol{\lambda}) = f_1(\boldsymbol{\psi}) + f_2(\boldsymbol{\mu})$ then the problem (15.14) can be reformulated in the setting of Theorem 15.1.1 as follows:

$$\text{find} \quad \mathbf{u} \in H \quad \text{such that} \quad k(\mathbf{u}, \mathbf{v}) = f(\mathbf{v}) \quad \text{for all} \quad \mathbf{v} \in H. \tag{15.16}$$

Based on Theorem 15.1.1 the following well-posedness result for the saddle point problem can be derived, cf. [55, 121], in which the conditions (15.3) and (15.4) on the bilinear form $k(\cdot,\cdot)$ are replaced by conditions on $\hat{a}(\cdot,\cdot)$ and $\hat{b}(\cdot,\cdot)$.

Theorem 15.3.1 *For arbitrary $f_1 \in V', f_2 \in M'$ consider the variational problem* (15.14). *Assume that the bilinear forms $\hat{a}(\cdot,\cdot)$ and $\hat{b}(\cdot,\cdot)$ are continuous and satisfy the following two conditions:*

$$\exists\, \beta > 0 : \ \sup_{\boldsymbol{\psi} \in V} \frac{\hat{b}(\boldsymbol{\psi}, \boldsymbol{\lambda})}{\|\boldsymbol{\psi}\|_V} \geq \beta \, \|\boldsymbol{\lambda}\|_M \quad \forall\, \boldsymbol{\lambda} \in M \quad \text{(inf-sup)}, \tag{15.17a}$$

$$\exists\, \gamma > 0 \ : \quad \hat{a}(\boldsymbol{\phi}, \boldsymbol{\phi}) \geq \gamma \, \|\boldsymbol{\phi}\|_V^2 \quad \forall\, \boldsymbol{\phi} \in V \quad \text{(V-ellipt.)}. \tag{15.17b}$$

Then the problem (15.14) *has a unique solution* $(\boldsymbol{\phi}, \boldsymbol{\lambda})$. *Moreover, the stability bound*

$$\left(\|\boldsymbol{\phi}\|_V^2 + \|\boldsymbol{\lambda}\|_M^2\right)^{\frac{1}{2}} \leq \frac{(\beta + 2\|\hat{a}\|)^2}{\gamma \beta^2} \left(\|f_1\|_{V'}^2 + \|f_2\|_{M'}^2\right)^{\frac{1}{2}}$$

holds. Hence the problem (15.14) *is well-posed.*

Remark 15.3.2 The *inf-sup* condition in (15.17a) is not only sufficient but also *necessary* for well-posedness of the saddle point problem (15.14). The condition on $\hat{a}(\cdot,\cdot)$ in (15.17b) is sufficient but not necessary. It turns out that the following two conditions for $\hat{a}(\cdot,\cdot)$ together are necessary and sufficient:

$$\exists\, \delta > 0 : \ \sup_{\boldsymbol{\psi} \in V_0} \frac{\hat{a}(\boldsymbol{\phi}, \boldsymbol{\psi})}{\|\boldsymbol{\psi}\|_V} \geq \delta \, \|\boldsymbol{\phi}\|_V \quad \text{for all} \quad \boldsymbol{\phi} \in V_0,$$

$$\forall\, \boldsymbol{\psi} \in V_0, \ \boldsymbol{\psi} \neq 0, \ \exists\, \boldsymbol{\phi} \in V_0 \ : \ \hat{a}(\boldsymbol{\phi}, \boldsymbol{\psi}) \neq 0,$$

with $V_0 := \left\{ \boldsymbol{\phi} \in V : \hat{b}(\boldsymbol{\phi}, \boldsymbol{\lambda}) = 0 \quad \text{for all} \quad \boldsymbol{\lambda} \in M \right\}$.

If $\hat{a}(\cdot,\cdot)$ is in addition assumed to be symmetric, then as in Theorem 15.2.2 there is a natural correspondence between the variational problem (15.14) and extrema of a functional:

Theorem 15.3.3 *Assume that the bilinear forms* $\hat{a}(\cdot,\cdot)$ *and* $\hat{b}(\cdot,\cdot)$ *are continuous and satisfy the conditions* (15.17). *In addition we assume that* $\hat{a}(\cdot,\cdot)$ *is* symmetric. *For arbitrary* $f_1 \in V'$, $f_2 \in M'$ *let* $(\boldsymbol{\phi}, \boldsymbol{\lambda})$ *be the unique solution of* (15.14). *Define the functional* $\mathcal{L} : V \times M \to \mathbb{R}$ *by*

$$\mathcal{L}(\boldsymbol{\psi}, \boldsymbol{\mu}) = \frac{1}{2}\hat{a}(\boldsymbol{\psi}, \boldsymbol{\psi}) + \hat{b}(\boldsymbol{\psi}, \boldsymbol{\mu}) - f_1(\boldsymbol{\psi}) - f_2(\boldsymbol{\mu}).$$

Then $(\boldsymbol{\phi}, \boldsymbol{\lambda})$ *is also the unique element in* $V \times M$ *for which*

$$\mathcal{L}(\boldsymbol{\phi}, \boldsymbol{\mu}) \leq \mathcal{L}(\boldsymbol{\phi}, \boldsymbol{\lambda}) \leq \mathcal{L}(\boldsymbol{\psi}, \boldsymbol{\lambda}) \quad \text{for all} \quad \boldsymbol{\psi} \in V, \ \boldsymbol{\mu} \in M \tag{15.18}$$

holds.

For a proof of this result we refer to the literature, e.g. [121]. The property in (15.18) explains why this type of variational equations are called "saddle point" problems.

The unknown $\boldsymbol{\lambda}$ in (15.14) can be eliminated, resulting in an equivalent formulation in which only the unknown $\boldsymbol{\phi}$ occurs. For this we introduce the following notation, for $f_2 \in M'$:

$$V_{f_2} := \left\{ \boldsymbol{\phi} \in V : \hat{b}(\boldsymbol{\phi}, \boldsymbol{\mu}) = f_2(\boldsymbol{\mu}) \quad \text{for all} \quad \boldsymbol{\mu} \in M \right\}.$$

We consider the variational problem: determine $\boldsymbol{\phi} \in V_{f_2}$ such that

$$\hat{a}(\boldsymbol{\phi}, \boldsymbol{\psi}) = f_1(\boldsymbol{\psi}) \quad \text{for all} \quad \boldsymbol{\psi} \in V_0. \tag{15.19}$$

The equivalence of this problem and the saddle point problem (15.14) is given in the following theorem.

Theorem 15.3.4 *Let the assumptions as in Theorem 15.3.1 be satisfied. Let* $(\boldsymbol{\phi}, \boldsymbol{\lambda})$ *be the unique solution of problem* (15.14). *Then* $\boldsymbol{\phi}$ *is the unique solution of the variational problem* (15.19).

Proof. For $\boldsymbol{\phi}$ we have $\hat{b}(\boldsymbol{\phi}, \boldsymbol{\mu}) = f_2(\boldsymbol{\mu})$ for all $\boldsymbol{\mu} \in M$, hence $\boldsymbol{\phi} \in V_{f_2}$. From (15.14a) and $\hat{b}(\boldsymbol{\psi}, \boldsymbol{\lambda}) = 0$ for all $\boldsymbol{\psi} \in V_0$ it follows that

$$\hat{a}(\boldsymbol{\phi}, \boldsymbol{\psi}) = f_1(\boldsymbol{\psi}) \quad \text{for all} \quad \boldsymbol{\psi} \in V_0,$$

and thus $\boldsymbol{\psi}$ solves the problem (15.19). Uniqueness of this solution follows using the ellipticity property (15.17b). □

Now we consider the *Galerkin discretization* of the saddle point problem formulated in (15.14). We introduce finite dimensional subspaces V_h and M_h:

$$V_h \subset V, \quad M_h \subset M.$$

The Galerkin discretization of the problem (15.14) is as follows: find $(\boldsymbol{\phi}_h, \boldsymbol{\lambda}_h) \in V_h \times M_h$ such that

$$\hat{a}(\boldsymbol{\phi}_h, \boldsymbol{\psi}_h) + \hat{b}(\boldsymbol{\psi}_h, \boldsymbol{\lambda}_h) = f_1(\boldsymbol{\psi}_h) \quad \text{for all} \quad \boldsymbol{\psi}_h \in V_h \tag{15.20a}$$

$$\hat{b}(\boldsymbol{\phi}_h, \boldsymbol{\mu}_h) = f_2(\boldsymbol{\mu}_h) \quad \text{for all} \quad \boldsymbol{\mu}_h \in M_h. \tag{15.20b}$$

For the discretization error we have the following result, cf. [55, 121, 106].

Theorem 15.3.5 *Consider the variational problem* (15.14) *and its Galerkin discretization* (15.20), *with continuous bilinear forms* $\hat{a}(\cdot,\cdot)$ *and* $\hat{b}(\cdot,\cdot)$ *that satisfy:*

$$\exists\, \beta > 0: \quad \sup_{\boldsymbol{\psi}\in V} \frac{\hat{b}(\boldsymbol{\psi},\boldsymbol{\lambda})}{\|\boldsymbol{\psi}\|_V} \geq \beta\, \|\boldsymbol{\lambda}\|_M \quad \forall\, \boldsymbol{\lambda} \in M, \tag{15.21a}$$

$$\exists\, \gamma > 0: \quad \hat{a}(\boldsymbol{\phi},\boldsymbol{\phi}) \geq \gamma\, \|\boldsymbol{\phi}\|_V^2 \quad \forall\, \boldsymbol{\phi} \in V, \tag{15.21b}$$

$$\exists\, \beta_h > 0: \quad \sup_{\boldsymbol{\psi}_h\in V_h} \frac{\hat{b}(\boldsymbol{\psi}_h,\boldsymbol{\lambda}_h)}{\|\boldsymbol{\psi}_h\|_V} \geq \beta_h\, \|\boldsymbol{\lambda}_h\|_M \quad \forall\, \boldsymbol{\lambda}_h \in M_h. \tag{15.21c}$$

Then the problem (15.14) *and its Galerkin discretization have unique solutions* $(\boldsymbol{\phi}, \boldsymbol{\lambda})$ *and* $(\boldsymbol{\phi}_h, \boldsymbol{\lambda}_h)$, *respectively. Furthermore the inequality*

$$\|\boldsymbol{\phi} - \boldsymbol{\phi}_h\|_V + \|\boldsymbol{\lambda} - \boldsymbol{\lambda}_h\|_M \leq C\big(\inf_{\boldsymbol{\psi}_h\in V_h} \|\boldsymbol{\phi} - \boldsymbol{\psi}_h\|_V + \inf_{\boldsymbol{\mu}_h\in M_h} \|\boldsymbol{\lambda} - \boldsymbol{\mu}_h\|_M\big)$$

holds, with $C = \sqrt{2}\big(1 + \gamma^{-1}\beta_h^{-2}(2\|\hat{a}\| + \|\hat{b}\|)^3\big)$.

Remark 15.3.6 The condition (15.21c) implies $\dim(V_h) \geq \dim(M_h)$. This can be shown by the following argument. Let $(\boldsymbol{\psi}_j)_{1\leq j\leq m}$ be a basis of V_h and $(\boldsymbol{\lambda}_i)_{1\leq i\leq k}$ a basis of M_h. Define the matrix $\mathbf{B} \in \mathbb{R}^{k\times m}$ by

$$\mathbf{B}_{ij} = \hat{b}(\boldsymbol{\psi}_j, \boldsymbol{\lambda}_i).$$

From (15.21c) it follows that for every $\boldsymbol{\lambda}_h \in M_h$, $\boldsymbol{\lambda}_h \neq 0$, there exists $\boldsymbol{\psi}_h \in V_h$ such that $\hat{b}(\boldsymbol{\psi}_h, \boldsymbol{\lambda}_h) \neq 0$. Thus for every $\mathbf{y} \in \mathbb{R}^k$, $\mathbf{y} \neq 0$, there exists $\mathbf{x} \in \mathbb{R}^m$ such that $\mathbf{y}^T\mathbf{B}\mathbf{x} \neq 0$, i.e., $\mathbf{x}^T\mathbf{B}^T\mathbf{y} \neq 0$. This implies that all columns of $\mathbf{B}^T$, and thus all rows of $\mathbf{B}$, are independent. A necessary condition for this is $k \leq m$. □

The first two conditions (15.21a) and (15.21b) are introduced in view of well-posedness of the given variational saddle point problem (15.14), cf. (15.17). The third condition (15.21c), which is called *discrete inf-sup condition*, is essential for the stability of the Galerkin discretization. Note that the constant C in the discretization error bound in Theorem 15.3.5 depends on β_h and that $C \to \infty$ if $\beta_h \downarrow 0$. The discrete inf-sup condition clearly depends on the specific pair of spaces (V_h, M_h) that is chosen.

15.4 A Strang lemma for saddle point problems

In this section we address the error analysis for the case that in the Galerkin discretization of a saddle point problem one uses a *perturbed* functional $\tilde{f}_1$ in the right-hand side. For the elliptic case this situation has first been analyzed by Strang in [227] and the error bound that he derived is often called the "first Strang lemma", cf. [71], Theorem 26.1. Here we present an analogon for the saddle point problem.

We consider the saddle point problem (15.14) with $f_2 = 0$, i.e.: find $(\boldsymbol{\phi}, \boldsymbol{\lambda}) \in V \times M$ such that

$$\hat{a}(\boldsymbol{\phi}, \boldsymbol{\psi}) + \hat{b}(\boldsymbol{\psi}, \boldsymbol{\lambda}) = f_1(\boldsymbol{\psi}) \quad \text{for all } \boldsymbol{\psi} \in V \tag{15.22a}$$

$$\hat{b}(\boldsymbol{\phi}, \boldsymbol{\mu}) = 0 \quad \text{for all } \boldsymbol{\mu} \in M. \tag{15.22b}$$

In the Galerkin discretization of this problem we use a perturbation $\tilde{f}_1 \in V_h'$ of f_1, i.e.: find $(\tilde{\boldsymbol{\phi}}_h, \tilde{\boldsymbol{\lambda}}_h) \in V_h \times M_h$ such that

$$\hat{a}(\tilde{\boldsymbol{\phi}}_h, \boldsymbol{\psi}_h) + \hat{b}(\boldsymbol{\psi}_h, \tilde{\boldsymbol{\lambda}}_h) = \tilde{f}_1(\boldsymbol{\psi}_h) \quad \text{for all } \boldsymbol{\psi}_h \in V_h \tag{15.23a}$$

$$\hat{b}(\tilde{\boldsymbol{\phi}}_h, \boldsymbol{\mu}_h) = 0 \quad \text{for all } \boldsymbol{\mu}_h \in M_h. \tag{15.23b}$$

For the perturbed problem the following generalization of Theorem 15.3.5 holds.

Theorem 15.4.1 *Let the assumptions of Theorem 15.3.5 be satisfied. Then the problem* (15.22) *and its perturbed Galerkin discretization* (15.23) *have unique solutions* $(\boldsymbol{\phi}, \boldsymbol{\lambda})$ *and* $(\tilde{\boldsymbol{\phi}}_h, \tilde{\boldsymbol{\lambda}}_h)$*, respectively. Furthermore the inequality*

$$\|\boldsymbol{\phi} - \tilde{\boldsymbol{\phi}}_h\|_V + \|\boldsymbol{\lambda} - \tilde{\boldsymbol{\lambda}}_h\|_M \leq C\big(\inf_{\boldsymbol{\psi}_h \in V_h} \|\boldsymbol{\phi} - \boldsymbol{\psi}_h\|_V + \inf_{\boldsymbol{\mu}_h \in M_h} \|\boldsymbol{\lambda} - \boldsymbol{\mu}_h\|_M\big) + \tilde{C}\|f_1 - \tilde{f}_1\|_{V_h'}$$

holds, with C *as in Theorem 15.3.5 and* $\tilde{C} = \gamma^{-1} + \beta_h^{-1}\big(\frac{\|\hat{a}\|}{\gamma} + 1\big)$.

Proof. Let $(\boldsymbol{\phi}_h, \boldsymbol{\lambda}_h)$ be the solution of the Galerkin discretization (15.20) with $f_2 = 0$. We introduce the notation $e_h := \boldsymbol{\phi}_h - \tilde{\boldsymbol{\phi}}_h \in V_h$, $\delta_h := \boldsymbol{\lambda}_h - \tilde{\boldsymbol{\lambda}}_h \in M_h$. From the triangle inequality and the result in Theorem 15.3.5 we get

$$\begin{aligned}\|\boldsymbol{\phi} - \tilde{\boldsymbol{\phi}}_h\|_V + \|\boldsymbol{\lambda} - \tilde{\boldsymbol{\lambda}}_h\|_M &\leq \|\boldsymbol{\phi} - \boldsymbol{\phi}_h\|_V + \|\boldsymbol{\lambda} - \boldsymbol{\lambda}_h\|_M + \|e_h\|_V + \|\delta_h\|_M \\ &\leq C\big(\inf_{\boldsymbol{\psi}_h \in V_h} \|\boldsymbol{\phi} - \boldsymbol{\psi}_h\|_V + \inf_{\boldsymbol{\mu}_h \in M_h} \|\boldsymbol{\lambda} - \boldsymbol{\mu}_h\|_M\big) \\ &\quad + \|e_h\|_V + \|\delta_h\|_M.\end{aligned}$$

It remains to prove that

$$\|e_h\|_V + \|\delta_h\|_M \leq \tilde{C}\|f_1 - \tilde{f}_1\|_{V_h'} \tag{15.24}$$

holds. Note that $\hat{b}(e_h, \boldsymbol{\mu}_h) = 0$ for all $\boldsymbol{\mu}_h \in M_h$. Taking $\boldsymbol{\psi}_h = e_h$ in (15.20a) and in (15.23a) results in $\hat{a}(\boldsymbol{\phi}_h, e_h) = f_1(e_h)$ and $\hat{a}(\tilde{\boldsymbol{\phi}}_h, e_h) = \tilde{f}_1(e_h)$, respectively. Hence,

$$\hat{a}(e_h, e_h) = f_1(e_h) - \tilde{f}_1(e_h)$$

holds. Ellipticity of $\hat{a}(\cdot,\cdot)$ implies $\|e_h\|_V^2 \le \gamma^{-1}\hat{a}(e_h, e_h) = \gamma^{-1}(f_1 - \tilde{f}_1)(e_h)$ and thus

$$\|e_h\|_V \le \frac{1}{\gamma}\|f_1 - \tilde{f}_1\|_{V_h'}. \tag{15.25}$$

From (15.20a) and (15.23a) it follows that

$$\hat{b}(\boldsymbol{\psi}_h, \delta_h) = -\hat{a}(e_h, \boldsymbol{\psi}_h) + (f_1 - \tilde{f}_1)(\boldsymbol{\psi}_h) \quad \text{for all} \quad \boldsymbol{\psi}_h \in V_h.$$

In combination with the discrete inf-sup property of $\hat{b}(\cdot,\cdot)$ this yields

$$\|\delta_h\|_M \le \beta_h^{-1} \sup_{\boldsymbol{\psi}_h \in V_h} \frac{\hat{b}(\boldsymbol{\psi}_h, \delta_h)}{\|\boldsymbol{\psi}_h\|_V} \le \beta_h^{-1}\big(\|\hat{a}\|\|e_h\|_V + \|f_1 - \tilde{f}_1\|_{V_h'}\big). \tag{15.26}$$

The results in (15.25) and (15.26) imply that the estimate (15.24) holds with $\tilde{C} = \gamma^{-1} + \beta_h^{-1}\big(\frac{\|\hat{a}\|}{\gamma} + 1\big)$. □

In view of our applications we formulate an easy corollary of this result in which the dependence of the discretization error bound on a scaling of the bilinear form $\hat{a}(\cdot,\cdot)$ is shown.

Corollary 15.4.2 Take $\mu > 0$ and consider the variational problem: find $(\boldsymbol{\phi}, \boldsymbol{\lambda}) \in V \times M$ such that

$$\begin{aligned} \mu\, \hat{a}(\boldsymbol{\phi}, \boldsymbol{\psi}) + \hat{b}(\boldsymbol{\psi}, \boldsymbol{\lambda}) &= f_1(\boldsymbol{\psi}) \quad \text{for all} \quad \boldsymbol{\psi} \in V \\ \hat{b}(\boldsymbol{\phi}, \boldsymbol{\mu}) &= 0 \quad \text{for all} \quad \boldsymbol{\mu} \in M. \end{aligned}$$

For $\tilde{f}_1 \in V_h'$ the perturbed Galerkin discretization is given by: find $(\tilde{\boldsymbol{\phi}}_h, \tilde{\boldsymbol{\lambda}}_h) \in V_h \times M_h$ such that

$$\begin{aligned} \mu\, \hat{a}(\tilde{\boldsymbol{\phi}}_h, \boldsymbol{\psi}_h) + \hat{b}(\boldsymbol{\psi}_h, \tilde{\boldsymbol{\lambda}}_h) &= \tilde{f}_1(\boldsymbol{\psi}_h) \quad \text{for all} \quad \boldsymbol{\psi}_h \in V_h \\ \hat{b}(\tilde{\boldsymbol{\phi}}_h, \boldsymbol{\mu}_h) &= 0 \quad \text{for all} \quad \boldsymbol{\mu}_h \in M_h. \end{aligned}$$

Let the assumptions as in Theorem 15.4.1 hold. From a scaling argument it follows that these problems have unique solutions and the error bound

$$\begin{aligned} \mu\|\boldsymbol{\phi} - \tilde{\boldsymbol{\phi}}_h\|_V + \|\boldsymbol{\lambda} - \tilde{\boldsymbol{\lambda}}_h\|_M \le C\,\big(\mu \inf_{\boldsymbol{\psi}_h \in V_h} \|\boldsymbol{\phi} - \boldsymbol{\psi}_h\|_V + \inf_{\boldsymbol{\mu}_h \in M_h} \|\boldsymbol{\lambda} - \boldsymbol{\mu}_h\|_M\big) \\ + \tilde{C}\|f_1 - \tilde{f}_1\|_{V_h'} \end{aligned}$$

holds, with the same constants C and $\tilde{C}$ as in Theorem 15.4.1.

15.5 Schur complement preconditioning for parameter dependent saddle point problems

In this section we present an abstract analysis of a Schur complement operator in Hilbert spaces. The results are from [189] and are not available in the standard literature on saddle point problems. Similar results are presented in [175]. Based on this framework one can derive Schur complement preconditioners for generalized Stokes equations that are robust with respect to variation of the parameter $\xi = \frac{1}{\Delta t}$, cf. (15.51) below. In Sect. 15.5.1 we present some preliminaries. In Sect. 15.5.2 we derive a Schur complement preconditioner. For part of the proofs we refer to [189]. In Sect. 15.5.3 we apply the abstract results to a concrete generalized Stokes equation.

15.5.1 Preliminaries

Let $H_1 \subset H_2$ and M be Hilbert spaces such that the identity $I : H_1 \to H_2$ is a continuous dense embedding, i.e., H_1 is dense in H_2 and $\|\boldsymbol{\phi}\|_{H_2} \leq c\|\boldsymbol{\phi}\|_{H_1}$ for all $\boldsymbol{\phi} \in H_1$. Assume bilinear forms $\hat{a} : H_1 \times H_1 \to \mathbb{R}$, $\hat{c} : H_2 \times H_2 \to \mathbb{R}$ and $\hat{b} : H_1 \times M$. Related to these bilinear forms we make the following assumptions. $\hat{a}(\cdot,\cdot)$ and $\hat{c}(\cdot,\cdot)$ are *symmetric* and the following *ellipticity, continuity* and *inf-sup* conditions hold with strictly positive constants γ_a, γ_b, γ_c:

$$\gamma_a\|\boldsymbol{\phi}\|_{H_1}^2 \leq \hat{a}(\boldsymbol{\phi},\boldsymbol{\phi}), \;\; \hat{a}(\boldsymbol{\phi},\boldsymbol{\psi}) \leq \Gamma_a\|\boldsymbol{\phi}\|_{H_1}\|\boldsymbol{\psi}\|_{H_1}, \;\; \boldsymbol{\phi},\boldsymbol{\psi} \in H_1, \tag{15.27a}$$

$$\gamma_c\|\boldsymbol{\phi}\|_{H_2}^2 \leq \hat{c}(\boldsymbol{\phi},\boldsymbol{\phi}), \;\; \hat{c}(\boldsymbol{\phi},\boldsymbol{\psi}) \leq \Gamma_c\|\boldsymbol{\phi}\|_{H_2}\|\boldsymbol{\psi}\|_{H_2}, \;\; \boldsymbol{\phi},\boldsymbol{\psi} \in H_2, \tag{15.27b}$$

$$\hat{b}(\boldsymbol{\phi},\boldsymbol{\lambda}) \leq \Gamma_b\|\boldsymbol{\phi}\|_{H_1}\|\boldsymbol{\lambda}\|_M, \quad \boldsymbol{\phi} \in H_1, \boldsymbol{\lambda} \in M, \tag{15.27c}$$

$$\gamma_b\|\boldsymbol{\lambda}\|_M \leq \sup_{\boldsymbol{\phi}\in H_1} \frac{\hat{b}(\boldsymbol{\phi},\boldsymbol{\lambda})}{\|\boldsymbol{\phi}\|_{H_1}}, \quad \boldsymbol{\lambda} \in M. \tag{15.27d}$$

For the analysis given below it is convenient to introduce corresponding linear mappings

$$\begin{aligned}
A &: H_1 \to H_1', && \langle A\boldsymbol{\phi},\boldsymbol{\psi}\rangle = \hat{a}(\boldsymbol{\phi},\boldsymbol{\psi}) && \text{for all } \boldsymbol{\phi},\boldsymbol{\psi} \in H_1,\\
C &: H_2 \to H_2', && \langle C\boldsymbol{\phi},\boldsymbol{\psi}\rangle = \hat{c}(\boldsymbol{\phi},\boldsymbol{\psi}) && \text{for all } \boldsymbol{\phi},\boldsymbol{\psi} \in H_2,\\
B &: M \to H_1', && \langle B\boldsymbol{\lambda},\boldsymbol{\psi}\rangle = \hat{b}(\boldsymbol{\psi},\boldsymbol{\lambda}) && \text{for all } \boldsymbol{\lambda} \in M,\ \boldsymbol{\psi} \in H_1.
\end{aligned}$$

To express the dependence of the duality pairing on the spaces used, we will also write, for example, $\langle C\boldsymbol{\phi},\boldsymbol{\psi}\rangle = \langle C\boldsymbol{\phi},\boldsymbol{\psi}\rangle_{H_2'\times H_2}$. To simplify the notation we write $\langle\cdot,\cdot\rangle$ if it is clear from the context which spaces are used in the duality pairing. The assumptions on the bilinear forms imply that

$$\gamma_a\|\boldsymbol{\phi}\|_{H_1} \leq \|A\boldsymbol{\phi}\|_{H_1'} \leq \Gamma_a\|\boldsymbol{\phi}\|_{H_1} \quad \text{for all } \boldsymbol{\phi} \in H_1, \tag{15.28a}$$

$$\gamma_c\|\boldsymbol{\phi}\|_{H_2} \leq \|C\boldsymbol{\phi}\|_{H_2'} \leq \Gamma_c\|\boldsymbol{\phi}\|_{H_2} \quad \text{for all } \boldsymbol{\phi} \in H_2, \tag{15.28b}$$

$$\gamma_b\|\boldsymbol{\lambda}\|_M \leq \|B\boldsymbol{\lambda}\|_{H_1'} \leq \Gamma_b\|\boldsymbol{\lambda}\|_M \quad \text{for all } \boldsymbol{\lambda} \in M. \tag{15.28c}$$

We also introduce the adjoint B' of B given by

$$B' : H_1 \to M', \quad \langle B' \boldsymbol{\psi}, \boldsymbol{\lambda} \rangle_{M' \times M} = \langle B \boldsymbol{\lambda}, \boldsymbol{\psi} \rangle_{H_1' \times H_1} \quad \text{for all} \quad \boldsymbol{\lambda} \in M, \ \boldsymbol{\psi} \in H_1.$$

Note that due to the symmetry of $\hat{a}(\cdot,\cdot)$ the operator $A : H_1 \to H_1'$ is selfadjoint: $\langle A\boldsymbol{\phi}, \boldsymbol{\psi} \rangle_{H_1' \times H_1} = \langle A\boldsymbol{\psi}, \boldsymbol{\phi} \rangle_{H_1' \times H_1}$ for all $\boldsymbol{\phi}, \boldsymbol{\psi} \in H_1$. The operator $C : H_2 \to H_2'$ is selfadjoint, too.

Consider the following parameter dependent variant of the saddle point problem in (15.14). Given $\tau \geq 0$ and $f_1 \in H_1'$, find $(\boldsymbol{\phi}, \boldsymbol{\lambda}) \in H_1 \times M$ such that

$$\begin{aligned} \hat{a}(\boldsymbol{\phi}, \boldsymbol{\psi}) + \tau \hat{c}(\boldsymbol{\phi}, \boldsymbol{\psi}) + \hat{b}(\boldsymbol{\psi}, \boldsymbol{\lambda}) &= f_1(\boldsymbol{\psi}) \quad \text{for all} \quad \boldsymbol{\psi} \in H_1 \\ \hat{b}(\boldsymbol{\phi}, \boldsymbol{\mu}) &= 0 \quad \text{for all} \quad \boldsymbol{\mu} \in M. \end{aligned} \tag{15.29}$$

The problem (15.29) can be rewritten in operator formulation: find $(\boldsymbol{\phi}, \boldsymbol{\lambda}) \in H_1 \times M$ such that

$$\begin{cases} A\boldsymbol{\phi} + \tau C\boldsymbol{\phi} + B\boldsymbol{\lambda} = f_1 \\ \qquad\qquad B'\boldsymbol{\phi} = 0. \end{cases} \tag{15.30}$$

From Theorem 15.3.1, with $V = H_1$ and $\hat{a}(\cdot,\cdot)$ replaced by $\hat{a}(\cdot,\cdot) + \tau\hat{c}(\cdot,\cdot)$, it follows that this problem has a unique solution. The *Schur complement operator*

$$S : M \to M', \quad S := B'(A + \tau C)^{-1} B$$

of this system is a *selfadjoint positive definite* operator. It defines a scalar product (and corresponding norm) on M:

$$\|\boldsymbol{\lambda}\|_S := \langle S\,\boldsymbol{\lambda}, \boldsymbol{\lambda} \rangle^{\frac{1}{2}} = \sup_{\boldsymbol{\phi} \in H_1} \frac{\langle B\boldsymbol{\lambda}, \boldsymbol{\phi} \rangle}{\langle (A + \tau C)\boldsymbol{\phi}, \boldsymbol{\phi} \rangle^{\frac{1}{2}}}, \qquad \boldsymbol{\lambda} \in M. \tag{15.31}$$

Remark 15.5.1 Consider the simple special case $\tau = 0$. Let $I_M : M \to M'$ be the Riesz isomorphism. From the properties of A and B it follows that

$$\gamma_b \Gamma_a^{-\frac{1}{2}} \|\boldsymbol{\lambda}\|_M \leq \|\boldsymbol{\lambda}\|_S \leq \gamma_a^{-\frac{1}{2}} \Gamma_b \|\boldsymbol{\lambda}\|_M \quad \text{for all} \quad \boldsymbol{\lambda} \in M,$$

holds. Hence, I_M is spectrally equivalent to S:

$$c_0 \langle I_M \boldsymbol{\lambda}, \boldsymbol{\lambda} \rangle \leq \langle S\boldsymbol{\lambda}, \boldsymbol{\lambda} \rangle \leq c_1 \langle I_M \boldsymbol{\lambda}, \boldsymbol{\lambda} \rangle \quad \text{for all} \quad \boldsymbol{\lambda} \in M,$$

with spectral constants $c_0 = \gamma_b^2 \Gamma_a^{-1}$, $c_1 = \gamma_a^{-1} \Gamma_b^2$.

In the analysis we use the concept of sums and intersections of vector spaces (cf. [34]). The idea of applying this concept in the analysis of Schur complement preconditioners is introduced in [174].

Let X, Y be compatible normed spaces, i.e., both X and Y are subspaces of some larger topological vector space Z. Then we can form their sum $X + Y$ and intersection $X \cap Y$. The sum $X + Y$ consists of all $\mathbf{z} \in Z$ such that

$\mathbf{z} = \mathbf{x}+\mathbf{y}$ with $\mathbf{x} \in X, \mathbf{y} \in Y$. The spaces $X \cap Y$ and $X + Y$ are normed vector spaces with norms

$$\|\mathbf{x}\|_{X\cap Y} = \left(\|\mathbf{x}\|_X^2 + \|\mathbf{x}\|_Y^2\right)^{\frac{1}{2}} \quad \text{for } \mathbf{x} \in X \cap Y, \tag{15.32}$$

$$\|\mathbf{z}\|_{X+Y} = \inf_{\mathbf{z}=\mathbf{x}+\mathbf{y}} \left(\|\mathbf{x}\|_X^2 + \|\mathbf{y}\|_Y^2\right)^{\frac{1}{2}} \quad \text{for } \mathbf{x} \in X,\ \mathbf{y} \in Y. \tag{15.33}$$

If X and Y are complete then both $X \cap Y$ and $X + Y$ are complete (Lemma 2.3.1 in [34]). A few properties that we will need further on are given in the following lemma. The space of bounded linear mappings $T : X \to Y$ is denoted by $\mathcal{L}(X, Y)$.

Lemma 15.5.2 *Let X_1, X_2 and Y_1, Y_2 be pairs of compatible normed vector spaces and let T be a linear mapping on $X_1 + X_2$ such that $T \in \mathcal{L}(X_1, Y_1) \cap \mathcal{L}(X_2, Y_2)$. Then $T:\ X_1 + X_2 \to Y_1 + Y_2$ is bounded and*

$$\|T\|_{X_1+X_2\to Y_1+Y_2} \leq \left(\|T\|_{X_1\to Y_1}^2 + \|T\|_{X_2\to Y_2}^2\right)^{\frac{1}{2}} \tag{15.34}$$

holds. If X_1 and X_2 are Hilbert spaces such that $X_1 \cap X_2$ is dense in both X_1 and X_2, then $(X_1 \cap X_2)' = X_1' + X_2'$ holds and

$$\|g\|_{(X_1\cap X_2)'} = \|g\|_{X_1'+X_2'} \quad \textit{for all} \quad g \in (X_1 \cap X_2)'. \tag{15.35}$$

In the remainder we assume $\tau > 0$. By τH_2 we denote the space H_2 with the scaled scalar product $\tau(\cdot,\cdot)_{H_2}$. Using the previous lemma we obtain the following equivalence result for the Schur complement norm in (15.31).

Theorem 15.5.3 *For all $\boldsymbol{\lambda} \in M$ we have*

$$\min\{\gamma_a, \gamma_c\}\|\boldsymbol{\lambda}\|_S^2 \leq \|B\boldsymbol{\lambda}\|_{H_1'+\tau^{-1}H_2'}^2 \leq \max\{\Gamma_a, \Gamma_c\}\|\boldsymbol{\lambda}\|_S^2. \tag{15.36}$$

Proof. For $\boldsymbol{\lambda} \in M$ we have

$$\|B\boldsymbol{\lambda}\|_{(H_1\cap\tau H_2)'} = \sup_{\phi\in H_1} \frac{\langle B\boldsymbol{\lambda}, \phi\rangle}{\left(\|\phi\|_{H_1}^2 + \tau\|\phi\|_{H_2}^2\right)^{\frac{1}{2}}}. \tag{15.37}$$

Due to the properties of A and C and the definition of $\|\cdot\|_S$ we get

$$\min\{\gamma_a, \gamma_c\}\|\boldsymbol{\lambda}\|_S^2 \leq \|B\boldsymbol{\lambda}\|_{(H_1\cap\tau H_2)'}^2 \leq \max\{\Gamma_a, \Gamma_c\}\|\boldsymbol{\lambda}\|_S^2 \quad \text{for all} \quad \boldsymbol{\lambda} \in M.$$

Now we apply the result in (15.35) to the case $X_1 = H_1, X_2 = \tau H_2$. Note that $H_1 \cap \tau H_2 = H_1$ (this should be understood as equality of sets) and that the intersection is dense in τH_2. Hence, we get

$$\|B\boldsymbol{\lambda}\|_{(H_1\cap\tau H_2)'} = \|B\boldsymbol{\lambda}\|_{H_1'+\tau^{-1}H_2'}$$

and thus the result is proved. □

We introduce a subspace W of M:

$$W = \left\{ \boldsymbol{\lambda} \in M : \sup_{\boldsymbol{\phi} \in H_1} \frac{\langle B\boldsymbol{\lambda}, \boldsymbol{\phi} \rangle}{\|\boldsymbol{\phi}\|_{H_2}} < \infty \right\} = \{ \boldsymbol{\lambda} \in M : B\boldsymbol{\lambda} \in H_2' \}. \tag{15.38}$$

(Recall that H_1 is dense in H_2). We define the following functional on W:

$$\|\boldsymbol{\lambda}\|_W := \sup_{\boldsymbol{\phi} \in H_2} \frac{\langle B\boldsymbol{\lambda}, \boldsymbol{\phi} \rangle}{\langle C\boldsymbol{\phi}, \boldsymbol{\phi} \rangle^{\frac{1}{2}}}. \tag{15.39}$$

The lemma below summarizes useful properties of W.

Lemma 15.5.4 *The following holds:*

$$\text{The identity } I : W \to M \text{ is a continuous embedding.} \tag{15.40a}$$

$$B(W) \text{ is a closed subspace of } H_2'. \tag{15.40b}$$

$$\| \cdot \|_W \text{ defines a norm and } (W, \| \cdot \|_W) \text{ is a Hilbert space.} \tag{15.40c}$$

Let $B_W' : H_2 \to W'$ *be the adjoint of* $B : W \to H_2'$, *i.e.,* $\langle B_W' \boldsymbol{\psi}, \boldsymbol{\lambda} \rangle_{W' \times W} = \langle B\boldsymbol{\lambda}, \boldsymbol{\psi} \rangle_{H_2' \times H_2}$ *for all* $\boldsymbol{\psi} \in H_2$, $\boldsymbol{\lambda} \in W$. *Define*

$$S_W : W \to W', \quad S_W := B_W' C^{-1} B. \tag{15.41}$$

The following holds:

$$(\boldsymbol{\lambda}, \boldsymbol{\mu})_W = \langle S_W \boldsymbol{\lambda}, \boldsymbol{\mu} \rangle_{W' \times W} \quad \text{for all } \boldsymbol{\lambda}, \boldsymbol{\mu} \in W. \tag{15.42}$$

Proof. Given in [189]. □

The Schur complement operator S_W in (15.41) has a simpler form than the Schur complement S since S_W does not involve A or τ. A relation between S and the operator S_W will be derived in Sect. 15.5.2.

The space $B(M)$ is a closed subspace of H_1'. In our analysis we will need the orthogonal projection

$$P : H_1' \to B(M), \quad (Pg, q)_{H_1'} = (g, q)_{H_1'} \quad \text{for all } q \in B(M).$$

The following lemma gives another characterization of this projection.

Lemma 15.5.5 *Let* $I_1 : H_1 \to H_1'$ *be the Riesz isomorphism, i.e.,* $\langle I_1 \boldsymbol{\phi}, \boldsymbol{\psi} \rangle = (\boldsymbol{\phi}, \boldsymbol{\psi})_{H_1}$ *for all* $\boldsymbol{\phi}, \boldsymbol{\psi} \in H_1$. *For* $f \in H_1'$ *let* $(\boldsymbol{\phi}, \boldsymbol{\lambda}) \in H_1 \times M$ *be the unique solution of*

$$\begin{aligned} I_1 \boldsymbol{\phi} + B\boldsymbol{\lambda} &= f \\ B' \boldsymbol{\phi} &= 0. \end{aligned}$$

Define the solution operator $S_1 : H_1' \to M$ *by* $f \to \boldsymbol{\lambda}$. *Then* $P = B\, S_1$ *holds.*

Proof. For arbitrary $f \in H_1'$ we have $B\, S_1 f = B\boldsymbol{\lambda} \in B(M)$ and for any $\boldsymbol{\mu} \in M$:

$$\begin{aligned} (f - B\boldsymbol{\lambda}, B\boldsymbol{\mu})_{H_1'} &= \langle I_1^{-1}(f - B\boldsymbol{\lambda}), B\boldsymbol{\mu} \rangle_{H_1 \times H_1'} = \langle \boldsymbol{\phi}, B\boldsymbol{\mu} \rangle_{H_1 \times H_1'} \\ &= \langle B' \boldsymbol{\phi}, \boldsymbol{\mu} \rangle_{M' \times M} = 0, \end{aligned}$$

and thus the result holds. □

15.5.2 Schur complement preconditioner

The space W is a subspace of M and thus the sum norm (15.33) on the space $M+\tau^{-1}W$ is given by

$$\|\boldsymbol{\lambda}\|_{M+\tau^{-1}W} = \inf_{\boldsymbol{\mu}\in W} \left(\|\boldsymbol{\lambda}-\boldsymbol{\mu}\|_M^2 + \tau^{-1}\|\boldsymbol{\mu}\|_W^2\right)^{\frac{1}{2}}. \tag{15.43}$$

From Theorem 15.5.3 it follows that $\|\boldsymbol{\lambda}\|_S$ is uniformly (w.r.t. τ) equivalent to $\|B\boldsymbol{\lambda}\|_{H_1'+\tau^{-1}H_2'}$. Using this, we now show that (under a certain assumption) $\|\boldsymbol{\lambda}\|_S$ is uniformly equivalent to $\|\boldsymbol{\lambda}\|_{M+\tau^{-1}W}$. We first consider $\|\boldsymbol{\lambda}\|_S \le c\|\boldsymbol{\lambda}\|_{M+\tau^{-1}W}$ and then $\|\boldsymbol{\lambda}\|_{M+\tau^{-1}W} \le c\|\boldsymbol{\lambda}\|_S$. We then show that $\|\boldsymbol{\lambda}\|_{M+\tau^{-1}W}$ satisfies $\|\boldsymbol{\lambda}\|_{M+\tau^{-1}W}^2 = \langle \tilde{S}\boldsymbol{\lambda},\boldsymbol{\lambda}\rangle$ with a feasible preconditioner $\tilde{S}$ of S.

Lemma 15.5.6 *Define* $\Gamma_s = \frac{\Gamma_b^2+\Gamma_c}{\min\{\gamma_a,\gamma_c\}}$. *For all* $\boldsymbol{\lambda}\in M$ *we have*

$$\|\boldsymbol{\lambda}\|_S^2 \le \Gamma_s\|\boldsymbol{\lambda}\|_{M+\tau^{-1}W}^2.$$

Proof. From Theorem 15.5.3 we get

$$\|\boldsymbol{\lambda}\|_S^2 \le \frac{1}{\min\{\gamma_a,\gamma_c\}}\|B\boldsymbol{\lambda}\|_{H_1'+\tau^{-1}H_2'}^2.$$

From (15.28b), (15.28c) and the definition of $\|\cdot\|_W$ we have

$$\|B\|_{M\to H_1'} \le \Gamma_b, \quad \|B\|_{W\to H_2'} \le \Gamma_c^{\frac{1}{2}},$$

and thus from (15.34) we obtain

$$\|B\boldsymbol{\lambda}\|_{H_1'+\tau^{-1}H_2'} \le (\Gamma_b^2+\Gamma_c)^{\frac{1}{2}}\|\boldsymbol{\lambda}\|_{M+\tau^{-1}W},$$

which completes the proof. □

For the inequality $\|\boldsymbol{\lambda}\|_{M+\tau^{-1}W} \le c\|\boldsymbol{\lambda}\|_S$, with a constant c independent of τ, we present an analysis which requires an assumption on the orthogonal projection $P: H_1' \to B(M)$.
This crucial assumption is as follows.

Assumption 15.5.7 Assume that $P: H_2' \to H_2'$ and that there exists a constant $c_P \ge 1$ such that

$$\|Pf\|_{H_2'} \le c_P\|f\|_{H_2'} \quad \text{for all} \quad f\in H_2'. \tag{15.44}$$

In [189] a slightly weaker, but more technical, assumption is used.

Lemma 15.5.8 *Let Assumption 15.5.7 hold. Define* $\gamma_s := \frac{\gamma_b^2\gamma_c}{c_P^2(\gamma_b^2+\gamma_c)\max\{\Gamma_a,\Gamma_c\}}$. *For all* $\boldsymbol{\lambda}\in M$ *we have*

$$\gamma_s\|\boldsymbol{\lambda}\|_{M+\tau^{-1}W}^2 \le \|\boldsymbol{\lambda}\|_S^2. \tag{15.45}$$

Proof. We outline the main arguments. For a complete proof we refer to [189]. From the definition of the norm $\|\cdot\|_W$ and properties of B we get

$$\|B^{-1}\|_{B(M)\to M} \leq \gamma_b^{-1}, \quad \|B^{-1}\|_{B(W)\to W} \leq \gamma_c^{-\frac{1}{2}},$$

and thus

$$\|B^{-1}g\|_{M+\tau^{-1}W} \leq (\gamma_b^{-2}+\gamma_c^{-1})^{\frac{1}{2}}\|g\|_{B(M)+\tau^{-1}B(W)} \quad \text{for all } \; g \in B(M).$$

Hence,

$$\frac{\gamma_b^2\gamma_c}{\gamma_b^2+\gamma_c}\|\boldsymbol{\lambda}\|^2_{M+\tau^{-1}W} \leq \|B\boldsymbol{\lambda}\|^2_{B(M)+\tau^{-1}B(W)} \quad \text{for all } \; \boldsymbol{\lambda} \in M.$$

Using Assumption 15.5.7 one can show that

$$\|B\boldsymbol{\lambda}\|^2_{B(M)+\tau^{-1}B(W)} \leq c_P^2\|B\boldsymbol{\lambda}\|^2_{H_1'+\tau^{-1}H_2'} \quad \text{for all } \; \boldsymbol{\lambda} \in M$$

holds. From Theorem 15.5.3 we have

$$\|B\boldsymbol{\lambda}\|^2_{H_1'+\tau^{-1}H_2'} \leq \max\{\Gamma_a, \Gamma_c\}\|\boldsymbol{\lambda}\|_S^2.$$

Hence the result in (15.45) holds. □

The results in these two lemmas imply the following main result.

Corollary 15.5.9 Suppose Assumption 15.5.7 holds. The following inequalities hold for any $\boldsymbol{\lambda} \in M$:

$$\gamma_s\|\boldsymbol{\lambda}\|^2_{M+\tau^{-1}W} \leq \|\boldsymbol{\lambda}\|_S^2 \leq \Gamma_s\|\boldsymbol{\lambda}\|^2_{M+\tau^{-1}W}. \tag{15.46}$$

It is not obvious how to use $\|\boldsymbol{\lambda}\|_{M+\tau^{-1}W}$ to construct a feasible preconditioner for the Schur complement S. We now address this issue.
Let $I_M : M \to M'$ be the Riesz isomorphism. Because the identity $I : W \to M$ is a continuous embedding we have $I_M(W) \subset W'$. The mapping $I_M : W \to W'$ is denoted by I_W (note that in general this is *not* the Riesz-isomorphism in W).

Theorem 15.5.10 *Define* $\tilde{S} : M \to M'$ *by*

$$\tilde{S} = I_M - I_M(I_W + \tau^{-1}S_W)^{-1}I_M,$$

with S_W *defined in* (15.41). *Then* $\tilde{S}$ *is selfadjoint and positive definite and*

$$\|\boldsymbol{\lambda}\|^2_{M+\tau^{-1}W} = \langle\tilde{S}\boldsymbol{\lambda}, \boldsymbol{\lambda}\rangle_{M'\times M} \quad \textit{for all} \;\; \boldsymbol{\lambda} \in M. \tag{15.47}$$

Proof. By assumption the operator $C^{-1} : H_2' \to H_2'$ is selfadjoint, therefore S_W and $\tilde{S}$ are selfadjoint as well.

With the help of elementary variational analysis we see that the infimum on the right-hand side in (15.43) is attained for $\tilde{\boldsymbol{\mu}} \in W$ that satisfies

$$(\tilde{\boldsymbol{\mu}} - \boldsymbol{\lambda}, \boldsymbol{\xi})_M + \tau^{-1}(\tilde{\boldsymbol{\mu}}, \boldsymbol{\xi})_W = 0 \quad \text{for all } \boldsymbol{\xi} \in W.$$

This can be reformulated in operator notation, using the definition of the W-scalar product and (15.42):

$$\langle I_M(\tilde{\boldsymbol{\mu}} - \boldsymbol{\lambda}) + \tau^{-1} S_W \tilde{\boldsymbol{\mu}}, \boldsymbol{\xi} \rangle_{W' \times W} = 0 \quad \text{for all } \boldsymbol{\xi} \in W. \tag{15.48}$$

Note that $I_M \boldsymbol{\lambda} \in M' \subset W'$ holds. The solution $\tilde{\boldsymbol{\mu}} \in W$ of (15.48) is given by

$$(I_W + \tau^{-1} S_W)\tilde{\boldsymbol{\mu}} = I_M \boldsymbol{\lambda},$$

and thus $\tilde{\boldsymbol{\mu}} = (I_W + \tau^{-1} S_W)^{-1} I_M \boldsymbol{\lambda}$. A straightforward computation yields

$$\|\boldsymbol{\lambda}\|^2_{M+\tau^{-1}W} = \|\boldsymbol{\lambda} - \tilde{\boldsymbol{\mu}}\|^2_M + \tau^{-1}\|\tilde{\boldsymbol{\mu}}\|^2_W = (\boldsymbol{\lambda} - \tilde{\boldsymbol{\mu}}, \boldsymbol{\lambda})_M = \langle I_M(\boldsymbol{\lambda} - \tilde{\boldsymbol{\mu}}), \boldsymbol{\lambda} \rangle.$$

Substituting $\tilde{\boldsymbol{\mu}} = (I_W + \tau^{-1} S_W)^{-1} I_M \boldsymbol{\lambda}$, we obtain (15.47). From (15.47) it follows that $\tilde{S}$ is positive definite. □

As a direct consequence of the results in Corollary 15.5.9 and Theorem 15.5.10 we obtain the following main result.

Corollary 15.5.11 Suppose Assumption 15.5.7 holds. The following inequalities hold for any $\boldsymbol{\lambda} \in M$:

$$\gamma_s \langle \tilde{S}\boldsymbol{\lambda}, \boldsymbol{\lambda} \rangle \le \langle S\boldsymbol{\lambda}, \boldsymbol{\lambda} \rangle \le \Gamma_s \langle \tilde{S}\boldsymbol{\lambda}, \boldsymbol{\lambda} \rangle. \tag{15.49}$$

In the setting of preconditioning one is interested in the *inverse* of the preconditioner. By a straightforward computation one can check that the inverse $\tilde{S}^{-1} : M' \to M$ of $\tilde{S}$ is given by

$$\tilde{S}^{-1} := I_M^{-1} + \tau S_W^{-1} = I_M^{-1} + \tau \big(B_W' C^{-1} B\big)^{-1}, \tag{15.50}$$

with B_W' the adjoint of $B : W \to H_2'$, cf. Lemma 15.5.4. In (15.50) $S_W^{-1} : W' \to W$ is considered as mapping $M' \to M$. As a final result in this section we give a simple criterion which is a useful sufficient condition for Assumption 15.5.7 to hold.

Lemma 15.5.12 *Let $S_1 : H_1' \to M$ be the solution operator from Lemma 15.5.5. Assume that there is a subspace $\widetilde{W} \subset M$ with norm $\|\cdot\|_{\widetilde{W}}$ such that both $S_1 : H_2' \to \widetilde{W}$ and $B : \widetilde{W} \to H_2'$ are bounded, i.e.,*

$$\|S_1 f\|_{\widetilde{W}} \le c_1 \|f\|_{H_2'} \quad \forall\, f \in H_2'\,, \quad \|B\boldsymbol{\lambda}\|_{H_2'} \le c_2 \|\boldsymbol{\lambda}\|_{\widetilde{W}} \quad \forall\, \boldsymbol{\lambda} \in \widetilde{W}.$$

Then Assumption 15.5.7 is fulfilled with $c_P = c_1 c_2$.

Proof. The proof immediately follows from $P = B\,S_1$ and

$$\|Pf\|_{H_2'} = \|B\,S_1 f\|_{H_2'} \le c_2\|S_1 f\|_{\tilde{W}} \le c_2 c_1 \|f\|_{H_2'} \quad \text{for all } f \in H_2'.$$

$\square$

15.5.3 Application to a generalized Stokes equation

We apply the abstract results presented in the previous section to a generalized Stokes equation of the form

$$\begin{aligned} \xi\mathbf{u} - \Delta\mathbf{u} + \nabla p &= \mathbf{g} \quad \text{in } \Omega, \\ \operatorname{div}\mathbf{u} &= 0 \quad \text{in } \Omega, \end{aligned} \tag{15.51}$$

in a bounded connected domain $\Omega \subset \mathbb{R}^3$, and with homogeneous Dirichlet boundary conditions for $\mathbf{u}$. For the weak formulation we use spaces

$$H_1 := H_0^1(\Omega)^3 =: \mathbf{H},\ H_2 := L^2(\Omega)^3 =: \mathbf{L}^2,\quad M = L_0^2(\Omega),$$

with scalar products

$$(\mathbf{u},\mathbf{v})_{H_1} := (\nabla\mathbf{u},\nabla\mathbf{v})_{L^2},\quad (\mathbf{u},\mathbf{v})_{H_2} := (\mathbf{u},\mathbf{v})_{L^2},\quad (p,q)_M := (p,q)_{L^2}.$$

The bilinear forms are defined by

$$\hat{a}(\mathbf{u},\mathbf{v}) := (\nabla\mathbf{u},\nabla\mathbf{v})_{L^2},\quad \hat{c}(\mathbf{u},\mathbf{v}) := (\mathbf{u},\mathbf{v})_{L^2},\quad \hat{b}(\mathbf{v},p) := -(p,\operatorname{div}\mathbf{v})_{L^2}.$$

With $\tau := \xi \ge 0$, the weak formulation is as follows: Find $(\mathbf{u},p) \in \mathbf{H}\times M$ such that

$$\left\{ \begin{aligned} \hat{a}(\mathbf{u},\mathbf{v}) + \tau\hat{c}(\mathbf{u},\mathbf{v}) + \hat{b}(\mathbf{v},p) &= (\mathbf{g},\mathbf{v})_{L^2} \quad \text{for all } \mathbf{v}\in\mathbf{H} \\ \hat{b}(\mathbf{u},q) &= 0 \quad \text{for all } q \in M. \end{aligned} \right. \tag{15.52}$$

Recall the inf-sup inequality:

$$\sup_{\mathbf{v}\in\mathbf{H}} \frac{(\operatorname{div}\mathbf{v},p)_{L^2}}{\|\nabla\mathbf{v}\|_{L^2}} \ge \gamma_b\|p\|_{L^2} \quad \text{for all } p \in M,$$

with $\gamma_b > 0$. Using this one easily verifies that the conditions in (15.27a)–(15.27d) are satisfied with $\gamma_a = \Gamma_a = \gamma_c = \Gamma_c = 1$, $\Gamma_b = \sqrt{3}$, $\gamma_b > 0$ the constant from the inf-sup inequality. For the operators A, B, B', C corresponding to the bilinear forms we use the (usual) notation

$$A =: -\Delta,\quad B := \nabla,\quad B' =: -\operatorname{div},\quad C =: I.$$

The result in Theorem 15.5.3 takes the form

$$(S\,p,p)_{L^2} = \|\nabla p\|^2_{\mathbf{H}^{-1}+\tau^{-1}\mathbf{L}^2} \qquad \text{for all } p \in L_0^2(\Omega)\ .$$

We now consider Assumption 15.5.7 and use the criterion in Lemma 15.5.12. Note that $-\Delta$ is the Riesz isomorphism $\mathbf{H} = H_0^1(\Omega)^3 \to \left(H_0^1(\Omega)^3\right)' =: \mathbf{H}^{-1}$. Thus for $\mathbf{f} \in \mathbf{H}^{-1}$ the solution $p = S_1\mathbf{f}$, with S_1 from Lemma 15.5.5, satisfies the weak formulation of the *stationary* Stokes problem:

$$\begin{aligned} -\Delta\mathbf{u} + \nabla p &= \mathbf{f} \quad \text{in } \Omega, \\ \operatorname{div}\mathbf{u} &= 0 \quad \text{in } \Omega, \\ \mathbf{u}|_{\partial\Omega} &= 0. \end{aligned} \tag{15.53}$$

In the following lemma it is shown *that H^2-regularity of this Stokes problem implies that Assumption 15.5.7 holds.*

Lemma 15.5.13 *Assume that the domain Ω is such that the Stokes problem (15.53) is H^2-regular, i.e., there is a constant c_R such that for any $\mathbf{f} \in \mathbf{L}^2(\Omega)$ the solution $(\mathbf{u}, p)$ is an element of $H^2(\Omega)^3 \times H^1(\Omega)$ and satisfies*

$$\|\mathbf{u}\|_{H^2(\Omega)} + \|\nabla p\|_{L^2} \leq c_R\|\mathbf{f}\|_{L^2}. \tag{15.54}$$

Then Assumption 15.5.7 is satisfied with $c_P = c_R$. Furthermore, we have $W = H^1(\Omega) \cap L_0^2(\Omega)$ and $\|p\|_W = \|\nabla p\|_{L^2}$.

Proof. We apply Lemma 15.5.12 with $\tilde{W} := H^1(\Omega) \cap L_0^2(\Omega)$ and norm $\|p\|_{\tilde{W}}^2 = (\nabla p, \nabla p)_{L^2}$. Due to the regularity assumption we have $\|S_1\mathbf{f}\|_{\tilde{W}} = \|\nabla p\|_{L^2} \leq c_R\|\mathbf{f}\|_{L^2}$. Furthermore, for $p \in \tilde{W}$ we have $\|Bp\|_{H_2'} = \|\nabla p\|_{L^2} = \|p\|_{\tilde{W}}$. Thus the assumptions in Lemma 15.5.12 hold with $c_1 = c_R$, $c_2 = 1$. It follows that Assumption 15.5.7 is fulfilled.

The definition (15.38) of W takes the form $W := \left\{ p \in L_0^2 : \nabla p \in \mathbf{L}^2 \right\}$. Thus $W = H^1(\Omega) \cap L_0^2(\Omega) = \tilde{W}$. Finally by the definition of the W-norm we have for $p \in \tilde{W}$:

$$\|p\|_W := \sup_{v \in H_2} \frac{\langle Bp, v\rangle}{\langle Cv, v\rangle^{\frac{1}{2}}} = \sup_{\mathbf{v} \in \mathbf{L}^2} \frac{(\nabla p, \mathbf{v})_{L^2}}{\|\mathbf{v}\|_{L^2}} = \|\nabla p\|_{L^2}.$$

□

Now consider the Schur complement of the generalized Stokes problem:

$$S = B'(A + \tau C)^{-1}B = -\operatorname{div}(-\Delta + \tau I)^{-1}\nabla. \tag{15.55}$$

We identify $L_0^2(\Omega)$ with its dual. Then $S : L_0^2(\Omega) \to L_0^2(\Omega)$ and $\langle\cdot,\cdot\rangle_{M' \times M} = (\cdot,\cdot)_{L^2}$.

If the stationary Stokes problem is H^2-regular the abstract theory in Sect. 15.5.2 can be applied, and we have a uniform equivalence result given in Corollary 15.5.11. This yields the following main result of this section.

Theorem 15.5.14 *Assume that the domain $\Omega \subset \mathbb{R}^3$ is such that the Stokes problem (15.53) is H^2-regular. Denote by $-\Delta_N^{-1} : L_0^2(\Omega) \to H^1(\Omega) \cap L_0^2(\Omega)$ the solution operator of the following Neumann pressure problem: Given $f \in L_0^2(\Omega)$ find $p \in H^1(\Omega) \cap L_0^2(\Omega)$ such that*

$$(\nabla p, \nabla q)_{L^2} = (f, q)_{L^2}, \quad \text{for all } q \in H^1(\Omega) \cap L_0^2(\Omega).$$

Define $\tilde{S}^{-1} = I - \tau \Delta_N^{-1}$. Then $\tilde{S}^{-1} : L_0^2(\Omega) \to L_0^2(\Omega)$ is selfadjoint and positive definite, and for all $p \in L_0^2(\Omega)$ and all $\tau \geq 0$ the following holds:

$$\gamma_s (\tilde{S}p, p)_{L^2} \leq (Sp, p)_{L^2} \leq \Gamma_s (\tilde{S}p, p)_{L^2}$$

with $\gamma_s = \frac{\gamma_b^2}{c_R^2(\gamma_b^2+1)}$, $\Gamma_s = 4$.

Proof. We apply Corollary 15.5.11. In the setting here we have $W = H_0^1(\Omega) \cap L_0^2(\Omega)$, $M = L_0^2(\Omega) = M'$. The mapping $\tilde{S} : M \to M$ is defined by, cf. (15.50), $\tilde{S}^{-1} = I_{L^2}^{-1} + \tau S_W^{-1} = I + \tau \big(B_W' C^{-1} B\big)^{-1}$. For $f \in M$ we have $w = S_W^{-1} f \in W$ iff

$$\begin{aligned}
& \langle B_W' C^{-1} B w, q \rangle_{W' \times W} = (f, q)_{L^2} \quad \forall q \in W \\
\Leftrightarrow\ & \langle Bq, C^{-1} B w \rangle_{L^2 \times L^2} = (f, q)_{L^2} \quad \forall q \in W \\
\Leftrightarrow\ & \langle \nabla q, I_{L^2}^{-1} \nabla w \rangle_{L^2 \times L^2} = (f, q)_{L^2} \quad \forall q \in W \\
\Leftrightarrow\ & (\nabla w, \nabla q)_{L^2} = (f, q)_{L^2} \quad \forall q \in W,
\end{aligned}$$

and thus S_W^{-1} is equal to the Neumann solution operator $-\Delta_N^{-1}$. Hence $\tilde{S}^{-1} = I + \tau S_W^{-1} = I - \tau \Delta_N^{-1}$. The values for the spectral bounds follow from Corollary 15.5.11 and from $\gamma_a = \Gamma_a = \gamma_c = \Gamma_c = 1$, $\Gamma_b = \sqrt{3}$ and $c_P = c_R$. □

References

1. H. Abels. On generalized solutions of two-phase flows for viscous incompressible fluids. *Interfaces and Free Boundaries*, 9:31–65, 2007.
2. H. Abels. On the notion of generalized solutions of viscous incompressible two-phase flows. In Y. Giga, H. Kozono, H. Okamoto, and Y. Shibata, editors, *Kyoto Conference on the Navier-Stokes Equations and their Applications*, volume B1, pages 1–19. RIMS Kokyuroku Bessatsu, 2007.
3. H. Abels. *Diffuse interface models for two-phase flows of viscous incompressible fluids.* MPI Leipzig, Lecture Notes nr. 36, 2008.
4. H. Abels. Existence of weak solutions for a diffuse interface model for viscous, incompressible fluids with general densities. *Comm. Math. Phys.*, 289(1):45–73, 2009.
5. H. Abels. On a diffuse interface model for two-phase flows of viscous, incompressible fluids with matched densities. *Arch. Rat. Mech. Anal.*, 194(2):463–506, 2009.
6. H. Abels and M. Wilke. Convergence to equilibrium for the Cahn-Hilliard equation with a logarithmic free energy. *Nonlinear Analysis*, 67:3176–3193, 2007.
7. H. Abels. Private communication, 2010.
8. R. A. Adams. *Sobolev Spaces.* Academic Press, 1975.
9. A. Alke and D. Bothe. 3D numerical modeling of soluble surfactant at fluidic interfaces based on the volume-of-fluid method. *Fluid Dynamics & Materials Processing*, 5:345–372, 2009.
10. A. Alke, D. Bothe, M. Kröger, and H.-J. Warnecke. VOF-based simulation of reactive mass transfer across deformable interfaces. *Progress in CFD*, 9:325–331, 2009.
11. A. Amar, S. Stapf, and B. Blümich. Internal fluid dynamics in levitated drops by fast magnetic resonance velocimetry. *Physical Review E (Statistical, Nonlinear, and Soft Matter Physics)*, 72:030201, 2005.
12. L. Ambrosio. Transport equation and Cauchy problem for BV vector fields. *Invent. Math.*, 158:227–260, 2004.
13. D. Anderson, G. McFadden, and A. Wheeler. Diffusive-interface methods in fluid mechanics. *Annu. Rev. Fluid Mech.*, 30:139–165, 1998.
14. N. Anderson and Å. Björck. A new high order method of regula falsi type for computing a root of an equation. *BIT*, 13:253–264, 1973.

15. R. Aris. *Vectors, tensors, and the basic equations of fluid mechanics.* Dover Publications, 1989.
16. D. Arnold, F. Brezzi, B. Cockburn, and L. Marini. Unified analysis of discontinuous Galerkin methods for elliptic problems. *SIAM J. Numer. Anal.*, 39:1749–1779, 2002.
17. E. Aulisa, S. Manservisi, R. Scardovelli, and S. Zaleski. Interface reconstruction with least-squares fit and split advection in three-dimensional cartesian geometry. *J. Comp. Phys.*, 225:2301–2319, 2007.
18. I. Babuška and J. Melenk. The partition of unity finite element method: Basic theory and applications. *Comp. Methods Appl. Mech. Engrg.*, 139:289–314, 1996.
19. I. Babuška and J. Melenk. The partition of unity method. *Int. J. Numer. Meth. Engrg.*, 40:727–758, 1997.
20. V. Badalassi, H. Ceniceros, and S. Banerjee. Computation of multiphase systems with phase field models. *J. Comp. Phys.*, 190:371–397, 2003.
21. R. Bank, A. Sherman, and A. Weiser. Refinement algorithms and data structures for regular local mesh refinement. In R. Stepleman, editor, *Scientific Computing*, pages 3–17. North-Holland, Amsterdam, 1983.
22. E. Bänsch. *Numerical methods for the instationary Navier-Stokes equations with a free capillary surface.* Habilitation thesis, University of Freiburg, 1998.
23. E. Bänsch. Finite element discretization of the Navier-Stokes equations with a free capillary surface. *Numer. Math.*, 88:203–235, 2001.
24. J. Barrett and C. Elliott. Fitted and unfitted finite-element methods for elliptic equations with smooth interfaces. *IMA Journal of Numerical Analysis*, 7:283–300, 1987.
25. P. Bastian. *Parallele adaptive Mehrgitterverfahren.* Teubner, Stuttgart, 1996.
26. P. Bastian, K. Birken, K. Johannsen, S. Lang, N. Neuß, H. Rentz-Reichert, and C. Wieners. UG – a flexible software toolbox for solving partial differential equations. *Comput. Vis. Sci.*, 1:27–40, 1997.
27. P. Bastian, K. Birken, K. Johannsen, S. Lang, V. Reichenberger, G. Wittum, and C. Wrobel. A parallel software-platform for solving problems of partial differential equations using unstructured grids and adaptive multigrid methods. In E. Krause and W. Jäger, editors, *High Performance Computing in Science and Engineering*, pages 326–339. Springer, Berlin, 1999.
28. M. Behr. Stabilized space-time finite element formulations for free-surface flows. *Comm. Numer. Meth. Engrg.*, 11:813–819, 2001.
29. M. Behr. Free-surface flow simulations in the presence of inclined walls. *Comp. Methods Appl. Mech. Engrg.*, 191:5467–5483, 2002.
30. T. Belytschko, N. Moës, S. Usui, and C. Parimi. Arbitrary discontinuities in finite elements. *Int. J. Numer. Meth. Engrg.*, 50:993–1013, 2001.
31. M. Benzi, G. Golub, and J. Liesen. Numerical solution of saddle point problems. *Acta Numerica*, 14:1–137, 2005.
32. M. Benzi and M. Olshanskii. An augmented Langrangian-based approach to the Oseen problem. *SIAM J. Sci. Comput.*, 28:2095–2113, 2006.
33. M. Benzi, M. Olshanskii, and Z. Wang. Modified augmented Langrangian preconditioners for the incompressible Navier-Stokes equations. *Int. J. Numer. Meth. Fluids*, 2010. DOI: 10.1002/fld.2267.
34. J. Bergh and J. Löfström. *Interpolation Spaces.* Springer, 1976.

35. E. Bertakis, S. Groß, J. Grande, O. Fortmeier, A. Reusken, and A. Pfennig. Validated simulation of droplet sedimentation with finite-element and level-set methods. *Chemical Eng. Science*, 65:2037–2051, 2010.
36. M. Bertalmio, L.-T. Cheng, S. Osher, and G. Sapiro. Variational problems and partial differential equations on implicit surfaces. *J. Comp. Phys.*, 174:759–780, 2001.
37. J. Bey. Tetrahedral grid refinement. *Computing*, 55:355–378, 1995.
38. J. Bey. *Finite-Volumen- und Mehrgitterverfahren für elliptische Randwertprobleme.* Advances in Numerical Methods. Teubner, Stuttgart, 1998.
39. J. Bey. Simplicial grid refinement: on Freudenthal's algorithm and the optimal number of congruence classes. *Numer. Math.*, 85:1–29, 2000.
40. B. Binks and T. Horozov. *Colloidal Particles at Liquid Interfaces.* Cambridge University Press, Cambridge, 2006.
41. D. Boffi. Stability of higher-order triangular Hood-Taylor methods for the stationary Stokes equations. *Math. Models and Methods in Appl. Sci. (M3AS)*, 4:223–235, 1994.
42. D. Boffi. Three-dimensional finite element methods for the Stokes problem. *SIAM J. Numer. Anal.*, 34:664–670, 1997.
43. D. Bothe, M. Koebe, K. Wielage, J. Prüss, and H.-J. Warnecke. Direct numerical simulation of mass transfer between rising gas bubbles and water. In M. Sommerfeld, editor, *Bubbly Flows: Analysis, Modelling and Calculation*, Heat and Mass Transfer. Springer, 2004.
44. D. Bothe, M. Koebe, K. Wielage, and H.-J. Warnecke. VOF-simulations of mass transfer from single bubbles and bubble chains rising in aqueous solutions. In *Proceedings 2003 ASME joint U.S.-European Fluids Eng. Conf.*, Honolulu, 2003. ASME. FEDSM2003-45155.
45. D. Bothe and J. Prüss. On the two-phase Navier-Stokes equations with Boussinesq-Scriven surface fluid. *J. Math. Fluid Mech.*, 12:133–150, 2010.
46. D. Bothe, J. Prüss, and G. Simonett. Well-posedness of a two-phase flow with soluble surfactant. In *Nonlinear Elliptic and Parabolic Problems*, volume 64 of *Progr. Nonlinear Differential Equations Appl.*, pages 37–61. Birkhäuser, Basel, 2005.
47. J. Brackbill, D. Kothe, and C. Zemach. A continuum method for modeling surface tension. *J. Comp. Phys.*, 100:335–354, 1992.
48. D. Braess. *Finite elements.* Cambridge University Press, Cambridge, second edition, 2001.
49. D. Braess, M. Dryja, and W. Hackbusch. A multigrid method for nonconforming FE-discretisations with application to non-matching grids. *Computing*, 63:1–25, 1999.
50. D. Braess and W. Hackbusch. A new convergence proof for the multigrid method including the V-cycle. *SIAM J. Numer. Anal.*, 20:967–975, 1983.
51. J. Bramble. *Multigrid Methods.* Longman, Harlow, 1993.
52. J. Bramble, J. Pasciak, and A. Vassilev. Analysis of the inexact Uzawa algorithm for saddle point problems. *SIAM J. Numer. Anal.*, 34:1072–1092, 1997.
53. L. Brenner, S.and Scott. *The Mathematical Theory of Finite Element Methods.* Springer, New York, second edition, 2002.
54. F. Brezzi and R. Falk. Stability of higher-order Hood-Taylor methods. *SIAM J. Numer. Anal.*, 28:581–590, 1991.
55. F. Brezzi and M. Fortin. *Mixed and Hybrid Finite Element Methods.* Springer, Berlin, 1991.

56. O. Bröker, M. Grote, C. Mayer, and A. Reusken. Robust parallel smoothing for multigrid via sparse approximate inverses. *SIAM J. Sci. Comput.*, 32:1395–1416, 2001.
57. M. Bruaset. *A Survey of Preconditioned Iterative Methods.* Longman, Harlow, 1995.
58. M. Burger. Finite element approximation of elliptic partial differential equations on implicit surfaces. *Comput. Vis. Sci.*, pages 87–100, 2008.
59. E. Burman. Consistent SUPG method for transient transport problems: stability and convergence. *Comp. Methods Appl. Mech. Engrg.*, 199:1114–1123, 2010.
60. J. Cahn and J. Hilliard. Free energy of a nonuniform system. I. interfacial free energy. *J. Chem. Phys.*, 28:688–699, 1958.
61. J. Cahouet and J.-P. Chabard. Some fast 3D finite element solvers for the generalized Stokes problem. *Int. J. Numer. Meth. Fluids*, 8(8):869–895, 1988.
62. M. Case, V. Ervin, A. Linke, and L. Rebholz. Improving mass conservation in FE approximations of the Navier-Stokes equations using continuous velocity fields: A connection between grad-div stabilization and Scott–Vogelius elements. Preprint 1510, Weierstrass Institute for Applied Analysis and Stochastics, 2010.
63. L. Cattabriga. Su un problema al contorno relativo al sistema di equazioni di Stokes. *Rend. Sem. Mat. Univ. Padov.*, 31:308–340, 1961.
64. Y. Chang, T. Hou, B. Merriman, and S. Osher. A level set formulation of Eulerian interface capturing methods for incompressible fluid flows. *J. Comp. Phys.*, 124:449–464, 1996.
65. Z. Chen and J. Zhou. Finite element methods and their convergence for elliptic and parabolic interface problems. *Numer. Math.*, 79:175–202, 1998.
66. J. Chessa and T. Belytschko. An extended finite element method for two-phase fluids. *ASME Journal of Applied Mechanics*, 70:10–17, 2003.
67. J. Chessa and T. Belytschko. Arbitrary discontinuities in space-time finite elements by level sets and X-FEM. *Int. J. Numer. Meth. Engrg.*, 61:2595–2614, 2004.
68. J. Chessa and T. Belytschko. A local space-time discontinuous finite element method. *Comp. Methods Appl. Mech. Engrg.*, 195:1325–1343, 2006.
69. A. Chorin. Flame advection and propagation algorithms. *J. Comp. Phys.*, 35:1–11, 1980.
70. P. Ciarlet. *The Finite Element Method for Elliptic Problems.* North-Holland, Amsterdam, 1978.
71. P. Ciarlet. Basic error estimates for elliptic problems. In P. Ciarlet and J.-L. Lions, editors, *Handbook of Numerical Analysis*, pages 17–351. North-Holland, Amsterdam, 1991.
72. M. Crandall, H. Ishii, and P. Lions. User's guide to viscosity solutions of second order partial differential equations. *Bull. Amer. Math. Soc.*, 27:1–67, 1992.
73. R. Croce, M. Griebel, and M. Schweitzer. Numerical simulation of bubble and droplet-deformation by a level set approach with surface tension in three dimensions. *Int. J. Numer. Meth. Fluids*, 62:963–993, 2009.
74. M. Crouzeix and P. Raviart. Conforming and non-conforming finite element methods for solving the stationary Stokes equations. *R.A.I.R.O. Anal. Numer.*, 7:33–76, 1973.

75. M. Dauge. Stationary Stokes and Navier-Stokes systems on two- or three-dimensional domains with corners. Part I: linearized equations. *SIAM J. Math. Anal.*, 20:74–97, 1989.
76. C. De Lellis. Ordinary differential equations with rough coefficients and the renormalization theorem of Ambrosio. *Seminaire Bourbaki*, 972:0–26, 2007.
77. G. de Rham. *Variétés Differentiables*. Hermann, 1960.
78. K. Deckelnick, G. Dziuk, and C. Elliott. Computation of geometric partial differential equations. *Acta Numerica*, pages 139–232, 2005.
79. K. Deckelnick, G. Dziuk, C. Elliott, and C.-J. Heine. An h-narrow band finite element method for elliptic equations on implicit surfaces. *IMA Journal of Numerical Analysis*, 30:351–376, 2010.
80. N. Deen, M. van Sint Annaland, and J. Kuipers. Direct numerical simulation of complex multi-fluid flows using a combined front tracking and immersed boundary method. *Chemical Engineering Science*, 64:2186–2201, 2009.
81. Y. Demay, A. Nouri, and F. Poupaud. An existence theorem for the multifluid Stokes problem. *Quart. of Appl. Math.*, 3:421–435, 1997.
82. A. Demlow. Higher-order finite element methods and pointwise error estimates for elliptic problems on surfaces. *SIAM J. Numer. Anal.*, 47:805–827, 2009.
83. A. Demlow and G. Dziuk. An adaptive finite element method for the Laplace-Beltrami operator on implicitly defined surfaces. *SIAM J. Numer. Anal.*, 45:421–442, 2007.
84. I. Denisova. Problem of the motion of two viscous incompressible fluids separated by a closed free interface. *Acta Appl. Math.*, 37:31–40, 1994.
85. I. Denisova and V. Solonnikov. Solvability of the linearized problem of a drop in a fluid flow. *J. Soviet Math.*, 56:2309–2316, 1991.
86. P. Deuflhard. *Newton Methods for Nonlinear Problems*. Springer, Berlin, 2004.
87. D. Di Pietro, S. Forte, and N. Parolini. Mass preserving finite element implementations of the level set method. *Appl. Numer. Math.*, 56(9), 2006.
88. W. Dijkhuizen, I. Roghair, M. van Sint Annaland, and J. Kuipers. DNS of gas bubbles behaviour using an improved 3D front tracking model-drag force on isolated bubbles and comparison with experiments. *Chemical Engineering Science*, 65:1415–1426, 2010.
89. R. DiPerna and P. Lions. Ordinary differential equations, transport theory and Sobolev spaces. *Invent. math.*, 98:511–547, 1989.
90. DROPS package for simulation of two-phase flows. `http://www.igpm.rwth-aachen.de/DROPS/`.
91. G. Duvaut and J. Lions. *Les Inéquations en Mécanique et en Physique*. Dunod, Paris, 1972.
92. G. Dziuk. Finite elements for the Beltrami operator on arbitrary surfaces. In S. Hildebrandt and R. Leis, editors, *Partial differential equations and calculus of variations*, volume 1357 of *Lecture Notes in Mathematics*, pages 142–155. Springer, 1988.
93. G. Dziuk. An algorithm for evolutionary surfaces. *Numer. Math.*, 58:603–611, 1991.
94. G. Dziuk and C. Elliott. Finite elements on evolving surfaces. *IMA J. Numer. Anal.*, 27:262–292, 2007.
95. G. Dziuk and C. Elliott. Surface finite elements for parabolic equations. *J. Comp. Math.*, 25:385–407, 2007.
96. G. Dziuk and C. Elliott. Eulerian finite element method for parabolic PDEs on implicit surfaces. *Interfaces and Free Boundaries*, 10:119–138, 2008.

97. G. Dziuk and C. Elliott. An Eulerian level set method for partial differential equations on evolving surfaces. *Comput. Vis. Sci.*, 13:17–28, 2008.
98. G. Dziuk and C. Elliott. L^2-estimates for the evolving surface finite element method. *SIAM J. Numer. Anal.*, 2011. to appear.
99. H. Elman. Preconditioning for the steady-state Navier-Stokes equations with low viscosity. *SIAM J. Sci. Comput.*, 20:1299–1316, 1999.
100. H. Elman, V. Howle, J. Shadid, R. Shuttleworth, and R. Tuminaro. Block preconditioners based on approximate commutators. *SIAM J. Sci. Comput.*, 27:1651–1668, 2006.
101. H. Elman, V. Howle, J. Shadid, D. Silvester, and R. Tuminaro. Least squares preconditioners for stabilized discretizations of the Navier-Stokes equations. *SIAM J. Sci. Comput.*, 30:290–311, 2007.
102. H. Elman, D. Silvester, and A. Wathen. *Finite Elements and Fast Iterative Solvers.* Oxford University Press, Oxford, 2005.
103. H. Elman and R. Tuminaro. Boundary conditions in approximate commutator preconditioners for the Navier-Stokes equations. *ETNA*, 35:257–280, 2009.
104. B. Engquist, A.-K. Tornberg, and R. Tsai. Discretization of Dirac delta functions in level set methods. *J. Comp. Phys.*, 207:28–51, 2005.
105. K. Eriksson and C. Johnson. Adaptive finite element methods for parabolic problems I: a linear model problem. *SIAM J. Numer. Anal.*, 28:43–77, 1991.
106. A. Ern and J.-L. Guermond. *Theory and Practice of Finite Elements.* Springer, New York, 2004.
107. M. Feistauer. *Mathematical Methods in Fluid Dynamics.* Longman Scientific & Technical, Harlow, 1993.
108. M. Feistauer, J. Felcman, and I. Straskraba. *Mathematical and Computational Methods for Compressible Flow.* Clarendon Press, Oxford, 2003.
109. M. Fortin and R. Glowinski. *Augmented Lagrangian Methods: Application to the Numerical Solution of Boundary-Value Problems*, volume 15 of *Studies in Mathematics and its Applications.* North-Holland, Amsterdam, 1983.
110. L. Franca and S. Frey. Stabilized finite element methods: II. the incompressible Navier-Stokes equations. *Comp. Methods Appl. Mech. Engrg.*, 99:209–233, 1992.
111. M. Francois, S. Cummins, E. Dendy, D. Kothe, J. Sicilian, and M. Williams. A balanced-force algorithm for continuous and sharp interfacial surface tension models within a volume tracking framework. *J. Comp. Phys.*, 213(1):141–173, 2006.
112. H. Freudenthal. Simplizialzerlegungen von beschränkter Flachheit. *Annals of Mathematics*, 43:580–582, 1942.
113. T. Fries and T. Belytschko. The generalized/extended finite element method: An overview of the method and its applications. *Int. J. Numer. Meth. Engrg.*, 84:253–304, 2010.
114. T. Fries and A. Zilian. The extended finite element method. *Int. J. Numer. Meth. Engrg.*, 2011. special issue.
115. S. Ganesan, G. Matthies, and L. Tobiska. On spurious velocities in incompressible flow problems with interfaces. *Comp. Methods Appl. Mech. Engrg.*, 196:1193–1202, 2007.
116. S. Ganesan and L. Tobiska. Finite element simulation of a droplet impinging a horizontal surface. In *Proceedings of ALGORITMY 2005*, pages 1–11, 2005.

117. S. Ganesan and L. Tobiska. Computations of flows with interfaces using arbitrary Lagrangian Eulerian method. In P. Wesseling, E. Onate, and J. Periaux, editors, *Proceedings of the ECCOMAS CFD conference*, Egmond aan Zee, The Netherlands, 2006. ISBN: 90-9020970-0.
118. S. Ganesan and L. Tobiska. A coupled arbitrary Lagrangian-Eulerian and Lagrangian method for computation of free surface flows with insoluble surfactants. *J. Comp. Phys.*, 228:2859–2873, 2009.
119. Y. Giga and S. Takahashi. On global weak solutions of the nonstationary two-phase Stokes flow. *SIAM J. Math. Anal.*, 25:876–893, 1994.
120. D. Gilbarg and N. Trudinger. *Elliptic Partial Differential Equations of Second Order*. Springer, Berlin, 1983.
121. V. Girault and P. Raviart. *Finite Element Methods for Navier-Stokes Equations*, volume 5 of *Springer Series in Computational Mathematics*. Springer, Berlin, 1986.
122. R. Glowinski. Numerical methods for fluids (part 3). In P. Ciarlet and J. Lions, editors, *Handbook of Numerical Analysis*, volume IX, pages 1–1083, Amsterdam, 2003. Elsevier.
123. J. Grande. *Non-stationary Incompressible Two-phase Flow Problems with Surfactant Transport - Analysis and Implementation of Numerical Methods*. Phd thesis, RWTH Aachen University, to appear 2011.
124. A. Greenbaum. *Iterative Methods for Solving Linear Systems*, volume 17 of *Frontiers in Applied Mathematics*. SIAM, Philadelphia, 1997.
125. S. Groß. Parallelisierung eines adaptiven Verfahrens zur numerischen Lösung partieller Differentialgleichungen. Diploma thesis (in german), IGPM, RWTH Aachen University, 2002.
126. S. Groß, V. Reichelt, and A. Reusken. A finite element based level set method for two-phase incompressible flows. *Comput. Vis. Sci.*, 9(4):239–257, 2006.
127. S. Groß and A. Reusken. Parallel multilevel tetrahedral grid refinement. *SIAM J. Sci. Comput.*, 26(4):1261–1288, 2005.
128. S. Groß and A. Reusken. An extended pressure finite element space for two-phase incompressible flows with surface tension. *J. Comp. Phys.*, 224:40–58, 2007.
129. S. Groß and A. Reusken. Finite element discretization error analysis of a surface tension force in two-phase incompressible flows. *SIAM J. Numer. Anal.*, 45(4):1679–1700, 2007.
130. M. Gurtin. *An Introduction to Continuum Mechanics*. Academic Press, New York, 1981.
131. M. Gurtin, D. Polignone, and J. Vinals. Two-phase binary fluids and immiscible fluids described by an order parameter. *Math. Models Methods Appl. Sci.*, 6:815–831, 1996.
132. W. Hackbusch. *Iterative Solution of Large Sparse Systems of Equations*, volume 95 of *Applied Mathematical Sciences*. Springer, New York, 1994.
133. W. Hackbusch. *Multi-Grid Methods and Applications*. Springer-Verlag, Berlin, second edition, 2003.
134. E. Hairer and G. Wanner. *Solving Ordinary Differential Equations II: Stiff and Differential-algebraic Problems*. Springer, New York, 1991.
135. A. Hansbo and P. Hansbo. An unfitted finite element method, based on Nitsche's method, for elliptic interface problems. *Comp. Methods Appl. Mech. Engrg.*, 191(47–48):5537–5552, 2002.

136. A. Hansbo and P. Hansbo. A finite element method for the simulation of strong and weak discontinuities in solid mechanics. *Comp. Methods Appl. Mech. Engrg.*, 193(33–35):3523–3540, 2004.
137. P. Hansbo. Nitsche's method for interface problems in computational mechanics. *GAMM-Mitt.*, 28(2):183–206, 2005.
138. M. Herrmann. A Eulerian level set/vortex sheet method for two-phase interface dynamics. *J. Comp. Phys.*, 203:539–571, 2005.
139. M. Herrmann. A balanced force refined level set grid method for two-phase flows on unstructured flow solver grids. *J. Comp. Phys.*, 227:2674–2706, 2008.
140. J. Hesthaven and T. Warburton. *Nodal Discontinuous Galerkin Methods: Algorithms, Analysis, and Applications.* Springer, New York, 2008.
141. J. Heywood and R. Rannacher. Finite element approximation of the nonstationary Navier-Stokes problem, II. stability of solutions and error estimates uniform in time. *SIAM J. Numer. Anal.*, 23:750–777, 1986.
142. J. Heywood and R. Rannacher. Finite element approximation of the nonstationary Navier-Stokes problem, III. smoothing property and higher order estimates for spatial discretization. *SIAM J. Numer. Anal.*, 25:489–512, 1988.
143. J. Heywood and R. Rannacher. Finite element approximation of the nonstationary Navier-Stokes problem, IV. error analysis for second-order time discretization. *SIAM J. Numer. Anal.*, 27:353–384, 1990.
144. C. Hirt and B. Nichols. Volume of fluid (VOF) method for the dynamics of free boundaries. *J. Comp. Phys.*, 39:201–22, 1981.
145. H. Hu, N. Patankar, and M. Zhu. Direct numerical simulations of fluid-solid systems using the arbitrary Lagrangian-Euler technique. *J. Comp. Phys.*, 169:427, 2001.
146. T. Hughes and L. Franca. A new finite element formulation for computational fluid dynamics: VII. the Stokes problem with various well-posed boundary conditions: symmetric formulations that converge for all velocity/pressure spaces. *Comp. Methods Appl. Mech. Engrg.*, 65:85–96, 1987.
147. S. Hysing. A new implicit surface tension implementation for interfacial flows. *Int. J. Numer. Meth. Fluids*, 51(6):659–672, 2006.
148. S. Hysing and S. Turek. The Eikonal equation: numerical efficiency vs. algorithmic complexity on quadrilateral grids. In *Proceedings of ALGORITMY 2005*, pages 22–31, 2005.
149. D. Jacqmin. Calculation of two-phase Navier-Stokes flows using phase-field modeling. *J. Comp. Phys.*, 155:96–127, 1999.
150. V. John. Higher order finite element methods and multigrid solvers in a benchmark problem for the 3D Navier-Stokes equations. *Int. J. Numer. Meth. Fluids*, 40:775–798, 2002.
151. V. John and L. Tobiska. Numerical performance of smoothers in coupled multigrid methods for the parallel solution of the incompressible Navier-Stokes equations. *Int. J. Numer. Meth. Fluids*, 33:453–473, 2000.
152. A. Johnson and T. Tezduyar. 3D simulation of fluid-particle interactions with the number of particles reaching 100. *Comp. Methods Appl. Mech. Engrg.*, 145:301, 1997.
153. C. Johnson. *Numerical Solution of Partial Differential Equations by the Finite Element Method.* Cambridge University Press, Cambridge, 1987.
154. D. Kay, P. Gresho, D. Griffiths, and D. Silvester. Adaptive time-stepping for incompressible flow; part II: Navier-Stokes equations. *SIAM J. Sci. Comput.*, 32:111–128, 2010.

155. J. Kim. A continuous surface tension force formulation for diffusive-interface models. *J. Comp. Phys.*, 204:784–804, 2005.
156. J. Kim and J. Lowengrub. Phase field modeling and simulation of three-phase flows. *Interfaces and Free Boundaries*, 7:435–466, 2005.
157. R. Kimmel and J. Sethian. Computing geodesic paths on manifolds. *Proc. Nat. Acad. Sci. USA*, 95(15):8431–8435, 1998.
158. D. Korteweg. Sur la forme que prennent les equations des mouvements des fluides si l'on tient compte des forces capillaires par des variations de densité. *Arch. Neer. Sci. Exactes Ser. II*, 6:1–24, 1901.
159. J. Krause and S. Margenov. *Robust Algebraic Multilevel Methods and Algorithms*, volume 5 of *Radon Series on Computational and Applied Mathematics*. De Gruyter, Landsberg, 2009.
160. O. Ladyzhenskaya. *Funktionalanalytische Untersuchungen der Navier-Stokesschen Gleichungen*. Akademie-Verlag, Berlin, 1965.
161. O. Ladyzhenskaya, V. Rivkind, and N. Ural'tseva. The classical solvability of diffraction problems. *Proc. Steklov Inst. Math.*, 92:132–166, 1966.
162. O. Ladyzhenskaya and N. Ural'tseva. *Linear and Quasilinear Elliptic Equations*. Academic Press, New York, 1968.
163. B. Lafaurie, C. Nardone, R. Scardovelli, S. Zaleski, and G. Zanetti. Modelling merging and fragmentation in multiphase flows with SURFER. *J. Comp. Phys.*, 113(1):134–147, 1994.
164. M. Larin and A. Reusken. A comparative study of efficient iterative solvers for generalized Stokes equations. *Numerical Linear Algebra with Applications*, 15:13–34, 2008.
165. W. Layton. *Introduction to the Numerical Analysis of Incompressible Viscous Flows*. SIAM, Computational Science & Engineering. SIAM, Philadelphia, 2008.
166. G. Legrain, N. Moës, and A. Huerta. Stability of incompressible formulations enriched with X-FEM. *Comp. Methods Appl. Mech. Engrg.*, 197:1835–1849, 2008.
167. J.-L. Lions. Sur les problemes mixtes pour certains systemes paraboliques dans les ouverts non cylindriques. *Annales de l'institut Fourier*, 7:143–182, 1957.
168. S. Lishchuk and I. Halliday. Effective surface viscosities of a particle-laden fluid interface. *Physical Review E*, 80:016306, 2009.
169. E. Loch and A. Reusken. On the accuracy of the level set SUPG method for approximating interfaces. Technical Report 319, IGPM, RWTH-Aachen, 2011.
170. J. Lopez and J. Chen. Coupling between a viscoelastic gas/liquid interface and a swirling vortex flow. *Journal of Fluids Engineering*, 120:655–661, 1998.
171. J. Lowengrub and L. Truskinovsky. Quasi-incompressible Cahn-Hilliard fluids and topological transitions. *R. Soc. Lond. Proc. Ser. A Math. Phys. Eng. Sci.*, 454:2617–2654, 1998.
172. E. Marchandise, P. Geuzaine, N. Chevaugeon, and J.-F. Remacle. A stabilized finite element method using a discontinuous level set approach for the computation of bubble dynamics. *J. Comp. Phys.*, 225(1):949–974, 2007.
173. E. Marchandise and J.-F. Remacle. A stabilized finite element method using a discontinuous level set approach for solving two phase incompressible flows. *J. Comp. Phys.*, 219(2):780–800, 2006.
174. K.-A. Mardal and R. Winther. Uniform preconditioners for the time dependent Stokes problem. *Numer. Math.*, 98:305–327, 2004.

175. K.-A. Mardal and R. Winther. Preconditioning discretizations of systems of partial differential equations. *Numerical Linear Algebra with Applications*, 2011. DOI:10.1002/nla.716.
176. M. Marion and R. Temam. Navier-Stokes equations: Theory and approximation. In P. Ciarlet and J. Lions, editors, *Handbook of Numerical Analysis*, volume VI, pages 503–689, Amsterdam, 1998. Elsevier.
177. P. Minev, T. Chen, and K. Nandakumar. A finite element technique for multifluid incompressible flow using Eulerian grids. *J. Comp. Phys.*, 187(1):255–273, 2003.
178. T. Misek, R. Berger, and J. Schröter. *Standard Test Systems for Liquid Extraction.* The Institution of Chemical Engineers, Warwickshire, second edition, 1985.
179. L. Modica. The gradient theory of phase transitions and the minimal interface criterion. *Arch. Rational Mech. Anal.*, 98(2):123–142, 1987.
180. N. Moës, M. Cloirec, P. Cartraud, and J.-F. Remacle. A computational approach to handle complex microstructure geometries. *Comp. Methods Appl. Mech. Engrg.*, 192:3163–3177, 2003.
181. N. Moës, J. Dolbow, and T. Belytschko. A finite element method for crack growth without remeshing. *Int. J. Numer. Meth. Engrg.*, 46:131–150, 1999.
182. M. Muradoglu and G. Tryggvason. A front-tracking method for computation of interfacial flows with soluble surfactants. *J. Comp. Phys.*, 227:2238–2262, 2008.
183. J. Nečas. *Les Méthodes Directes en Théorie des Équations Elliptiques.* Masson, Paris, 1967.
184. T. Nguyen. *Numerical Methods for Mass Transport Equations in Two-phase Incompressible Flows.* Phd thesis, RWTH Aachen University, 2009. www.igpm.rwth-aachen/DROPS.
185. P. Nithiarasu. An arbitrary Lagrangian-Eulerian (ALE) formulation for free surface flows using the characteristic-based split (CBS) scheme. *Int. J. Numer. Meth. Fluids*, 12:1415–1428, 2005.
186. J. Nitsche. Über ein Variationsprinzip zur Lösung von Dirichlet-Problemen bei Verwendung von Teilräumen, die keinen Randbedingungen unterworfen sind. *Abh. Math. Sem. Univ. Hamburg*, 36:9–15, 1971.
187. W. Noh and P. Woodward. A simple line interface calculation. In A. van de Vooran and P. Zandberger, editors, *Proceedings, Fifth International Conference on Fluid Dynamics.* Springer, 1976.
188. A. Nouri and F. Poupaud. An existence theorem for the multifluid Navier-Stokes problem. *J. Differential Equations*, 122:71–88, 1995.
189. M. Olshanskii, J. Peters, and A. Reusken. Uniform preconditioners for a parameter dependent saddle point problem with application to generalized Stokes interface equations. *Numer. Math.*, 105(252):159–191, 2006.
190. M. Olshanskii and A. Reusken. On the convergence of a multigrid method for linear reaction-diffusion problems. *Computing*, 65:193–202, 2000.
191. M. Olshanskii and A. Reusken. Convergence analysis of a multigrid method for a convection-dominated model problem. *SIAM J. Numer. Anal.*, 42:1261–1291, 2004.
192. M. Olshanskii and A. Reusken. A Stokes interface problem: stability, finite element analysis and a robust solver. In P. Neittaanmäki et al., editors, *Proceedings of European Congress on Computational Methods in Applied Sciences and Engineering, ECCOMAS*, 2004.

193. M. Olshanskii and A. Reusken. Analysis of a Stokes interface problem. *Numer. Math.*, 103:129–149, 2006.
194. M. Olshanskii and A. Reusken. A finite element method for surface PDEs: matrix properties. *Numer. Math.*, 114:491–520, 2009.
195. M. Olshanskii, A. Reusken, and J. Grande. An Eulerian finite element method for elliptic equations on moving surfaces. *SIAM J. Numer. Anal.*, 47:3339–3358, 2009.
196. E. Olsson and G. Kreiss. A conservative level set method for two phase flow. *J. Comp. Phys.*, 210:225–246, 2005.
197. A. Onea, M. Wörner, and D. Cacuci. A qualitative computational study of mass transfer in upward bubble train flow through square and rectangular mini-channels. *Chemical Engineering Science*, 64(7):1416–1435, 2009.
198. S. Osher and R. Fedkiw. *Level Set Methods and Dynamic Implicit Surfaces.* Applied Mathematical Sciences,Vol. 153. Springer, Berlin, Heidelberg, New York, 2003.
199. C. Paige and M. Saunders. Solutions of sparse indefinite systems of linear equations. *SIAM J. Numer. Anal.*, 12(4):617–629, 1975.
200. A. Paschedag and M. Wegener. Three-dimensional simulations of mass transfer at single droplets. In *Proceedings of the Fifth International Conference on Computational Fluid Dynamics in the Process Industries*, 2006.
201. L. Payne and H. Weinberger. An optimal Poincaré inequality in convex domains. *Arch. Rat. Mech. Anal.*, 5:280–292, 1960.
202. J. Peters, V. Reichelt, and A. Reusken. Fast iterative solvers for discrete Stokes equations. *SIAM J. Sci. Comput.*, 27(2):646–666, 2005.
203. S. Pijl, A. Segal, C. Vuik, and P. Wesseling. A mass-conserving level-set method for modelling of multi-phase flows. *Int. J. Numer. Meth. Fluids*, 47:339–361, 2005.
204. J. Pilliod and E. Puckett. Second-order accurate Volume-Of-Fluid algorithms for tracking material interfaces. *J. Comp. Phys.*, 199:465–502, 2004.
205. J. Qian, Y.-T. Zhang, and H.-K. Zhao. Fast sweeping methods for Eikonal equations on triangular meshes. *SIAM J. Numer. Anal.*, 45(1):83–107, 2007.
206. A. Quarteroni and A. Valli. *Numerical Approximation of Partial Differential Equations*, volume 23 of *Springer Series in Computational Mathematics.* Springer-Verlag, Berlin, Heidelberg, 1994.
207. R. Rannacher. On the numerical solution of the incompressible Navier-Stokes equations. *Z. Angew. Math. Mech.*, 73(9):203–216, 1993.
208. L. Rayleigh. On the theory of surface forces - II. compressible fluids. *Phil. Mag.*, 30:209–220, 1892.
209. A. Reusken. Analysis of an extended pressure finite element space for two-phase incompressible flows. *Comput. Vis. Sci.*, 11:293–305, 2008.
210. A. Reusken. Introduction to multigrid methods for elliptic boundary value problems. In J. Grotendorst, N. Attig, and S. Bügel, editors, *Simulation Methods in Molecular Sciences.* Forschungszentrum Jülich, NIC Series Vol. 42, 2009.
211. H.-G. Roos, M. Stynes, and L. Tobiska. *Numerical Methods for Singularly Perturbed Differential Equations — Convection-Diffusion and Flow Problems*, volume 24 of *Springer Series in Computational Mathematics.* Springer-Verlag, Berlin, second edition, 2008.
212. T. Rusten and R. Winther. A preconditioned iterative method for saddlepoint problems. *SIAM J. Matrix Anal. Appl.*, 13(3):887–904, 1992.

213. Y. Saad. *Iterative Methods for Sparse Linear Systems.* SIAM, Philadelphia, PA, USA, 2003.
214. R. Scardovelli and S. Zaleski. Interface reconstruction with least-square fit and split Eulerian-Lagrangian advection. *Int. J. Numer. Meth. Fluids*, 41:251–274, 2003.
215. J. Schlottke, E. Dülger, and B. Weigand. A VOF-based 3D numerical investigation of evaporating, deformed droplets. *Progress in Computational Fluid Dynamics*, 9:426–435, 2009.
216. J. Schlottke and B. Weigand. Direct numerical simulation of evaporating droplets. *J. Comp. Phys.*, 227:5215–5237, 2008.
217. C. Schwab. *p and hp–Finite Element Methods.* Clarendon Press, Oxford, 1998.
218. L. Scott and M. Vogelius. Norm estimates for a maximal right inverse of the divergence operator in spaces of piecewise polynomials. *RAIRO, Model. Math. Anal. Numer.*, 19:111–143, 1985.
219. L. Scriven. Dynamics of a fluid interface. Equations of motion for Newtonian surface fluids. *Chem. Eng. Sci.*, 12:98–108, 1960.
220. J. Sethian. A fast marching level set method for monotonically advancing fronts. *Proc. Nat. Acad. Sci. USA*, 93:1591–1595, 1996.
221. J. Sethian. Theory, algorithms, and applications of level set methods for propagating interfaces. *Acta Numerica*, 5:309–395, 1996.
222. J. Sethian. *Level Set Methods and Fast Marching Methods.* Cambridge University Press, 1999.
223. Y. Shibata and S. Shimizu. On a resolvent estimate of the interface problem for the Stokes system in a bounded domain. *J. Differential Equations*, 191:408–444, 2003.
224. D. Silvester and A. Wathen. Fast iterative solution of stabilised Stokes systems. II. Using general block preconditioners. *SIAM J. Numer. Anal.*, 31(5):1352–1367, 1994.
225. J. Slattery, L. Sagis, and E.-S. Oh. *Interfacial Transport Phenomena.* Springer, New York, second edition, 2007.
226. A. Smolianski. *Numerical Modeling of Two-Fluid Interfacial Flows.* Phd thesis, University of Jyvaskyla, 2001.
227. G. Strang. Variational crimes in the finite element method. In A. Aziz, editor, *The Mathematical Foundations of the Finite Element Method with Applications to Partial Differential Equations*, pages 689–710. Academic Press, New-York, 1972.
228. G. Strang and G. Fix. *An Analysis of the Finite Element Method.* Prentice-Hall, Englewood Cliffs,NJ, 1973.
229. M. Sussman. A second order coupled level set and volume-of-fluid method for computing growth and collapse of vapor bubbles. *J. Comp. Phys.*, 187:110–136, 2003.
230. M. Sussman and E. Fatemi. An efficient interface preserving level set redistancing algorithm and its application to interfacial incompressible fluid flow. *SIAM J. Sci. Comput.*, 20(4):1165–1191, 1999.
231. M. Sussman, P. Smereka, and S. Osher. A level set approach for computing solutions to incompressible two-phase flow. *J. Comp. Phys.*, 114:146–159, 1994.
232. N. Tanaka. Global existence of two-phase nonhomogeneous viscous incompressible fluid flow. *Comm. in Partial Differential Equations*, 18:41–81, 1993.

233. K. Teigen, X. Li, J. Lowengrub, F. Wang, and A. Voigt. A diffusion-interface approach for modelling transport, diffusion and adsorption/desorption of material quantities on a deformable interface. *Comm. Math. Sci.*, 7:1009–1037, 2009.
234. K. Teigen, P. Song, J. Lowengrub, and A. Voigt. A diffusive-interface method for two-phase flows with soluble surfactants. Technical report, Department of Mathematics, TU-Dresden, 2010.
235. R. Temam. *Navier-Stokes Equations*, volume 2 of *Studies in Mathematics and its Applications.* North-Holland publishing company, Amsterdam, 1984.
236. R. Temam. *Infinite-dimensional Dynamical Systems in Mechanics and Physics.* Springer, New York, 1988.
237. R. Thatcher. Locally mass-conserving Taylor-Hood elements for two- and three-dimensional flow. *Int. J. Numer. Meth. Fluids*, 11(3):341–353, 1990.
238. V. Thomee. *Galerkin Finite Element Methods for Parabolic Problems.* Springer, Berlin, 1997.
239. L. Tobiska and R. Verfürth. Analysis of a streamline diffusion finite element method for the Stokes and Navier-Stokes equations. *SIAM J. Numer. Anal.*, 33:673–688, 1996.
240. A.-K. Tornberg and B. Engquist. A finite element based level-set method for multiphase flow applications. *Comput. Vis. Sci.*, 3:93–101, 2000.
241. A.-K. Tornberg and B. Engquist. Numerical approximations of singular source terms in differential equations. *J. Comp. Phys.*, 200:462–488, 2004.
242. U. Trottenberg, C. Oosterlee, and A. Schüller. *Multigrid.* Academic Press, London, 2001.
243. G. Tryggvason, B. Bunner, A. Esmaeeli, D. Juric, N. Al-Rawahi, W. Tauber, J. Han, S. Nas, and Y.-J. Jan. A front-tracking method for the computations of multiphase flow. *J. Comp. Phys.*, 169:708–759, 2001.
244. S. Turek. *Efficient Solvers for Incompressible Flow Problems: An Algorithmic and Computational Approach*, volume 6 of *Lecture Notes in Computational Science and Engineering.* Springer, Berlin, Heidelberg, 1999.
245. UG, software tool for the numerical solution of partial differential equations on unstructured meshes in 2D/3D.
`http://atlas.gcsc.uni-frankfurt.de/~ug/`.
246. S. Unverdi and G. Tryggvason. A front-tracking method for viscous, incompressible multi-fluid flows. *J. Comp. Phys.*, 100:25–37, 1992.
247. J. Van der Waals. The thermodynamic theory of capillarity under the hypothesis of a continuous density variation. *J. Stat. Phys.*, 20:197–244, 1893.
248. H. Versteeg and W. Malalasekra. *An Introduction to Computational Fluid Dynamics: The Finite Volume Method, 2nd ed.* Prentice Hall, London, 2007.
249. J. Wang, P. Lu, Z. Wang, C. Yang, and Z.-S. Mao. Numerical simulation of unsteady mass transfer by the level set method. *Chemical Engineering Science*, 63:3141–3151, 2008.
250. M. Wegener, M. Fevre, Z. Wang, A. Paschedag, and M. Kraume. Marangoni convection in single drop flow - experimental investigations and 3D-simulations. In M. Sommerfeld, editor, *Proceedings of the International Conference on Multiphase Flows (ICMF), Leipzig, Germany*, 2007.
251. P. Wesseling. *An Introduction to Multigrid Methods.* Wiley, Chichester, 1992.
252. P. Wesseling. *Principles of Computational Fluid Dynamics.* Springer, Berlin, 2000.

253. G. Wittum. Linear iterations as smoothers in multigrid methods : Theory with applications to incomplete decompositions. *Impact Comput. Sci. Eng.*, 1:180–215, 1989.
254. G. Wittum. Multigrid methods for Stokes and Navier-Stokes equations. Transforming smoothers – algorithms and numerical results. *Numer. Math.*, 54:543–563, 1989.
255. G. Wittum. On the robustness of ILU-smoothing. *SIAM J. Sci. Stat. Comp.*, 10:699–717, 1989.
256. J. Wloka. *Partial Differential Equations.* Cambridge University Press, Cambridge, 1987.
257. J. Xu. Iterative methods by space decomposition and subspace correction. *SIAM Review*, 34:581–613, 1992.
258. J. Xu and Y. Zhu. Uniform convergent multigrid methods for elliptic problems with strongly discontinuous coefficients. *Mathematical Models and Methods in Applied Sciences*, 18(1):77–105, 2008.
259. J.-J. Xu and H.-K. Zhao. An Eulerian formulation for solving partial differential equations along a moving interface. *J. Sci. Comput.*, 19:573–594, 2003.
260. H. Yserentant. Old and new convergence proofs of multigrid methods. *Acta Numerica*, pages 285–326, 1993.
261. E. Zeidler. *Nonlinear Functional Analysis and its Applications, II/A.* Springer, New York, 1990.
262. S. Zhang. Successive subdivisions of tetrahedra and multigrid methods on tetrahedral meshes. *Houston J. Math.*, 21(3):541–556, 1995.
263. S. Zhang. A new family of stable mixed finite elements for the 3D Stokes equations. *Math. Comp.*, 74(250):543–554, 2005.
264. S. Zhang. On the $P1$ Powell-Sabin divergence-free finite element for the Stokes equations. *J. Comp. Math.*, 26:456–470, 2008.
265. S. Zhang. Quadratic divergence-free finite elements on Powell-Sabin tetrahedral grids. Preprint 10/2008,6/2009, Department of Mathematics, University of Delaware, 2009, www.math.udel.edu/ szhang/.
266. H.-K. Zhao. A fast sweeping method for Eikonal equations. *Mathematics of Computation*, 74(250):603–627, 2004.

Index

MIX
Papier aus verantwortungsvollen Quellen
Paper from responsible sources
FSC® C105338

If you have any concerns about our products,
you can contact us on
ProductSafety@springernature.com

In case Publisher is established outside the EU,
the EU authorized representative is:
Springer Nature Customer Service Center GmbH
Europaplatz 3, 69115 Heidelberg, Germany

Printed by Libri Plureos GmbH
in Hamburg, Germany